TOURS SYMPOSIUM ON NUCLEAR PHYSICS III

TOURS SYMPOSIUM ON NUCLEAR PHYSICS III

Tours, France   September 1997

EDITORS
M. Arnould
Universite Libre de Bruxelles, Belgium

M. Lewitowicz
GANIL, France

Yu. Ts. Oganessian
FLNR-JINR, Russia

M. Ohta
H. Utsunomiya
T. Wada
Konan University, Japan

AIP CONFERENCE
PROCEEDINGS 425

American Institute of Physics   Woodbury, New York

Editor:

Hiroaki Utsunomiya
Department of Physics
Konan University
8-9-1 Okamoto
Higashinada, Kobe 658
JAPAN

E-mail: hiro@konan-u.ac.jp

Authorization to photocopy items for internal or personal use, beyond the free copying permitted under the 1978 U.S. Copyright Law (see statement below), is granted by the American Institute of Physics for users registered with the Copyright Clearance Center (CCC) Transactional Reporting Service, provided that the base fee of $15.00 per copy is paid directly to CCC, 222 Rosewood Drive, Danvers, MA 01923. For those organizations that have been granted a photocopy license by CCC, a separate system of payment has been arranged. The fee code for users of the Transactional Reporting Service is: 1-56396-749-9/ 98 /$15.00.

© 1998 American Institute of Physics

Individual readers of this volume and nonprofit libraries, acting for them, are permitted to make fair use of the material in it, such as copying an article for use in teaching or research. Permission is granted to quote from this volume in scientific work with the customary acknowledgment of the source. To reprint a figure, table, or other excerpt requires the consent of one of the original authors and notification to AIP. Republication or systematic or multiple reproduction of any material in this volume is permitted only under license from AIP. Address inquiries to Office of Rights and Permissions, 500 Sunnyside Boulevard, Woodbury, NY 11797-2999; phone: 516-576-2268; fax: 516-576-2499; e-mail: rights@aip.org.

L.C. Catalog Card No. 97–70290
ISBN 1-56396-749-9
ISSN 0094-243X
DOE CONF- 9709162

Printed in the United States of America

CONTENTS

Preface xiii
Organizing Committee xv

SUPER-HEAVY ELEMENTS

EXPERIMENTAL DEVELOPMENTS

GSI Experiments on the Synthesis of Superheavy Elements 3
    F. P. Heßberger, S. Hofmann, V. Ninov, P. Armbruster, H. Folger,
    G. Münzenberg, Ch. Stodel, A. Lavrentev, A. G. Popeko, A. V. Yeremin,
    M. E. Leino, and S. Saro
Fusion Reactions and Experimental Approaches to the Synthesis
of Superheavy Nuclei 16
    A. V. Yeremin, V. K. Utyonkov, Yu. Ts. Oganessian
The RNB Project in Japanese Hadron Facility and Possible Use
of Neutron-Rich Beam for the Study of Superheavy Nuclei 29
    T. Nomura

THEORETICAL DEVELOPMENTS

Fusion of Massive Nuclei and Synthesis of Superheavy Elements
in the Framework of the DNS Concept 41
    E. A. Cherepanov, G. G. Adamian, V. V. Volkov, and N. V. Antonenko
Competition Between Complete Fusion and Quasifission
in Reactions with Heavy Nuclei 51
    N. V. Antonenko, G. G. Adamian, W. Scheid, and V. V. Volkov
Diffusion Model for the Synthesis of Superheavy Elements 61
    Y. Aritomo, K. Okazaki, T. Wada, M. Ohta, and Y. Abe

DECAY OF SUPERHEAVY ELEMENTS

Stability and Production of Superheavy Nuclei 75
    P. Möller and J. R. Nix
Superheavy Nuclei in Selfconsistent Nuclear Models 85
    M. Bender, K. Rutz, T. Bürvenich, T. Schilling, J. A. Maruhn,
    W. Greiner, and P. G. Reinhard
Decay Properties of Superheavy Nuclei 97
    R. Smolańczuk
Simulation for Fusion and Fusion-Fission Dynamics in Heavy Nuclei 107
    K. Okazaki, Y. Aritomo, T. Tokuda, T. Wada, M. Ohta, K. Hatogai,
    and Y. Abe

FUSION-FISSION DYNAMICS

EXPERIMENTAL ADVANCES

Fission of Exotic Nuclei 113
    K.-H. Schmidt, J. Benlliure, A. Heinz, S. Steinhäuser,
    C. Böckstiegel, A. Grewe, H.-G. Clerc, M. de Jong, A. R. Junghans,
    J. Müller, and M. Pfützner

**Production Cross-Section of Very Neutron-Rich Nuclei
in Relativistic Projectile Fission of ^{238}U** 126
    C. Engelmann, P. Armbruster, H. Geissel, A. Heinz, C. Kozhuharov,
    G. Münzenberg, W. Schwab, K. Sümmerer, B. Voss, F. Ameil,
    M. Bernas, C. Donzaud, C. Stéphan, L. Tassan-Got, S. Czajkowski,
    Ph. Dessagne, Ch. Miehé, Z. Janas, M. Pfützner, and C. Böckstiegel

**Evolution of Fission Lifetime with Temperature: A Straightforward
Measurement by the Blocking Technique** 134
    M. Morjean, J. Galin, F. Goldenbaum, E. Lienard, B. Lott, A. Péghaire,
    Y. Périer, M. Chevallier, D. Dauvergne, R. Kirsch, J. C. Poizat,
    J. Rémillieux, C. Cohen, G. Prevot, D. Schmaus, D. Jacquet, J. Dural,
    and M. Toulemonde

**Angular Momentum Dependence of Pre-Scission Particle Multiplicity
in Medium Mass Systems** 142
    K. Yuasa-Nakagawa, T. Nakagawa, K. Yoshida, S. Yamaji, K. Furutaka,
    K. Matsuda, Y. Futami, X. Liu, S. M. Lee, D. X. Jiang, Y. Aoki, J. Kasagi,
    T. Suomijärvi, W. Q. Shen, T. Wada, and Y. Abe

FUSION OF MASSIVE NUCLEI

**Investigation of the Extra-Extra-Push by Pre-Scission Neutron
Measurements with DEMON** 155
    G. Rudolf for the DEMON Collaboration

Fusion Dynamics of Massive Nuclei 164
    F. A. Ivanyuk, G. I. Kosenko, Yu. Ts. Oganessian, and V. V. Pashkevich

Multi-Dimensional Langevin Approach to the Fusion of Massive Nuclei 171
    T. Tokuda, K. Okazaki, T. Wada, M. Ohta, and Y. Abe

Time Scale of Fission Process of Cold Heavy Nuclei 179
    A. A. Goverdovski and Yu. B. Ostapenko

FISSION MODES

Observation of Fission Modes in Heavy Ion Induced Reactions 189
    M. G. Itkis, N. A. Kondratiev, E. M. Kozulin, Yu. Ts. Oganessian,
    V. V. Pashkevich, I. V. Pokrovsky, V. S. Salamatin, A. Ya. Rusanov,
    L. Calabretta, C. Maiolino, K. Lukashin, C. Agodi, G. Bellia,
    G. G. Chubarian, B. J. Hurst, D. O'Kelly, R. P. Schmitt, F. Hanappe,
    E. Liatard, A. Huck, and L. Stuttgé

New Fission Mode of the ^{252}Cf Spontaneous Fission Obtained
with Modern HPGE Detectors 202
      A. V. Daniel, G. M. Ter-Akopian, Yu. Ts. Oganessian, G. S. Popeko,
      J. H. Hamilton, J. Kormicki, A. V. Ramayya, B. R. S. Babu, T. Ginter,
      S. J. Zhu, W.-C. Ma, J. Rasmussen, M. A. Stoyer, I. Y. Lee, S. Asztalos,
      S. Y. Chu, K. E. Gregorich, A. O. Macchiavelli, M. F. Mohar, S. G. Prussin,
      J. Kliman, M. Morhac, J. D. Cole, R. Aryaeinejad, Y. K. Dardenne,
      and M. Driger

Fission Mode Study for Low-Energy Fission of Light Actinide Elements 212
      H. Baba

Vlasov Treatment of Spontaneous Fission and Sub-Barrier Fusion 222
      A. Iwamoto, V. Kondratyev, and A. Bonasera

SUB-BARRIER FUSION

Fusion Barrier Distributions - What Have We Learned? 233
      D. J. Hinde and M. Dasgupta

Fusion Reactions of Deformed Nuclei Near Coulomb Barriers 249
      H. Ikezoe, T. Ikuta, S. Mitsuoka, T. Kuzumaki, L. Jun, Y. Nagame,
      I. Nishinaka, and K. Tsukada

Anharmonic Phonon Excitations in Subbarrier Fusion Reactions 259
      K. Hagino, N. Takigawa, and S. Kuyucak

PHYSICS WITH EXOTIC NUCLEI

Mass and Nuclear Moment Measurement with High and Low
Energy RIB's 271
      W. Mittig

Gamma Spectroscopy with Low and High Energy Radioactive Beams 279
      W. Gelletly

Study of Deformation in Light Neutron-Rich Nuclei 290
      D. Guillemaud-Mueller

The September 1997 Status of SPIRAL Project and Future Developments 300
      N. Alamanos

Inelastic Proton Scattering of Unstable Nuclei 305
      F. Maréchal, Y. Blumenfeld, S. Hirzebruch, J. H. Kelley, J. A. Scarpaci,
      T. Suomijärvi, A. Azhari, D. Bazin, J. A. Brown, M. Fauerbach,
      T. Glasmacher, P. F. Mantica, D. J. Morissey, M. Steiner, P. D. Cottle,
      J. K. Jewell, K. W. Kemper, L. A. Riley, and S. Ottini

Double Giant Resonance States 315
      H. Kurasawa and T. Suzuki

ASTRONUCLEAR PHYSICS

THERMONUCLEAR REACTIONS I

New Experimental Approaches to Quests in Static Burning .... *
    C. Rolfs (contribution not received)
Reactions with Radioactive Beams: Direct Measurements. 327
    W. Galster
A Plan for ^{4}He(^{12}C, ^{16}O) γ Experiment 337
    K. Sagara
Laboratory Electron Screening .... *
    C. Rolfs (contribution not received)
Time Scale for Non-Resonant Breakup of ^{7}Li Over the Gamow
Energy Region. 343
    H. Utsunomiya, Y. Tokimoto, K. Osada, T. Yamagata, M. Ohta,
    Y. Aoki, K. Hirota, K. Ieki, Y. Iwata, K. Katori, S. Hamada, Y.-W. Lui,
    and R. P. Schmitt

THERMONUCLEAR REACTIONS II

New Experiments for the Breakout Off the Hot-CNO Cycle. 355
    S. Kubono, X. Liu, T. Miyachi, M. Kurokawa, K. I. Hahn, P. Strasser,
    S. H. Park, J. C. Kim, C. H. Lee, Y. Fuchi, S. C. Jeong, H. Kawashima,
    M. H. Tanaka, S. Kato, C. S. Lee, J. H. Lee, T. Minemura, T. Motobayashi,
    P. D. Parker, T. Shimoda, M. Smith, H. Utsunomiya, and M. Yasue
Nuclear Astrophysics with Intermediate-Energy RI Beams. 362
    T. Motobayashi
Information About the ^{12}C(α,γ) ^{16}O Reaction From the β-Delayed
Proton Decay of ^{17}Ne 372
    J. D. King, J. C. Chow, A. C. Morton, R. E. Azuma, N. Bateman, C. Iliadis,
    L. Buchmann, M. Dombsky, E. Gete, U. Giesen, K. P. Jackson, T. Shoppa,
    R. N. Boyd, J. M. D'Auria, T. Davinson, A. Shotter, W. Galster, and G. Roy
Coulomb Dissociation of ^{8}B at 254 MeV/u 382
    N. Iwasa, K. Sümmerer, F. Boue, G. Surowka, P. Senger, T. Baumann,
    H. Geissel, M. Hellström, P. Koczon, F. Laue, A. Ozawa, E. Schwab,
    W. Schwab, A. Surowiec, E. Grosse, F. Uhlig, A. Förster, H. Oeschler,
    C. Sturm, A. Wagner, J. Speer, B. Kohlmeyer, B. Blank, C. Marchand,
    M. S. Pravikoff, S. Czajkowski, R. Kulessa, W. Walus, T. Motobayashi,
    T. Teranishi, and M. Gai
Astronuclear Physics with Coulomb Dissociation 389
    S. Typel

THERMONUCLEAR REACTIONS III

Pulsed keV Neutrons for Nuclear Astrophysics and Recent
Results of (n,γ) Reactions of Light Nuclei 399
    Y. Nagai, T. Shima, T. Kii, T. Baba, K. Takaoka, S. Naito,
    A. Tomyo, T. Takahashi, Y. Nobuhara, M. Kinoshita, M. Igashira,
    and T. Ohsaki

Contraints for s-Process Scenarios: Neutron Capture Studies
in the Lanthanide Region 408
    F. Käppeler and K. Wisshak

Recent Progress in Theoretical Nuclear Astrophysics 418
    P. Descouvemont

The Cross Section of the Neutron Capture Reaction $^{13}C(n,\gamma)\,^{14}C$ 428
    H. Herndl, R. Hofinger, and H. Oberhummer

Radiative Neutron Captures by Exotic Nuclei 436
    S. Goriely

NON-THERMAL REACTIONS AND NEUTRINO ASTROPHYSICS

Spallation Reactions in Extraterrestrial Matter 447
    R. Michel

Nuclear Data for Gamma Ray Astronomy and the Cosmochemistry
of Isotopic Anomalies 457
    A. Coc and M.-G. Porquet

The Gamma-Ray Line Emission of Orion 465
    M. Cassé, E. Vangioni-Flam, and S. T. Scully

OTHER BASIC NUCLEAR DATA FOR ASTROPHYSICS

The ETFSI Mass Formula - Recent Developments 475
    J. M. Pearson, R. C. Nayak, F. Tondeur, A. Mamdouh, M. Rayet,
    and I. N. Borzov

Beta-Decay Rates: Towards A Self-Consistent Approach 485
    I. N. Borzov, S. Goriely, and J. M. Pearson

β-Decay Rates - The Semi-Gross Theory 495
    T. Tachibana, M. Yamada, and H. Nakata

STELLAR EVOLUTION AND NUCLEOSYNTHESIS

Nucleosynthesis in Low- and Intermediate-Mass Stars: An Overview 507
  N. Mowlavi
Supernova Nucleosynthesis 517
  M. Hashimoto, S. Nagataki, K. Sato, and S. Yamada
The Pre-Supernova Evolution of Massive Stars and Concomitant
Nucleosynthesis 526
  G. Meynet

BINARY STAR EVOLUTION AND THE HOT MODES OF H-BURNING

Nuclear Uncertainties and Their Role in Nova Nucleosynthesis 539
  J. José and M. Hernanz
X-Ray Bursts 551
  R. E. Taam
Nucleosynthesis at the Proton Drip Line - A Challenge
for Nuclear Physics 559
  H. Schatz, J. Görres, M. Wiescher, L. Bildsten, T. Rauscher,
  and F.-K. Thielemann

CHEMICAL EVOLUTION OF THE GALAXY

Origin and Evolution of the Light Elements Li, Be, and B 571
  M. Cassé, E. Vangioni-Flam, and S. T. Scully
The "Stellar Yields - Galactic Chemical Evolution" Connection:
A Re-Analysis 581
  S. Sandrelli, A. Visco, and M. Tosi
Heavy Elements Abundances in Metal-Poor Stars 592
  P. Magain, E. Jehin, C. Neuforge, and A. Noels

THE SYNTHESIS OF THE A>56 NUCLIDES IN REALISTIC MODEL STARS

The s-Process Efficiency in Massive Stars 605
  M. Rayet and M. Hashimoto
The ^{187}Re - ^{187}Os Cosmochronometry - The Latest Developments 616
  K. Takahashi
The p-Process in Exploding Massive Stars 626
  M. Arnould, M. Rayet, and M. Hashimoto

CLOSING

Theoretical Astronuclear Physics in Tours: A Brief Summary 637
    M. Arnould

Schedule 641
Scientific Program 642
List of Participants 647
Author Index 657

PREFACE

Tours Symposium on Nuclear Physics III was held on the basis of invitation at Hôtel de l'Univers in Tours, France, from September 2 through 5, 1997. Sixty two speakers were invited from fourteen countries over the world. Eighty nine scientists including post-docs and Ph.D. students gathered for this symposium.

This symposium was devoted to three fields of nuclear physics : (1) synthesis of superheavy elements, including studies of fusion-fission dynamics and sub-barrier fusion ; (2) physics with exotic nuclei ; and (3) astronuclear physics. We thank the members of the Organizing Committee for their excellent works on setting up topics and selecting invited speakers. We are grateful to all speakers and participants for their vital contributions which determined the value of this symposium.

We received strong support from many organizations, representatives and personnel which was indispensable in leading this symposium to a great success. In particular, we thank Mr Jean DELANEAU, Président du Conseil Général d'Indre-et-Loire ; Professeur Norihiko NAKAN-ISHI, Président of Konan University ; Professeur Daniel GUERREAU, Directeur du GANIL ; Mr Tamaki FUJIE, Directeur du Lycée Collège Konan de Touraine ; Mrs Monique BEX, Ingénieur Documentaliste, Chargée de Mission Information Scientifique du GANIL ; Mrs Brigitte ESPAZE, Chargée de Mission, Agence de Développement de la Touraine ; Mrs Miki CLEORA, Secrétaire du Lycée Collège Konan de Touraine ; the HIRAO Foundation ; and Mitsubishi Electric Co. Inc.

M. Arnould
M. Lewitowicz
Yu. Ts. Oganessian
M. Ohta
H. Utsunomiya
T. Wada

ORGANIZING COMMITTEE

Y. Abe (YITP, Japan)
M. Arnould (ULB, Belgium)
D. Guerreau (GANIL, France)
A. Iwamoto (JAERI, Japan)
M. Lewitowicz (GANIL, France)
G. Münzenberg (GSI, Germany)
T. Nomura (IPNS, Japan)
Yu. Ts. Oganessian (FLNR-JINR, Russia)
M. Ohta (Konan, Japan)
B. Remaud (IRESTE, France)
H. Utsunomiya (Konan, Japan)
T. Yamagata (Konan, Japan)
T. Wada (Konan, Japan)

HOST INSTITUTE

Konan University

SUPPORTED BY

Conseil Général d'Indre-et-Loire
GANIL
Lycée-Collège Konan de Touraine
The Hirao Foundation

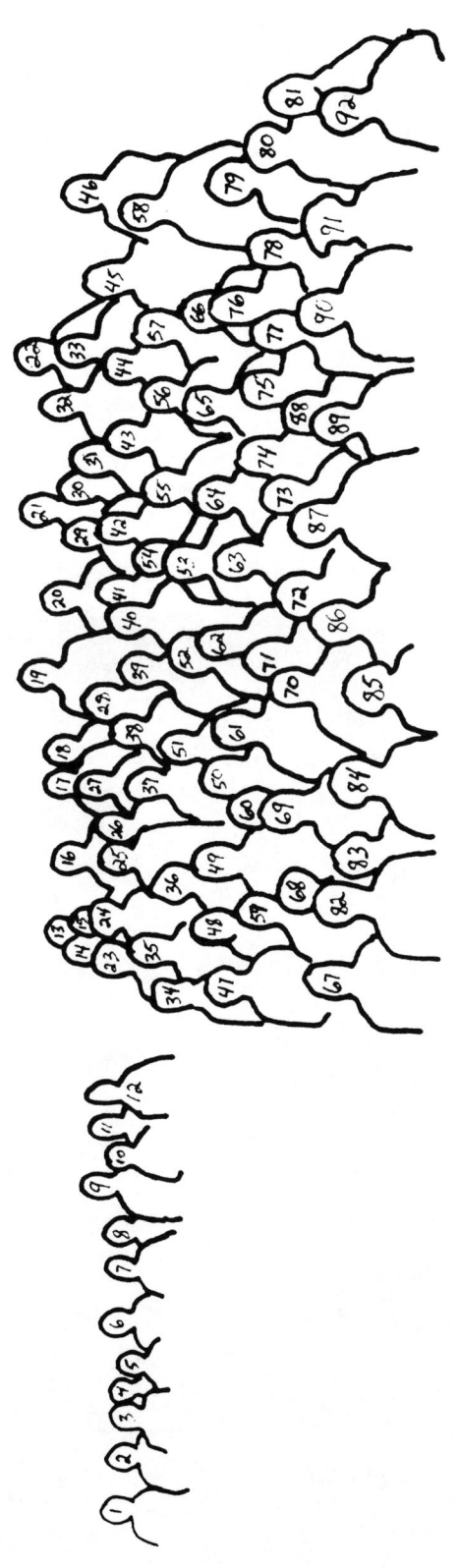

ID of figures on the photograph

1. K. Sagara
2. K. Yonehara
3. K. Takahashi
4. Y. Yamamoto
5. **M. Böhnke-Takahashi**
6. M. Rayet
7. B. Rayet
8. R. Michel
9. M. Arnould
10. C. Angulo
11. F. Käppeler
12. P. Descouvemont
13. M. Mizuno
14. E. Tachibana
15. R. Shoda
16. T. Tachibana
17. A. Coc
18. G. Meynet
19. A. Eremin
20. Y. Nagai
21. R. Smolanczuk
22. P. Möller
23. H. Kato
24. N. Takahashi
25. A. Daniel
26. G. Münzenberg
27. N. Antonenko
28. J. M. **Pearson**
29. F. **Marechal**
30. M. Hashimoto
31. N. Iwasa
32. K. Hagino
33. J. José
34. R. Morikawa
35. T. Tokuda
36. G. Kosenko
37. T. Motobayashi
38. A. Goverdovski
39. R. Taam
40. M. Bender
41. I. Borzov
42. C. Engelmann
43. S. Sandrelli
44. D. Hinde
45. W. Gelletly
46. H. Herndl
47. Y. Aritomo
48. M. Shibutani
49. K. Okazaki
50. K. Osada
51. F. P. Heßberger
52. H. Ikezoe
53. G. Rudolf
54. W. Galster
55. M. Morjean
56. K. **Ohta**
57. Y. Tanaka
58. H. Schatz
59. H. Hirose
60. A. Fukushima
61. Y. Arimoto
62. J. **King**
63. G. Chubaryan
64. T. Suzuki
65. Y. Tokimoto
66. Miss. Ohta
67. S. Typel
68. T. Wada
69. S. Kubono
70. C. Rolfs
71. Y. Abe
72. **R. King**
73. E. Uegaki
74. K.-H. Schmidt
75. M. Nomura
76. K. Hotagai
77. A. Iwamoto
78. T. Nomura
79. **M. Cleora**
80. B. Espaze
81. M.-L. Seguin
82. H. Utsunomiya
83. E. Utsunomiya
84. M. Ohta
85. Mrs. Ohta
86. S. Rolfs
87. Yu. Ts. Oganessian
88. **K. Iwamoto**
89. K. Yuasa-Nakagaea
90. A. Yokoyama
91. Y. Suzuki
92. E. Cherepanov

EXPERIMENTAL DEVELOPMENTS

GSI Experiments on the Synthesis of superheavy Elements

F.P. Heßberger[1], S. Hofmann[1], V. Ninov[1], P. Armbruster[1], H. Folger[1],
A. Lavrentev[2], M.E. Leino[3], G. Münzenberg[1], A.G. Popeko[2], S. Saro[4],
Ch. Stodel[1] and A.N. Yeremin[2]

[1]*Gesellschaft für Schwerionenforschung mbH, Planckstraße 1, D-64291 Darmstadt, Germany*
[2]*Flerov Laboratoy of Nuclear Reactions, JINR, 141980 Dubna, Russia*
[3]*University of Jyväskylä, Department of Physics, P.O. Box 35, FIN-40351 Jyväskylä, Finland*
[4]*Comenius University, Dept. of Nuclear Physics, Mlynska Dolina F1, SK-84215 Bratislava, Slowakia*

Abstract. Evaporation residue production was investigated at SHIP in cold fusion reactions of Pb- and Bi- target nuclei with projectiles of elements between Ti (Z=22) and Se (Z=34) leading to compound nuclei Z_{CN} = 104-116. The isotopes ^{269}110, ^{271}110, ^{272}111, and ^{277}112 of the elements Z=110, Z=111 and Z=112 were unambiguously identified for the first time in bombardments of ^{208}Pb, ^{209}Bi with ^{62,64}Ni and ^{70}Zn. Excitation functions for ^{50}Ti + ^{208}Pb and ^{58}Fe + ^{208}Pb were measured with high precision, three new spontaneous fission (sf) activities ^{253}104, ^{254}104, ^{258}106 were identified. A small α -decay branch of the even-even nucleus ^{256}104 ($b_\alpha \approx$ 0.003) was confirmed, allowing to estimate mass excesses Δmc^2 for N-Z = 48 nuclei up to ^{264}Hs (Z=108). An analysis of the α-decay chains observed in a bombardment of ^{209}Bi with ^{58}Fe projectiles showed evidence for an isomeric state in ^{266}Mt (Z=109). We further report on an attempt to produce element 116 and a second isotope of element 112 by the reactions ^{82}Se + ^{208}Pb and ^{68}Zn + ^{208}Pb, respectively.

INTRODUCTION

The search for superheavy elements, predicted close to the double magic nucleus ^{298}114(1) - more recent theoretical results are found in (2,3) - was a substantial motivation for the construction of the UNILAC and the velocity filter SHIP (4) at GSI in Darmstadt. Concerning the possibilities to reach the 'island of superheavy elements' in the beginning of the experimental work at SHIP in 1976 only one method seemed open: to jump across the 'sea of instability'. Although this method was tempting, it contained severe uncertainties. Decay properties of nuclei in the intended region, such as decay modes, decay energies and half-lives, were not known and only could be estimated on the basis of predicted mass excesses, shell effects, fission barriers etc., and thus were extremely uncertain. The same held, mutans mutandis, for production cross sections predicted using fusion - evaporation codes optimized to reproduce data in the region of known elements. Experiments, performed at SHIP, to produce superheavy elements in bombardments of ^{170}Er with ^{136}Xe or ^{238}U with ^{65}Cu (5), as well as by the reaction ^{48}Ca + ^{248}Cm (6) did not show positive results.

It turned out to be more successful to approach the heavier elements step by step. Following the concept of 'cold' fusion of lead or bismuth - targets with medium heavy projectiles like ^{40}Ar or ^{50}Ti, first applied successfully by Oganessian et al. (7), we succeeded to produce and identify about 25 new isotopes with atomic numbers from Z = 98 up to Z=112. Mutual influencing of experimental results and theoretical calculations allowed a better understanding of their stability, while measured excitation functions allowed a reliable empirical extrapolation of optimum bombarding energies and cross - sections for 1n - deexcitation channels.

In the present paper we predominantly will report on the results of a series of experiments performed between 1994 and 1997 (8) - (14) and briefly discuss preliminary results on an attempt to produce a second isotope of element 112 by the reaction ^{68}Zn + ^{208}Pb and an isotope of element 116 in bombardments of ^{208}Pb with ^{82}Se. For details on the experimental set-up we refer to (8).

EXCITATION FUNCTIONS

Complete fusion reactions appear as most successful method for the production of transactinide nuclei. The formation cross section of a specific nuclide in a given reaction, however, is strongly dependent on the excitation energy E^* of the compound nucleus, according to the relation $E^* = E_{CM} + Q$ (where E_{CM} denotes the energy in the center-of-mass system and Q the Q-value of the reaction), and thus on the bombarding energy $E_{lab} = (m_p + m_t)/m_t \times E_{CM}$. Since maximum production cross sections are decreasing rapidly with increasing atomic numbers as shown in fig.1, the choice of the optimum E_{lab} is crucial for the production of heaviest nuclei. We decided to determine it empirically by extrapolating values obtained from measured excitation functions.

Results for bombardments of ^{208}Pb, ^{209}Bi targets with ^{50}Ti, ^{58}Fe and ^{64}Ni - projectiles are presented in fig. 2. Excitation energies were calculated using experimental mass excesses published by Audi and Wapstra (15) and values predicted by Myers and Swiatecki (16). They were computed for the center of the target using energy losses of the projectiles according to (17). A clear trend to lower excitation energies for the maxima of the 1n- deexcitation channels, which turned out to be the most promising reactions for the production of heaviest nuclei, are indicated. Comparing the reactions leading to even-even compound nuclei we find by fitting Gaussians to the mesured data (fig. 2a, b, d) the cross section maximum for ^{50}Ti + ^{208}Pb at $E^* = (15.6 \pm 0.1)$ MeV. It decreases to $E^* = (13.6 \pm 0.2)$ MeV for ^{58}Fe + ^{208}Pb and to $E^* = (12.4 \pm 0.5)$ MeV for ^{64}Ni + ^{208}Pb. The errors indicated cover both, the uncertainty of the projectile energy measurement and of the fitting procedure. The widths of the 1n - deexcitation functions are rather narrow. From fits of Gaussians to the measured data we obtain values of (4-5) MeV (FWHM).

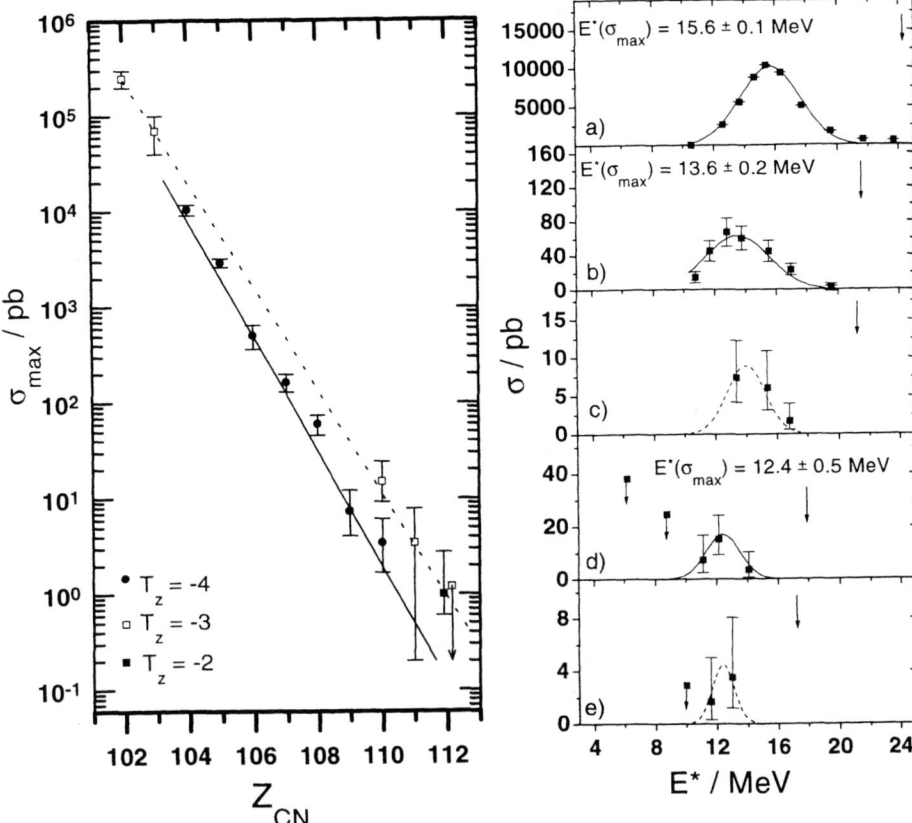

FIGURE 1. Systematics of maximum 1n cross sections for heavy element production in cold fusion reactions. $T_z = -4$: ^{50}Ti, ^{54}Cr, ^{58}Fe, ^{62}Ni - projectiles, $T_z = -3$: ^{48}Ca, ^{64}Ni, ^{68}Zn - projectiles, $T_z = -2$: ^{70}Zn - projectile. The lines are to guide the eye.

FIGURE 2. Excitation functions for 1n - deexcitation channels. a) ^{50}Ti + ^{208}Pb, b) ^{58}Fe + ^{208}Pb, c) ^{58}Fe + ^{209}Bi, d) ^{64}Ni + ^{208}Pb, e) ^{64}Ni + ^{209}Bi. The Bass model barrier for each reaction is marked by an arrow. The dashed lines are to guide the eye, the full lines are the results of fitting Gaussians to the data.

Another important question in this context is, to which extent the extrapolations for even - even compound nuclei can be used for systems leading to odd-mass compound nuclei by replacing ^{208}Pb targets by ^{209}Bi. Although no complete measured excitation functions are availible for the latter reactions, the present data do not indicate significant differences. More certain information we expect from a detailed measurement of the 1n-deexcitation function for ^{50}Ti + ^{209}Bi → ^{258}105 + 1n, which is planned in autumn 1997.

IDENTIFICATION AND DECAY PROPERTIES OF THE NEW ELEMENTS Z = 110, 111, 112

The Isotope $^{269}110$

A linear extrapolation of the optimum excitation energies for the production of $^{257}104$ and ^{265}Hs$_{108}$ (see fig. 1) resulted in an 'optimum' value of $E^* = 12.3$ MeV for the production of $^{269}110$ by the reaction ^{62}Ni + ^{208}Pb. In a run of 12 days, from November 8th to November 20th, 1994, where a total projectile dose of 2.2×10^{18} was collected, we observed four α-decay chains which were attributed to the isotope with the mass number 269 of the new element 110 (8). The assignment was based on the observation that the α - decays directly preceeded the well established α- decay chain of ^{265}Hs and thus have to origin from the α - mother $^{269}110$. From the measured decay data we obtained a mean decay energy of $E_\alpha = 11.112 \pm 0.020$ MeV and a half-life of $T_{1/2} = 170^{+160}_{-60}$ µs, the production cross section was $\sigma = 3.5^{+2.7}_{-1.8}$ pb.

The Isotope $^{271}110$

Since it is well established in the region of transfermium nuclei that more neutron rich projectiles lead to higher formation cross sections, one could expect for the combination ^{64}Ni + ^{208}Pb also a higher evaporation residue (ER) cross section than for ^{62}Ni + ^{208}Pb. In a succeeding experiment from November 22th to November 30th, 1994, and December 18th to December 22th, 1994, the ER production by the reaction ^{64}Ni + ^{208}Pb was investigated at $E^* = (8-13)$ MeV. Nine α - decay chains observed in this experiment could be attributed to $^{271}110$ (10). A maximum cross section of $\sigma = 15^{+9}_{-6}$ pb was measured at $E^* = 12.1$ MeV. The observed decay chains can be subdivided into three groups (disregarding the one chain where the α-particle attributed to $^{271}110$ escaped the detectors and only a ΔE signal was registered): five events decayed with a mean energy $E_\alpha = 10.738$ MeV, two events with $E_\alpha = 10.682$ MeV. The time differences $\Delta t_{ER-\alpha 1}$ between the implantation of the ER and the succeeding α -decays lay between 0.2 ms and 4.4 ms, resulting in a half-life $T_{1/2} = 1.1^{+0.6}_{-0.3}$ ms. The third group existed of only one decay chain starting with $E_\alpha = 10.709$ MeV. A time distance $\Delta t_{ER-\alpha 1} = 81$ ms was registered. Because of the significantly longer lifetime, it seems justified to assign this transition to a second level of $^{271}110$. The results on the first two groups can be used for a first qualitative discussion on the ground state level of $^{271}110$. Recent theoretical calculations (18) predict levels with spin and parity $1/2^+$, $3/2^+$, $7/2^+$, $9/2^+$, $11/2^-$ in the energy interval $\Delta E < 200$ keV above the ground-state. The hindrance factors for the 10.738 MeV and 10.682 MeV transitions are calculated as 3.4 and 6.2, respectively, their intensity ratio is 5/2, the energy difference $\Delta E = 56 \pm 7$ keV. These values are typical for favoured transitions to the band head and the first rotational level. Using the energy difference and the low lying levels and spins proposed for ^{267}Hs, the α- decay daughter of $^{271}110$, we obtain inertia parameters $A = \hbar^2/2\Theta = \Delta E/(2\times(I+1))$ of A = 5.1, 6.2 and 11.2 keV for the spins I = 9/2, 7/2 and 3/2, respectively. The first two values of A are in good agreement with data known from other deformed heavy nuclei, wheras A = 11.2 keV is unusually high. The remaining levels $9/2^+[615]$ and $7/2^+[613]$ arise from the i11/2 subshell. Theory predicts the $9/2^+[615]$ Nilsson orbital closer to the Fermi surface in $^{271}_{161}110$. Therefore, we tentatively assign the 10.738 MeV transition to the decay of the $9/2^+[615]$ state into the analog Nilsson state in ^{267}Hs, and the 10.682 MeV transition into the first rotational level $11/2^+$ (10).

The Isotope $^{272}111$

The positive results in the synthesis of element 110 in the reactions ^{62,64}Ni + ^{208}Pb encouraged us to start an attempt to produce an isotope of element 111 by the reaction ^{64}Ni + ^{209}Bi. In an experimental run from December 1st to December 18th, we used three bombarding energies. Using the predicted mass excess of (16) for the compound nucleus $^{273}111$ excitation energies of 10.0 MeV, 11.6 MeV, and 13.0 MeV are obtained. Projectile doses of 1.0×10^{18} (at $E^* = 10.0$ MeV), 1.1×10^{18} (at $E^* = 11.6$ MeV), and 1.1×10^{18} (at $E^* = 13.0$ MeV), were collected. While no decay chain that could be attributed to $^{272}111$ was registered at $E^* = 10.0$ MeV, one event was observed at $E^* = 11.6$ MeV, and two events at $E^* = 13.0$ MeV (9), referring to a cross section of $\sigma = 3.5^{+4.6}_{-2.3}$ pb. Only one decay chain was complete, i.e. it consisted of five successive α -

TABLE 1. α-decay chains attributed to $^{272}111$

Isotope	Chain 1	Chain 2	Chain 3
$^{272}111$	$E_\alpha = 0.533$ MeV (escape) $\Delta t = 3600$ μs	$E_\alpha = 4.612$ MeV (escape) $\Delta t = 696$ μs	$E_\alpha = 10.820$ MeV $\Delta t = 2042$ μs
^{268}Mt	$E_\alpha = 10.259$ MeV $\Delta t = 71$ ms	$E_\alpha = 10.097$ MeV $\Delta t = 171$ ms	$E_\alpha = 10.221$ MeV $\Delta t = 72$ ms
^{264}Bh	$E_\alpha = 9.475$ MeV $\Delta t = 98$ ms	$E_\alpha = 9.618$ MeV $\Delta t = 334$ ms	$E_\alpha = 9.621$ MeV $\Delta t = 1452$ ms
$^{260}105$	$E_\alpha = 0.873$ MeV (escape) $\Delta t = 1.97$ s	$E_\alpha = 9.146$ MeV $\Delta t = 0.953$ s	$E_\alpha = 9.200$ MeV $\Delta t = 0.573$ s
^{256}Lr			$E_\alpha = 8.463$ MeV $\Delta t = 66.3$ s

decays with full energy deposit in the detector (table 1). On the whole, the assignment was based on the fact that decay chains of evaporation residues produced in other energetic possible deexcitation channels (γ, p, α) do not show the observed properties and on the agreement of energy and life-time of the decays α_4 and α_5 with literature values for ^{260}105 (E = 9.0 - 9.2 MeV, $T_{1/2}$ = 1.52 s) (19,20) and ^{256}Lr (E = 8.3-8.6 MeV, $T_{1/2}$ = 31 s) (21).

The Isotope ^{277}112

In an experimental run from January, 26th to March, 7th, 1996 we searched for element 112 using the projectile - target combination ^{70}Zn + ^{208}Pb. A total projectile dose of 3.4 × 10^{18} was collected. According to our systematics on optimum excitation energies a bombarding energy according to E^* = 10.1 MeV was chosen. Two decay chains which could be attributed to ^{277}112 were observed (Fig.3), the resulting production cross section was σ = 1.0 $^{+1.3}_{-0.6}$ pb (11). The most striking result, however, is the significant difference in the decay energies and lifetimes of the daughter isotope ^{273}110 of E_α = 9.73 MeV, Δt = 170 ms (chain 1) and E_α = 11.08 MeV, Δt = 110 μs (chain 2). Due to the large differences in the lifetimes the two transitions must be assigned to different levels in ^{273}110. Half-life calculations result in values of $t_{1/2,calc}$ = 175 ms and of $t_{1/2,calc}$ = 67 μs, respectively, both for Δl = 0 transitions. Since these values compare well with the measured lifetimes we conclude that both transitions are unhindered and connect analogous states in the mother and daughter nuclei. As a first attempt for an explanation the systematics of Q_α - values (as derived from the maximum decay - energies observed) are displayed in fig. 4. 'Chain 1' indicates the continuation of a trend of decreasing Q_α - values with increasing neutron numbers, as also evident for Z = 104 to Z = 108, which

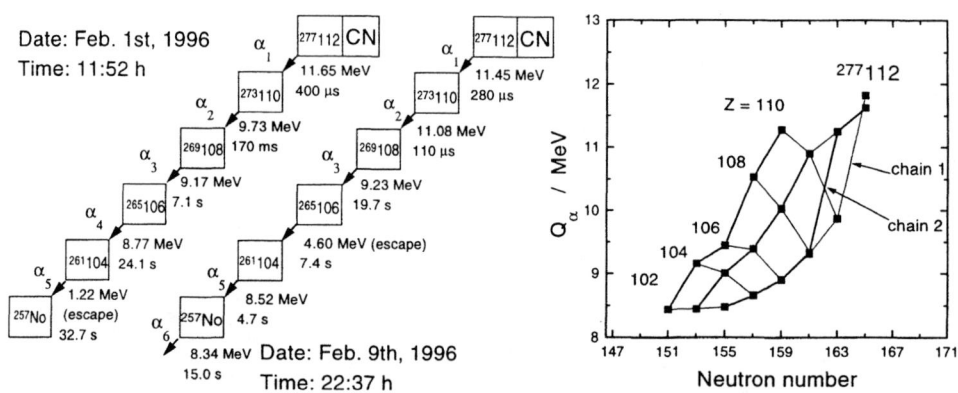

FIGURE 3. The two α- decay chains observed during an irradition of ^{208}Pb with ^{70}Zn projectiles and attributed to ^{277}112.

[11], Reprinted with permission of Springer-Verlag.

FIGURE 4. Systematics of measured Q_α - values. The thick lines connect members of α- decay chains starting with ^{269,271}110 and ^{277}112

[11], Reprinted with permission of Springer-Verlag.

can be understood as the consequence of a general trend of an increasing ratio m(Z,A)/m(Z-2,A-4) with decreasing neutron number. 'Chain 2' indicates a break of this trend and a local minimum at N=161, which can be interpreted as the consequence of a local minimum in (ΔE_{micr}(Z,A)-ΔE_{micr}(Z-2,A-4)), i.e. the difference in the ground-state shell correction energies of the mother- and daughter nucleus. Such minima are expected when neutron shells are crossed. From this side it could be a hint for a neutron shell at N = 162 (22). On the other hand, the energy difference of ΔE = 0.20 MeV between the two observed α - transitions attributed to ^{277}112 may suggest that two parallel running chains, one through the ground state, and the other one through a high energy isomeric state, are already initiated by the reaction process, populating two different levels in ^{277}112, which both have similar halflives. On the basis of the present data, definite conclusions are not possible, however. So we will not regard our results as a proof for a shell crossing and we will for the moment present only one value of $T_{1/2}$ = 240^{+430}_{-90} µs obtained from the two events.

SPONTANEOUS FISSION PROPERTIES

Theoretical calculations using refined models based on the Nilsson-Strutinsky approach presently predict two areas of enhanced nuclear stability in the region Z=104 to Z=120: one around $^{273}_{163}$110 (23) or $^{270}_{162}$108 (22) with microscopic ground state shell corrections of $\Delta E_{micr} \approx$ -(7-8) MeV and a maximum deformation of $\beta_2 \approx$ 0.22 and one around $^{292}_{178}$114 with $\Delta E_{micr} \approx$ -8 MeV, forming an island of spherical superheavy nuclei, located slightly below the predicted double magic nucleus $^{298}_{184}$114. As a consequence enhanced stability against sf is expected in these regions. In fig.5 we have displayed calculated (24) and experimental fission half-lives T_{sf} for isotopes Z $\geq$ 102. The local maximum of T_{sf} around the deformed neutron shell at N=152 known for elements Z $\leq$ 102 is vanished at Z=104. This effect was explained by the decrease of the outer fission barrier below the groundstate (25) and is also indicated by a strong increase of the barrier curvature energy $\hbar\omega$ (26). For the Z = 104 - isotopes we rather observe at N=152 a plateau than a maximum in T_{sf}. Below N=152 a sharp decrease is present: $^{254}_{150}$104 produced in the course of these experiments by ^{206}Pb(^{50}Ti,2n)^{254}104, decays by sf with $T_{1/2}$ = (23±3) µs (14), which is a factor of $\approx$300 lower than that of the neighbouring even - even nucleus $^{256}_{152}$104. A further steep decrease towards lower neutron numbers is indicated by the observation of a 48 µs sf activity in an irradiation of ^{204}Pb with ^{50}Ti, which was attributed to ^{253}104 (14). The half-life is roughly five orders of magnitude lower than that of the neighbouring odd-even nucleus ^{255}104. Since sf of odd-even nuclei is hindered by a factor of typically 1000 compared to neighbouring even - even nuclei, for ^{252}104 a value of T_{sf} < 0.1 µs can be expected. At the neutron rich side a steep increase for N > 156 is indicated: $T_{1/2} = (2.1\pm0.2)$ s recently was published by Lane et al. (27) for $^{262}_{158}$104, which roughly is a factor of 25 higher than the old value of $T_{1/2} = (47\pm5)$ ms (28). At Z = 106 sf was observed unambiguously only for two isotopes with N ≤ 158, $^{260}_{154}$106 ($T_{1/2} = 7.2^{+4.8}_{-2.7}$ ms) (29) and $^{258}_{152}$106 ($T_{1/2} = 2.9^{+1.3}_{-0.7}$ ms) produced in the course of these experiments by the reaction ^{209}Bi(^{51}V,2n)^{258}106. The heaviest even-even nucleus of element 106, $^{266}_{160}$106 (30) was identified by its α-decay. An sf branch of this isotope seems to be established, but the reported value of $T_{sf} = 55^{+463}_{-33}$ s (31) still suffers from large error bars. On the basis of these results T_{sf} of this isotope, conservatively, may be estimated at least a factor of 2000 higher than that of the element 106 isotopes at N= 152,154. The isotope $^{264}_{154}$Hs originally was identified by one observed α-α-sf correlation chain (32). In the recent SHIP experiments one more α- decay chain containing the full α-energy of ^{264}Hs ($E_\alpha = (10.48\pm0.02)$ MeV) and two sf events attributed to this isotope were observed (13). Although these data are still rather scanty, an sf branch of $b_{sf} \approx 0.5$,

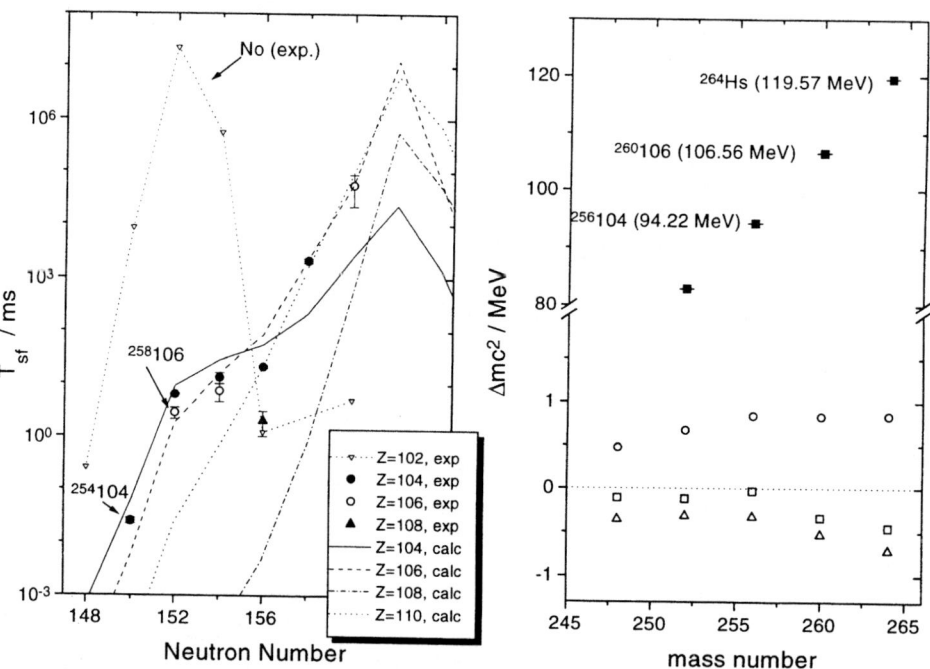

FIGURE 5. Comparison of calculated (22) and experimental fission half-lives for even-even nuclei Z = 104 to Z = 110.

FIGURE 6. Comparison of experimental and calculated mass excesses. Upper part: ■ Experimental mass excesses. Lower part: $(\Delta m(exp) - \Delta m(calc))c^2$ according to □ (22), ○ (23), △ (16).

resulting in $T_{sf} \approx 2$ ms, is indicated, about a factor of 10 lower than the value predicted in (24). As a consequence of the stabilisation against sf α- decay is predicted as the dominating decay mode for isotopes in the region Z=106 to 114 and N=152 to 162 (24).

MASS EXCESSES FOR ^{256}104, ^{260}106, ^{264}Hs

α-decay studies provide more information about nuclear structure than sf. Q_α - values from ground-state to ground-state transitions can be used to calculate mass excesses Δmc^2 from decaying nuclei if those of the daughter products are known. In case of very low production rates this procedure can be applied unambigouosly only to even-even nuclei, since only for these nuclei transitions having the highest energy and intensity can be regarded as ground state transitions. At Z = 104 sf is the dominating decay mode for the even - even isotopes. Only for ^{256}104 a small α - branch has been reported so far (33), allowing a direct connection of the masses of the T_z = -24 isotopes of Z ≥ 104 to masses of lighter elements. This is the only mass measurement we have for elements Z ≥104 so far. From the recent experiments we obtained values b_α = 0.0032±0.0017 and E_α = (8790±20) keV for ^{256}104 (14). From the latter value we obtain Δmc^2 = 94.22 MeV (^{256}104), Δmc^2 = 106.56 MeV (^{260}106), and Δmc^2 = 119.57 MeV (^{264}Hs). In fig.6 we compare them to recent mass predictions. Smolanczuk and Sobiczewski (22) as well as Myers and Swiatecki (16) predict similar values, their Δmc^2 -values are higher than the experimental ones, the differences are ≤0.7 MeV, while Möller et al. (23) predict lower Δmc^2 -values with differences of roughly one MeV. In all three cases a slight trend to increasing differences is visible.

α - DECAY OF ^{266}Mt

In addition to the three α-decay chains observed in our first experiments to synthesize element 109 (34,35) twelve more α-decay chains, attributed to decays of ^{266}Mt$_{109}$, were observed in a recent bombardment of ^{209}Bi with ^{58}Fe. α-decays of E = (10.4-11.4) MeV were followed by decays of E = (9.7-10.5) MeV, which represents the energy range of α- particles attributed to ^{262}Bh$_{107}$ and ^{262m}Bh$_{107}$ (36), in eleven cases. These correlations are displayed in fig. 6. The bars represent an energy range $E_{\alpha,mean}$ ± 65 keV, i.e. taking into account the energy deviation of a single event from the mean energy. One single event of E_α = 11.739 MeV was followed ($\Delta t_{ER-\alpha}$ = 7.8 ms) by an event of E_α = 6.4 MeV, probably an α-particle emitted by ^{262}Bh, whose energy was not completly registered. Experimental half-lives for the daughter-decays are $T_{1/2} = 73^{+40}_{-19}$ ms for the events in the interval E_α = (9.7-10.15) MeV (^{262}Bh), and $T_{1/2} = 9^{+7}_{-3}$ ms for those in E_α = (10.15-10.45) MeV (^{262m}Bh), both in agreement with the values reported for ^{262}Bh and ^{262m}Bh (36). Differences for the ^{266}Mt - decays correlated to ^{262}Bh or ^{262m}Bh are evident:

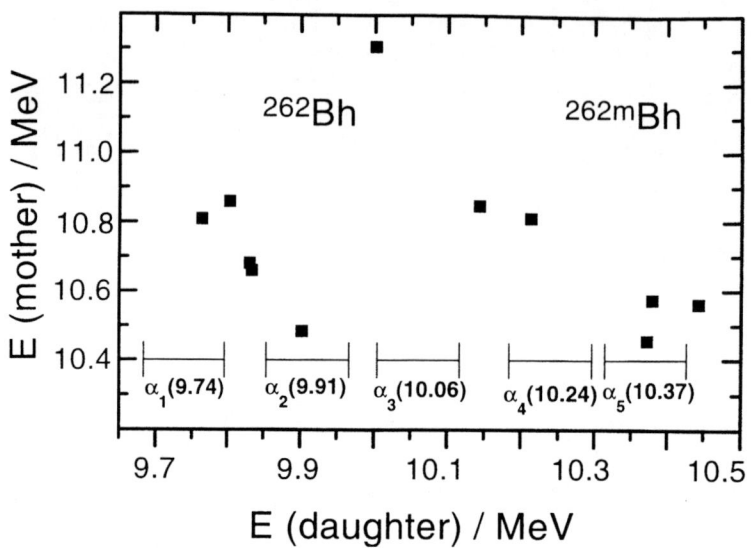

FIGURE 10. α - α - correlation plot for ^{266}Mt decays followed by decays of ^{262,262m}Bh.

a) Leaving aside the single event at E_α = 11.307 MeV, we obtain a 'mean energy' of E = (10.72 ± 0.13) MeV for the events followed by α-decays of ^{262}Bh, and E = (10.60 ± 0.13) MeV for those followed by ^{262m}Bh.

b) Half-lives of $T_{1/2}$ = 0.7 $^{+0.4}_{-0.2}$ ms and $T_{1/2}$ = 1.2 $^{+1.0}_{-0.4}$ ms for events followed by ^{262}Bh and ^{262m}Bh, respectively, are obtained.

These observations indicate that possibly two different levels of ^{266}Mt are populated by the reaction process. A preliminary assignment of the decay energies is presented in table 2. The predominantly similar decay energies indicate that the decays of the ^{266}Mt - level with $T_{1/2}$ = 0.7 ms populate levels in ^{262}Bh below the isomeric state, while those with $T_{1/2}$ = 1.2 ms populate the isomeric state ^{262m}Bh or levels above it. The low hindrance factors (HF) for the events E<11 MeV indicates that they represent 'favoured' decays, while the strongly hindered E_α =11.306 MeV decay probably reprents the grond-state transition ^{266}Mt(gs) → ^{262}Bh(gs). Due to the relatively large Δt = 7.8 ms, the α-

TABLE 2. Decay properties of ^{266}Mt and ^{266m}Mt (preliminary data)

^{266}Mt ($T_{1/2}$ = 0.7 $^{+0.4}_{-0.2}$ ms)			^{266m}Mt ($T_{1/2}$ = 1.2 $^{+1.0}_{-0.4}$ ms)		
$E_{\alpha, mean}$/ MeV	Σevents	HF	$E_{\alpha, mean}$/ MeV	Σevents	HF
11.306	1	≈460	10.814	1	≈35
10.837	3	≈13	10.569	2	≈5
10.671	2	≈8	10.456	1	≈5
10.484	1	≈6	11.739 (tentative)	1	>5000

decay of $E_\alpha = 11.739$ MeV is tentatively assigned to the ^{266m}Mt (1.2 ms) - level; it probably may represent the transition ^{266m}Mt $\rightarrow$ ^{266}Bh(gs). To confirm these interpretations and to improve the quality of the data, additional experiments are necessary, however.

ATTEMPTS TO SYNTHESIZE ELEMENTS Z ≥ 112 BY ^{68}Zn + ^{208}Pb $\rightarrow$ ^{266}112* AND ^{82}Se + ^{208}Pb $\rightarrow$ ^{290}116*

Using the combination ^{70}Zn + ^{208}Pb for synthesis of element 112 a compound nucleus with N=166 is produced, i.e. we overshoot the region of predicted maximum ground state stabilisation around N=162. It thus seems tempting to try a reaction leading closer to the predicted neutron shell at N=162 in order to profit from an enhanced ground state stability. Of course such a procedure may be under discussion. It is known that at increasing E^* ground state shell effects are washed out with a decay constant of typically 18.5 MeV (37). From this point an enhancement of the survival probability by higher ground state shell effects seems questionable. We have investigated this possibility using the combination ^{68}Zn + ^{208}Pb leading to the compound nucleus $^{266}_{164}$112. In an experimental run from March, 27th, 1997 to April, 21st, 1997, we used two projectile energies E_{lab} = 341.4 MeV and E_{lab} = 339.3 MeV. Beam doses of 2.3×10^{18} (341.4 MeV) and 2.4×10^{18} (339.3 MeV) were collected. In a preliminary analysis of the data, no decays that could be attributed to an isotope of element 112, preferrably ^{275}112 (1n deexcitation channel), were observed. The cross section limits are 1.2 pb for both energies (fig. 8).

The most striking result of the 1994 - experiments was the proof that the E^* - value at

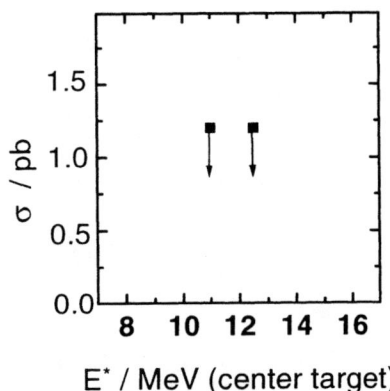 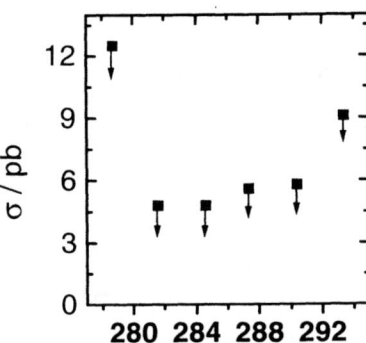

FIGURE 8. Upper limits for evaporation residue - production by the reaction ^{68}Zn + ^{208}Pb $\rightarrow$ ^{266}112*. The cross section limits refer to the observation of one event with a probability of 0.68.

FIGURE 9. Upper limits for evaporation residue production by the reaction ^{82}Se + ^{208}Pb $\rightarrow$ ^{290}116*. The fusion barrier according to (37) is marked by an arrow. The cross section limits refer to the observation of one event with a probability of 0.68.

which the maximum of the 1n - channel occurs gradually decreases from element 104 (^{50}Ti + ^{208}Pb) to 110 (^{64}Ni + ^{208}Pb), although an increase of the extra-push energy from $\Delta E = 24$ MeV to $\Delta E = 112$ MeV (38) is predicted. This trend runs parallel to a decrease of E^* at the fusion barrier, calculated e.g. according to the Bass model (39). Together with the trend of the cross sections a tempting perspective was obvious at that time: since the trend to lower E^* - values at the fusion barrier is preserved towards higher Z-values, ER production via γ-deexcitation seemed feasible. Especially for $E^* < B_n$ (neutron binding energy) one might speculate on increasing cross sections. A suitable reaction seemed ^{82}Se + ^{208}Pb → ^{290}116*, where $E^* = 6.1$ MeV (23) - 2.4 MeV (22) at the fusion barrier (39). We investigated this reaction in an experiment from November 9th to December 11th, 1995. The beam energy was varied between $E_{CM} = 293$ MeV and 278 MeV and covered an excitation energy range from $E^* = 0$ MeV up to $E^* = 8.9$ MeV according to (22) or 12.6 MeV (23). In consideration of a lower mass excess of the compound nucleus ^{290}116 than predicted we conservatively chose our lowest bombarding energy at $E_{CM} = 278$ MeV, for which (22,23) already predict negative excitation energies. No event that could be attributed to an isotope of element 116 was observed. Upper limits for ER production are displayed in fig.9.

REFERENCES

1. Meldner, H., *Arkiv för fysik* **36**, 593-598 (1967)
2. Möller, P., Nix, J.R., *J. Phys. G. Part. Phys* **20**, 1681-1747 (1994)
3. Cwiok, S., Sobiczewski, A., *Z. Phys. A* **342**, 203-213, (1992)
4. Münzenberg, G., Faust, W., Hofmann, S., Armbruster, P., Güttner, K., Ewald, H., *Nucl. Instrum. Methods* **161**, 65-82 (1979)
5. Münzenberg, G., Armbruster, P., Faust, W., Hofmann, S., Reisdorf, W., Schmidt, K.-H., Valli, K., Ewald, H., Güttner, K., Clerc, H.G., Lang, W., *GSI Jahresbericht 1977* **GSI-J-1-78**, 75 (1978)
6. Agarwal, Y.K., Armbruster, P., Hofmann, S., Heßberger, F.P., Münzenberg, G., Poppensieker, K., Reisdorf, W., Schmidt, K.-H., Schneider, J.R.H., Schneider, W.F.W., Vermeulen, D., Ghiorso, A., Leino, M., Moody, K.J., *GSI Scientific Report 1983* **GSI-84-1**, 79 (1984)
7. Oganessian, Yu.Ts.; editors: Harney,H.L., Braun-Munzinger. P., Gelbke, C.K., *Lecture Notes in Physics Vol. 33,* Berlin, Heidelberg, New York: Springer, 1974, pp. 221-252
8. Hofmann, S., Ninov, V., Heßberger, F.P., Armbruster, P., Folger, H., Münzenberg, G., Schött, H.-J., Popeko, A.G., Yeremin, A.V., Andreyev, A.N., Saro, S., Janik, R., Leino, M., *Z. Phys. A* **350**, 277-280 (1995)
9. Hofmann, S., Ninov, V., Heßberger, F.P., Armbruster, P., Folger, H., Münzenberg, G., Schött, H.-J., Popeko, A.G., Yeremin, A.V., Andreyev, A.N., Saro, S., Janik, R., Leino, M., *Z. Phys. A* **350**, 281-282 (1995)
10. Hofmann, S., *GSI - Nachrichten 02-95*, 4-15 (1995)
11. Hofmann, S., Ninov, V., Heßberger, F.P., Armbruster, P., Folger, H., Münzenberg, G., Schött, H.-J., Popeko, A.G., Yeremin, A.V., Saro, S., Janik, R., Leino, M., *Z. Phys. A* **354**, 229-230 (1996)
12. Hofmann, S., Heßberger, F.P., Ninov, V., Armbruster, P., Münzenberg, G., Stodel, C., Popeko, A.G., Yeremin, A.V., Saro, S., Leino, M., *Z. Phys. A* (in press)
13. Ninov, V. et al. to be published
14. Heßberger, F.P., Hofmann, S., Ninov, V., Armbruster, P., Folger, H., Münzenberg, H., Schött, H.J., Popeko, A.G., Yeremin, A.V., Andreyev, A.N., Saro, S., submitted to *Z. Phys. A*
15. Audi, G., Wapstra, H., *Nucl. Phys. A* **595**, 409-480 (1995)
16. Myers, W.D., Swiatecki, W.J., *Nucl. Phys. A* **601**, 141-167 (1996)

17. Hubert, F., Bimbot, R., Gauvin, H., *Atomic Data and Nuclear Data Tables* **46**, 1-213 (1990)
18. Cwiok, S., Hofmann, S., Nazarewicz, W., *Nucl. Phys.* A **573**, 356-394 (1994)
19. Ghiorso, A., Nurmia, M., Eskola, K., Harris, J., Eskola, P., *Phys. Rev. Lett.* **24**, 1498-1503 (1970)
20. Bemis,Jr., C.E., Dittner, P.F., Silva, R.J., Hahn, R.L., Tarrant, J.R., Hunt, L.D., Hensley, D.C., *Phys. Rev. C* **16**, 1146-1157 (1977)
21. Eskola, K., Eskola, P., Nurmia, M., Ghiorso, A., *Phys. Rev. C* **4**, 632-642 (1971)
22. Smolanczuk, R., Sobiczewski, A.,"Shell effects in the properties of Heavy and Superheavy Nuclei" in *Proceedings of the EPS Cnference on Low Energy Nuclear Dynamics*, St. Petersburg (Russia), April 18-22, 1995, World Scientific, Singapore, New Jersey, London, Hong Kong, 1995, pp. 313-320 and private communication 1995
23. Möller, P., Nix, J.R., Myers, W.D., Swiatecki, W.J., *Atomic Data and Nuclear Data Tables* **59**, 185-381 (1995)
24. Smolanczuk, R., Skalski, J., Sobiczewski, A., *Phys. Rev. C* **52**, 1871-1879 (1995)
25. Oganessian, Yu.Ts., Demin, A.G., Iljinov, A.S., Tretyakova, S.P., Pleve, A.A., Penionzhkevich, Yu.E., Ivanov, M.P., Tretyakov, Yu.P., *Nucl. Phys.* A **239**, 157-171 (1975)
26. Heßberger, F.P., Münzenberg, G., Hofmann, S., Armbruster, P., Agarwal, Y.K., Reisdorf, W., Poppensieker, K., Schmidt, K.-H., Schneider, J.R.H., Schneider, W.F.W., Schött, H.J., Sahm, C.-C., Vermeulen, D., Thuma,B., *Journal of Less- Common Metals* **122**, 445-451 (1986)
27. Lane, M.R., Gregorich, K.E., Lee, D.M., Mohar, M.F., Hsu, M., Kacher, C.D., Kadkhodayan, B., Neu, M.P., Stoyer, N.J., Sylwester, E.R., Yang, J.C., Hoffman, D.C., *Phys. Rev. C* **53**, 2893-2899 (1996)
28. Somerville, L.P., Nurmia, M.J., Nitschke, J.M., Ghiorso, A., Hulet, E.K., Lougheed, R.W., *Phys. Rev. C* **31**, 1801-1815 (1985)
29. Münzenberg, G., Hofmann, S., Folger, H., Heßberger, F.P., Keller, J., Poppensieker, K., Quint, B., Reisdorf, W., Schmidt, K-H., Schött, H.J., Armbruster, P., Leino, M.E., Hingmann, R., *Z. Phys.* A **322**, 227-235 (1985)
30. Lazarev, Yu.A., Lobanov, Yu.V., Oganessian, Yu.Ts., Utyonkov, V.K., Abdullin, F.Sh., Buklanov, G.V., Gikal, B.N., Iliev, S., Mezentsev, A.N., Polyakov, A.N., Sedykh, I.M., Shirokovsky, I.V., Subbotion, V.G., Sukhov, A.M., Tsyganov, Yu.S., Zhuchko, V.E., Lougheed, R.W., Moody, K.J., Wild, J.F., Hulet, E.K., McQuaid, J.H., *Phys. Rev. Lett.* **73**, 624-627 (1994)
31. Schädel, M., Brüchle, W., Schausten, B., Schimpf, E., Jäger, E., Wirth, G., Günther, R., Kratz, J.V., Paulus, W., Seibert, A., Thörle, P., Trautmann, N., Zauner, S., Schumann, D., Andrassy, M., Misiak, R., Gregorich, K.E., Hoffman, D.C., Lee, D.M., Sylwester, E.R., Nagame, Y., Oura, Y., to be published in *Radiochim. Acta* (in press) and *GSI-Preprint-96-56* (1996)
32. Münzenberg, G., Armbruster, P., Berthes, G., Folger, H., Heßberger, F.P., Hofmann, S., Keller, J., Poppensieker, K., Quint, A.B., Reisdorf, W., Schmidt, K.-H., Schött, H.-J., Sümmerer, K., Zychor, I., Leino, M.E., Hingmann, R., Gollerthan, U., Hanelt, E., *Z. Phys.* A **328**, 49-59 (1987)
33. Heßberger, F.P., Münzenberg, G., Hofmann, S., Reisdorf, W., Schmidt, K.H., Schött, H.J., Armbruster, P., Hingmann, R., Thuma, B., Vermeulen, D., *Z. Phys.* A **321**, 317-327 (1985)
34. Münzenberg, G., Armbruster, P., Heßberger, F.P., Hofmann, S., Poppensieker, K., Reisdorf, W., Schneider, J.R.H., Schneider, W.F.W., Schmidt, K.-H., Sahm, C.-C., Vermeulen, D., *Z. Phys.* A **309**, 89-90 (1982)
35. Münzenberg, G., Hofmann, S., Heßberger, F.P., Folger, H., Ninov, V., Poppensieker, K., Quint,A.B., Reisdorf, W., Schött, H.-J., Sümmerer, K., Armbruster, P., Leino, M.,E., Ackermann, D., Gollerthan, U., Hanelt, E., Morawek, W., Fujita, Y., Schwab, T., Türler, A., *Z. Phys.* A **330**, 435-436 (1988)
36. Münzenberg, G., Armbruster, P., Hofmann, S., Heßberger, F.P., Folger, H., Keller, J.G., Ninov, V., Poppensieker, K., Quint, A.B., Reisdorf, W., Schmidt, K.-H., Schneider, J.R.H., Schött, H.-J., Sümmerer, K., Zychor, I., Leino, M.E., Ackermann, D., Gollerthan, U., Hanelt, E., Morawek, W., Vermeulen, D., Fujita, Y., Schwab, T., *Z. Phys.* A **333**, 163-175 (1989)
37. Ignatyuk, A.V., Smirenkin, G.N., Tishin, A.S., *Sov. Journal Nucl. Phys* **21**, 255-257 (1975)
38. Swiatecki, W.J., *Nucl. Phys.* A **376**, 275-291 (1982)
39. Bass, R., *Phys. Rev. Lett.* **39**, 265-268 (1977)

Fusion Reactions and Experimental Approaches to the Synthesis of Superheavy Nuclei

A.V. Yeremin, V.K. Utyonkov and Yu.Ts. Oganessian

*Flerov Laboratory of Nuclear Reactions, JINR,
141 980 Dubna, Moscow region, Russia*

Abstract. The question whether the asymmetric actinide based heavy ion reactions could be used for the synthesis of heavy ($Z \geq 106$) nuclides is essential from the point of view of the study of limitation on fusion, it is also important in such reactions new nuclides close to the magic number N=162 can be produced. Thus as the problem of a hindrance to fusion still remains unsolved the high excitation energy of the compound nucleus looks to be an obvious obstacle to using these reactions.

Using the gas–filled recoil separator [1] and electrostatic recoil separator VASSILISSA [2] installed at the beam lines of the U–400 heavy ion cyclotron of the FLNR JINR we investigated the fusion reactions leading to 102, 103, 104, 105 and heaviest isotopes of the 106, 108 and 110 elements.

The analysis of the measured cross–sections did not reveal any evidence of a hindrance to fusion at the ion bombarding energy close to the Coulomb barrier. ^{48}Ca + ^{232}Th $\rightarrow$ ^{280}110*, ^{48}Ca + ^{238}U $\rightarrow$ ^{286}112*, ^{48}Ca + ^{244}Pu $\rightarrow$ ^{292}114* appear to be the best reactions from the point of view of their cross–sections.

INTRODUCTION

At the last two decades for the synthesis of the heaviest transfermium elements so called "cold" fusion reactions with led and bithmuth targets were succesfully used [3,4]. In the eighties, experiments on the synthesis of heavy elements in "cold" fusion reactions led to the discovery of the new elements with Z = 107, 108 and 109 and at the 1994 – 1996 with the use of velocity filter SHIP the heaviest ones with atomic numbers 110 – 112 were synthesised [5-7]. The advantage of these reactions is that only slightly excited ($E^* \approx 10$–20 MeV) compound nuclei at bombarding energies close to the fusion barrier are produced. This is a consequence of the double magic structure of ^{208}Pb. Low–excitation energies were considered to be the reason for the survival of fragile heavy nuclei against prompt fission.

The possibilities for the so called "hot" fusion reactions with the use of U – Cf targets were estimated even more pessimistic. The asymmetric systems fuse with higher probabilities at the Bass barrier [8], but, at the same time, the excitation energies of the compound nuclei increase from 10–20 MeV, which is typical for "cold" fusion, to 40 - 50 MeV. This increase of the excitation energy reduces the survival probability of heavy compound nuclei.

via the complete fusion reactions between ^{232}Th, ^{238}U and ^{244}Pu targets and ^{48}Ca projectiles delivered by the FLNR JINR U400 cyclotron.

FUSION REACTIONS

The only method succesful up to now for the production of new elements in the transfermium region is the complete fusion of heavy ions. As the projectile energy to overcome the Coulomb fusion barrier generally exceeds the Q-value nessesary for compound nucleus formation, the last one is excited and has to dissipate energy by particle emission to come to the ground state. The strong competition between fusion probability on the one hand and survival probability (written by us in terms of Γ_n and Γ_{tot}) on the other hand, leads to comparatively sharp excitation functions with a halfwidth of only 5 MeV to 10 MeV.

Two types of target–projectile combinations have been used succesfully for the synthesis of the transfermium elements: actinide targets from thorium to californium with corresponding beams from sulfur to carbon or targets near lead and beams from zinc to titanium. More symmetric target–projectile combinations have been tried but were unsuccessful in heavy element synthesis. Excitation energies at the Coulomb barrier for actinide reactions are near 40 – 50 MeV whereas with lead, bithmus targets typical excitation energies are 15 – 20 MeV. Correspondingly the compound nuclei need four – five or one – two evaporation steps to dissipate the excitation energy. The production cross sections for both types of reactions leading to the elements with Z $\geq$ 102 are of the same order, "hot" fusion reactions have cross sections smaller not more that one order of magnitude. It is surprising as one would expect the "cold" fusion–evaporation cross sections to be much enhanced. The explanation was found in our recent study of the neutron–to–total width ratios for highly excited heavy nuclei. We obtained the ratios $< \Gamma_n/\Gamma_{tot} > = 0.4$–$0.6$ at the compound nucleus excitation energy E* > 40 MeV [15]. One can conclude that the obtained values of $< \Gamma_n/\Gamma_{tot} >$ at high excitation energies signify that the main losses in the yields of transcurium ER formed in heavy ion "hot" fusion reactions arise at the final steps of deexcitation cascades.

The difference in the cross sections for "hot" fusion reactions roughly coincides with the ratio of the reduced de Broglie wavelength of the dinuclear systems. Larger cross sections for the "cold" fusion reactions could be explained by double magic (or close to it) structure of the target nuclei which allows for compound nucleus to come to ground state after evaporation of one (or two) neutrons.

Various theoretical and semi-empirical calculations were carried out to choose optimum target-projectile-energy combinations for the production of the heaviest nuclei via the (HI, xn) reactions [16,17].

We tried to approach this problem on the basis of systematic trends in the experimental cross section values at maxima of excitation functions for highly fissile and highly excited nuclei. The data obtained by us [18,19] and the other results for the actinide-based systems (see most of refs in [16–18]) were used in the analysis.

Many authors (see for example review papers [9,10] and refs. therein) compared the formation cross sections of evaporation residues (ER's) obtained in "hot" and "cold" fusion reactions and the possibilities of the synthesis of transfermium elements. The "hot" fusion reactions followed by evaporation of 4 neutrons from compound system were studied in the synthesis of Z = 102 – 106 elements with using of actinide targets; experimental data on 1n-evaporation cross-sections were obtained in more symmetric target-projectile combinations leading to Z = 102-109 nuclei. This comparison showed a considerable advantage for producing nuclei with Z>104 via "cold" fusion reactions. But the results obtained in our experiments [11] allowed a conclusion that the 4n-evaporation reaction is somewhat suppressed in these cases whereas the 5n- and may be even 6n- reaction channels seem to be more promising for heavy-element production. Partially this conclusion had a support at the recent experiments in which heaviest isotopes of element 108 and 110 were synthesised via 5 neutron evaporation channel in the reactions $^{34}S+^{238}U$ [12] and $^{34}S + ^{244}Pu$ [13].

In addition we could explain the rising interest to the "hot" fusion reactions leading to the production of transfermium nuclides by a number of reasons. Particularly one can mention the follows:

- decay properties of new nuclides with $Z \geq 106$, which are impossible to obtain in the reactions of "cold" synthesis (for example, nuclides with the number of neutrons close to N = 162);

- the promising results of measuring the number of neutrons emitted before fission which enables to hope for the experimentally accessible cross section values of the ER's formation;

- the general interest to the study of the heavy ion fusion mechanism and to the formation of heaviest nuclei, which ground state stability against fission is caused by shell corrections only.

PROPERTIES OF THE HEAVIEST ELEMENTS

The stability of heavy nuclei is governed by nuclear shell structure whose influence is dramatically amplified near closed proton and neutron shells. Beyond the spherical shells Z = 82 and N = 126, the stability of nuclei diminishes rapidly with increasing Z until the transuranium region, where this trend is altered due to the influence of shell gaps in single - particle level spectra near Z = 100 and N = 152 that appear here at deformed shapes and provide the unusual stability of ^{252}Fm against spontaneous fission (SF).

Since the mid-1960's, nuclear theory has been predicting with increasing confidence the next spherical shells be located at Z = 114 and N = 178 - 184. Numerous attempts to synthesize spherical superheavies were undertaken, making use of different experimental methods and various types of nuclear reactions. Complete fusion reactions followed by evaporation of several neutrons from the excited composite system look the most promising to reach this goal, especially fusion reactions with the neutron - rich projectile ^{48}Ca, which allows the most close approach to the expected neutron shell N = 178 - 184. Yet, no evidence of the superheavy elements production has been presented until now.

More recently, it was realized that this region of spherical superheavy nuclides might be connected by a "peninsula" of stability to the edge of the known heaviest elements. This far - reaching conclusion was based on the predicted existence of the deformed proton and neutron shell closures near Z = 108 and N = 162.

In 1993 - 1995, a series of experiments was carried out designed to provide a direct and decisive test of the theoretical predictions regarding the existence of the new shell closures in the vicinity of Z = 108 and N = 162. Prior to that experiments, no evidence was available to make a definite conclusion about these predictions. The only exception was the 5-ms nuclide ^{262}No showing a hint of unexpected stability against SF at N = 160.

The experiments performed at Dubna by employing the gas-filled recoil separator have resulted in the discovery of the new heavy nuclides ^{262}104, ^{265}106, ^{266}106, ^{267}108 and ^{273}110. The ground-state decay properties that were established for ^{266}106 and ^{262}104, by using the ^{248}Cm + ^{22}Ne reaction, revealed a large enhancement in their stability as compared to that of nuclides with lower Z or N values; an addition of four protons or four neutrons increases the stability against SF or alpha decay by a factor of 3000. The only explanation for this fact can be the approach to a nearby proton and neutron shell closures. Another test of the theory was the observation of a decrease in stability for nuclide with Z, N beyond the predicted magic numbers that would allow the exact Z, N localizations of the new shell closures. The alpha - decay energy measured for neutron - rich ^{273}110, produced by ^{244}Pu + ^{34}S reaction, provided indication that a neutron shell closure probably exists and is located at N = 162. The N = 162 shell closure appears much weaker than the spherical shell N = 126, but seems at least comparable in strength to the deformed shell N = 152 that is in good agreement with the recent theoretical predictions.

In the same period in ^{62,64}Ni + ^{208}Pb experiments conducted at GSI(Darmstadt) by employing the separator for heavy - ion reaction products (SHIP) the new light isotopes of the element 110 with masses 269 and 271 were identified. In the experiment performed at the beginning of 1996 year in "cold" fusion reaction ^{70}Zn + ^{208}Pb the new even - Z element ^{277}112 was discovered; the nuclides ^{273}110 and ^{269}108 were observed as the α - decay daughter of ^{277}112.

The most recent and the most sensitive complex experiment aimed to synthesize element 116 via the reaction ^{82}Se + ^{208}Pb $\rightarrow$ ^{290}116* at GSI (Darmstadt) was performed at the end of 1995 [14]. No event that could be attributed to an isotope of element 116 was observed. Upper limit for evaporation residue production is equal to 5 pb.

To check the hypothetical existence of nuclear shell closures at Z = 114 and N = 178 - 184, one of the fundamental predictions of modern nuclear theory, a set of experiments will be performed, aiming at the production of heaviest nuclides with Z = 110, 112 and 114 (N = 166-167, 170-171 and 174-175, respectively,)

The analysis of the measured cross-sections did not reveal any evidence of a hindrance to fusion at the ion bombarding energy close to the Coulomb barrier. A modified version of the statistical code ALICE [20] described rather well all the existing data for fusion reactions for which the effective fissility parameter x_{eff} ranged from 0.68 (^{12}C+^{248}Cm) to 0.81 (^{34}S+^{244}Pu). On these grounds, we extrapolated the calculations to the fusion reactions leading to Z=110 – 114 nuclides with the neutron numbers N ≥ 162 (see figure 1). These extrapolations demonstrate the possibility of the experiments aimed at the production and investigation of these nuclides. ^{48}Ca + ^{232}Th → ^{280}110*, ^{48}Ca + ^{238}U → ^{286}112*, ^{48}Ca + ^{244}Pu → ^{292}114* appear to be the best reactions from the point of view of their cross-sections. For the last – mentioned reactions the extrapolation yields the cross-section values at between 1 and 10 pb for the 3n and 0.2 – 1 pb for the 4n reaction channels. Some improvements of the detector modules of the gas-filled separator and separator VASSILISSA will enable us to carry out these experiments during a reasonable time (1 – 1.5 months for one experiment).

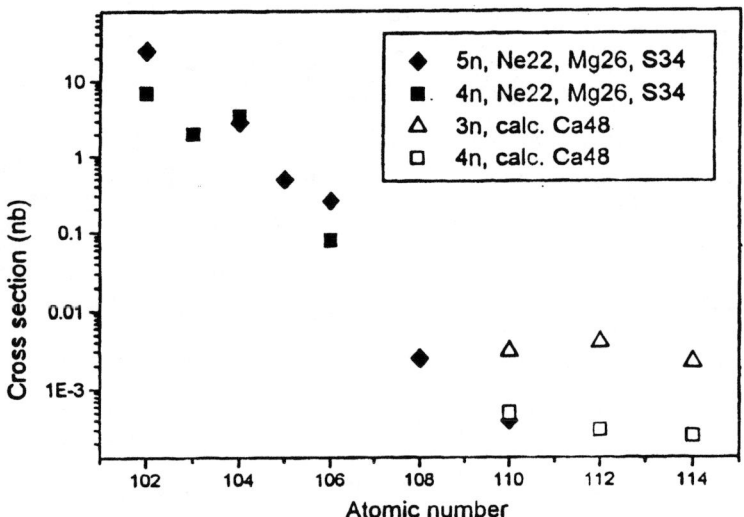

FIGURE 1. Experimental and calculated production cross sections for actinide based fusion evaporation reactions.

EXPERIMENTAL APPROACHES

Complete fusion reactions followed by evaporation of few (2-5) neutrons from the excited composite system are to be used to produce the new superheavy nuclides. The targets of ^{232}Th, ^{238}U and the heaviest available isotope ^{244}Pu are preferable to be used in a number of experiments with ^{48}Ca ions. Intense beams of ^{48}Ca projectiles are to be delivered by the JINR U400 cyclotron with new ECR ion source. The consumption of ^{48}Ca material will be not more than 1 mg per hour resulting beam intensity onto separators targets not smaller than 2×10^{12} pps.

In the number of experiments with ^{48}Ca ions the first will be with the use of ^{232}Th target. It is planned to perform this experiment with internal beam of Ca ions, because the intensity of the internal beam is factor of 3 higher than external one. This experiment will include the use of fast chemistry for extraction of the fraction containing 106 element nuclei from ^{232}Th target material and the off line measurements of the correlation chains from α decay of ^{268}106 and SF of ^{264}104. Theory predicts [21,22] for the ^{268}106, the granddaughter of ^{276}110 - result of the 4n evaporation channel after fusion of ^{48}Ca and ^{232}Th, the halflive against α and SF decay more than 2 hours. In this case it will be possible to use fast chemistry and to perform after that off line measurements at the geometry close to 4π. The fraction extracted after chemistry will be deposited onto 30 μg/cm^2 thick carbon foil and will perform 10 μg/cm^2 thick and 20 mm in diameter source. Carbon foils with that sources will be placed in between two silicon detectors (30 mm in diameter each) at special vacuum chambers. Total registration efficiency for α particles and SF fragments will be close to 90 %. Altogether 8 vacuum chambers with pairs of silicon detectors will be used in this experiment. In each chamber the measuring time will be 1 - 3 days. The suppression factors for long lived transfer reactions products are expected to be more than 10^6. Thickness of ^{232}Th target will be 8 mg/cm^2, more than range of evaporation residues of 110 element but less than ranges of transfer reaction products. Selectivity of the chemical methods will allow to suppress the fraction of U isotopes, probable candidates for long lived background products, by factor of more than 10^5. This experiment will need 250 hours of beam time and will show the level of ER formation cross sections with actinide target and ^{48}Ca ions.

The next planned experiment will be the study of the reaction ^{48}Ca + ^{238}U $\rightarrow$ ^{282}112 + 4n. The measurement of the formation cross section of this isotope will show a possibility to continue this line with ^{244}Pu target. It is planned to perform this experiment with the use of kinematic separator VASSILISSA (see figure 2) installed on extracted beam line of U400 cyclotron.

The principle component of the VASSILISSA facility is a system consisting of three electrostatic dipoles (condensers), which accomplish the spatial separation of the trajectories of recoil nuclei, multinucleon transfer reaction products and beam particles by virtue of differences in their energy and ionic charges.

Recoil nuclei are deflected by 8° in the first electrostatic dipole to enter the second dipole aperture whereas the full energy projectiles pass through the first dipole almost unperturbed and then are stopped in the Faraday cup. Further separation of recoil nuclei from scattered projectiles and other background particles takes place in the second and third dipoles. Electric rigidity is the same for three dipoles. The design goal for its maximum value was 2.5 MV. The focusing system of VASSILISSA consists of two triplets of wide aperture magnetic quadrupoles. The first triplet located just behind the target focuses the ER's knocked out of the

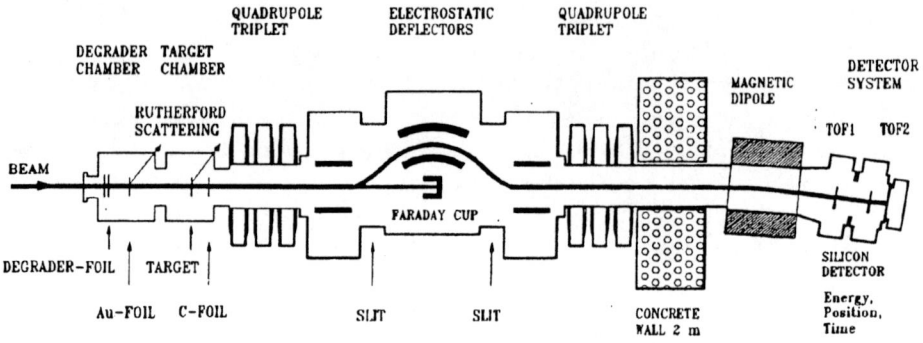

FIGURE 2. Schematic view of the VASSILISSA separator

target and shapes their flux into a quasiparallel beam. The second triplet follows the electric dipoles and serves for collecting the ER's onto the focal plane detectors. An additional dipole magnet provides, after the separator, an $8°$ deflection for the ER's and gives an additional background suppression by factor of 10 – 50 for the scattered beam projectiles.

The thicknesses of separator targets are typically (150–500) $\mu g/cm^2$. In the case of high beam intensity (10^{12}–$10^{13}s^{-1}$) the target mounted on a rotating wheel is used. The rotation of the target wheel is synchronized with the cyclotron beam pulses. Recoil nuclei emerging from the target are accepted by the separator within a solid angle of 15 msr ($\approx \pm 4°$ in the vertical and horizontal directions). VASSILISSA tolerates the energy and charge deviations of up to $\pm 15\%$. The distance from the target to the focal plane is about 12 m.

The detector system consisting of two (start and stop) time–of–flight detectors and an array of silicon detectors has been developed and installed in the separator focal plane. The value of 99.95% has been achieved for the detection probability of such recoil nuclei by making use of a single timing detector. After passing the time–of–flight detectors, recoil nuclei are implanted in an array of detectors assembled on the basis of position–sensitive silicon strips. The detector array consists of five identical 16 – strip silicon wafers. The active area of one of the silicon strip detectors is 60×60 mm^2. In the case of the stop detector each strip is position sensitive in the vertical direction with a resolution of 300 μm between α decays of the α decay chain. The optimum energy resolution is 20 keV for α's of a ^{241}Am source. Four wafers are mounted in the backward hemisphere facing the stop detector. They will measure escaping α's or fission fragments with a solid angle of 85 % of 2 π. In case of the backward detectors the strips do not have the position resolution and each four neighboring strips are connected galvanicly so that 16 energy sensitive segments will be formed.

The fraction of the ER's passing through the separator entrance aperture ranged from 25% to 75% for the targets with the optimal thickness dependending on the reaction asymmetry. The corresponding separation efficiencies are approximately equal to 5 - 6% for Ne, 10 - 15 % for Al, 20 - 30% for Ar and 30 - 40% for Ti and Ca beams. The losses are caused by the energy and charge selection in VASSILISSA.

Compared with the beam intensity on the separator target, the flow of low energy beam–like ions passing through the time–of–flight detectors is suppressed by the factor of 10^8–10^{10} depending on the projectile – target mass ratio. The count rate on the Si detectors is reduced by an additional factor of 10 – 30 because of the absorption of the scattered ions in the time–of–flight detector foils and the foil is installed before the Si detectors. Due to the measurement of the recoil time–of–flight and energy we additionally suppressed this background by the factor of ≈ 100. The overall background suppression ranges from 10^{11} to 10^{13} being dependent on the projectile to target mass ratio and also on the tuning of the U–400 beam. The full energy beam ions were not detected in the focal plane of the separator in all our experiments, and on this basis the suppression factor to be not worse than 10^{19} was estimated.

The described conditions allows, with the use of the Si detector array, to establish the time and position correlations between the signals of recoil nuclei and there α decay for short–lived, $T_{1/2} = (5\ \mu s - 10\ s)$, ER's. The α–α correlations occurring as a result of α–decay chains can be observed for the time intervals ranging from 2 μs to (50–500) s. The attainable upper limit depends on the α energy and count rate. Such correlations are used each time when a new nucleus is identified or excitation functions are measured for the known nuclei produced with a cross section of ≤ 1 nb. Long lived nuclei with higher production cross sections are recognized unambiguously due to their known α–lines.

The rotating target wheel, the parameters of the separation efficiency and suppression factors, as well as the separator detection and electronic systems are capable of sustaining the heavy ion beam currents up to 3 pμA. This enables to carry out the actinide target based experiments on the synthesis of transfermium nuclei expecting the observation of a single mother-daughter $\alpha - \alpha -$ correlation event per day for the reaction cross section of 10 - 50 pb.

The experiment ^{48}Ca + ^{244}Pu is planned to be carried out by using the gas-filled recoil separator (GFRS) (see figure 3). ER's recoiling out of the target are separated in flight from beam particles and various transfer reaction products by the gas-filled dipole magnet followed by the quadrupole doublet. The separator is filled with hydrogen at a pressure of about 1 Torr. A 0.5 μm "exit" mylar window separates the pentane filled detection module from the gas media of the separator. The field B of the separator's dipole magnet is adjusted to center the quasi-Gaussian distribution of ER's on the focal plane detector in the horizontal direction.

The separation time is determined by the time of flight of the produced nuclei through the separator and is about one microsecond for the ^{244}Pu + ^{48}Ca reaction. The collection efficiency for the Z = 114 recoils at the separator's focal plane detector could be 50 – 60 % for this reaction. Under these conditions, the separator provides the primary beam suppression by a factor up to 10^{16} - 10^{18}; various transfer

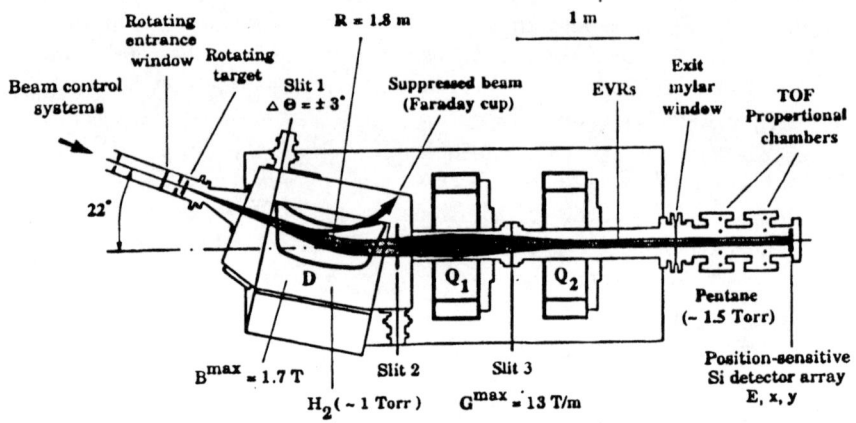

FIGURE 3. Schematic view of the gas-filled separator

reaction products are suppressed by a factor of 10^4-10^8, depending on the reaction channel.

Separated complete fusion products pass through a time-of-flight measurement system composed of two multiwire proportional chambers operated at low pressure (1.5 Torr of pentane) to determine their speed and are implanted into a 40×120 mm^2 position sensitive silicon detector array. The 12 strip PSD array is composed of three 40×40 mm^2 passivated boron implanted planar silicon (PIPS) detectors, with each detector having four 40 mm high × 9.7-mm wide strips. With each detected energy event, the strip number, TOF information, the time in μs from the beginning of each beam pulse to either α/implant or SF events, and the running time in 0.1-ms intervals are also recorded. New nuclides are identified by measuring correlations in energy, time and position to establish genetic links between their implantation in a position sensitive detector and subsequent alpha decay followed by alpha decays of descendant nuclides. To increase detection efficiency of the alpha particles which, due to their low implantation depth, escape the focal plane detector, the lather will be surrounded with an array of eight strip 40× 40 mm^2 detectors. New detection system of the "well" type will allow to increase alpha detection efficiency from 50 % to 85 % and to determine total kinetic energy of the spontaneous fission fragments, as well.

The separator has highly reliable target module which makes it possible to use beams of ^{22}Ne ions with intensities up to 1.5× 10^{13} pps, ^{34}S and ^{48}Ca ions with intensities up to (5-10) × 10^{12} pps. Previous experiments demonstrate that the separator can be continuously exploited in the course of long-term irradiations during thousands of hours. Electrodeposition technique has been developed to produce durable large-area (10-35 cm^2) rotating targets of highly radioactive materials, such as ^{244}Pu or ^{248}Cm. These targets withstand extreme total irradiation doses - up to $(2\text{-}3) \times 10^{19}$ beam particles.

RESULTS OF THE TEST EXPERIMENTS

The experimental conditions encountered at the synthesis of transfermium elements are rather difficult. This work sets the highest requirements for the experimental apparatus and accelerator. Subnanobarn and picobarn values of the production cross sections of the nuclei lead to the requirement of a high beam intensity and long (more than one week) irradiation time. During these long term experiments the highest possible background suppression factors and transport efficiency of the experimental set-up should be maintained. The goal of our test experiments was first – to obtain data on the suppression factors of the unwanted reaction products for all experimental methods including irradiations with internal beam of U400 and chemistry for extraction of the fraction containing 106 element nuclei from ^{232}Th target material and second – to measure transportation efficiency of heavy ER's to the focal plane detectors of the recoil separators. The parameters of recoil separators are known (see [1,2]), but we performed the experiments to test set – ups after all improvements.

The experiment with internal U400 ^{40}Ar beam and ^{232}Th target was performed with the following conditions. ^{40}Ar beam intensity was 1.6×10^{12} ions/sec, beam energy at the entrance of the target 226 MeV, integral flux – 2.5×10^{16} ions. ^{232}Th target size was $25\times8\times4.2$ mm$^2\cdot\mu$, the mass of the target was 9.85 mg. After the irradiation the fast chemistry (≈ 1 hour) was performed for extraction of the fraction, which should contain 106 element nuclei, from ^{232}Th target material containing all descendant nuclei of ^{232}Th α decay and transfer reaction products. The source from this fraction, deposited onto thin (30 μg/cm^2) carbon foil, was prepared and off line measurements were performed. An example of the obtained α spectra is presented at the figure 4. Mainly, we are interested in the α energy region from 7.5 MeV to 8.1 MeV, because the α particle energy for ^{268}106, the granddaughter of ^{276}110 formed in the reaction ^{48}Ca + ^{232}Th $\to$ ^{276}110 +4n, is predicted to be 7.9 MeV [23].

The results of the off line measurements of the α activity from the source after chemical separation are presented at the Table 1. The suppression factors after chemical extraction and separation were obtained $K_{Bi} \approx 100$ (for Bi), $K_{Pb} \approx 400$ (for Pb) and $K_{Th} \approx 2\cdot10^4$ (for Th). To improve suppression factors from Pb and Bi at the second experiment an additional extraction stage was performed. The suppression factors were obtained $K_{Bi} \geq 3\cdot10^4$ (for Bi) and $K_{Th} \geq 3\cdot10^5$ (for Th).

An extracted ^{40}Ar beam from U400 was used for the test experiments with the use of recoil separators. The beam intensity on the separators targets (10 mm in diameter) was typically $(0.6 - 1) \times 10^{12}s^{-1}$. The beam energy was controlled by measuring the energy of the ions scattered at $30°$ in a thin (200 $\mu g/cm^2$) gold foil. Foils of Al and Ti were used as degraders to vary smoothly the beam energy. The typical energy spread of bombarding ions was $\approx (1 - 1.5)\%$.

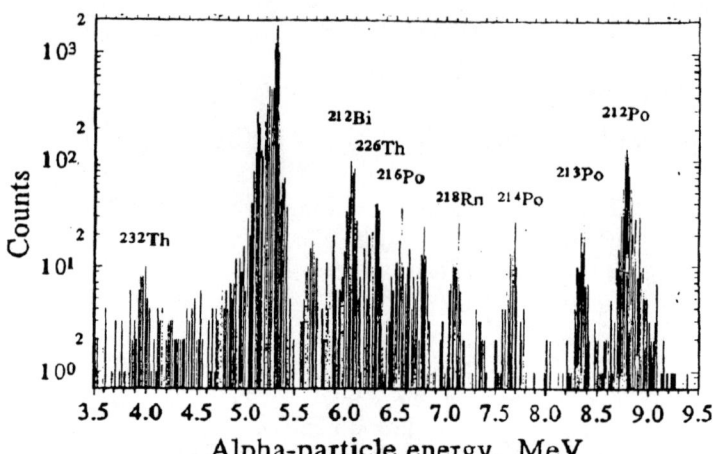

FIGURE 4. Off line α spectuma registered after chemical separation (18 hours)

TABLE 1. The results of the off line measurements. E_α, MeV – energy region, R – detected nuclide, S – mother nuclide, $T_{1/2}$ – half-live, N_m – total number of the registered α particles from daughter nuclei, A_m – measured α activity of daughter nuclei (at the beginning of the measurements), N_t – total number of nuclei at the source after chemical extraction.

E_α, MeV	R	S	$T_{1/2}$	N_m	A_m	N_t
8.52 – 9.20	^{212}Po	^{212}Bi	49^{+18}_{-12} min	240	2.7 min^{-1}	2000
		^{212}Pb	$12.5^{+4.3}_{-3.1}$ min	660	0.7 min^{-1}	5500
		^{228}Th			0.8 h^{-1}	165000
8.10 – 8.52	^{213}Po	^{213}Bi	48^{+7}_{-5} min	130	1.9 min^{-1}	680
		^{225}Ac	$11.2^{+2.9}_{-2.1}$ d	300	0.9 h^{-1}	1600
7.48 – 7.82	^{214}Po	^{226}Th	24^{+18}_{-9} min	30	0.6 min^{-1}	150
		^{226}Ac	29^{+12}_{-8} min	190	4.6 h^{-1}	1200
		^{230}U			2.1 h^{-1}	8000

In the case of separator VASSILISSA integral flux of the bombarding ions on the ^{208}Pb (0.5 mg/cm^2) and ^{232}Th (0.2 mg/cm^2) targets was 1.7×10^{16} and 1.8×10^{16} respectively. Beam energy at the middle of the targets was 198 MeV and 215 MeV respectively. The suppression factors from transfer reactions products (^{211,212}At, ^{211}Po, ^{211}Bi) were achieved better than 10^4. Separation efficiency for ER's was ≈ 30 %. In the case of Th target only few events of α decay of ^{211}Po were detected. Counting rate of all events for total focal plane detector area was ≈ 100 Hz in the case of Pb target and only 5 – 10 Hz in the case of Th target.

In the case of GFRS integral flux of the bombarding ions on the ^{207}Pb (0.4 mg/cm^2) and ^{232}Th (0.37 mg/cm^2) targets was 0.8×10^{16} and 1.2×10^{16} respectively. Beam energy at the middle of the targets was 196 MeV and 208 MeV respectively. Separation efficiency for ER's measured in reaction ^{207}Pb(^{40}Ar,3n)^{244}Fm was estimated to be ≈ 40 %. The results of the irradiation of ^{232}Th targets are presented in Table 2. τ_{tot} is time interval between detected events for ER's or α's for total focal plane detector area, τ_{pos} is the same, but with taking into account the position information for the events.

TABLE 2. The results of the irradiation of ^{232}Th targets at GFRS and VASSILISSA.

Event	GFRS		VASSILISSA	
	τ_{tot}	τ_{pos}	τ_{tot}	τ_{pos}
ER's				
(7–16) MeV	1 – 3 min	4 – 10 h	30 – 50 s	16 – 25 h
$\alpha_{beam-on}$				
(10–11.5) MeV	22 s	1.2 h	10 – 20 s	5 – 10 h
(9–10) MeV	16 s	0.9 h	5 – 10 s	2 – 5 h
$\alpha_{beam-off}$				
($\geq$ 7.5) MeV	400 s	22 h	$\approx$ 1000 s	
SF	$\geq$ 2 h	$\geq$ 400 h	$\geq$ 10 h	

CONCLUSION

Among all the target - projectile combinations available at present, complete fusion reactions induced by ^{48}Ca ions on heavy actinide targets, such as ^{232}Th, ^{238}U and ^{244}Pu, provide the closest approach to the neutron number expected for the spherical superheavy elements. The complete fusion reactions of ^{48}Ca projectiles with heavy transactinide targets result in about 30 MeV excitation energy of the compound nuclei. This is 20 MeV lower than the corresponding values for compound nuclei produced in the reactions ^{34}S + ^{238}U, ^{244}Pu, in which the isotopes ^{267}108 and ^{273}110 have been synthesized after evaporation of 5 neutrons. In the ^{48}Ca case, this can mean a higher probability of 3n evaporation channel, which can increase production cross sections of corresponding nuclei by several orders of magnitude compared to 5n channel, due to lower fission losses in deexcitation process.

A cardinal advantage of the proposed experiments with ^{48}Ca ions will be their high sensitivity reaching 0.5 pb level of cross sections or even less. In the ^{48}Ca experiments, improvements of sensitivity will be provided by higher ^{48}Ca beam intensities (2×10^{12} pps compared to 5×10^{11} pps in previous experiments) and due to increase of integral beam flux by a factor of more than 100. The collection efficiency of the separators planned to be used at these experiments will be 40% - 60%, α particle detection efficiency will be 80% - 90% with new detection systems.

ACKNOWLEDGMENTS

This work is performed partially under the financial support of Russian Foundation for Basic Research, contracts N 96–02–17209 and N 96–02–17377.

REFERENCES

1. Yu.A. Lazarev et. al., "The Dubna gas-filled recoil separator: a facility for heavy element research." in *Proceedings of the Int. School - Seminar on Heavy Ion Physics*, Dubna, 1993 (JINR E7-93-274, Dubna, 1993) Vol. 2, pp. 497–502.
2. A.V. Yeremin et. al., *Nucl. Instr. & Meth.* **A350**, 608–617 (1994).
3. Münzenberg G., *Rep. Prog. Phys.* **51**, 57–104 (1988).
4. Armbruster P., *Ann. Rev. Nucl. Part. Sci.* **35**, 135–194 (1985).
5. Hofmann S. et. al., *Z. Phys.* **A350**, 277–280 (1995).
6. Hofmann S. et. al., *Z. Phys.* **A350**, 281–282 (1995).
7. Hofmann S. et. al., *Z. Phys.* **A354**, 229–230 (1995).
8. Bass R., *Nucl. Phys.* **A231**, 45–63 (1974).
9. Armbruster P., "On the production of superheavy elements and the limitations to go beyond." in *Proceedings of: Scuola Int. di Fisica "Enrico Fermi", Course "Trends in Nuclear Physics"*, Varenna 1987, Italy.; GSI preprint 87-59 (1987)
10. Oganessian Yu.Ts., "Reactions leading to the synthesis of heavy elements." in *Proceedings of Int. Symp. Nikko 91 "Towards a Unified Picture of Nuclear Dynamics"*, Japan, 1991, (New York: AIP) (1991) pp. 465–487.
11. Yeremin A.V. et. al., "Synthesis of Heavy Nuclei In Asymmetric Target–Projectile Combinations." in *Proceedings of Int. School-Seminar on Heavy Ion Physics*, Dubna, 10–15 May 1993, JINR E7-93-274, vol. 1, pp. 109–118.
12. Lazarev Yu. A. et. al., *Phys. Rev. Lett.* **75**, 1903–1906 (1995).
13. Lazarev Yu. A. et. al., *Phys. Rev.* **C54**, 620–625 (1996).
14. Heßberger F.P., "GSI experiments on the synthesis of superheavy elements." in *Proceedings of the Int. Workshop XXIV on Gross Properties of Nuclei and Nuclear Excitations*, Hirschegg, Austria, January 15–20, 1996, pp. 1–10.
15. Sagaidak R.N. et. al., "Fission–evaporation competition in excited uranium and fermium nuclei." Submitted to *J. Phys. G: Nucl. Part. Phys.*
16. Iljinov A.S. et. al., *J. Phys. G: Nucl. Phys.* **9**, 931 (1983).
17. Reisdorf W. and Schädel M., *Z. Phys.* **A343**, 47 (1992).
18. Andreyev A.N. et. al., *Z. Phys.* **A345**, 389–394 (1993).
19. Andreyev A.N. et. al., *Z. Phys.* **A344**, 225–226 (1992).
20. B.I. Pustylnik, "Estimation of the production cross sections of heavy nuclei in "cold" and "hot" fusion reactions." in *Proceedings of the Int. Conf. on Dynamical Aspects of Nuclear Fission*, Častá - Papiernička, Slovak Republic, Aug. 30 – Sept. 4, 1996, pp. 121–128.
21. Smolańczuk R., Skalski J. and Sobiczewski A., *Phys. Rev* **C52**, 1871–1880 (1995).
22. Möller P. and Nix J.R., *J. Phys. G: Nucl. Part. Phys.* **20**, 1681–1747 (1994).
23. Smolańczuk R., Skalski J. and Sobiczewski A., "Masses and half-lives of superheavy elements." in *Proceedings of the Int. Workshop XXIV on Gross Properties of Nuclei and Nuclear Excitations*, Hirschegg, Austria, January 15–20, 1996, pp. 35–42.

The RNB Project in Japanese Hadron Facility and Possible Use of Neutron-Rich Beam for the Study of Superheavy Nuclei

Toru Nomura

Institute of Particle and Nuclear Studies, High Energy Accelerator Research Organization, Tanashi Branch, 3-2-1 Midori-cho, Tanashi, Tokyo, 188 Japan

Abstract. We first describe briefly a radioactive nuclear beam (RNB) facility based on the isotope separator on-line and post-accelerator scheme planned in Japanese Hadron Project. In this facility, various radioactive nuclear species produced in 3 GeV proton-induced reactions will be accelerated through heavy-ion linacs in three stages, the maximun output energy in each stage being 0.17, 1.05 and 6.5 MeV/nucleon, respectively. Secondly, we discuss the feasibility of the use of neutron-rich RNB for experimental study of more neutron-rich superheavy nuclei than those presently known. It is shown that the increase of the survival probability of neutron-rich compound nuclei can possibly compensate for a difficulty arising from expected weak intensities of the secondary-beams. In addition, cold-fusion-like reactions as well as possible enhancement of near-barrier fusion cross sections that can become more prominent by use of neutron-rich beams are discussed.

1. INTRODUCTION

Firstly I will report on our future plan of the radioactive nuclear beam (RNB) facility, which is being planned in the so-called Japanese Hadron Facility (JHF) (1,2) and nicknamed as Exotic Nuclear Arena (E-arena) (1,3). In the present plan of the E-arena, various radioactive nuclear species produced in 3 GeV proton-induced reactions will be separated with an isotope separator on-line (ISOL), and the selected nuclei will be injected into heavy-ion linacs to be accelerated up to 6.5 MeV/nucleon. The present status of the short-lived nuclear beam facility already constructed as "Research and Development" of the E-arena is also briefly described.

Secondly I am going to make a few comments on the possible use of neutron-rich nuclear beams available by the E-arena for production of superheavy elemnts (SHE). At present, fusion reactions induced by neutron-rich nuclear beam are supposed to be an unique way of producing more neutron-rich superheavy nuclei than presently known, that are more close to the neutron magic number N=184. A main interest of this part concerns a question whether a difficulty arising from the expected weak beam intensity of neutron-rich beams can be overcome with possible enhancement of the survival probability of neutron-rich compound nuclei.

2. EXOTIC NUCLEI ARENA

The JHF is presently being proposed by "High Energy Accelerator Research Organization" (KEK), which is a new national institute in Japan realized in April, 1997 as a consequence of the reorganinization of National Laboratory for High Energy Physics, Institute for Nuclear Study and Meson Science Center, University of Tokyo.

The JHF is based on a high-intensity proton accelerator complex comprising a 200 MeV linac, a 3 GeV booster synchrotron and a 50 GeV main ring as shown in Fig. 1. Their designed average beam intensities are 400 µA, 200 µA and 10 µA, respectively. The research fields of JHF are classified into four categories called arenas; K-arena for intermediate energy nuclear and particle physics, E-arena producing low-energy RNB mainly for nuclear physics, M-arena for muon physics and N-arena for high-intensity pulsed neutron source for material science. The facility will be constructed at the present site of KEK by utilizing its existing infrastructure as much as possible. For instance, the 3 GeV booster will be installed at the existing tunnel of the 12 GeV PS at KEK. It is intended that its construction be completed in 5 years after full approval.

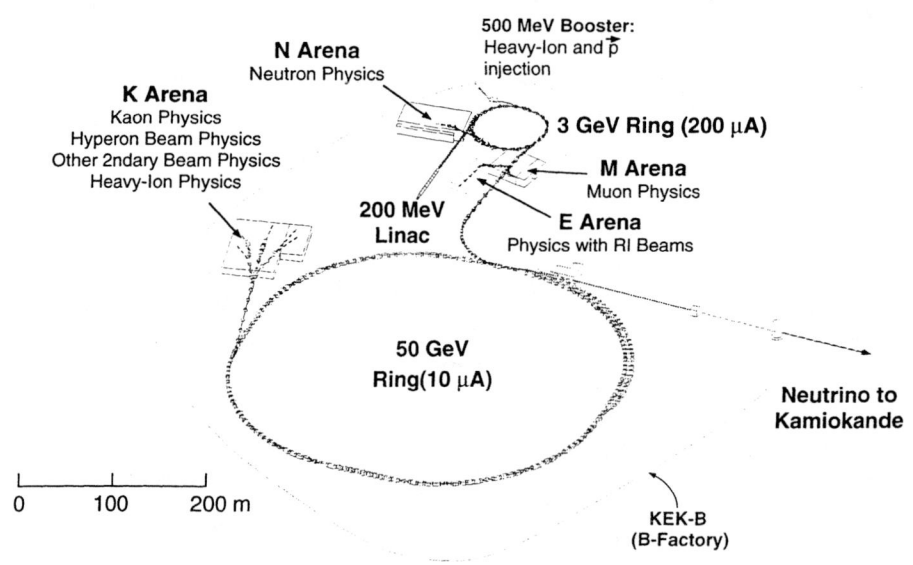

FIGURE 1. Schematic layout of JHF.

The E-arena will utilize a part of proton beam supplied from the 3-GeV booster ring for producing a wide spectrum of radioactive nuclear species via spallation, multi-fragmentation and/or fission processes of a proper thick target. They will thermally diffuse to the target surface and desorb from it, before being ionized in an ion source, pre-accelerated and formed into a beam that can be mass-separeated in an isotope-separator on-line. They can be accelerated in three stages through heavy-ion linacs, split-coaxial RFQ linac (SCRFQ) and two Interdigital-H linacs (IH1 and IH2) as shown in Fig. 2, the maximum output energy in each stage being 0.17, 1.05 and 6.5 MeV/nucleon, respectively. The variable energy can be realized between 0.17 and 6.5 MeV/nucleon by changing the applied RF voltages and/or its phases (4). The injection energy from the ISOL to the SCRFQ linac is 2 keV/nucleon, while the minimum charge-to-mass (q/A) ratio of radioactive ions that can be accelerated has been chosen to be 1/30. Some charachteristics of the E-arena are listed in Table 1.

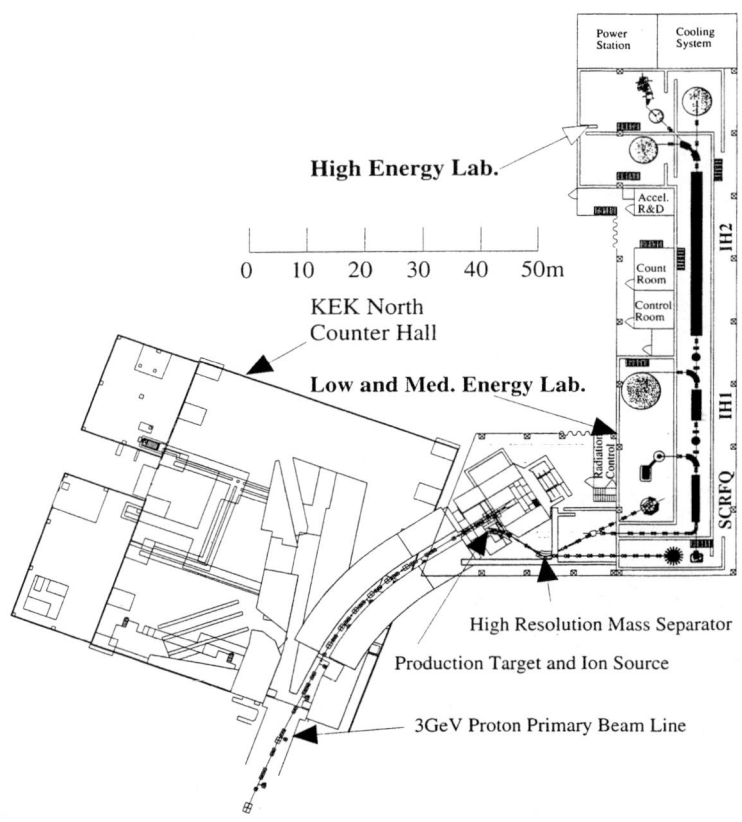

FIGURE 2. Plan view of the E-arena.

TABLE 1. Some characteristics of the E-arena

Primary beam	3 GeV Protons, 10 μA
ISOL	M/ΔM=9,000
Split-Coaxial RFQ (SCRFQ)	
RF	25.5 MHz
Duty Fctor	30-100 % depending on values of q/A
q/A	≥1/30
Injection energy	2 keV/nucleon
0utput energy	170 keV/nucleon
Interdigital-H linac 1 (IH1)	
RF	51 MHz
Duty Factor	100 %
q/A	≥1/10
Output energy	170-1050 keV/nucleon (variable)
Interdigital-H linac 1 (IH2)	
RF	102 Mhz
Duty Factor	100 %
q/A	≥1/6
Output energy	1.05-6.5 MeV/nucleon (variable)

One of the big merits of the use of 3 GeV protons as the primary beam is that we are able to use a very thick production target. The energy loss of 3 GeV protons through a 10^{24} atoms/cm^2 thick ^{40}Ca traget, for instance, is about 100 MeV, so that the transportation of the primary beam to the beam dump after the passage of the thick target can be done without big difficuluties. To be realistic, the maximum intensity of the primary beam is chosen to be 10 μA in the early stage of the E-arena. This limitation comes mainly from various radiation safety problems, and if everything works safely, we have a possibility of increasing the primary-beam intensity by small modification of the facility. When we bombard a target of the above thickness with 10 μA beam, then the yield of nuclei produced with 10 μb cross section is 6×10^8 atoms/s. According to the semi-empirical formula on spallation cross sections given in ref. (5), such nuclei correspond to those several nucleons off the stability line in the region of light nuclides, while they are 10-15 nucleons far from stability in the case of medium-mass nuclides.

It should be mentioned that we have already constructed the first stage of the E-arena, that is, ISOL, SCRFQ and IH1 linacs, at the Institute for Nuclear Study, University of Tokyo, which has become, due to the reorganization mentioned before, Tanashi Branch of KEK since the beginning of April, 1997. They have been installed by a K=68 cyclotron existing at KEK-Tanashi, which can supply, for example, protons up to 45 MeV, α particles up to 68 MeV, and various relatively light heavy ions up to about 10 MeV/nucleon. The injection energy and the minimum q/m ratio are the same as described above. The variability of the IH1 output energy has already been realized. When the JHP is approved, the whole system of this prototype facility will be moved from the present site to the 3 GeV booster ring in the JHF, and will be used as a part of the E-arena.

3. STUDY OF SHE BY NEUTRON-RICH BEAM

Figure 3 shows a partial nuclear chart showing the region of Z≥102 together with the predicted SHE region having large fission barriers calculated for the ground states of nuclei (6). Thanks to beautiful experiments recently carried out at GSI in Germany and at Dubna in Russia (7), elements up to Z=112 are known today. However, as is seen from this figure, the maximum neutron number of known nuclei is 165, being about 20 neutrons less than the predicted neutron magic number N=184, although the maximum atomic number is already close to the proton magic number Z=114. This is a natural consequence of the fact that the experiments so far done are based on fusion of stable beam nucleus and stable or nearly stable target nucleus, except for the case of ^{48}Ca-induced reactions used a few times in the past (8). This difficulty cannot be avoided essentially before neutron-rich unstable nuclear beam of several MeV/u becomes available in future. The RNB facilities based on the ISOL and post-accelerator scheme is suited for realizing such neutron-rich beams with relatively high intensity. Today such a facility is under construction at GANIL in France, in which heavy ion beams of about 100 MeV/nucleon will be used as primary beam to produce unstable nuclei, and will have capability to study a new region of SHE. To my knowledge, however, the E-arena is more suited for the above purpose, since it will utilize 3 GeV protons as

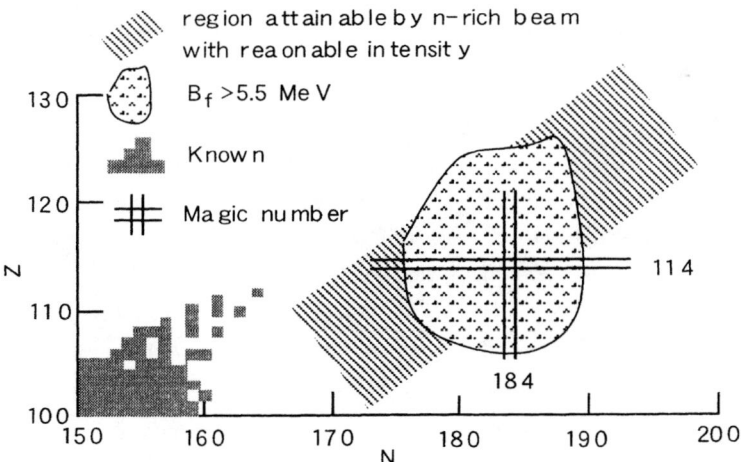

Figure 3. Partial nuclear chart in the region of Z≥102. The region indicating the calculated fission barriers is taken from ref. (6).

primary beam, facilitating the use of a very thick production target with higher primary beam intensity to gain large yield.

However, it is of course true that the beam intensity of such neutron-rich beams must be very much limited anyway. The usual beam intensity of stable heavy-ion beams generally used at GSI and Dubna experiments is, roughly speaking, 0.5-2 pµA. When we want to study heavy nuclei of which the production cross section is 1 pb, and assume that we use a target of 10^{18} atoms/cm^2 thickness, the production rate with 1 pµA beam is about 0.5 counts per day. Therefore, you might think that the use of neutron-rich beam for production of SHE would be unrealistic. We have a hope, however, that this may not be the case for the following reasons: (i) The production cross section of residual heavy nuclei having more neutrons must become much larger than the case of neutron deficient residual nuclei presently known. This is a consequence of the increase of the so-called survival probability due to the increase of the fission barrier and of the decrease of neutron binding energy. (ii) Possible enhancement of the near-barrier fusion cross sections may take place in the case of neutron-rich beams, as pointed out by several authors (9). (iii) The excitation energies of the compound nuclei formed by neutron-rich beams are often lower than those by the corresponding stable heavy-ion beam, which facilitates cold-fusion-like reactions even with the use of actinide targets like Uranium. Let us consider a little more details of these facts below.

3.1 Survival Probability

We first note that the zero-th approximation of production cross sections for evaporation residues in the (HI,xn) type fusion reactions can be written by

$$\sigma_{xn} \approx \sigma_{fus}(\Gamma_n/\Gamma_f)^x \qquad (1),$$

where σ_{fus} is fusion cross section, and Γ_n/Γ_f is ratio of neutron emission width to fission width averaged over all the deexcitation processes of a compound nucleus. Figure 4 shows empirical values of Γ_n/Γ_f deduced from the measured production cross section of various (HI,xn) type reactions, and plotted versus the neutron number (N) of the last evaporating nuclei. It is clear that Γ_n/Γ_f increases with the increase of N. This is generally true, because Γ_n/Γ_f can be written in the first approximation as

$$\Gamma_n/\Gamma_f \propto \exp(B_f - B_n)/T \qquad (2),$$

where B_f is fission barrier, B_n neutron binding energy, and T nuclear temperature. In the case of Fm isotopes, which have the most data points, the increase Γ_n/Γ_f amounts to 10 times for the increase of 7 neutrons ($\Delta N=7$), while it is 14 times more for $\Delta N=10$. In the case of ΔN=7, the increase of the survival probability reaches 10^4 and 10^5 for the (HI,4n) and (HI,5n) reaction channels, respectively, while it becomes even 10^6 for the (HI,5n) channel when ΔN=10.

Because we have so far had no experience of accelerating neutron-rich beams in the energy region of interest, the practical intensity of the neutron-rich beam is hard to estimate. However, according to the empirical formula (6) of spallation cross sections of nuclei corresponding to ΔN=7, it is by 10^5 less than the corresponding stable beam even in the most favorate case. If it is realized, the increase of the survival probability almost cancels the decrease of the beam intensity. When we are allowed to extrapolate the same trend to the SHE region near Z=114 and N=184, we have a hope that the decrease of the secondary neutron-rich beam intensity can nearly be cancelled out by the increase of the survival probability.

3.2 Possible Enhancement of the Near-Barrier Fusion Cross Section

Since the survival probability of the compound nucleus decreses very rapidly with the increase of its excitation energy, we usually want to realize the "cold" compound nucleus as much as possible. Therefore, it is ideal to make the compound nucleus at the lowest bombarding energies with reasonably large cross sections. In this sense, the study of the near-barrier fusion is very important.

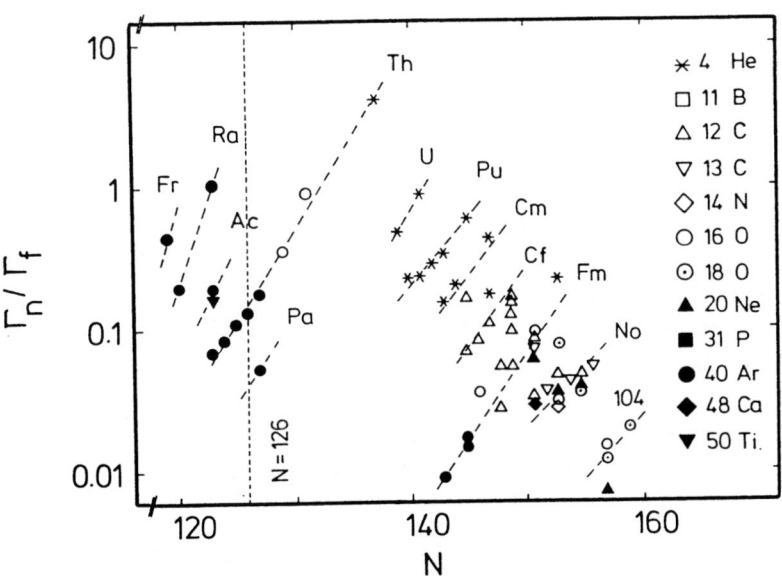

Figure 4. Systematics of effective values of the neutron-to-fission width ratio averaged over evaporation cascade of four neutrons. Taken from ref. (10) (slightly modified).

Here I would like to point out the following hopeful features of the neutron-rich beam; (i) decrease of the effective Coulmb barrier at the contact point of two colliding nculei, and (ii) possible neck formation triggered by weakly bound neutrons.

The reason for the former is clear. To have an idea how much the corresponding decrease is expected, let us compare the Coulomb barrier for the ^{62}Cr + ^{238}U (RNB) reaction with that for the ^{52}Cr + ^{238}U (stable H.I.) reaction. When they are evaluated, for simplicity, by $V_{CB} = 1.02 \times Z_p Z_t/(A_p^{1/3}+A_t^{1/3})$ MeV, in which Z_p (Z_t) and A_p (A_p) are atomic and mass numbers of projectile (target) nucleus, respectively, the difference is around 5 MeV. A consequence of this effect together with the Q-value for compound-nucleus formation will be discussed in subsection 3.3.

The latter possible effect mentioned above is interesting, and I believe it is worthwhile to study also for the fusion mechanism. Stelson (9) pointed out that one of the origins for the enhancement of near-barrier fusion cross sections observed in heavy-ion reactions is that the neck formation between the two colliding nuclei lowers the fusion barrier. He also pointed out strong correlation of the experimental data with possibilities of neutron flow occurring from one nucleus to another at 1-2 fm larger than the mean barrier distance. This is due to the lowering of the potential barrier between the two nuclei, as indicated in Fig. 5, when they come close to each other. It suggests that the neck formation is triggered by the flow of least bound neutrons. In other words, weakly bound neutrons may be able to work as glue for fusion. If it is true, this effect must become more evident in the case of neutron-rich beams, which have weakly bound neutrons. It should be noted, however, that this is not the case for neutron-hallo or neutron-skin nuclei, because radial wave functions of such hallos or skins distribute too much widely. Neutron-rich nuclei having the neutron binding energies a few MeV less than the usual one seems to be ideal.

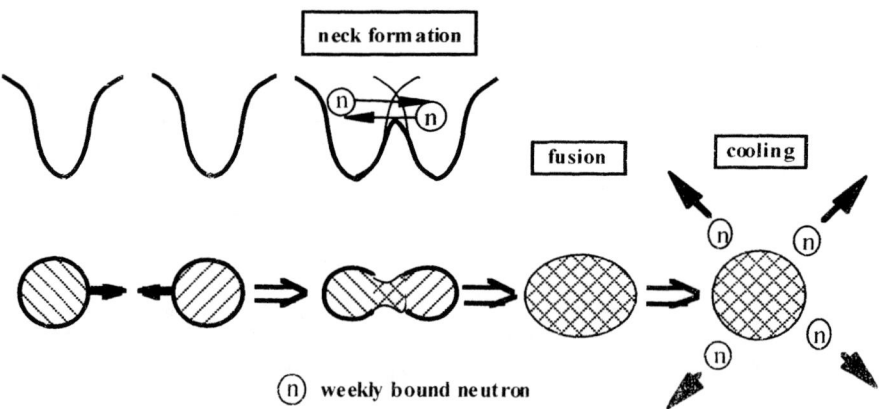

Figure 5. Schematic view of two colliding nuclei to show the neck formation triggered by less bound neutrons.

3.3 Cold-Fusion-Like Reactions

Table 2 compares some calculated excitation energies (U_{CN}) of compound nuclei formed by neutron-rich RNB with those by the most neutron-rich stable heavy-ion beams. The values of U_{CN} are evaluated at bombarding energies equal to the Coulomb barriers estimated in the same way as written in the previous subsection. It should be noted that this Coulomb barrier is approximately equal to the interaction barrier calculated by the Bass model (12). Except for the case of zinc-induced reactions, the actinide targets are chosen, so that the compound nuclei have sufficently large neutron numbers. In the case of RNB we have calculated U_{CN} in each target for beams with various neutron numbers up to 9 neutrons more than the corresponding stable beam, and is shown the case that yields the minimum value of U_{CN}.

As seen in this table, the values of U_{CN} in the case of neutron-rich beam induced reactions are always smaller than those in the case of stable heavy-ion beams. This is a consequence of the fact that the Coulomb barrier is always lower in the case of the neutron-rich beam as mentioned in the previous subsection, while in most cases the Q-values for the compound-nucleus formation are also somewhat smaller in the neutron-rich beam. In the cases of the actinide targets, the difference in excitation energy is 5.5-8.6 MeV, which roughly corresponds to the energy carried away by one neutron emission, resulting in the considerably large increase of the survival probability. Quite remarkable is the ^{66}Fe+^{232}Th reaction, which yields U_{CN}=23.1 MeV; it can be regarded as a more or less cold-fusion type reaction realized in the case of the actinide targets.

TABLE 2. Calculated Excitation Energies of Some Compound Nuclei at Bomarding Energies Equal to the Coulomb Barrier

Beam	Target	CN[a]	Q_{CN}[b] (MeV)	V_{CB}[c] (MeV)	U_{CN}[d] (MeV)
^{40}Ar(stable)	^{248}Cm	^{288}114$_{174}$	-138.0	181.7	43.7
^{47}Ar(RNB)	^{248}Cm	^{295}114$_{181}$	-143.1	178.2	35.1
^{54}Cr(stable)	^{238}U	^{292}116$_{176}$	-193.2	225.7	32.5
^{60}Cr(RNB)	^{238}U	^{298}116$_{182}$	-195.7	222.7	27.0
^{58}Fe(stable)	^{232}Th	^{290}116$_{174}$	-207.9	238.3	30.4
^{66}Fe(RNB)	^{232}Th	^{298}116$_{182}$	-211.2	234.3	23.1
^{70}Zn(stable)	^{208}Pb	^{278}112$_{166}$	-242.4	249.8	7.4
^{78}Zn(RNB)	^{208}Pb	^{286}112$_{174}$	-242.5	246.0	3.5

[a] Compound nucleus.
[b] Relevant Masses of unknown nuclei are taken from ref. (11).
[c] Estimated in the same way as described in subsection 3.2 in the text.
[d] Excitation energy of the compound nucleus.

Also very interesting is the case of the ^{78}Zn+^{208}Pb reaction, in which the calculated value of U_{CN} is only 3.5 MeV. It indicates a possibility of making a really cold compound nucleus, which makes radiative capture processes (0n channel) quite probable by adopting proper bombarding energies. Perhaps, some resonance-like fusion reactions might be also possible.

References

1. For example, see " Proposal of the Japanese Hadron Facility", KEK, 1997.
2. For instance, see Proc. of Int. Workshop on " Physics with 50 GeV PS", INS, Univ. of Tokyo, Japan, December 1995, JHP-Supplement-18, April, 1996, eds. Fukuda, T., Hamagaki, H., and S. Nagamiya, S.
3. For the early version of the E-arena, see, for instance, Nomura, T., Nucl. Instr. Meth. **B70**, 407 (1992).
4. Tomizawa, M., et al., Proc. of the 5th European Particle Accelerator Conf., 1996, p. 780; also see Arai, S., et al, Nucl. Instr. Meth. **B70**, 414 (1992).
5. Rudstam, G., Z. Naturforsch. **21A**, 1027 (1966); Silberberg. R., and Tsao, C.H., Naval Research Laboratory, Washington DC, NRL Report 7593 (1973).
6. Cwiok, S. and Sobiczewski, A., Z. Phys. **A342**, 203 (1992).
7. For example, see Hofmann, S., et al., Z. Phys. **A350**, 277 (1995); Hofmann, S. et al., Z. Phys. **A350**, 281 (1995); Hofmann, S. et al., Z. Phys. **A354**, 229 (1996); Lazarev, Yu A., et al.. Phys. Rev. **C54**, 620 (1996).
8. For example, see Armbruter P., et al., Phys. Rev. Lett. **54**, 406 (1985).
9. For example, see Stelson, P.H., Phys. Lett. **B205**, 190 (1988), and references therein.
10. Treatise on Heavy-Ion Science, **Vol. 4**, ed. D.A. Bromeley, Chapter 1 by Oganessian, Y.T., and Lazarev, Y.A., p. 77.
11. Moller P., et al., Atomic Data and Nucl. Data Tables, **59**, 185 (1995).
12. Bass, R., Nucl. Phys. **A231**, 45 (1974).

THEORETICAL DEVELOPMENTS

Fusion of Massive Nuclei and Synthesis of Superheavy Elements in the Framework of the DNS Concept

E.A.Cherepanov[*], G.G.Adamian[*], N.V.Antonenko[**], V.V.Volkov[*]

[*]*Flerov Laboratory of Nuclear Reactions, Joint Institute for Nuclear Research,
141980 Dubna, Moscow region, Russia*
[**]*Justus-Liebig-Universitat, Giessen, Germany*

Abstract. The dinuclear system (DNS) concept of formation of compound nuclei has been applied to analysis the conditions necessary to synthesize superheavy elements (SHE). For elements 110-114 the inner fusion barriers have been calculated. Thus, it has become possible to estimate the optimal collision kinetic energy. Using the model of competition between complete fusion and quasi-fission, the formation probability of a compound nucleus in element 110-114 synthesis reactions has been calculated.

INTRODUCTION

In planning and preparing experiments, aimed at the synthesis of SHE, experimentalists need theoretical estimations of: 1) the properties of the SHE being synthesized - their lifetimes and radioactive decay modes; 2) the optimal bombarding energy for the chosen reaction and the expected SHE formation cross sections. In the literature (see, for example, [1]) is given the necessary information on the first point. The macroscopic dynamical model of Swiatecki (MDM) [2] has been used lately as a basis for the theoretical analysis of complete fusion of massive nuclei. However, the predictions of MDM concerning the conditions of the synthesis of elements 110-114 were found to contradict the experimental data. MDM required a large excess (many tens of MeV) of the collision kinetic energy E above the entrance Coulomb barrier B_{Bass} (extra-extra push, E_{xx}). However, the synthesis of elements 110, 111 and 112 was achieved at collision energies even lower than B_{Bass}. Also, MDM does not allow to estimate SHE-production cross sections. For instance, for the reaction $^{110}Pd + ^{110}Pd \rightarrow ^{220}U$ the formation probability of the compound nucleus ^{220}U according to MDM was estimated to be about four orders of magnitude higher than its experimental value [3].

The standard statistical calculations of production cross sections of 104, 108 and 110 elements in different reactions (see Fig. 1) demonstrate the increase of discrepancy with experimental data for more heavy elements.

Our earlier investigations of the process of forming a compound nucleus have shown that the suggested in FLNR (JINR) concept of the dinuclear system (DNS-concept) [4] gives the most realistic representation of the complete fusion process. This fact inspired us to apply the DNS-concept to the analysis of SHE-synthesis reactions. The obtained results are shown in the present report.

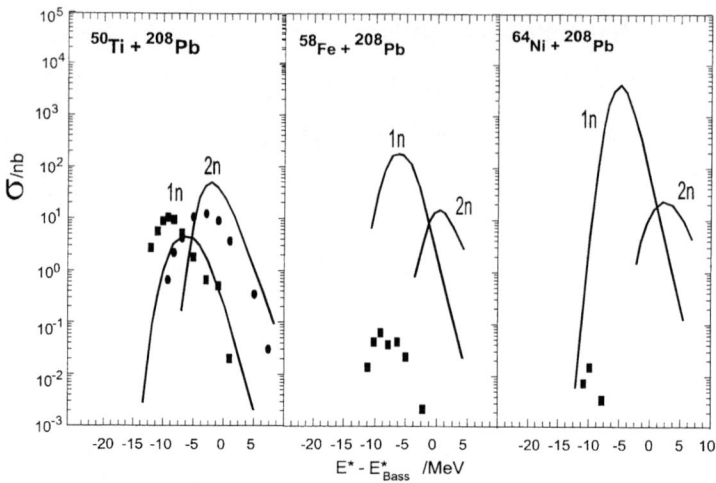

FIGURE 1. The comparison experimental data [7,8] (points) and theoretical calculations [13] for reactions which lead to produce 104, 106 and 110 elements.

BASIC ASSUMPTIONS OF THE DNS-CONCEPT

The motivation of the DNS-concept, the comparison of the DNS-concept and MDM have already been presented in ref.[3,4]. Therefore, here we are going to point out only the basic assumptions of the DNS-concept, which are used in the analysis of the SHE-synthesis reactions. According to the DNS-concept the complete fusion process proceeds in the following way. On the capture stage, after full dissipation of the collision kinetic energy, a dinuclear system is being formed (well known from deep inelastic transfer reactions). Complete fusion is an evolution in which the nucleons of one nucleus sequentially, shell after shell, are transferred to the second nucleus. All the way to forming the compound nucleus, the nuclei of the DNS keep their individuality -- this is consequence their shell structure. The main characteristic of the DNS, determining its evolution, is the potential energy, $V(Z,L)$, of the system. Here, Z is the atomic number of one of the nuclei in the DNS, L is the spin of the system. Also,

$$V(Z,L) = B_1 + B_2 - B_{cn} + V(R^*,L), \qquad (1)$$

where B_1, B_2 and B_{cn} are the mass excess of the DNS nuclei and compound nucleus, $V(R)$ - the nucleus-nucleus potential. It incorporates the nuclear, Coulomb and centrifugal potentials:

$$V(R,L) = V_n(R) + V_c(R) + V_{rot}(L), \qquad (2)$$

where R is the distance between the centers of the nuclei in the DNS. In calculating $V(R,L)$, the DNS was assumed to have the shape of two slightly overlapping spheres. R^* is the value of R, at which the DNS is to be found at the bottom of the "pocket" in the potential $V(R,L)$. The nuclear potential $V_n(R)$ was calculated using the double folding method [5]. The Coulomb potential $V_c(R)$ took account of the partial overlapping of the DNS nuclei. The centrifugal potential was calculated for the case of the rigid body moment of inertia. The excitation energy, E*, of the DNS was assumed to be equal to the difference between the collision kinetic energy E and the potential energy of the system V(Z,L) :

$$E^* = E - V(Z,L). \qquad (3)$$

PECULIARITIES OF FUSION OF TWO MASSIVE NUCLEI WITHIN THE DNS CONCEPT

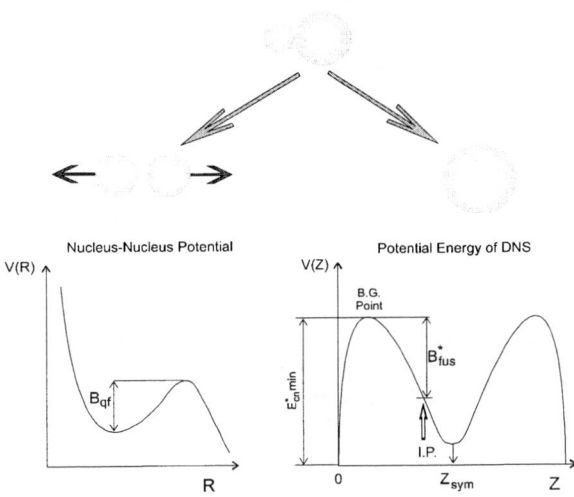

FIGURE 2. Schematic illustration of two ways of evolution for the initial dinuclear system

In Fig.2 The curve of potential energy (L is fixed) V(Z) exhibits two minima: the first one (Z=0) - corresponding to complete fusion and the second one $Z_{sym}=(Z_1+Z_2)/2$ - corresponding to the formation of a symmetric DNS. In the latter, the Coulomb repulsion between the nuclei reaches a maximum value and the system easily decays into two, close in mass, nuclei, viz. quasi-fission takes place. If the injection point of the reaction is situated to the right of the maximum in V(Z,L) (the Businaro-Gallone (BG) point), the initial DNS may follow either of the two possible evolution paths: to a larger or smaller charge asymmetry. On the way to a symmetric configuration, the DNS meets no obstacles. On the contrary, on the way to the compound nucleus the DNS has to overcome the potential barrier, B^*_{fus}, which is equal to the difference in V(Z,L) at the B.G. point and at the injection point of the reaction. The appearance of the inner fusion barrier B^*_{fus} [3] is due to the endothermic character of the process of nucleon rearrangement in the massive DNS, which leads the system to the compound nucleus. The energy necessary for this rearrangement is supplied from the DNS excitation energy E^*. The formation of the compound nucleus is not possible, if the DNS excitation energy is less than the value of B^*_{fus}. Thus, the relation $E^* = B^*_{fus}$ determines the minimum value of the excitation energy of the DNS and the corresponding collision kinetic energy E, which allow complete fusion with compound nucleus formation take place.

Another important characteristic of the fusion of massive nuclei, which manifest itself only in the DNS concept, is the competition between complete fusion and quasi-fission. Due to the statistical character of the exchange of nucleons between the DNS nuclei, a certain probability exists that either the system reaches and overcomes the B.G. point, which leads to the formation of a compound nucleus, or the system reaches a symmetric configuration from where it undergoes quasi-fission. The more symmetric the reaction, the higher the inner fusion barrier B^*_{fus} which has to be overcome by the DNS on its way to the compound nucleus and, also, the stronger is the quasi-fission channel (see Fig.2).

APPLICATION OF THE DNS CONCEPT TO THE ANALYSIS OF SHE-SYNTHESIS REACTIONS

Modification of the calculation of V(Z,L)

The synthesis of SHE is usually performed at projectile energies leading to excitation energies of the compound nucleus as low as possible. This ensures higher survival probability after deexcitation. As can be seen from fig.2, the main "heating" of the compound nucleus takes place during the descent of the DNS from the B.G. point. It is at this evolution stage that the greater part of the system's potential energy is transformed into thermal excitation. However, whether the DNS will reach the state of a compound nucleus or will undergo quasi-fission is determined already when approaching the B.G. point. At the same time, exactly at this evolution stage the DNS excitation energy is lowest. Thus, one can say, that during the most important stage to complete fusion, the DNS is in a cold state. This peculiarity in the evolution of the DNS in SHE-synthesis reactions required some modifications in calculating the potential energy V(Z,L). Instead of the liquid-drop mass values in relation (1), real (taken from tables) masses were used. Account was taken also of the deformation of the DNS heavy nuclei formed during its evolution to the compound nucleus. The deformation was considered for the ground state, while the large axis orientation corresponded to the minimum of the DNS potential energy.

Inner fusion barriers B^*_{fus} for reactions used to synthesize elements 110, 112, 114 and 116

It was shown in above that the minimum excitation energy E^* necessary to realize complete fusion of massive nuclei, and the corresponding collision energy E depend on the inner fusion barrier B^*_{fus} (see. Fig.2), the height of which is determined by the shape of the potential energy V(Z,L) and the reaction injection point.

Fig.3 represents the potentials V(R) and potential energy V(Z,L) for the DNS, formed in the cold fusion reaction leading to element 110: $^{208}Pb + ^{62}Ni \rightarrow ^{270}110(-1n) \rightarrow ^{269}110$. It can be seen that taking account of the deformation of the heavy nucleus in the DNS is of great importance when estimating the value of B^*_{fus} - the latter can be lowered by as much as 15~MeV.

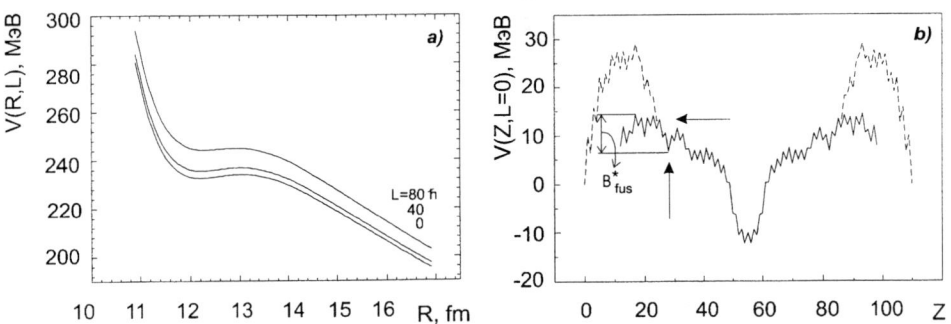

FIGURE 3. a) $^{62}Ni + ^{208}Pb = ^{270}110$ reaction input potential as a function of the distance between the centers of interacting nuclei and angular momentum, b) for the same combination of nuclei, given is the DNS' potential energy. Bold curve represents the potential energy calculated with allowance made for the deformation of the DNS' heavy partner in the ground state. Dotted curve gives the calculations for spherical nuclei. The vertical arrow points to the position of the reaction input point. The horizontal arrow points to the position of the excitation energy at which the synthesis of element 110 was carried out.

The potential energy $V(Z,L)$ was calculated for the case of zero angular momentum, $L=0$. In SHE-synthesis reactions, 10-15 partial waves contribute to the formation of compound nucleus. For this reason, the shape of $V(Z,0)$ is well enough representative. As can be seen from Fig.2b, in the mentioned reaction the inner fusion barrier B^*_{fus} amounts to 8~MeV. After overcoming it, the nucleus $^{270}110$ will has excitation energy of about 18~MeV. Experiment has shown [7] that, at the maximum yield of $^{269}110$-nuclei, the excitation energy of the compound nucleus was 13~MeV.

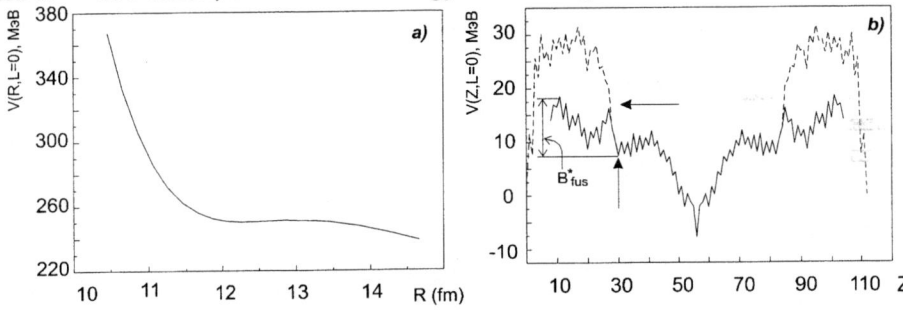

FIGURE 4. The designations are the same as in Fig.3, but the $^{70}Zn + ^{208}Pb = ^{278}112$ reaction is considered.

Fig.4 represents $V(R)$ and $V(Z,0)$ for the DNS, formed in the cold fusion reaction $^{208}Pb + ^{70}Zr \rightarrow ^{278}112(-1n) \rightarrow ^{277}112$, leading to element 112. In this case, also agreement is achieved between the calculated value of B^*_{fus} and the compound nucleus excitation energy, determined in the experiment [8].

In GSI (Darmstadt), an attempt was made to synthesize element 114 using the reaction $^{208}Pb + ^{82}Sc \rightarrow ^{290}116 \; (-\alpha) \rightarrow ^{286}114$ [9].

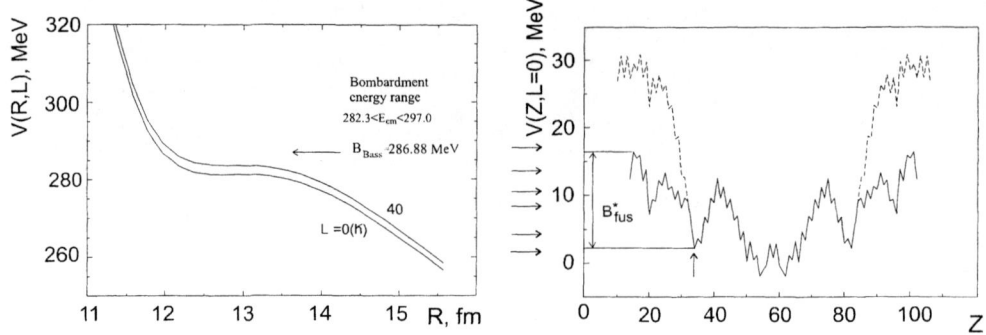

FIGURE 5. The same as in Fig.3, but the ^{82}Se+ ^{208}Pb= ^{290}116 reaction is considered. The horizontal arrows in Figure (on the right) point to the values of excitation energies at which the synthesis of element 116 was attempted.

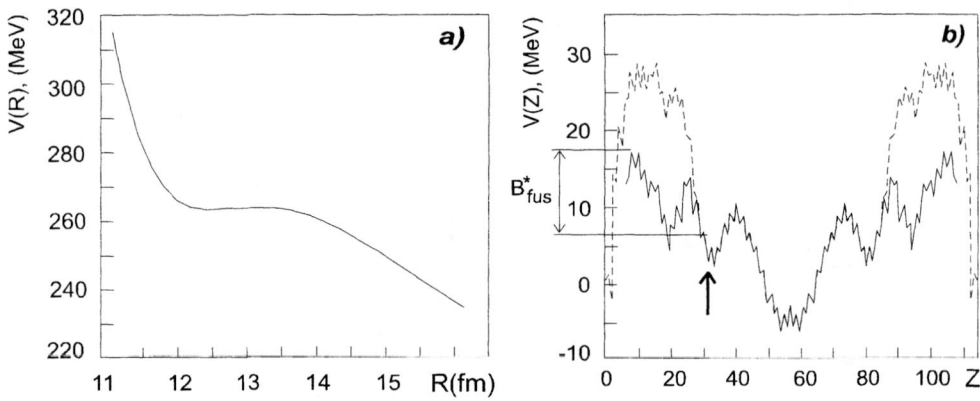

FIGURE 6. Input potential and the DNS' potential energy for the ^{76}Ge + ^{208}Pb= ^{284}114 reaction. All the designations are the same as in Fig.3.

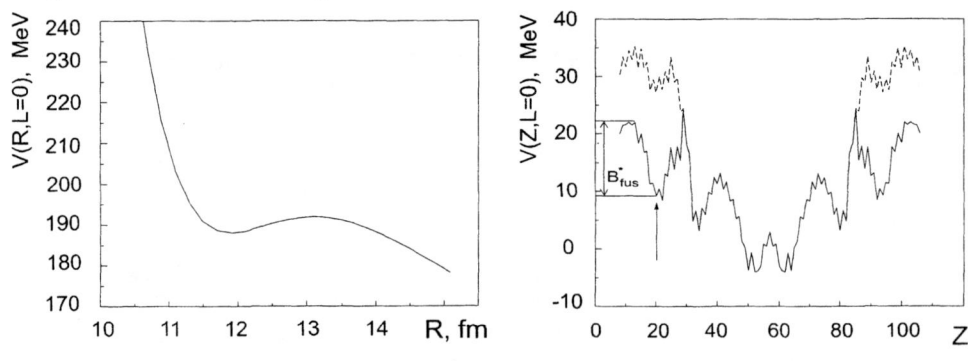

FIGURE 7. Input potential and the DNS' potential energy for the ^{48}Ca + ^{244}Pu= ^{292}114 reaction. All the designations are the same as in Fig.3.

Fig.5 shows the potentials V(R) and V(Z,O) for the DNS, formed in the above reaction. The horizontal arrows indicate six values of the excitation energy of the compound nucleus $^{290}116$, which were obtained at different incident energies of the ^{82}Sc-projectiles. It can be seen that in four cases the excitation energy of the compound nucleus was insufficient to overcome B^*_{fus} and it was hardly possible to expect its formation.

Fig.6 and 7 present V(R) and V(Z,O) for the two different reactions to be used for the synthesis of element 114. On the basis of these data, one can estimate both E^* and E, necessary to realize complete fusion of the participating nuclei.

The Minimum of the Excitation Energy of Heavy Compound Nuclei

According to the DNS-concept the minimum excitation energy of the compound nucleus coincides with the height of the Bussinaro-Gallone point. It means that the minimum of the excitation energy of the compound nucleus is determined by the shape of the potential energy curve.

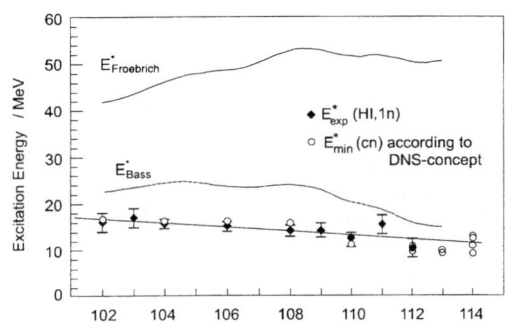

FIGURE 8. Excitation energy of compound nucleus of the 102-112 elements in cold fusion reactions (HI,1n) (theory and experiment [10]).

In the calculations potential energy V(Z,0) the real masses taken from the tables were used for the DNS nuclei. The deformation of the DNS nuclei was taken into account. The deformation of the heavy nucleus was taken in the ground state, the deformation of the light nucleus - in the 2^+ state.

Fig.8 demonstrates the experimental data for the excitation energy of the compound nuclei of elements from 102 to 112 produced in cold fusion reactions and the minimal excitation energy calculated within the framework of the DNS-concept. The calculated data based on the surface frictional model [13] are also indicated.

The Role of Quasi-Fission in the Reactions of Synthesis of SHE

The DNS concept gives the following relation for the formation cross section of evaporation residues in reactions between massive nuclei:

$$\sigma_{ER} = \sigma_c \cdot P_{cn} \cdot W_{sur}, \qquad (4)$$

where σ_c is the capture cross section, P_{cn} - the probability of forming a compound nucleus, W_{sur} - the probability of survival of the compound nucleus during its deexcitation. In the advent to the center of spherical SHE (Z=114, N=184), the shell corrections to the liquid-drop fission barrier increase. For this reason, the factor W_{sur}, reflecting the competition between neutron evaporation and fission, should not strongly decrease. Obviously, the same holds true for σ_c. The angular momentum L_{Bf}, at which the fission barrier vanishes, will also increase with increasing the fission barrier. However, it is L_{Bf} which determines the set of partial waves which will contribute to the capture cross section σ_c. It is difficult to estimate P_{cn}, since there exist no methods for its calculation. An model was suggested by the present authors, which took into account the competition between complete fusion and quasi-fission in reactions between massive nuclei.

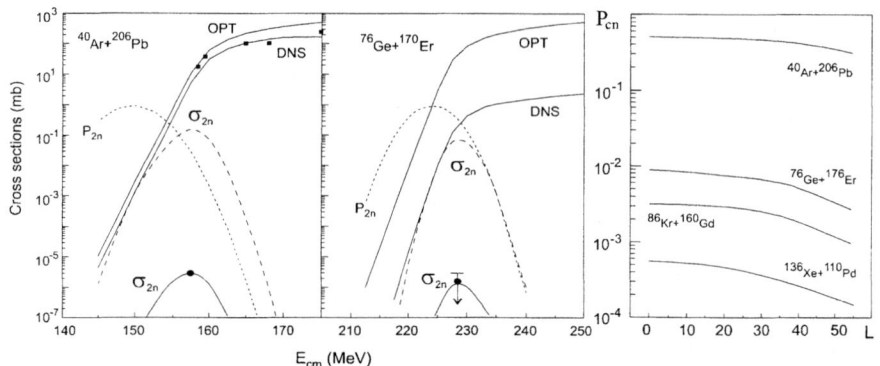

FIGURE 9. Comparison the experimental data (points) and calculations for two different combinations (left) of colliding nuclei resulting in the production the same ^{246}Fm compound nucleus. On the right – probability of compound nucleus formation for different reactions as a function of angular momentum.

In the our model certain simplifications of the DNS evolution process have been introduced [14]. From any configuration the DNS may pass to the neighboring in Z and N configuration only. This means that one proton and one or two neutrons are transferred from one nucleus to the other. The cluster transfer is excluded. The probability of the nucleon transfer is proportional to the DNS level densities in the neighboring configuration.

For the calculation of the probability of a proton transfer from one nucleus to the other in the DNS, we have used the expression from ref. [11] when the macroscopic transition probabilities can be written in terms of the microscopic one $\lambda_{zz'}$ and of the level densities of the macroscopic states ρ_z:

$$P_{zz'} = \lambda_{zz'} \cdot \rho_z, \qquad (5)$$

The level densities can be written in terms of the potential energy of the dinuclear system $\rho_z=\rho(E-V(Z,L))$, where E is the energy of the system. According to [11] $\lambda_{zz'} = \lambda_0 (\rho_z \cdot \rho_{z'})^{-1/2}$, where λ_0 is a constant. Finally for probabilities of capture of proton P^- and left of one P^+ we can write:

$$P^+ = \left\{1+\exp\left[\frac{V(Z+1,L)-V(Z-1),L)}{2T}\right]\right\}^{-1}, \quad P^- = \left\{1+\exp\left[\frac{V(Z-1)-V(Z+1,L)}{2T}\right]\right\}^{-1} \qquad (6)$$

where $T=(E/a)^{1/2}$, and a is a parameter of level densities.

Knowing these relative ($P^+ + P^- = 1$) probabilities (6), using random numbers we emulate the direction of motion of the DNS (either in the direction of the B.G. point or in the direction of the system symmetry point). Repeating this procedure as many times as necessary, we obtain the wanted statistics for our calculations. The ratio of the number of trajectories, which have led to complete fusion, and the total number of events gives us the fusion probability P_{cn} characterizing the competition between quasi-fission and fission during the evolution of the DNS.

A large number of the DNS trajectories in the Z and A space reduce to one trajectory which goes along the bottom of the potential energy valley. The calculation of the DNS evolution process is carried out using the Monte-Carlo method for different angular moment L. It is assumed that the DNS which crosses over the maximum $V(Z,L)$ goes irreversibly into the complete fusion channel. The DNS which has reached the symmetric configuration irreversibly proceeds into the quasi-fission channel. The model was tested in the calculation of the production cross section of ^{244}Fm in four reactions with different charge and mass asymmetries (see Fig.9).

Fig.9 demonstrate a powerful influence of quasi-fission on P_{cn} in symmetric nuclear reactions. Using this model it was possible to reproduce the experimental values of the production cross section for ^{244}Fm in the reactions with ^{40}Ar and ^{76}Ge ions [6].

In the framework of this model, the quantity P_{cn} was calculated for the reactions leading to the synthesis of elements 102-114. The results of these calculations are shown on the Fig.10.

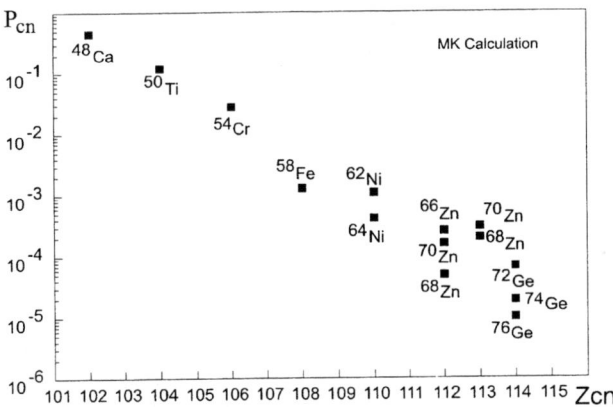

FIGURE 10. Probability of compound nucleus formation for different reactions as a function of Z of compound nuclei. Targets were ^{208}Pb and ^{209}Bi. The bombardment particles are indicate on the figure.

As can be seen from the Fig.10, in going from element 110 to element 112, the value of P_{cn} changes by a factor of 10 which is close to the experimentally obtained ratio of the formation cross sections of these elements: 15 and 2~pb. The significant decrease of P_{cn} for the 116-element synthesis reaction makes it possible to understand the main reason for the failure to synthesize in the performed experiment. On the basis of this analysis, a conclusion can be drawn that, in "cold" fusion reactions, quasi-fission is the main factor determine the decrease of SHE-formation cross sections with increasing their atomic number.

CONCLUSIONS

The DNS concept was used in the analysis of reactions leading to the synthesis of elements 110 and 114. The attempt to synthesize element 116 was also analyzed. The DNS concept allows to carry out calculations of the inner fusion barrier B^*_{fus} for synthesis reactions, which in turn gives an estimation of the energy threshold for complete fusion and minimum of excitation energy of compound nuclei. The DNS concept allows to estimate the competition between complete fusion and quasi-fission in SHE synthesis reactions, which gives the probability P_{cn} of forming a compound nucleus after capture. In "cold" fusion reactions, the quasi-fission is the main factor determining the decrease of SHE- production cross sections when increasing the atomic number of the synthesized SHE.

As the next step of developing our model of the competition of complete fusion and quasi-fission will be take into account the possibility of a moving DNS on the R-coordinate during its evolution into the compound nucleus. By the another words, we want to include the possibility competition of quasi-fission and fusion on each step of evolution of DNS. To compare with available experimental data on synthesis SHE it is necessary to estimate the value of W_{sur} (4) which defines a competition between fission and evaporation of particles from excited compound nucleus. It will be done by the authors of this report in nearest future.

REFERENCES

1. Smolanczuk R., Skalski J., Sobiczewski A. *Phys. Rev.* **C52,** 1871 (1995); Smolanczuk R. *Phys. Rew.* **C56**, 812-824 (1997).
2. Swiatecki W.J., *Phys. Scripta* 24, 113 (1981); Bjornholm S. and Swiatecki W.J., *Nucl. Phys.* **A391**, 471 (1982); Blocki J.P., Feldmeier H, Swiatecki W.J. *Nucl. Phys.*, **A459**, 145 (1986).
3. Antonenko A.V., Cherepanov E.A., Nasirov A.K., Permjakov V.P. and Volkov V.V., *Phys. Lett.* **B319**, 425 (1993); ibid. *Phys. Rev.* **C51**, 2635 (1995).
4. Volkov V.V. *Izvestiya AN USSR ser. fiz.* **50,** 1879 (1986).
5. Adamian G.G., Antonenko N.V., Jolos R.V., Ivanova S.P., Melnikova O.I., *Int. Journal of Mod. Phys.* **E5**, n1, 191-216 (1996).
6. Gaggeler H., Sikkeland T., Wirth G., Bruchle W., Bogl W., Franz G., Herrmann G., Kratz J.V., Schaedel M., Summerer K. and Weber W. *Z.Phys.*, **A316**, 291 (1984).
7. Hofmann S., Ninov V., Hesberger F.P., Armbruster P., Folger H., Muenzenberg G., Schott H.J., Popeko A.G., Eremin A.V., Andreev A.N., Saro S., Janic R., Leino M.. *Z.Phys.* **A350**, 277 (1995); ibid *Z.Phys.* **A359**, 281 (1995).
8. Hofmann S., Ninov V., Hesberger F.P., Armbruster P., Folger H., Muenzenberg G., Schott H.J., Popeko A.G., Eremin A.V., Saro S., Janic R., Leino M. *Z.Phys.* **A354**, 229 (1995)
9. Hofmann S., private communication.
10. Popeko A.G. "Untersuchung der Fusionsreaktionen $^{16}O+^{208}Pb$ und $^{86}Kr+^{136}Xe$. Report at Deutsche Physicalische Gesselschaft, 1997, p.13.
11. Moretto L.G., Sventek J.S. Phys. Lett. <u>58B</u>, p. 26 (1975).
12. Frobrich P. *Phys. Report,* 116 , 337 (1984).
13. Pustylnik B.I. "Estimation of the production cross sections of heavy nuclei in "cold" and "hot" fusion reactions". *Proceedings of 3 Int. Conf. Dynamical Aspects of Nuclear Fission,* Slovak Republic, August 30 – September 4, 1996, p.121-128.
14. Cherepanov E.A., Volkov V.V., Antonenko N.V., Nasirov A.K. "Complete fusion of massive nuclei in frame of DNS-concept and macroscopic dynamical model". *In Proceedings of the II Int. Symp. Heavy Ion Physics and its applications,* 29 August – 1 September , 1995, pp.272-282.

Competition between Complete Fusion and Quasifission in Reactions with Heavy Nuclei

N.V. Antonenko*, G.G. Adamian*†, W. Scheid* and
V.V. Volkov†

*Institut für Theoretische Physik der Justus–Liebig–Universität,
D-35392 Giessen, Germany
†Joint Institute for Nuclear Research, 141980 Dubna, Russia

Abstract. A model based on the dinuclear system concept is suggested for the calculation of the competition between complete fusion and quasifission in reactions with heavy nuclei. The fusion rate through the inner fusion barrier in mass asymmetry is found by using the Kramers-type expression. The calculated cross sections for the heaviest nuclei are in a good agreement with the experimental data. The experimentally observed rapid fall-off of the cross section of the cold fusion with increasing charge number Z of the compound nucleus is explained.

INTRODUCTION

The competition between the complete fusion and quasifission processes occurs in the reactions with massive nuclei at bombarding energies smaller than 15 MeV/nucleon. In these reactions the quasifission channel dominates and leads to a strong reduction of few orders of magnitude of the fusion cross section [1]. In order to reach superheavy elements and the island of stability at $Z = 114$ and $N = 178 - 184$ [2], two heavy nuclei must fuse. At the GSI (Darmstadt) the elements with $Z =110$, 111 and 112 were recently synthesized in cold fusion reactions [3]. The heaviest isotope of the element with $Z = 110$ was produced in the FLNR, JINR (Dubna) [4]. The nucleus with $Z = 110$ was also produced in LBL (Berkeley) [5]. The next important step is the synthesis of the elements with $Z =113$ and 114 by using both Pb-based [6] and actinide-based reactions [7].

The new model suggested in [1] yields a good agreement between the theoretical predictions and experimental data on the fusion of heavy nuclei. Within this model the evaporation residue cross section can be written as

$$\sigma_{ER}(E_{\text{c.m.}}) = \sum_{J=0}^{J_{max}} \sigma_c(E_{\text{c.m.}}, J) P_{CN}(E_{\text{c.m.}}, J) W_{sur}(E_{\text{c.m.}}, J). \quad (1)$$

The capture cross section σ_c defines the transition of the colliding nuclei over the Coulomb barrier and the formation of the dinuclear system (DNS) when the kinetic energy is transformed into the excitation energy of the DNS. The value of J_{max} corresponds to $E_{\text{c.m.}}$ and is smaller than the limit value of J for the compound nucleus formation. The probability of complete fusion P_{CN} depends on the competition between the complete fusion and quasifission processes after the capture stage in the DNS. The surviving probability W_{sur} estimates the competition between fission and neutron evaporation in the excited compound nucleus.

In the DNS model [1,8] the fusion process is assumed as a transfer of nucleons from the light nucleus to the heavy one. The DNS evolves as a diffusion process in the mass asymmetry degree of freedom $\eta = (A_1 - A_2)/A$ to the compound nucleus (A_1 and A_2 are the mass numbers of the nuclei and $A = A_1 + A_2$). Evolving to the compound nucleus the DNS should overcome the inner fusion barrier B^*_{fus} in the mass asymmetry degree of freedom. The top of this barrier (the Businaro-Gallone point at $\eta = \eta_{BG}$) coincides with the maximum of the DNS potential energy as a function of η. We assume that complete fusion occurs after the DNS overcomes this inner barrier. In the DNS-concept the value of B^*_{fus} represents a hindrance for complete fusion of the initial DNS with $|\eta_i| < |\eta_{BG}|$. Besides the motion in η a diffusion process in the variable of the relative distance R between the DNS nuclei occurs leading to a decay of the DNS which we denote as quasifission. For quasifission, the DNS should overcome the potential barrier B_{qf} which coincides with the depth of the pocket in the nucleus-nucleus potential. The energy required to overcome the fusion and quasifission barriers is contained in the excitation energy of the DNS.

MODEL AND RESULTS

The application of transport equations to describe the DNS evolution in heavy ion collisions is well known. Since the fusion process looks like the DNS evolution in the DNS-concept [8], the Fokker-Planck equation (FPE) can be used to describe the diffusion processes in the collective variables R and η characterizing the DNS [1,9]. The neck degree of freedom that is important in the macroscopic dynamical model is not considered at all in Ref. [1]. As follows from our analysis in Ref. [9], in the DNS the neck size is close to the one obtained by a simple superposition of the frozen density tails of the nuclei.

The collective Hamiltonian of the DNS is

$$H_{coll} = \frac{\mu_{RR} \dot{R}^2}{2} + \frac{\mu_{\eta\eta} \dot{\eta}^2}{2} + U(R, \eta, J), \quad (2)$$

where μ_{RR} and $\mu_{\eta\eta}$ are the mass parameters, $U(R,\eta,J)$ the potential energy of the DNS depending on R, η, and the angular momentum J of the system. For $|\eta| < |\eta_{BG}|$, $\mu_{R\eta} \ll \sqrt{\mu_{RR}\mu_{\eta\eta}}$. Since the fusion of massive nuclei with not too large η is of our interest here, we can omit the nondiagonal term of the kinetic energy.

The value of $U(R,\eta,J)$ in (2) is calculated as

$$U(R,\eta,J) = B_1 + B_2 + V(R,J) - [B_{12} + V'_{rot}(J)], \qquad (3)$$
$$V(R,J) = V_{Coul}(R) + V_n(R) + V_{rot}(R,J),$$

where B_1, B_2, and B_{12} are the binding energies of the fragments and compound nucleus, and V_n, V_{Coul}, and V_{rot} are the nuclear, Coulomb, and centrifugal parts of the nucleus-nucleus potential $V(R,J)$, respectively. The liquid-drop and realistic binding energies are used in our calculations for large and small excitation energies of the DNS, respectively. The isotopic composition of the nuclei forming the DNS is chosen with the condition of the N/Z-equilibrium in the system. The value of the $U(R,\eta,J)$ in (3) is normalized to the energy of the rotating compound nucleus by $B_{12}+V'_{rot}$. The method for the calculation of the nucleus-nucleus potential was described in [10]. The minimal excess of the kinetic energy, $\Delta E_{min} = E_{c.m.}^{min} - V(R_b)$, above the entrance barrier in $V(R)$, at which fusion becomes possible in our model, is compared with the experimental data [11] in Table 1. We set $\Delta E_{min} = B^*_{fus}-B_{qf}$ for $B^*_{fus}-B_{qf} \geq 0$ and $\Delta E_{min} = 0$ for $B^*_{fus}-B_{qf} < 0$. It is seen that the prediction of ΔE_{min} in our model is in a good agreement with ΔE_{min} obtained from the experiment.

TABLE 1. Experimental and calculated minimal values of ΔE (ΔE_{min}), at which fusion is possible.

System	ΔE_{min} [MeV] exp. [11]	ΔE_{min} [Mev] our model
^{40}Ar+^{206}Pb	-0.5±3	0
^{76}Ge+^{170}Er	10±5	8
^{86}Kr+^{160}Gd	≥15.7	11.5
^{110}Pd+^{136}Xe	≥23.5	15
^{96}Zr+^{124}Sn	6.5±3	5

The calculated driving potential $U(R_m,\eta) = U(\eta)$ as a function of η for the reaction ^{54}Cr+^{208}Pb is presented in Fig. 1 for $J = 0$. For a given η, the value R_m coincides with the position of the minimum of the potential pocket in $V(R,\eta)$. For the 1n reactions (small excitation energies of the DNS), the deformation effects are taken into account in the calculation of the potential energy. For the heavy nuclei in the DNS, which are deformed in the ground state, the parameters of deformation are taken from Ref. [12]. The light nuclei of the DNS are assumed to be deformed only if the energies of their 2^+ states are smaller than 1.5 MeV. As known from experiments on subbarrier fusion

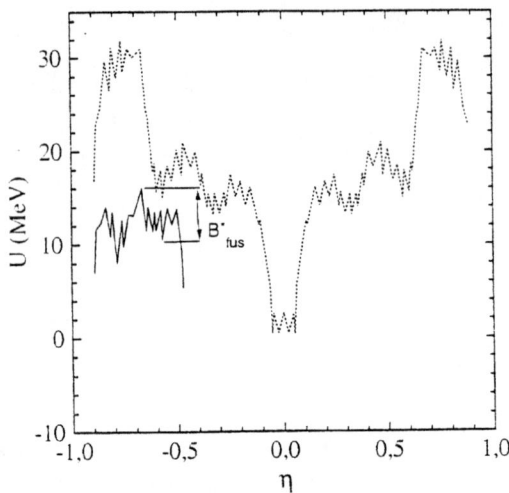

FIGURE 1. Dependence of the potential energy of the DNS on η for the ^{54}Cr+^{208}Pb reaction ($|\eta_i| = 0.59$). The calculated results with and without deformation of the DNS nuclei are presented by solid and dashed lines, respectively. The result with deformation is presented for η-values which are of interest to calculate B^*_{fus}.

of lighter nuclei, these states are easily populated. The relative orientations of nuclei in the DNS follows the minimum of the potential energy during the evolution in η. We find that the values of P_{CN} calculated with the deformation of both nuclei in the DNS are practically the same as the ones calculated previously in [1] where a deformation only in the heavy nucleus of the DNS was taken into account. However, a deformation of both nuclei in the DNS yields a good agreement with the experimental excitation energies of the compound nucleus for the 1n reactions.

With the value of the inner fusion barrier, the optimal excitation energy of the compound nucleus is calculated as $E^*_{CN} = U(R_m, \eta_i) + B^*_{fus}$ ($U(R_m, \eta_i)$ is the energy of the initial DNS). Therefore, the optimal kinetic energy is $E_{c.m.} = E^*_{CN} - Q$. Note that all considered collisions occur above the calculated entrance barrier ($\Delta E > 0$). For smaller and larger excitation energies, the evaporation residue cross sections in the 1n fusion reaction decrease due to the decrease of the values of P_{CN} and W_{sur} in (1). The calculated optimal values of E^*_{CN} (see Fig. 2) are in a good agreement with the experimental data on the 1n fusion reactions used in the production of the heaviest elements with Z =102–112 [6]. The macroscopical model [13] overestimate the minimal values of E^*_{CN} in the fusion reactions. The model [13] predicts E^*_{CN} between 50 and 300 MeV. With the model [14] the values of E^*_{CN} are estimated to be about 40–50 MeV. The use of the Bass potential overestimates the experimental value of E^*_{CN} by 5–7 MeV [6].

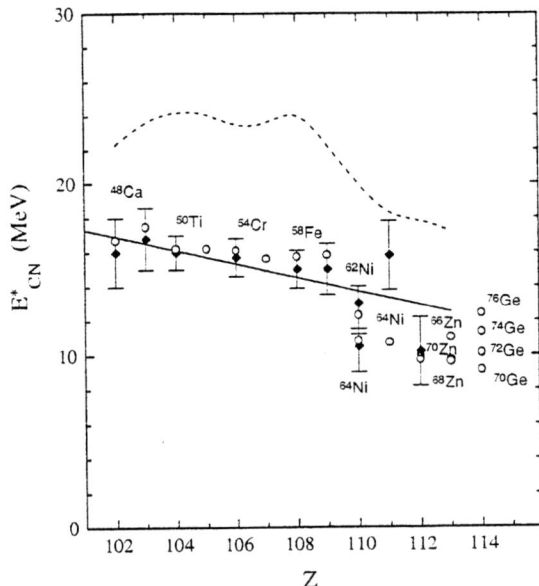

FIGURE 2. Optimal excitation energies of the compound nucleus for the 1n Pb-based reactions. The nuclei with even and odd Z are produced with ^{208}Pb and ^{209}Bi targets, respectively. The experimental data are shown by solid diamonds. The projectiles are indicated. The experimental point for $Z = 112$ is shown for ^{70}Zn as a projectile. The calculated results are depicted by open circles for different projectiles. The values of E^*_{CN} obtained with the Bass potential are presented by the dashed line [6]. The solid line is drawn to guide the eye.

In order to obtain the complete fusion probability

$$P_{CN} \approx \int_{\eta_{BG}}^{\infty} P(R,\eta,t_0)d\eta, \qquad (4)$$

we solve the Fokker-Planck equation for the distribution function $P(R,\eta,t)$ within the global momentum approach [1]. The value of t_0 determines the half-life of the DNS which means a maximum interaction time of about $t_{int} = (3-4)t_0$. The data obtained with this function (Fig. 3) are in agreement with the ones calculated in the approach based on the Kramers-type expression for the fusion rate through the inner fusion barrier of the DNS. The values of P_{CN} obtained here are smaller than ones in [10] where other method of the calculation is used and the value of the interaction time is overestimated.

The leakage of probability through the inner fusion barrier in η is defined by the rate $\lambda_\eta(t)$ at $\eta = \eta_{BG}$. Then we obtain

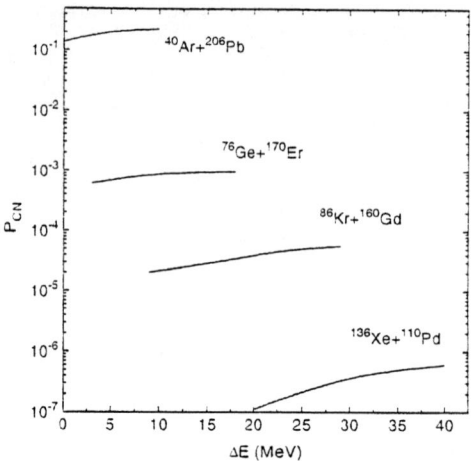

FIGURE 3. Dependence of the fusion probability on ΔE for the reactions leading to the ^{246}Fm compound nucleus. The calculations were done by using the solution of the equations for the first and second moments of the distribution function.

$$P_{CN} = \int_0^{t_{int}} \lambda_\eta(t)dt \approx \frac{\lambda_\eta^{Kr}}{\lambda_R^{Kr} + \lambda_\eta^{Kr}}. \quad (5)$$

Here, λ_i^{Kr} are asymptotic values of the fusion or quasifission rate $\lambda_i(t)$ at the corresponding barriers. In (5), we neglect the transient time τ_i ($i = R, \eta$), which is the time when the value of $\lambda_i(t)$ reaches its asymptotic value λ_i^{Kr}, because $\tau_i \ll 1/\lambda_i^{Kr}$, $\tau_R \approx \times 10^{-21}$s and $\tau_\eta \approx 2 \times 10^{-21}$s. This is fulfilled in the reactions considered. Since the estimated t_{int} in our calculations is about 10^{-20}s, we can use Kramers-type quasistationary formulas. Using the results of Ref. [16] we obtained in [1] the approximate expression for $\lambda_\eta(t)$.

To obtain the asymptotic fusion and quasifission rates λ_i^{Kr} ($i = R, \eta$), we use the formalism elaborated in Ref. [17]. We approximate the expression for the quasistationary rate λ_i^{Kr} over a multidimensional potential barrier with a Kramers-type formula

$$\lambda_i^{Kr} = \frac{1}{2\pi} \frac{\omega_i^2}{\sqrt{\omega_i^{B_R}\omega_i^{B_\eta}}} \left(\sqrt{\left[\frac{(\Gamma/\hbar)^2}{\omega_i^{B_R}\omega_i^{B_\eta}}\right]^2 + 4} - \frac{(\Gamma/\hbar)^2}{\omega_i^{B_R}\omega_i^{B_\eta}} \right)^{1/2} \exp\left[-\frac{B_i}{\Theta}\right]. \quad (6)$$

Here, B_i ($i = R, \eta$) defines the height of the fusion ($B_\eta = B_{fus}^*$) or quasifission ($B_R = B_{qf}$) barriers. The possibility to apply the Kramers-type expression to relatively small barriers ($B_i/\Theta > 0.5$) was demonstrated in [18]. The local thermodynamic temperature Θ is calculated with the expression $\Theta = \sqrt{E^*/a}$, where $a = A/12$ MeV^{-1} and E^* is the DNS excitation energy. In Eq. (6), the

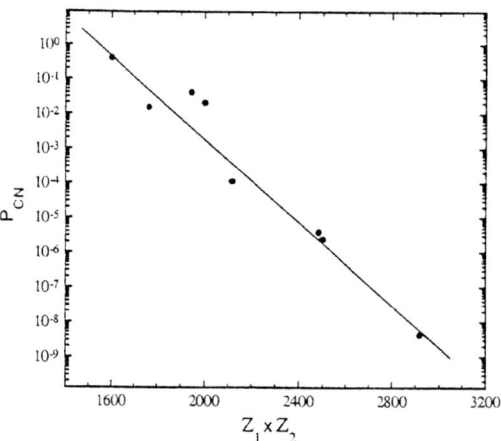

FIGURE 4. Calculated values (solid points) of the fusion probability as a function of $Z_1 \times Z_2$ at $J = 0$, $\Gamma = 2$ MeV and $E^* = 30$ MeV. The solid line is drawn to guide the eye.

frequencies $\omega_i^{B_j}$ ($j = R, \eta$) of the inverted harmonic oscillators approximate the potential in the variables $i = R, \eta$ on the tops of the barriers B_j, and ω_i are the frequencies of the harmonic oscillators approximating the potential in the same variables for the initial DNS. Since the oscillator aproximation of the potential energy surface is good for the reaction considered, we neglected the nondiagonal components of the curvature tensors in (6).

The values of P_{CN} were calculated for the reactions ^{90}Zr+^{90}Zr, ^{100}Mo+^{100}Mo, ^{110}Pd+^{110}Pd, ^{136}Xe+^{136}Xe, ^{86}Kr+^{136}Xe, ^{110}Pd+^{136}Xe ^{96}Zr+^{124}Sn and ^{124}Sn+^{124}Sn with $\Gamma = 2$ MeV, $J = 0$ and the DNS excitation energy 30 MeV. The values of P_{CN} as a function of $Z_1 \times Z_2$ (Z_1 and Z_2 are the charge numbers of the colliding nuclei) are presented in Fig. 4. One can see the exponential decrease of the fusion probability with increasing $Z_1 \times Z_2$ in the symmetric and almost symmetric reactions. Therefore, the experimentally observed [19] rapid fall-off of the fusion cross sections with increasing $Z_1 \times Z_2$ is simply explained in our model.

The calculated values of P_{CN} for the $1n$ Pb-based reactions are presented in Fig. 5. These values are in agreement with the ones extracted from experimental data [3,6,20]. The decrease of P_{CN} in (1) by about four orders of magnitude with Z increasing from 104 to 112 explains the observed rapid fall-off of the evaporation residue cross sections. The factors σ_c and W_{sur} do not strongly change with Z for the cases considered. The fusion probability strongly decreases with decreasing mass asymmetry of the initial DNS (increasing Z for the Pb-based reactions) because the inner fusion barrier B^*_{fus} increases and the quasifission barrier B_{qf} decreases (the Coulomb repulsion increases). For example, in the ^{76}Ge+^{208}Pb→^{284}114 reaction the estimated value of σ_{ER} is

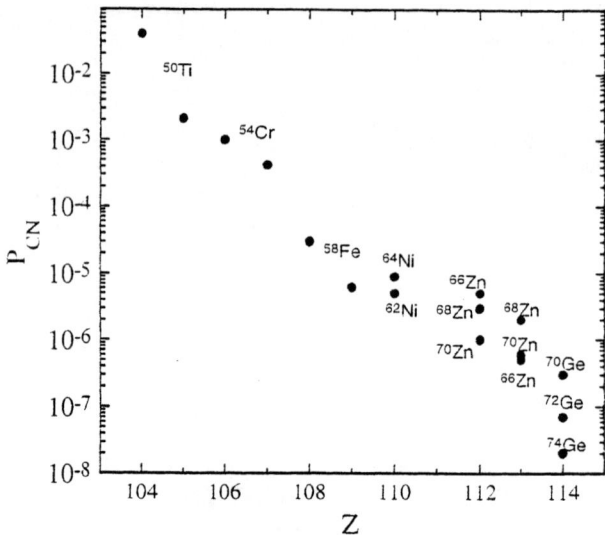

FIGURE 5. Calculated fusion probability P_{CN} for different $1n$ Pb-based reactions. The projectiles are indicated. The excitation energies in the calculations are taken the same as in Fig. 2.

near the limit of present measurements. The probability to obtain the nucleus with $Z = 116$ in the ^{82}Se+^{208}Pb reaction is smaller than this limit. From our analysis it follows that the fusion of symmetric combinations ($\eta_i = 0$) for the synthesis of the heaviest elements yields smaller cross sections.

Using the data in Fig. 5, we can explain the smaller yield of the nucleus with $Z = 110$ in the reaction with ^{62}Ni than the one with ^{64}Ni. The fusion probability in the reactions with ^{66,68}Zn is larger than the one with ^{70}Zn. However, W_{sur} in the reaction with ^{70}Zn can be larger than W_{sur} in the reactions with other Zn isotopes because of the smaller neutron separation energy in ^{278}112. It could be that the increase of P_{CN} is compensated by a decreasing W_{sur} in the reactions with the lighter isotopes. In addition, to obtain the same values of σ_c for the reactions with ^{70}Zn and ^{68}Zn, the excitation energy in the reaction with ^{68}Zn should be larger by 2-3 MeV than the one with ^{70}Zn. We note that in the reactions used for the production of the heaviest elements all factors in (1) become important.

In order to calculate the evaporation residue cross sections in the $1n$ Pb-based reactions, we use values of surviving probabilities $W_{sur}(E^*_{CN}) \approx \Gamma_n/\Gamma_f$ (the values given refer to an angular momentum of zero) which are few times larger than the ones estimated in [19], but smaller than the values from the analysis of $4n$ reactions [20–22]. In accordance with the experimental data and shell-model calculations [2] the value of Γ_n/Γ_f increases slightly for $Z = 108$

because of the shell closures in the vicinity of $N = 162$. Since for larger Z the neutron separation energies and fission barriers are almost the same, we took the same value of Γ_n/Γ_f for these nuclei. The value of Γ_n/Γ_f is sensitive to shell effects and excitation energies and has to be studied in further details. The calculated values of σ_{ER} (Table 2) are in a good agreement with the known experimental data [3,6,19,21,22].

TABLE 2. Calculated [th.] and experimental [exp.] evaporation residue cross sections for the $1n$ Pb-based reactions.

Reactions	E^*_{CN} (MeV)	P_{CN}	σ_c (mb)	W_{sur}	σ_{1n} [th.]	σ_{1n} [exp.]
^{50}Ti+^{208}Pb →^{258}104	16.1	3×10^{-2}	5.3	1×10^{-4}	16 nb	10 nb
^{54}Cr+^{208}Pb →^{262}106	16.0	9×10^{-4}	4.6	2×10^{-4}	0.8 nb	0.5 nb
^{58}Fe+^{208}Pb →^{266}108	15.5	3×10^{-5}	4.0	6×10^{-4}	72 pb	70 pb
^{64}Ni+^{208}Pb →^{272}110	10.7	1×10^{-5}	3.4	6×10^{-4}	20 pb	15 pb
^{70}Zn+^{208}Pb →^{278}112	9.8	1×10^{-6}	3.0	6×10^{-4}	1.8 pb	1 pb
^{70}Zn+^{209}Bi →^{279}113	9.8	4×10^{-7}	2.9	6×10^{-4}	0.7 pb	

In the recent experiments the present limit of the heaviest element production in the cold fusion has been reached. More asymmetric combinations of the colliding nuclei than in the Pb-based reactions (the initial DNS is near or behind the top of the inner fusion barrier in mass asymmetry) can be used to extend the production of superheavy elements. According to our model one should take targets heavier than Pb. For the actinide-based reactions with the projectiles like ^{48}Ca, ^{34,36}S, the fusion probability is much larger than in Pb-based reactions. This effect can compensate the increase of the fission of the compound nucleus due to a higher excitation energy which corresponds to the $3n - 4n$ channels. Our calculated cross section for the $4n$ reaction ^{48}Ca+^{244}Pu→^{292}114 is about 1 pbarn.

SUMMARY

Our model gives simple explanation of so called hindrance of the fusion in the entrance channel of the reactions used for the production of new superheavy nuclei. It should be applied for the further analysis of the experimental data.

ACKNOWLEDGMENTS

We thank Dr. E.A.Cherepanov, Dr. A.K.Nasirov (Dubna), Dr. S.Hofmann and Prof. G.Münzenberg (Darmstadt) for fruitful discussions. The author (N.V.A.) is grateful to the Alexander von Humboldt-Stiftung for the financial support. This work was supported in part by DFG.

REFERENCES

1. Antonenko, N.V., Cherepanov, E.A., Nasirov, A.V., Permjakov, V.B., and Volkov, V.V., *Phys. Lett.* **B319**, 425 (1993); *Phys. Rev.* **C51**, 2635 (1995); Adamian, G.G., Antonenko, N.V., and Scheid, W., *Nucl. Phys.* **A618**, 176 (1997).
2. Smolanczuk, R., and Sobiczewski, A., in *Proc. Int. Conf. on Low Energy Nuclear Dynamics*, St.Petersburg, 1995, eds. Oganessian, Yu. *et al.*, Singapore: World Scientific, 1995, p. 313.
3. Hofmann, S. *et al.*, *Z. Phys.* **A350**, 277 (1995); **A350**, 281 (1995); **A354**, 229 (1996).
4. Lazarev, Yu.A. *et al.*, *Phys. Rev.* **C54**, 620 (1996).
5. Ghiorso, A. *et al.*, *Phys. Rev.* **C51** R2293 (1995).
6. Hofmann, S., in *Proc. Int. Conf. on Low Energy Nuclear Dynamics*, St.Petersburg, 1995, eds. Oganessian Yu. *et al.*, Singapore: World Scientific, 1995, p. 305; Popeko, A.G., in *Proc. Int. Workshop on New Ideas on Clustering in Nuclear and Atomic Physics*, Rauischholzhausen, 1997 (in print).
7. Oganessian, Yu.Ts., in *Proc. Int. Conf. on Nuclear Physics Structure of Vacuum and Elementary Matter*, Wilderness, 1996, eds. Stöcker, H. *et al.*, Singapore: World Scientific, 1997, p. 11.
8. Volkov, V.V., *Izv. AN SSSR ser. fiz.* **50**, 1879 (1986); in *Proc. of the 6th Intern. Conf. on Nuclear Reaction Mechanisms*, Varenna, 1991, ed. Gadioli, E., Varenna: Ricerca Scientifica, 1992, p.39.
9. Adamian, G.G., Antonenko, N.V., Jolos, R.V., and Scheid, W., *Nucl. Phys. A* (in print) (1997).
10. Adamian, G.G. *et al.*, *Int. J. Mod. Phys.* **E5**, 191 (1996).
11. Gaggeler, H. *et al.*, *Z. Phys.* **A316**, 291 (1984).
12. Raman, S. *et al.*, *At. Data and Nucl. Data Tables* **36**, 1 (1987).
13. Swiatecki, W.J., *Phys. Scripta* **24**, 113 (1981); Bjornholm, S., and Swiatecki, W.J., *Nucl. Phys.* **A391**, 471 (1982).
14. Fröbrich, P., *Phys. Rep.* **116**, 337 (1984).
15. Volkov, V.V. *et al.*, *Nuova Cimento* (in print) (1997).
16. Bhatt, K.H., Grange, P., and Hiller, B., *Phys. Rev.* **C33**, 968 (1986).
17. Fröbrich, P., and Tillack, G.R., *Nucl. Phys.* **A540**, 353 (1992); Weidenmüller, H.A., and Zhang, Jing-Shang, *J. Stat. Phys.* **34**, 191 (1984).
18. Gonchar, I.I., and Kosenko, G.I., *Sov. J. Nucl. Phys.* **53**, 133 (1991).
19. Armbruster, P., *Ann. Rev. Nucl. Part. Sci.* **35**, 135 (1985).
20. Cherepanov, E.A., Iljinov, A.S., and Mebel, M.V., *J. Phys.* **G9** 653 (1983).
21. Oganessian, Yu.Ts., Lazarev, Yu.A. in *Treatise on Heavy-Ion Science*, ed. D.A. Bromley, v.4, New York: Plenum Press, 1985, p. 3.
22. Münzenberg, G., *Rep. Prog. Phys.* **51**, 57 (1988).

Diffusion model for the synthesis of superheavy elements

Y. Aritomo, K. Okazaki, T. Wada, M. Ohta and Y. Abe[†]

Department of Physics, Konan University, Okamoto 8-9-1, Kobe 658, Japan
[†] *Yukawa Institute for Theoretical Physics, Kyoto University, Kyoto 606-01, Japan*

Abstract. The fusion-fission process for synthesizing superheavy elements is studied on the basis of the dissipative dynamics. We use a multi-dimensional Langevin equation for the first stage where the reaction system evolves from the contact configuration of two incident nuclei to the later time when the complete dissipation of the initial relative kinetic energy is accomplished, and a two-dimensional Smoluchowski equation for the following stage. The evaporation residue cross sections for superheavy elements have been shown to have an optimum value at a certain initial energy, due to the balance between the diffusibility for fusion at high temperature and the restoration of the shell correction energy against fission at low temperature. The isotope dependence is also discussed.

I INTRODUCTION

The synthesis of superheavy elements is a long-standing important subject in nuclear physics [1]. Recent several years, the synthesis of new heavy elements has been performed by the so called *cold fusion* reaction, and the findings of the heaviest elements 110, 111 and 112 are reported by GSI group with the cross section of several picobarns region [2,3]. The reaction mechanism is not well understood, but cross sections are empirically inferred to much smaller for heavier elements. We, therefore, propose to explore another way, so called *warm fusion* or *hot fusion* reaction.

As is well-known, in heavy systems around $Z \sim 80$, the trajectory calculations with friction [4] was very useful for the explanation of the extra- or extra-extra-push energy. In superheavy region, however, the mean trajectory calculations are not suitable, because mean trajectories can not reach the spherical shape region and around due to the strong dissipation [5]. However, the extremely small part of distribution can be found there due to fluctuation. Therefore, it is important to take into account the fluctuating part from the mean trajectory. It becomes necessary to solve a full dissipative dynamics, or a fluctuation-dissipation dynamics with the Kramers (Fokker-Planck) equation or with the Langevin equation.

Here, we divide the whole dynamical process into two stages; the stage (I) is that of before the kinetic energy of relative motion dissipates, and the stage (II) is that of after the kinetic energy of relative motion completely dissipates. In the preliminary analysis [6–8], to study the fusion-fission dynamics in the stage (II) by the diffusion model, we employ the one-dimensional Smoluchowski equation in the elongation degree of freedom which is a strong friction limit of Fokker-Planck equation, and we take into account the temperature dependent shell correction energy. We pointed out a new possibility of synthesizing the superheavy element with relatively *hot fusion* reaction.

In the present study, we discuss quantitatively about the evaporation residue cross section of superheavy elements by a more realistic model of multi-dimensional dynamics assuming the symmetric incident channel. In the stage (I), we employ the two-dimensional Langevin equation including the elongation and necking degrees of freedom. Then in the stage (II), by taking the results of the previous stage as the initial condition, we calculate the evaporation residue cross section of superheavy elements using the two-dimensional Smoluchowski equation in the same parameter space.

In Section 2, we discuss about the fusion mechanism in very heavy nuclei and we introduce the thermal fluctuation around the mean trajectory. In Section 3, the importance of the temperature dependence of the shell correction energy is discussed. We explain our framework in Section 4. In Section 5, we present the results of our calculations and show the excitation function for producing superheavy elements, and discuss the isotope dependence in connection with the neutron separation energies. Summary is given in Section 6.

II CHARACTERISTIC OF FUSION IN MASSIVE NUCLEI

We discuss about the fusion mechanism in massive nuclei. Figure 1 shows the potential energy surface of liquid drop model (LDM) in the nuclear deformation space. Nuclear shapes are described with two-center potential parametrization by Maruhn and Greiner [10]. The horizontal axes denote z_0 (the distance between the centers of two potential wells) and the vertical axes denote δ (the deformation of fragments); $z_0 = \delta = 0$ corresponds to spherical compound nucleus (marked by circles). Solid curves denote the scission lines (zero-neck line). Di-nucleus region and mono-nucleus region are separated by this scission line. Saddle points are marked by crosses. Dashed curves denote the ridge lines. If the colliding partners are in the right-hand region of the ridge line, they automatically go to fission. On the other hand, if they are in the left-hand region, they go to fusion.

First, for light mass systems (Fig.1a for $Z = 50$), the contact point (marked by a square) is inside the ridge line, therefore the colliding partners go down the potential slope, and automatically go to the spherical region. Fusion occurs with large probability. This situation is described successfully by the Bass model [16].

Then, for heavy mass system (Fig.1b for $Z = 80$), the contact point is outside of the ridge line. Therefore, to go to the spherical region, it needs extra energy to go over the extra barrier. In this mass region, if we put an extra energy, mean trajectory overcomes this barrier. This situation is described by the extra push model [9].

But for superheavy mass system (Fig.1c for $Z = 114$), the situation drastically change. The contact point is far from the ridge line. Therefore, almost all of the colliding partners go down the potential slope, and go to fission immediately. Even if we give any extra energy, it can not across the ridge line. Then, we turn our attention to the fluctuation around the mean trajectory.

The fluctuation comes from the coupling of collective shape degree of freedom to nucleonic degree of freedom. Dissipation can be expressed by friction force which

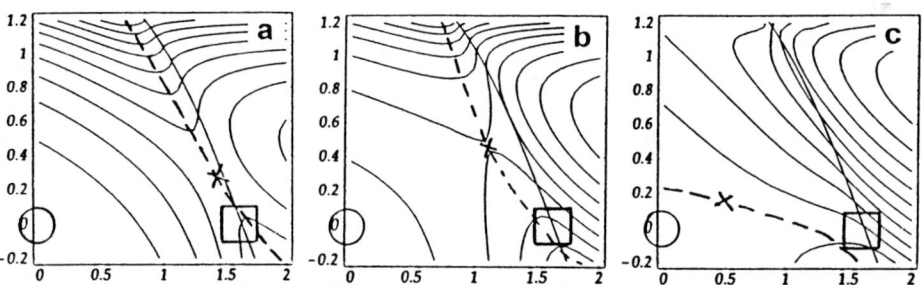

FIGURE 1. Macroscopic energy surface in nuclear deformation space. Figure 1a is for $Z=50$, Fig. 1b for $Z=80$, and Fig. 1c for $Z=114$. Abscissa denotes the separation between two potential centers and ordinate denotes the fragment deformation. Scission line (zero-neck radius) is denoted by the solid curve and ridge is given by the dashed line. Symbols are given in the text.

transfer the energy from collective motion to nucleonic motion. To the contrary, fluctuation can be expressed by random force which transfer the energy in the opposite direction. Due to these forces, nuclear shapes fluctuate around the mean trajectory and some trajectories can overcome the extra barrier with very small probability. In superheavy mass region, it is essential to take into account the fluctuation, therefore we have to deal with fluctuation-dissipation dynamics. Fluctuation-dissipation dynamics can be treated with the Langevin equation or with the Fokker-Planck equation.

III SHELL CORRECTION ENERGY

In superheavy mass region, another very important effect is shell correction energy. In this region, there are no fission barrier in LDM potential, but the shell correction energy is very large. The shell correction energy is predicted to be as large as 10 MeV for the case of double magic nucleus $Z=114$ and $N=184$. Figure 2 shows the potential energy surface in the nuclear deformation space for this nucleus. The LDM energy surface is shown in Fig.2a, and the potential energy including the shell corrections is shown in Fig.2b. The stabilization by the shell correction energy is essential, and the shell correction energy depends on temperature. When the temperature is high, the shell correction energy disappears (Fig.2a), but when the temperature is low, it recovers (Fig.2b). Thus, it is essential to take into account the temperature dependence (excitation energy dependence) of the shell correction energy in estimating the fission-evaporation competition.

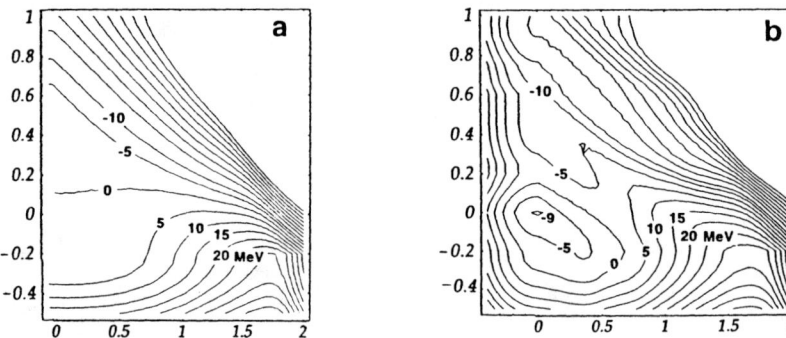

FIGURE 2. Energy surface in nuclear deformation space for the superheavy element 114. The LDM energy surface is shown in Fig.2a, and the potential energy including the shell corrections is shown in Fig.2b. Abscissa denotes the separation between two potential centers and ordinate denotes the fragment deformation.

IV MODEL

We divide the whole fusion-fission process into two stages as mentioned in Introduction. Here, more details for both stages are presented. In this work, to calculate the Langevin equation in the first stage, we adopt the two-dimensional nuclear deformation space assuming the symmetric incident channel with the two-center parametrization [10]. The neck parameter ϵ is fixed to 1.0 in the present calculation, deformation parameters δ_1 and δ_2 of the colliding nuclei are taken to be equal, i.e., $\delta_1 = \delta_2 = \delta$, and we set the asymmetry parameter $\alpha=0$. We treat z_0 (distance between two potential centers) and δ as the two collective parameters to be described by the Langevin equation. About the process in the first stage, the details can be found in Ref. [18].

As the probabilities reaching the spherical shape and around is extremely small in the case of superheavy elements synthesis, the treatment with the Langevin equation is not practical. Therefore, in the second stage after the complete dissipation of the initial kinetic energy, we use the Smoluchowski equation. The Smoluchowski equation is the strong friction limit of the Fokker-Planck equation, and the evolution of the probability distribution $P(q, l; t)$ in collective coordinate space is assumed to follow the Smoluchowski equation;

$$\frac{\partial}{\partial t}P(q,l;t) = \left(\gamma^{-1}\right)_{ij}\left[\frac{\partial}{\partial q_i}\left\{\frac{\partial V(q,l;t)}{\partial q_j}P(q,l;t)\right\} + T\frac{\partial^2}{\partial q_i \partial q_j}P(q,l;t)\right]. \quad (1)$$

$V(q, l; t)$ is the potential energy, and the angular momentum of the system is expressed by l. $\gamma_{ij}(q)$ is dissipation tensors which is calculated by the wall-and-window one-body dissipation model [13]. T is the temperature of the compound nucleus calculated from the excitation energy as $E^* = aT^2$ with a denoting the level density parameter of Töke and Swiatecki [11]. The temperature dependent shell correction energy is added to the macroscopic potential energy,

$$V(q,l;t) = V_{\text{macro}}(q,l) + V_{\text{shell}}(q)\Phi(t), \quad (2)$$

where V_{macro} is the potential energy of the liquid drop model and V_{shell} is the shell correction energy at $T = 0$.

The temperature dependence of the shell correction energy is extracted from the free energy calculated with single particle energies [14]. The temperature dependent factor $\Phi(t)$ in Eq. (2) is parameterized as;

$$\Phi(t) = \exp\left(-\frac{aT^2(t)}{E_d}\right), \qquad (3)$$

following the work by Ignatyuk et al. [15]. The shell-damping energy E_d is chosen as 20 MeV. The cooling curve $T(t)$ is calculated by the statistical model code SIMDEC [14].

The Smoluchowski equation is solved with the initial condition which is the results of the previous stage. With the temperature decreasing by neutron evaporation, the LDM-potential changes due to the restoration of shell correction energy. The potential energy surface $V_{\text{macro}}(q, l = 0) + V_{\text{shell}}(q)\Phi(T = 0)$ is shown in Fig. 2b. We can see that the nucleus which stays around the spherical shape is very stable.

The evaporation residue cross section is defined as the probability which is left inside the fission barrier in the final stage of the cooling process and is proportional to the quantity $d(T_0, l; t)$ at $t = \infty$;

$$d(T_0, l; t) = \int_{\text{inside saddle}} P(q, l; t) dq. \qquad (4)$$

Here, T_0 is the initial temperature to be calculated with the incident kinetic energy. The evaporation residue cross section σ_{EV} is calculated as;

$$\sigma_{EV} = \frac{\pi\hbar^2}{2\mu_0 E_{cm}} \sum_l (2l+1) d(T_0, l; t = \infty), \qquad (5)$$

where μ_0 denotes the reduced mass in the entrance channel and E_{cm} the incident energy in center-of-mass frame.

V NUMERICAL RESULTS

Now, we present the excitation function calculated by the present dynamical model, taking as an example the reaction forming the superheavy nucleus with the doubly closed shell, i.e., the reaction $^{149}_{57}\text{La} + ^{149}_{57}\text{La} \rightarrow {}^{298}114$. According to our calculation using the Langevin equation, the initial relative kinetic energy is found to dissipate completely around $t_{\text{diss}} = 0.5 \times 10^{-21}$s. Therefore, it is valid to change the calculation method from using the Langevin equation to the Smoluchowski equation at this time.

The time dependent feature of the probability $d(T_0, l = 10; t)$ are plotted in Fig. 3 for five different initial temperatures $T_0 = 0.68, 0.79, 0.96, 1.11$ and 1.24 MeV (which, of course, correspond to excitation energy equal to 15., 20., 30., 40. and 50. MeV, respectively). They all increase in the beginning and later on start to decrease or stay almost constant, which is easily understood by the diffusion picture. Up to the time of $(10 \sim 20) \times 10^{-21}$ s, the probability in the region of the compact configuration is supplied by diffusion from the contact region and its yield increases rapidly. But during that time, the main part of the probability has descended down the slope of the potential and the supply ceases.

For high temperature such as 1.24 MeV, the main part of the initial distribution goes down the potential slope, but a part of this probability overcomes the ridge line by diffusion. Therefore, the probability d increases rapidly, and the peak value is very large due to the large diffusion coefficient $D = T/\gamma$. However, due to the low fission barrier, the probability accumulated in the compact configuration area diffuses back over the fission barrier. The probability d decreases drastically, and finally the probability remains this region is small. On the other hand, for low temperature such as 0.68 MeV, the small probability overcomes the ridge line, due to the small diffusion coefficient $D = T/\gamma$. As the fission barrier, however, is high, most of the probability reached around the spherical region survives against fission.

The height of the peak around 15×10^{-21}s is essentially determined by the diffusibility into the compact configuration area, while the decrease from the peak value to the final yield at $t_\infty = 2000 \times 10^{-21}$s is determined by how fast the shell

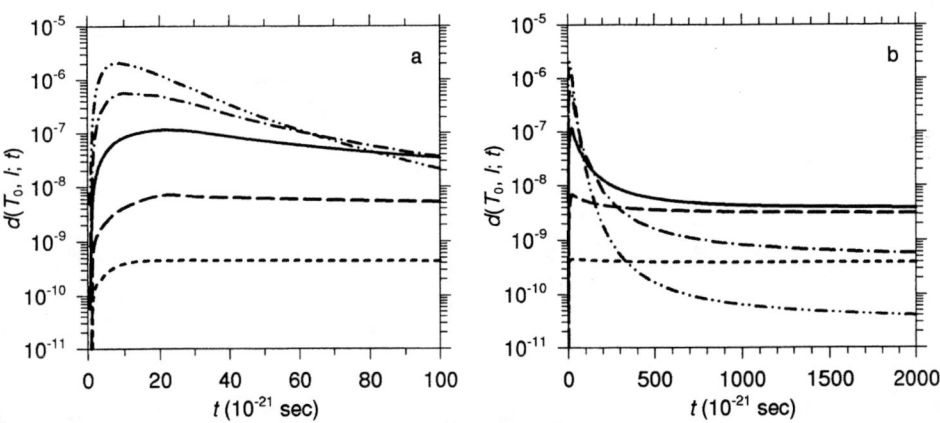

FIGURE 3. Time evolution of the probability density in the compact configuration region $d(T_0, l = 10; t)$ for five initial temperatures; $T_0 = 0.68$ (short-dashed), 0.79 (long-dashed), 0.96 (solid), 1.11 (dot-dashed), and 1.24 MeV (dot-dot-dashed). Figure 3a is for short time behavior and Fig. 3b for long time behavior.

correction energy is restored to give rise to sufficient barrier height. Thus, the final yield surviving in the compact configuration area is determined by two factors: the diffusibility depending on the temperature and the restoration of the shell correction energy. Due to the competition between high fusion and low fission, the optimum value of the probability d at $t = 2000 \times 10^{-21}$ s can be obtained at intermediate temperature.

Then, we calculate the excitation function of evaporation residue cross section σ_{EV} by summing up the partial wave according to Eq. (5). The excitation function of σ_{EV} for the ^{149}La + ^{149}La → ^{298}114 reaction is shown in Fig. 4 by solid line. The excitation function shows a bell shape having a maximum at excitation energy about 25 MeV. For this reaction system, the Bass potential barrier height [16] is 325 MeV in the center of mass frame and corresponds to $E^* = 10$ MeV in the compound nucleus. It should be emphasized, above the Bass barrier the optimum cross section can be realized in this reaction system and thereby can be observed experimentally.

Then, we compare the results calculated by two-dimensional model with those by one-dimensional model which we used in the previous analysis. In that analysis, we solved the one-dimensional Smoluchowski equation along the fission path [6–8]. In Fig. 4, the result with the one-dimensional model is shown with dashed line. In the present improved calculation, the essential feature is the same as in the previous work.

Next interesting point is the result on the isotope dependence of the cross section. Neutron rich isotopes are considered to obtain a large residue cross section, because they generally have large neutron decay widths and thereby favored in the survival probability Γ_n/Γ_f.

We calculated the evaporation residue cross section for a series of Z=114 isotopes from N=176 to 184 with the one-dimensional model [6–8]. We used different cooling curves for each isotope while we neglected the isotope dependence of the energy surface. The energy surface for N=184 was used for all cases. Figure 5 shows the calculated evaporation residue cross section for $^{145,146,147,148,149}_{57}$La + $^{145,146,147,148,149}_{57}$La → ^{290,292,294,296,298}114 reactions as functions of initial excitation energy. The isotope dependence of evaporation residue cross section is found to be very strong. Larger evaporation residue cross section is seen in neutron rich nucleus as is expected. The excitation function changes from the monotonically decreasing to the bell-shape one with the increase of the neutron number.

The neutron separation energy depends on the neutron number. The theoretical neutron separation energies [17] averaged over 4 neutron emissions $<B_n>$ for the corresponding composite systems, $^{290,292,294,296,298}114$, are 6.4, 6.1, 5.8, 5.4, and 5.0 MeV, respectively. When the neutron separation energy is small, the cooling due to the neutron evaporation is fast. In this case, due to the restoration of the shell correction energy, fission barrier arises rapidly to prevent the fission decay of compound nucleus. Therefore, we can observe the competition between large diffusibility at high temperature and high survival probability at low temperature. On the other hand, when the neutron separation energy is large, the cooling is slow. Therefore, fission barrier grows slowly. In this case, we will obtain monotonically decreasing excitation function. We can see that the cooling speed is very important in this dynamical mechanism.

In Fig.6, a map of the neutron separation energy averaged over four successive neutron emissions $<B_n>$ is displayed. The numerical numbers associated with lines are $<B_n>$ values. Naturally neutron-rich isotopes have smaller B_n's than neutron-deficient ones. Three triangles indicate isotopes which GSI succeeded in synthesizing by the *cold fusion*. $<B_n>$'s are 7 ~ 8 MeV, so there in no hope for quick cooling which gives rise to the enhancement in higher energies. On the

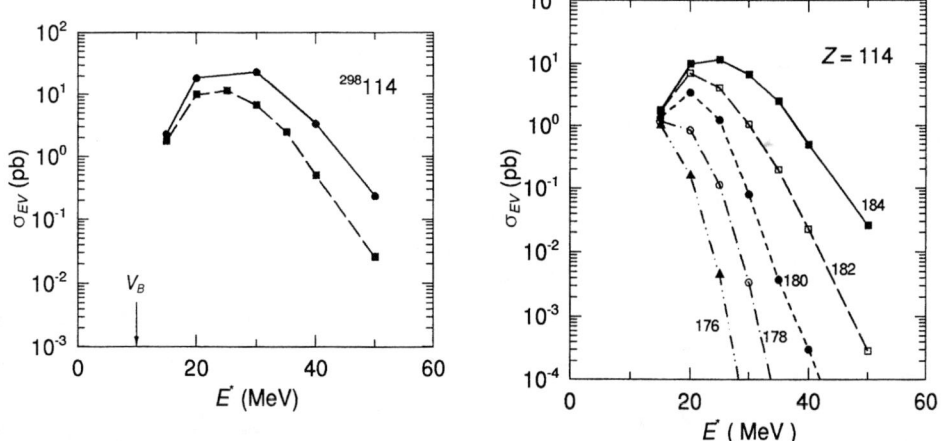

FIGURE 4. (left) Excitation function of the evaporation residue cross section for $^{149}_{57}$La + $^{149}_{57}$La → $^{298}114$ reaction. The corresponding Bass potential barrier is indicated by the arrow. The result with the present model is shown by the solid line, while the dashed line denotes the result with the one-dimensional model.

FIGURE 5. (right) Isotope dependence of the excitation function of the evaporation residue cross section for $Z=114$. Figures denote neutron numbers.

other hand, the dots indicate those which are investigated in the present work with $<B_n>$ equal to 5 ~ 6 MeV. It is expected that even more neutron rich isotopes are more favorable for the enhancement of residue cross sections. Thus, an exploration of experimental feasibility to reach the neutron rich side of the superheavy elements is an extremely interesting and urgent subject.

VI SUMMARY

A diffusion model which takes into account dynamical evolution of a distribution including statistical fluctuations in the deformation parameter space is shown to be a necessary and an appropriate way to describe fusion-fission process. We use the two-dimensional Langevin equation for the first stage, from the contact of two incident nuclei to the complete dissipation of the initial relative kinetic energy, and apply the two-dimensional Smoluchowski equation for the following stage.

With the model, it is shown for the synthesis of superheavy elements that there exists the optimum temperature or the excitation energy of compound system due to the competition between the diffusibility for fusion and the restoration of the shell correction energy against fission. Roughly speaking, the optimum temperature

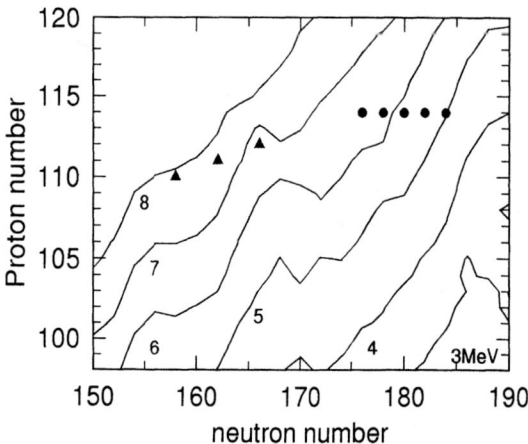

FIGURE 6. A map of the neutron separation energy averaged over four successive neutron emissions $<B_n>$. The numerical numbers associated with lines are $<B_n>$ values. Three triangles indicate isotopes which GSI succeeded in synthesizing by *cold fusion*. The dots indicate those which are investigated in the present paper.

is around the restoration temperature of the shell correction energy. In the present symmetric system, the maximum cross section of several tens picobarns is obtained around $E^* = 20\sim30$ MeV. There still exists a possibility of *hot* or *warm fusion* reaction for synthesize of the superheavy elements. These characteristic feature are the same as the results obtained by the previous one-dimensional analysis. The absolute value of the cross section, of course, depends on the friction coefficient, which has not been well determined yet.

It is found that the position of the optimum energy strongly depends on the time scale of the cooling by neutron evaporation, *i.e*, the existence of the bell-shape peak in the cross section depends on isotope. Larger cross section is obtained for neutron-rich system because of the faster cooling. The shape of the excitation function of the evaporation residue cross section changes from a monotonically decreasing one to a bell-shape one as the number of neutron increases.

REFERENCES

1. Y.T. Oganessian and Y.A. Lazarev, *Treatise on Heavy-Ion Science* ed. by D.A. Bromley (Plenum, 1985) pp.3-251; G. Münzenberg, Rep. Prog. Phys., **51** (1988) 57.
2. S. Hofmann et al., Z.Phys. **A350** (1995) 277.
3. S. Hofmann et al., Z.Phys. **A354** (1996) 229.
4. J. P. Blocki, H. Feldmeier and W. J. Swiatecki, Nucl. Phys. **A459** (1986) 145.
5. S. Bjørnholm and W. J. Swiatecki, Nucl. Phys. **A391** (1982) 471.
6. Y. Aritomo, T. Wada, M. Ohta, and Y. Abe, Phys. Rev., **C55** (1997) R1011.
7. T. Wada, Y. Aritomo, M. Ohta and Y. Abe, Nucl. Phys. **A616** (1997) 446c.
8. Y. Abe, Y. Aritomo, T. Wada and M. Ohta, J. Phys. in print. N.V. Antonenko et al., Phys. Lett., **B319** (1993) 425; C.E. Aguiar et al., Nucl. Phys., **A517** (1990) 205.
9. W.J. Swiatecki, Phys. Scripta **24** (1981) 113; W.J. Swiatecki, Nucl. Phys., **A376** (1982) 275.
10. J. Maruhn and W. Greiner, Z. Phys., **251** (1972) 431; K. Sato, A. Iwamoto, K. Harada, S. Yamaji, and S. Yoshida, Z. Phys., **A288** (1978) 383.
11. J. Töke and W.J. Swiatecki, Nucl. Phys., **A372** (1981) 141.
12. H.J. Krappe, J.R. Nix, and A.J. Sierk, Phys. Rev., **C20** (1979) 992.
13. J.R. Nix and A.J. Sierk, Nucl. Phys., **A428** (1984) 161c.
14. M. Ohta, Y. Aritomo, T. Tokuda and Y. Abe, Proc. of Tours Symp. on Nuclear Physics II (World Scientific, Singapore, 1995) p.480.
15. A.V. Ignatyuk, G.N. Smirenkin and A.S. Tishin, Sov. J. Nucl. Phys., **21** (1975) 255.
16. R. Bass, *Nuclear Reactions with Heavy Ions* (Springer, 1980).
17. P. Möller et al., Atomic Data and Nuclear Data Tables, **59** (1995) 185.
18. T. Tokuda, Talk in Tours Symposium on Nuclear Physics III, Tours, 1997.

DECAY OF SUPERHEAVY ELEMENTS

Stability and Production of Superheavy Nuclei

Peter Möller[*,†] and J. Rayford Nix[†]

[*] *P. Moller Scientific Computing and Graphics, Inc., P. O. Box 1440, Los Alamos, New Mexico 87544, USA*
[†] *Theoretical Division, Los Alamos National Laboratory, Los Alamos, New Mexico 87545, USA*

Abstract. Beyond uranium heavy elements rapidly become increasingly unstable with respect to spontaneous fission as the proton number Z increases, because of the disruptive effect of the long-range Coulomb force. However, in the region just beyond $Z = 100$ magic proton and neutron numbers and the associated shell structure enhances nuclear stability sufficiently to allow observation of additional nuclei. Some thirty years ago it was speculated that an island of spherical, relatively stable superheavy nuclei would exist near the next doubly magic proton-neutron combination beyond ^{208}Pb, that is, at proton number $Z = 114$ and neutron number $N = 184$. Theory and experiment now show that there also exists a rock of stability in the vicinity of $Z = 110$ and $N = 162$ between the actinide region, which previously was the end of the peninsula of known elements, and the predicted island of spherical superheavy nuclei slightly southwest of the magic numbers $Z = 114$ and $N = 184$. We review here the stability properties of the heavy region of nuclei.

Just as the decay properties of nuclei in the heavy region depend strongly on shell structure, this structure also dramatically affects the fusion entrance channel. The six most recently discovered new elements were all formed in cold-fusion reactions. We discuss here the effect of the doubly magic structure of the target in cold-fusion reactions on the fusion barrier and on dissipation.

INTRODUCTION

It was predicted more than 30 years ago that the next doubly magic nucleus beyond $^{208}_{82}$Pb$_{126}$ is ^{298}114$_{184}$ and that nuclei on an island of superheavy nuclei in its vicinity have half-lives of up to perhaps several billion years [1–7]. In Fig. 1 we show the calculated enhancement to binding due to microscopic effects [8] for nuclei throughout the periodic system. The two nuclei ^{266}Pb and ^{238}Hf are so near the neutron drip line that one can anticipate that they will never be observable. Of the remaining doubly magic regions only the recently observed region near ^{272}110 consists of nuclei deformed in their ground states. The only spherical doubly magic region that realistically remains to be observed is therefore the island of spherical superheavy elements currently predicted to occur near ^{292}114. Although no nuclei on this superheavy island have been observed so far, six new elements with proton numbers $Z = 107$–112 have been discovered between this superheavy island and the previously heaviest-known elements at the edge of the actinide region. The heaviest of the new elements are localized in the vicinity of a rock of *deformed* shell-stabilized nuclei near proton number $Z = 110$ and neutron number $N = 162$. We show in Fig. 2 the location of the currently heaviest known nucleus ^{277}112 and its observed α-decay products superimposed on calculated microscopic corrections for nuclei throughout the periodic system. In the following we discuss what we have learned experimentally and theoretically about (1) the stability properties of the heaviest nuclei and (2) the cold-fusion reaction mechanism that has been so essential in the discovery of the heaviest elements. A few aspects of hot fusion will also be reviewed.

I MODELS

In the study of nuclear structure a substantial number of different models are used, many of which are discussed at this conference. The significance of these models and their relation to each other is made more clear if we observe that it is possible to put the nuclear models used into one of the groups below:

1. Models where the physical quantity of interest is given by an expression such as a polynomial or a more general algebraic expression. The parameters are usually determined by adjustments to experimental data. Normally models of this type describe only a single nuclear property. No nuclear wave functions are obtained in these models. An example of a model of this type is the original Bethe-Weizsäcker macroscopic mass model. A nice

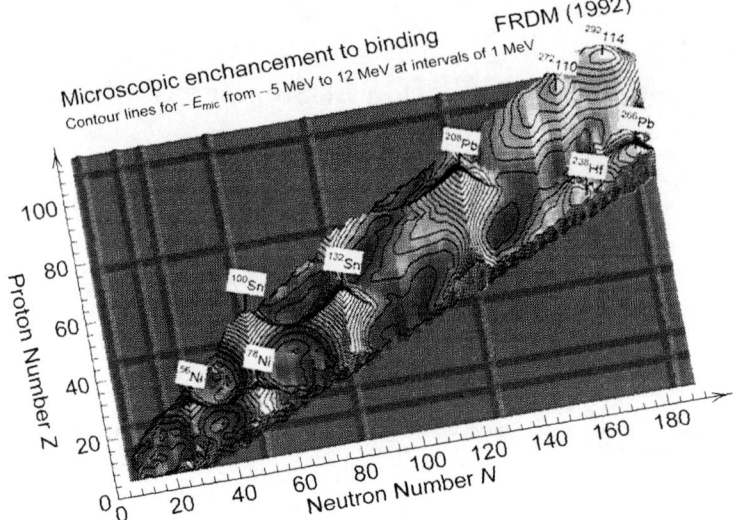

FIGURE 1. Calculated microscopic enhancement to binding for nuclei throughout the periodic system.

feature of this macroscopic mass model is that it can be generalized, without any additional parameters, to describe macroscopic fission barriers. This generalization property is unusual for models in this category.

2. Models that are claimed to be based on microscopic equations with realistic two-body interactions but which utilize so many approximations that the end result is again some algebraic expression with parameters that are adjusted to experimental data. No microscopic differential equations are actually solved. Examples of models of this type are the Duflo mass model [9] and the fermion dynamical-symmetry model [10] which is applied to nuclear mass calculations and some other calculations.

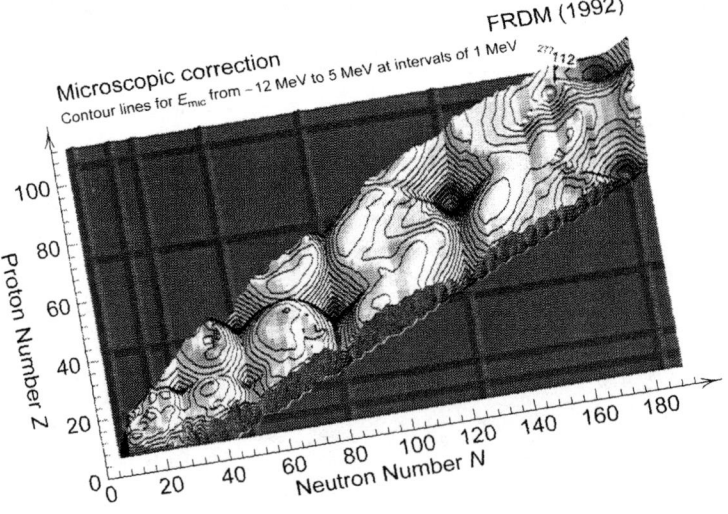

FIGURE 2. The heaviest known element $^{277}112$ and its α-decay products as dots superimposed on calculated microscopic corrections for nuclei throughout the periodic system.

3. Models that use an effective nuclear interaction and *actually solve* the resulting microscopic quantum-mechanical equations, for example a Schrödinger or a Dirac equation. The solutions provide *nuclear wave functions* which allow a vast number of nuclear properties to be modeled within *a single* framework. Most models of this type that are currently used fall into two subgroups depending on the type of wave function and the type of microscopic interaction used:

   3a Single-particle models that use a simple central potential with additional residual interactions such as the two-body pairing interaction. The Schrödinger equation is solved in a single-particle approximation and additional two-body interactions are treated in the BCS, Lipkin-Nogami, or RPA approximations, for example. Wave functions are obtained, which allows a large number of nuclear-structure features to be predicted, such as transition rates within or between rotational bands or beta-decay transition rates.

   For the calculation of the total nuclear energy it is not possible in the single-particle model to obtain the nuclear ground-state energy as $E = <\Psi_0|H|\Psi_0>$, where Ψ_0 is the ground-state nuclear wave function. To obtain the nuclear potential energy as a function of shape one combines the single-particle model with a macroscopic model, which leads to the macroscopic-microscopic model in which the energy is calculated as a sum of a microscopic correction obtained from calculated single-particle levels by use of the Strutinsky method and a macroscopic energy.

   3b Hartree-Fock-type models in which the postulated effective interaction is of two-body type and the wave function is an antisymmetrized Slater determinant. In other respects, these models have many similarities to those in **3a**, with the exception that it is possible to obtain the nuclear ground-state energy as $E = <\Psi_0|H|\Psi_0>$.

We sometimes hear proponents of models of type **3b** refer to models of type **3a** as "macroscopic." This is clearly inaccurate since

a both models solve microscopic wave equations. The difference between the two approaches lies in the type of interaction and type of wave function used.

b to date most if not all important new insight into *microscopic* nuclear structure has been provided by models in category **3a**. For example, reasonably accurate ground-state deformations and masses, nuclear level structure, including spins and parities, and the mass asymmetry of the fission saddle point were first obtained in this approach. The first two model categories are only able to parameterize or polynomialize data by use of expressions that normally contain a vast number of parameters and consequently reproduce experimental

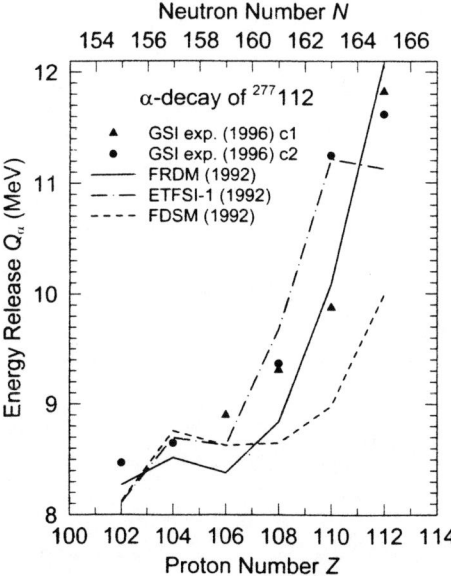

FIGURE 3. Comparison between energy releases Q_α obtained in the FRDM (1992), ETFSI-1 (1992) model, and FDSM (1992) and recent experimental data for the heaviest known element. Two different decay chains were observed experimentally.

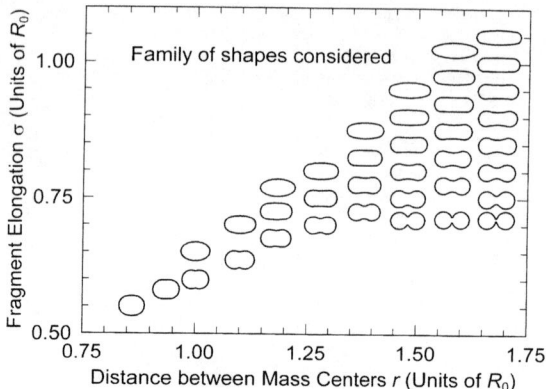

FIGURE 4. Nuclear shapes leading from the ground state to elongated and compact scission shapes. The corresponding potential-energy surface for ^{264}Fm is shown in Fig. 5. The quantity R_0 is the radius of the original spherical nucleus.

data used in the adjustment well, but fail catastrophically for data not used in the adjustments or for new data. Models in category **3b** are expected to *in principle* be more accurate than models in category **3a**, because the wave function is more realistic and more realistic effective-interactions can be used. However, two problems remain today: what effective two-body interaction is *realistic*, in the sense that it will yield more accurate results than the well-studied and well-optimized single-particle effective interactions, and what are the optimized parameter values of such a realistic two-body interaction?

Let us further emphasize that there is no "correct" model in nuclear physics. Modeling of nuclear physics involves simplifying the true forces and equations with the goal to obtain a formulation that can be solved in practice, but that "retains the essential features" of the true system under study. How to obtain an effective force from the true

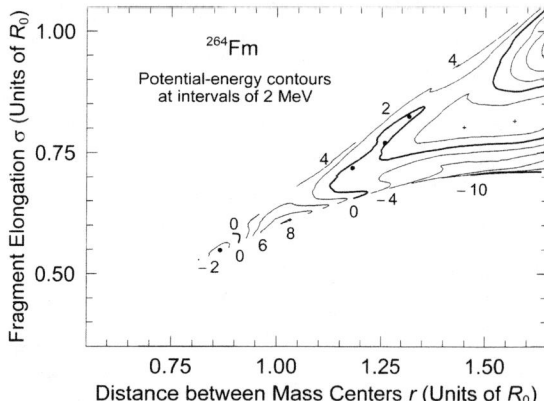

FIGURE 5. Potential-energy surface leading from the ground state to elongated and compact scission shapes. The corresponding shapes are shown in Fig. 4.

force is not well-defined. What we mean by "retains the essential features" depends on the circumstances. Simply speaking, it means that it retains sufficiently much of the true system that we can learn something from the simplified model. Bearing this in mind, it is clear that the microscopic single-particle models have been enormously successful over the years, since we have learned so much from them.

In our brief overview here we will discuss what has been learned about the stability of the heaviest elements from calculations based on a macroscopic-microscopic model with a realistic diffuse-surface folded-Yukawa single-particle potential as a starting point for the microscopic term and a generalized droplet model with a Yukawa-plus-exponential potential for the nuclear energy in the macroscopic term. The model has been described in sufficient detail elsewhere [8,11–14]. For a review of other results we refer to the many other interesting talks at this conference and to a recent review [15].

II STABILITY AT THE END OF THE PERIODIC SYSTEM

Whereas predictions of an island of superheavy elements were made already 30 years ago, the existence of a "rock" of relatively long-lived *deformed* neutron-deficient shell-stabilized superheavy nuclei in the vicinity of $Z = 110$ and $N = 162$ has been a subject of study only for about 15 years or so. In our first global nuclear mass calculation in 1981 [13,14], which was limited to 4023 nuclei, and which did not reach the neutron and proton drip lines, part of this rock is nevertheless clearly visible in the tabulated microscopic corrections. In Fig. 1 we see that the island of superheavy elements predicted in the mid 1960s is not isolated from the relatively stable actinide region. Instead, a stabilizing peninsula extends from the spherical superheavy region towards the actinide region. On this peninsula two stabilizing ridges corresponding to $N = 152$ and $N = 162$ are clearly visible.

For additional clarity we have included in Figs. 1 and 2 only even-even nuclei so that odd-even staggering is removed. The ridge at $N = 152$ has long been connected to the unusually long spontaneous-fission half-lives of ^{250}Cf, ^{252}Fm, and ^{254}No [16–18].

The peninsula structure extending from the spherical superheavy island is somewhat similar to the smaller peninsula extending to the southwest from the doubly magic ^{208}Pb. On this peninsula the most prominent shell-stabilized ridges occur at $N = 102$ and $N = 108$, but are less developed than the ridges in the heavy region.

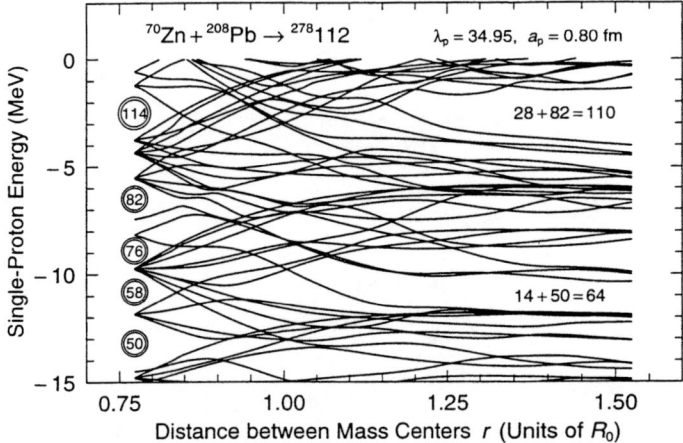

FIGURE 6. Proton single-particle level diagram for merging nuclei in an asymmetric heavy-ion collision leading to the heaviest known nucleus. The intersecting-sphere parameterization is used for the entire path from touching spheres to a single sphere. Thus, the level-diagram path does not pass exactly through the calculated ground-state shape. The asymmetric configurations in the entrance channel lead to a mixing of states with odd and even parity. The magic-fragment gaps associated with the initial entrance-channel configuration remain far inside the touching point, to about $r/R_0 = 1.15$, somewhat outside the maximum in the fusion barrier.

In Fig. 2 we have superimposed the location of the most recently discovered heavy element $^{277}112$ and its α-decay daughters [19] on calculated microscopic corrections obtained in the FRDM (1992) [8]. The α-decay Q-values are plotted in Fig. 3 and compared to three theoretical calculations [8,10,20,21]. Two experimental decay chains have been observed. These decay chains provide for the first time the half-life of a nucleus, ^{271}Hs, at the center of the predicted rock of stability. The measured half-life of about 10 s is in excellent agreement with the half-life of 630 s predicted by the FRDM (1992) [20]. The agreement between these experimental observations and theoretical predictions confirms the predictive powers of current nuclear-structure models and represents a triumph for nuclear physics.

III THE COLD-FUSION ENTRANCE CHANNEL

The six heaviest-known elements were all produced in cold-fusion reactions with doubly magic $^{208}_{82}\text{Pb}_{126}$ or near-doubly magic $^{209}_{83}\text{Bi}_{126}$ targets. The cold-fusion reaction has long been thought to enhance heavy-element evaporation-residue cross sections primarily because it leads to compound nuclei of low excitation energy, which enhances de-excitation by neutron emission relative to fission. Higher excitation energies would lead to higher fission probabilities. However, the evaporation-residue cross section is the product of the cross section for compound-nucleus formation and the probability for de-excitation by neutron emission. One may therefore ask if cold fusion *also* enhances the cross section for compound-nucleus formation. Because of the low excitation energies in the entrance channel, the large negative shell correction associated with target nuclei near the doubly magic ^{208}Pb should be almost fully manifested at touching and slightly inside touching.

Nuclei near ^{258}Fm have already provided important insight into fragment shell effects in symmetric fission and fusion configurations [15,22,23]. At ^{258}Fm fission becomes symmetric with a very narrow mass distribution, the kinetic energy of the fragments is about 35 MeV higher than in the asymmetric fission of ^{256}Fm, and the spontaneous-fission half-life is 0.38 ms for ^{258}Fm compared to 2.86 h for ^{256}Fm. These features are well understood in terms of the macroscopic-microscopic model. Shell effects associated with division into fragments near ^{132}Sn lower the fusion valley at touching by about 20 MeV in the most favorable case relative to that in macroscopic model. This fragment shell effect remains important far inside the touching point and results in fission into the fusion valley with very

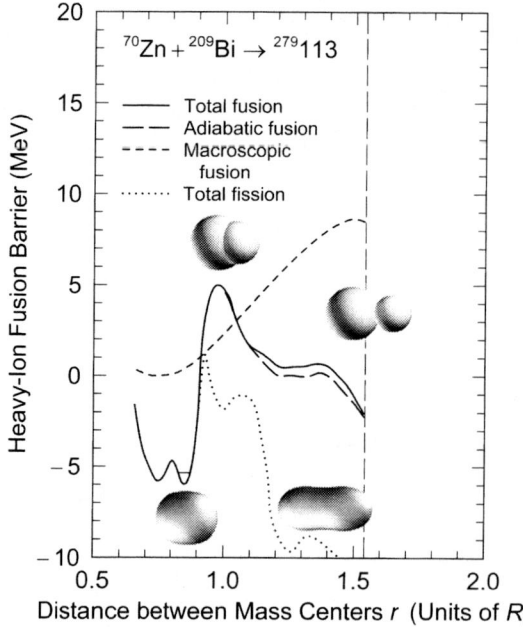

FIGURE 7. Total, adiabatic, and macroscopic fusion barriers for the cold-fusion reaction $^{70}\text{Zn} + {}^{209}\text{Bi} \rightarrow {}^{279}113$ and fission barrier corresponding to spontaneous fission from the ground state.

compact *cold* fragments for several fissioning nuclei in the vicinity of ^{258}Fm. The maximum effect of fragment shells in symmetric fission occurs in the fission of the hypothetical nucleus ^{264}Fm. In Fig. 4 we show a set of shapes leading from the nuclear ground state towards elongated scission configurations in the upper right part of the figure and to a compact scission configuration in the lower right part of the figure. The corresponding potential-energy surface calculated in the FRDM (1992) is shown in Fig. 5. The deep valley to the lower right, corresponding to cold fission into spherical fragments or to cold fusion of two ^{132}Sn nuclei, is very prominent and extends to the inner fission saddle near the ground state. The deformation coordinates r and σ correspond to the distance between the centers of mass of the two parts of the system and to the sum of the root-mean-square extensions along the symmetry axis of the mass of each half of the system about its center of mass, respectively.

We now show that shell effects are also very important in the fusion entrance channel in cold-fusion heavy-ion reactions, which usually involve *asymmetric* projectile-target combinations. In Fig. 6 we show calculated proton single-particle levels for merging, intersecting spheres in terms of the r shape coordinate for the reaction ^{70}Zn + ^{208}Pb → ^{278}112. This represents the reaction employed to reach the heaviest nucleus known thus far. We note that the magic-fragment gap combination 28 + 82 = 110 remains far inside the touching point, up to about $r/R_0 = 1.15$. The quantity R_0 is the radius of the spherical compound system. Because of the stability of the fragment gaps during the merging of the two nuclei, excitation should be minimal until late in the fusion process, which should favor evaporation-residue formation. These results are in excellent agreement with the results of calculations related to the symmetric fission of nuclei near ^{258}Fm into symmetric spherical fragments near ^{132}Sn.

To quantitatively study the effect of the persistent magic-fragment gaps on the fusion barrier as the heavy ions merge, we have calculated the fusion barrier for intersecting spheres for the proposed reaction
$$^{70}\text{Zn} + ^{209}\text{Bi} \rightarrow {}^{279}113$$
which is shown in Fig. 7. Just inside the peak in the fusion barrier at about $r/R_0 = 1.0$ we have switched from the intersecting-sphere parameterization to Nilsson's perturbed-spheroid ϵ parameterization so that we accurately obtain the energy of the ground state. Such a switch is not carried out in the calculation of the level diagram, so the level-diagram path does not pass exactly through the calculated ground-state shape. The calculated ground-state shape is indicated in Fig. 7. We also show the touching configuration and one intersecting-sphere configuration at $r/R_0 = 1.0$, near the maximum in the fusion barrier. The dotted line shows the calculated fission barrier, for which we considered ϵ_2, ϵ_4, and ϵ_6 shape distortions. The effect of mass asymmetry on the fission barrier is expected to be small. The fusion barrier in the macroscopic FRDM without any shell effects is given by the short-dashed line. The touching configuration is indicated by a thin vertical long-dashed line. The thicker long-dashed line is the calculated adiabatic barrier without any specialization energy. Despite the fairly high spin 9/2 of the ^{209}Bi ground state the calculated specialization energy is quite low. The adiabatic curve is shown only from touching to about

Fusion configurations for a spherical projectile

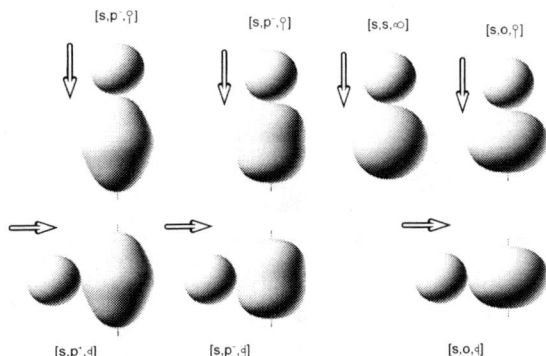

FIGURE 8. Seven touching configurations for heavy-ion collisions with spherical projectiles and general targets.

$r/R_0 = 1.0$. However, the specialization energy is quite low also between $r/R_0 = 1.0$ and the ground-state shape. For the cold-fusion reactions that have resulted in the formation of the elements $Z = 107$–112, our calculations show that the incident energy resulting in maximum 1n cross section corresponds to an energy just a few MeV above the calculated total fusion barrier.

IV SUPERHEAVY ELEMENTS BY HOT-FUSION REACTIONS

When hot-fusion reactions are used to produce heavy elements, these reactions usually involve a deformed actinide target and a spherical, light projectile. The combination of a spherical projectile and deformed target is illustrated for various representative target deformations in Fig. 8. For such configurations the fusion barrier is not one-dimensional, but two-dimensional [24,25]. More generally, both the projectile and targets may be deformed. We illustrate the case of prolate, negative-hexadecapole targets and projectiles in Fig. 9. In this case the full characterization of the potential between the deformed projectile and the deformed target leads to a four-dimensional potential-energy surface. However, major features of this multi-dimensional potential-energy surface can be obtained from a calculation of the barrier for the five limiting configurations shown in Fig. 9.

As limiting orientations we consider only situations where the projectile center is on the x, y, or z axis of the target and orientations of the projectile where the projectile symmetry axis is either parallel to or perpendicular to the target symmetry axis. Since we restrict ourselves to axial symmetry, configurations with the projectile center located on the x or y axis are identical. If the projectile is located in the equatorial region of the target it can be oriented in three major orientations, and if it is located in the polar region it can be oriented in two major orientations. Thus, for a particular projectile-target deformation combination there are five possible limiting configurations. Because compact touching configurations are thought to favor compound-nucleus formation, a particularly favorable configuration could be the equatorial-cross configuration, which is the configuration shown to the right in the bottom row of Fig. 9. For the prolate, negative-hexadecapole targets and projectiles shown in this figure we call this configuration the "hugging" configuration. Another close-approach configuration would involve an oblate target and an oblate projectile in a polar-parallel configuration. The fusion-barrier configurations in deformed heavy-ion collisions are discussed in greater detail in Ref. [25].

To reach element $Z = 114$ the reaction ^{48}Ca + ^{244}Pu → ^{292}114 is being considered. In Fig. 10 we show our calculated two-dimensional fusion potential-energy surface for this reaction. The target is centered at the origin. The fusion potential is shown for locations (ρ, z) of the center of the projectile, where ρ is the distance from the symmetry axis z. Since the projectile center is some distance away from the target surface when the target and

Fusion configurations of deformed nuclei

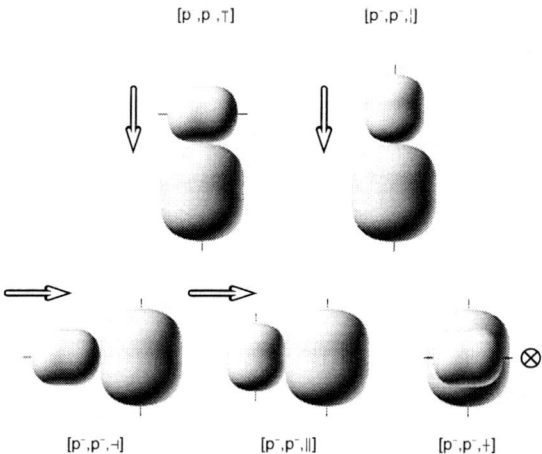

FIGURE 9. Five touching configurations for heavy-ion collisions with prolate, negative-hexadecapole projectiles and targets.

projectile surfaces touch, the energy function that we use is not defined inside the gray area for the separated projectile-target configuration. It is immediately clear from the figure that there is a substantial difference between the polar barrier height of 193.3 MeV and the equatorial barrier height of 208.0 MeV.

We have earlier argued that compact hugging configurations are desirable for evaporation-residue formation [25]. That is, targets and projectiles should have positive values of ϵ_4 (negative hexadecapole moments) so that waistlines develop. This permits a close approach in the equatorial-cross orientation. Suitable targets would then be nuclei in the rare-earth region starting at approximately $Z = 68$. Corresponding projectiles would then be in the range $Z = 42$–50. One may, for example, consider a reaction with a spherical projectile: ^{124}Sn + ^{176}Yb → ^{300}120, for which the equatorial barrier is 378.2 MeV. This means that only two neutrons would be emitted for collisions at the barrier. An example of a true hugging configuration with negative-hexadecapole target and projectile shapes is ^{114}Cd + ^{180}Hf → ^{294}120. Here, the *hugging* barrier is 381.8 MeV and the number of evaporated neutrons is four. The distance between mass centers for this configuration at touching is 11.08 fm. Reactions with oblate projectiles exhibit interesting features. An example is ^{116}Cd + ^{180}Hf → ^{296}120, for which the polar-parallel barrier is 365.6 MeV, corresponding to a 2n evaporation process for collisions at the barrier. The touching distance is 11.91 fm for this orientation. For the equatorial-transverse orientation the touching distance is only 10.49 fm, but the barrier is 389.3 MeV for this orientation, corresponding to the emission of five neutrons.

The cold-fusion process between spherical projectiles and targets has led to the discovery of six new elements. Some features of the cold-fusion process are well understood today but other important aspects remain to be explored. In particular, we need to understand how the evaporation-residue cross section depends on the projectile and target species and on the reaction energy. The exploration of reactions between spherical or deformed projectiles and deformed targets leading to heavy elements also presents a fascinating challenge for the future. Here we need to understand also the influence of deformation and relative orientation of the target and projectile on the evaporation-residue cross section. Also, in this case one can find fairly cold reactions leading to the evaporation of relatively few neutrons. In the future, some aspects of these reactions will perhaps be explored in a radioactive-ion-beam facility with polarized targets.

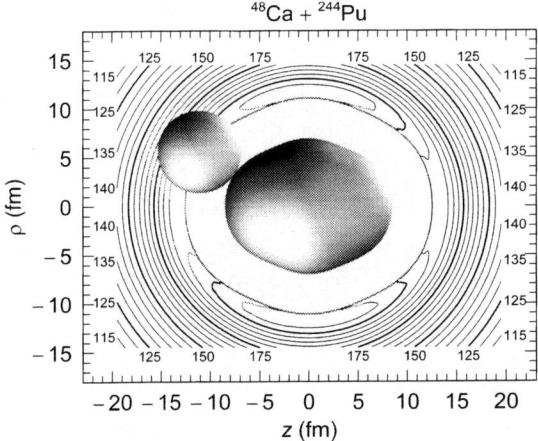

FIGURE 10. Calculated potential-energy surface in units of MeV for the reaction ^{48}Ca + ^{244}Pu. The energy in the medium-gray area outside the target nucleus in the center was not calculated, because the points in this region correspond to points inside the touching configuration. Note the ridge with saddle points and peaks around the target.

References

1) H. W. Meldner, unpublished (1965).

2) H. W. Meldner, Proc. Int. Symp. on why and how to investigate nuclides far off the stability line, Lysekil, 1966, Ark. Fysik **36** (1967) 593.

3) W. D. Myers and W. J. Swiatecki, Ark. Fys. **36** (1967) 343.

4) S. G. Nilsson, J. R. Nix, A. Sobiczewski, Z. Szymański, S. Wycech, C. Gustafson, and P. Möller, Nucl. Phys. **A115** (1968) 545.

5) S. G. Nilsson, C. F. Tsang, A. Sobiczewski, Z. Szymański, S. Wycech, C. Gustafson, I.-L. Lamm, P. Möller, and B. Nilsson, Nucl. Phys. **A131** (1969) 1.

6) J. R. Nix, Ann. Rev. Nucl. Sci. **22** (1972) 65.

7) M. Brack, J. Damgaard, A. S. Jensen, H. C. Pauli, V. M. Strutinsky, and C. Y. Wong, Rev. Mod. Phys. **44** (1972) 185.

8) P. Möller, J. R. Nix, W. D. Myers, and W. J. Swiatecki, Atomic Data Nucl. Data Tables **59** (1995) 185.

9) J. Duflo, Nucl. Phys. **A576** (1994) 29.

10) X.-L. Han, C.-L. Wu, D. H. Feng, and M. W. Guidry, Phys. Rev. **C45** (1992) 1127.

11) M. Bolsterli, E. O. Fiset, J. R. Nix, and J. L. Norton, Phys. Rev. **C5** (1972) 1050.

12) P. Möller and J. R. Nix, Nucl. Phys. **A229** (1974) 269.

13) P. Möller and J. R. Nix, Nucl. Phys. **A361** (1981) 117.

14) P. Möller and J. R. Nix, Atomic Data Nucl. Data Tables **26** (1981) 165.

15) P. Möller and J. R. Nix, J. Phys. G: Nucl. Part. Phys. **20** (1994) 1681.

16) W. J. Swiatecki, Phys. Rev. **100** (1955) 937.

17) J. Randrup, C. F. Tsang, P. Möller, S. G. Nilsson, and S. E. Larsson, Nucl. Phys. **A217** (1973) 221.

18) J. Randrup, S. E. Larsson, P. Möller, S. G. Nilsson, K. Pomorski, and A. Sobiczewski, Phys. Rev. **C13** (1976) 229.

19) S. Hofmann, N. Ninov, F. P. Heßberger, P. Armbruster, H. Folger, G. Münzenberg, H. J. Schött, A. G. Popeko, A. V. Yeremin, S. Saro, R. Janik, and M. Leino, Z. Phys. **A354** (1996) 229.

20) P. Möller, J. R. Nix, and K.-L. Kratz, Atomic Data Nucl. Data Tables **66** (1997) 131.

21) Y. Aboussir, J. M. Pearson, A. K. Dutta, and F. Tondeur, Atomic Data Nucl. Data Tables **61** (1995) 127.

22) P. Möller, J. R. Nix, and W. J. Swiatecki, Nucl. Phys. **A469** (1987) 1.

23) P. Möller, J. R. Nix, and W. J. Swiatecki, Nucl. Phys. **A492** (1989) 349.

24) P. Möller and A. Iwamoto, Nucl. Phys. **A575** (1994) 381.

25) A. Iwamoto, P. Möller, J. R. Nix, and H. Sagawa, Nucl. Phys. **A596** (1996) 329.

Superheavy nuclei in selfconsistent nuclear models

M. Bender[a], K. Rutz[a,b], T. Bürvenich[a], T. Schilling[a],
P.-G. Reinhard[c,d], J. A. Maruhn[a,d], and W. Greiner[a,d]

[a] *Institut für Theoretische Physik, Universität Frankfurt,*
Robert-Mayer-Str. 10, D-60325 Frankfurt am Main, Germany.

[b] *Gesellschaft für Schwerionenforschung mbH,*
Planckstr. 1, D-64291 Darmstadt, Germany.

[c] *Institut für Theoretische Physik, Universität Erlangen,*
Staudtstr. 7, D-91058 Erlangen, Germany.

[d] *Joint Institute for Heavy-Ion Research, Oak Ridge National Laboratory,*
P. O. Box 2008, Oak Ridge, TN 37831, U.S.A.

Abstract. The shell structure of superheavy nuclei is investigated within various parametrizations of relativistic and nonrelativistic nuclear mean-field models. The heaviest known even-even nuclei are used as a benchmark to estimate the predictive value of the models. From that starting point, spherical and deformed shell closures in the superheavy region are searched.

I INTRODUCTION

The possible existence of islands of shell-stabilized superheavy nuclei has been an inspiring problem in heavy-ion physics for almost three decades. Recent experiments at GSI [1] and Dubna [2] brought innovations by producing isotopes at and in the vicinity of the *deformed* doubly magic nucleus $^{270}_{162}\text{Hs}_{108}$, as theoretically verified in macroscopic-microscopic models [3,4]. The ultimate goal remains the spherical doubly magic superheavy nucleus $^{298}_{184}114$ which was predicted in the earliest macroscopic-microscopic investigations [5,6] and confirmed in more recent models of this type [3,4].

It is the aim of this contribution to scan a wide region of superheavy nuclei for occurrence of spherical and deformed magic shells within the framework of the relativistic mean-field model (RMF) [7] and within the nonrelativistic Skyrme–Hartree–Fock (SHF) approach [8]. The extrapolation towards superheavy nuclei challenges the predictive power of nuclear structure models. The macroscopic-microscopic method requires preconceived knowledge about the expected densities and single-particle potentials. But this knowledge fades away when stepping into new regions where strong polarization effects may occur. The selfconsistent nuclear models, on the other hand, determine potentials and deformations unambiguously. These models are nowadays capable of describing all known nuclei from ^{16}O on with satisfying quality by fixing a handful of model parameters [7,9,10]. There remain, however, several loosely fixed aspects in these parametrizations which amplify as uncertainties in extrapolations, e.g., to nuclei near the drip line [11] or to superheavy nuclei as discussed here.

II THE FRAMEWORK

In view of the uncertainties in extrapolations, we consider a broad selection of parametrizations with about comparable quality concerning normal nuclear properties but differences in less well determined aspects (spin-orbit, isovector properties).

For the nonrelativistic SHF calculations we consider the parametrizations SkM* [12], SkI1 [10], SkP [13], and SLy6 [9] which all employ the standard form but differ in bias. The force SkP uses effective mass $m^*/m = 1$ and is designed to allow a selfconsistent treatment of pairing. The other forces all have smaller effective masses around $m^*/m = 0.7$–0.8. The force SkM* was first to deliver acceptable incompressibility and fission properties. The force SLy6 stems from an attempt to cover properties of pure neutron matter together with normal nuclear ground state properties. The force SkI1 stems from a recent systematic fit already embracing data from exotic nuclei. The forces SkI3 and SkI4 are fitted exactly as SkI1 but using a variant of the Skyrme parametrization where the spin-orbit force is complemented by an explicit isovector degree-of-freedom [10]. They are designed to overcome the different isovector trends of spin-orbit coupling between conventional Skyrme forces and the RMF. SkI3 contains a fixed isovector part exactly analogous to the RMF, whereas SkI4 is adjusted allowing free variation of the isovector spin-orbit force. The modified spin-orbit force has a strong effect on the spectral distribution in heavy nuclei and thus even more influence for the predictions of superheavy nuclei. For the RMF we consider the parametrizations NL–Z [14], PL–40 [15], NL–SH [16], and TM1 [17]. The force NL–Z aims at a best fit to nuclear ground state properties for the standard nonlinear ansatz [7]. The force PL–40 is a similar fit, but with a stabilized form of the scalar nonlinear selfcoupling. It shares most properties with NL–Z, as the good reproduction of ground state properties. But PL–40 is somewhat more appropriate in the regime of small densities at the outer nuclear surface. The force NL–SH also employs the standard ansatz, but was adjusted with a bias to isotopic trends. Finally, the force TM1 includes a nonlinear selfcoupling of the vector field

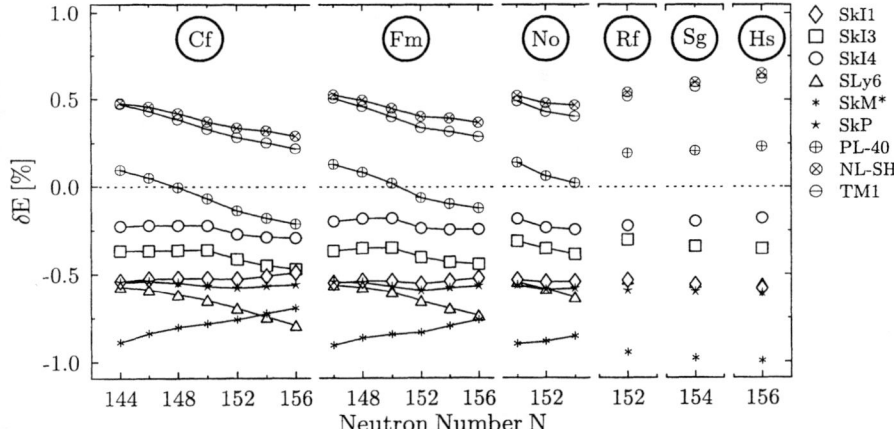

FIGURE 1. Relative error of the binding energy for the isotope chains of the heaviest known even–even nuclei, calculated in an axially deformed representation with the forces as indicated.

as well, and is fitted in the same way as NL–SH. The features of the forces subject to our investigation are discussed in more detail in [18].

In both, SHF and RMF, the pairing correlations are treated in the BCS scheme using a delta pairing force $V_{\text{pair}} = V_{p/n} \delta(\boldsymbol{r}_1 - \boldsymbol{r}_2)$. The strengths V_p for protons and V_n for neutrons are adjusted for every mean-field parametrization separately to the pairing gaps in chains of semimagic nuclei [19].

The numerical procedure solves the coupled SHF and RMF equations on a grid in coordinate space with the damped gradient iteration method [20].

III COMPARISON FOR EXPERIMENTALLY KNOWN SUPERHEAVY NUCLEI

The question is how all these parametrizations which are comparable for stable nuclei perform when extrapolating to the new area of superheavy nuclei. A first answer comes from comparison of the ground states from deformed mean-field calculations with the recently measured, heaviest known even-even nuclei which are located between the deformed neutron shell closure at $N = 152$ and a region of enhanced stability in the vicinity of $^{270}_{162}\text{Hs}$, for which the occurrence of a doubly deformed shell closure is predicted within macroscopic–microscopic models [3,4].

Figure 1 shows the relative error $\delta E = (E_{\text{calc}} - E_{\text{exp}})/E_{\text{exp}}$ between the calculated and the experimental value of the binding energy for the isotope chains of these nuclei. The experimental binding energies are taken from [21]. Negative values of the error indicate under-bound nuclei. Although all forces

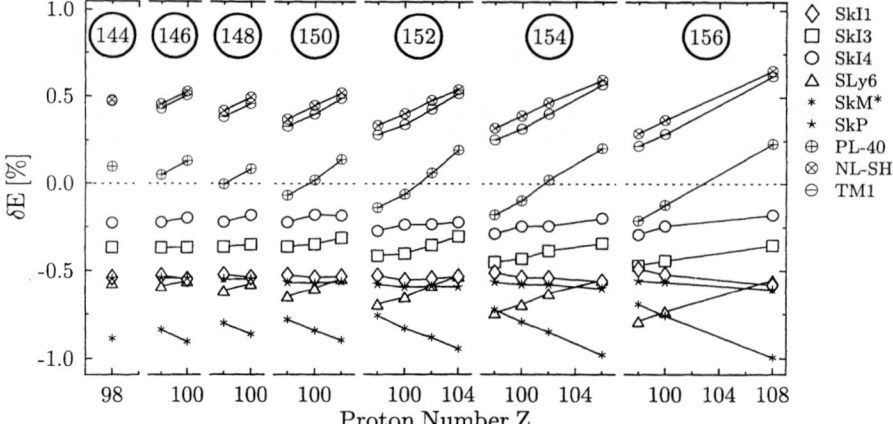

FIGURE 2. The same as in figure 1, but for isotone chains with the neutron number as indicated.

show acceptable quality in these extrapolated results, there are clearly visible differences in the performance.

The Skyrme forces with the old standard spin-orbit coupling lead all to somewhat under-bound superheavy nuclei. The isovector-extended spin-orbit coupling in SHF (the forces SkI3 and SkI4) produces much better binding energies confirming once again that there is some truth in the relativistic isovector mix of the spin-orbit coupling. This is corroborated by the equally good results for the RMF parametrizations NL–Z and PL–40. The results from NL–Z are so close to those of PL–40, that we have displayed only one case. Although NL–Z and PL–40 show the smallest deviation from the experimental values for the considered set of nuclei, the absolute value of the bindings energy rises not fast enough with increasing neutron number, a feature that all RMF parameter sets share. Figure 2 shows the same data as figure 1, but plotted for isotone chains. The behavior is much similar to the previous case of isotopic chains.

At second glance at figures 1 and 2, however, we see significant differences in the isotopic and isotonic trends. Most SHF forces produce flat curves which means that the error has no trend and remains about the same throughout the whole region (a curious exception is SLy6 where the mismatch in trends is surprising because this force had been fitted with emphasis on neutron rich nuclei). All RMF parametrizations, on the other hand, show strong isotopic and isotonic trends in the errors of the binding energies. Even the attempt with NL–SH to adjust particularly exotic nuclei [16] did not cure these problems. This deviation of the trends hints at an essential difference in the isovector properties between SHF and RMF, yet to be understood.

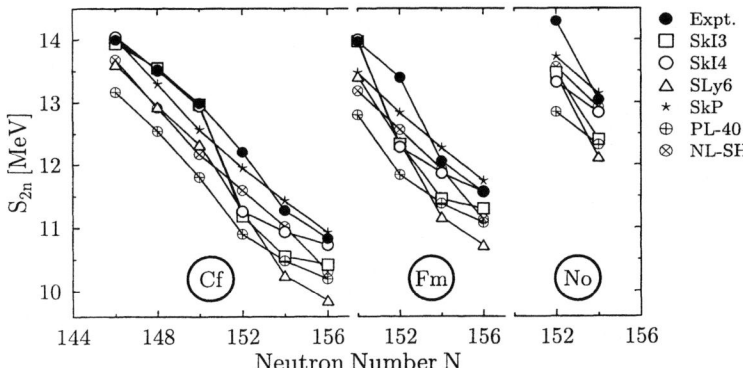

FIGURE 3. Two–neutron separation energy S_{2n} in units of MeV for the experimentally known superheavy even–even nuclei.

Isotopic and isotonic trends of the binding energy are better quantified in terms of the two–nucleon separation energies, i. e.,

$$S_{2n}(N,Z) = B(N,Z) - B(N-2,Z) \quad . \tag{1}$$

Figure 3 shows the two–neutron separation energies S_{2n} of the heaviest even–even nuclei. The SHF models are generally closer to the data than the RMF models. That is the same feature as seen from the slopes in figure 1. Drawing the energy differences, however, reveals more details. Note the jump of the S_{2n} between $N=150$ and 152 for most of the SHF models except SkP, which indicates a shell closure at $N=150$. The RMF models, on the other hand, show no shell closure in that region at all, and the data hint such a closure rather at $N=152$, that becomes more pronounced with increasing charge number. This detailed quantity is thus a very critical test of models.

IV SPHERICAL MAGIC SHELLS IN LARGER SUPERHEAVY NUCLEI

The most interesting feature for even larger systems is the possible occurrence of new spherical doubly magic nuclei. We have seen that a sudden jump in the two-nucleon separation energies (1) indicates a a shell closure. An even better signal is provided by the second difference of binding energies, the two-nucleon gaps

$$\begin{aligned}\delta_{2p}(N,Z) &= 2B(N,Z) - B(N,Z-2) - B(N,Z+2) \quad , \\ \delta_{2n}(N,Z) &= 2B(N,Z) - B(N-2,Z) - B(N+2,Z) \quad , \end{aligned} \tag{2}$$

which show a pronounced peak for magic numbers. It is to be noted that the amplitude of shell effects decreases with increasing system size, due to the increasing level density.

We now explore very large systems where magic shells occur at spherical shape and can thus use the much simpler spherical calculations. Figure 4 shows the proton and neutron gaps for the chosen forces and for a large variety of superheavy even–even nuclei. Like for the comparison with the experimental data in section III, the results from NL–Z are so close to those of PL–40 that we have displayed only one case. In the following discussion we will consider the black squares (standing for the largest gaps) as indicators of a shell closure. The left column of figure 4 shows the proton gaps δ_{2p}. The isotopes of $Z=120$ have the most pronounced proton gaps in most cases. For SkI4 is $Z=114$ the preferred case, and SkM* and SkP favor $Z=126$. The right column of figure 4 shows the neutron gaps δ_{2n}. The dominant shell closure is $N=184$ which appears for all forces except for NL–SH and TM1. The forces PL–40 and NL–Z, on the other hand, deliver an alternative (and preferred) choice with $N=172$, which appears to be also the dominant shell closure for NL–SH and TM1.

The most interesting species are, of course, the doubly magic systems. These require a simultaneous occurrence of a large shell gap for the protons as well as for the neutrons. Such a coincidence is not trivial, as we see from the many forces where it cannot be found, namely SkI1, SkI3, SLy6, and TM1. The remaining parametrizations do predict doubly magic nuclei, however, at different places. The forces SkP and SkM* predict $Z=126$, $N=184$. The relativistic PL–40 parametrization predicts $Z=120$, $N=172$, whereas the nonrelativistic SkI4 prefers $Z=114$, $N=184$. Thus even our two preferred forces, SkI4 and PL–40, make conflicting predictions.

One sees in figure 4 that the proton shell closures for a given Z can change with varying neutron number N, and similarly the neutron shell closures can vary with changing proton numbers. A vivid example is the $Z=120$ shell computed with SkI1, which starts with closure, looses that property with increasing neutron number, and regains it later. In figure 5 we show the single proton spectra for this case. The quantity of interest is the shell gap at $Z=120$. Minimal relative changes of the single proton levels produce a regime of higher level density at the Fermi surface around $N=184$, the neutron number where the proton shell gap is lowest, see figure 4.

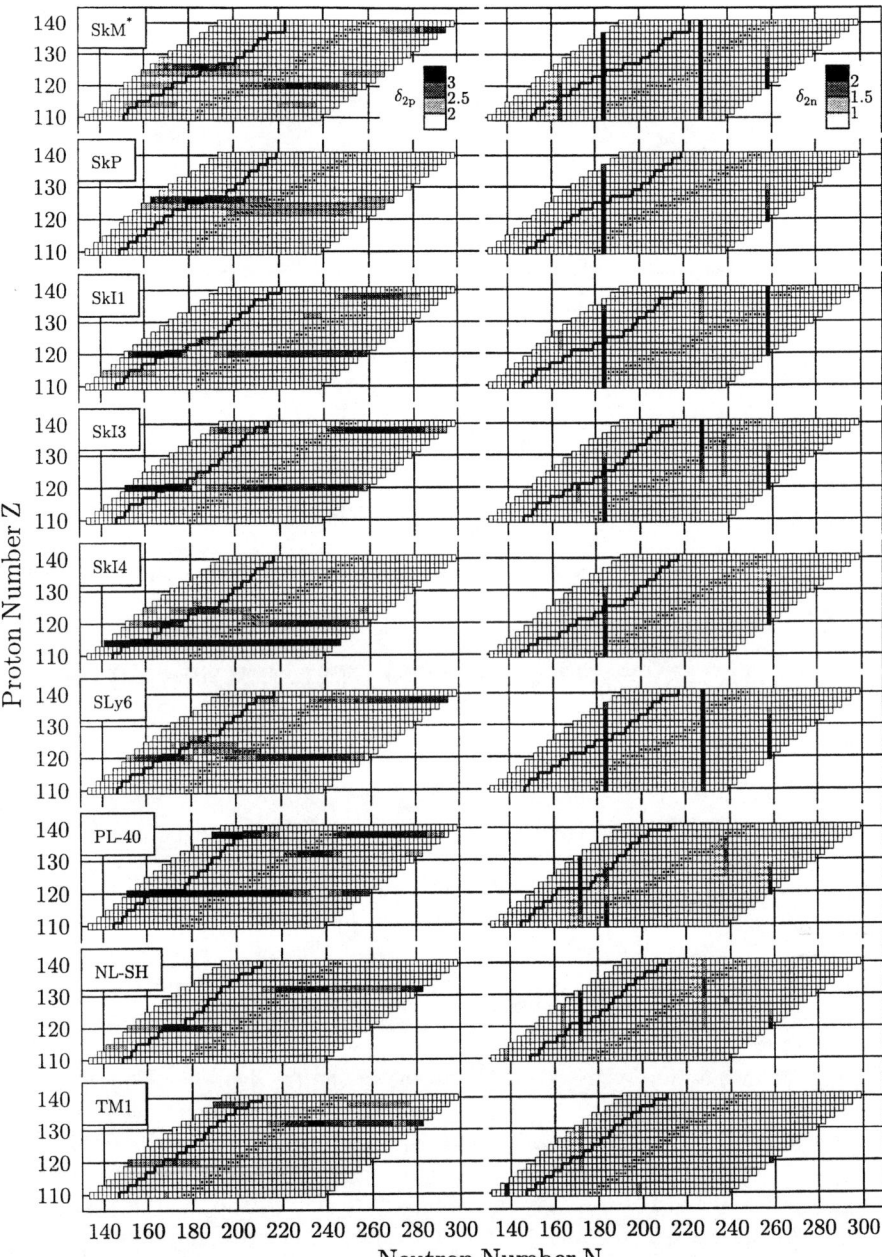

FIGURE 4. Grey scale plots of proton gaps (left column) and neutron gaps (right column) in the N-Z plane for spherical calculations with the forces as indicated. Nuclei that are stable with respect to β decay and the two-proton drip-line are emphasized.

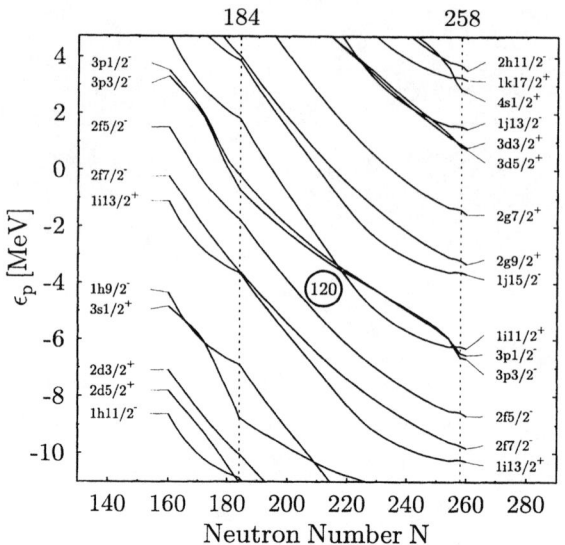

FIGURE 5. The single proton levels near the Fermi energy for the isotopes of $Z=120$ versus the neutron number, computed with SkI1.

V GROUND–STATE DEFORMATIONS OF SUPERHEAVY NUCLEI

Sphericity is only trivially granted for the doubly magic nuclei. In all other cases, only a full deformed calculation can answer the question of the appropriate ground state properties. Thus we have explored systematically the superheavy nuclei in the range up to $Z=128$ and $N=190$ with axially symmetric deformed mean-field calculations. The selection of forces has been limited to SkI4 and PL–40 which perform best in the region of known superheavy elements, see section III. A similar study for the forces SkP and SLy7 can be found in [22], for NL–SH in [23]. Figure 6 shows the two–proton gaps δ_{2p} (left column), two–neutron gaps δ_{2n} (middle column) and the corresponding relative quadrupole moment β_2 (right column) in the N–Z plane. The dimensionless relative quadrupole deformation is defined as $\beta_2 = 4\pi \langle r^2 Y_{20}\rangle/(3Ar_0^2)$ with $r_0 = 1.2\,A^{1/3}$ fm and provides a more immediate geometrical understanding than the quadrupole moment as such. We recognize first the shell closures at $Z=120$, or $Z=114$ respectively, which we had seen in the spherical calculations. But it is interesting to note that deformation can set on quickly when moving away from the doubly magic case. For example, the spherical proton shell closure at $Z=114$ for SkI4 disappears in an area of massive deformation at neutron numbers smaller than $N=178$, and the spherical neutron shell closure at $N=172$ for PL–40 vanishes immediately at proton numbers smaller

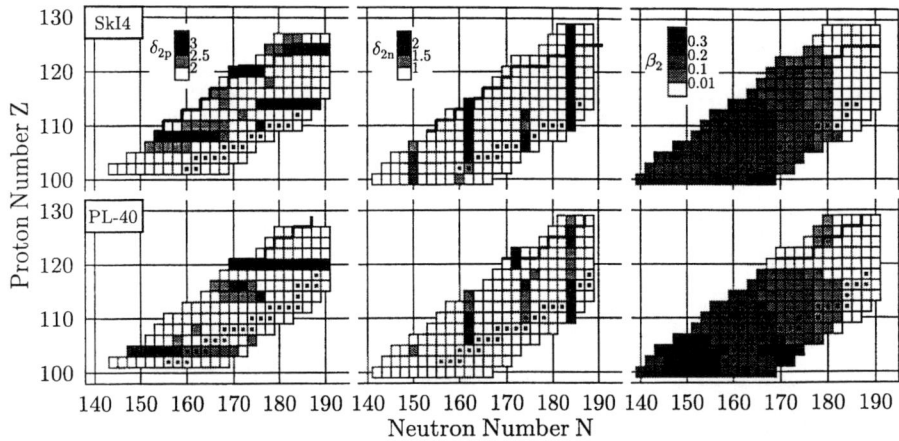

FIGURE 6. Two–proton gaps δ_{2p} (left column), two–neutron gaps δ_{2n} (middle column) and the corresponding relative quadrupole momenta β_2 (right column) in the N–Z plane, resulting from axially deformed calculations with the forces SkI4 and PL–40. The grey scales are the same like in figure 4.

than the magic $Z=120$, where the the nuclei are massively deformed in the ground state.

An interesting feature is the occurrence of deformed shell closures which appear here at smaller Z and N, namely $Z=108$ for SkI4 or $Z=104$ for PL–40 and $N=162$, and a softer closure at $N=150$ for SkI4 as already discussed in section III. These are located deep in the region of massive β_2 deformation. Like for the spherical proton shell closures, SkI4 and PL–40 give conflicting predictions also here. While PL–40 predicts $Z=104$ and $Z=116$, SkI4 prefers $Z=108$, leading to a deformed doubly magic ^{270}Hs in accordance with the macroscopic–microscopic models [3,4].

VI THE POTENTIAL ENERGY SURFACE OF ^{264}HS

In the previous section, we have looked at the ground state deformation. The full potential energy surface (PES) is also of interest as it allows to estimate the stability against spontaneous fission and the fusion path for the synthesis of these nuclei. As an example, Figure 7 shows the PES of ^{264}Hs, computed with SkI4 and a quadrupole constraint. The higher multipole deformations have been left free to adjust themselves to the minimum configuration, see e. g. [24].

The PES shows significant deviations from the fission barriers of actinides. There is no fission isomer and the second barrier vanishes. The fission path will follow the reflection symmetric solution, that gives a much narrower barrier

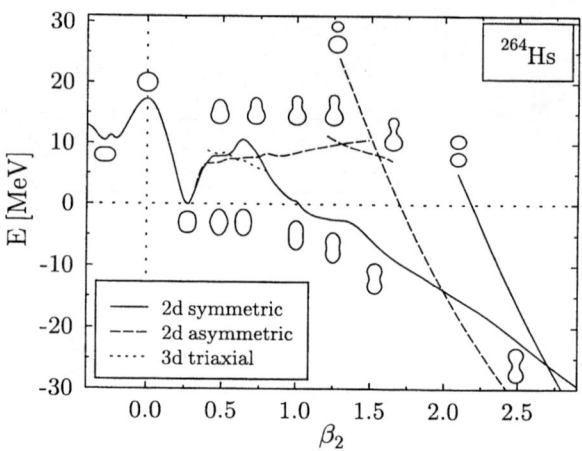

FIGURE 7. Valleys in the PES of ^{264}Hs, calculated with SkI4. The various shapes along the paths are indicated by the contours of the density at $\rho_0 = 0.07\,\text{fm}^{-3}$.

than the asymmetric solution. Although the first barrier has similar width and height as the first barrier of typical actinide nuclei, the absence of the second barrier will lower the lifetime against spontaneous fission dramatically. Like in the actinide region, the first barrier is a bit lowered if one allows for triaxial configurations. The reflection asymmetric solution does not lower the overall barrier, but it coexists far inside the barriers. It connects the asymptotically separated combination ^{210}Po + ^{54}Cr with the ground state, and corresponds to the fusion path. This combination of projectile and target differs only slightly from the experimentally successful choice ^{207}Pb(^{58}Fe, n)^{264}Hs [25]. A calculation in the RMF using PL–40 gives qualitatively the same results.

VII CONCLUSIONS

We have investigated the description of superheavy nuclei in the framework of relativistic and nonrelativistic nuclear mean-field models. A comparison with the binding and the separation energy of the heaviest experimentally measured even-even nuclei shows a clear preference for the standard relativistic forces NL–Z, PL–40, and relativistically corrected Skyrme forces SkI4 and SkI3. There are, however, significant differences in the isotopic and isotonic trends where the RMF have generally more problems to adjust to the experimental values. This is due to the rigidity of the RMF in the isovector channel. The most subtle and detailed quantities are the proton and neutron shell gaps. The different parametrizations make significantly different predictions for those shell gaps. The experimentally hinted soft shell gap at $N = 152$ is not reproduced by any model in this study. The shell structure of superheavy nuclei is thus a very critical probe for all mean-field models. New experimental information on superheavy nuclei will disclose the weaknesses of present approaches and help to improve on the models.

ACKNOWLEDGMENTS

The authors would like to thank S. Hofmann, G. Münzenberg, and D. Habs for many valuable discussions. This work was supported by Bundesministerium für Bildung und Forschung (BMBF), by Deutsche Forschungsgemeinschaft (DFG), by Gesellschaft für Schwerionenforschung (GSI), and by Graduiertenkolleg Schwerionenphysik. The Joint Institute for Heavy Ion Research has as member institutions the University of Tennessee, Vanderbilt University, and the Oak Ridge National Laboratory; it is supported by the members and by the Department of Energy through Contract No. DE-FG05-87ER40361 with the University of Tennessee.

REFERENCES

1. Hofmann, S., Ninov, V., Hessberger, F. P., Armbruster, P., Folger, H., Münzenberg, G., Schött, H. J., Popeko, A. G., Yeremin, A. V., Andreyev, A. N., Saro, S., Janik, R., and Leino, M., *Z. Phys.* **A350**, 277 (1995), *Z. Phys.* **A350**, 281 (1995), and *Z. Phys.* **A354**, 229 (1996).
2. Lazarev, Yu. A., Lobanov, Yu. V., Oganessian, Yu. Ts., Utyonkov, V. K., Abdullin, F. Sh., Polyakov, A. N., Rigol, J., Shirokovsky, I. V., Tsyganov, Yu. S., Iliev, S., Subbotin, V. G., Sukhov, A. M., Buklanov, G. V., Gikal, B. N, Kutner, V. B., Mezentsev, A. N., Subotic, K., Wild, J. F., Lougheed, R. W., and Moody, K. J., *Phys. Rev.* **C54**, 620 (1996).
3. Patyk, Z., and Sobiczewski, A., *Nucl. Phys.* **A533**, 132 (1991).
4. Möller, P., and Nix, J. R., *Nucl. Phys.* **A549**, 84, (1992); *J. Phys.* **G 20**, 1681, (1994).
5. Mosel, U., and Greiner, W., *Z. Phys.* **222**, 261 (1969).
6. Nilsson, S. G., Tsang, C. F., Sobiczewski, A., Szymanski, Z., Wycech, S., Gustafson, C., Lamm, I.-L., Möller, P., and Nilsson, B., *Nucl. Phys.* **A131**, 1 (1969).
7. Reinhard, P.-G., *Rep. Prog. Phys.* **52**, 439 (1989).
8. Quentin, P., and Flocard, H., *Ann. Rev. Nucl. Part. Sci.* **28**, 523 (1978).
9. Chabanat, E., Bonche, P., Haensel, P., Meyer, J., and Schaeffer, R., preprint, 1996.
10. Reinhard, P.-G., and Flocard, H., *Nucl. Phys.* **A584**, 467 (1995).
11. Nazarewicz, W., Dobaczewski, J., Werner, T. R., Maruhn, J. A., Reinhard, P.-G., Rutz, K., Chinn, C. R., Umar, A. S., and Strayer, M. R., *Phys. Rev.* **C53**, 740 (1996).
12. Bartel, J., Quentin, P., Brack, M., Guet, C., and Håkansson, H.-B., *Nucl. Phys.* **A386**, 79 (1982).

13. Dobaczewski, J., Flocard, H., and Treiner, J., *Nucl. Phys.* **A422**, 103 (1984).
14. Rufa, M., Reinhard, P.-G., Maruhn, J. A., Greiner, W., and Strayer, M. R., *Phys. Rev.* **C38**, 390 (1989).
15. Reinhard, P.-G., *Z. Phys.* **A 329**, 257 (1988).
16. Sharma, M. M, and Ring, P., *Phys. Rev.* **C45**, 2514 (1992).
17. Sugahara, Y., Toki, H., *Nucl. Phys.* **A579**, 557 (1994).
18. Rutz, K., Bender, M., Bürvenich, T., Schilling, T., Reinhard, P.-G., Maruhn, J. A., and Greiner, W., *Phys. Rev.* **C56**, 238 (1997).
19. Bender, M., Reinhard, P.-G., Rutz, K., and Maruhn, J. A., (in preparation).
20. Blum, V., Lauritsch, G., Maruhn, J. A., and Reinhard, P.-G., *J. Comp. Phys.* **100**, 364 (1992).
21. Audi, G., and Wapstra, A. H., *Nucl. Phys.* **A595**, 409 (1995).
22. Ćwiok, S., Dobaczewski, J., Heenen, P.-H., Magierski, P., and Nazarewicz, W., *Nucl. Phys.* **A611**, 211 (1996).
23. Lalazissis, G. A., Sharma, M. M., Ring, P., and Gambhir, Y. K., *Nucl. Phys.* **A608**, 202, (1996).
24. Rutz, K., Maruhn, J. A., Reinhard, P.-G., and Greiner, W., *Nucl. Phys.* **A590**, 680 (1995).
25. Münzenberg, G., Hofmann, S., Folger, H., Heßberger, F. P., Keller, J., Poppensieker, K., Quint, B., Reisdorf, W., Schmidt, K.-H., Schött, H.-J., Armbruster, P., Leino, M. E., and Hingmann, R., *Z. Phys.* **A317**, 235 (1984),

Decay Properties of Superheavy Nuclei

Robert Smolańczuk

Sołtan Institute for Nuclear Studies
Hoża 69, PL-00-681 Warszawa, Poland

Abstract. Theoretical results on α-decay energies, α-decay half-lives, dynamical fission barriers, as well as spontaneous-fission half-lives, for both deformed and spherical superheavy atomic nuclei, are presented and discussed. The calculations are based on the macroscopic-microscopic model and are performed in a multidimensional deformation space describing axially-symmetric nuclear shapes.

INTRODUCTION

The main aim of the present study is to present new theoretical results on decay properties of the hypothetical spherical superheavy nuclei [1]. We consider α-decay and spontaneous-fission properties bacause majority of spherical superheavy atomic nuclei are situated on the nuclear chart in, or close to, the area of β-stability and, therefore, α decay and spontaneous fission are the main decay modes for these nuclei. We should keep in mind, however, that the dominance of β decay may appear for those spherical superheavy nuclei which are located outside the area of β-stability and which have very large α-decay and spontaneous fission half-lives [2]. The area of β-stability for even-even superheavy nuclei is also presented in this paper.

We perform the calculations for even-even spherical superheavy nuclei with $Z = 104 - 120$ [1]. We use the model [3-5] which had some success in reproducing and predicting decay properties of deformed superheavy nuclei with $Z = 104 - 114$. For completeness, we show our results obtained earlier [4,6] for the latter ones. Although the calculations are performed for even-even systems, we also draw certain conclusions for odd-A and odd-odd superheavy nuclei. This is possible because the spontaneous-fission half-life increases considerably due to the effect of an unpaired nucleon [7] while the α-decay half-life is much less sensitive to this effect [8]. Moreover, we determine the boundaries of the region of superheavy nuclei which are expected to live long enough to be detected after the synthesis in a present-day experimental setup.

DESCRIPTION OF THE CALCULATION

All details of the calculation are given in Ref. [1]. Here, we only present its main points.

The potential energy and its dependence on deformation is calculated by means of the macroscopic-microscopic method. The Yukawa-plus exponential model [9] is taken for the macroscopic energy and the microscopic energy is obtained by using the Strutinsky method [10]. The single-particle levels, from which the microscopic energy is calculated, are obtained by the diagonalization of the Woods-Saxon single-particle hamiltonian [11] in the deformed harmonic oscillator basis. The Woods-Saxon potential with the univesal variant of parameters [11] is used.

The probability of the tunneling of the potential energy barrier by a nucleus along the fission trajectory and, consequently, the spontaneous-fission half-life, is calculated by means of the semiclasical WKB approximation. The fission trajectory is found in a dynamical way [12,13], i.e., it is determined by the minimization of the action integral which describes the penetration of a nucleus through the potential energy barrier in a multidimensional deformation space. The resistance of a nucleus against shape changes along any trajectory in a multidimensional deformation space is described by the effective inertia which is dependent on both the tensor of inertia and a shape of the trajectory. The tensor of inertia is calculated within the cranking model (with the inclusion of pairing) [14] and, therefore, it is sensitive to the shell structure of a nucleus. In the dynamical calculation of the spontaneous-fission half-lives the effective inertia besides the potential energy influences significantly the shape of the fission trajectory through the action integral. This results in a shorter dynamical fission trajectory than the static one, along which the potential energy is minimal, and, consequently, to a smaller spontaneous-fission half-life obtained in the dynamical calculation [3].

In the calculations of nuclear masses and, consequently, α-decay energies, we use the following values of the volume-asymmetry parameter κ_V, the charge-asymmetry parameter c_a and the constant a_0:

$$\kappa_V = 1.990 , \quad c_a = 0.572 \text{ MeV} , \quad a_0 = 11.0 \text{ MeV} . \tag{1}$$

These parameters appear in the macroscopic part of the mass formula and were refitted to 77 experimentally known masses [15] of even-even nuclei with $Z \geq 82$ and $N \geq 126$. This readjustment was necessary because of the use of the larger (four-dimensional) deformation space than the ones used in other macroscopic-microscopic calculations [9,16,17].

The α-decay energy Q_α for each nucleus is obtained by subtracting from its theoretical mass the theoretical mass of the daughter nucleus and the experimentally known mass of the α-particle. The the α-decay half-life T_α is calculated by means of the Viola and Seaborg formula [8] with parameters [6,5,1]:

$$a = 1.81040 , \quad b = -21.7199 , \quad c = -0.26488 , \quad d = -28.1319 . \tag{2}$$

They were adjusted to 58 even-even nuclei with $Z > 82$ and $N > 126$ for which both T_α and Q_α are measured [18]. Previous adjustments of a, b, c and d were done for a much smaller number of nuclei [8,19].

RESULTS AND DISCUSSION

Equilibrium Deformation Energy

Figure 1 shows a contour map of the equilibrium deformation energy E_{def} calculated for even-even nuclei with $Z = 82 - 120$ [1]. This quantity is the gain in nuclear energy due to the deformation of a nucleus. We consider the nuclei with $E_{\text{def}} \gtrsim 2$ MeV as well deformed. Most of nuclei with $Z = 82 - 120$ shown in Figure 1 are, or are expected to be, well deformed. We also predict that the superheavy nuclei synthesized so far are well deformed. The transitional ($E_{\text{def}} \lesssim 2$ MeV) and spherical ($E_{\text{def}} \approx 0$) superheavy nuclei are situated around the doubly magic nucleus $^{298}114_{184}$. The area of spherical superheavy nuclei shown in Figure 1 is restricted to the ones for which we calculate the α-decay and spontaneous-fission half-lives larger than 0.1 μs. This value is one order of magnitude less than the smallest half-life possible to measure after the synthesis of a superheavy nucleus in a present-day experimental setup.

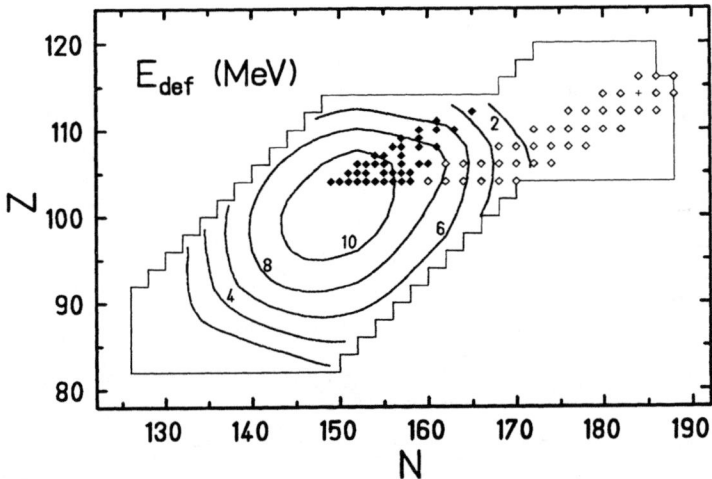

FIGURE 1. Contour map of the equilibrium deformation energy E_{def} for even-even nuclei with the proton number $Z = 82 - 120$ [1]. Numbers at contour lines give energy in MeV. The energy difference between neighboring contour lines is equal to 2 MeV. Full rhomb-shaped symbols denote the deformed superheavy nuclei synthesized so far. Even-even β-stable superheavy nuclei are indicated by open rhomb-shaped symbols and the β-stable doubly magic $^{298}114_{184}$ - by a cross.

One can see that the heaviest superheavy nucleus produced so far, $^{277}112_{165}$ [20], is located on the nuclear chart very close to the area of spherical superheavy nuclei. This means that the region od deformed superheavy nuclei touches the area of the hypothetical spherical superheavy nuclei. In other words we expect that both deformed and spherical superheavy nuclei constitute the continuation of the peninsula of known nuclei.

The area of β-stable even-even superheavy nuclei [1] is also shown in Figure 1. It contains pairs of nuclei with two smallest masses obtained for each isobaric chain of even-even nuclides with $A \geq 264$. However, it may happen for some isobaric chains of even-even nuclei that for more than two nuclides with the smallest masses the neighboring odd-odd isobars will have larger masses. In such a case the region of β-stable even-even superheavy nuclei will be a little larger than the one indicated in Figure 1.

Shell Effects

For both deformed and spherical superheavy nuclei large shell effects are obtained which stabilize these nuclei. It is clearly seen in Figure 2 which gives a contour map of the total shell correction energy E_{sh} [1]. This quantity is defined as the difference between the total potential energy for the equilibrium shape and the macroscopic part of the potential for the spherical shape. Since the spherical shape is the equilibrium point in the macroscopic (i.e., without shell structure) model, the total shell correction E_{sh} is the gain in energy of a nucleus due to its shell structure, including the effect of pairing.

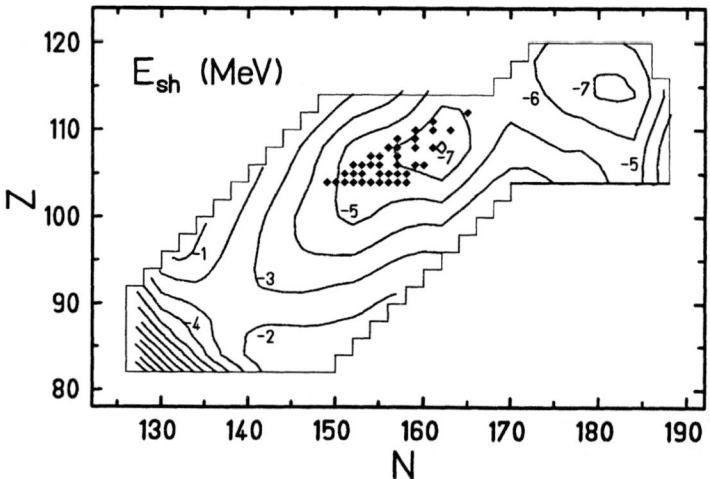

FIGURE 2. Contour map of the total shell correction energy E_{sh} for even-even nuclei with $Z = 82 - 120$ [1]. Numbers at contour lines give energy in MeV. The energy difference between neighboring contour lines is equal to 1 MeV. Rhomb-shaped symbols denote the deformed superheavy nuclei synthesized so far.

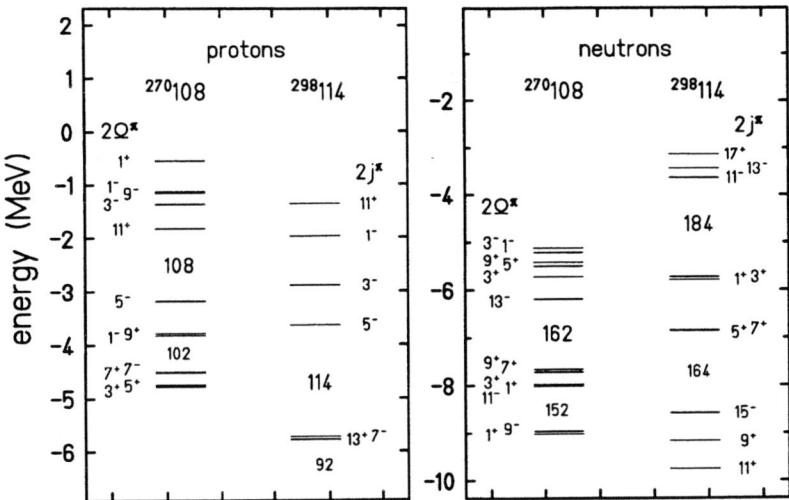

FIGURE 3. Proton and neutron single-particle-energy spectra calculated for the deformed $^{270}108_{162}$ ($^{270}\text{Hs}_{162}$) and spherical $^{298}114_{184}$ doubly magic superheavy nuclei [1]. At each single-particle level the absolute value of the projection of spin Ω, multiplied by two, on the symmetry axis, together with state's parity π ($^{270}108_{162}$) or the state parity π with the spin j, multiplied by two ($^{298}114_{184}$), are indicated.

Two minima of E_{sh} in the region of superheavy nuclei are shown in Figure 2 [1]. Both of them have similar depth exceeding slightly 7 MeV. They are obtained at the deformed $^{270}108_{162}$ ($^{270}\text{Hs}_{162}$) and for nuclei close to the spherical doubly magic $^{298}114_{184}$. These minima indicate the creation of large gaps at the Fermi level in the single particle spectra and, consequently, large nuclear shells which are shown in Figure 3 [1]. The energy gaps, between $Z = 114$ and 115 in the proton spectrum and between $N = 184$ and 185 in the neutron spectrum of the spherical nucleus $^{298}114_{184}$, are both equal to 2.1 MeV. Similar energy gaps, both equal to 1.4 MeV, are obtained between $Z = 108$ and 109 in the proton spectrum, and between $N = 162$ and 163 in the neutron spectrum for the not yet observed deformed nucleus $^{270}108_{162}$ ($^{270}\text{Hs}_{162}$). These features of the latter nucleus cause its increased stability against α decay and spontaneous fission and, therefore, we call this nucleus "deformed doubly magic" as distinct from the "traditional" doubly magic $^{298}114_{184}$, which is expected to be spherical.

α-Decay Energy and α-Decay Half-Life

Increased stability against α decay for nuclei with the deformed neutron magic number $N = 162$ is demonstrated in Figure 4 [1]. In this figure both the α-decay energy Q_α and the α-decay half-life T_α for elements $Z = 100 - 120$ are given as a function of neutron number N.

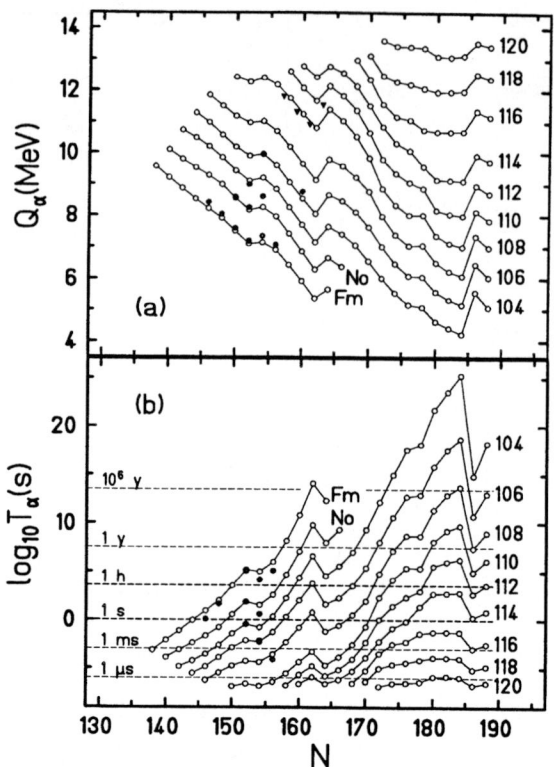

FIGURE 4. (a) Calculated alpha-decay energy Q_α, in MeV, as a function of neutron number N, for elements $100-120$ [1]. Experimental values taken from Refs. [18,21–25] are indicated by full circles, for even-even nuclei, and full triangles, for odd-N isotopes of the element 110.
(b) Calculated alpha-decay half-life T_α, in seconds, as a function of neutron number N, for elements $100-120$ [1]. Experimental values for even-even nuclei are indicated by full circles [18,26–28].

The effect of the deformed neutron shell is seen as minima of Q_α and corresponding maxima of T_α for particular elements at $N = 162$. This effect is weaker than that of the spherical neutron shell at $N = 184$. The effect of the deformed proton shell at $Z = 108$ manifests itself for nuclei with $N \approx 162$ as a larger gap between the curves describing Q_α versus N for $Z = 108$ and 110 in comparison to the gaps between other pairs of neighboring curves. This effect is comparable to that of the spherical proton shell at $Z = 114$ for nuclei with $N \approx 184$.

All measured α-decay energies for even-even nuclei [18,21] are shown in Figure 4(a) together with the α-decay energies for recently discovered odd-N isotopes of element 110, $^{269}110_{159}$ [22], $^{271}110_{161}$ [23] and $^{273}110_{163}$ [24]. In this figure, we also show the α-decay energy from a report [25] on the possible production of the isotope $^{267}110_{157}$. In one of the two α-decay chains observed at GSI-Darmstadt after the synthesis of the nucleus $^{277}112_{165}$ [20], the α-decay energy measured for its daughter nucleus, which is $^{273}110_{163}$, is very close to the value obtained in the joint Dubna-Livermore experiment [24]. The α-decay energies measured for odd-N isotopes of the element 110 with $N \approx 162$ show the same tendency as Q_α calculated for even-even isotopes of this element. This seems to confirm that the deformed neutron shell appears exactly at $N = 162$.

Fission Barrier and Spontaneous-Fission Half-Life

Increased stability against spontaneous fission for nuclei around $^{270}108_{162}$ (^{270}Hs$_{162}$) and $^{298}114_{184}$ is demonstrated in Figure 5 which gives a contour map of the calculated dynamical fission-barrier height B_f^{dyn} for even-even superheavy nuclei [1]. The highest barriers are obtained for nuclei around the minimum of E_{sh} created by the deformed shells at $Z = 108$ and $N = 162$, as well as for spherical nuclei around the minimum at $Z = 114$ and $N = 184$. Although the barriers calculated for the spherical nuclei are about 2 MeV lower than the barriers obtained for the deformed ones, they are considerably broader. For example, the barrier obtained for $^{298}114_{184}$ is by about 35 % broader than the one for $^{270}108_{162}$ (^{270}Hs$_{162}$). Due to the larger broadness of the barrier together with the larger effective inertia obtained for spherical superheavy nuclei they are considerably more stable against spontaneous fission in comparison to the deformed ones. It is shown in Figure 6 where the logarithm of the spontaneous-fission half-life T_{sf} for even-even superheavy nuclei is given [1]. One can see the local maxima of T_{sf} at $N = 162$ which are a consequence of the deformed neutron shell at this number. The much higher maxima of T_{sf} are obtained at the spherical neutron magic number $N = 184$. For the spherical doubly magic nucleus $^{298}114_{184}$ we obtain the value of $4.4 \cdot 10^5$ y for T_{sf} which is much larger than the values of the order of 1s - 1h calculated for deformed superheavy nuclei very close to the nucleus $^{270}108_{162}$ (^{270}Hs$_{162}$).

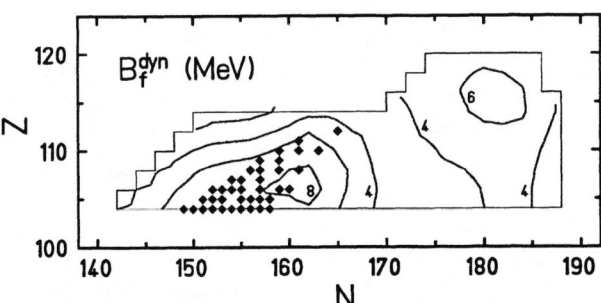

FIGURE 5. Contour map of the calculated dynamical fission-barrier height B_f^{dyn} for even-even nuclei with the atomic number $Z = 104 - 120$ [1]. Numbers at contour lines give energy in MeV. The energy difference between neighboring contour lines is equal to 2 MeV. Rhomb-shaped symbols denote the deformed superheavy nuclei synthesized so far.

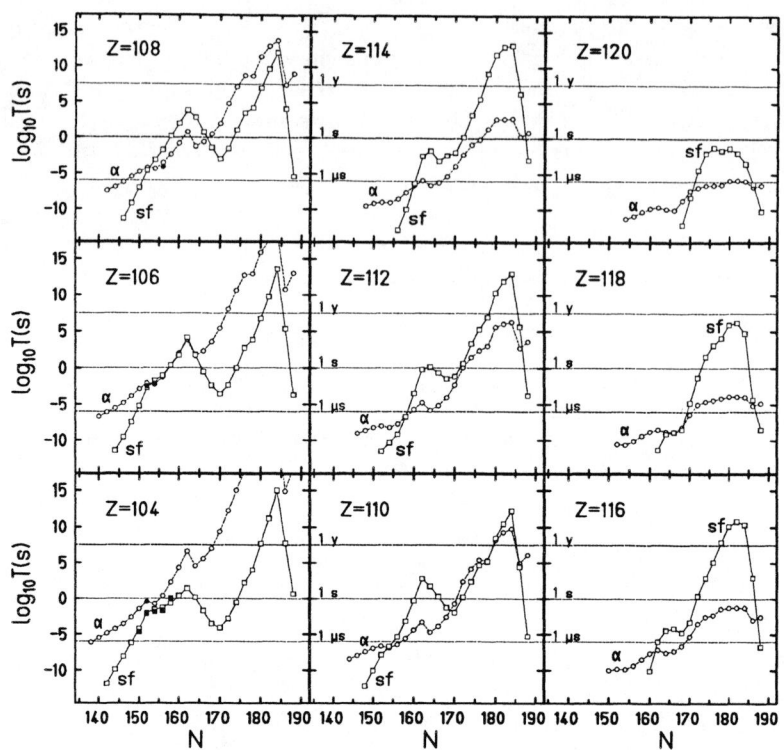

FIGURE 6. Dependence of logaritm of the calculated spontaneous-fission (sf) and alpha-decay (α) half-lives, given in seconds, on neutron number N, for elements 104 – 120 [1]. Experimental values [21,26–30] are indicated by full symbols.

CONCLUSIONS

The α-decay half-lives obtained in the present calculation [1] are smaller than 1 μs for even-even nuclei with $Z > 120$. The α-decay half-lives for odd-odd and odd-A isotopes of the element 121 may be comparable or even a little larger than those calculated for the neighboring even-even isotopes of the element 120 due to the effect of an unpaired nucleon. This means that the element 121 is probably the heaviest one which may be detected at present, if it would be synthesized.

The comparison of α-decay and spontaneous-fission half-lives shown in Figure 6 leads to the conclusion that α decay dominates for even-even deformed superheavy nuclei with $Z > 106$ and, roughly speaking, even-even spherical ones with $Z > 110$. This conclusion also holds for odd-A and odd-odd superheavy nuclei due to the effect of an unpaired nucleon.

The total half-lives obtained for many β-stable even-even isotopes of the elements $Z = 104 - 114$ are larger than 1 second which is the lower limit for radiochemical investigations [31]. We expect that the even-even nuclei with larger proton number live shorter than 1 second. This means that due to the effect of an unpaired nucleon the investigations of chemical properties of elements up to $Z = 115$ should be posible, if their spherical isotopes were synthesized.

According to the results given in Figure 6, many superheavy nuclei might be stored for a long time, if they were produced in experiment. For example, the total half-life calculated for the β-stable nucleus $^{292}110_{182}$ is equal to 51 years. The discovery of the long-living isotopes of the superheavy elements would open unique possibilities for chemistry and atomic physics which have at disposal large theoretical and experimental basis.

ACKNOWLEDGMENTS

The author is thankful to Janusz Skalski for many valuable discussions. Support by the Polish Committee for Scientific Research (KBN) through Grant No. 2 P03B 156 08 is gratefully acknowledged.

REFERENCES

1. Smolańczuk, R., *Phys. Rev.* **C56**, 812 (1997).
2. Tondeur, F., *Z. Phys.* **A288**, 97 (1978).
3. Smolańczuk, R., Klapdor-Kleingrothaus, H. V., and Sobiczewski, A., *Acta Phys. Pol.* **B24**, 685 (1993).
4. Smolańczuk, R., Skalski, J., and Sobiczewski, A., *Phys. Rev.* **C52**, 1871 (1995).
5. Smolańczuk, R., Skalski, J., and Sobiczewski, A., "Masses and half-lives of superheavy elements," in *Proceedings of the International Workshop XXIV on Gross Properties of Nuclei and Nuclear excitations "Extremes of Nuclear Structure"*, Hirschegg, Austria, 1996, edited by Feldmeier, H., Knoll, J., and Nörenberg, W. (GSI, Darmstadt, 1996), pp. 35-42.
6. Smolańczuk, R., and Sobiczewski, A., "Shell effects in the properties of heavy and superheavy nuclei," in *Proceedings of XV Nuclear Physics Divisional Conference "Low Energy Nuclear Dynamics"*, St. Petersburg, Russia, 1995, edited by Oganessian, Yu., Kalpakchieva, R., and von Oertzen, W. (World Scientific, Singapore, 1995), pp. 313-320.
7. Bjørnholm, S., and Lynn, J. E., *Rev. Mod. Phys.* **52**, 725 (1980).
8. Viola Jr., V. E., and Seaborg, G. T., *J. Inorg. Nucl. Chem.* **28**, 741 (1966).
9. Möller, P., and Nix, J. R., *At. Data Nucl. Data Tables* **26**, 165 (1981).
10. Strutinsky, V. M., *Nucl. Phys.* **A95**, 420 (1967); **A122**, 1 (1968).
11. Ćwiok, S., Dudek, J., Nazarewicz, W., Skalski, J., and Werner, T., *Comput. Phys. Commun.* **46**, 379 (1987).
12. Pauli, H. C., *Phys. Reports* **7C**, 35 (1973); *Nukleonika* **20**, 601 (1975).
13. Baran, A., Pomorski, K., Lukasiak, A., and Sobiczewski, A., *Nucl. Phys.* **A361**, 83 (1981).
14. Brack, M., Damgaard, J., Jensen, A. S., Pauli, H. C., Strutinsky, V. M., and Wong, C. Y., *Rev. Mod. Phys.* **44**, 320 (1972).

15. Audi, G., and Wapstra, A. H., *Nucl. Phys.* **A565**, 1 (1993).
16. Patyk, Z., and Sobiczewski, A., *Nucl. Phys.* **A533**, 132 (1991).
17. Ćwiok, S., Hofmann, S., and Nazarewicz, W., *Nucl. Phys.* **A573**, 356 (1994).
18. Buck, B., Merchant, A. C., and Perez, S. M., *At. Data Nucl. Data Tables* **54**, 53 (1993).
19. Sobiczewski, A., Patyk, Z., and Ćwiok, S., *Phys. Lett.* **B224**, 1 (1989).
20. Hofmann, S., Ninov, V., Hessberger, F. P., Armbruster, P., Folger, H., Münzenberg, G., Schött, H. J., Popeko, A. G., Yeremin, A. V., Saro, S., Janik, R., and Leino, M., *Z. Phys.* **A354**, 229 (1996).
21. Lazarev, Yu. A., Lobanov, Yu. V., Oganessian, Yu. Ts., Utyonkov, V. K., Abdullin, F. Sh., Buklanov, G. V., Gikal, B. N., Iliev, S., Mezentsev, A. N., Polyakov, A. N., Sedykh, I. M., Shirokovsky, I. V., Subbotin, V. G., Sukhov, A. M., Tsyganov, Yu. S., Zhuchko, V. E., Lougheed, R. W., Moody, K. J., Wild, J. F., Hulet E. K., and McQuaid, J. H., *Phys. Rev. Lett.* **73**, 624 (1994).
22. Hofmann, S., Ninov, V., Hessberger, F. P., Armbruster, P., Folger, H., Münzenberg, G., Schött, H. J., Popeko, A. G., Yeremin, A. V., Andreyev, A. N., Saro, S., Janik, R., and Leino, M., *Z. Phys.* **A350**, 277 (1995).
23. Hofmann, S., *GSI-Nachrichten* 02-95, 4 (1995).
24. Lazarev, Yu. A., Lobanov, Yu. V., Oganessian, Yu. Ts., Utyonkov, V. K., Abdullin, F. Sh., Polyakov, A. N., Rigol, J., Shirokovsky, I. V., Tsyganov, Yu. S., Iliev, S., Subbotin, V. G., Sukhov, A. M., Buklanov, G. V., Gikal, B. N., Kutner, V. B., Mezentsev, A. N., Subotic, K., Wild, J. F., Lougheed, R. W., and Moody, K. J., *Phys. Rev.* **C54**, 620 (1996).
25. Ghiorso, A., Lee, D., Somerville, L. P., Loveland, W., Nitschke, J. M., Ghiorso, W., Seaborg, G. T., Wilmarth, P., Leres, R., Wydler, A., Nurmia, M., Gregorich, K., Czerwinski, K., Gaylord, R., Hamilton, T., Hannink, N. J., Hoffman, D. C., Jarzynsky, C., Kacher, C., Kadkhodayan, B., Kreek, S., Lane, M., Lyon, A., McMahan, M. A., Neu, M., Sikkeland, T., Swiatecki, W. J., Türler, A., Walton, J. T., and Yashita, S., *Phys. Rev.* **C51**, R2293 (1995).
26. Hessberger, F. P., Münzenberg, G., Hofmann, S., Reisdorf, W., Schmidt, K. H., Schött, H. J., Armbruster, P., Hingmann, R., Thuma, B., and Vermeulen, D., *Z. Phys.* **A321**, 317 (1985).
27. Münzenberg, G., Hofmann, S., Folger, H., Hessberger, F. P., Keller, J., Poppensieker, K., Quint, B., Reisdorf, W., Schmidt, K. H., Schött, H. J., Armbruster, P., Leino, M. E., and Hingmann, R., *Z. Phys.* **A322**, 227 (1985).
28. Münzenberg, G., Armbruster, P., Berthes, G., Folger, H., Hessberger, F. P., Hofmann, S., Poppensieker, K., Reisdorf, W., Quint, B., Schmidt, K. H., Schött, H. J., Sümmerer, K., Zychor, I., Leino, M. E., Gollerthan, U., and Hanelt, E., *Z. Phys.* **A324**, 489 (1986).
29. Hessberger, F. P., Hofmann, S., Ninov, V., Armbruster, P., Folger, H., Münzenberg, G., Popeko, A. G., Yeremin, A. V., Andreyev, A. N., Saro, S., and Leino, M., *GSI Scientific Report 1995* (Report GSI 96-1, Darmstadt 1996), p.9.
30. Somerville, L. P., Nurmia, M. J., Nitschke, J. M., Ghiorso, A., Hulet E. K., and Lougheed, R. W., *Phys. Rev.* **C31**, 1801 (1985).
31. Schädel, M., et al., *GSI Scientific Report 1995* (Report GSI 96-1, Darmstadt 1996), p. 10.

SIMULATION FOR FUSION AND FUSION-FISSION DYNAMICS IN HEAVY NUCLEI

K. Okazaki, K. Hatogai[†], Y. Aritomo, T. Tokuda, T. Wada, M. Ohta and
Y. Abe[††]

Department of Physics, Konan University, Okamoto 8-9-1, Kobe 658, Japan
[†]*Konan Education and Research Center of Information Science, Konan University, Okamoto, Kobe 658 Japan*
[††]*Yukawa Institute for Theoretical Physics, Kyoto University, Kyoto 606-01, Japan*

Abstract. Starting from the contact configuration of colliding nuclei, the evolution of the nuclear shape is simulated in two-dimensional deformation space fixing the asymmetry by means of the Langevin equation under the influence of the one body dissipation. It is quantitatively shown that, in heavy mass region, the asymmetry of the entrance channel plays an important role to get an optimum fusion probability in connection with the extra barrier to be overcome. It is necessary to take into account this effect together with the excitation energy at Bass barrier to consider the optimum target-projectile combination for the synthesis of super heavy elements.

I INTRODUCTION AND OUR METHOD

On the analysis of fusion evaporation residue cross sections in heavy mass region beyond $Z \sim 100$, a very few theoretical investigation is found since the middle of 1980's when the extra-push model was studied extensively [1-3].

From the experimental point of view, it is clear that the fusion hindrance becomes more and more distinguished when the atomic number of compound nucleus goes beyond 80 [4]. These features are explained clearly in the talk of this symposium by using the potential energy surface in the deformation space, i.e. by the situation whether the contact configuration of entrance channel is located inside the conditional saddle point or not [5,6].

It is also pointed out that the analysis by mean trajectories in the deformation space looses its meaning when the atomic number of compound nucleus increases beyond $90 \sim 100$ due to strong dissipation of relative kinetic energy in the entrance channel [5,6]. The proposed scenario to achieve a breakthrough on this problem is that in the first step, Langevin equation is solved from the contact stage of colliding nuclei until the complete dissipation of the relative kinetic energy is realized, and in the next step the diffusion equation such as the Smoluchowski equation is applied in the multi-dimensional deformation parameter space [5-8]. This is an expedient simulation method for an extremely hindered process.

For the estimation of the evaporation residue cross section in very heavy mass region, another essential point in this scenario is that we have to take into account the temperature dependent shell correction energy which is restored in the cooling process due to neutron evaporation and stabilizes the compound nucleus against fission.

In the poster presentation, we displayed the time evolution of the distribution function in two-dimensional deformation space calculated with the Langevin equation, and demonstrated how the contact configuration evolves to the fusion area or to the valley of fission. These results are useful in considering the initial condition of the diffusion equation in the following stage of the evolution.

II RESULTS AND DISCUSSION

We presented the animation showing how the initial probability located at the contact configuration evolves in the deformation space and how the extra push energy works to enhance the fusion probability in the region $80 < Z < 100$ in the symmetric case. It was also presented, however, the fusion is a very rare event in spite of any strong extra push force in the region beyond $Z \sim 100$.

Here, we show two useful figures to think over from experimental point of view, which kind of target and projectile combination is suitable to form compound nucleus in superheavy region. The relative position between the contact configuration and the conditional saddle point is drawn in Fig. 1 for three different atomic number $Z = 80, 90$ and 100. The contour lines in each panels show the equi-potential energy in the two center model [9]. The abscissa indicates the distance of the center-of-mass of two fissioning fragments and the ordinate indicates the asymmetry of the fragments. The Businaro-Gallone point (BG) is indicated by the plus, the solid line shows the scission line on which the contact configuration is included, and the unconditional saddle point (SD) is marked by the solid circle.

The Businaro-Gallone point and the unconditional saddle point are connected by the straight broken line. The value of asymmetry parameter α_{cr} where the solid line intersect with the broken line can be defined as a critical asymmetry. Above this critical value α_{cr}, the fusion hindrance will become less important. The critical value α_{cr} approaches to the Businaro-Gallone point when the fissility parameter x goes to unity, which is shown in Fig. 2. The fusion hindrance becomes more appreciable the less asymmetric parameter becomes below α_{cr} when the fissility is fixed.

Therefore, we can make the following remarks. As can be seen from Fig. 1, when the asymmetry increases, the potential energy for the contact configuration on the solid line increases with respect to the ground state energy of spherical nuclei($\alpha = 0, Z_0 = 0$). Since the contact configuration corresponds to the top of the Bass barrier, the excitation energy for the compound nucleus produced by this barrier energy increases from a few MeV to $\sim 50 MeV$(around BG point) when the asymmetry increases. The excitation energy beyond $30 \sim 40 MeV$ affects to reduce the survival probability exponentially. So, the competition between the less fusion hindrance and the high excitation energy at large asymmetry is expected, and the optimum target-projectile combination will exist slightly below the critical value α_{cr}.

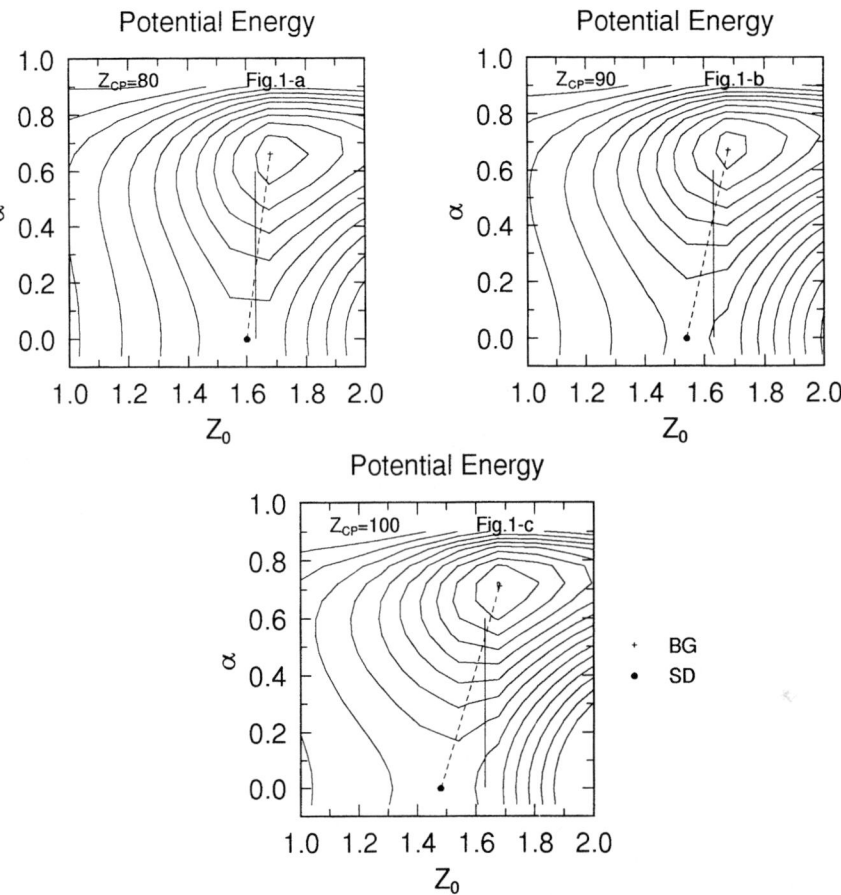

Fig. 1-a,b,c The contour lines show the equi-potential energy in deformation space with the two center parameterization [9]. The Businaro-Gallone point(BG) is indicated by the plus, the solid line shows the scission line on which the contact configuration is included, and the unconditional saddle point(SD) is marked by the solid circle. The atomic number of compound nucleus $Z_{\rm CP}$ is indicated in each panel.

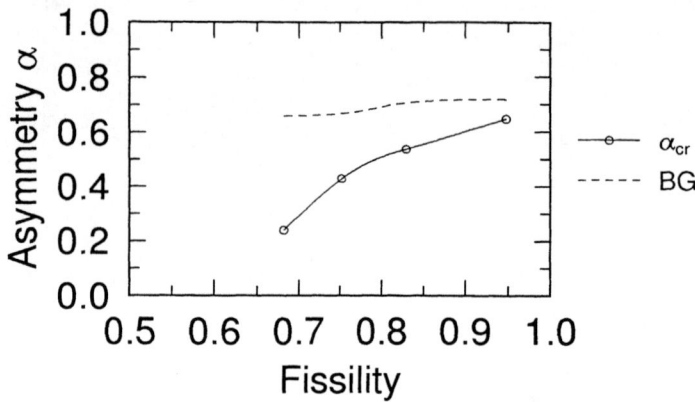

Fig. 2 The fissility parameter dependence of the critical asymmetry α_{cr} and the Businaro-Gallone point. Above the critical asymmetry, the fusion hindrance becomes less important.

REFERENCES

1. W.J. Swiatecki, Phys. Scripta **24** (1981) 113.
2. S. Bjørnholm and W.J. Swiatecki, Nucl. Phys. **A391** (1982) 471.
3. J.P. Blocki and H. Feldmeier, Nucl. Phys. **A459** (1986) 145.
4. K.-H. Schmidt and W. Morawak, Rep. Prog. Phys., **54** (1994) 949.
5. Y. Aritomo, talk presented in this symposium
6. T. Tokuda, talk presented in this symposium
7. Y. Aritomo, T. Wada, M. Ohta, and Y. Abe, Phys. Rev. **C55** (1997) R1011.
8. T. Wada, Y. Aritomo, T. Tokuda, M. Ohta, and Y. Abe, Nucl. Phys. **A616** (1997) 446c.
9. J. Maruhn and W. Greiner, Z. Physik **251** (1972) 431

EXPERIMENTAL ADVANCES

Fission of exotic nuclei

K.-H. Schmidt[a], S. Steinhäuser[b], C. Böckstiegel[b], A. Grewe[b],
J. Benlliure[a], H.-G. Clerc[b], A. Heinz[a], M. de Jong[b],
A. R. Junghans[b], J. Müller[b], M. Pfützner[c]

[a]*Gesellschaft für Schwerionenforschung m.b.H., Planckstraße 1, 64291 Darmstadt, Germany*

[b]*Technische Hochschule Darmstadt, Institut für Kernphysik,
Schloßgartenstraße 9, 64289 Darmstadt, Germany*

[c]*Institute of Experimental Physics, University of Warsaw,
Ul Hoza 69, 00-381 Warszawa, Poland*

Abstract. A new experimental technique for nuclear-fission studies has been developed at GSI, Darmstadt. Relativistic secondary projectiles are produced by fragmentation of a 1 A GeV ^{238}U primary beam and identified in nuclear charge and mass number. The giant resonances are excited by electromagnetic interactions in a secondary lead target, and fission from excitation energies around 11 MeV is induced. The fission fragments are identified in nuclear charge, and their velocity vectors are determined. Nuclear-charge distributions and total kinetic energies have been determined for a number of neutron-deficient actinides and preactinides which were not accessible with conventional techniques. The characteristics of multimodal fission of nuclei around ^{226}Th are systematically investigated and related to the influence of shell effects on the potential energy and on the level density between saddle point and scission. A systematic view on the large number of elemental yields measured gave rise to a new interpretation of the enhanced production of even elements in nuclear fission and allowed for a new understanding of pair breaking in fission.

INTRODUCTION

Nuclear fission provides unique information on the reordering of nucleons in a large-scale collective motion. The signatures of shell structure and pairing correlations show up in fission from low excitation energies. They have general implications on the influence of shell structure on nuclear dynamics and on the viscosity of cold nuclear matter. The use of secondary beams gives access to a large new field of fisssioning systems by overcoming restrictions of conventional experimental techniques.

Shell structure

Separate components appear in the yields and in the kinetic-energy distributions due to the shell structure in the potential-energy landscape in the highly deformed fissioning system. These "fission modes" are identified with valleys in the potential energy in the direction of elongation due to shell effects (1, 2). There exist models which relate the characteristics of the fission modes to the properties of the scission configuration (3). Others expect a decisive influence of the shell effects at the saddle point (4, 5).

Studies on low-energy fission were restricted to about 80 fissioning nuclei up to now. They represent only about 10 percent of all known nuclei with Z above 82. The fission of most of them can rather well be described by a superposition of two mass-asymmetric modes: standard I with a spherical heavy fragment around the spherical 82 neutron shell and standard II with a deformed heavy fragment around the deformed 86 neutron shell. In particular at higher excitation energy, a third mode appears, the superlong, mass-symmetric component. Only for a very limited number of nuclei around ^{259}Fm, a very specific symmetric mode shows up with a narrow mass distribution and exceptionally high kinetic energies (6 - 8) which is interpreted as the formation of two spherical fission fragments near the doubly magic ^{132}Sn (9 - 12). Near ^{227}Ac triple-humped mass distributions have been measured (13 - 21), and around ^{208}Pb symmetric fission prevails (22). In the present work, we perform a systematic study of the transition from mass-symmetric fission around ^{208}Pb to mass-asymmetric fission above ^{233}U. An overview on this transition is not yet available due to the lack of long-lived isotopes, necessary as target material for conventional fission experiments.

Pairing correlations

In low-energy fission, the production of fission fragments with an even number of protons and neutrons was found to be generally enhanced. This was interpreted as a measure for the probability that completely paired single-particle configurations are preserved up to the scission point. If fission starts from a completely paired superfluid configuration or from a configuration with a definite number of quasiparticle excitations, the amount of enhancement of even splits is connected to the dissipation in the fission process which couples intrinsic excitations and collective motion. Thus, it is a unique source of information on the viscosity of cold nuclear matter.

Since the complete identification of fission fragments in nuclear charge and mass number is very difficult, even-odd effects in the yields as a function of proton or neutron number only exist for a very limited number of systems. In contrast to mass yields, which are available more frequently, the elemental yields are particularly interesting, since they are not modified by the evaporation of neutrons from the excited fission fragments. In the available data, the even-odd effect in elemental yields for different fissioning systems varies as a function of fissility. In addition, for some of the systems, a variation as a function of the charge split is found. This has been interpreted as a corresponding variation of the dissipated energy (23).

With the new experimental technique applied in the present work, elemental yields can be determined with high quality for a large number of fissioning systems. Therefore, the study of pairing correlations in the fission process is another subject of interest followed here.

EXPERIMENT

The secondary-beam facility of GSI Darmstadt offers unique possibilities to provide secondary beams of neutron-deficient actinides and preactinides produced by fragmentation of ^{238}U. Within the intensity limits given by the primary-beam intensity and the fragmentation cross sections (24, 25), nuclear charge und mass number of the secondary projectiles can freely be selected by accordingly tuning the fragment separator (26).

In the present experiment, isotopically identified secondary beams of a number of neutron-deficient preactinides and actinides were produced by fragmentation from a 1 A GeV ^{238}U primary beam in a 657 mg/cm^2 beryllium target. The target was coated downstream with a 212 mg/cm^2 niobium layer to increase the fraction of fully stripped fragments. The fragment separator was operated as a momentum-loss achromat (27), using an intermediate degrader of 3.5 g/cm^2 aluminum. For the isotopical identification, scintillation detectors (28) were used to measure the positions of the fragments at the center and at the exit of the fragment separator as well as their time-of-flight..

In order to study the fission properties of the secondary projectiles, excited states in the desired energy range had to be populated, and the fission products had to be detected. The specific properties of the secondary beams do not allow to use excitation mechanisms usually applied for low-energy fission studies, like the capture of thermal neutrons. Instead, dedicated experimental methods had to be developed. Since the secondary-beam intensities only reach hundred per second or less for a specific isotope, it is crucial to ensure a high secondary reaction rate and a large detection efficiency. As excitation mechanism, electromagnetic interactions with a heavy target material have been chosen. They essentially populate the giant dipole resonance with a mean energy only a few MeV above the fission barrier with cross sections as high as a few barn.

The experimental setup behind the fragment separator is sketched in figure 1. As secondary target we used a stack of lead foils with a total thickness of 3 g/cm^2 mounted in a gas-filled chamber which acts as a subdivided ionization chamber (active target). With this device it is possible to determine the lead foil in which fission took place and to discriminate against fission induced in the scintillator. The average energy of the secondary projectiles in the lead target was about 420 A MeV. The differential energy loss of each fission fragment was measured separately with an horizontally subdivided ionization chamber (twin MUSIC). In order to correct the energy loss for the velocity dependence, the time-of-flight of the fission fragments was measured by means of a (1m × 1m) scintillator wall.

Figure 2 shows the nuclear-charge response of the experimental set up for fission fragments after electromagnetic-induced fission of ^{226}Th. Due to the high center-of-mass energies, an excellent charge resolution is achieved. Events stemming from reactions at lower impact parameters with nuclear contact were suppressed. For details of the analysis procedure see refs. (29, 30).

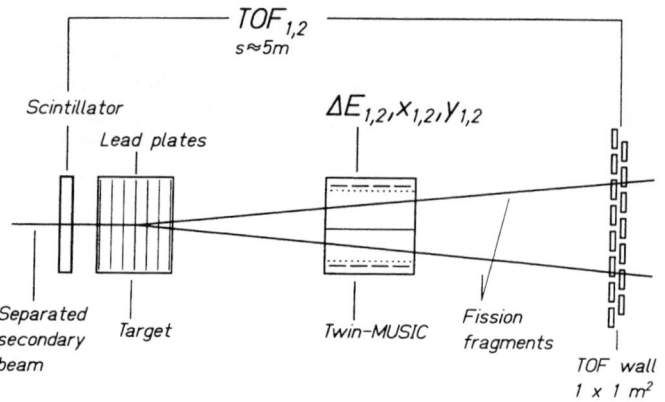

FIGURE 1. Schematic drawing of the set up for the fission experiment mounted behind the fragment separator

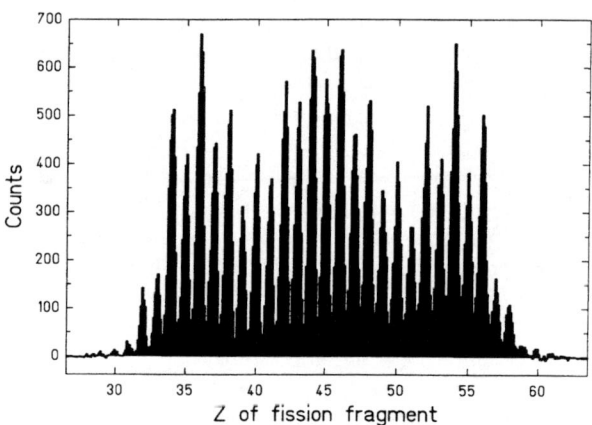

FIGURE 2. Nuclear-charge response of the twin MUSIC with velocity correction applied as obtained for the fission fragments after electromagnetic excitation of ^{226}Th secondary projectiles.

RESULTS AND DISCUSSION

In the present experiment, the elemental yields and the total kinetic energies of a series of neutron-deficient preactinides and actinides from ^{205}At to ^{234}U have been determined. The elemental yields covering the transition from symmetric fission at ^{221}Ac to asymmetric fission at ^{233}U are shown in figure 3. In the transitional region, around

^{226}Th, triple-humped distributions appear, demonstrating comparable weights for symmetric and asymmetric fission.

Fission modes

The weights of the two fission components were quantitatively determined by fitting three Gaussian curves to the charge-yield distributions, where the width of the symmetric component was fixed to 9.5 charge units (FWHM). This value was deduced from isotopes fissioning purely symmetrically. The ratio of symmetric to asymmetric fission was then determined by the ratio of the areas of the Gaussians describing the data.

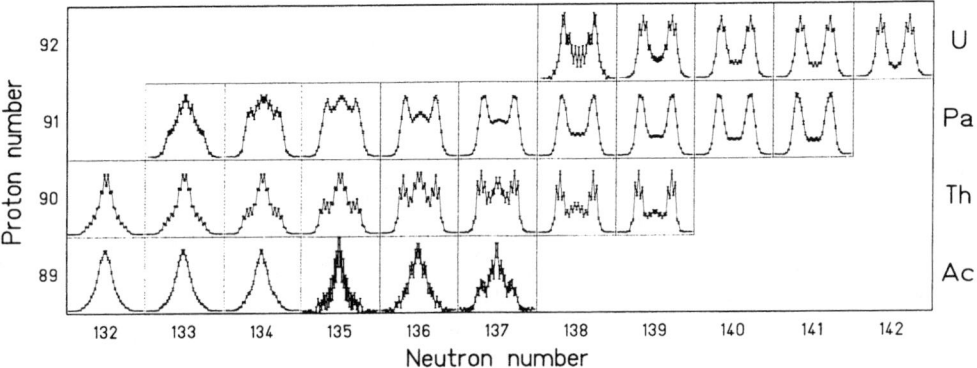

FIGURE 3. Measured fission-fragment charge distributions from ^{234}U to ^{221}Ac are shown on a chart of the nuclides.

The results of this procedure are shown in figure 4. For all nuclei with A < 220, the symmetric component has been found to prevail. The transition is rather smooth, and the weights of the two fission components have been found to scale with the mass of the fissioning nucleus.

In detail, the charge-yield distributions and the total kinetic energies of ^{233}U, ^{228}Pa, ^{226}Th, ^{223}Th, ^{219}Ac, and ^{214}Ra are shown in figure 5. For the uranium and thorium isotopes, a strong even-odd structure is observed for both the asymmetric and the symmetric component. The gross structural effects observed in the charge yields are different from those showing up in the total kinetic energies. From ^{233}U to ^{223}Th, the weight of the asymmetric fission component decreases, while the enhancement of the total kinetic energies (with respect to the expectation of the macroscopic model) for fission with a neutron number of the heavy fragment around N = 82 is preserved. For the lighter fissioning nuclei, another double-humped structure appears in the enhancement of the total kinetic energies while the charge distributions stay single humped. In all cases, the total kinetic energies follow rather closely the structure of the Q values as a function of the charge split.

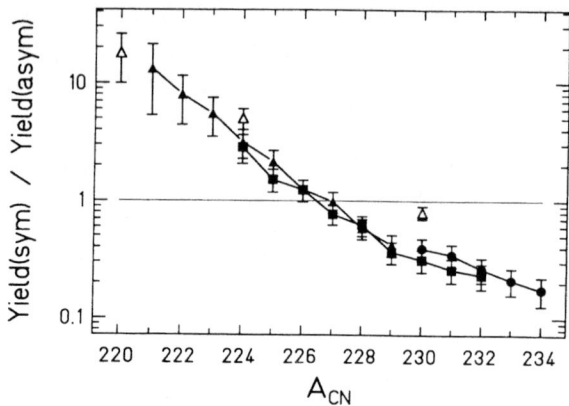

FIGURE 4. Intensity ratios of the symmetric and the asymmetric fission components in the transitional region as a function of mass number. The full triangles (squares, circles) correspond to thorium (protactinium, uranium) isotopes. The open symbols for ^{220}Th and ^{224}Th measured by Itkis et al. (21) and for ^{230}Th measured by Unik et al. (15) are included.

In a simultaneous description of elemental yields and total kinetic energies, it was tried to reproduce these data with the assumption of two fission modes, a symmetric and an asymmetric one. Each mode was represented by a Gaussian contribution in the yields and a specific scission-point configuration. Position and width of the Gaussians in nuclear charge as well as the tip distance of the scission configuration were treated as free parameters for each component. For the fissioning system ^{228}Pa, figure 6 (upper part) shows that the yields can be well reproduced, but a satisfactory fit of the total kinetic energy is not possible. Only when two asymmetric modes with comparable weights are introduced in addition to the symmetric mode, a good description of both, the elemental yields and the total kinetic energies is achieved, see lower part of figure 6. The parameters, in particular the relative heights of the total kinetic energies, attributed to the individual modes, roughly coincide with the expectations for the three most intense modes known from heavier fissioning systems.

FIGURE 5. (on next page) Measured elemental yields (left part) and average total kinetic energies (right part) as a function of the nuclear charge measured for fission fragments of several fissioning nuclei. Only statistical errors are given. The total kinetic energies are subject to an additional systematic uncertainty of 2 %, common to all data (30). In addition, for each charge split the maximum Q (ground-state to ground-state) value determined by varying the neutron numbers of the fission fragments is shown. Arrows indicate the positions of neutron (N = 50, 82) and proton shells (Z = 50). The positions of the neutron shells are calculated from the proton numbers by assuming an unchanged charge density (UCD). Finally, the expectations of a macroscopic scission-point model are shown (dashed lines) based on the Coulomb repulsion of the fragments at scission including a quadrupole deformation of $\beta = 0.625$ and a tip distance of 2 fm.(3).

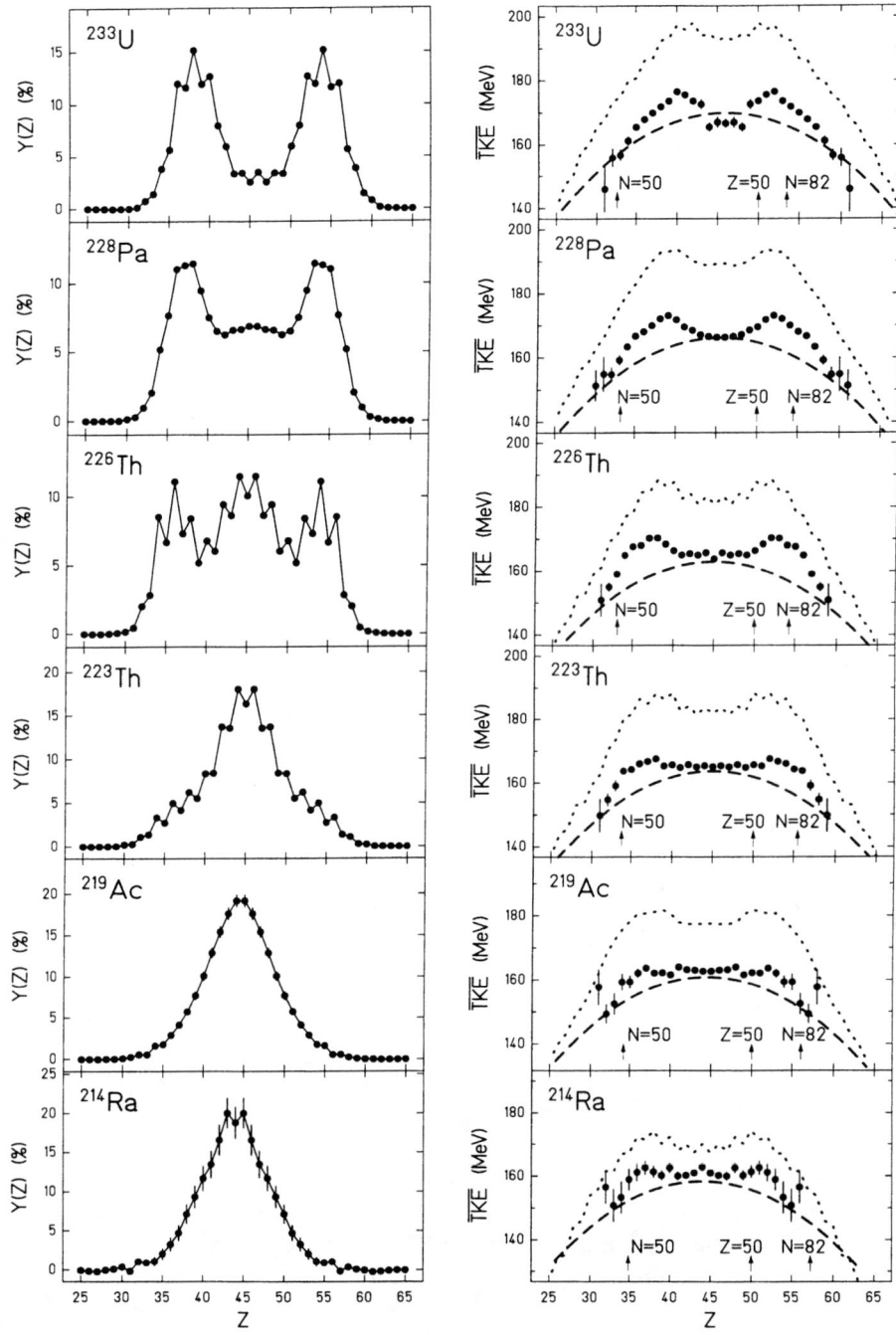

FIGURE 5.

This finding also agrees with the results of Itkis et al. (21) who investigated fission of neutron-deficient actinides from appreciably higher excitation energies induced by subbarrier heavy-ion fusion reactions.

Unfortunately, the dispersion of the total kinetic energy could not be deduced in the secondary-beam experiment due to the limited resolution. Therefore, the relative weights of the two asymmetric modes are not well defined from the present data. Nevertheless, it is clear that the triple-humped fission-fragment distributions are composed of strong contributions of three modes. This is true not only for ^{228}Pa but also for neighbouring nuclei showing triple-humped charge-yield distributions.

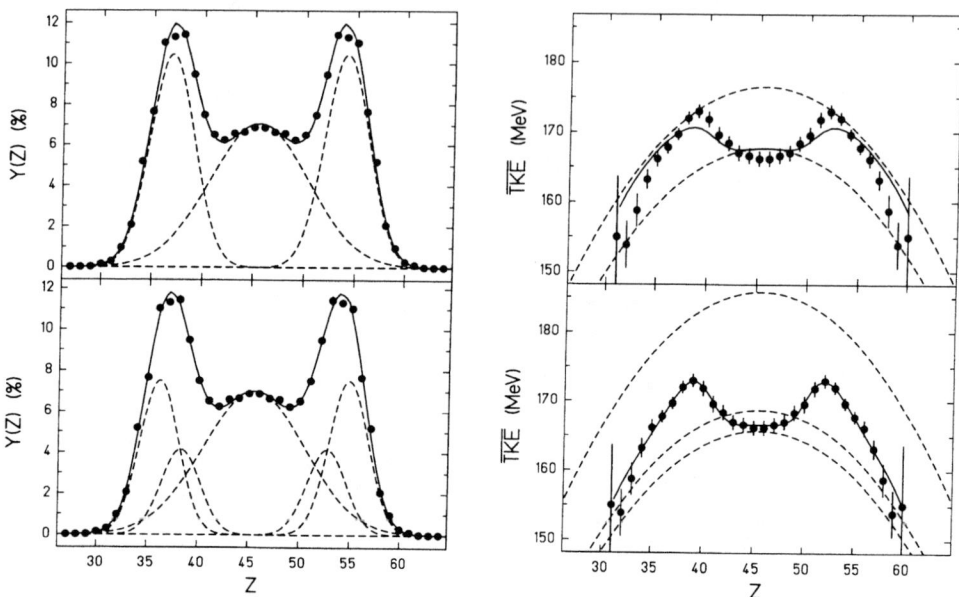

FIGURE 6. Elemental yields and total kinetic energies for electromagnetic-induced fission of ^{228}Pa. In addition to the data points, fits with two fission modes (upper part) and with three fission modes (lower part) are shown as full lines. The yields and the total kinetic energies of the individual contributions are shown separately as dashed lines.

The data on elemental yields support the idea, stated by Itkis et al. (31), that the weights of the fission modes are principally determined by an interplay of the neutron shells at $N = 82$ and $N = 86$ with the liquid-drop potential. A quantitative formulation of this idea has been developed by A. Grewe (32) by relating the charge distributions to the density of transition states in the vicinity of the outer fission barrier. The total kinetic energies, however, seem to be closely related to the ground-state properties of the fission fragments. This finding agrees with the expectation that the total kinetic energies are essentially determined by the shell effects in the scission-point configuration.

Pair breaking in fission

The excellent nuclear-charge resolution of the present experiment allowed to determine the even-odd structure in the elemental yields for a large number of fissioning systems. Figures 7 and 8 display the experimental results for four fissioning nuclei. In the lower part, the corresponding local even-odd effect D(Z) of the charge distributions is shown which is defined as (33):

$$D(Z+3/2) = 1/8 \, (-1)^{Z+1} \{- \ln Y(Z) + 3 \ln Y(Z+1) - 3 \ln Y(Z+2) + \ln Y(Z+3)\} \quad (1)$$

Herein, Y(Z) are the elemental yields. Thus, D(Z+3/2) is a quantity measured over four consecutive yields. centered at (Z+3/2). This quantity describes the local deviation from a Gaussian-like distribution.

The local even-odd effect shows a striking result for ^{220}Ac and ^{228}Pa: For these odd-Z fissioning nuclei, the values of D(Z) tend to be positive for the light fragments and negative for the heavy fragments. Absolute values up to more than 0.2, that means by 20 % enhanced production yields of even elements, are found for the extremely asymmetric light charge splits. This behaviour of the local even-odd effect is found to be essentially the same for the two nuclei, although their charge distributions show very different characteristics.

For the even-Z fissioning nuclei ^{226}Th and ^{233}U, a completely different behaviour of the local even-odd effect is observed. As expected, the average value does not vanish. However, there is a strong variation from a minimum value near symmetry to an increase of about a factor of two in the extremely asymmetric charge splits. Again, this behaviour proves to be rather general, independent of the gross structure of the charge distribution. This finding goes in line with previous observations in ^{235}U(n_{th},f) (34) and ^{249}Cf(n_{th},f) (35).

The observation of even-odd effects in odd-Z fissioning nuclei and the variation of the even-odd effect as a function the charge split in both odd-Z and even-Z fissioning nuclei can be interpreted with the assumption that unpaired protons are distributed to the nascent fragments according to the available phase space. Any intrinsic excitations induced during the descent from saddle to scission are assumed to be carried by unpaired nucleons. Thus, these nucleons populate excited single-particle states. On the way to scission, the single-particle states of the fissioning system evolve to states of the individual fragments. A quantitative formulation of this approach, based on the density of quasiparticle excitations as formulated by Strutinsky (36), is given in ref. (37). As an essential result, the local proton even-odd effect for a given number N of unpaired protons is given by the following relation:

$$D(Z) = (1 - 2 \, Z/Z_{CN})^N \quad (2)$$

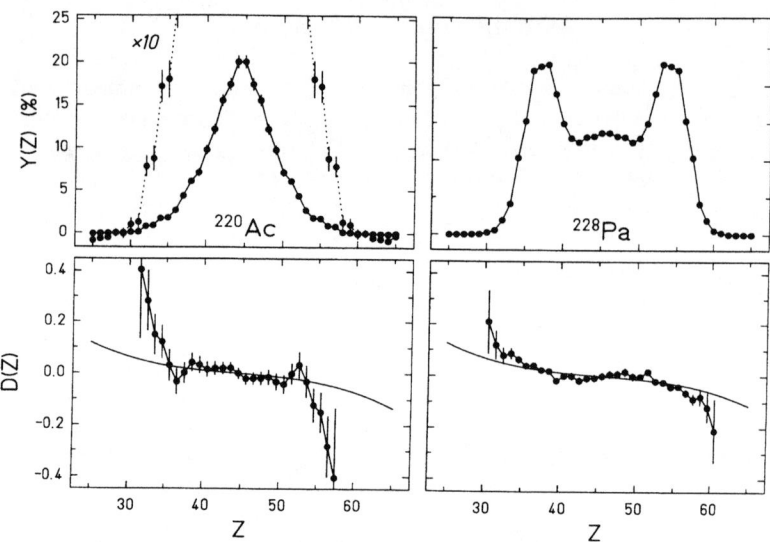

FIGURE 7. In the upper part, the charge distributions Y(Z) of ^{220}Ac and ^{228}Pa after electromagnetic-induced fission are shown. The charge yields are normalized to 200 %. The lower part displays the corresponding local even-odd effect D(Z) as a measure of the local deviation from a Gaussian-like distribution. The line represents a model prediction with a statistical contribution included (see text).

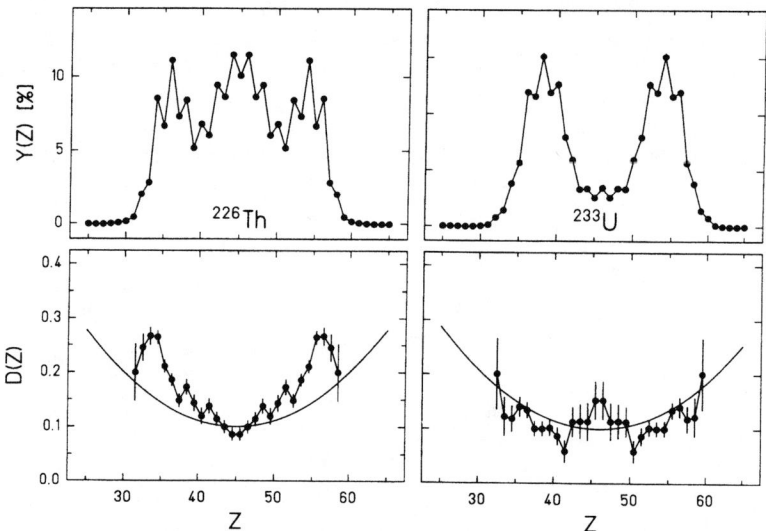

FIGURE 8. In the upper part, the charge distributions Y(Z) of ^{226}Th and ^{233}U after electromagnetic-induce fission are shown. The charge yields are normalized to 200 %. The lower part displays the corresponding local even-odd effect D(Z) as a measure of the local deviation from a Gaussian-like distribution. The line represents a model prediction with a statistical contribution included (see text).

where Z is the nuclear charge number of the fission fragment and Z_{CN} the nuclear charge number of the fissioning nucleus. In this expression, shell effects in the single-particle level scheme are disregarded. The data shown in figure 8 can be reproduced fairly well when a fraction of 10 % for the fully paired configuration and a fraction of 90 % for one proton pair broken are assumed. In the odd-Z fissioning systems, shown in figure 7, a fraction of 10 % refers to one unpaired proton and a fraction of 90 % refers to three unpaired protons.

This model at least partly explains the strongly enhanced even-odd effect observed previously in the asymmetric tails of the nuclear-charge distributions after fission of the even-Z nuclei ^{236}U (34) and ^{250}Cf (35) without assuming particularly low intrinsic excitation energies in extremely asymmetric fission. Also the well known decrease of the even-odd effect with increasing fissility parameter Z_{CN}^2/A_{CN} observed between ^{232}Th and ^{252}Cf can partly be attributed to a decreasing statistical contribution, because the asymmetry parameter $(Z_{heavy} - Z_{light})/Z_{CN}$ decreases from ^{232}Th to ^{252}Cf., due to the stationary position of the heavy component in the mass distributions near A = 138.

The new findings reveal that the enhanced production of fission fragments with even proton numbers cannot directly be interpreted as the probability that no quasiparticle excitations up to the scission point are induced. The relation between the even-odd effect and the intrinsic excitation energy at scission is more complex than assumed previously.

CONCLUSION

A new experimental technique using secondary beams opened new possibilities for low-energy fission studies. First experiments provided the elemental yields and the total kinetic energies for the fission of a large number of short-lived radioactive nuclei.

The transition from symmetric to asymmetric fission around mass number 227 seems to be caused by the competition between strong neutron shells corresponding to fragments with 82 and 86 neutrons which favour the asymmetric modes (standard I and standard II) and the liquid-drop potential which favours symmetric fission. While the relative strengths of these modes vary drastically as a function of mass number of the fissioning system, leading to very different charge distributions, the scission configurations corresponding to these modes seem to be rather universal.

The new understanding of the even-odd effect in fission deduced from the present data will allow to extract more reliable information on the viscosity of cold nuclear matter. It seems that strong variations of the observed even-odd effect in the elemental yields for different systems and as a function of asymmetry for a specific system are attributed to a great part to a statistical contribution to this quantity rather than to a variation of the intrinsic excitation energy.

REFERENCES

1. V. V. Pashkevich, Nucl. Phys. **A169** (1971) 275
2. U. Brosa, S. Grossmann, A. Müller, Phys. Rep. **197** (1990) 167
3. B. D. Wilkins, E. P. Steinberg, R. R. Chasman, Phys. Rev. **C14** (1976) 1832
4. G. A. Kudyaev, Yu. B. Ostapenko, G. N. Smirenkin, Yad. Fiz. **45** (1987) 1534 (Sov. J. Nucl. Phys. **45** (1987) 951)
5. G. A. Kudyaev, Yu. B. Ostapenko, E. M. Rastopchin, Yad. Fiz. **47** (1988) 1540 (Sov. J. Nucl. Phys. **47** (1988) 976)
6. E. K. Hulet, J. F. Wild, R. J. Dougan, R. W. Lougheed, J. H. Landrum, A. D. Dougan, P. A. Baisden, C. M. Henderson, R. J. Dupyzk, R. L. Hahn, M. Schädel, K. Sümmerer, G. Bethune, Phys. Rev. Lett. **56** (1986) 313
7. E. K. Hulet, J. F. Wild, R. J. Dougan, R. W. Lougheed, J. H. Landrum, A. D. Dougan, P. A. Baisden, C. M. Henderson, R. J. Dupzyk, R. L. Hahn, M. Schädel, K. Sümmerer, G. Bethune, Phys. Rev. **C40** (1989) 770
8. D. C. Hofmann, Nucl. Phys. **A502** (1989) 21c
9. P. Möller, J. R. Nix, Proc. Symp. on Phys. and Chem. of Fission, Rochester 1973, vol 1, (Vienna, IAEA, 1974) pp 329
10. M. G. Mustafa, U. Mosel, H. W. Schmitt, Phys. Rev. **C7** (1973) 1519
11. V. V. Pashkevich, Nucl. Phys. **A477** (1988) 1
12. S. Cwiok, P. Rozmej, A. Sobiczewski, Z. Patyk, Nucl. Phys. **A491** (1989) 281
13. R. C. Jensen, A. W. Fairhall, Phys. Rev. **109** (1958) 942
14. H. C. Britt, H. E. Wegner, J. C. Gursky, Phys. Rev. **129** (1963) 2239
15. J. P. Unik, J. R. Huizenga, Phys. Rev. **B134** (1964) 90
16. E. Konecny, H. W. Schmitt, Phys. Rev. **172** (1968) 1213
17. E. Konecny, H.-J. Specht, J. Weber, Phys. Lett. **B45** (1973) 329
18. H. J. Specht, Rev. Mod. Phys. **46** (1974) 773
19. H. J. Specht, Phys. Scripta **A10** (1974) 21
20. J. Weber, H. C. Britt, A. Gavron, E. Konecny, J. B: Wilhelmy, Phys. Rev. **C13** (1976) 2413
21. M. G. Itkis, Yu. Ts. Oganessian, G. Chubarian, V. N. Okolovich, G. N. Smirenkin, in Nuclear Fission and Fission-product Spectroscopy, H. Faust and G. Fioni (eds.), ILL Grenoble (1994) pp. 77
22. M. Itkis, V. N. Okolovich, A. Ya. Rusanov, G. N. Smirenkin, Sov. J. Nucl. Phys. **41** (1985) 544
23. H. Nifenecker, G. Mariolopoulos, J. P. Bocquet, Z. Phys. **A308** (1982) 39
24. H.-G. Clerc, M. de Jong, T. Brohm, M. Dornik, A. Grewe, E. Hanelt, A. Heinz, A. R. Junghans, C: Röhl, S. Steinhäuser, B. Voss, C. Ziegler, K.-H. Schmidt, S. Czajkowski, H. Geissel, H. Irnich, A. Magel, G. Münzenberg, F. Nickel, A. Piechacyek, C. Scheidenberger, W. Schwab, K. Sümmerer, W. Trinder, M. Pfützner, B. Blank, A. Ignatyuk, G. Kudyaev, Nucl. Phys. **A590** (1995) 785
25. A. R. Junghans et al., to be published
26. H. Geissel, P. Armbruster, K.-H. Behr, A. Brünle, K. Burkard, M. Chen, H. Folger, B. Franczak, H. Keller, O.Klepper, H. Langenbeck, F. Nickel, F. Pfeng, M. Pfützner, E. Roeckl, K. Rykaczewski, I. Schall, D. Schardt, C. Scheidenberger, K.-H. Schmidt, A. Schröter, T. Schwab, K. Sümmerer, M. Weber, G. Münzenberg, T. Brohm, H.-G. Clerc, M. Fauerbach, J.-J. Gaimard, A. Grewe, E. Hanelt, B. Knödler, M. Steiner, B. Voss, J. Weckenmann, C. Ziegler, A. Magel, H. Wollnik, J.P. Dufour, Y. Fujita, D. J. Vieira, B. Sherrill, Nucl. Instr. Meth. **B70** (1992) 286
27. K.-H. Schmidt, E. Hanelt, H. Geissel, G. Münzenberg, J.-P. Dufour, Nucl. Instr. Meth. **A260** (1987) 287-303
28. B. Voss, T. Brohm, H.-G. Clerc, A. Grewe, E. Hanelt, A. Heinz, M. de Jong, A. R. Junghans, W. Morawek, C. Röhl, S. Steinhäuser, C. Ziegler, K.-H. Schmidt, K.-H. Behr, H. Geissel, G. Münzenberg, F. Nickel, C. Scheidenberger, K. Sümmerer, A. Magel, M. Pfützner, Nucl. Instr. Meth **A364** (1995) 150

29. A. Grewe, S. Andriamonje, C. Böckstiegel, T. Brohm, H.-G. Clerc, S. Czajkowski, E. Hanelt, A. Heinz, M. G: Itkis, M. de Jong, M. S: Pravikoff, K.-H. Schmidt, W. Schwab, S. Steinhäuser, K. Sümmerer, B. Voss, Nucl. Phys. **A614** (1997) 400
30. C. Böckstiegel, S. Steinhäuser, J. Benlliure, H.-G. Clerc, A. Grewe, A. Heinz, M. de Jong, A. R. Junghans, J. Müller, K.-H. Schmidt, Phys. Lett. **B398** (1997) 259
31. M. G. Itkis, V. N. Okolovich, A Ya, Rusanov, G. N. Smirenkin, Sov. J. Part. Nucl. **19** (1988) 301
32. A. Grewe, thesis, Techn. Hochschule Darmstadt, 1997, unpublished
33. B: L. Tracy, J. Chaumont, R. Klapisch, J. M. Nitschke, A. M. Poskanzer, E. Roeckl, C. Thibault, Phys. Rev. **C5** (1972) 222
34. J. L. Sida, P. Armbruster, M. Bernas, J. P. Bocquet, R. Brissot, H. R. Faust, Nucl. Phys. **A502** (1989) 233c
35. R. Hentzschel, H. R. Faust, H. O. Denschlag, B. D. Wilkins, J. Gindler, Nucl. Phys. **A571** (1994) 427
36. V. M. Strutinsky, Contr. to Intern. Conf. on Nucl. Phys. (Paris 1958), pp. 617
37. S. Steinhäuser, J. Benlliure, C. Böckstiegel, H.-G. Clerc, A. Heinz, A. Grewe, M. deJong, A. R. Junghans, J. Müller, M. Pfützner, K.-H. Schmidt, submitted to Nucl. Phys. A

Production cross-section of very neutron-rich nuclei in relativistic projectile fission of ^{238}U

C. Engelmann[1], F. Ameil[2], P. Armbruster[1], M. Bernas[2],
C. Böckstiegel[6], S. Czajkowski[3], Ph. Dessagne[4], C. Donzaud[2],
H. Geissel[1], A. Heinz[1], Z. Janas[5], C. Kozhuharov[1], Ch. Miehé[4],
G. Münzenberg[1], M. Pfützner[5], W. Schwab[1], C. Stéphan[2],
K. Sümmerer[1], L. Tassan-Got[2] and B. Voss[1]

[1] *Gesellschaft für Schwerionenforschung, Planckstr. 1, 64291 Darmstadt, Germany*
[2] *IPN Orsay, BP 1, IN2P3 CNRS France*
[3] *CENBG, Bordeaux-Gradignan, IN2P3 CNRS France*
[4] *CRN Strasbourg, IN2P3 CNRS France*
[5] *Warsaw University, Poland*
[6] *Institut für Kernphysik, Schlossgartenstr. 9, 64289 Darmstadt*

Abstract. Nuclear fission of ^{238}U projectiles at 750·A MeV, impinging on a beryllium target, was used to produce about 60 new neutron-rich isotopes. Most of them were identified in the region of chromium to technetium, including the doubly magic nucleus $^{78}_{28}$Ni. Due to the kinematics of the fission reaction at relativistic energy, the fragment separator (FRS) at the GSI, Darmstadt allows to separate fission fragments produced after fission of the projectile and to identify both, their mass and their nuclear charge. Indications for a low-energy fission process can be found. The isotopic evolution of production cross sections is presented and compared between two targets, Be and Pb.

I KINEMATICS OF FISSION OF RELATIVISTIC PROJECTILES

At relativistic energies, fission of ^{238}U projectiles can be induced either by electro-magnetic exitation or by nuclear collisions. In case of a low atomic number of the target, fission is mainly induced by nuclear collisions whereas for targets with high atomic numbers fission after electro-magnetic interaction is the dominating process. In this experiment a Be-target was used to investigate fission induced at low exitation energies and to produce n-rich isotopes,

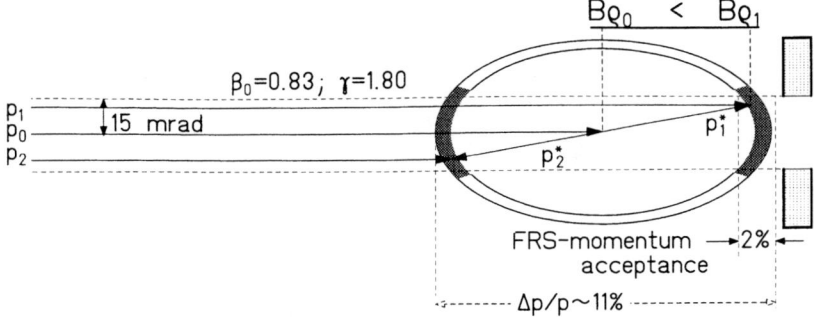

FIGURE 1. kinematics of relativistic projectile fission

using the fact that low-energy fission after nuclear collisions is covering 4% [1], fission in total is covering 30% [2], of the total reaction cross section. The kinematics of fission of relativistic ^{238}U projectiles can be described in two steps. Independent of the exitation process itself, the exited ^{238}U nucleus de-excites via neutron evaporation or fission. In binary fission the fission-fragments are emitted nearly isotropical back-to-back with equal momentum $|\vec{p_1^*}| = |\vec{p_2^*}|$. The kinetic-energy of the fission-fragments varies with the partition of the fissioning nucleus and the masses of the fragments. For the element Zn a fission-velocity of $\beta_f \approx 0.05$ is estimated. The momenta of one given fission fragment are filling a spherical shell in the phase-space, the radius and the thickness of the shell are given by the fission-process itself. In the laboratory-system the incident ^{238}U-beam energy is E/A=750·A MeV, resulting in a beam velocity of $\beta_0 \approx 0.83$ and a magnetic rigidity $B\rho_0 = 12.08$ Tm. The momenta of the fission-fragments are to be added relativistically. The result of the conversion is a cone based upon an ellipsoidal shell of momenta, as presented in Fig.1. All fission-fragments, emitted into 4π in the projectile-system, are now focussed into this cone in forward direction, the opening angle of the cone is dependent on the element. For the element Zn an opening angles of $2 \cdot \Theta_{Zn} = 74$ mrad is calculated. Due to the high velocity of the fragments, all nuclei are fully stripped, $q = Z$, enabling a unambiguous identification of the nuclear charge. The magnetic rigidity of the fission fragments allows to focus on a specific A/Z ratio, using $q \cdot B\rho = p = M \cdot v$, $q = Z$, $M \equiv A\gamma$, resulting in $B\rho \approx A/Z \cdot \beta\gamma$. When only regarding fission-fragments emitted in forward direction in the center of mass system with a well defined fission velocity β_f, $B\rho_1 > B\rho_0$ (see Fig.1), higher magnetic rigidity corresponds to a higher A/Z ratio of more neutron-rich isotopes. Combining these kinematics with the ±15 mrad angular acceptance and the two percent momentum acceptance of the fragment separator, two groups of fission-fragments could be transmitted: Either those fragments emitted in beam direction or those emitted in backward direction in the center of mass (see Fig.1) with a difference in momentum of $\delta p/p \approx 11\%$ between these two groups. Only fission can produce fragments with velocities

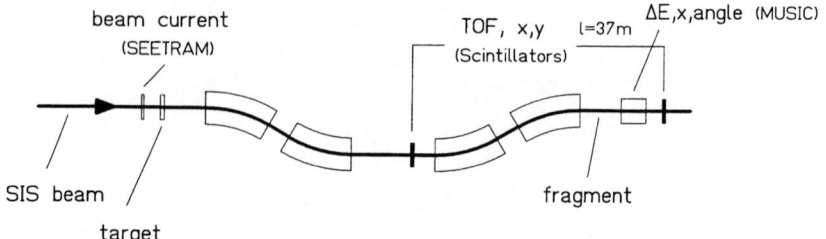

FIGURE 2. setup of the GSI-fragment separator in the standard achromatic mode

and A/Z-ratios higher than those of the primary beam, therefore a selection of the fragments emitted in forward direction is resulting in a low background of events of other nuclear reactions.

II SETUP OF THE EXPERIMENT AND ISOTOPE IDENTIFICATION

Extracted at 750·A MeV from the heavy-ion synchrotron SIS, the ^{238}U-ions are crossing the secondary-electron transmission detector SEETRAM to determine the beam-current. The average beam intensity was $5 \cdot 10^7$ ions per spill with a rate of one spill every four seconds. After hitting the 1006 mg/cm^2 Be-target, the first selection in terms of $B\rho$ is done with two dipole magnets in the first half of the spectrometer. In the central focal plane the fragments are passing a position-sensitive plastic scintillation detector to determine their momentum and to start a time-of-flight measurement. After a second analysis in $B\rho$ in the second half of the separator the selected fragments are crossing a multiple-sampling ionization-chamber to measure the energy-loss ΔE, the angle and the position of their trajectories. The stop signal of the time-of-flight is taken at the end of the FRS, after a flight-path of 37 m. The final identification is done event-by-event, using the information on $B\rho$, ΔE, ToF and trajectory. Adding up all these information, a scatter-plot of energy-loss ΔE in the ionization chamber versus the A/Z- ratio can be obtained (see Fig.3a). Each single island in the spectra belongs to one specific isotope, clearly separated from its neighbors. The whole plot corresponds to one fixed setting of the separator at $B\rho = 1.16 \cdot B\rho_0$. Several isotopes are labeled to give the orientation. The drawn line corresponds to the border of known isotopes in the year 1992, as given in Ref. [3,4]. All isotopes at the right hand side of this line are identified for the first time, including the doubly-magic nucleus ^{78}Ni$_{50}$ [5]. In Fig.3b the mass distribution of the elements Ni, Cu and Zn are obtained by projecting the single elements on the A/Z-axis. The mass-spectra are plotted with logarithmic scales to show the very low background and the achieved mass-resolution of 260.

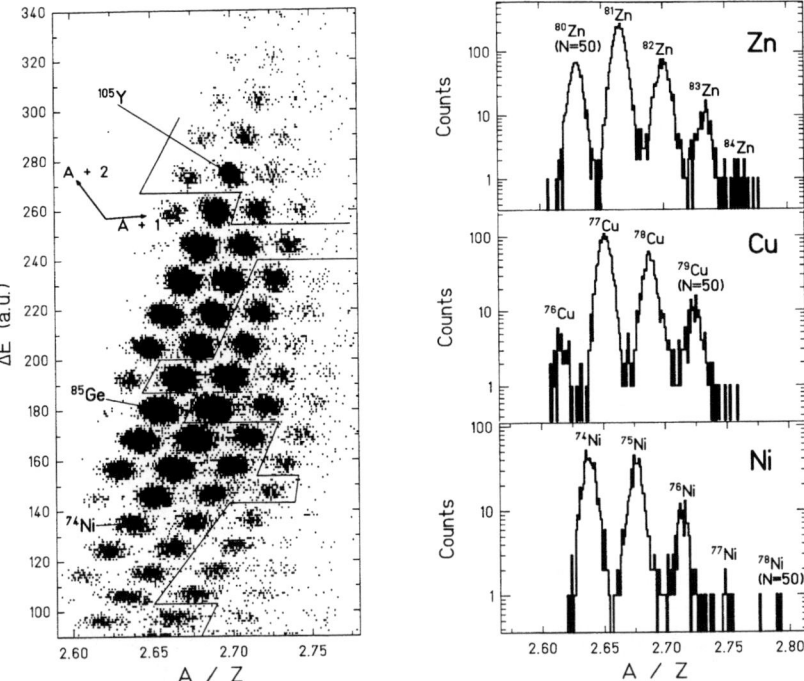

FIGURE 3. a) isotope plot of new isotopes found at a setting of $B\rho = 1.16 \cdot B\rho_0$; b) mass spectra obtained by projecting the scatter-plot on the A/Z axis for the elements Ni, Cu, Zn

III VELOCITY DISTRIBUTION OF THE ISOTOPES AND PRODUCTION CROSS SECTIONS

The production cross section of each single isotope can be extracted after accounting for the velocity distribution of the fission-fragments. The momentum spread due to the fission-process and the cut in momentum by the separator are adding up to a total momentum width of $\Delta p/p = 0.8\%$ [1]. This is smaller than the momentum acceptance of 2% of the FRS and is not explaining the experimental velocity distribution as shown in Fig.4. The main contribution to the velocity distribution is caused by the difference in energy-loss between ^{238}U and the fission-fragments in a thick target. The reaction-probability is constant all along the path of the projectile in the target. The kinetic energy distribution of the fission-fragments is centered around the mean value corresponding to a fission-reaction taking place in the middle of the target. The width of the distribution is given by two limits, fission directly at the beginning of the target or fission at the far end of the target. With increas-

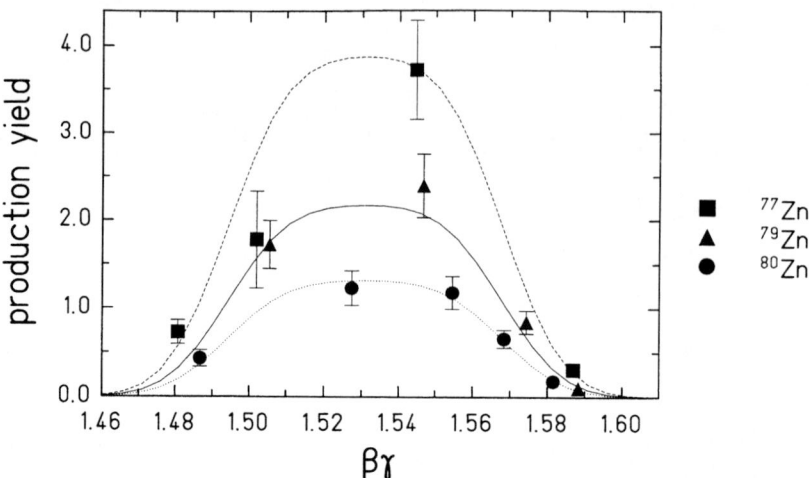

FIGURE 4. measured velocity distribution of ^{77,79,80}Zn in arbitrary scale and result of the fitting procedure

ing difference in atomic number between the projectile and the fragment the target location straggling grows up to 4.8% for Zn and can be determined by energy-loss calculations. This large momentum width is the cause of the large spread in magnetic rigidity of one isotope, resulting in a transmission of about five isotopes in one setting of the FRS instead of two expected from a sharp 2% momentum acceptance. The final momentum distribution can be described by a Gaussian curve folded with a rectangle of a calculated width. This is chosen as a fit-function (see Ref. [6]) to describe all isotopes of a given element simultaneously (see Fig.4). The total kinetic energy and the target location straggling of the fission-fragments are both dependent on the atomic number, the mass dependence can be neglected. Therefore all curves of one element have a similar shape (see Fig.4) and the different resulting normalizations of the fit-function are giving the different isotopic production-yields. The production cross sections are obtained by correcting the yields for the transmission of the FRS, for the beam intensity and the target thickness. The transmission is calculated by the Monte-Carlo simulation MOCADI [7], including the total kinetic energy distribution of fission-fragments known from thermal-neutron induced fission. The transmission increases from 1.6% for Zn up to 2.3% for Zr directly reflecting the decrease in fission velocity with increasing atomic number. The resulting distribution of production cross sections as a function of the neutron-number is shown in Fig.5 for all elements between Ca and Nb. For 22 elements with eight to twelve isotopes each, production cross sections are measured over a range of seven orders of magnitude. The lines are drawn only to connect the data points. For the lighter isotopes a small neutron odd-even effect can be seen, whereas for the heavier isotopes a

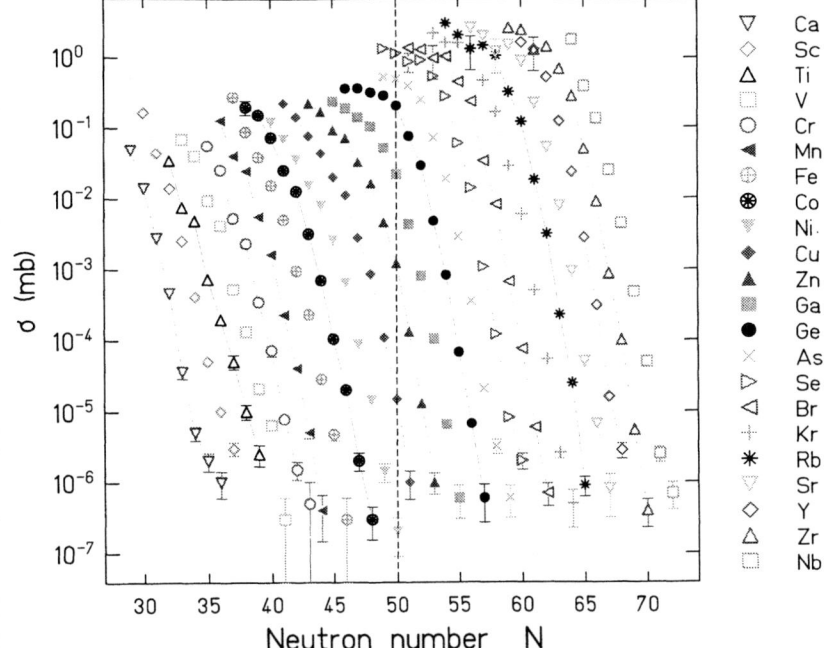

FIGURE 5. production cross sections in 750·A MeV ^{238}U+Be projectile fission

steep, structure-less exponential decrease is found. The isotopic distributions of Ni, Cu, Zn, Ga and Ge are reaching or crossing the closed neutron-shell of 50. An enhancement of production is observed for Zn, Ga and Ge, but the large statistical errors for the most exotic isotopes prevent to conclude the same for ^{79}Cu and ^{78}Ni. Normalized to the number of target atoms, the production cross sections can be compared directly between the two targets. The total low-energy fission cross sections is 2.1 barn for Pb, compared to 0.115 barn for Be [1]. This huge difference is due to the large electro-magnetic cross section of the exitation of the giant resonances, when using the Pb target. The average exitation energy in this case is approximately 12 MeV, compared to a broad distribution of $0 \leq E^* \leq 100$ MeV after nuclear reactions. The smaller low-energy fission cross section in the Be target is compensated by the number of target atoms per square centimeter, which is about a factor of 20 higher than for the Pb target. Two examples are shown in Fig.6. The first one is the production of Zr/Te, the most abundant elements in asymmetric fission of Uranium. For the most abundant isotopes, ^{100}Zr,^{134}Te, the production rate is clearly higher when using the Pb-target. When looking at very neutron-rich isotopes, the production rates are the same inside the error bars. The gain in total low-energy fission cross section in the Pb-target is compensated by the number of target atoms and the broader mass distribution of fission at higher

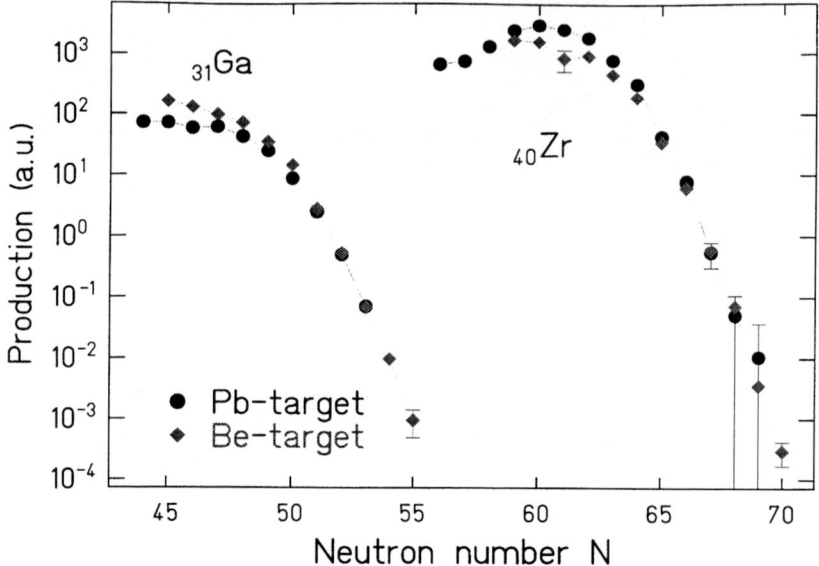

FIGURE 6. production of fission fragments in 750·A MeV ^{238}U projectile fission on Pb and Be targets

exitation energies. The second example is the element Ga, this is the lightest element for which a distribution of production cross section is measured with the Pb target. In the maximum of the distribution now the best production is obtained with the Be target, while for the more neutron-rich isotopes there is no decision possible with the present error bars. In table 1 are shown the production rates of the pair ^{134}Te,^{100}Zr, representing the maximum of the possible production rate for fission fragments, the doubly magic nuclei ^{132}Sn and ^{78}Ni and of ^{76}Ga. They are calculated with the present beam intensity and with the intensity after the first upgrade of the accelerator.

IV SUMMARY

Using relativistic ^{238}U beams allows to produce, separate and identify very neutron-rich isotopes produced in fission. Between Z= 28 and Z= 35 in the chart of nuclei the astrophysical r-process path of element-synthesis is touched or crossed. Together with the first experiment of projectile fission on a Pb-target [4] more than 110 new isotopes have been identified. Even with the present beam-intensities several experimental investigations of these very exotic nuclei are possible e.g. beta-decay lifetime measurements after implantation or search for isomers to get access to spectroscopic informations. With the intensity upgrade of the SIS in 1997 a factor of 15 in increase of the production rates is expected, allowing the investigation of even more exotic isotopes

target	σ (mb)	present intensity		intensity upgrade	
		ions/pulse	ions/s	ions/pulse	ions/s
^{100}Zr, ^{134}Te					
Pb	86	1500	250	22600	3800
Be	2.5	750	125	11200	1800
^{132}Sn					
Pb	15	540	90	8100	1300
^{76}Ga					
Pb	2	24	4	360	50
Be	0.3	54	9	800	130
^{78}Ni					
Be	0.3 nb	0.5/day		7/day	

TABLE 1. production rates of fission fragments on Pb and Be targets

and is opening the possibility to do mass measurement, β-γ coincidence after implantation or γ-spectroscopy after Coulomb-exitation of the more abundant fission-fragments.

REFERENCES

1. Armbruster, P. et al., *Z. Phys.* **A 355** (1996) 191
2. Hesse, M. et al., *Z. Phys.* **A 355** (1996) 69
3. Antony, M.S. et al., chart of the nuclides *CRN-Strasbourg* (1992)
4. Bernas, M. et al., *Phys. Lett.* **B 331** (1994) 19
5. Engelmann, C. et al., *Z. Phys.* **A 352** (1995) 351
6. Donzaud, C. *et al.*, submitted to em Z. fur Phys. A
7. Geissel, H. et al., *Nucl. Instr. Meth.* **A 282** (1989) 247

Evolution of fission lifetime with temperature: a straightforward measurement by the blocking technique

M. Morjean*, M. Chevallier†, C. Cohen‡, D. Dauvergne†,
J. Dural§, J. Galin*, F. Goldenbaum*, D. Jacquet¶, R. Kirsch†,
E. Lienard*, B. Lott*, A. Péghaire*, Y. Périer*, J.C. Poizat†,
G. Prevot‡, J. Rémillieux†, D. Schmaus‡, M. Toulemonde§

*GANIL DSM/CEA, IN2P3/CNRS ,BP 5027, 14076 Caen Cedex 5, France
†Institut de Physique Nucléaire de Lyon, IN2P3/CNRS, Univ. Cl. Bernard,
43 Bd. 11 Novembre 1918, F-69622 Villeurbanne Cedex, France
‡GPS,2 place Jussieu, 75251 Paris Cedex 05, France
¶Institut de Physique Nucléaire d'Orsay,BP 1, F-91406 Orsay Cedex, France
§CIRIL, BP 5133, 14040 Caen Cedex, France

Abstract. The blocking patterns in a single crystal have been used to measure fission lifetimes as a function of the excitation energy in the ^{238}U+^{28}Si reactions at 24 A.MeV. The neutron multiplicity measured on 4π has been used to infer the excitation energy of the fissioning nuclei. The fission lifetimes measured for uranium-like nuclei with temperatures up to 3 MeV are longer than 10^{-19}s, a much larger value than those inferred from previous measurements by less direct techniques.

INTRODUCTION

The dynamics of fission has been widely studied in order to get information on the nature of the nuclear dissipation processes [1]. Many experimental approaches have been followed up to now in order to determine the evolution of the fission lifetime with excitation energy. At low excitation energies, fission lifetimes have been inferred in a very direct way from the blocking technique in a single crystal [2], leading, for fission induced by low energy neutrons on a uranium target [3], to lifetimes in good agreement with the predictions of the statistical model of Bohr and Wheeler. Unfortunately, as soon as the excitation increases in the fissioning nuclei, the fission lifetimes decrease by orders of magnitude and reach values out of the range covered by the blocking technique as used before [4,5]. More indirect methods had then to be used at higher excitation energies [6,7], mostly based on the measurement of pre- and post-scission neutron, charged particle or γ ray multiplicities. These experiments concluded that the lifetimes were decreasing steadily at higher excitation energies, but very different values are inferred, depending on the probe considered or on assumptions performed in the data analysis [8].

In the present paper, the results of an experiment in which the limit of the blocking technique has been lowered by more than one order of magnitude are reported, allowing thus the measurement of fission lifetimes down to 10^{-19}s. This limit could be pushed away taking advantage of the large velocity of the fissioning nuclei produced in reactions induced on a single silicon crystal by the high quality uranium beam available at GANIL at 24 MeV per nucleon. Furthermore, the excitation energy was inferred from the associated neutron multiplicity measured in a solid angle close to 4π sr, making it possible the determination, from the very same experiment, of lifetimes in a very large range of temperatures of the fissioning nuclei.

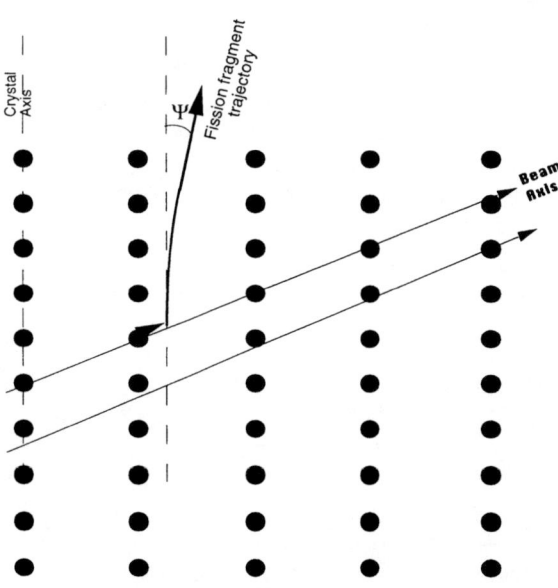

FIGURE 1. Principle of the blocking technique.

EXPERIMENTAL PROCEDURE

The principle of time measurements by the blocking technique is shown in figure 1. Fission fragments emitted in a direction close to an axis (or plane) of a single crystal are deflected by the high repulsive field of the atomic string (or plane). This so-called blocking effect leads to dips in the fission fragment angular distribution in the direction of the axis or planes of a single crystal. The closer to the axis the emitted fragment, the stronger the field is. The shorter the fission lifetime, the deeper the dips in the angular distribution are. This is illustrated in figure 2 where the angular distributions with respect to the < 110 > axis of a silicon crystal have been simulated for various lifetimes of an uranium nucleus. In these simulations, the uranium nucleus recoils at 24 MeV/nucleon at 7 degrees with respect to the crystal axis and a fission fragment with Z=40 is emitted in a direction close to this axis. A clear evolution of the dip is observed for lifetimes between 10^{-17} and 10^{-19}s, but the thermal vibrations of the atomic strings and planes prevent time discrimination for lifetimes shorter than 10^{-19}s.

A ^{28}Si single crystal with an effective thickness 5.8 μm was mounted on a goniometer inside a reaction chamber and bombarded by a 24 MeV/nucleon ^{238}U beam. The two coincident fission fragments of the projectile-like nuclei were identified in charge by two 5x5 cm^2 telescopes, each consisting of two silicon strip detectors (150 μm and 500 μm in depth, respectively). One of the telescopes was located at 3 meters downstream from the target, covering angles around 7 degrees, and the second one at only 20 cm from the target, in the same plane, on the opposite side to the beam axis, covering angles between -5 and -13 degrees. The telescope located at 3 meters from the target was aimed at the detection of one of the fission fragments with an angular resolution of the order of 10^{-3} degrees. The atomic numbers up to Z=50 were separated by both telescopes and an extrapolation towards higher Z values permitted a reconstruction of the charge of any primary nuclei before fission.

A third telescope, identical to the one located at 7 degrees, was also set at 3 meters from the target but was covering angles around one degree, inside the grazing angle, in order to detect elastically scattered uranium nuclei. The Rutherford scattering process is supposed to be an instantaneous process (as compared to the time scale for fission) and the behaviour of the dip observed in the angular distribution of elastically scattered nuclei provides a reference for both the quality of the crystal and of the beam. Any variation during the experiment of the shape of this dip would have been an indication of beam radiation damages, but no significant evolution has been observed with a beam intensity of about 10^7 ions per second during 8 days.

All the experimental set-up was surrounded by ORION [9], a 4π neutron detector that provides event-by-event the neutron multiplicity associated with a nuclear reaction. The excitation energy can then be inferred from the neutron multiplicity after correction for the detection efficiency that was calculated

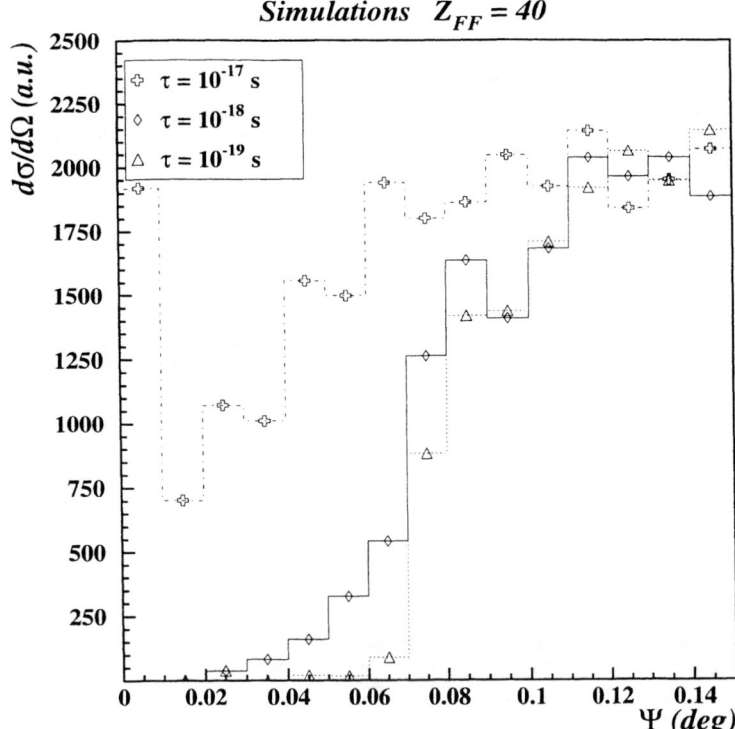

FIGURE 2. Simulations of fission fragment angular distributions. An uranium nucleus has a recoil velocity of 6.2 cm/ns at 7 degrees from the $<110>$ axis of a single silicon crystal and emits a fission fragment with Z=40 around the axis of the crystal. The angles are presented with respect to the $<110>$ axis of the crystal.

from Monte Carlo simulations [10] to be of the order of 50% for neutrons from the projectile-like nuclei. The correlation between neutron multiplicity and thermal excitation energy has been calculated with the Monte Carlo statistical code PACE for an uranium nucleus that does not undergo fission. A rough correction was then applied on the data in order to take into account the neutrons gained during the fission process itself.

RESULTS

The blocking patterns observed with the telescope located at 7 degrees when two fission fragments have been detected in coincidence are shown in figure 3 for ten bins of associated neutron multiplicity. The dips corresponding either to the $<110>$ axis or to various atomic planes can be easily identified:

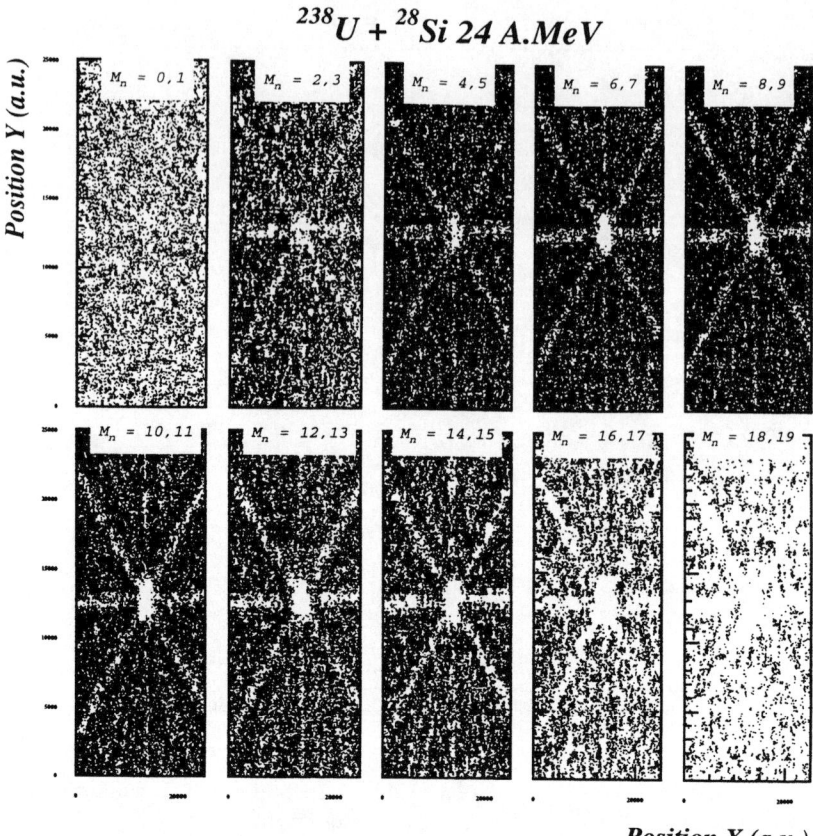

FIGURE 3. Blocking patterns observed for fission fragments.

within the covered angular range of about 1 degrees, the fission yield is almost constant except in the regions where the blocking effect is dominant. Even for the lowest and highest multiplicity bins, despite the poor statistics, the blocking effect is visible. The dips become deeper and deeper when the neutron multiplicity increases, suggesting that that the fission lifetime becomes shorter and shorter at high excitation energy.

In order to quantify the information on the lifetime, the blocking ratio $\mathcal{R}$ has been defined as the ratio between the yield inside the dip corresponding to the $<110>$ axis (integrated up to half the constant average value observed far from the axis) and the yield integrated over the whole detection range. The evolution of $\mathcal{R}$ with excitation energy is presented in figure 4. It decreases

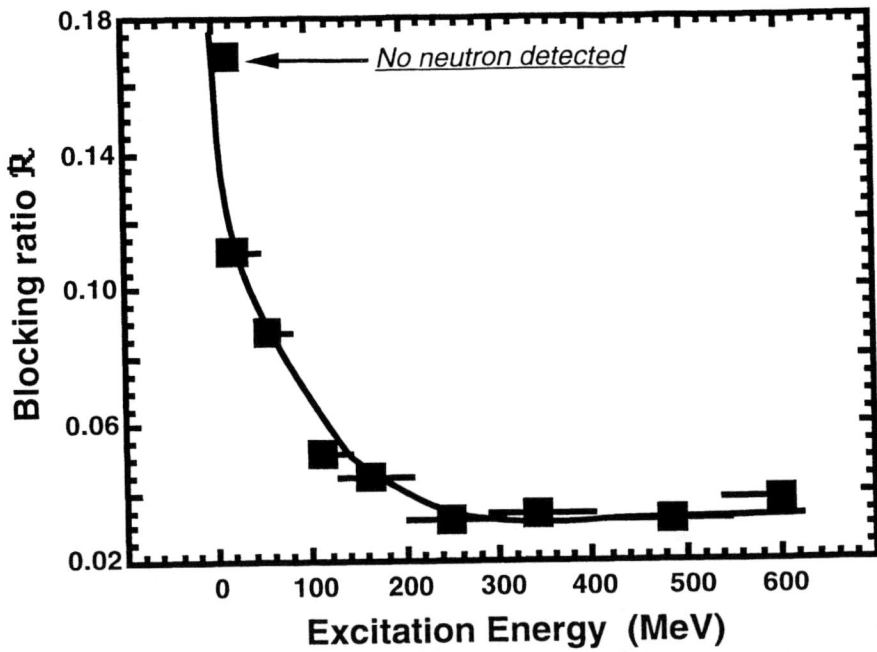

FIGURE 4. Evolution of the blocking ratio with excitation energy.

rapidly when the excitation energy E* increases up to about 250 MeV, and is nearly constant for larger E* values. Decreasing values of $\mathcal{R}$ indicate that the fission lifetime has been long enough to permit the projectile-like nucleus to leave, before it undergoes fission, the region in which a smearing of the atomic potential arises due to thermal vibrations.

Figure 5 presents the values of $\mathcal{R}$ simulated for the < 110 > axis assuming, for the fission process, an exponential decay time with an average time τ. A continuous decrease of $\mathcal{R}$ is observed for τ values larger than 10^{-19}s and the smearing due to thermal vibrations occurs only for smaller τ values. From a comparison with figure 4, it can thus be concluded that the limit for time discrimination, 10^{-19}s, is not yet reached for excitation energies smaller than 250 MeV, whereas it is most likely reached for larger excitations. The values of $\mathcal{R}$ deduced from simulations depend on assumptions on the actual shape of the fission time distribution (assumed exponential in figure 5) and on the effect of post-scission neutron emission on the fission fragment trajectories (neglected in figure 5). Whatever the assumptions included in simulations, the average lifetimes are found to decrease from a few 10^{-18}s at excitation energies of a

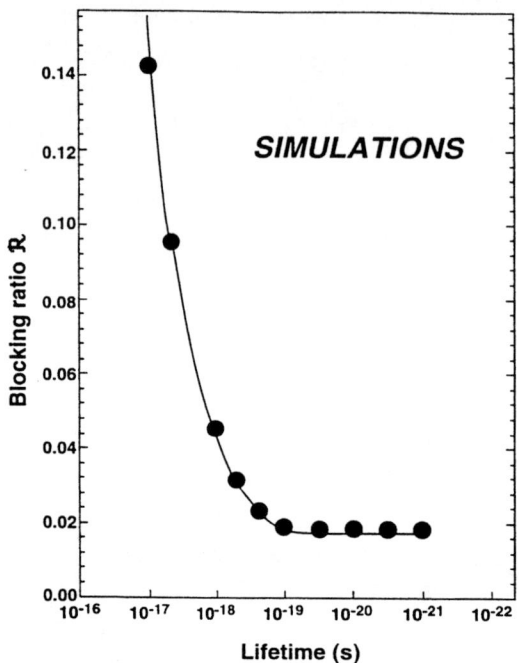

FIGURE 5. Results of simulations for the evolution of the blocking ratio with the lifetime.

few tenths of MeV down to a few 10^{-19}s at 250 MeV. These lifetimes are larger by at least one order of magnitude than the ones inferred from other techniques [6–8], indicating that these latters, contrary to the present one, are only sensitive to the shortest part of the fission time distribution.

CONCLUSION

The fission lifetime evolution with excitation energy has been inferred in a very simple way from the blocking effect magnitude in a single crystal. The limit for time discrimination has been pushed away, as compared to previous experiments using the same technique, down to about 10^{-19}s taking advantage of the large velocity of the fissioning projectile-like nucleus in reactions induced by uranium projectiles accelerated at 24 MeV per nucleon.

Once the excitation energy is larger than a few tenths of MeV, the fission lifetime of uranium-like nuclei is found to be in the range of 10^{-18}s, decreasing when the excitation energy increases up to about 250 MeV where it reaches

10^{-19}s, the actual limit for time discrimination in the present experiment. Such times are much larger than the ones inferred from other techniques. This discrepancy seems to arise from the long part of the time distribution that has an important weight on the average lifetime and that can only be probed by the blocking technique.

ACKNOWLEDGMENTS

The authors are grateful to C. Spitaels for the set-up of the position sensitive silicon detectors. They would also like to acknowledge the help of R. Beunard, G. Fremont, Y. Georget, J. Moulin, J.L. Vignet and A. Vigot before and during the experiment.

REFERENCES

1. D. Hilscher and H. Rossner, *Ann. Phys. Fr.* **17**, 471 (1992)
2. W.M. Gibson, *Ann. Rev. Nucl. Sci.* **25**, 465 (1975)
3. J.U. Andersen et al., *Nucl. Phys.* **A241**, 317 (1975)
4. J.U. Andersen et al., *Phys. Rev. Lett.* **36**, 1539 (1976)
5. J.S. Forster et al., *Nucl. Phys.* **A464**, 497 (1987)
6. D.J. Hinde et al., *Phys. Rev.* **C45**, 1229 (1992)
7. P. Paul and M. Thoennessen, *Ann. Rev. Part. Nuc. Sci.* **44**, 65 (1994)
8. K. Siwek-Wilczynska et al., *Phys. Rev.* **C51**, 2054 (1995)
9. J. Galin and U. Jahnke, *J. Phys. G* **20**, 1105 (1994)
10. J. Poitou and C. Signarbieux, *Nucl. Inst. Meth.* **114**, 113 (1974)

Angular momentum dependence of prescission particle multiplicity in medium mass systems

K. Yuasa–Nakagawa[a,1], T. Nakagawa[a], K. Yoshida[a],
K. Furutaka[b,2], K. Matsuda[b,3], Y. Futami[c,4], X. Liu[c], Y. Aoki[d],
S. M. Lee[c], J. Kasagi[d], D. X. Jiang[c,5], T. Suomijärvi[e],
W. Q. Shen[f], T. Wada[g], S. Yamaji[a] and Y. Abe[h]

[a] The Institute of Physical and Chemical Research(RIKEN),
Hirosawa 2-1, Wako, Saitama 351-01, Japan
[b] Department of Physics, Tokyo Institute of Technology, Meguro, Tokyo 152, Japan
[c] Institute of Physics, University of Tsukuba, Tsukuba, Ibaraki 305, Japan
[d] Laboratory of Nuclear Science, Tohoku University, Sendai, Miyagi 982, Japan
[e] Institute de Physique Nucléaire, 91406 Orsay Cedex, France
[f] Shanghai Institute of Nuclear Research, Chinese Academy of Science,
Shanghai 201800, China
[g] Department of Physics, Konan University, Kobe, Hyogo 658, Japan
[h] Yukawa Institute of Theoretical Physics, Kyoto University, Kyoto 606-01, Japan

Abstract. We measured γ–ray multiplicity and light charged particle (LCP) multiplicity simultaneously in coincidence with binary decay fragments in the reactions of ^{84}Kr (10.6 and 8.5 MeV/nucleon) + ^{27}Al and ^{58}Ni (10.0 MeV/nucleon) + ^{56}Fe. We observed the prescission time and charge distribution of heavy fragments as a function of the angular momentum of the system applying the measured γ–ray multiplicity as an angular momentum filter. The deduced prescission time becomes smaller as the angular momentum of the system increases. For ^{84}Kr + ^{27}Al, the fusion–fission reaction seems to occur dominantly only at the angular momentum $\sim 70\ \hbar$, which is concluded from the dependence of prescission proton multiplicity on the total kinetic energy (TKE). The charge distribution of binary decay fragments changes from the symmetric mass division to asymmetric one with increasing the γ–ray multiplicity.

---

1) E-mail address: KEIKO@RIKAXP.RIKEN.GO.JP
2) Present address: Japan Atomic Energy Research Institute, Tokai, Ibaraki 319-11, Japan
3) Present address: Mitsubishi Electric Co., Tokyo, Japan
4) Present address: National Institute of Radiological Science, Anagawa, Chiba 260, Japan
5) On leave from Department of Technical Physics, Pekin University, Beijing, China

INTRODUCTION

In the heavy–ion induced reactions we observe various reaction processes according to energies, masses of the systems and angular momenta brought into the systems. In the reaction processes binary decay process is considered to give an excellent field to study the fundamental properties of nuclear matter such as the viscosity or the friction in the collective motions. For the purpose the detection of prescission light particles and γ–rays in coincidence with binary decay fragments gives new information on the reaction mechanism including its time scale. Numerous experimental results are reported especially for heavy mass systems $A \sim 200$ [1-4]. The careful study of prescission multiplicities has concluded that the main binary decay is a slow process, because the experimental prescission multiplicity exceed the prediction of the standard statistical model calculation.

In case of heavy mass systems ($A \sim 200$) for the study of nuclear dissipation using the measured particle multiplicity it is essential to separate the pre–saddle component from the post–saddle component in the analysis [2].

For the mass region $A \sim 100$, in contrast to $A \sim 200$, the pre–saddle component is dominant in case of the fusion–fission reaction, then it is suitable to observe the physical quantities at the pre–saddle point. In this mass region, however, the reaction mechanism itself is very complex. We observe the fusion–fission reactions and deep–inelastic collisions superposed each other at high angular momentum state. It should be at first made to distinguish the fusion–fission reaction from another reaction processes like quasi–fission processes, deep inelastic collisions (DICs) etc. to study the basic characteristics of nuclear matter using binary decay processes.

In the heavy ion induced reactions it is well known that the angular momentum brought into the system plays an essential role in the classification of its decay processes. Because the measurements of γ–ray multiplicities have been proven to be a good technique for determining the intrinsic angular momentum of compound nuclei [5], we additionally paid attention to γ–ray multiplicities and applied it as an angular momentum filter to classify the reaction processes.

We have utilized the 3π phoswich detector system and measured light charged particle (LCP) and γ–ray multiplicities simultaneously in coincidence with binary decay fragments. Such measurements must give us more detailed information on the reaction and allow us to select the pure fusion–fission reactions and to observe the physical quantities as a function of the angular momentum.

In this report, we present the results of the dependence of prescission time and charge distribution on angular momentum of the system and the dependence of particle multiplicity on total kinetic energy (TKE), and discuss about

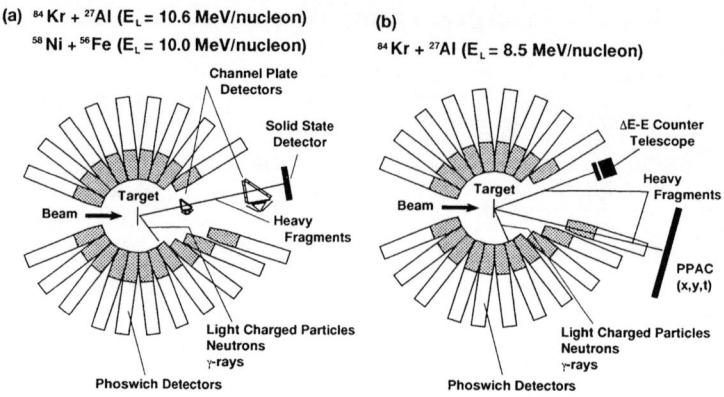

FIGURE 1. Schematic drawing of the experimental set–up in the large scattering chamber ASCHRA. (a) for ^{84}Kr(10.6 MeV/nucleon) + ^{27}Al and ^{58}Ni + ^{56}Fe. (b) for ^{84}Kr + ^{27}Al at 8.5 MeV/nucleon.

the reaction mechanism of medium mass systems from these experimental results.

EXPERIMENT

The experiment was performed using the large scattering chamber (ASCHRA) in the RIKEN Accelerator Research Facility. Figure 1 shows the schematic drawing of the experimental set–up. The multiplicities of protons and α–particles were measured in coincidence with a binary decay fragment in the reactions ^{84}Kr + ^{27}Al at the incident energies of 8.5 and 10.6 MeV/nucleon and ^{58}Ni + ^{56}Fe at 10.0 MeV/nucleon. The thicknesses of the self–supporting Al and Fe targets were 600 μg/cm^2 and 1.0 mg/cm^2, respectively.

In the experiment of ^{84}Kr(10.6 MeV/nucleon) + ^{27}Al and ^{58}Ni + ^{56}Fe, heavy fragments were measured by a time–of–flight (TOF) counter telescope which consists of two channel plate detectors [6] and a large solid state detector ($\Delta\Omega \sim 4$ msr). This telescope was placed at 10 degree from the beam direction. The flight path between two channel plate detectors was 35 cm and the typical time resolution was 300 ps.

For ^{84}Kr(8.5 MeV/nucleon) + ^{27}Al heavy fragments were detected by three ΔE-E counter telescopes and a PPAC. Each telescope consisted of two solid state detectors. The thickness of the ΔE and E detectors were 30 μm and 1 mm, respectively. The telescopes were placed at 10, 20 and 30 degree from the beam direction.

Figure 2 shows the contour plots of the reaction products, the left panel

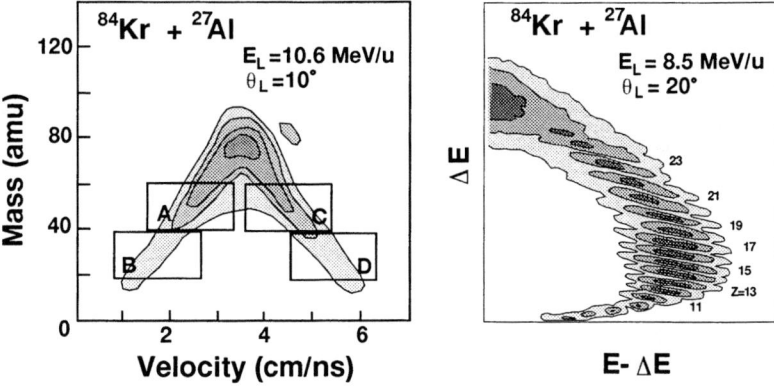

FIGURE 2. The contour plots of the reaction products detected in the time–of–flight counter telescope for the reaction of ^{84}Kr(10.6 MeV/nucleon) + ^{27}Al (left) and in the ΔE–E counter telescope for ^{84}Kr(8.5 MeV/nucleon) + ^{27}Al (right). The mass and velocity gates used in the analysis are also shown.

is the case of the reaction ^{84}Kr + ^{27}Al at the incident energy of 10.6 MeV/nucleon, which were detected in the time–of–flight counter telescope. The right panel is for the incident energy of 8.5 MeV/nucleon, measured by the ΔE–E counter telescope at 20 degree in the laboratory system.

Light charged particles (LCPs) and γ–rays were detected event by event using a 3π–phoswich detector system. The system consists of 120 phoswich detectors and covers the angular range between 10° to 160° in the laboratory system. Each detector has a thin plastic scintillator (NE102A, 100 or 200 μm in thickness) and a thick BaF$_2$ crystal (180 mm long). By this combination it is possible to identify not only LCPs and also γ–rays and neutrons [7]. The time of flight which was derived from the time difference between an RF signal from the cyclotron and a timing signal of a phoswich detector was also measured for the identification of these particles.

Figure 3 shows the contour plots of the light particles detected in a phoswich detector. Light particles are clearly identified by the combination of detected energy and time–of–flight.

RESULTS AND DISCUSSION

To select the fully damped binary decay fragment we made gates on two dimensional plots of heavy fragments. For ^{84}Kr(10.6 MeV/nucleon) + ^{27}Al we selected the regions A and C of the left panel of fig. 2. For ^{58}Ni + ^{56}Fe the region B1 and C of fig. 1(c) in ref. [8] were used. In the case of ^{84}Kr (8.5

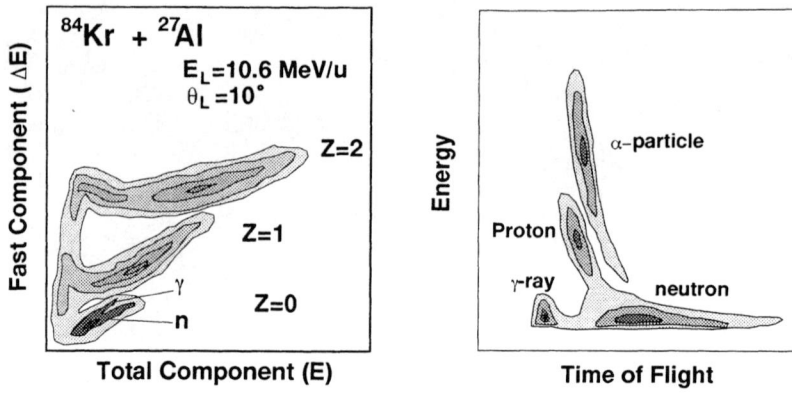

FIGURE 3. The contour plots of the light particles detected in a phoswich detector for the reaction ^{84}Kr(10.6 MeV/nucleon) + ^{27}Al. The left shows the total component(energy) vs. fast component(ΔE) of the phoswich detector. The right is energy vs. time–of–flight.

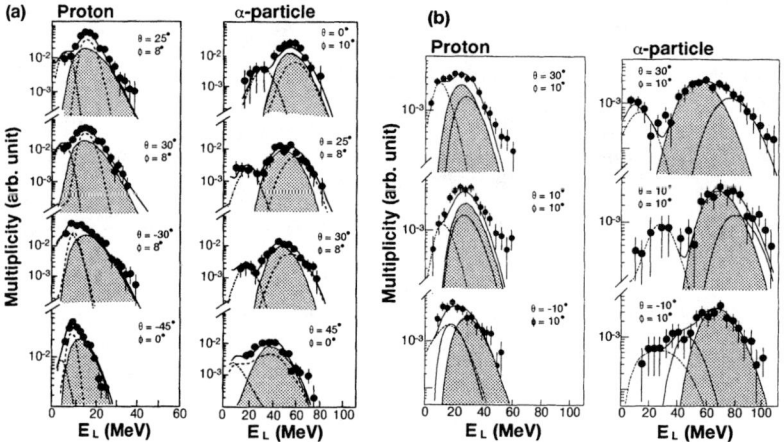

FIGURE 4. Measured energy spectra of LCPs (solid circles) with the fitting results. (a) for the reaction of ^{58}Ni + ^{56}Fe and (b) for ^{84}Kr (8.5 MeV/nucleon) + ^{27}Al. The angle ϕ corresponds to the angle between the direction of the particle emission and spin axis of the compound nucleus. Two postscission components from the detected and undetected fragments are shown by dashed and dotted–dashed lines, respectively. Total spectra is shown by continuous lines. The hatched regions correspond to prescission components.

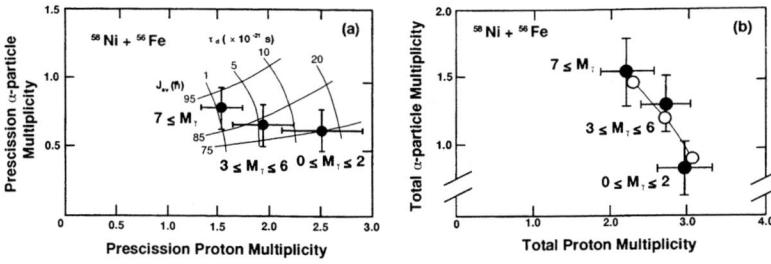

FIGURE 5. For the system ^{58}Ni + ^{56}Fe the experimental prescission α-particle and proton multiplicities, M_α^{pre} and M_p^{pre}, plotted two-dimensionally by closed circles for three gates of γ-ray multiplicity. The solid lines are results of the calculation with a statistical model code PACE2 for various fission delay times (τ_d) and angular momenta ($J_{av}(\hbar)$). (b) Same as (a), but of the total α-particle and proton multiplicities: M_α^{tot} and M_p^{tot}. Closed circles show the experimental results, whereas open circles are calculated ones with the statistical model code feeding back the angular momentum and pre-scission time obtained from fig. 5(a).

MeV/nucleon) + ^{27}Al, $20 \leq Z \leq 25$ events were used.

For these events we made a further selection in terms of γ-ray multiplicity to extract information of the angular momentum state of the compound nucleus. The selected bins are $0 \leq M_\gamma \leq 2$, $3 \leq M_\gamma \leq 6$ and $7 \leq M_\gamma$. For the events selected by mass and velocity of heavy fragments and γ-ray multiplicity, the energy spectra of the protons and α-particles were fitted by the moving source analysis in which three sources were assumed, i.e., a composite system before scission and two fully accelerated fragments after scission. We have fitted the energy spectra of protons and α-particles over the whole angular range (the detail of the fitting procedure, see [9]), and finally extracted the pre- and post-scission LCP multiplicities for each γ-ray multiplicity region. Figure 4(a) and (b) show the typical examples of fitting results. (a) shows the case of ^{58}Ni + ^{56}Fe and (b) for ^{84}Kr(8.5 MeV/nucleon) + ^{27}Al.

Angular momentum selection and prescission time determination

We tried to extract the prescission time and angular momentum for each γ-ray multiplicity region from the obtained prescission and total proton and α-particle multiplicities(M_p^{pre}, M_p^{tot}, M_α^{pre} and M_α^{tot}). We made a statistical calculation using the code PACE2 [10]. We calculated the total and prescission proton and α-particle multiplicities changing the fission delay time and angular momentum. We used the excitation energy $E^* = 160$ and 200 MeV for

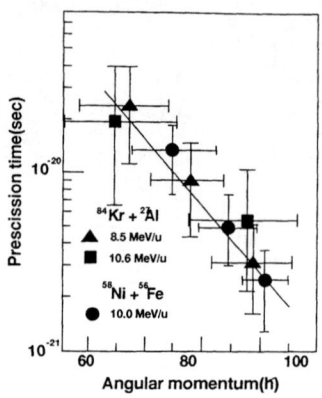

FIGURE 6. Plot of the pre–scission times as a function of the angular momentum of the system. Closed circles, triangles and squares shows the results of ^{58}Ni + ^{56}Fe, ^{84}Kr + ^{27}Al at the incident energy of 8.5 and 10.6 MeV/nucleon, respectively.

^{84}Kr + ^{27}Al at the incident energies of 8.5 and 10.6 MeV/nucleon, respectively. For ^{58}Ni + ^{56}Fe, we used the excitation energy E* = 205 MeV, which is about 10 % lower than the initial value [11]. The used level density parameter is $a = A/8$. $a_f/a_n = 1.0$ was used. It is necessary to reproduce the ratio of proton and α–particle multiplicities by the calculation, because the ratio of proton and α–particle evaporation is strongly dependent on the angular momentum of the system at the unique excitation energy.

The experimental M_p^{pre} and M_α^{pre} are two–dimensionally plotted with the calculated results (solid lines) in fig. 5(a), M_p^{tot} and M_α^{tot} in fig. 5(b), for the case of ^{58}Ni + ^{56}Fe. It is clearly shown in fig. 5(b) that the ratio M_α^{tot}/M_p^{tot} increases as the γ–ray multiplicity increases. From the comparison between the calculated results and experimental data we have determined the angular momentum and prescission time for each γ–ray multiplicity bin.

The obtained prescission time of two reactions are plotted as a function of angular momentum of the system in fig. 6. It is clearly seen that the prescission time decreases as the angular momentum of the system increases.

Fusion–fission and quasi–fission process

It is reported that the prescission neutron multiplicity is essentially independent of the total kinetic energies of fragments (TKE) for fusion–fission reactions, however, the dependence in quasi–fission reactions is completely different, that is, the prescission neutron multiplicity decreases as the TKE increases [1,11,12]. According to this result we used the TKE to select the

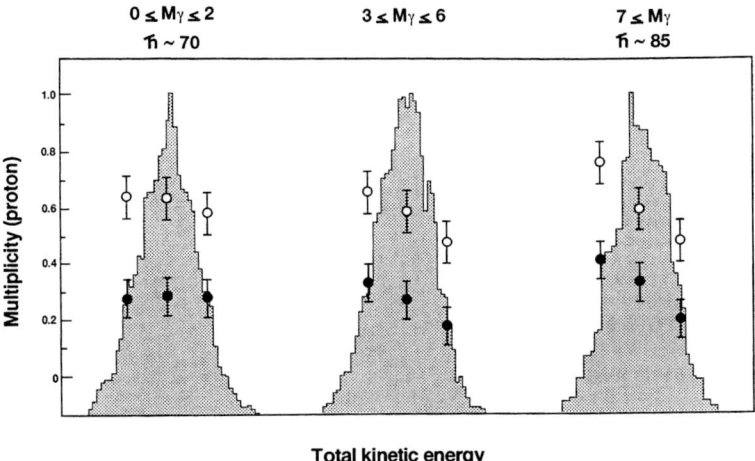

FIGURE 7. The dependence of proton multiplicity on TKE of binary decay fragments for three γ-ray multiplicity regions. Open and closed circles show total and prescission proton multiplicities, respectively.

pure fusion–fission reactions. For the three γ-ray multiplicity regions, we have additionally divided the TKE of fragments into three parts (low, medium and high). For each TKE part, we have extracted the prescission LCP multiplicity by moving source fit. Figure 7 shows the results of this analysis for the reaction of ^{84}Kr + ^{27}Al at the incident energy of 8.5 MeV/nucleon. For the low angular momentum state, the prescission proton multiplicity (closed circles) seems constant against the TKE, however, for the larger angular momentum state (medium and high), the prescission proton multiplicity decreases as the TKE increases.

From these results, for the angular momentum $\sim 70\,\hbar$ the pure fusion–fission reaction is dominant and for the higher angular momentum the quasi–fission is dominant. The extracted prescission time is $\sim 2 \times 10^{-20}$ s for the pure fusion–fission processes.

Charge distribution

We have also observed the charge distribution of heavy fragments as a function of angular momentum state. The γ-ray multiplicity was divided into 4 regions ($0 \leq M_\gamma \leq 1$, $2 \leq M_\gamma \leq 3$, $4 \leq M_\gamma \leq 6$ and $7 \leq M_\gamma$). Cross sections of fragments were extracted for each region. Figure 8 shows the obtained

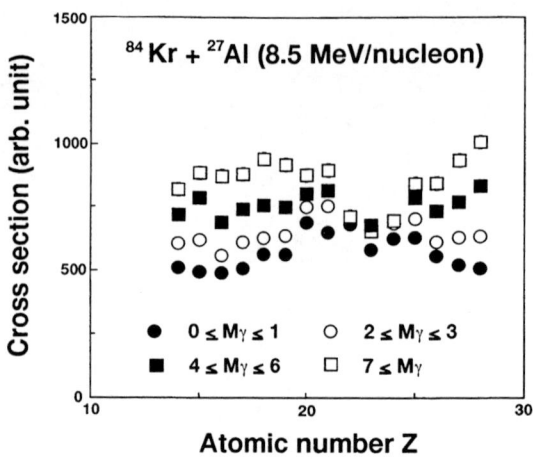

FIGURE 8. The charge distributions of fragments obtained in the reaction of ^{84}Kr(8.5 MeV/nucleon) + ^{27}Al with four γ-ray multiplicity gate. Closed and open circles, closed and open squares correspond to the γ-ray multiplicity, $0 \leq M_\gamma \leq 1$, $2 \leq M_\gamma \leq 3$, $4 \leq M_\gamma \leq 6$ and $7 \leq M_\gamma$, respectively.

charge distributions of fragments for the reaction ^{84}Kr + ^{27}Al at the incident energy of 8.5 MeV/nucleon. The charge distribution changes drastically: for the low γ-ray multiplicity the charge distribution has a bump at the mass ~ 22, with increasing γ-ray multiplicity the charge distribution changes from the symmetric mass division to the asymmetric one, indicating the transition from fusion–fission process to quasi–fission process.

SUMMARY AND CONCLUSION

We made an experiment in the systems of ^{84}Kr (10.6 and 8.5 MeV/nucleon) + ^{27}Al and ^{58}Ni (10.0 MeV/nucleon) + ^{56}Fe. We measured γ-ray multiplicity and LCP multiplicity simultaneously in coincidence with binary decay fragments in event–by–event mode. From the results of the system ^{84}Kr + ^{27}Al at two incident energies the prescission time deduced from the observed multiplicities is found to become smaller exponentially from 2×10^{-20} s to 3×10^{-21} s with increasing the angular momentum from 65 to 90 $\hbar$.

For ^{84}Kr (8.5 MeV/nucleon) + ^{27}Al the dependence of prescission proton multiplicity on the TKE of the fragments shows that the pure fusion–fission reaction is dominant only for the angular momentum $\sim 70~\hbar$ and for the higher angular momentum the quasi–fission process is dominant. The charge distribution of heavy fragments also supports this conclusion, because it changes drastically from symmetric distribution to asymmetric one as the angular momentum of the system increases.

ACKNOWLEDGMENTS

We would like to thank the crew of RIKEN Accelerator Research Facility for their excellent machine operation during the experimental course of work. One of the authors (K.Y.N.) appreciates the financial support by Special Researcher's Basic Science Program of the Institute of Physical and Chemical Research (RIKEN).

REFERENCES

1. Hilscher, D., and Rossner, H., *Ann. Phys. Fr.* **17**, 471 (1992) and references therein.
2. Lestone, J., P., *et al*, Nucl. Phys. **A559**, 277 (1993).
3. Benrachi, F., *et al*, Phys. Rev. C **48**, 2340 (1993).
4. Paul, P., and Thoennessen, M., *Annu. Rev. Nucl. Part. Sci.* **44**, 65 (1994).
5. Schröder, W., U., and Huizenga, J., R., *Treatise on Heavy-Ion Science*, New York: Prenum, 1984, Vol. 2, Sect. 3.6., pp. .
6. Mizota, T., *et al*, Nucl. Instr. Meths. **A305**, 125 (1991).
7. Futami, Y., *et al*, Nucl. Instr. Meths. **A326**, 513 (1993).
8. Yuasa-Nakagawa, K., *et al*, Phys. Lett. **B351**, 77 (1995).
9. Yuasa–Nakagawa, K., *et al*, Phys. Rev. C **53**, 997 (1996).
10. Gavron, A., Phys. Rev. C **21**, 230 (1980).
11. Hinde, D., J., *et al*, Phys. Rev. C **45**, 1229 (1992).
12. Rossner, H., *et al*, Phys. Rev. C **43**, 2434 (1991).

FUSION OF MASSIVE NUCLEI

Investigation of the Extra-extra-push by Pre-scission Neutron Measurements with DEMON

Gérard Rudolf* for the Demon collaboration*,†

*Institut de Recherches Subatomiques, IN2P3-CNRS/Université Louis Pasteur
BP 28, F-67037 Strasbourg Cedex (France)
†ISN, IN2P3-CNRS/Université Joseph Fourier, Grenoble (France)
Université Libre de Bruxelles, Bruxelles (Belgium)
Flerov Laboratory of Nuclear Reactions, JINR, Dubna (Russia)
Vinca Institute, Belgrade (Yugoslavia)
LPC, IN2P3-CNRS/ISMRA et Université de Caen (France)

Abstract. The objective of this talk is to present a simple method to calculate pre- and post-scission neutron multiplicities in the frame of the Bass model. This method is of particular interest for very heavy systems for which an extra-extra-push is supposed to hinder fusion. The multiplicities calculated by the model are compared to published data covering a broad range of projectile and target masses, and to more recent ones obtained with the help of the Demon detector and addressing specifically the existence of the extra-extra-push.

INTRODUCTION

The fusion of two nuclei occurs as soon as the bombarding energy exceeds the fusion barrier, which, for light or intermediate mass nuclei, is generally quite close to the interaction barrier. For heavy nuclei, a hindrance to fusion appears. Indeed, it has been shown [1] that the threshold for the production of symmetric fragments can be several tens of MeV higher than the interaction barrier. The two nuclei may come to contact, interact strongly through a deep inelastic reaction, but do not fuse even for head-on collisions.

Deep inelastic collisions are essentially explained by the high angular momentum brought in the system, so that repulsive forces overcome the attractive ones and prevent the system from fusion. The hindrance to fusion in the case of head-on collisions between heavy nuclei can be explained by an increase, with respect to the Bass model, [2] of the Coulomb force, or by a loss of the kinetic energy before the system reaches the entrance saddle point. The first explanation relies on a dynamical deformation due to the strong mutual electric field [3], the second on the effect of friction for systems where the saddle point configuration is more compact than the contact point [4].

Symmetric fragmentation appears only once the kinetic energy is increased by the so-called "extra-push" which compensates either of the above mentionned effects. This extra-push has of course a direct implication on the minimum excitation energy at which one can expect to fuse heavy nuclei with sizeable cross-section.

But symmetric fragmentation is not a signature for fusion-fission, since other mechanisms such as fast-fission or quasi-fission can produce equal mass fragments without the formation of a compound nucleus. Both mechanisms are explained by the lifetime of the di-nuclear system which can be long enough to equilibrate all degrees of freedom. Dynamical calculations have suggested that an even higher bombarding energy, corresponding to the extra-extra-push, is needed to overcome these mechanisms and reach true fusion.

The extra-extra-push is difficult to measure. A few experiments have tried to distinguish fast-fission from fusion-fission on the basis of fragment angular distributions [5]. The values which were obtained for the extra-extra-push seem too high and contradict the cross-sections which lead to the synthesis of the heaviest elements.

A possible new way to measure the extra-extra-push has been provided by the discovery that, during the fission process, the path from the equilibrated shape to scission lasts long enough to allow many neutrons to be emitted by the compound nucleus before it separates in two fragments [6]. The deconvolution of neutron angular distributions into pre- and post-scission distributions allows to establish a clock [7] on the fusion process. Accordingly, it should allow to distinguish fast-fission from fusion-fission.

THE MODEL

Pre-scission neutron multiplicities ν_{pre} and post-scission ones ν_{post} have generally been analysed in the frame of the statistical model [6]. However, in its original version this model underpredicts strongly ν_{pre}. It has to be modified by introducing a delay during which fission is blocked. This delay allows the compound nucleus to evaporate particles. When the experimental and calculated multiplicities match, one has determined the delay to fission. In the case of the ^{20}Ne + ^{nat}Re system, for example, a Gemini calculation predicts about ν_{pre}=3 and ν_{post}=8 if the delay is 10^{-20}s (please note that we define ν_{post} as the total number of neutrons emitted by both fragments) . For a delay of 10^{-19}s, these numbers become 7 and 4, respectively. The measured values of 5.6 and 4.8 correspond to $4*10^{-20}$s which is considered as the delay time for fission [8].

Such a model has to be adapted to take into account the effect of an possible extra-extra-push. As a first step, I prefer here to use a more simple model based on the Bass model [2]. Since it has been shown that fission fragments are produced almost cold at the scission point, independent on the bombarding energy, we make the assumption that post-scission evaporation is due to a small excitation energy E_{excff} remaining at the scission point plus the deformation energy stored in the fragments. Moreover, we admit that this deformation energy E_{evapff} is given by the difference between the saddle-point potential energy for two undeformed spheres $E_{spheres}$, and the Viola energy E_{Viola}. Thus

$$E_{evapff} = E_{spheres} - E_{Viola} + E_{excff} \qquad (1)$$

$E_{spheres}$ is calculated with the expressions given in ref. [9], which are based on phenomenological radii and a proximity potential.

We make moreover the extreme assumption that no excitation energy is evaporated by the combined system in the case of fast-fission. In the case of fusion-fission, the energy available for pre-scission evaporation is

$$E_{evapcn} = E_{cm} + Q_{gg} - E_{Viola} - E_{evapff} \qquad (2)$$

i.e. the total available excitation energy minus E_{evapff}. The multiplicity of evaporated neutrons is given in all cases by

$$\nu = E_{evap}/\Delta E_{evap} \qquad (3)$$

where

$$\Delta E_{evap} = e_{bind} + 2t + a_p E_{exc} \qquad (4)$$

$$t = (aE_{exc}/A)^{1/2} \qquad (5)$$

and E_{exc} is the excitation energy of the given nucleus before it undergoes evaporation. The binding energy e_{bind} for one neutron and the density parameter a have been taken equal to 8 MeV throughout the range of systems studied in this paper. The last term in the expression of ΔE_{evap} is introduced to take the charged particle evaporation into account. A constant value of 0.019 has been taken for a_p.

Measurements of ν_{pre} are generally triggered by the detection of symmetric fragments. Therefore, one may mix events corresponding to fast-fission and fusion-fission. For head-on collisions, the transition between the two mechanisms occurs when the kinetic energy overcomes the extra- extra-push. For a given l-value, one has to take into account the centrifugal energy and therefore the transfer of angular momentum. Thus, ν_{pre} is given by:

$$\nu_{pre} = \frac{\sigma_{fusion}}{\sigma_{fusion} + \sigma_{fast-fission}} \frac{E_{evapcn}}{\Delta E_{evapcn}} \qquad (6)$$

where

$$\sigma_{fusion} = (1 - f_{lfusion})^{-2} \frac{\pi r^2}{E_{cm}} [E_{cm} - B_{fus} - E_{push}] \qquad (7)$$

$$\sigma_{fast-fission} = (1 - f_{lfast-fission})^{-2} \frac{\pi r^2}{E_{cm}} [E_{epush} - E_{push}] \qquad (8)$$

The values of the extra-push E_{push} and of the extra-extra-push E_{epush} have been taken from the systematics of ref. [5]. The only free parameters are therefore:

- E_{excff} which is the total excitation energy of the fission fragments at the scission point. In the same manner as the delay time in the case of the statistical model, E_{excff} can be tuned so that the predicted multiplicities correspond to the measured ones.
- $f_{lfusion}$ and $f_{lfast-fission}$ which give the rate of angular momentum transfered from the relative motion to the intrinsic degrees of freedom. These parameters can vary between 0 and a maximum given by the Bass model [2].

Two different hypothesis have been examined [10]:

- $f_{lfusion}$ and $f_{lfast-fission}$ are equal. This hypothesis corresponds to a picture where the system follows a given path, experiences some energy loss and momentum transfer, and according to the balance between the remaining kinetic energy and the potential energy fuses or re-separates.
- $f_{lfusion}$ is equal to 0 and $f_{lfast-fission}$ is equal to the maximum corresponding to a sticking configuration. It means that two different paths exist, one without friction leading to fusion and the other with friction leading to fast-fission. This hypothesis seems not to agree with the dynamical picture developped in ref [4]. However, it has the merit to lead to a better agreement with the data [10], and has been used in the following sub-section.

COMPARISON WITH DATA

With extra-extra-push

Let us first use the model with the values for the extra-extra-push given in ref. [5] and with E_{excff} equal to 0. Applied to the ^{20}Ne + ^{187}Re system, the

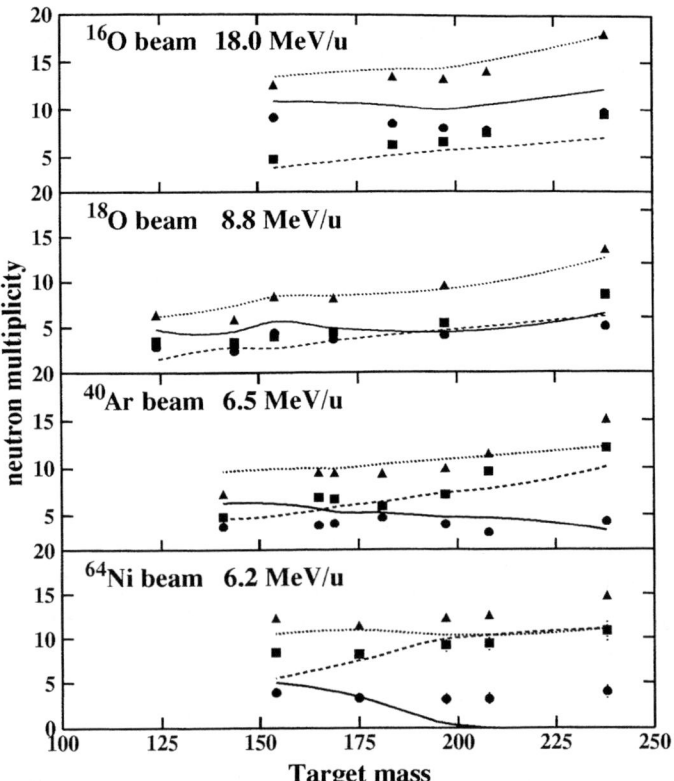

FIGURE 1. Experimental and calculated neutron multiplicities: ν_{tot} (triangles and dotted curve), ν_{pre} (circles and continuous curve), and ν_{post} (squares and dashed curve). Please note that in this paper $\nu_{tot}=\nu_{pre}+\nu_{post}$.

model predicts $\nu_{pre}=6.7$ and $\nu_{post}=4.4$. These values correspond to $9*10^{-20}$s in the Gemini calculation [8] i.e. to a very long time at which the decay process is almost finished. Fig.1 shows a comparison between the data covering a broad range of projectile and target masses [11] and the predictions of the model. Since E_{excff} has been taken equal to 0, the model overestimates ν_{pre} in most cases. However, for the ^{64}Ni induced reactions, it underestimates it strongly. This is due to the extra-extra-push which is not overcome by a system like ^{64}Ni + ^{208}Pb at 6.2 MeV/u. In opposition to the model, the observed values for ν_{pre} depend little on the target mass, and seem to exceed reasonable expectations for evaporation by the combined system in a fast-fission process. This is a first hint that the concept of the extra-extra-push is not valid.

Another study has been performed more recently on the ^{58}Ni + ^{208}Pb and

^{64}Ni + ^{208}Pb systems, for which the excitation functions of ν_{pre} and ν_{post} have been measured by the Demon collaboration [12]. Fig. 2 shows the comparison between the model and the data. A fair agreement is obtained for the ^{58}Ni

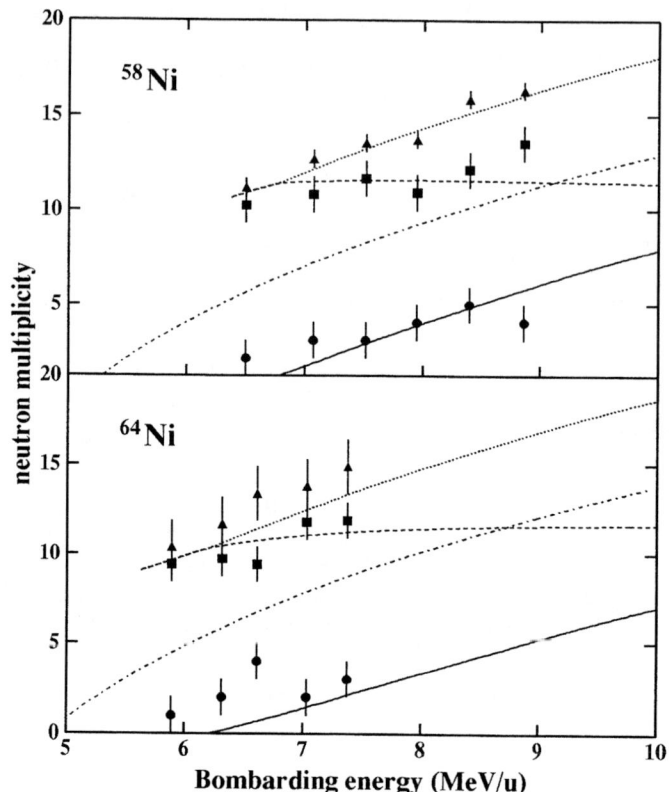

FIGURE 2. Same as Fig.1. The dashed-dotted curve indicates the total number of neutrons evaporated in a deep-inelastic reaction.

induced reaction. At the two lowest bombarding energies, the measured ν_{pre} values exceed somehow the calculated ones, but this can be explained by the fact that fast-fission can be accompanied by evaporation from the combined system. More astonishing is the result obtained at the highest energy. An increase of ν_{post} and a decrease of the ν_{pre} is not expected by the model. The phenomenon is more pronounced in the case of the ^{64}Ni induced reaction, where two regimes seem to be observed. The model predicts correctly the higher energy points. But the three lower energy points deviate clearly from the expectations.

Without extra-extra-push

Let us now use the model with different parameters:

- we suppose that the extra-extra-push does not exist, or better said that E_{epush} is equal to E_{push}. Equation 8 shows that the cross-section for fast-fission is then equal to 0, which means that as soon as the system has overcome the extra-push, it fuses to a compound nucleus. Equation 6 is then much simplified, and we need not to make any assumption about the transfer of angular momentum.

- the excitation energy of the fission fragments at the scission point E_{excff} is taken equal to 25 MeV, regardless of the system

Applied to the ^{20}Ne + ^{187}Re system, the predictions of the model are now 5.2 for ν_{pre} and 6 for ν_{post}. These values are closer to the experimental ones measured for the ^{20}Ne + ^{nat}Re system and correspond to 3.5 10^{-20}s in the Gemini calculation [8].

Fig.3 compares the predictions of the model to the data of ref. [11]. The over-all agreement is greatly improved. The strongest deviations occur for the lightest and the heaviest targets. Light systems are not very far from the Businaro-Gallone point, at which the model may no more be relevant. Especially astonishing is the relatively good agreement for the ^{64}Ni induced reactions, when compared to the preceding calculation (see Fig. 1).

Fig.4 compares the predictions of the model with this parameter set to the Demon data. The model predicts now correctly the three lower energy points of the ^{64}Ni induced reaction. Of course, there is now a definite disagreement for the ^{58}Ni induced reaction, and for the two highest points of the ^{64}Ni induced one.

As shown by Fig. 2, these points seem to correspond better to the hypothesis that the extra- extra-push exists and that E_{excff} is equal to 0. However, it would be difficult to understand why an extra-extra-push could appear only at high bombarding energy in the case of the ^{64}Ni induced reaction. Moreover, we have seen already that the hypothesis we had to make on the angular momentum transfer is not quite founded. Considering the good overall agreement illustrated by Fig. 3, it sems more reasonable to suppose that extra-extra-push does not exist and that in some cases E_{excff} is higher than 25 MeV. Fig.5 shows the result without extra-extra-push and whith E_{excff}=70 MeV. The agreement is now quite good with nearly all points which are off the model in Fig. 4.

CONCLUSION

It is possible to calculate pre- and post-scission neutron multiplicities through a very simple model based on the Bass model. The predictions have

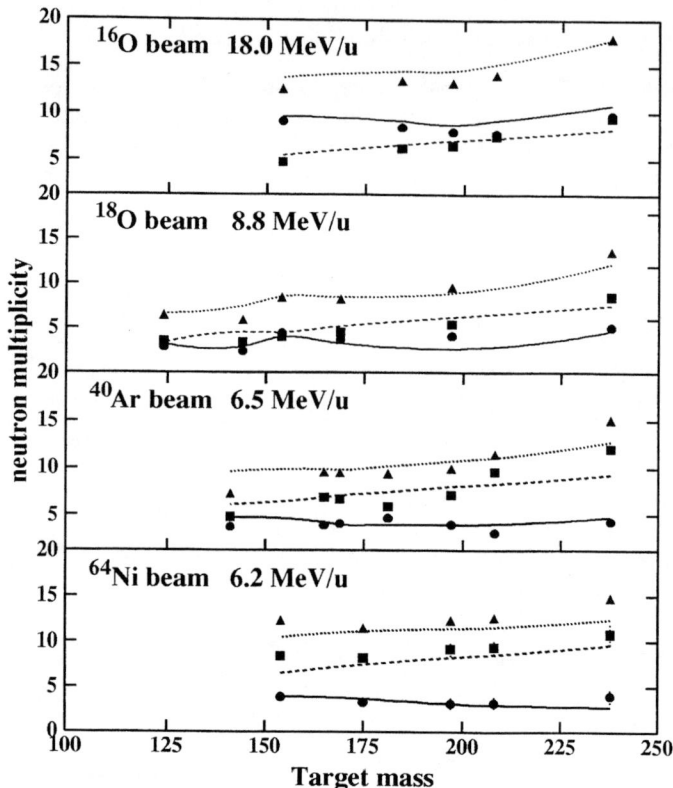

FIGURE 3. Same as Fig. 1, but here $E_{epush}=E_{push}$ and $E_{excff}=25$ MeV

been compared to data spaning a broad range of projectile and target masses. With one single set of parameters, they agree with most of the data to within less than one unit with the assumptions that:

- $E_{epush}=E_{push}$, which means that no extra-extra-push is needed.

- E_{excff}, which is the total excitation energy of the fragments at the scission point, is equal to 25 MeV.

In the case of the two reactions leading to element 110, only the low energy points for the ^{64}Ni + ^{208}Pb reaction agree with these parameters. For the higher energies and for the ^{58}Ni induced reaction, a second regime has been evidenced, corresponding to $E_{excff}=70$ MeV. Detailed statistical calculations will be needed to decide if this observation can be interpreted as an indication

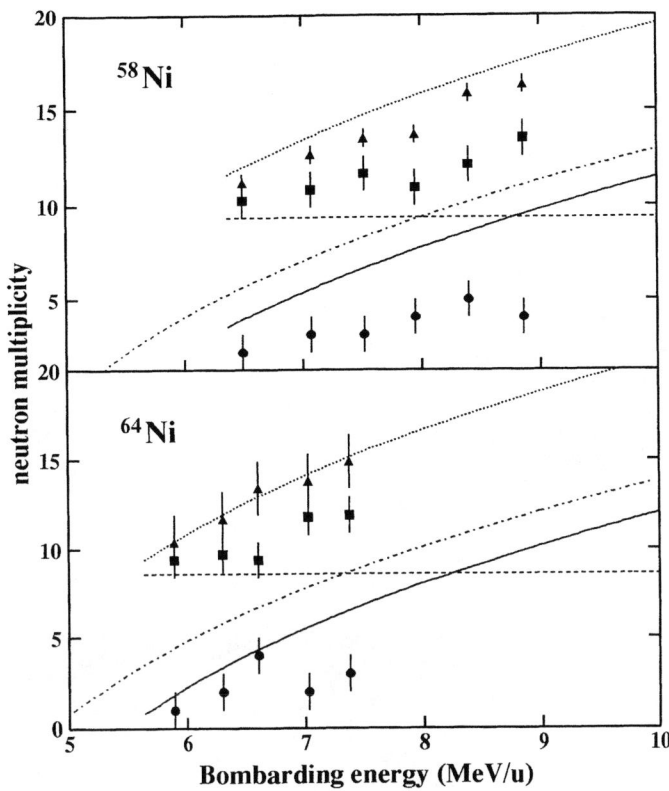

FIGURE 4. Same as Fig.3.

that the delay time for fission of some very heavy nuclei at moderately high excitation energy is shorter than suggested by systematics. The difference between the ^{58}Ni and ^{64}Ni induced reactions may mean that not only the mass and the excitation energy play a role in the shortening of the lifetime of the nucleus against fission, but also the A/Z ratio.

REFERENCES

1. Sann, H., et al., *Phys. Rev. Lett.* **47**, 1248 (1981).
2. Bass, R., *Nucl. Phys.* **A231**, 45 (1974).
3. Blocki, J.P., *Nucl. Phys.* **A459**, 145 (1986)
4. Frobrich, P., *Phys. Lett.* **B215**, 36 (1988)
5. Shen, W.Q., et al., *Phys. Rev.* **C36**, 115 (1987)
6. Hinde, D.J., et al., *Nucl. Phys.* **A452**, 550 (1986)

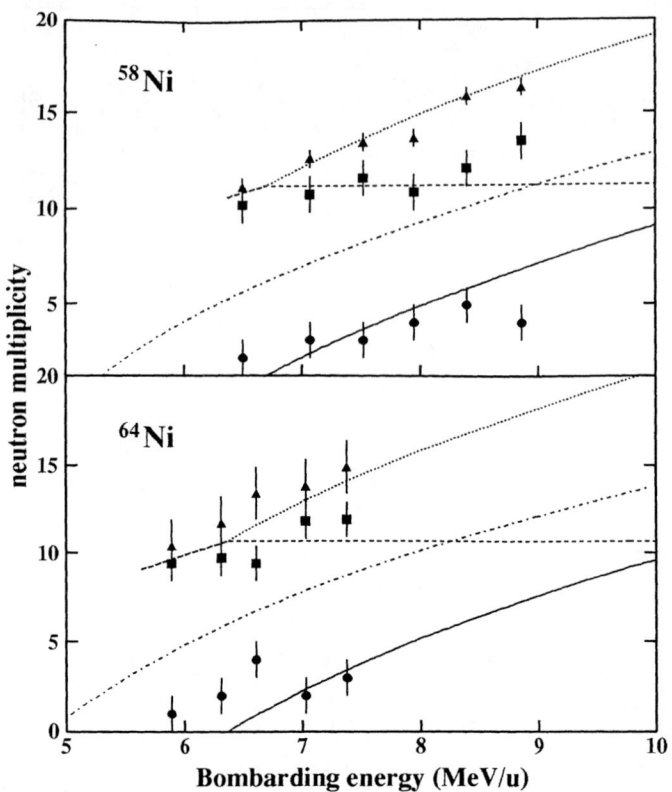

FIGURE 5. Same as Fig.1, but $E_{epush}=E_{push}$ and $E_{excff}=70$ MeV

7. Hilscher, D., et al., *Phys. Rev. Lett.* **62**, 1099 (1989)
8. Tilquin, I., PhD work in progress, UCLN, Louvain-la Neuve
9. Toke, J., et al., *Nucl. Phys.* **A440**, 327 (1985)
10. Rudolf, G., "Investigation of fission time scales in various mass regions by neutron spectra measurements", presented at the Conference on Advances in Nuclear Physics and Related Areas, Thessaloniki, Greece, July 8 - 12, 1997.
11. Hinde, D.J., et al., *Phys. Rev.* **C45**, 1229 (1992).
12. Donadille, L., PhD work in progress, ISN Grenoble.

Fusion Dynamics of Massive Nuclei

F.A. Ivanyuk[1], G.I. Kosenko[2], Yu.Ts. Oganessian[2], V.V. Pashkevich[3]

[1] Institute for Nuclear Research, 252028, Kiev-28, Ukraine
[2] Flerov Laboratory of Nuclear Reactions JINR, 141980 Dubna, Russia
[3] Bogoliubov Laboratory of Theoretical Physics JINR, 141980 Dubna, Russia

Introduction.

The permanent interest of scientists in the synthesis of superheavy elements and in the search for the island of stability of superheavy elements is a stimulating factor for a further investigation of the properties of the nuclear system in the process of compound-nucleus formation. This paper deals with the dependence of the probability of compound-nucleus formation on the distance between the ions approaching in the entrance channel. After the system has been formed in the absence of relative motion it has only inner excitation for some time during which the system remains unchanged. We shall try to answer the following question: "In what way does the probability of compound-nucleus formation vary as a function of the excitation energy stored in the system and as a function of the distance between the ions before translational motion stops?" In other words, the question is whether or not complete fusion occurs if the initial energy is not enough for complete fusion to take place before the translational motion of the ions stops. What will happen if the system will stop at the barrier vertex or a little before reaching it? A similar problem has been investigated by a Japanese group [1] who used a one-dimensional model. Our colleagues studied the influence of the excitation energy on the probability of compound-nucleus formation with the starting configuration corresponding to two touching spheres. In our case we consider the starting configuration in the region of the saddle point.

Model.

To solve the problem formulated we used the following Langevin equations

$$\dot{q}_i = \mu_{ij} p_j, \quad \dot{p}_i = -\frac{1}{2} p_j p_k \frac{\partial \mu_{jk}}{\partial q_i} - \frac{\partial F}{\partial q_i} - \gamma_{ij} \mu_{jk} p_k + \vartheta_{ij} \xi_j, \qquad (1)$$

where $\vec{q}=(\rho,h), \vec{p}=(p_\rho, p_h)$ are vectors of the collective coordinates and conjugate momenta, F(q) is free energy, m_{ij} ($\|\mu_{ij}\|=\|m_{ij}\|^{-1}$) and γ_{ij} are the inertial and frictional tensors, and $\vartheta_{ij}\xi_j$ is a random force. We note that ξ is a random quantity (value) having the properties $<\xi>=0, <\xi_i\xi_j>=2\delta_{ij}$. The amplitudes of the stochastic force ϑ_{ij} are related to the diffusion tensor D_{ij} as follows

$$D_{ij} = \xi_{ik}\xi_{kj}, \qquad (2)$$

In turn, D_{ij} satisfies the Einstein relation

$$D_{ij} = T\gamma_{ij}. \qquad (3)$$

Here T is the nuclear temperature which can be determined according to the formula

$$T = \left\{\frac{d\ln\rho(U)}{dU}\right\}^{-1}, \qquad (4)$$

with parameters taken from [2]. Here ρ is the single-particle level-density. The calculations have been carried out using two parametrizations of the nuclear shape. The first one was taken from [3]. In this case the liquid-drop model [4] was employed for potential energy calculations. The inertial and frictional tensors were determined by the Werner-Wheeler method [5]. The nuclear viscosity was assumed to be a one-body quantity and calculated according to the "wall-and-window" formula. The other parametrization was based on Cassini ovaloids [6]. In this parametrization the inertial and frictional tensors were calculated by using the linear response theory, as in [7]. In figs.1 one can see the friction tensor as functions of the coordinates and temperature. It is clearly seen that at a low excitation energy the frictional tensors has many peaks. The surface of $\gamma_{\varepsilon\varepsilon}$-component friction tensors is very uneven as a result of the quasi-crossing of the single-particle levels. As the temperature increases the peaks become almost smooth and the absolute values of the frictional parameter increase. In coming closer to the scission point the inertial and frictional tensors increase for the same value of temperature (or energy) and reach their maximum values in the region of neck occurrence. At this point they begin to decrease. In this case the potential energy was calculated taking the shell correction into account, i.e.

$$V(\vec{q}) = V(\vec{q})_{LDM} + V(\vec{q})_{shell} f(T), \qquad (5)$$

^{216}Th, $\gamma_{\varepsilon\varepsilon}$

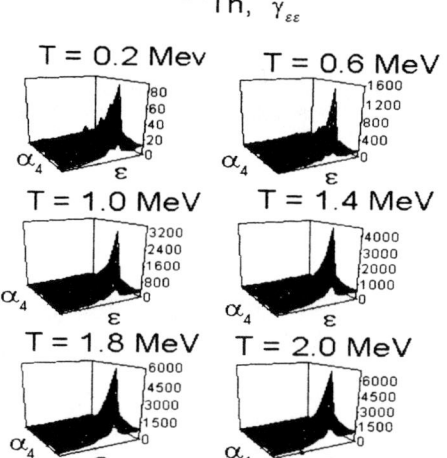

Fig.1 $\gamma_{\varepsilon\varepsilon}$ component of the friction tensor

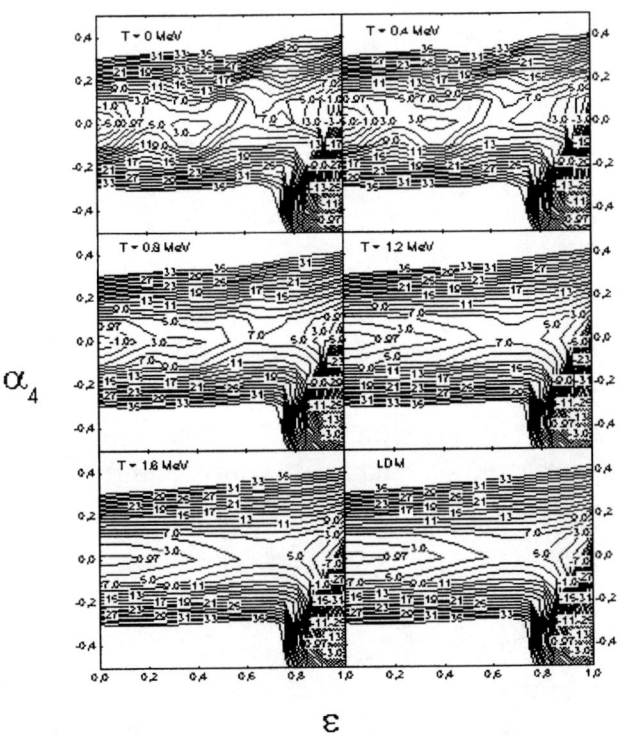

^{216}Th, deformation energy

Fig.2

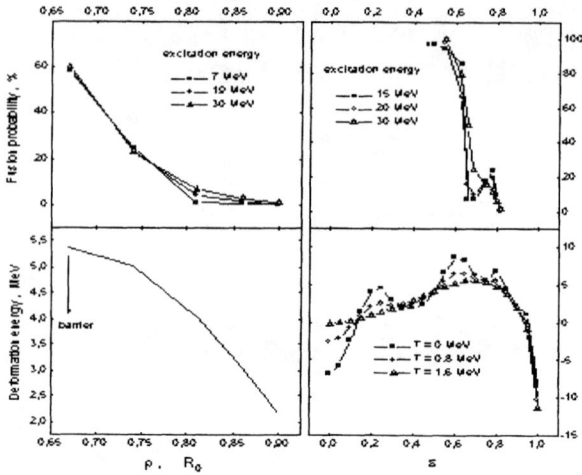

fig.3

where $f(T) = \exp(-\gamma a T^2)$. Here a is the level density parameter, $\gamma = 0.070 MeV^{-1}$ in accordance with [3]. The temperature dependence of the potential energy is presented in fig. 2. The shell effect decreases with increasing temperature and the potential energy slightly differs from the liquid-drop value already at $T = 1.6 MeV$

Results.

The results of the calculation performed are presented in figs. 3. The data for the first and the second (taking the shell structure into account) parametrizations are given in the right and the left panels, respectively. Let us first turn to the left-hand panel in which the results obtained using the $\{c,h,\alpha\}$ parametrization are shown. The dependence of the potential energy on the distance between the centres of mass of the future fragments, $\rho(c)$, is shown in lower parts of the figures. In the upper part one can see the probability of compound-nucleus formation as a function of ρ. This calculation has been performed for three values of the initial excitation energy. In one case the excitation energy is above the barrier by 4-5 MeV. In another case the difference reached 20 MeV. In the situation where the starting point lies on the barrier vertex the probability of compound-nucleus formation is close to 50-60 %. As we move the starting point farther from the barrier vertex the compound-nucleus probability decreases. If the system stops at a point lying 3 MeV below the barrier vertex the probability of complete fusion does not exceed 3%. We note that the initial excitation energy influences the result very slightly.

For the second-parametrization calculation (left-hand panel) taking the shell structure effect into account we observe another situation. The lower part of the figure shows the potential energy surface along the $\alpha_4 = 0$ line for different temperatures. The shell structure is seen at zero temperature whereas at a temperature of 1.6 MeV the picture becomes practically a liquid-drop one. In this case the compound-nucleus formation probability strongly depends on the initial excitation energy. If the excitation energy exceeds the shell barrier as little as by 3-4 MeV, the shell structure of the potential energy surface affects the probability of compound-nucleus formation. Initially the particle is situated on the lower potential energy surface. However, as soon as the fluctuation mechanism begins to act, a part of the inner energy is transformed to the collective one. This fact leads to the occurrence of the shell structure and the particle loses its velocity. If the initial excitation energy is of the order of 30 MeV the probability of compound-nucleus formation behaves as in the liquid-drop case. In this case the shell structure effect almost does not manifest itself. The probability of complete fusion drops abruptly as the starting point is displaced from the barrier vertex.

The calculations carried out allow us to draw the following conclusions. First, as a result of fluctuations the compound-nucleus can be formed even if the system does not reach the barrier vertex. Second, the complete fusion probability considerably depends on the fact at which point the nascent system stops in the entrance channel. At low excitation energies the probability of compound-nucleus formation depends on the shell structure of the nucleus. Therefore, the situations in which the excitation energy exceeds the shell barrier by 10-15 MeV are most preferable.

References.
[1] Wada T et al. // 2rd Japanese-Italian Symp. 1995, RIKEN; Talk at Russian-German Workshop, Dubna, 12-14 Septemer,1996.
[2] Iljinov A.S. et al. // Nucl.Phys. A543 (1992) 517.
[3] Brack M. et al. //Rev.Mod.Phys. 44 (1972) 320.
[4] Myers W.D., Swiatecki W.J. // Ark.Fys. 36 (1967) 343.
[5] Davies K.T.R. et al. // Phys.Rev. C13 (1976) 2385.
[6] Pashkevich V.V. // Nucl.Phys. A169 (1971) 275.
[7] Ivanyuk F.A. et al. // Phys.Rev. C55 (1997) 1730; Preprint RIKEN-AFNP-230, Juli 1996.

MULTI-DIMENSIONAL LANGEVIN APPROACH TO THE FUSION OF MASSIVE NUCLEI

T. Tokuda, K. Okazaki, T. Wada, M. Ohta and Y. Abe[†]

Department of Physics, Konan University, Okamoto 8-9-1, Kobe 658, Japan
[†] *Yukawa Institute for Theoretical Physics, Kyoto University, Kyoto 606-01, Japan*

Abstract. We have investigated the fusion-fission process of massive nuclei using the multi-dimensional Langevin equation and the statistical model. The compound nucleus formation probability has been obtained as a function of extra energy above incident barrier. The excitation function of evaporation residue cross section for three systems; ^{100}Mo+^{100}Mo,^{100}Mo+^{110}Pd, and ^{110}Pd+^{110}Pd, were calculated using the survival probability against fission obtained with the statistical model and were compared with experimental data. Our results reproduce the phenomenon of fusion hindrance for the fusion reactions of massive nuclei in the case of nearly symmetric projectile-target combinations.

I INTRODUCTION

We are investigating the fusion reaction of massive nuclei in which heavy compound nucleus ($Z > 80$) is formed. This study is an important step for understanding the formation of very heavy nucleus, for example the super heavy element(SHE); $Z = 114$. Estimating the evaporation residue cross section of the SHE is our major theme [1]. To understand the dynamical process of the synthesis of very heavy nucleus, the information on the early stage of fusion reaction is very important. On this stage, as we mention below, the energy dissipation and fluctuation play important roles to determine whether colliding nuclei fuse each other or reseparate. Thus, we use the Langevin equation which can treat the dissipation and fluctuation, to investigate fusion-fission process of massive nuclei.

In the fusion reaction of massive nuclei with nearly symmetric projectile-target combinations, there exists a problem of fusion hindrance, which means that colliding nuclei need extra energy above their incident Coulomb barrier to fuse each other. To understand this phenomenon, Bjørnholm and Swiatecki and their collaborators investigated this extra energy as the extra-extra push energy [2,3]. In their model, the collective motion of nucleus is described as a classical trajectory of a particle moving on the potential energy surface in nuclear deformation space with frictional force which is based on the dissipation of collective kinetic energy to internal nucleonic thermal energy. The extra-extra push energy is defined as the minimum kinetic energy above the Coulomb barrier needed to reach fusion.

However, if we investigate the fusion-fission dynamics more precisely, we should introduce the thermal fluctuation. Then, the energy flow occurs not only from collective motion to nucleonic motion, but also from nucleonic motion to collective motion. The collective motion of nucleus is described as a trajectory of a Brownian particle moving on the potential energy surface in nuclear deformation space. Thus, we investigate the fusion-fission dynamics by calculating the Brownian motion with the multi-dimensional Langevin equation. Studies in this direction have been performed by Aguiar et al. [4]. They calculated the heavy-ion fusion probability for some systems, for example; ^{100}Mo+^{100}Mo,^{104}Ru+^{104}Ru, and so on. They indicated the utility of the inclusion of fluctuations on description of very dissipative processes.

In the present work, we investigated the fusion reactions for three nearly mass-symmetric systems; ^{100}Mo+^{100}Mo,^{100}Mo+^{110}Pd, and ^{110}Pd+^{110}Pd. We calculated the compound nucleus formation probability with the three-dimensional Langevin equation. We employed three degrees of freedom as parameters for describing the nuclear shape; distance between two fragments, fragment deformation, and mass asymmetry. Besides, we calculated the survival probabilities against fission with the statistical model [5]. From these results, we obtained the evaporation residue cross sections and compared them with experimental data.

The paper is organized as follows. In Sect. 2, we give the ingredients of the model and explain the extra energy with the potential energy surface in nuclear deformation space. The results of our calculations are presented in Sect. 3 with discussions about the fusion reactions of massive nuclei. In Sect. 4, we summarize the present work.

II MODEL

For describing the nuclear shape, we adopt the three-dimensional nuclear deformation space with two-center parameterization [6,7]. The neck parameter ϵ is fixed to 1.0 in the present calculation, deformation parameters δ_1 and δ_2 of the colliding nuclei are taken to be equal, i.e., $\delta_1 = \delta_2 = \delta$. We treat z_0 (distance between two potential centers) and δ and the mass asymmetry parameter α ($\alpha = (A_1 - A_2)/(A_1 + A_2)$, where A_1 and A_2 are the mass numbers of two fragments) as the three collective parameters to be described by the Langevin equation.

The potential energy of deformation is calculated as the sum of a generalized surface energy [8] and Coulomb energy. Figure 1 shows the potential energy surface in the nuclear deformation space for three systems of mass symmetric fragments combinations ($\alpha = 0$). Abscissas denote z_0 and ordinates denote δ; $z_0 = \delta = 0$ corresponds to a spherical compound nucleus (marked by circles). Solid curves denote the scission line (zero-neck line) which includes incident touching configuration and dashed curves denote the ridge passing through the fission saddle point (marked by crosses). Figure 1a shows the energy surface for $Z = 50$ system. In this case, the contact configuration (marked by a square) locates inside the ridge. Thus, the colliding nuclei automatically form a spherical compound nucleus when they have enough energy to go over the incident Coulomb barrier. For heavy nuclei (Fig. 1b for $Z = 80$ and Fig. 1c for $Z = 100$), the contact configuration is outside of the ridge. Thus, the colliding nuclei need extra energy to go over the ridge to form spherical compound nucleus. This extra energy corresponds to the phenomenological extra push or extra-extra push energy.

In the present framework, the collective motion of nucleus is described as a trajectory of a Brownian particle which is started from the contact configuration and moving in the nuclear deformation space. The Brownian motion is calculated by the multi-dimensional Langevin equation.

The multi-dimensional Langevin equation is given in the following form,

$$\frac{dq_i}{dt} = \left(m^{-1}\right)_{ij} p_j,$$
$$\frac{dp_i}{dt} = -\frac{\partial V}{\partial q_i} - \frac{1}{2}\frac{\partial}{\partial q_i}\left(m^{-1}\right)_{jk} p_j p_k - \gamma_{ij}\left(m^{-1}\right)_{jk} p_k + g_{ij} R_j(t), \qquad (1)$$

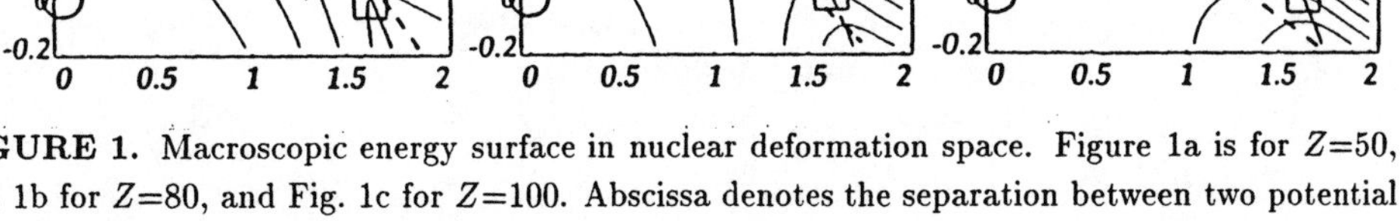

FIGURE 1. Macroscopic energy surface in nuclear deformation space. Figure 1a is for $Z=50$, Fig. 1b for $Z=80$, and Fig. 1c for $Z=100$. Abscissa denotes the separation between two potential centers and ordinate denotes the fragment deformation. Scission line (zero-neck radius) is denoted by the solid curve and ridge is given by the dashed line. Symbols are given in the text.

with summation from 1 to N ($N=3$) over repeated indices; $V(q)$ is the potential energy, $m_{ij}(q)$ and $\gamma_{ij}(q)$ are the shape-dependent collective inertia and dissipation tensors, respectively. The normalized random force $R_i(t)$ is assumed to be a white noise, i.e., $<R_i(t)>=0$ and $<R_i(t_1)R_j(t_2)>=2\delta_{ij}\delta(t_1-t_2)$. The strength of the random force g_{ij} is given by $\gamma_{ij}T = g_{ik}g_{jk}$ under a generalization of the fluctuation-dissipation theorem, where T is the temperature of the compound nucleus calculated from the excitation energy as $E_x = aT^2$ with a denoting the level density parameter of Töke and Swiatecki [9]. Hydrodynamical inertia tensor is adopted with the Werner-Wheeler approximation for the velocity field, and the wall-and-window one-body dissipation [10] is adopted for the dissipation tensor. Excitation energy of the composite system E_x is calculated for each trajectory as,

$$E_x = E_0 - \frac{1}{2}\left(m^{-1}\right)_{ij} p_i p_j - V(q), \qquad (2)$$

where E_0 is given as $E_0 = E_{cm} - Q$ with Q denoting the Q-value of this reaction and E_{cm} the incident energy in center-of-mass frame. At $t=0$, each trajectory starts from the contact configuration; $z_0=1.63R_0$, $\delta=0$, and α corresponding to the projectile-target combination, with the initial velocity in the z_0 direction, where R_0 denotes the radius of the spherical compound nucleus. In present work, the composite system has no E_x at $t=0$. E_x grows as the initial relative kinetic energy between the projectile and the target nucleus dissipates, and the dissipation almost achieves at the time corresponding to the order of 10^{-22}s.

The numerical method to solve the equation is given in Ref. [11]. Time development is obtained by iterations with a small time step, which was taken as $0.01\hbar$MeV in the present work. In order to obtain good accuracy, we expanded the equation to the second order in the time step.

III RESULTS

Figure 2 shows the compound nucleus formation probabilities(CNFP) for the nearly mass-symmetric systems; ^{100}Mo+^{100}Mo, ^{100}Mo+^{110}Pd, and ^{110}Pd+^{110}Pd, as functions of the center of mass energy E_{cm}, obtained by solving the Langevin equation. In this figure, each system has two lines; a thin line and a thick line. Thin lines are obtained by using the normal value of the one-body friction. Thick lines are obtained by using the three times of the normal one. It is known that the extra energy for fusion reaction mentioned above and also the CNFP are very sensitive to the friction coefficient especially around the contact configuration and near the scission line [4]. This is because of the fact that the Brownian particle moves on that region at the early stage of fusion reaction. However, the friction coefficient in this region is poorly known, compared with that around the fission saddle point. This is the reason for the existence of two lines for each system in Fig. 2. For comparison between our results and experimental data of evaporation residue cross section σ_{ER} which is shown later, we used the values denoted by the thick lines. It can be seen that the CNFP grows slower for heavier system compared with lighter case, and that the typical value is smaller for heavier case. By comparing thin and thick lines, we can conclude that the CNFP is actually very sensitive to the strength of the friction; larger friction gives smaller CNFP. When we define the extra energy as the energy where the CNFP is equivalent to 0.5 in Fig. 2, our results are consistent with experimental measurement. The extra energy strongly depends on the difference between the potential energies at the ridge and that at the contact configuration. We estimate the energy difference on the one-dimensional potential energy surface at $\delta = 0$ and α corresponding to the contact configuration, which are 0.4MeV, 0.8MeV, and 1.5MeV for the three systems; ^{100}Mo+^{100}Mo, ^{100}Mo+^{110}Pd, and ^{110}Pd+^{110}Pd, respectively. The extra energies estimated from Fig. 2 are 16MeV,

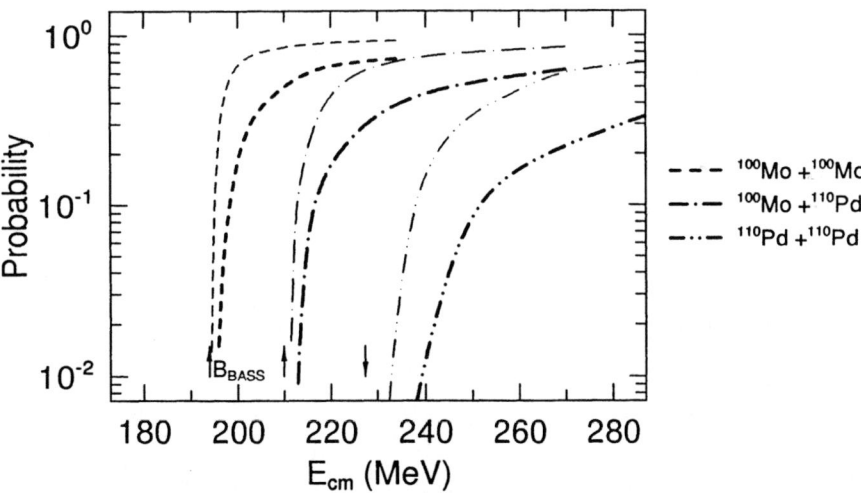

FIGURE 2. Compound nucleus formation probability(CNFP) as functions of E_{cm} for three systems; ^{100}Mo+^{100}Mo, ^{100}Mo+^{110}Pd, and ^{110}Pd+^{110}Pd. Thin lines denote the results calculated by using the normal value of the one-body friction, and thick lines by using the three times of that. The Bass barriers(B_{Bass}) are marked by arrows.

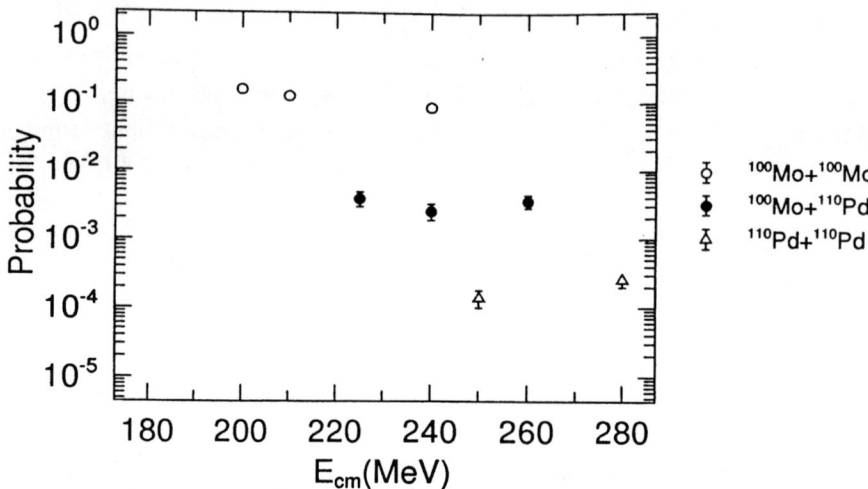

FIGURE 3. Survival probability against fission versus E_{cm} calculated by SIMDEC.

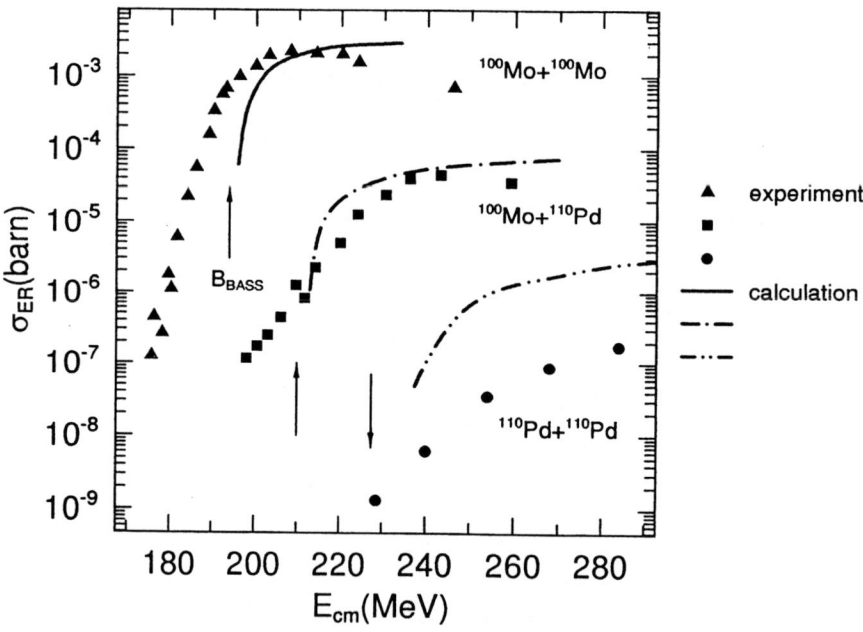

FIGURE 4. Evaporation residue cross section as functions of E_{cm}. Experimental data for each system are presented by solid triangles, solid squares, and solid circles respectively. The Bass barriers(B_{Bass}) are marked by arrows.

35MeV, and 96MeV for the three systems respectively. The extra energy increases by several tens of MeV, while the energy difference increases only by 1MeV. It can be expected that the extra energy becomes too large to achieve the fusion reaction as a main process for very heavy nucleus such as the SHE in the case of nearly symmetric projectile-target combinations. These results indicate the existence of fusion hindrance.

To estimate σ_{ER}, we calculated the survival probability against fission by simulating the competition between fission and particle evaporations with the statistical model code SIMDEC [5]. Figure 3 shows the results for the systems; ^{100}Mo+^{100}Mo, ^{100}Mo+^{110}Pd, and ^{110}Pd+^{110}Pd. The order of the survival probability becomes much smaller for heavier system. From these results; CNFP, survival provability, and the geometrical fusion cross section which is also obtained by SIMDEC, we estimated σ_{ER}. Figure 4 shows the results of our calculations and experimental data [12]. Our results show good agreements with the data, in particular for the systems of ^{100}Mo+^{100}Mo and ^{100}Mo+^{110}Pd. However, we have some discrepancies between calculated values and experimental data. It can be seen that we underestimate the σ_{ER} around the energy corresponding to the Bass barrier(B_{Bass}) [13] because of the simplicity of our model, for example we didn't include the effect of quantum barrier penetration. For ^{110}Pd+^{110}Pd reaction, the gradient of the calculated value is roughly in agreement with experimental data, though there is the discrepancy of about one order of magnitude. This seems due to the ambiguities in the statistical calculation for calculating the survival probability, for example the fission barrier height of the system and so on.

Although our results have some discrepancies with experimental data, taking into account some ambiguities and simplicities of our model, these results are satisfactory.

IV SUMMARY

We have investigated the fusion-fission process of massive nuclei for nearly symmetric projectile-target systems; ^{100}Mo+^{100}Mo, ^{100}Mo+^{110}Pd, and ^{110}Pd+^{110}Pd, with the multi-dimensional Langevin equation and the statistical model calculation. We obtained the compound nucleus formation probabilities by calculating the Brownian motions, which correspond to the time evolution of nuclear collective motions, in nuclear deformation space with the three-dimensional Langevin equation. Furthermore we calculated the survival probabilities against fission with the statistical model calculation. From these results, we obtained the evaporation residue cross sections and compared them with experimental data. Our results are in fair agreement with the data and give the interpretation to the phenomenon of the fusion hindrance which appears in the fusion reaction of nearly mass-symmetric massive nuclei. There remains some discrepancies between our results and experimental data, which shows the necessity for further work, in particular in near barrier region and in heavier mass case.

As a next step, we have to consider microscopic friction coefficients, for example that is based on the linear response theory [14], for more realistic calculations near the scission line in nuclear deformation space. In addition to this, asymmetry dependence of fusion dynamics is important to investigate the possibility of the synthesis of the SHE. Studies in these directions are now in progress.

REFERENCES

1. Y. Aritomo, T. Wada, M. Ohta, and Y. Abe, Phys. Rev., **C55** (1997) R1011; Y. Aritomo, talk presented in this symposium.
2. S. Bjørnholm, W.J. Swiatecki, Nucl. Phys., **A391** (1982) 471.
3. J.P. Blocki, H. Feldmeier and W.J. Swiatecki, Nucl. Phys., **A459** (1986) 145.
4. C.E. Aguiar, V.C. Barbosa, R. Donangelo and S.R. Souza, Nucl. Phys., **A491** (1989) 301; C.E. Aguiar, V.C. Barbosa and R. Donangelo, Nucl. Phys., **A517** (1990) 205.
5. M. Ohta, Y. Aritomo, T. Tokuda and Y. Abe, Proc. of Tours Symp. on Nuclear Physics II (World Scientific, Singapore, 1995) p.480.
6. J. Maruhn and W. Greiner, Z. Phys., **251** (1972) 431.
7. K. Sato, A. Iwamoto, K. Harada, S. Yamaji, and S. Yoshida, Z. Phys., **A288** (1978) 383.
8. H.J. Krappe, J.R. Nix, and A.J. Sierk, Phys. Rev., **C20** (1979) 992.
9. J. Töke and W.J. Swiatecki, Nucl. Phys., **A372** (1981) 141.
10. J.R. Nix and A.J. Sierk, Nucl. Phys., **A428** (1984) 161c.
11. Y. Abe, C. Grégoire, and H. Delagrange, J. Phys.(Paris), **47** (1986) C4-329.
12. W. Morawek, PhD Thesis Inst. Kernphysik, TH Darmstadt,(1991).
13. R. Bass. Proc. Symp. on Deep-inelastic and fusion reactions with heavy ions (Berlin, 1979) (Springer, Berlin) p. 281.
14. S. Yamaji, H. Hofmann and R. Samhammer, Nucl. Phys., **A475** (1988) 487

Time Scale of Fission Process of Cold Heavy Nuclei

Goverdovski A.A., Ostapenko Yu.B.

Fission Dynamics Laboratory, Institute of physics and Power Engineering, Obninsk, Kaluga Region, Russia, 249020

Abstract. Fission fragments mass-kinetic energy correlations have been investigated for uranium-235 thermal fission. Pronounced shifts of mass components were observed and then used for extraction of prescission neutrons multiplicities v_{PRE}. It was shown that those neutrons are emitted from fission modes corresponding to the asymmetric mass splits far from double magic mode. Fragment mass and kinetic energy dependence of v_{PRE} indicated different nuclear viscosity of fissile system on whole stage of system's descent from saddle point to scission one. As a result, fission time depends on mass split and it is determined by first third part of the descent. Reduced viscosity coefficient β for cold fissioning nucleus is much higher than for hot nucleus.

INTRODUCTION

Cold fissioning nuclei are very interesting objects for detailed investigations of both statistical and dynamical features of so large-scale collective motion like nuclear fission. The time of the process is a direct reflection of nuclear viscosity and therefore any information sensitive to the energy dissipation in fission can be used for the analysis. One of the methods widely used in this field is so called "neutron clock" which gave very important and essential part of the information about fission time. Traditionally it is using for hot nuclei in fusion-fission reactions. In those cases the initial excitation energy of the system is much higher than the excitation due to viscosity. Really only time is affecting particle emission probabilities before scission. In addition, shell, pairing effects can not be studied in the respect to fission time. Completely another situation could be reached in low energy neutron induced fission of heavy nuclei. Not so long ago it was shown that in fission of uranium-235 by thermal neutrons very small part of prompt neutrons can be emitted before scission (1). Effect exists in fission valleys far from double-magic one. It was used here in careful analysis of neutron width to get descent time value. For this calculations we used level density model developed for super-fluid nucleus with shell, pairing and collective effects (2). Very essential part of the analysis was the assumption that in cold fission thermal equilibrium inside the highly deformed fissioning system can not be reached. Detailed description of data processing and data analysis is given below. Comparison with other experimental and theoretical results is done.

ANALYSIS OF EXPERIMENTAL MASS SPECTRA

The experiment was done with Frisch gridded ionization chamber in which thin layer of ^{235}U was deposited and irradiated by neutrons from T(p,n) reaction moderated in polyethylene block. To produce fast neutrons the accelerator KG-2,5 of IPPE, Obninsk was used. Fragments masses were determined with well known double-energy method. Details of experimental procedure and fragments spectroscopy can be found in our work (1) together with technical details. Here we'll only present main equations to explain carefully the method we used for determination of prescission neutrons multiplicities v_{PRE} versus mass and kinetic energies of fission fragments.

One can get fragment masses $m_{L,H}$ (L-light, H-heavy) after determination of the final energies of the fission fragments $E_{L,H}$ including all of corrections. Sum of unchanged masses ($m^*_L + m^*_H = A = 236$) is equal to the mass of compound nucleus and:

$$m^*_{L,H} = \frac{AE_{H,L}}{E_L + E_H} \quad (1)$$

Prompt neutron emission will change mass spectrum, but for provisional masses μ determined directly by 2E method the effective shift will be rather small:

$$\Delta m_L = \mu - m^*_L = A\chi(E_{L,H})(\frac{v_L}{m_L} - \frac{v_H}{m_H}) \quad (2)$$

$$\chi(E_{L,H}) = \frac{E_L E_H}{(\frac{E_L m^*_L}{m_L} + \frac{E_H m^*_H}{m_H})(E_L + E_H)} \quad (3)$$

Provisional mass distributions corrected on neutron emission effects can be considered as good estimations of initial fragment mass distributions. But in present work we are mainly interesting in small shifts in mass spectra. To observe them we have to go to high energy region where so-called cold fragmentation spectrum demonstrates pronounced structure associated with mass channels or Brosa-valleys on the potential energy surface of fissioning nucleus. Two dimensional data matrix around high energies far from average values is presented in Fig. 1. One can see very pronounced structure of the yield surface landscape. Some kind of shift by 1-2 a.m.u. of the mass component

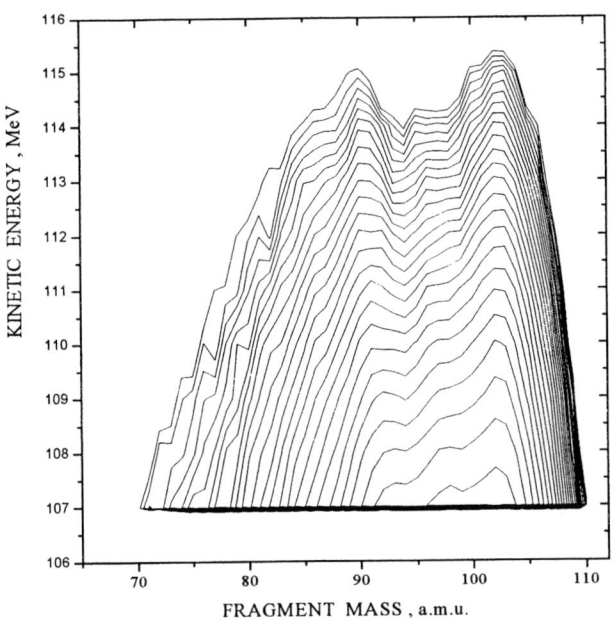

FIGURE 1. Two dimensional fission fragments mass-energy spectrum.

centered around m_L = 90 a.m.u. takes place while the position of so-called Standard-I (1) component is stable: $<m_L>$ = 102 ± 0.2 a.m.u. Figure 2 illustrates this situation in more detail. Here, locations of two main mass components are presented Vs light fragment kinetic energy. Solid lines show positions expected from the assumption that prompt neutrons emission from the fragment only exists. In our case kinetic energy is a specific measure of the descent length just before a rupture of two-center figure on two parts. This degree of freedom can be converted into total excitation energy of the system as difference between reaction Q-value and total kinetic energy of both fragments. From this point of view the data presented in Fig. 2 can be interpreted as positions of mass components on different stages of the descent to scission where the total excitation energy of both future fragments can be quite different. The last point connects with viscous properties of the system. In principle they can depend on deformation, internal temperature and mass asymmetry of fissile nucleus. Coming back to Fig. 2 one can say, that we have to find additional source of mass shift because neutron emission from fully accelerated fission fragments are not able to explain visible effect for component centered at m_L = 90 a.m.u. especially comparing with double magic fission valley (m_L / m_H = 102/134). The simplest and obvious source is neutron emission with changing the compound system mass. There are two possibilities. One of them is neutron emission from uranium-236 before penetration of fission barrier. Since

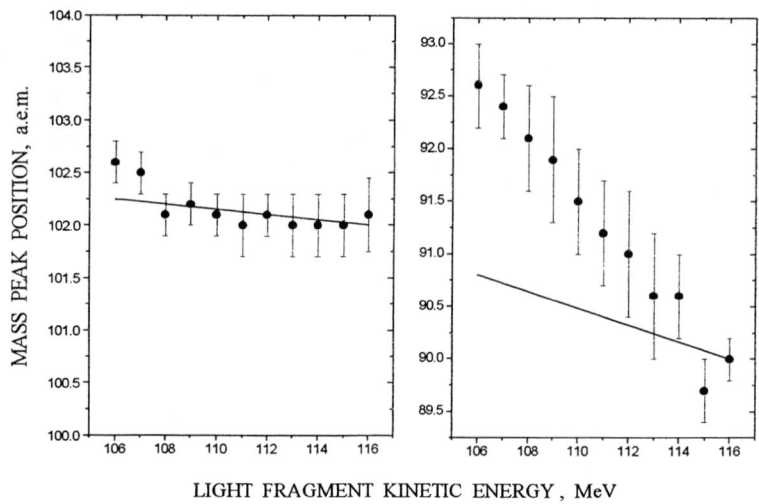

FIGURE 2. Positions of main mass components peaks Vs kinetic energy

the energy of incident neutrons is thermal or 1 MeV, any transition to the first barrier well will do following fission process practically impossible. The second possibility is the neutron emission during the motion along the fission valleys after saddle point but before scission one (or at least not later than fragments will be accelerated enough to change mass spectra with the sensitivity determined by kinematics). By definition, corresponding neutrons will be prescission neutrons. Fission fragments mass spectra are more sensitive to these neutrons according to following relation $\Delta m_L = v_{PRE} \, m_L / A$. Obviously, the effect is essentially positive and it does not depend on difference of neutron multiplicities from light and heavy fragments. Since the biggest effect was observed for very asymmetric mass component all following operations will be done only for it. Fig. 3 shows neutron multiplicity extracted from data of Fig.2. Three data sets were used in analysis. There are data for thermal neuron induced fission of ^{235}U from our experiment and from work by Knitter and co-workers made with analogous experimental set-up in Geel, Belgium (3). The third data set is for 1-MeV neutrons. One can see very strong correlation between neutron binding energy and growth of v_{PRE}.

CALCULATION OF FISSION TIME AND DISCUSSION

For determination of fission time scale or at least time of descent from saddle to scission we have to calculate total neutron width Γ_n of excited nucleus with definite N,Z numbers. Obviously, this nucleus is low excited, practically cold and therefore structural effects have to be taken into account. Formal scheme of Γ_n computation is well known

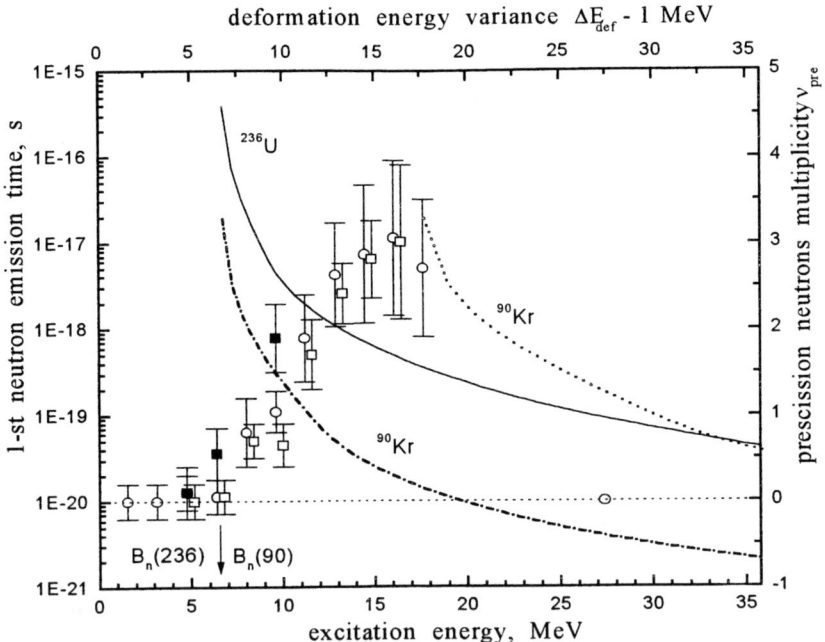

FIGURE 3. Prescission neutrons multiplicities determined from fragment mass spectra shifts for thermal neutron induced fission of ^{235}U : O - present work; □ - Knitter e.a. (3); and ■ -uranium fission by 1 MeV neutrons (present work). Neutron binding energy for ^{236}U and ^{90}Kr is indicated. Dot line - is zero level of ν_{PRE} since for the average kinetic energy and therefore high yields prescission neutrons were not observed yet. Curves with captures ^{236}U and ^{90}Kr are results of first neutron emission time where lowest curve is if all excitation energy is concentrated on Kr- nucleus only.

and is widely used in practice (2). In addition, fortunately, all parameters needed for calculations of nuclear level density are well known from nuclear reactions studies. Theoretical background is well developed and we came in our work along the lines of the corresponding models (2). One difficulty was in this way. Assuming ^{236}U as a source of the neutrons in question one has to choose parameters for highly deformed nucleus just after saddle configuration. Variations of main parameters like shell correction δW, correlation function Δ_0 and level density parameter α in realistic region gave in principle quite stable results influenced mainly by excitation energy of the nucleus. Some results of calculations are given in Fig.3. Axis of excitation energy means our first and very strong assumption that all free energy determined as (Q-TKE) is concentrated in internal excitation of the fissile system at least on early stage of the descent from the barrier top. In nature it's not true of course, but here we need fission time in first approximation. It will be only low limit of time scale. On the other hand experimental ν_{PRE} - curve comes up near the neutron binding energy of uranium. Therefore our assumption is not so unrealistic. The calculations were made for first

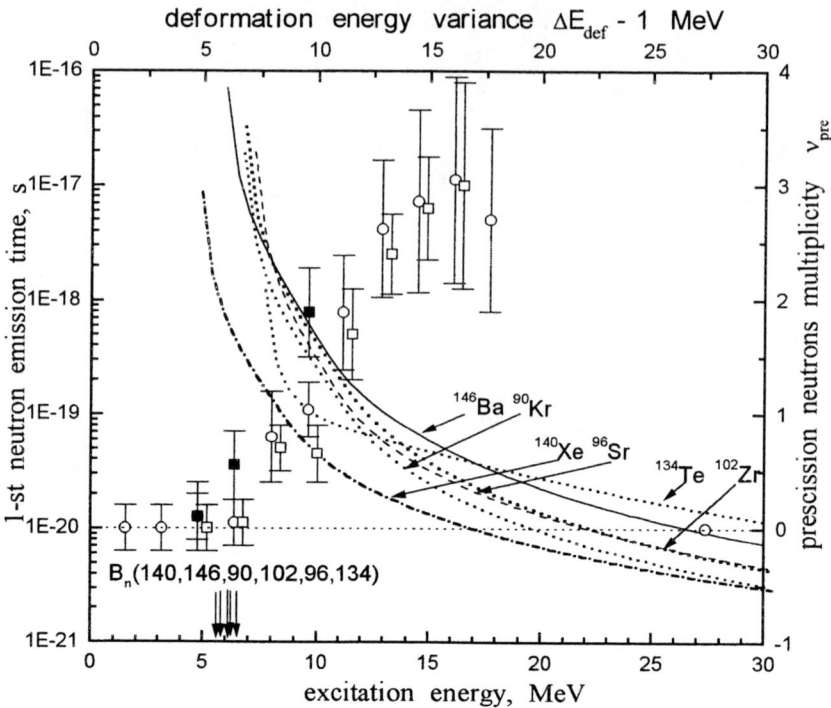

FIGURE 4. Same as in Fig.3 but calculated curves are for fission fragments indicated.

neutron emission only. Emission of second neutron after cooling of the nucleus needs much more time. But even within indicated kind of assumption emission time from uranium as common emitter is too long to be real - 10^{-17} s. Here we oriented on the data for hot nuclei (4). Any possible variations of model parameters could not change quantitatively and qualitatively the results.

Another source of prescission neutrons can be associated with future fission fragments. Most probable fragment pair in fission valley centered around m_L = 90 a.m.u. is ^{90}Kr/^{146}Ba. The effect of ν_{PRE} was observed for very high total kinetic energies where all free energy has to be concentrated in excitation energy to do neutron emission possible. It means that the fragments born just after the neutron emission had equilibrium deformations (no energy for deformation), and further computations are based on well known statistical model parameters (2). Figure 4 demonstrates results of calculations for all interesting mass splits. One can see that the time needed for neutron emission is approximately the same in all of fission valleys. Lack of neutrons from double magic valley of ^{102}Zr/^{134}Te is due to lack of excitation energy needed for neutron emission and therefore this fission mode is almost unviscous. Completely another situation is realized in far asymmetric fission with ^{90}Kr/^{146}Ba fragment split.

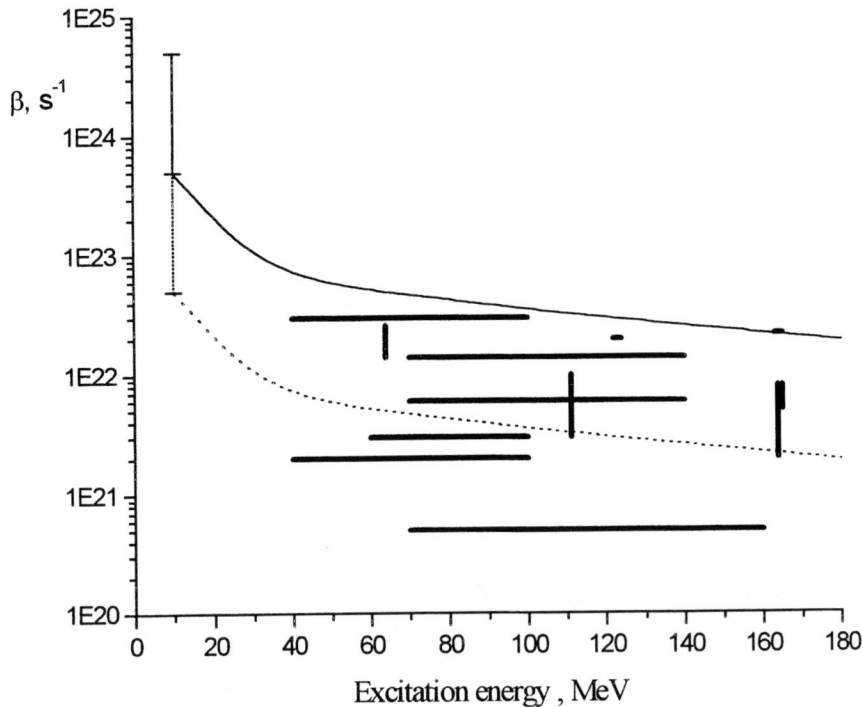

FIGURE 5. Reduced dissipation coefficient β determined for different excitation energies of fissioning nuclei. Thick lines indicate area of change of β for hot nuclei (4). Solid and dot lines are extrapolations made under model assumption of $1/T^2$-kind of temperature dependence of dissipation coefficient β. First is if neutrons are emitting from uranium. Second – is for light fragment of $^{90}Kr/^{146}Ba$ split.

One can see that the motion of fissile compound along the bottom of corresponding valley of the potential energy surface is very slow, viscous motion accompanied by prescission emission of neutrons. This takes place during the initial stage of the descent where the yields are low. Having in mind that the contribution of fission mode centered around fragment pair $^{90}Kr/^{146}Ba$ to final mass spectrum of uranium fission process is rather low, the prescission emission can be characterized as a rare event in cold nuclear matter existing not far from fission barrier top. This could explain why nobody in the world observe prescission neutrons in spontaneous fission process. Explanation is obvious: in spontaneous fission a system does not reach compact configurations corresponding to fission barrier top and therefore its motion is relatively fast so that descent time is not enough for neutron emission.

It's interesting and important to compare our data with data from other experiments. The subject of the comparison could be reduced coefficient of dissipation β deduced from experimental values of fission times. In Fig.5 we present this kind of comparison. Here we have used two β values corresponding to different descent times and therefore to different principal assumptions. First is that neutrons are emitting from uranium. Second – is for light fragment of ^{90}Kr/^{146}Ba fragment pair. Extrapolations are made under model assumption of $1/T^2$-kind of temperature dependence of dissipation coefficient β. One can see that the second case is much more preferable than first one. But in principle both cases can be used in future data analysis. The biggest problem here is in complex explanation of all data on low energy nuclear fission. If the first case is correct too, one of possible hard question is why does nucleus have so strong memory as it remembers input channel properties which is observable in fission fragments angular distributions ? We need more information to answer this question.

CONCLUSIONS

Now we know that the dissipation process in fissile system depends on mass-split and temperature. The motion from top of partial fission barrier along the separate fission valley of high mass asymmetric system is slow and very viscous which does possible the emission of neutrons before scission. This effect has to be investigated in detail for many fissile systems and their temperatures. The systems with pronounced multi-component structure of mass distribution looks like the best objects of studies.

ACKNOWLEDGMENTS

The work was supported by President of Russian Federation (grant #96-15-96938).

REFERENCES

1. Goverdovski, A.A., *Sov.J.Nucl.Phys.* **55**, 2033-2038 (1992).
2. Ignatyuk, A.V., *Statistical properties of atomic nuclei*, Moscow: ATOMIZDAT, 1983, p.p.10-49.
3. Knitter, H.-H., *Nuclear Physics* **A462**, 85-106 (1987).
4. Hilscher, D., and Rossner H., *Annales de Phys. Fr.* **17**, 471-490 (1992).

FISSION MODES

OBSERVATION OF FISSION MODES IN

HEAVY ION INDUCED REACTIONS

M.G. Itkis[1], N. A. Kondratiev[1], E. M. Kozulin[1], Yu. Ts. Oganessian[1],
V. V. Pashkevich[1], I.V. Pokrovsky[1], V. S. Salamatin[1], A.Ya. Rusanov[2],
L. Calabretta[3], C. Maiolino[3], K. Lukashin[3], C. Agodi[3], G. Bellia[3],
G. G. Chubarian[4], B. J. Hurst[4], D. O'Kelly[4], R. P. Schmitt[4], F. Hanappe[5],
E. Liatard[6], A. Huck[7], L. Stuttgé[7]

[1]FLNR, JINR, Dubna, Russia 141980; [2]INP, Alma-Ata, Kazakhstan; [3]LNS-INFN, 57 Corso Italia, I-95100, Catania, Italy; [4]Cyclotron Institute, Texas A&M University, TX 77843, USA; [5]ULB, PNTPM CP229, av. F.D. Roosewelt, B1050, Brussels, Belgium; [6]ISN, 53 avenue des Martyrs, 38026, Grenoble, Cedex, France; [7]CRN, 23, rue du Loess, 67037 Strasbourg, Cedex 2, France

Abstract. The fission of the systems ^{220,224,226}Th was investigated by measuring the mass-energy distributions of the fission fragments. The corresponding excitation energies at the saddle point, E^*_{sp}, ranged from 16 to 40 MeV. As E_{sp}^* decreases, an asymmetric mass component becomes visible on the predominately symmetric mass distribution. The contribution of the asymmetric mode is characterized by the total yield ratio Y_s/Y_a, which decreases rapidly for the heavier isotopes of thorium. This behavior of Y_s/Y_a is in qualitative agreement with theoretical calculations. For all isotopes studied, the subtracted asymmetric fission component, $Y_a = Y_t - Y_s$, exhibits a complex structure, actually showing two components, $Y_a = Y_{a1} + Y_{a0}$, which have average masses $M_{a1} = 132$ and $M_{a0} = 140$.

INTRODUCTION

Numerous experimental studies have provided a rather clear picture of the phenomenology of fission properties as a function of excitation energy and nucleonic composition: (a) Near the valley of beta stability, spontaneous and low-energy fission of the actinides from Th to Cf generally display asymmetric mass distributions; (b) The tendency towards mass symmetric fission increases rapidly with excitation energy, E^*. When $E^* \geq 50$ MeV, the mass distributions become close to Gaussians. These observations have generally been attributed to the disappearance of shell effects in the heavy fragments with $Z \approx 50$ and $N \approx 82$.

At the same time, it is well known that dramatic changes occur in the mass and energy distributions of certain spontaneously fissioning Fermium isotopes and some other transuranic elements [1,2]. For the isotopes ^{258,259}Fm and ^{260}Md, fission produces narrow symmetric mass distributions and fragments with relatively high kinetic energies. Only small changes in the nucleonic composition radically alter this picture.

Dramatic changes in the symmetric/asymmetric character of the fission probabilities can be accounted for by theories, which incorporate a micro-macroscopic approach including the effects of nuclear shell structure on the formation of the fragments. Detailed calculations of the nuclear potential energy surface indicate the possibility of a co-existence of various (at least two) distinct fission modes. There is a standard path which, leads to asymmetric fission fragments and a shorter path, which leads to two nearly symmetric fragments. These different sequences in the shape evolution result in bimodal fission phenomena [3].

Multimodal fission phenomena have also been observed in the Pb-At region in proton and alpha induced reactions [4], as earlier predicted [5]. Unfortunately, with light charged particles it is not possible to extend this work into the largely unstudied At-Th region due to the absence of suitable targets. This region is especially interesting for studying fission dynamics as a whole and the nature of fission modes in particular. Here dramatic changes are expected to occur in the fission properties as a function of the nucleonic composition and the excitation energy. The transition form symmetric to asymmetric fission along with the accompanying changes in the barrier height and the saddle to scission times can be expected to strongly influence the fission mass and energy distributions, the fission cross sections, the fragment γ-ray multiplicities as well as the pre- and post-fission neutron multiplicities.

A new strategy for probing the At-Th region was developed several years ago utilizing heavy ion beams in near- and sub-barrier fusion-fission reactions. The first experiments were conducted 4 years ago in the FLNR at the JINR in Dubna [6]. This work has been continued at the LNS-INFN in Catania, the INS at Grenoble, and the Cyclotron Institute at Texas A&M University. Similar experiments have been performed at GSI to investigate Coulomb fission reactions of secondary radioactive beams [7].

The current paper focuses on the results of studies of the multimodal structure of the fission mass and energy distributions of the neutron deficient nuclei ^{220}Th, ^{224}Th and ^{226}Th formed via ^{16}O and ^{18}O reactions at near and sub-barrier energies. Additional aspects of the work including γ-ray multiplicities and pre- and post-fission neutron multiplicities will be presented in a future when the analysis has been completed.

EXPERIMENTAL PROCEDURES

The main problem in studying low energy fission ($20 < E^* < 30$ MeV) in the above mentioned transition region using heavy ion beams is the very small fusion-fission cross section near the Coulomb barrier. By choosing the appropriate target-projectile combination, one can achieve the "lowest possible excitation energy," $E^* = E_{cm} + Q$, in the compound system when $E_{cm} \approx B_C$ where Q is the fusion Q-value. This is, of course, the strategy used in cold fusion reactions for producing heavy nuclei [8].

As an illustration, Fig. 1 shows the fission excitation function for the reaction

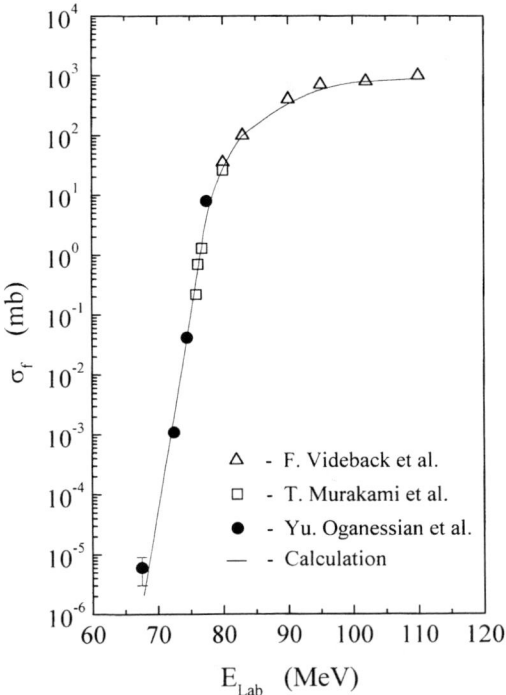

FIGURE 1. Fission excitation function for the system ^{208}Pb + ^{16}O. Calculations and • - [9], □ - [10], Δ - [11].

^{208}Pb(^{16}O,f)^{224}Th. The range of bombarding energies which produce compound systems with E* = 20 - 30 MeV corresponds to 72 < E_p(^{16}O) < 83 MeV. Over this energy interval, the fission cross section varies from 10^{-3} to 10^2 mb, placing severe constraints on the types of measurements which can be successfully conducted.

The experiments were carried out using time-of-flight spectrometer DEMAS and his following modification [12]. This spectrometer was based on large area parallel-grid avalanche counters (PGAC) that provided a reliable and precise system for measuring mass and energy distributions of correlated fragments. A schematic of the spectrometer is shown in Fig.2. The time-of-flight spectrometer consists of four PGAC's ("stop" detectors) with 200x300 mm^2 active area each and two small parallel plate avalanche counters (PPAC's) with areas of 30x40 mm^2 ("start" detectors). Each arm had a flight path about 40 cm. The spectrometer was calibrated with fission fragments from a thin ^{252}Cf source. The measured peak-to-valley ratio for the reconstructed masses is greater than 20, indicating a mass resolution ΔM/M ≈ 3%.

The angular distributions, time-of-flight and energy spectra of neutrons were measured using eight elements of the DEMON array [14]. The γ-ray multiplicities were

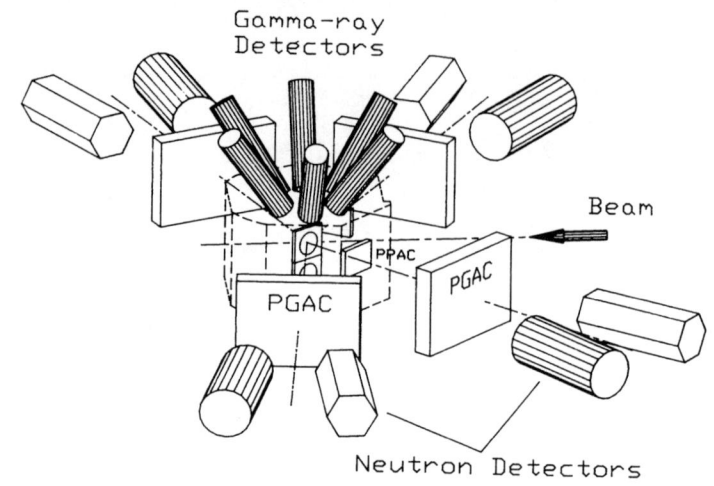

FIGURE 2. Schematic of the experimental setup (see text).

measured with an array of six 63x63 mm NaI detectors arranged symmetrically around the target.

RESULTS AND ANALYSIS

The characteristics of the reactions studied are summarized in Tab. 1. This table lists the entrance channels, the bombarding energies, E_i, the compound nucleus excitation energies, E^*, the fission barriers, E_f, the saddle point excitation energies, E^*_{sp}, the average total kinetic energies, $<TKE>$, the TKE dispersions, σ^2_{TKE}, and the

TABLE 1. Parameters For the Various Fission Reaction.

Reaction	E_i, MeV	E^* MeV	E_f MeV	E^*_{sp} MeV	TKE MeV	σ^2_{TKE} MeV^2	σ^2_M amu^2	$\Sigma Y_s/\Sigma Y_a$
^{208}Pb+^{18}O	78	26.1	6.9	19.2	165.4±1.1	137±6	280±15	2.8±0.5
	75	23.3		16.4	165.2±2.0	150±11	291±22	2.4±0.8
^{208}Pb+^{16}O	108	53.8	7.2	46.6	162.4±1.0	137±9	224±15	-
	85	32.4		25.2	164.0±1.1	113±7	193±13	8.0±1.0
	77	25.0		17.8	163.6±1.2	98±6	187±12	4.8±0.5
	75	23.2		15.9	164.7±1.4	94±6	202±13	3.3±0.4
^{204}Pb+^{16}O	108	53.6	8.8	46.8	162.1±1.0	151±10	236±15	-
	85	34.3		25.5	163.1±1.2	118±7	188±12	7.6±0.8
	77	26.9		18.1	163.6±1.2	102±6	166±11	6.0±0.7
	75	25.0		16.2	163.7±1.4	103±6	162±11	6.3±1.0

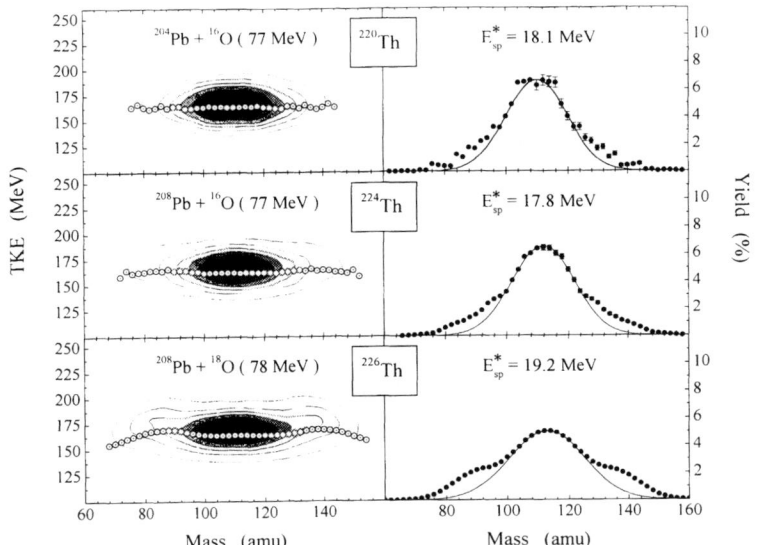

FIGURE 3. (Left) Contour plots showing the distribution of fission events in the TKE-mass plane for the reactions $^{204,\,208}$Pb + ^{16}O (77 MeV), ^{208}Pb + ^{18}O (78 MeV). The open circles indicate the average TKE-values. (Right) Corresponding mass yield distributions for the reactions above. The curves reflect fits to the symmetric fission component.

mass dispersions, σ^2_M. The last column is related to the relative ratios of different fission modes (see below).

Figure 3 shows contour plots of fission events in the TKE versus mass plane for the systems ^{220}Th, ^{224}Th and ^{226}Th along with the projected total mass distributions (normalized to 200%). The average kinetic energies as a function of fragment mass are indicated by the circles superimposed on the contour diagrams. These plots clearly show that at similar excitation energies there is an increasing tendency to fission asymmetrically as the N of the compound system increases.

Mass and kinetic energy distributions for ^{224}Th at different bombarding energies are shown on Fig. 4. Evidence for the presence of different fission modes is also readily apparent in the mean kinetic energies as a function of fragment mass. According to the macroscopic model [15], the mean TKE is expected to exhibit a parabolic mass dependence. This behavior is clearly indicated at the highest excitation energies, ≈ 46 MeV. However, at the lowest bombarding energies, deviations from this smooth parabolic dependence as large as 15 MeV are visible in the mass range 132 - 140.

The data for Y(M), TKE(M), and σ^2_{TKE}(M) suggest the presence of two asymmetric modes for M > 130 designated by a_0 and a_1. These appear to peak around the mean masses $M_{a1} = 132$ and $M_{a0} = 140$. To estimate the contribution from each of the three modes (the symmetric mode and two asymmetric modes), the total mass distributions were analyzed assuming that each component is described by a Gaussian.

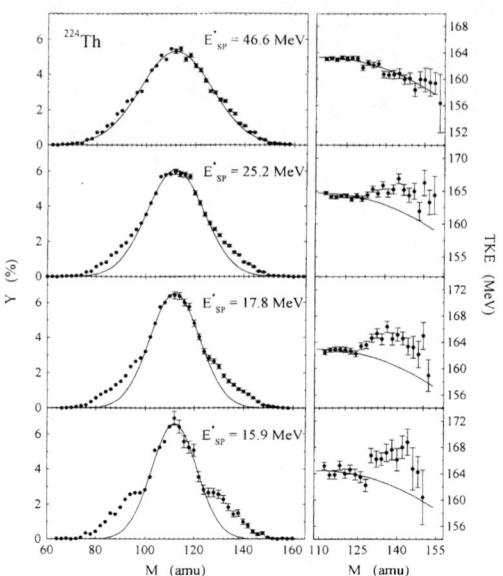

FIGURE 4. (Left) Mass yield distributions for fission following the reaction ^{208}Pb + ^{16}O at four different bombarding energies. (Right) Corresponding average TKE-values as a function of fission fragment mass.

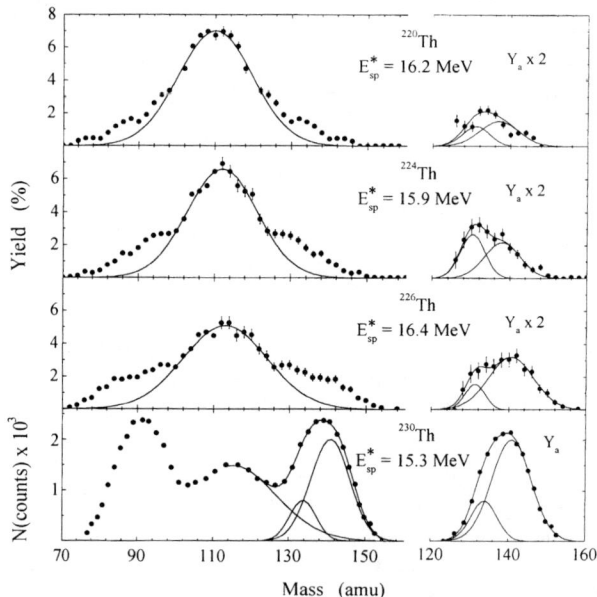

FIGURE 5. Mass distributions for fission of ^{220,224,226}Th and ^{230}Th [16] at excitation energies $E^*_{sp} \approx 16$ MeV. The curves depict fits used to decompose the asymmetric component into the contributions from the fission modes a_o and a_1 (see text).

Thus, the total mass distributions, Y(M), normalized on 200% were subjected to least square fits with a total of five Gaussians.

The fits to the total and asymmetric portions of the mass distribution for ^{220}Th, ^{224}Th, ^{226}Th and ^{230}Th [16] for $E^*_{sp} \approx 16$MeV are shown in Fig. 5. The left part of the figure presents the total mass distributions. The fits to the asymmetric portions of the mass distributions are shown on the right. The curves show the yield distributions of $Y_{a1}(M)$ and $Y_{a0}(M)$, which are centered about M = 132 and 140, respectively, and their sum. It is interesting to note that the Y_{a1} component decreases in yield as the compound system becomes more neutron deficient.

The last column of Tab. 1 shows the Y_s^t/Y_a^t ratio from the five-component fits to the ^{220,224,226}Th mass distributions. The dependence of ratio Y_s^t/Y_a^t on the excitation energy of the fissioning nuclei at the saddle point is shown in Fig. 6. Additional results are also shown from analyses of the mass distributions for other Th isotopes studied in reactions induced with neutrons, γ-quanta, and ^{4}He ions [17-23].

The values of Y_s^t/Y_a^t for all nuclei studied at the same excitation energy, $E^*_{sp} \approx 16$ MeV, are shown in insert in Fig. 6. The main point of this figure is to demonstrate the strong dependence of Y_s^t/Y_a^t on the neutron number of the fissioning nucleus.

Using the Bohr-Wheeler transition state formalism, one can calculate the ratio of the yields, $Y_i^i(E^*)$, or, equivalently, the ratio of the average fission widths, $\Gamma_f^i(E^*)$, for the distinct two fission modes:

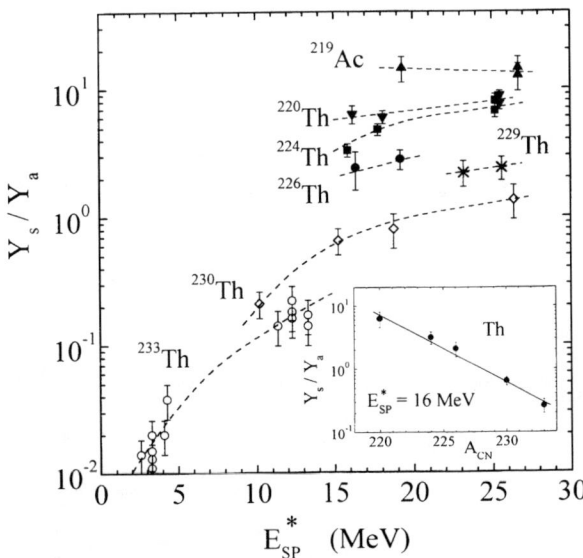

FIGURE 6. The ratio of the total yields of symmetric to asymmetric fission for a wide range of Th isotopes and ^{219}Ac as a function of the excitation energy at the saddle point. ✶ - [16], ◇ - [16, 17], ○ - [18-23], ▲, ▼, ■, ● - present work. (Insert) The same ratio as a function of compound nucleus mass for fixed excitation energy.

$$\frac{Y_a^t(E^*)}{Y_s^t(E^*)} = \frac{\Gamma_f^a(E^*)}{\Gamma_f^s(E^*)} = \frac{\int_0^{E^*-E_f^a} \rho_a(E^* - E_f^a - E)dE}{\int_0^{E^*-E_f^s} \rho_s(E^* - E_f^s - E)dE}. \quad (1)$$

The level density can be written as:

$$\rho_i(E^*) = K^i(E^*)\rho_{in}^i(E^*), \quad (2)$$

where $K^i(E^*)$ is the total collective enhancement to the level density and $\rho_{in}^i(E^*)$ is the intrinsic level density. As an approximation one can assume that

$$\rho_{in}^i(E^*) \propto \exp(\frac{E^*}{T}), \quad \text{and} \quad K^i(E^*) \propto K_{rot}^i(E^*), \quad (3)$$

where T is the temperature (assumed constant) and $K_{rot}^i(E^*)$ is the rotational enhancement to the level density. Equation 3 neglects possible differences between the vibrational coefficients in $K = K_{rot} * K_{vib}$ for the fission modes, but does account for the strong dependence of K_{rot} on the type of nuclear symmetry. For constant E_{sp}^* one obtains the relation

$$\frac{Y_a^t}{Y_s^t} \approx \frac{K_{rot}^a}{K_{rot}^s} \exp\left(\frac{E_f^s - E_f^a}{T}\right), \quad \text{or equivalently,} \quad \ln\frac{Y_a^t}{Y_s^t} \approx C_1 + C_2(E_f^s - E_f^a), \quad (4)$$

where $C_1 = \ln(K_{rot}^a/K_{rot}^s)$ and $C_2 = T^{-1}$.

In deriving eq. 4, it is not necessary that the level densities follow eq. 3. The same result is obtained using the Fermi gas model provided the excitation energy of the system is larger that the difference $E_f^s - E_f^a$, which is expected to hold in the current situation. It should also be noted that angular momentum effects in the level densities have been neglected. This is not expected to have much influence on the analysis.

From the above, one sees that a value of $E_f^s - E_f^a$ can be deduced from the experimental yield ratio Y_a^t/Y_s^t. The values of $K_{rot}^a/K_{rot}^s = e^{C_1}$ convey information on the symmetry of the nuclear saddle point shape. Theoretical values of K_{rot} for different nuclear symmetries are taken from refs. [25-29].

The values of $E_f^s - E_f^a$ deduced from eq.4 are given in the Tab. 2 for various values of C_1. One quickly concludes that the values of $E_f^s - E_f^a$, deduced from experiment are compatible with parameter $C_1 = -1.9$. This corresponds to pear shaped Th nuclei in their ground states.

Dramatic effects are observed in the mass distributions when they are gated on different TKE values. Figure 7 shows the energy gated differential mass distributions for ^{224}Th and ^{226}Th at compound nucleus excitation energies of $E^* = 26$ MeV. It is readily apparent in the case of ^{226}Th that changing the total kinetic energy TKE from 146 MeV to 194 MeV produces strong transformations in the mass distributions.

TABLE 2. Comparison between experimental and theoretical differences in fission barriers for symmetrical and asymmetrical fission.

Compound Nucleus	$\left(E_f^s - E_f^a\right)^{exp}$, MeV			$\left(E_f^s - E_f^a\right)^{theor}$, MeV
	$C_1 = -1.9$	$C_1 = 0.7$	$C_1 = -3.3$	
^{220}Th	-0.30	-2.57	+1.30	-1.0
^{224}Th	+0.80	-2.20	+2.40	+0.9
^{226}Th	+1.15	-1.95	+2.75	+1.2
^{230}Th	+2.60	-0.40	+4.20	+2.9
^{232}Th	+3.20	+0.60	+4.80	+3.1
^{233}Th	+3.70	+1.10	+5.30	-

Thus, if the N of the fissioning nucleus is changed by just two neutrons, one observes a transition from symmetric to asymmetric fission when the fission excitation energy, Q-TKE, approaches 0.

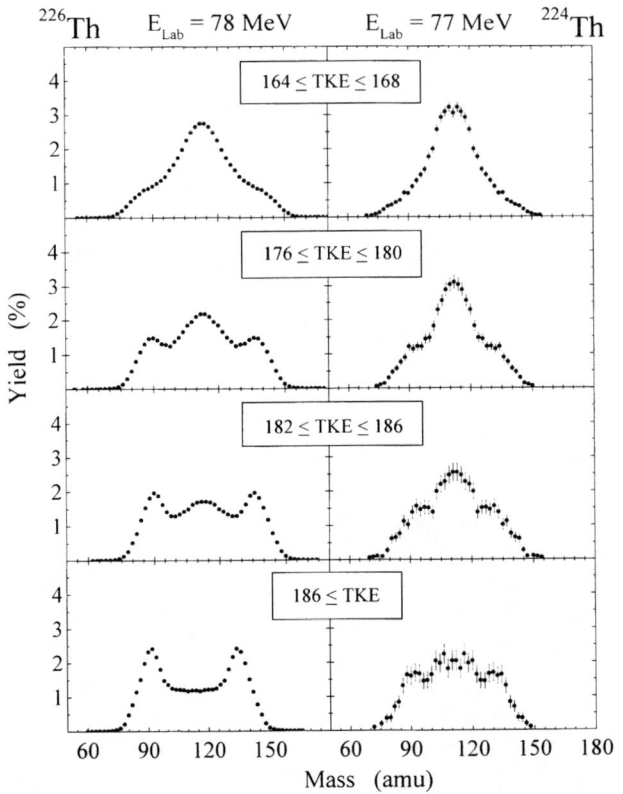

FIGURE 7. Mass yield distributions for ^{226}Th (Left) and ^{224}Th (Right) for four TKE windows.

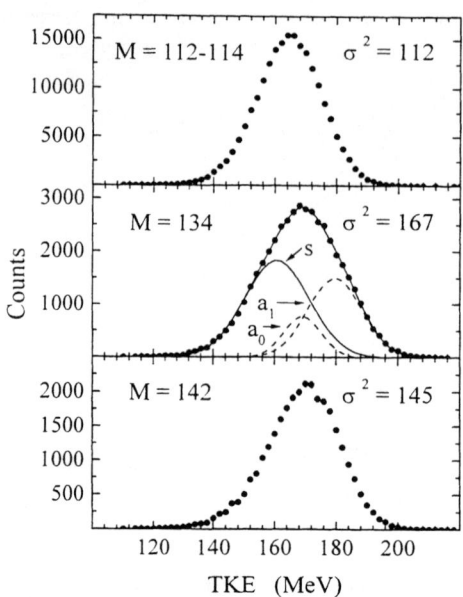

FIGURE 8. Energy spectra of fragments from the fission of ^{226}Th for three mass cuts. (Center) The curves show the results of fits to the three components. Energy spectra for all symmetric fission s (solid curve), and for the two asymmetric fission modes a_0 and a_1 (dashed curves).

Further understanding of the characteristics of the fission modes can be achieved by analyzing the differential total kinetic energy spectra at different mass asymmetries. Figure 8 shows the experimental spectra, Y(TKE), for ^{226}Th (E_i=78MeV) for three different mass regions: M=A/2=112-114, M=134, and M=142. Both the average total kinetic energy, and the corresponding dispersion, σ_{TKE}, vary with mass cut. When one of the fission modes is dominant (i.e., symmetric fission for M=112-114, or asymmetric fission for a_0 at M=142), the spectra are essentially Gaussians. The energy dispersions σ_{TKE} are also smaller than those seen in the spectra for M=134 where there are comparable contributions from all the fission modes (s, a_0 and a_1).

The TKE spectra for M=132 and 134 have been fitted with three Gaussians. Both the dispersion and the area for one of the Gaussians had a fixed value. Since the $\sigma^2_{TKE}(M)$ for symmetric fission does not depend on fragment mass, it was assumed that σ^2_{TKE} (M=A/2), which has been obtained from the fit to the TKE spectrum with M=112-114. The relative contribution of Y_s^t to Y^t was determined from the decomposed mass yield. The results of the analysis are illustrated in the middle panel of Fig. 8. This shows the experimental energy spectra, the simulated distributions for the symmetric fission mode (solid curve), and difference spectra (dashed curves), which should correspond to both the asymmetric modes a_0 and a_1. A similar analysis was carried out for M=132. These observations again provide strong evidence for independent fission modes.

DISCUSSION AND RESULTS

The potential energy surfaces for the various Th isotopes have been calculated as a function of the mass asymmetry. Cassinian ovals [5] were used to describe the nuclear shapes. In the calculations, shell corrections were taken into account using spherical harmonics up to 7th order. For a given shape, a Woods-Saxon single-particle potential was generated as a function of the distance $l(\vec{r})$ from the origin to the surface [30].

The methods used to calculate the single-particle spectrum [30,31], the shell corrections [32], the liquid drop parameters [33], and the potential [34] are the same as those described in ref. 3. In some cases, the surface diffuseness was taken into account using the "Yukawa plus exponential" model [35] with the parameters given in ref. 36. The results of calculations for ^{220,226}Th are shown in Fig. 9. Here the potential energy is given as a function of the elongation parameter α and the mass asymmetry parameter α_3. At each point on the surfaces, a minimization was carried out with respect to the parameters α_4-α_7. The surfaces are completely symmetrical with respect to α_3, so only one of the two asymmetric valleys is presented here. Because of the choice of scales, the ground state, the second minimum, and the intervening barrier (usually called A) are not visible in Fig. 9. For all three systems, a symmetric valley develops at $\alpha \approx 0.7$. This valley runs almost parallel to the asymmetric valley. The occurrence of these two valleys obviously influences the final mass asymmetry. For example, in ^{220}Th one sees that as the system evolves towards scission the potential energy surface favors a shift to the valley with smaller mass asymmetry. In ^{226}Th the two valleys are separated by

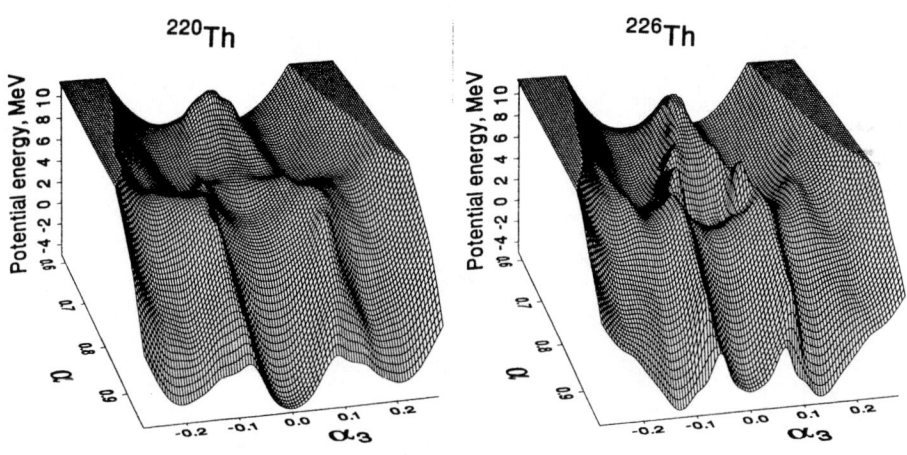

FIGURE 9. Calculated potential energy as a function of the elongation parameter, α, and the asymmetry variable, α_3, for the fission of (a) ^{220}Th, (b) ^{224}Th and (c) ^{226}Th.

only a very small ridge. In the same nucleus, another shallow valley develops at still larger asymmetries when the system is close to scission. This valley could be populated from the primary asymmetric valley by passing over the very thin ridge that separates them. This ridge, apparently due to the shell effects, should quickly disappear as the excitation energy is increased.

The same type of potential energy surface calculations was performed for all the thorium isotopes from ^{220}Th and ^{232}Th. All of the surfaces show the same basic valley structure. The differences lie in the quantitative details of the surfaces, i.e., the depths of the asymmetric and symmetric valleys, the height of the third fission barrier (E_C), the energy at the exit of the asymmetric valley, and the energy of the pass (E_S) between the asymmetric and the symmetric valleys. The latter quantity is expected to play a crucial role in determining the ratio of asymmetric to symmetric fission.

SUMMARY

The mass and the energy spectra of the fission fragments of the neutron deficient isotopes ^{220,224,226}Th have been investigated using near- and sub-barrier fusion reactions. The data gathered in this work provide a detailed view of the evolution of the mass and the energy distributions with the excitation energy and the nucleonic composition of the fissioning system.

The measured mass and energy distributions of the fission fragments show that the ratio of symmetric to asymmetric fission depends very strongly on the neutron number of the Th nucleus. These distributions also exhibit features associated with multimodal fission phenomena.

Calculations of the fission potential energy surfaces have been presented. The predicted features of these surfaces are in good agreement with the observed experimental features.

The experimental and the theoretical work described in this paper clearly demonstrate the utility of near- and sub-barrier, heavy-ion fusion reactions in exploring low energy fission phenomena.

ACKNOWLEDGMENTS

This was supported by the Russian Fundamental Research Foundation, under grant 96-02-17743A, and the US Department of Energy, under grant DE-FG03-93ER40773.

REFERENCES

1. Hullet E.K et al., *Phys. Rev. Let.* **56**, 313 (1986).
   Hullet E.K. et al., *Phys. Rev.* **C 40**, 779 (1989).

2. Hoffman D., Proc. of the 4th Int. Symposium on the Phys. and Chem. Of Fission, Julich, Germany, 1979, **1**, 275 (IAEA, Vienna, 1980).
3. Möller P., Nix R.J. and Swiatecki W.J, *Nucl. Phys.* **A 492**, 349 (1989); Pashkevich V.V., *Nucl. Phys.* **A 477**, 1 (1988);
4. Itkis M.G., Okolovich V.N.,Rusanov A.Ya. , Smirenkin G.N., *Z. Phys.* **A 320**, 433 (1985); Itkis M.G., Okolovich V.N., Smirenkin G.N., *Nucl. Phys.* **A 502**, 243 (1989) 243;
5. Pshkevich V.V., *Nucl. Phys.* **A 169**, 275 (1971).
6. Itkis M.G. et al., JINR Rapid Commun. 3 (1994) 19;
7. Shmidt K.-H. et al., *Phys. Lett.* **B325**, 313 (1994).
8. Oganessian Yu.Ts. et al, In Proc. of the Robert A. Welch Conf. On Chemical Research, Houston, 159, Texas, 1990.
9. Oganessian Yu.Ts. et al., *JINR Rapid Commun.*, **1(75)**, 123 (1996).
10. Murakami T. et al.,, *Phys. Rev.* **C 34**, 1353 (1986).
11. Videback F., Goldstein R.B., Grodzins L., and Steadman S.G., *Phys. Rev.* **C 15**, 954 (1977).
12. Chubarian G.G. et. al., *Sov. Journal of Nuclear Phys.* **56(3)**, 3 (1993).
13. Whetstone S.L., *Phys. Rev.* **131**, 1232 (1963).
14. Tilquin I. et al., *Nucl. Instr. And Meth. In Phys. Res.* **A 365**, 446 (1995).
15. Cohen S., Swiatecki W.J., *Ann. Phys.* **22**, 406 (1963).
16. Unik J.P., Huizenga J.R., *Phys. Rev.* **134**, B390 (1964).
17. Britt H.C., Wegner H.E., Gursky J.C., *Phys. Rev.* **129**, 2239 (1963).
18. Pfeiffer E., *Z. Phys.* **240**, 403 (1970).
19. Dubrovina S. M. et al., *Sov. Journal of Nuclear Phys.* **17**, 470 (1973).
20. Gonnenwein F., Pfeiffer E., *Z. Phys.* **207**, 209 (1967).
21. Glendenin L.E., *Phys. Rev.* **C 22**, 152 (1980).
22. Vorobjeva V.G., Djachenko N.P., Kuzminov B.D., In Proc. of the All Union Conf. On Neutron Physics, Kiev, USSR, April 1977, **Part 3**, 183, Atomenergoizdat, Moscow, 1997.
23. Holubarsch W, Pfiffer E., Gonnenwein F., *Nucl. Phys.* **A 171**, 631 (1971).
24. Speht H.J., *Nucleonica*, **20**, 717 (1975).
25 Børnholm S., Bohr O., Mottelson B.R., In Proc. of Int. Symp. Phys. and Chem. of Fission, Rochester, 1973, **1**, 361 (IAEA, Vienna, 1974).
26. Ignatyuk A.V., Istekov K., Smirenkin G.N., *Sov. Journal of Nuclear Phys.* **29**, 875 (1979).
27 Kudjaev G. A., Ostapenko Yu.B., Svirin M.I., Smirenkin G.N., *Sov. Journal of Nuclear Phys.* **47**, 1540 (1988).
28. Rastopchin E., Svirin M., Smirenkin G.N., *Sov. Journal of Nuclear Phys.* **55**, 310 (1992).
29. Strutinsky V.M., *Sov. Journal of Nuclear Phys.* **1**, 821 (1965).
30. Damgaard J., Pauli H.C., Pashkevich V.V., Strutinsky V.M., *Nucl. Phys.* **A 135**, 432 (1969).
31. Pashkevich V.V. and Strutinsky V.M., *Sov. Journal of Nuclear Phys.* **9**, 56 (1969).
32. Strutinsky V.M., *Nucl. Phys.* **A 95**, 420 (1967).
33. Myers W.D. and Swiatecki W.J., *Avk. Fys.* **36**, 343 (1967).
34. Ivanova S.P., Komov A.L., Malov L.A., Soloviev V.G., *Particles and Nuclei* **7(2)**, 450 (1976).
35. Krappe H.J., Nix J.R. and Sierk A.J., *Phys. Rev.* **C 20**, 992 (1979).
36. Sierk A.J., *Phys. Rev.* **C 33**, 2039 (1986).
37. Pashkevich V.V., In Proc. of the XV EPS Nuclear Physics Divisional Conference on Low Energy Nuclear Dynamics. April 1995, St. Petersburg, Russia, 161, World Scientific, Singapore, 1996.

New Fission Mode of the ^{252}Cf Spontaneous Fission Obtained with Modern HPGE Detectors

A.V. Daniel[1], G.M. Ter-Akopian[1], J.H. Hamilton[2], Yu.Ts. Oganessian[1],
J. Kormicki[2], G.S. Popeko[1], A.V. Ramayya[2], W.-C. Ma[3], B.R.S. Babu[2],
T. Ginter[2], S.J. Zhu[2], J. Rasmussen[4], M.A. Stoyer[4], I.Y. Lee[4], S. Asztalos[4],
S.Y. Chu[4], K.E. Gregorich[4], A.O. Macchiavelli[4], M.F. Mohar[4],
S.G. Prussin[4], J. Kliman[5], M. Morhac[5], J.D. Cole[6],
R. Aryaeinejad[6], Y.K. Dardenne[6], M. Driger[6]

[1]Flerov Laboratory of Nuclear Reactions, JINR, Dubna 141980, Russia
[2]Department of Physics and Astronomy, Vanderbilt University, Nashville, TN 37235
[3]Department of Physics, Mississippi State University, Mississippi State, MS 39762
[4]Lawrence Berkeley National Laboratory, Berkeley, CA 94720
[5]Institute of Physics, Slovak Academy of Sciences, Bratislava, Slovak Republic
[6]Idaho National Engineering Laboratory, Idaho Falls, ID 83415

Abstract. The data of Independent yields of secondary fission fragment pairs (emerging after prompt neutron emission from primary fragment pairs) obtained by detecting coincidences between γ rays following the spontaneous fission of ^{252}Cf have been expanded. Our approach to estimate characteristics of the primary fragments pairs (mass and excitation energy distributions) by unfolding the yields of secondary fragment pairs is discussed. Mew model parameters were introduced and results are presented here. The new results confirmed our old assumption that in case of Mo-Ba charge split the two fission modes differing with average total kinetic energy <TKE> on ~36 MeV are realized.

INTRODUCTION

Recently a new approach to the low energy nuclear fission investigation has been suggested and realized in Refs. (1,2,3). It is based on the high energy resolution measurements of the γ-ray coincidences following the nuclear fission events. Its realization is possible with the large experimental setups based on the Compton suppressed high purity Ge detector arrays (GAMMASPHERE, EUROGAMM, GASP and their prototypes). We applied the approach to study the ^{252}Cf spontaneous fission and succeeded in the observation of new kind experimental data. These are the yields of correlated fission fragment pairs. The conventional charge and mass distributions of fission fragments inferred from our results are in a good agreement with the known data. Prompt neutron multiplicity distributions and rough estimations of the total kinetic energy (TKE) distributions were obtained from the yields of fragment pairs for the different charge split of the ^{252}Cf. Strong enhancement of the 7-10 neutron emission was observed for the Mo-Ba charge split (1).

Mass and excitation energy distributions of primary Ru-Xe, Mo-Ba and Zr-Ce fission fragments were deduced by applying the unfolding procedure to the experimental yields of secondary fragment pairs (3,4). For the Ru-Xe and Zr-Ce charge splits, the experimental data were well fitted by assuming one fission mode with <TKE> values close to the value of <TKE> known for the ^{252}Cf spontaneous fission. In case of Mo-Ba charge split a successful fit has been obtained only with the assumption that, in addition to the "normal" fission mode, a second fission mode with a lower value of <TKE> of 153 MeV contributes to this charge split. The difference of the <TKE> values makes about 36 MeV. Details and various modifications of the unfolding procedure are discussed here. All of them confirmed he assumption about the existence of two fission modes in case of Mo-Ba charge split.

EXPERIMENTAL PROCEDURES AND RESULTS

Two independent experiments were carried out with the hermetically closed ^{252}Cf sources. The first one was done at the Holifield Heavy Ion Research Facility with the Oak Ridge Compton Suppression Spectrometer System (20 working detectors), and the second experiment was done at the GAMMASPHERE (36 working detectors).

In the both cases the same procedure of data treatment was used. A two dimensional matrix of γ-γ coincidences (4096x4096 channels) was created from the initial data by selecting the γ-ray coincidences occurring within 200 ns. One of the main problem to extract reliable data from this matrix is a complicated form of background. It consists of a smooth part and a part that has the form of ridges parallel to the axes. The search for γ-γ coincidence peaks and estimations of their areas were carried out with the use of our code M2 (5). In our approach, the full two-dimensional matrix was divided in small regions, and each region was separately fitted with a two-dimensional function describing all components (real peaks and various types of background) of the two-dimensional spectrum. Examples of the M2 code application to a simulated γ-γ coincidence spectrum and to the real one are shown in Fig. 1,2 respectively. In both cases the plots are presented in the same scale to emphasize the real level of the various types of background. In Fig. 1a one can see the false peaks (x_1-y_2, x_2-y_1 and x_3-y_1) originated only from the intersections of the background ridges that are separately shown in Fig. 1c,d. In Fig. 2 (part of the real γ-γ coincidences spectrum) one can see a large background ridge that goes parallel to E2 axis. This ridge was well subtracted, as it can be seen in Fig. 2b. The existence of such ridges is caused by coincidences of full absorption pulses with the pulses from the incomplete energy release of other γ-rays.

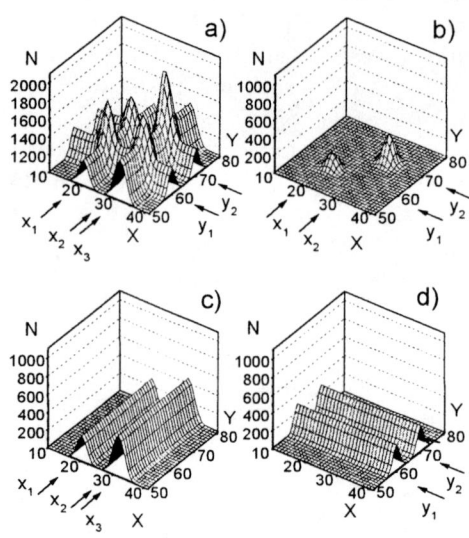

FIGURE 1. This is an example of the simulated γ-γ coincidence spectrum - a, the spectrum obtained by the background subtraction with the M2 code - b, and the background ridges - c, d.

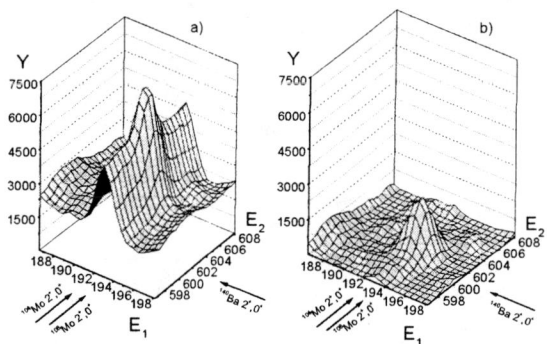

FIGURE 2. Original two-dimension γ-γ coincidence spectrum and the spectrum obtained after background subtraction are shown in a small region. The region is located around the coincidence peaks associated with $2^+ \rightarrow 0^+$ γ transition for pairs ^{104}Mo-^{140}Ba and ^{108}Mo-^{140}Ba.

The yields of the pairs involving both fragments with even A were deduced from the areas of the peaks originating from the coincidences of the $2^+ \rightarrow 0^+$ γ transitions in both fragments. In the cases when odd A fragments occurred in fragment pairs, the areas of the peaks originating from the coincidences between γ transitions to their ground states were summed up to obtain the yields. The relative yields of the fragment pairs were obtained after corrections for the known detection efficiency and internal conversion probability. Additional corrections needed for a number of fragment pairs were discussed in (3).

The relative yields of fragment pairs were transformed into the independent yields by a normalization procedure. Firstly the relative yields of fission fragments were obtained by making a convolution of the relative yields of fragment pairs. The obtained data were normalized to the independent yields known from the literature (6) for some particular fission fragments of ^{252}Cf. In fact, we used the experimental values of the independent yields of ^{135,136,138}Xe, ^{140,142,144,146}Ba, ^{144,146,148,150}Ce compiled in (6). For one charge split a number of normalization coefficients were calculated and these coefficients were averaged with the weights that were equal to the reciprocal values of their errors.

The yields of the fragment pairs allowed us also to obtain the multiplicity distributions of prompt neutrons by summing up the yields of the pairs that were responsible for the same total number of evaporated neutrons. A number of significant values missed in the experimental yields of the fragment pairs were added by using the two-dimensional B-spline approximation of the primary data. A similar procedure was used to estimate the total kinetic energy distributions. The obvious relation was used in this case:

$$TKE = M_{Cf} - \left(M_{LF} + M_{HF} + n \cdot M_n + \sum E_i + E_{LF} + E_{HF}\right), \qquad (1)$$

where E_{LF} and E_{HF} are the residual parts of the excitation energies of light and heavy fragments retained after the neutron evaporation; E_i are the energies of the evaporated neutrons; n is the total number of neutrons (n = A_{Cf}-A_{LF}-A_{HF}). The real distributions of E_i, E_{LF} and E_{HF} are not known, therefore the simplest hypotheses were used. It was supposed that a simple fission spectrum (Maxwellian) describes the energy distributions of neutrons E_i and a random distribution between zero and the neutron binding energy is enough to describe the distributions of E_{LF} and E_{HF}. As a result of using these hypotheses we can talk about the estimation of the TKE distributions only. The results of this procedure are shown in Fig. 3.

The neutron multiplicity distributions (on exclusion of the Mo-Ba distribution) were well fitted with Gaussian curves. The Mo-Ba distribution could not be described by a single Gaussian because of the excess of 7-10 neutron events. We explained this result in (4) by attracting the hypothesis that the two distinct fission modes for the same charge and mass asymmetry exist for the Mo-Ba charge split. The second fission mode shows a very low value of <TKE> and a very narrow mass distribution of primary fragments. Now it is confirmed by the TKE distributions, to some extent. The Mo-Ba TKE distribution shown in Fig. 3e demonstrates a long tail in the region of low energy which is absent in other distributions in Fig. 3d and 3f.

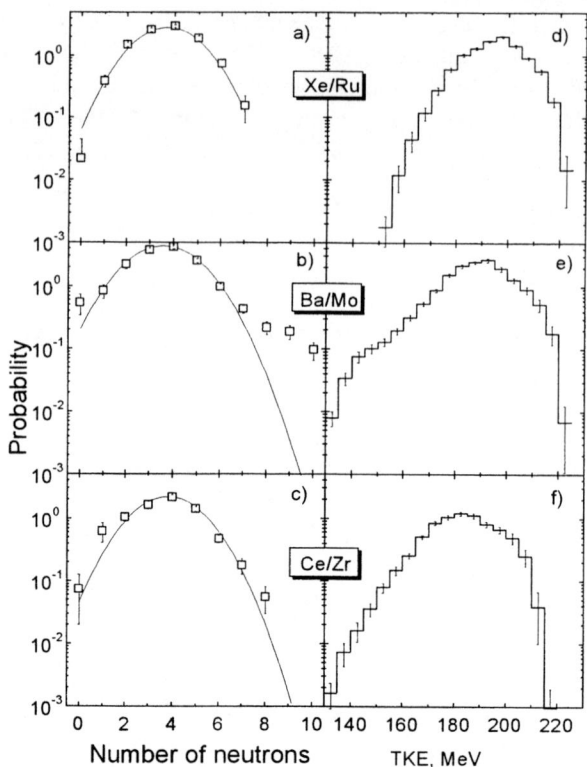

FIGURE 3. Multiplicity distributions of prompt neutrons (a-c) and TKE distributions (d-f) for three charge splits of ^{252}Cf. Solid lines in Fig. a-c show the Gaussian fit applying to the data.

CHARACTERISTICS OF THE PRIMARY FRAGMENT PAIRS

The experimental yields of fission fragment pairs discussed here originate from de-excitation of primary fission fragments, and carry information about the mass and excitation energy distributions of primary fission fragments of the fixed charge splits $Y(A_L, E_L^*, A_H, E_H^* | Z_L, Z_H)$. Two possible ways to use such data are the following. One way is to test the ability of different theories to predict the primary fragment distributions which reproduce the experimental data after applying a statistical code for modeling the de-excitation of primary fragments. The second way, which we suggested and realized in (4) for the first time, is to extract the primary fragment distributions by unfolding the experimental data.

Unfolding Procedure

The experimental yields of secondary fission fragment pairs created after neutron evaporation $Y^{\exp}(A'_L, A'_H | Z_L, Z_H)$ can be connected to the yields of primary fission fragment pairs $Y(A_L, A_H | Z_L, Z_H)$ by the following equations,

$$Y^{calc}(A'_L, A'_H | Z_L, Z_H) = \sum_m \sum_{(A_L, A_H)} Y(A_L, A_H | Z_L, Z_H) \times I_L \times I_H, \quad (2)$$

$$I_L = \int F(E^*_L, A_L) \times P_n(E^*_L, A_L) \times \delta(A_L - A'_L - n) dE^*_L, \quad (3)$$

$$I_H = \int F(E^*_H, A_H) \times P_n(E^*_H, A_H) \times \delta(A_H - A'_H - n) dE^*_H, \quad (4)$$

where $F(E^*, A)$ - is the excitation energy distribution of the primary fission fragment; $P_n(E^*, A)$ - is the probability of evaporation of n neutrons from a primary fragment $n = A - A'$. Equation (2) involves two summations: the first one is done over all the primary fragment pairs (A_L, A_H) of a given charge split, and the second is introduced to take into account possible contributions of more than one fission mode.

The unfolding procedure is based on the least squares method and implied minimization of the form

$$\chi^2 = \sum \left[\frac{Y^{\exp}(A'_L, A'_H | Z_L, Z_H) - Y^{calc}(A'_L, A'_H | Z_L, Z_H)}{\sigma_{\exp}(A'_L, A'_H | Z_L, Z_H)} \right]^2, \quad (5)$$

where, separately for each charge split, the summation is done over all the fragment pairs of a fixed charge split for which the yields were obtained in the experiment.

To make the task feasible we should to assume that the primary fragment mass and excitation energy distributions can be described by simple forms with a small number of free parameters. As in (3,4), here we confined ourself with the consideration of only gaussian forms for all the distributions resulting in inherent two free parameters for each of them. In order to reduce the total number of free parameters additional assumptions were made.

First it was assumed, that for a particular charge split and at a fixed fission mode, the mean value of $<TKE>$ and its variance σ_{TKE} are the same for different primary fragment pairs. These assumption and the following relations between the parameters of our model

$$Q_{fiss} = <TKE> + <E^*_L> + <E^*_H>, \quad (6)$$

$$\sigma^2_{TKE} = \sigma^2_{E^*_L} + \sigma^2_{E^*_H} \quad (7)$$

allowed us to reduce the number of free parameters. Here $<E^*_{L(H)}>$ and $\sigma_{E^*_{L(H)}}$ are the fragment mean excitation energy values and their variances. Equations (6) and (7) are valid in the framework of the model underlying Eq. (2) and the assumption of the Gaussian form for the primary fragment excitation energy distribution.

Second, we imposed the additional boundary conditions

$$\frac{\sigma^2_{E^*_H}}{\sigma^2_{E^*_L}} = \frac{<E^*_H>}{<E^*_L>} \tag{8}$$

for the dispersions of the excitation energies of the light and heavy fragments as defined above.

In (3,4) for each charge split, the mean excitation energies $<E^*_H>$ were searched for nine heavy primary fragments centered around the mean mass fragment. Here we replace the Eq. (8) by an assumption, that for a particular charge split and at a fixed fission mode, $\sigma_{E^*_L}, \sigma_{E^*_H}$ are the same for different primary fragment pairs. It is more physical assumption that introduced only one additional free parameter. Also we began to check possible relations between values of $<E^*_H>$ and $<E^*_L>$ that allow to reduce the nine free parameters $<E^*_H>$ to two or tree ones. We started with the simple relation

$$\frac{<E^*_L>}{<E^*_H>} = P1 + P2 \times (A_H - <A_H>). \tag{9}$$

Neutron Evaporation

Realization of the unfolding procedure demands the precise calculations of the $P_n(E^*, A)$ values for the all primary fission fragments. The calculations must be done in a wide range of the excitation energy E^* from the neutron binding energy and up to the highest possible excitation energy of fragments. The calculations were carried out, both in the present paper and in (3,4), with the use of the GNASH code (7) that implements the Hauser-Feshbach theory. In both cases Ignatyuk level density was used. The neutron transmission coefficients are provided to the GANSH code from an external file. These data were calculated here with the coupled channel deformed optical model code ECIS (8), at variance with the calculations of (3,4) that had been carried out with the spherical optical model code SCAT2 (9). Also a new global optical model neutron potentials were used. These were the potentials of Walter and Guss (10) with the surface imaginary term modified by Yamamuro (11). In fact one can not see a serious difference in the results of two systematic calculations shown on Fig. 4.

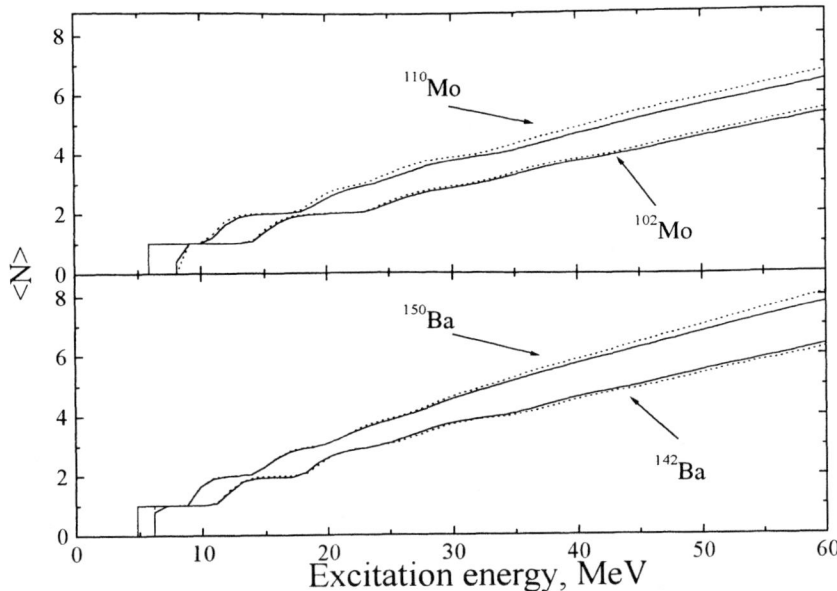

FIGURE 4. Mean number of evaporated neutrons vs the excitation energy of fragments. Solid line present the new results and dotted lines presents the results of (3,4).

Practically the values of $P_n(E^*, A)$ obtained with the new calculations were used. But we believe that the use the global optical model neutron potentials for the case of neutron-rich fragments demands its future revision.

Fission Modes

The unfolding procedure can be applied only if the extensive data for the yields of the secondary fission fragments pairs of the same charge split are known. Experimental yields for three charge splits Ru-Xe, Mo-Ba, and Zr-Ce presented in (3) are notable for their extensive data. Separately for each of these charge splits the unfolding procedure was applied. At first it appeared "natural" to carry this out by assuming that only a single fission mode contributes to each splits. In fact, taking only one fission mode (later referred to as *Mode 1*), the data for the Ru-Xe and Zr-Ce fragment pairs were successfully fitted (3). However, no reasonable solutions could be obtained with the single fission mode for the pattern of the Mo-Ba yields. This was to be expected, as the Mo-Ba charge split has strongly enhanced high neutron multiplicity yields that are not seen in the charge splits of Ru-Xe and Zr-Ce.

An excellent fit to the Mo-Ba yields were obtained by assuming that two distinct fission modes (*Mode 1* and *Mode 2*) contribute to the formation of the primary Mo-Ba fission fragments (3,4). The ratio of *Mode 1/Mode 2* intensities is 14 and other parameters are presented in (3,4).

We applied the new assumptions about relations between the model parameters (Eq. 9) to fit the data for the Mo-Ba fragment pairs (3) extended to add six new experimental points. Also the points corresponding to the zero neutron emission were excluded. The best result obtained with the two fission modes is shown on Fig. 5. Characteristics of these two fission modes extracted from the calculations are presented in Table 1.

TABLE 1. Characteristics of two fission modes.

	$<A_H>$	σ_A	$<TKE>$	σ_{EH}	σ_{EL}	P1	P2
Mode 1	145.5	1.3	188.9	6.7	5.6	0.98	0.02
Mode 2	145	0	153.7	8.6	6.1	0.70	0.

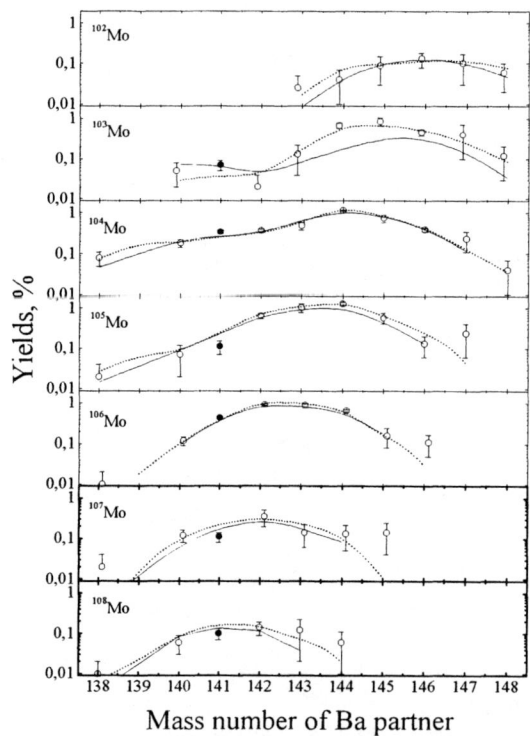

FIGURE 5. Points are our experimental yields of fission fragment pairs. Solid curves show the results of the unfolding procedure with new assumptions discussed in text and dotted curves show the results presented in (3,4).

CONCLUSIONS

The yields of fission fragment pairs provide the direct information that allows one to investigate the excitation energy of primary fragments and derive their mass distributions. The fact, that this can be done really, was proved for the Mo-Ba charge split of ^{252}Cf. The existence of two distinct fission modes in this split was confirmed. It was shown, that in spite of the model variations, one can not avoid the contribution of the *Mode 2* in the yields of fission fragment pairs in the case of Mo-Ba charge split.

ACKNOWLEDGMENTS

Two of the author (G.M.T.-A. and A.V.D.) would like to express appreciation for the hospitality and financial support received during their stays at Vanderbilt University and Oak Ridge National Laboratory. This research used resources of the National Energy Research Scientific Computing Center, which is supported by the Office of Energy Research of the U.S. Department of Energy. This work was supported in part at Vanderbilt University and Idaho National Engineering Laboratory by the U.S. Department of Energy under Grant and contract Nos. DE-FG05-88ER40407 and DE-AC07-76ID01570 respectively.

REFERENCES

1. Ter-Akopian G. M., et al, *Phys. Rev. Lett.* 73, 1477-1480 (1994)
2. Hamilton J. H., et al, *Prog. Part. Nucl. Phys.* 35, 635-704, (1995)
3. Ter-Akopian G. M., et al, *Phys. Rev. C*, 55, 1146-1161 (1997).
4. Ter-Akopian G. M., et al, *Phys. Rev. Lett.* 77, 32-35 (1996).
5. Daniel A. V., Popeko G. S. and Ter-Akopian G. M., "*Possibilities of the γ-γ coincidences analysis,*" JINR Comm. P10-97-109, Dubna, 1997 (in Russian).
6. England T. R. and Rider B. F., "*Evaluation and Compilation of Fission Product Yields,*" LANL report LA-UR-94-3106, 1994.
7. Young P. G. and Arthur E. D., "*GNASH: A Preequilibrium, Statistical Nuclear Model Code for Calculation of Cross Sections and Emission Spectra,*" LANL report LA-6947, 1977.
8. Raynal J., "*Optical Model and Coupled-Channel Calculations in Nuclear Physics,*" International Atomic Energy Agency report, IAEA SMR-9/8, 1970.
9. Bersillon O., "*SCAT2 - A Spherical Optical Model Code,*" in Progress Report of the Nuclear Physics Division, Bruyeres-le-Chatel 1977, CEA-N-2037, p.111, 1978.
10. Walter R. L., and Guss P. P., "A Global Optical Model for Neutron Scattering for A>53 and 10 MeV<E<80 MeV," in *Proc. Int. Conf. Nuclear Data for Basic and Applied Science, Santa Fe, N.M., May 13-17, 1985,* Gordon and Breach Science Pub., Inc. 1986, pp.1079-1090.
11. Yamamuro N. "Nuclear Cross Section Calculations with a Simplified-Input Version of ELIESE-GNASH Joint Program," in *Proc. Int. Conf. on Nuclear Data for Science and Technology, May 30 - June 3, 1988, Mito, Japan,* LA-UR 88-1618, Ed. S. Igarasi, Saikon Publ. Co., Ltd., 1988, pp. 489-492.

Fission Mode Study for Low-Energy Fission of Light Actinide Elements

Hiroshi Baba

Department of Chemistry, Graduate School of Science, Osaka University,
1-1 Machikaneyama, Toyonaka, Osaka 560, Japan

Abstract. The double-velocity and double-energy measurements for thermal-neutron induced fissions of ^{235}U and ^{233}U and spontaneous fission of ^{252}Cf were carried out to extract correlation data among primary and secondary fragment masses, kinetic energy and neutron multiplicity event by event. The resulting data were discussed from a point of view of the multi-mode fission mechanism. Then, a new approach by means of a deformation parameter of fragments was devised to elucidate the correlation of fission configuration and fission mode.

1. INTRODUCTION

The multi-mode random neck-rupture model was proposed by Brosa et al.[1-3] in order to explain fission characteristics. The model requires mainly four scission configurations, namely, "superlong", "standard", "superasymmetry" and "supershort", for the fission in the actinide region. Which one of them becomes prevailing is claimed to depend on the fissioning nucleus. In the case of ^{236}U, the mass asymmetric standard and symmetric superlong shapes appear and the former further splits into two subgroups, the less asymmetric standard I and the main asymmetric component named standard II.

According to the model prediction, several works have tried to ascribe low-energy fission observables to this multi-mode fission mechanism. Weigmann et al.[4] compare the mass-yield distribution of thermal-neutron fission of ^{235}U with that of resonance fission obtained with 19.3 eV neutrons and observe enhancement of the latter yield in the mass range from 127 to 137 u and suppression of the yield just next to the mass region, which is attributed to the multi-mode fission. Knitter et al.[5] decompose the mass-yield distribution of the thermal-neutron fission of ^{235}U into two components by means of the symmetric Gaussian distribution and try to interpret the fragment kinetic energy distribution versus fragment mass in terms of thus deduced proportions of standard I and II. Weber et al.[6] also conclude the appearance of different fission-fragment mass asymmetries for specific nuclear excitation in their study of ^{236}U(e,e'f) and (γ, f) reactions supporting the feasibility of the multi-mode fission mechanism.

The most persuading work for the mechanism is probably the one given by Wagemans et al.[7,8] in the form of the double-energy measurement of a series of spontaneous fission of plutonium isotopes. They demonstrate that the peak of the mass-yield distribution systematically transfers from 141 to 135 as the isotopic mass increases from 236 to 242, which implies the existence of two groups, and that the fragment kinetic energy distribution is decomposed into two corresponding components.

Thus, the existence of two mass asymmetric groups besides the symmetric mode is considered certain in the thermal-neutron-induced and spontaneous fissions of uranium and plutonium isotopes. However, there are still some reservations to be cleared in this subject. For instance, one must find out if it is allowed to deconvolute the inclusive mass-yield distribution into two symmetric Gaussians to assign them to standard I and II as Knitter et al.[5] attempted. Further, we do not feel that the bifurcation in the path from saddle to scission is proved to appear in the asymmetric mode despite of the elaborate work by Wagemans et al.

The purposes of the present work are then to find whether correlation of the fission observables that we measured for thermal-neutron induced fissions of ^{235}U and ^{233}U and spontaneous fission of ^{252}Cf is well explained as the superposition of several fission modes the model predicts, and to elucidate the origin of the modes through the analysis. Besides that, It was also attempted to deduce deformation parameters of the fission fragments and to clarify the correlation of the scission configuration and the fission modes.

2. EXPERIMENTAL

The measurements were carried out at Super-Mirror Neutron-Guide-Tube of Kyoto University Reactor. The ^{235}U or ^{233}U target was bombarded in a vacuum chamber by the thermal neutron beam from the guide tube (beam flux was about 6×10^7 n/s), and fission fragment's velocities and kinetic energies were measured. ^{235}U target was prepared in the form of thin collodion film containing uranyl di-benzoyl methane. The thickness of ^{235}U target was 80 μg U/cm^2. ^{233}U target was prepared by electrodeposition on Nickel foil of 10 μg/cm^2. The thickness of ^{233}U target was about 100μg U/cm^2.

To measure velocity and kinetic energies of the complementary fission fragments simultaneously, two start and two stop detectors were used. For start detectors, plastic scintillators combined with photomultipliers were employed. The plastic scintillator was made of about 25μg/cm^2 thick NE102A film. For stop detectors, 900 mm^2 Silicon Surface Barrier Detectors, SSBD, (ORTEC 900F) were used.

The time resolution of this system was below about 250 ps and the mass resolution for fission fragment was evaluated to be 2 to 3 u. About 16,000 fission events were observed in 110 hours for the ^{235}U fission, while 18,000 events were acquired for the ^{233}U fission. The measurement for spontaneous fission of ^{252}Cf was also performed off-beam with the same system for comparison and about 100,000 events were observed for it. The ^{252}Cf target was prepared by eletrodeposition on Nickel foil of 0.1 μm in thickness and its source intensity was 150 kBq.

Primary fragment masses and kinetic energies were estimated from velocities of complementary fragments with the assumption that prompt neutrons were emitted isotropically after scission. The secondary mass number m was evaluated using Schmitt's prescription[9]. The number of prompt neutron ν was obtained from the difference between the primary mass m' and the secondary mass m event by event.

3. RESULTS AND DISCUSSION

3.1. Effective Distance of Fragments at Scission Point

In order to find an appropriate constraint for the deconvolution, the effective distance D_{eff} between the two charge centers at the moment of scission was then deduced from the obtained TKE by assuming the scission configuration consisting of two spheres connected with a thin neck, namely, D_{eff} is given by

$$D_{\text{eff}} = \frac{Z_\text{H} Z_\text{L} e^2}{TKE}, \quad (1)$$

where Z is the charge of the fragment and suffices H and L designate heavy and light fragments, respectively. Here, we further assumed the uniform charge distribution Z_{UCD}, since the deviation of the charge from Z_{UCD} is so small that the effect is negligible in Eq.(1).

The kinetic energy of the light fragment is kept constant over the asymmetric mode as has already been known among neutron-induced and spontaneous fissions of actinides. [10] This constancy in the fragment kinetic energy is considered to represent the characteristic of the prevailing mode, "standard II", of the asymmetric mass division since the expected "standard I" component should be localized in a rather narrow mass range around the double closed shell, 132 u, and then utilized as a constraint of the deconvolution. That is, the effective distance in a given mass division of the principal asymmetric mode was calculated following the discussion of Ref.[11] with the formula:

$$D_{\text{eff}}^{\text{II}} = \frac{1.44 Z_F^2 \eta^2 (1-\eta)}{E_{\text{k,L}}}, \qquad (2)$$

where η is the mass fraction of the heavy fragment, A_H/A_F with the fissioning mass A_F.

One consequence of Fig. 1 is that one can not expect a constant D_{eff} value for the asymmetric (standard II) mode as Knitter *et al.*[5] assumed in their multimode analysis of thermal-neutron fission of ^{235}U. This means the random neck-rupture model[1-3] fails at this point.

Another consequence of Fig. 1 is appearance of a systematic deviation of $D_{\text{eff}}^{\text{II}}$ from D_{eff} deduced from the first moment of the total kinetic energy distribution for $A_H \geq 130$ u. Such a systematic deviation was, however, observed very small even in the case of ^{235}U and hardly detected for ^{233}U fission. Hence, we must conclude the content of the so-called standard I is very small, if any, much smaller than expected ever. The total content of the minor component was found to be 3.1% for ^{236}U and 12.8% for ^{234}U.

3.2. Two-Dimensional Analysis of Mass-TKE Distribution

Based on the multi-mode fission mechanism suggested by Brosa[1-3], Siegler *et al.* showed that the theory could reproduce the correlation between the total kinetic energy and mass of fission fragments by means of the two-dimensional analysis of the yield distribution.[12] We tried the same approach for spontaneous fission of ^{252}Cf and thermal neutron induced fission of ^{233}U and ^{235}U to investigate the feasibility for various fissioning systems according to the prescription given in Ref.[12].

The ^{252}Cf(sf) data were satisfactorily reproduced with four components as predicted by the multi-mode theory, just as well as the case of ^{237}Np(n_{th},f) by Siegler *et al.*[12] as shown in Fig. 2. On the contrary, the agreement between the observed data and calculated with two components according to the prediction of the theory was very poor in the cases of ^{233}U(n_{th},f) and ^{235}U(n_{th},f). The most striking change in ^{234}U or ^{236}U data is the non-symmetric overall distribution in the mass axis. Since such anomalous shape was not observed in the ^{252}Cf data measured under practically the same condition, it should not be brought about due to inadequate experimental setup.

The situation was not much improved even when we introduced a skewed Gaussian function for a mass distribution of each component for the two thermal-neutron fissions. Then, we allowed one more component to fairly satisfactorily reproduce the observed distribution as exemplified for ^{233}U(n_{th},f) in Fig. 2. However, we intend to further investigate if thus obtained agreement is true consequences from the multi-mode mechanism or merely of a superficial nature.

3.3. Scission Configuration of Thermal-Neutron Fissions

The excitation energy given to a fission fragment is released as neutron and γ-ray emission until the nuclide gets settled in the ground state. Hence the excitation energy is able to be evaluated from the multiplicity and kinetic energies of the prompt neutrons and the γ-ray energy. With an assumption that all the excitation energy is used as deformation of the fission fragment, the deformation energy is equal to the summation of kinetic and binding energies of prompt neutrons and the γ-ray energy plus the ground-state deformation energy of the product nucleus as follows;

$$E_{\text{def}} = E_{\text{def}}^0(m_i) + \sum_{k=0}^{\nu_i} \{E_n(m_i) + E_{b,n}(m_i)\} + E_\gamma, \qquad (3)$$

where E_{def}^0 is the ground-state deformation energy of the secondary nucleus[13], E_n the kinetic energy of prompt neutron, $E_{b,n}$ the neutron binding energy[14], and E_γ the energy of γ-ray. The value of E_n is taken to be 2 MeV which is the averaged value of kinetic energies of prompt neutron measured by Ref.[15], and E_γ is taken to be half the neutron binding energy from the statistical consideration.

The deformation rate β was then deduced by equation given by Myers and Swiatecki[16] for a prorate spheroid with the eccentricity ε;

$$E_{\text{def}}(m) = 4\pi r_A^2 r_0 \left\{ \frac{\arcsin\varepsilon + \varepsilon\sqrt{1-\varepsilon^2}}{2\varepsilon \sqrt[6]{1-\varepsilon^2}} - 1 + 2x \left[\frac{\sqrt[3]{1-\varepsilon^2}}{2\varepsilon} \ln\left(\frac{1+\varepsilon}{1-\varepsilon}\right) - 1 \right] \right\}, \qquad (4)$$

where x is the fissility parameter, $r_A = 1.2A^{1/3}$ and $r_0 = 0.9$ MeV· fm^{-2}. The deformation parameter β is related to ε as $\beta = 2/3\sqrt{4/5\pi}\varepsilon$.

The results clearly reveal that there are two types of the scission configuration in the thermal neutron induced fission of ^{235}U and ^{233}U. One configuration consists of a pair of deformed light and heavy fragments (type-I), and the other is constructed with a deformed light fragment and a spherical heavy fragment (type-II). The yields of type-I and II configurations were 64.1% and 35.9% for ^{235}U(n_{th},f), and 58.9% and 41.1% for ^{233}U(n_{th},f), respectively.

Observed fission events can be sorted to type-I and type-II by means of the deformation parameter β of the heavy fragment. The mass and total kinetic energy (TKE) distributions are parted then to two distributions as illustrated in Figs. 3 and 4. Peak positions of the type-II mass distributions in Fig. 3, 136 and 139, correspond to Z=53, 55 and N=83, 84, respectively. Hence, the spherical neutron shell at N=82 is considered to play an important role in the formation of the type-II configuration. Likewise, the type-I scission configurations may be related to the deformed neutron shell at N=86. It is conceivable that the peak position of TKE distribution for type-I is higher than for type-II (Fig. 4) since the deformed heavy fragments may make the distance between the centers of fragments longer than the spherical heavy fragments and consequently lower the Coulomb repulsion.

The first moment of the kinetic energy distribution of a given fragment is shown in Fig. 5 as a function of fragment mass for respective configurations for the same systems. The average kinetic energy of the light fragments of type-II is higher than that of type-I by about 3%. It follows that for a given number of the light fragment, momentum conservation requires a smaller momentum to be possessed by the complementary heavy fragment of type-I than that of the corresponding type-II heavy fragment by about 3%. Therefore, the mass of the relevant heavy fragment of type-I can not be equal to that of type-II fragment, but must actually be slightly smaller than the latter.

This implies that the fissioning mass of the type-I component must be smaller than that of the type-II, probably by one, and consequently the excitation energy attributed to deformation should be considered to be spent at least partly for neutron emission in the case of the type-I component. Furthermore, nearly half the difference of the most probable total kinetic energy between the two components by 10 MeV comes from the difference of the most probable mass division between them so that the expected difference in the scission configuration should not be so large as the multi-mode fission mechanism predicts.

This may be a rather awkward consequence since it is hardly conceivable that an ordinary type of second chance fission takes place in thermal neutron fission. However, this is a straightforward conclusion one must accept if one admits the present results.

Recent studies of cold fission indicate non-trivial probabilities of barrier-penetrating type of fission even in thermal neutron fission. It has been also reported that some amount of pre-fission neutron emission is observed among thermal neutron fission.

Therefore, a sort of ternary fission may possibly the answer to the present findings, in which a neutron is emitted from the neck part at the moment of scission that does not participate in the mass split into two fragments.

In conclusion, the double-velocity and double-energy measurements for thermal-neutron induced fissions of ^{235}U and ^{233}U and spontaneous fission of ^{252}Cf were carried out to extract correlation data among primary and secondary fragment masses, kinetic energy and neutron multiplicity event by event. Then the data analysis based on the multi-mode fission mechanism was attempted for the above systems. It was found that the analysis does not give a satisfactory result for all the cases studied and that there are several observed results that the multi-mode fission mechanism fails to give a satisfactory explanation.

REFERENCES

1. Brosa, U., Grossmann, S., and Müller, A.: Z. Phys. **A325**, 241 (1986).
2. Brosa, U., Grossmann, S., and Müller, A.: Z. Naturforsch. **41a**, 1341 (1986).
3. Brosa, U., Grossmann, S., and Müller, A.: Proc. XVIth Int. Symp. on nuclear physics, Gaussig, 1986.
4. Weigmann, H., Knitter, H.-H., and Hambsch, F.-J.: Nucl. Phys., **A502** 177c (1989).
5. Knitter, H.-H., Hambsch, F.-J., Budtz-Jörgensen, C., and Theobald, J. P.: Z. Naturforsch. **42a**, 786 (1987).
6. Weber, Th., Wilke, W., Emrich, H. J., Heil, R. D., Kihm, Th., Kneissl, U., Knöpfle, K. T., Seemann, U., Steiper, S., and Ströher, H.: Nucl. Phys. **A502**, 279c (1989).
7. Wagemans, C., Schillebeeckx, P., and Deruytter, A.: Nucl. Phys. **A502**, 287c (1989).
8. Schillebeeckx, P., Wagemans, C., Deruytter, A. J., Berthélémy, R.: Nucl. Phys. **A545**, 623 (1992).
9. Schmitt, H. W., Neiler, J. H., and Walter, F. J.: Phys. Rev. **141**, 1146 (1966).
10. Hyde, E. K.: *The Nuclear Properties of the Heavy Elements, vol.III, Fission Phenomena*, Prentice-Hall Inc., Englewood Cliffs, 1964, p.190.
11. Baba, H., Saito, T., Takahashi, N., Yokoyama, A., Miyauchi, T., Mori, S., Yano, D., Hakoda, T., Takamiya, K., Nakanishi, K., Nakagome, Y.: J. Nucl. Sci. Tech. (Tokyo), in press (1997).
12. Siegler, P., Hambsch, F. -J., Oberstedt, S., and Theobald,J. P.: Nucl. Phys. **A594**, 45 (1995).
13. Möller, P., and Nix, J. R.: Atom. Data and Nucl. Data Tables **26**, 165 (1981).
14. Wapstra, A. H., and Audi,G : Nucl. Phys. **A432**, 1 (1985).
15. Leachman, R. B.: Proc. Int. Conf. Peaceful Uses At. Energy **2**, United Nations, New York, 1956, p.193.
16. Myers, W.D., and Swiatecki, W.J.: Nucl. Phys. **81**, 1 (1966).

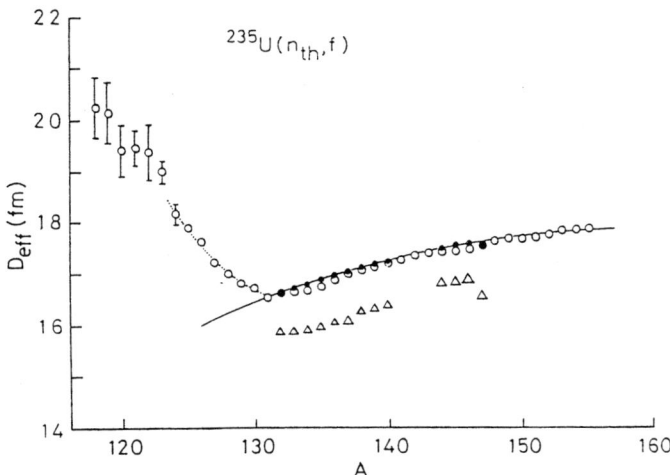

FIGURE 1. Effective charge separation D_{eff} (open circles) plotted versus heavy fragment mass A_H for thermal-neutron fission of ^{235}U. The solid curve represents the expected effective separation for the main asymmetric component (standard II). Triangles give the effective separations for the minor asymmetric component deduced by deconvolution with constraint of fixing the separation for the main component at the value represented with solid circles. The dotted line is given to guide the eye for the deviation from the expected trend.

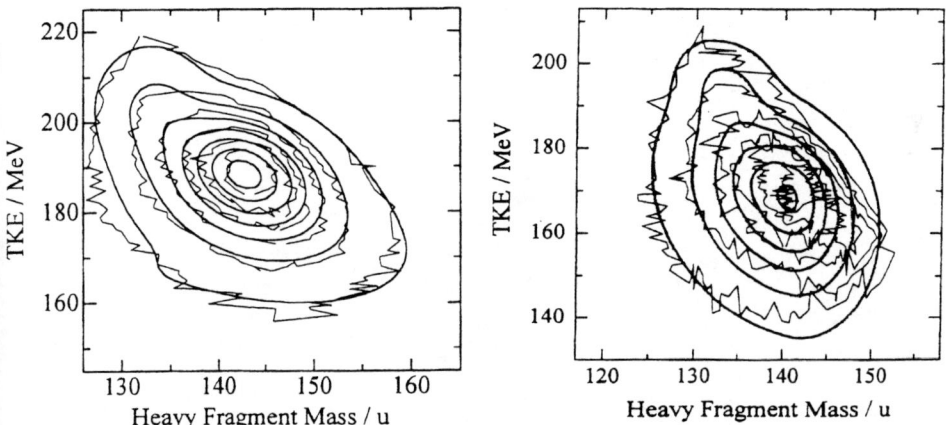

FIGURE 2. Two-dimensional (mass number and TKE) distributions of ^{252}Cf(sf) (left) and ^{233}U(n_{th},f) (right).

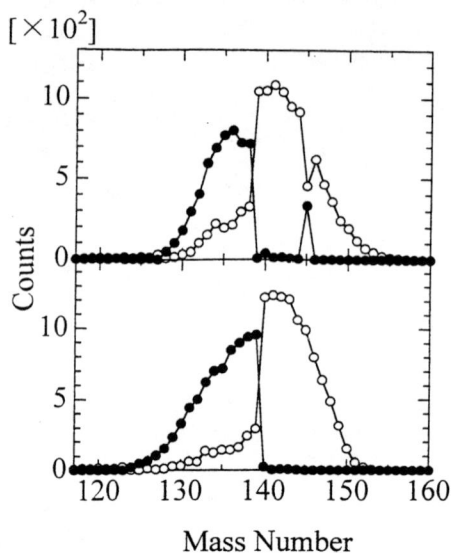

FIGURE 3. Mass distributions sorted by the magnitude of deformation for ^{235}U(n_{th},f) (top) and ^{233}U(n_{th},f) (bottom). Open and closed circles represent cases for type-I and type-II scission configurations, respectively. See text for detail.

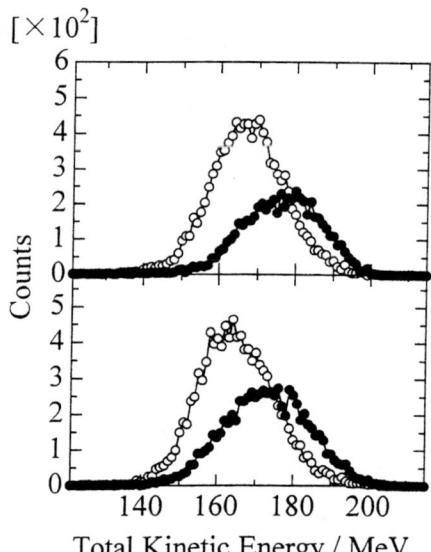

FIGURE 4. Total kinetic distributions sorted by the magnitude of deformation for ^{235}U(n_{th},f) (top) and ^{233}U(n_{th},f) (bottom). Open and closed circles represent cases for type-I and type-II scission configurations, respectively. See text for detail.

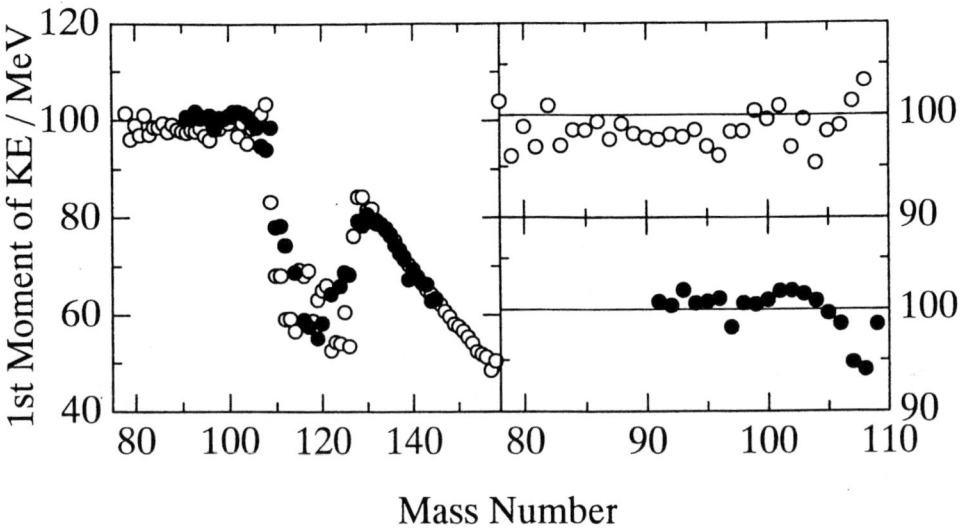

FIGURE 5. First moment of the kinetic energies of fragments sorted by the magnitude of deformation for ^{235}U(n_{th},f). The whole figure (left) and magnified parts on light fragments for respective configurations (right) are shown. Open and closed circles represent cases for type-I and type-II scission configurations, respectively. See text for detail.

Vlasov Treatment of Spontaneous Fission and Sub-barrier Fusion

A. Iwamoto[†], V. Kondratyev[†] [1] and A. Bonasera[‡]

[†] *Advanced Science Research Center, Japan Atomic Energy Research Institute Tokai, Naka-gun, Ibaraki, 319-11 Japan*
[‡] *Laboratorio Nazionale del Sud - Istituto Nazionale di Fisica Nucleare via S. Sofia 44, I-95123 Catania, Italy*

Abstract.
We present a model towards the many-body description of spontaneous fission and sub-barrier fusion based on the semiclassical Vlasov equation and the Feynman path integral method. We define suitable collective variables from the Vlasov solution and use the imaginary time technique for the dynamics below the Coulomb barrier. Internal excitations in fission process and the effect of particle transfer in fusion process are discussed.

INTRODUCTION

Many-body description of nuclear dynamics at the energies below the Coulomb barrier is still an open problem. A recently proposed method [1] based on the Vlasov equation (VE) and the Feynman path integral technique is expected to represent an approach towards a microscopic description of tunnelling phenomena in nuclear processes. This method follows the early attempts of Negele and coworkers [2] who used Time Dependent Hartree-Fock (TDHF). Because of numerical difficulties this model has not been applied to the very heavy system which fissions spontaneously. Thus it is useful to apply some simplified approaches, like VE which contains main features of many-body system. Because of the semiclassical nature of VE some genuine quantum effects contained in TDHF might be lost. However, such a treatment should be rather useful for quantitative understanding of some general properties of many-body dynamics. The preliminary studies of spontaneous fission of heavy nuclei and sub-barrier fusion of light nuclei [1] show that it is really a promising approach to describe the tunneling process with particle degree of freedom.

---

[1] on leave from: *Institute for Nuclear Research, 47, Pr.Nauki, Kiev, 252028 Ukraine*

We encountered, however, some difficulties which are inherent to many body dynamics. One is related to fission process where some kind of internal excitation inevitably arises. It is because of the mean- field nature of Vlasov model and restriction of the rotational symmetric shape assumed in the calculation. Another is related to the sub-barrier fusion where it was found that to deduce an effective potential energy between two colliding nuclei, the particle exchange between them play role especially at small inter-nuclear distance. This effect becomes prominent as the mass number increases and what is interesting is the effect acts inversely to the scattering of below-barrier and one of the above-barrier. In this report, we will discuss critically on these dificulties which appear in many-body model of spontaneous fission and sub-barrier fusion.

THE MODEL

The starting point is the Vlasov equation:

$$\frac{\partial f(\mathbf{r},\mathbf{p},t)}{\partial t} + \{h(\mathbf{r},\mathbf{p},t), f(\mathbf{r},\mathbf{p},t)\} = 0, \qquad (1)$$

where $f(\mathbf{r},\mathbf{p},t)$ is the one-body distribution function, $\{..\}$ is the Poisson bracket, $h(\mathbf{r},\mathbf{p},t) = T + U(r)$ is the hamiltonian with the kinetic energy T and the mean-field $U(r)$ which depends on density ρ and contains the Coulomb and surface terms.

The VE is solved by means of the test particles (TP) method [3,4] :

$$f(\mathbf{r},\mathbf{p},t) = \frac{1}{N} \sum_{i=1}^{N \times A_T} \delta(\mathbf{r} - \mathbf{r}_i(t))\delta(\mathbf{p} - \mathbf{p}_i(t)), \qquad (2)$$

where N is the number of TP per nucleon and A_T the total mass number. Substituting eq.(2) in eq.(1) we get the classical equations of motions for $\mathbf{r}_i$ and $\mathbf{p}_i$ of all the TP under the influence of the mean-field $U(r)$. In order to discuss the dynamics below the barrier we first define the collective variables. Here we follow the approach of refs. [1] and [5].

The most important degrees of freedom for fusion or fission, for a case of well separated fragments, are the the relative motion of the nuclei or of the fission fragments (FF) R and its conjugate momentum P. For more compact configurations, deformation and the size of the neck connecting the fragments become important [5,6]. In order to take into account these features we define the following degree of freedom:

$$\{\tilde{\mathbf{R}}(\tilde{\mathbf{P}})\} = \int_A d\mathbf{r}d\mathbf{p}\{\mathbf{r}(\mathbf{p})\}f(\mathbf{r},\mathbf{p},t) - \int_B d\mathbf{r}d\mathbf{p}\{\mathbf{r}(\mathbf{p})\}f(\mathbf{r},\mathbf{p},t), \qquad (3)$$

where A and B are the regions having $z < z_{plane} - r_N$ and $z > z_{plane} + r_N$ respectively. Here z is the coordinate along the fissioning motion, r_N is the neck radius and z_{plane} is the position of the plane which separates two fragments. The z_{plane} coincides with the C.M. of the total system for symmetric fission and by changing the plane position, we could describe asymmetric fission as well [1].

The physical meaning of our choices is rather simple. For compact shapes, eq.(3) gives somehow a measure of the degree of deformation of the nucleus. For spherical shapes $r_N = R$ (where R is the radius of the fissioning nucleus) the number of particles located in the tips (along the z-axis) are almost equal to zero (the system has a diffuse surface). When the system starts to elongate, $r_N < R$, only those particles which are located in the tips of the deformed nucleus give a contribution to the collective degree of freedom. After the scission point $r_N = 0$ and $\tilde{\mathbf{R}} \to \mathbf{R}$ ($\tilde{\mathbf{P}} \to \mathbf{P}$) i.e. the collective degrees of freedom become the relative distance and momentum as it should be.

Taking the time derivative of eq.(3) and using the Vlasov eq. (1) we get the classical Hamilton equations of motion for the collective variables $\tilde{\mathbf{R}}$ and $\tilde{\mathbf{P}}$. These equations are valid for the dynamics above the barrier.

To describe the dynamics below the barrier, we use the Wick transformation: $t \to i\tau$, i.e. we change the time from real to imaginary (cf. [1,2]). It is straightforward to show that the Hamilton's equation of motion for the (imaginary) $\tilde{\mathbf{P}}$ changes sign. To realize this procedure in the microscopic calculations, we impose a special force to each test particle from the time when the collective momentum becomes zero and takes a maximum value, i.e., we change the force acting on each TP [1] as

$$\frac{\partial \mathbf{p}_i}{\partial t} = -(\nabla U - 2\xi_{A(B)}) \qquad (4)$$

where the newly added term $\xi_{A(B)} = \frac{1}{A(B)}\mathbf{F}_{A(B)}$ is the normalized composition of the forces acting on a TP belonging to fragment A(B). At this stage the FF are pushed away from each other by the inverted collective force. We continue to solve the Vlasov eq.(1) with the modified force given by eq.(4) until the time when the collective momentum becomes zero again. From that time on, the motion becomes classically allowed, and thus we solve the real time Vlasov eq.(1) again up to scission and beyond.

FISSION

The definition of collective coordinate in fission process is not so simple like in the case of heavy-ion reaction. The original parent nucleus has a quite compact shape and the collective degree of freedom is no more the relative distance and momentum. It is also expected that one needs more than 1-degree of freedom. In order to take into account this fact we defined the degrees of freedom as in eq.(3) where the neck radius is involved. Our problem is still one dimensional which was chosen to simplify the calculation but with this choice, we can satisfy the following basic conditions for the collective deformation energy of the nucleus: (1)at spherical ground state(in case of no quantum effect), the collective force is zero,(2)the collective force becomes negative and its strength increases from the ground state to reach a maximum and then becomes zero again at the saddle deformation,(3)after the saddle deformation, the collective force becomes positive and increase,(4) reaching finally to the pure Coulomb force for separated fragments.

The choice of the relative distance and its momentum between two fragments doesn't satisfy the above given basic feature of the fission collective force. We tried also this choice for the numerical calculations and it resulted too big action, typically 10 and 20 times larger than data! The reason for this unphysical results is rather simple. By an inadequate choice of the collective force, the system when it enters into the sub-barrier region, is suddenly separated with a very strong force. The volume conservation is broken completely with this choice and it means the fission happens through the inadequate path, where internal state is highly excited and thus the barrier becomes very high.

An example of the system evolution for ^{252}Cf fission is given in the second reference in [1] and we will not show it again. (Caution should be paid that the scales of axes in Fig.1 of this reference should be multiplyed by factor 1.5!) We see that the system has some quadrupole deformation when entering the imaginary time evolution. Then, because of the rather strong collective force two fragments start to deform and finally separate. Scission occurs almost at the second turning point for this heavy element. The final value of the kinetic energy of the fission fragments are in rather good agreement to the Viola systematics. This quantity appears to be very sensitive to the choice of the surface term. We found that the action value is roughly a factor 2 larger for this nucleus than data and does not show the shell fluctuations. We performed calculations for seven parent nuclei and the result is mostly the same, that is, the final kinetic energies come close to the data but the action value is roughly two times larger than the experimentally determined values. [1]

With respect to this overestimation of the action integral, we point out one difficulty of this model calculation. It is related to the symmetry of the model space of Vlasov calculation. In our case, there is no effect which brakes the rotational symmetry of the system. Therefore, the component of the angular momentum parallel to the molecular axis is conserved between the parent nucleus and final two fragments. Because the radius of the parent nucleus and those of the daughter nuclei differ, the daughter nuclei cannot be in their ground state but are excited. This is a well known effect which is inherent to the mean-field approach and happens also in classical system as well as the quantum system. [7] Our model suffers this difficulty, too, and this is an important origin of the overestimation of the action integral. A question arises with this problem:For heavy system like actinide nuclei, this kind of internal excitation is estimated rather large, typically several MeV per particle. Why such a big excitation causes only a factor of two change of action value? The reason is that near the exit point of the tunneling process, the potential energy of the system is already decreasing rapidly. Especially in our case, the numerical calculation shows an already separated configuration for two fragments at barrier exit point. The distance of this point really increases as a result of a large intrinsic excitation compared with the ground-state nucleus. But the action value is an integration of the momentum in the barrier measured from the ground state energy level and thus this value doesn't change much although the excitation energy is big.

This difficulty should be solved by introducing the residual interaction which can break the symmetry of the system. Solution of this problem together with the introduction of the quantum effect is the next step for the development of the fission theory.

SUB-BARRIER FUSION

As another example of application of our method we discuss the heavy-ion fusion reaction at beam energies above and below the Coulomb barrier. In this case, as far as we calculate the overcome of a fusion barrier or the tunneling through it, the overlap of two colliding nuclei is assumed not to be very large. Thus we use the definition of collective variable eq.(3) in which we approximately put $r_N = 0$. This is a good approximation when two nuclei are apart but as they approach each other, we find an effect which comes from the particle exchange between the dividing plan.

To discuss this effect, we assume that the phase-space distribution function which characterizes the system has the following form,

$$f(\mathbf{r},\mathbf{p}) = f_A(\mathbf{r}, \mathbf{p} - \mathbf{P}/2) + f_B(\mathbf{r}, \mathbf{p} + \mathbf{P}/2) \qquad (5)$$

with $\mathbf{P} = \{0, 0, P\}$, f_A and f_B are generally distorted distribution functions of nuclei A and B with the center of masses at rest. We showed the case of symmerric reaction case for simplicity. Using this parametrization we obtain the following equation of motion for the collective momentum

$$\begin{aligned}\dot{P} &= \int d\mathbf{r}d\mathbf{p}|\mathbf{P}|\partial_t f(\mathbf{r},\mathbf{p}) = \int d\mathbf{r}d\mathbf{p}|\mathbf{P}|[-\{h,f\}] \\ &\approx \int d\mathbf{r}\ n(\mathbf{r})[sign(z)\partial_z U[n] - 2\delta(z)\{\Pi_{zz}/m + P^2/m\}] \\ &= F^l + F^{nl} = F\end{aligned} \qquad (6)$$

where $n(\mathbf{r}) = \int d\mathbf{p} f(\mathbf{r},\mathbf{p})$ and $\Pi_{zz} = <p_z^2>_f /n(\mathbf{r})$ denotes the diagonal component of the static pressure tensor. The first term in the square brackets of the integrand in the second line of eq. (6) give the local part (F^l) of the collective force, while the time-reversible non-local part (F^{nl}) of the collective force is caused by non-equilibrium nucleon exchange. The latter is related to the coordinate-dependent collective mass [8]. In contrast to the "randomization hypothesis" approximation (i.e. zero nuclear response time), the eq. (6) does not contain the frictional force.

In the case of a head on collision we can calculate the potential barrier corresponding to the '*static*' interaction between two nuclei as

$$V_{int}(R) = | \int_R^\infty dR \cdot F^l | . \qquad (7)$$

In the following we will also consider the effective potential barrier

$$V_{eff}(R) = | \int_R^\infty dR \cdot F | \qquad (8)$$

accounting for the non-local components of the collective force caused by the dynamical effect.

In the energy region below the barrier we employ the Feynman path integration technique within the microscopic treatment. For a given path connecting the entrance and exit channels of fusion reaction the probability for sub-barrier process is obtained from the equations

$$T_0(E) = (1 + \exp\{2\mu_r S/\hbar\})^{-1}, \quad S = \int \mathbf{P}^I \, d\mathbf{R}^I . \qquad (9)$$

where corresponding to the self-consistent evolution along the path in classically forbidden region action S is determined by the respective collective coordinate $\mathbf{R}^I$ and momentum $\mathbf{P}^I$.

The classically forbidden region for the collective variables becomes accessible after the so-called Wick transformation $t \to i\tau$, where τ is real. Changing simultaneously $P \to -iP^I$ we get the following equation of motion for $\{\mathbf{R}, \mathbf{P}\}$ imaginary time evolution

$$\frac{d\mathbf{R}^I}{d\tau} = \frac{\mathbf{P}^I}{m}; \quad \frac{d\mathbf{P}^I}{d\tau} = \mathbf{F}^I \approx -\mathbf{F}^l + \mathbf{F}^{nl} . \qquad (10)$$

From this eqs. we see that the path leading to sub-barrier fusion is associated with dynamics in inverted potential barrier (7) (i.e. negative sign of the local collective force), while the respective nonlocal component remains unchanged. This feature of the fusion path can be also obtained going through the standard steps of WKB approximation applied for the collective degrees of freedom when accounting for the variation of the respective inertia parameter [8]. As we will see below such a property results in a lower effective barrier for imaginary time propagation (i.e. V_{eff}^I given by (8) with the collective force $\mathbf{F}^I$ from the eq. (10)) as compared to the associated potential barrier, and therefore in an enhancement of the fusion cross section.

Using the test particle method, we can easily identify the local force F^l as the normal potential force acting on test particles and the non-local force F^{nl}

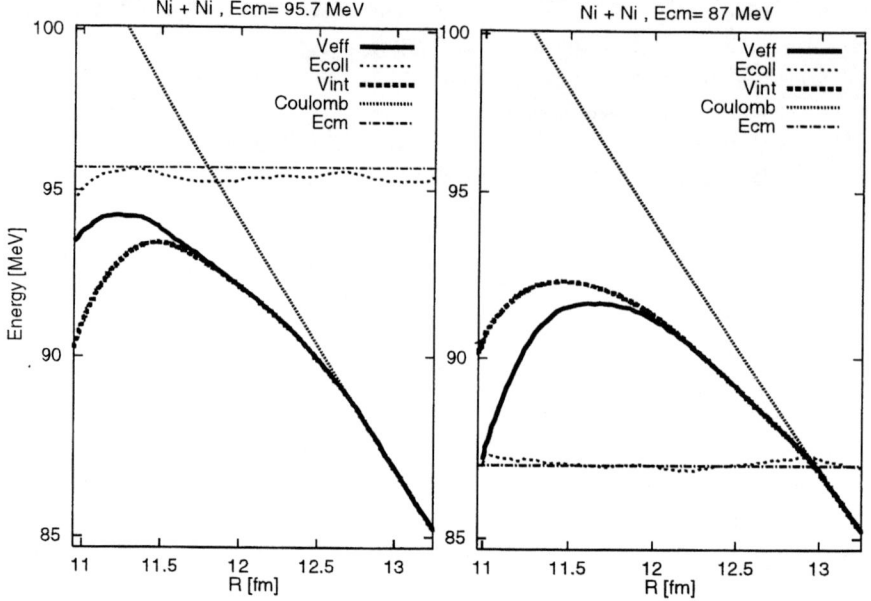

FIGURE 1. Local and effective potentials for the fusion of $^{58}Ni + ^{58}Ni$. The left panel is for above-barrier incident energy and right panel is for sub-barrier energy.

as the force arising from the nucleon exchange between the dividing plane. At this stage, it became clear the procedure to impose a Wick transformation to test particle dynamics, i.e., to change the signature of the local force and the non-local force is kept unchanged. In this way, we can calculate the evolution of the tunneling process microscopically.

Fig.1 shows the potential for the fusion reaction for $^{58}Ni + ^{58}Ni$. Left panel shows the potential for incident energy of 95.7 MeV, which is several MeV above the barrier. The right panel is for incident energy of 87 MeV, which is several MeV below the barrier. In this figure, the solid lines are the effective potentials and the bold dotted lines are the local potentials. In these calculations, we used the Skyrme-type interaction of the compressibility K=200MeV, Yukawa-type interaction with the range 1.4fm, Coulomb force and the symmetry potential. From this figure, it is seen that the local potential is the main part and it reproduced the height of the barrier obtained from experiment rather well. We can see that the local potential barrier for sub-barrier energy is lower than for the above-barrier case. We think this effect originates from the coupled-channel effect discussed frequently in the sub-barrier reactions. In our case, various kinds of the coupling to intrinsic excitation are included simultaneously and in this respect, there must be difference between ours and the standard coupled-channel model. On the other hand, the dynamical potential behaves very differently for above-barrier and sub-barrier. This part of the potential acts to increase the potential for the above-barrier case but it lowers the potential for the sub-barrier case. This feature is expected from the equation of motion for the collective variables and it is definitely ascertained in our microscopic calculations. This kind of behavior is almost equivalent to the model of changing collective mass parameter, if it decreases as two nuclei comes closer. This general behavior is consistent with the phenomenological treatment of the non-local effect discussed in [9] and in [10].

CONCLUSION

In this contribution we have presented an approach to nuclear dynamics based on the Vlasov equation and the Feynman path integral method. We have discussed the application of the model to spontaneous fission and sub-barrier fusion. A reasonable reproduction of different observables has been obtained [1]. The method is rather flexible and it can give a starting point of the tunneling phenomena of a many-body system.

In case of spontaneous fission, we found that the method gives a reasonable description of the time evolution of the system in the tunneling region. However, the definition of the collective variable is not unique and further study is necessary to put the model on more firm ground. The spurious excitation caused by the restriction of the model is one of the most urgent task to solve. For the complete description of the fission process, it is clear that we need to include fully the quantum effect which should appear as a shell effect.

In case of sub-barrier fusion between two heavy ion, the definition of the collective variable is not a matter of the biggest problem. In this report, we discussed mainly on the non-local force effect which originate from the nucleon exchange between two nuclei when two come close together. Using Vlasov dynamics, we showed clearly the importance of this nonlocal effect even in the sub-barrier reaction and showed the relation of the force and the particle exchange process between the dividing plane. In addition, we showed that the force acts inversely for above-barrier and sub-barrier reactions. This feature was expected from the macroscopic model but a microscopic foundation of it was achieved with the Vlasov dynamics.

It is true that there remains so many theoretical problems to be solved but nevertheless this approach seems to be a very promising way to clarify the tunneling process of many-particle system.

Acknowledgments

Two of the authors (A.B. and V.N.K.) are indebted to the research group for Hadron Transport Theory, JAERI, for warm hospitality and financial support. This work is supported in part by Japan STA program.

REFERENCES

1. A. Bonasera and V.N. Kondratyev, Phys. Lett. **B339** (1994)207; A. Bonasera and A. Iwamoto, Phys. Rev. Lett. **78** (1997) 187; and A. Bonasera, V.N. Kondratyev and A. Iwamoto, J.Phys. **G23** (1997) no.10 in press
2. J. Negele, Physics Today **38** (1985) 24; Nucl. Phys. **A502** (1989) 371c and references therein.
3. G.F. Bertsch and S. Das Gupta, Phys. Rep. **160** (1988) 190; A. Bonasera, F. Gulminelli and J. Molitoris, Phys. Rep. **243** (1994) 1.
4. C. Wong, Phys. Rev. **C17** (1978) 1832.
5. A. Bonasera, G.F. Bertsch and E. El Sayed, Phys. Lett. **B141** (1984) 9; A. Bonasera, Nucl. Phys. **A439** (1985) 353 ; Phys.Rev. **C34** (1986) 740; and Nuovo Cimento **A99** (1988) 435.
6. P. Möller, J.R. Nix, J. Phys. **G20** (1994) 1681, and references therein.
7. G. Bertsch and A. Bulgac, private communication
8. A. Iwamoto, Z. Phys. A **349** (1994) 265.
9. D. Galetti and M.A. Cândido Ribeiro, Phys.Rev. **C50** (1994) 2136.
10. R. Dutt and T. Sil, Phys.Rev. **C54** (1996) 319.

SUB-BARRIER FUSION

Fusion Barrier Distributions – What Have We Learned?

D.J. Hinde, M. Dasgupta

Department of Nuclear Physics, Research School of Physical Sciences and Engineering, The Australian National University, Canberra ACT 0200, Australia

Abstract.
The study of nuclear fusion received a strong impetus from the realisation that an *experimental* fusion barrier distribution could be determined from precisely measured fusion cross–sections. Experimental data for different reactions have shown in the fusion barrier distributions clear signatures of a range of nuclear excitations, for example the effects of static quadrupole and hexadecapole deformations, single– and double–phonon states, transfer of nucleons, and high–lying excited states. The improved understanding of fusion barrier distributions allows more reliable prediction of fusion angular momentum distributions, which aids interpretation of fission probabilities and fission anisotropies, and understanding of the population of super–deformed bands for nuclear structure studies. Studies of the relationship between the fusion barrier distribution and the extra–push energy should improve our understanding of the mechanism of the extra–push effect, and may help to predict new ways of forming very heavy or super–heavy nuclei.

I INTRODUCTION

The concept of a distribution of fusion barrier heights can be most easily demonstrated in the case of fusion of nuclei with a static deformation [1–3]. In a classical picture, different orientations of the deformed target nuclei are encountered during the collision, giving rise to a distribution of barrier heights, some lower and some higher than the fusion barrier for spherical nuclei.

Such a classical picture is not expected to be valid for other modes which can be excited, such as surface vibrations (phonons) and nucleon transfers. The quantum-mechanical effects of such couplings can be most simply visualised in the eigenchannel approximation, where it was shown [4] that the effect of couplings to the elastic channel is to replace the single fusion barrier by a distribution of barriers of different heights (energies), some of which are necessarily lower than the height of the single barrier without couplings.

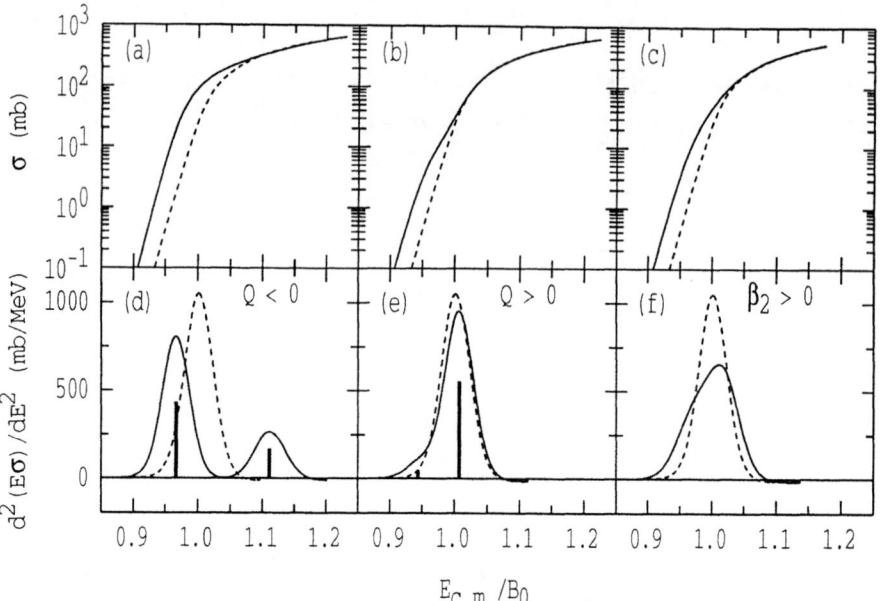

FIGURE 1. Calculated excitation functions and barrier distributions for a single-barrier (dashed lines) compared with three coupling schemes involving coupling to a negative Q-value channel (a) and (d), a positive Q-value channel (b) and (e), and coupling associated with a nucleus with a permanent quadrupole deformation (c) and (f). The cross–sections, plotted against the ratio of the energy to the average barrier B_0, are essentially identical at low and high energies. The type of coupling is more easily seen in the lower part of the figure than in the excitation functions themselves. (Adapted from Ref. 7).

This provided a mechanism to explain the observed enhancement [5,6] of experimental data at sub-barrier energies, over calculations based on penetration of a single barrier. However since coupling to any channels gives rise to enhancement, an educated guess was required as to the type of coupling to be included in a given reaction.

This point is illustrated in Fig. 1, where three calculated excitation functions for reactions involving different coupling schemes are shown. These are coupling to a negative Q-value channel in Fig. 1(a), a positive Q-value channel in Fig. 1(b), and coupling associated with a deformed nucleus in Fig. 1(c). The discrete barriers (in the eigenchannel approximation) associated with the first two cases are shown in Fig. 1(d) and (e) by the thick vertical lines whose lengths indicate the probability of encountering the barrier; the third coupling scheme gives a continuous distribution of barrier heights associated with different orientations of the deformed nucleus. The enhancements (cross-sections) at the lower energies are essentially identical in the three cases, despite the differences in the nature of couplings. There are significant differences in the excitation functions at higher energies, but the type of coupling involved is not immediately apparent from the excitation functions themselves.

It was realised by Rowley et al. [8] that an experimental distribution of fusion barrier heights in a reaction can be extracted *directly* from a fusion excitation function. In a classical sharp cut-off model,

$$E\sigma = \pi R^2 \sum_\alpha w_\alpha (E - B_\alpha), \tag{1}$$

where E is the energy, σ is the fusion cross-section, R is the fusion radius, w_α and B_α are the weight and barrier energy respectively for the channel index α. In this case, $d^2(E\sigma)/dE^2$ returns the original discrete barrier distribution as a set of delta functions, centered at the barrier energies and weighted by the w_α [8]. When quantum mechanical penetration of the barriers is permitted, the cross-sections vary smoothly in the vicinity of each barrier and $d^2(E\sigma)/dE^2$ becomes continuous, the delta functions being replaced by near-Gaussian functions [8].

The calculated excitation functions in Fig. 1 have been treated by this method, resulting in the functions shown by the full lines in the lower panels. They are very different for the three cases, and reveal directly where the barrier strength lies; in Fig. 1(d) and (e) the solid curves are centred around the calculated discrete barrier positions (thick lines) and the heights of the peaks reflect the weights of the barriers. In the third case the solid curve coincides with the barrier distribution expected for a deformed nucleus. These are of course simple cases, and in reality several different types of coupling may be involved in a given reaction, complicating the situation. Nonetheless, the ability to see directly how barrier strength is distributed is useful in understanding the reaction process.

The function $d^2(E\sigma)/dE^2$ will henceforth be referred to as the barrier distribution, even though it is recognised that the concept of a barrier distribution is only strictly valid in limiting cases.

The experimental data are measured at discrete energies, so the function $d^2(E\sigma)/dE^2$ is extracted using a point difference formula. For consistency, when comparing experimental barrier distributions with theory, the theoretical distribution is evaluated from the calculated excitation function in an identical manner. The energy step used to evaluate the curvature is typically 2-3 MeV, since the fwhm of a single barrier is typically 2 MeV, and larger steps, though giving a better-defined curvature, suppress structure in the barrier distribution.

II STATIC DEFORMATION

Would the appealing concept of an experimental, measured fusion barrier distribution actually prove to be valid in the real world?

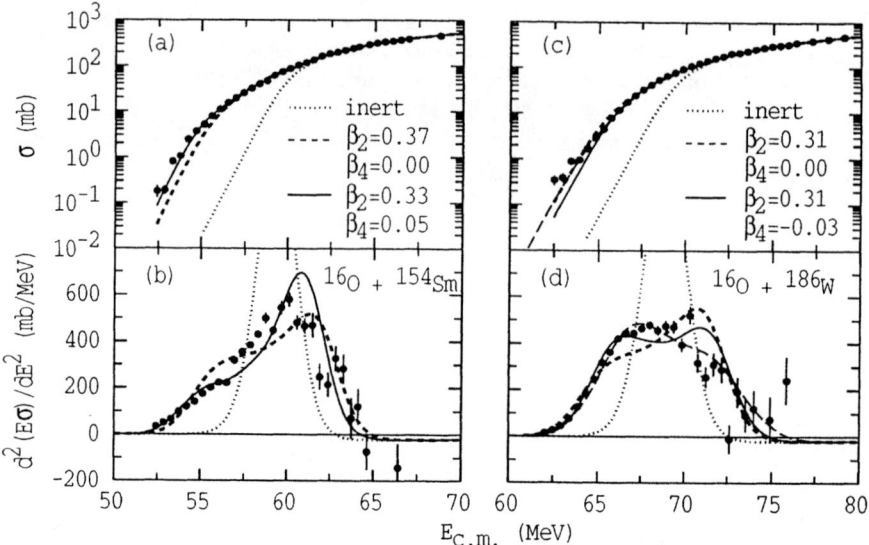

FIGURE 2. Comparison of the measured fusion excitation functions and the extracted barrier distributions for the indicated reactions. The dotted curves are the result of single-barrier penetration calculations. The dashed lines show the calculations with the inclusion of the quadrupole deformations of ^{154}Sm and ^{186}W; the solid lines include hexadecapole deformations. The differences in the shape of the barrier distribution is essentially due to the different sign of the hexadecapole deformations for the two nuclei. Inclusion of small transfer and phonon couplings results in the long-dashed lines (adapted from Ref. 9).

To test this experimentally required data measured with much higher precision than had previously been deemed necessary, since the second derivative of the data must be taken. To obtain a meaningful fusion barrier distribution with energy step sizes as indicated requires data with a standard deviation of $\sim 1\%$, at least at energies above the mean barrier.

The first reaction to be measured with the intent of determining the fusion barrier distribution was ^{16}O + ^{154}Sm [9], chosen because ^{16}O was expected to be inert, and ^{154}Sm has a well-known static deformation, and the fusion reaction could be treated semi-classically. Following this measurement, the reaction ^{16}O + ^{186}W was studied [10]. The excitation functions are shown in Fig. 2(a) and (c), and the barrier distribution in Fig. 2(b) and (d), where an energy step size of 2.3 MeV was used to determine the curvature of $E\sigma$.

The dotted curves are the results of calculations for a single barrier, which describe the data very poorly. The inclusion of the static quadrupole deformations of the ^{154}Sm and ^{186}W nuclei in the calculations (thick dashed lines) dramatically improves the agreement with experiment, producing wide barrier distributions. Comparing the calculated and experimental excitation functions, one could be led to believe that they agree very well. The barrier distributions clearly show a different picture: significant disagreements occur, which are not the same for the two reactions. This shows that the data are sensitive to nuclear properties other than the quadrupole deformation.

The ^{154}Sm and ^{186}W nuclei are known to have significant hexadecapole deformations, which are similar in magnitude but are expected to differ in sign; the former has a positive β_4 and the latter a negative one. The result of including the hexadecapole deformations in the calculations are shown by the solid lines, which agree remarkably well with the experimental results. It is clear that the differences in the shape of the calculated (and experimental) barrier distributions for the two systems is mainly due to the differences in the sign of the β_4 value. Some minor discrepancies between the calculated and experimental barrier distributions are still present. As discussed in Ref. [7]; inclusion of small phonon and transfer couplings allows even better reproduction of the data, as shown by the long dashed lines.

The results of these measurements exceeded all expectations, showing that fusion excitation functions are not only very sensitive to quadrupole deformations but also to the sign of small hexadecapole deformations. They also showed that for these statically deformed nuclei, the function $d^2(E\sigma)/dE^2$ does indeed return a sensible representation of the barrier distribution *expected* to be present.

III VIBRATIONAL NUCLEI

The concept of a fusion barrier distribution is strictly valid for coupling to states with zero excitation energy (eigenchannel approximation) and thus it was not clear whether couplings to vibrational states with typical excitation energies of a few MeV will give rise to the simple forms of barrier distributions predicted in Ref. [4]. The reactions ^{16,17}O + ^{144}Sm were chosen to investigate this question, as the coupling scheme was expected to be relatively simple. The use of ^{17}O was to check if the distributions are sensitive to the additional neutron in the projectile.

A The ^{16,17}O + ^{144}Sm reactions

The measured [11] excitation functions along with the extracted barrier distributions for the ^{16,17}O + ^{144}Sm reactions are shown in the upper and lower panels of Fig. 3. Calculations assuming no coupling (dotted lines) are inconsistent with the data. The experimental barrier distributions for both reactions are dominated by a two–peaked structure. closely resembling that of Fig. 1(d), a typical case of coupling to −Q value channels. The most important channels associated with excitation of ^{144}Sm are expected to involve those states with the largest B(Eλ)↑ values, *i.e.* the first 2^+ and 3^- states.

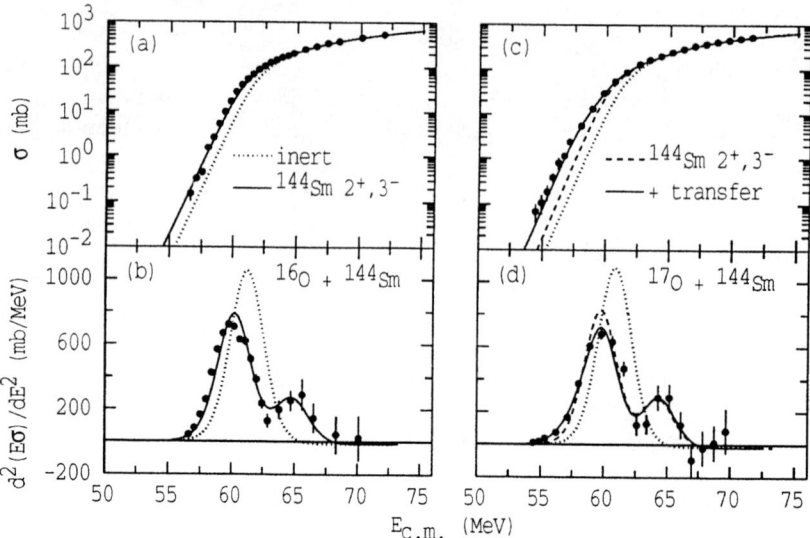

FIGURE 3. Comparison of the measured excitation functions (a and c) and extracted barrier distributions (b and d) for ^{16}O + ^{144}Sm and for ^{17}O + ^{144}Sm. Coupling to the 2^+ and 3^- states of ^{144}Sm is adequate to describe the ^{16}O data as shown by the solid lines in the left panels. These couplings are however not sufficient for the ^{17}O data (dashed lines in the right panels), which requires additional coupling to a neutron transfer channel with positive Q-value (solid lines).

Considering firstly the ^{16}O reaction, including the 2^+ and 3^- states of ^{144}Sm in the calculation gives a good representation [7] of the data, as shown in Fig. 3(a) and (b).

In the reaction with the ^{17}O projectile, the data at higher energies are very similar to those for ^{16}O, but at low energies the cross-sections for the ^{17}O reaction are more than four times higher than those for ^{16}O. A similar calculation to that described above, is shown by the dashed lines in Fig. 3(c) and (d), failing to fit the low energy data. The barrier distribution shows that this calculation gives too much strength around the main peak, and fails to reproduce the observed low energy tail, indicating a need for additional coupling. An increased cross-section results with any form of extra coupling, as was shown in Fig. 1, but the type of coupling required is not immediately apparent in the excitation function. A low energy tail in the barrier distribution can be obtained by coupling to a $+Q$-value channel (see Fig. 1(e)), indicating coupling to transfer channels may be responsible. Unlike the ^{16}O–induced reaction, the neutron stripping reaction with ^{17}O has a positive Q-value of 2.6 MeV, and is the obvious candidate for inclusion in the calculations. Including couplings to this transfer channel give an extremely good representation of the data [7]

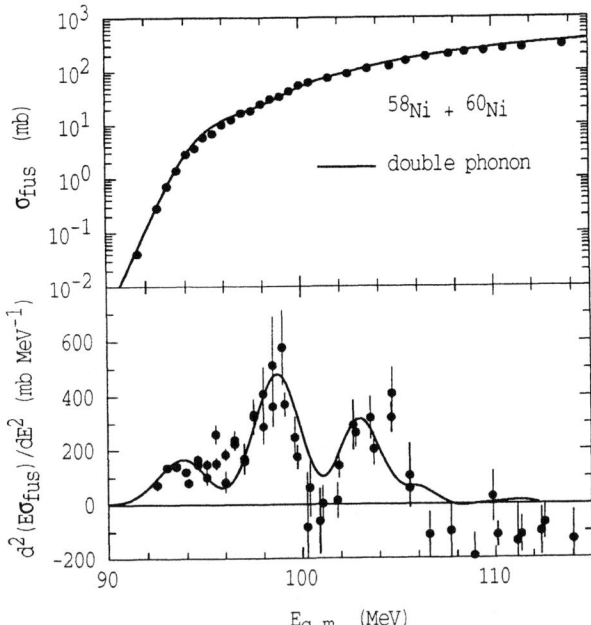

FIGURE 4. Measured fusion excitation function and extracted barrier distribution ^{58}Ni + ^{60}Ni. A calculation including up to mutual double quadrupole excitations in each nucleus describes the data well.

as shown by the solid line in the right panels of Fig. 3.

The excellent agreement between the theoretical calculations and the experimental results strongly supports the barrier distribution picture, even when the excitation energies are not very close to zero as in the eigenchannel picture, thus indicating that this technique can potentially be used as a tool to give a better understanding of more complex reaction processes. These experiments unambiguously demonstrated the effects on fusion of couplings to inelastic and transfer channels. Since these experiments, measurements for the reactions ^{32,36}S + ^{110}Pd and ^{40}Ca + ^{46,48,50}Ti have been studied at other laboratories [12,13] and have also demonstrated similar effects on the barrier distributions of couplings to transfer channels.

B The ^{58}Ni + ^{60}Ni reaction

Simple barrier distributions as exhibited by the ^{16,17}O + ^{144}Sm reaction are not typical. A number of reactions involving vibrational nuclei have been studied, and typically the barrier distributions exhibit structure indicating more complex processes are occurring during fusion, as exemplified in the reaction discussed below.

Fusion following the reaction ^{58}Ni + ^{60}Ni was measured [14] to investigate the influence of elastic ($Q = 0$) two neutron transfer. However, it was found that the barrier distribution showed a beautiful three-peaked structure (see Fig. 4) which could not be reproduced by coupling to neutron-pair transfer and/or coupling to simple 1-phonon states in the Ni nuclei. It was shown [14] that couplings not only to the 1-phonon states, but more importantly coupling from them to the 2-phonon states in each nucleus (including mutual excitations) was necessary to obtain a good representation of the data shown in Fig. 4. This was the first time the effects of such complex couplings were clearly identified as affecting the fusion process.

C The ^{16}O ,^{28}Si + ^{208}Pb reactions

Difficulties in the interpretation of measured fission fragment anisotropies [15], and thus the deduced fusion mean square angular momenta for the reaction ^{16}O + ^{208}Pb, prompted the precise re-measurement of the fusion excitation function at near- and sub-barrier energies [16]. Subsequently, measurements have been made for reactions of ^{28}Si [17], ^{19}F and ^{32}S on ^{208}Pb. Analysis of the latter two reactions is not yet complete, however comparison of the ^{16}O ,^{28}Si + ^{208}Pb data with calculations of the type which fitted the ^{16}O + ^{144}Sm reaction show that the data cannot be well reproduced by including only coupling to single phonon states (the strongest being the 3^- state) in ^{208}Pb. Calculations including double phonon states gave improved agreement. Analysis of the other reactions, and further exploration of the parameter space of the calculations is proceeding, to determine more precisely the role played by ^{208}Pb in the fusion process.

This is important, as a good understanding of the influence of ^{208}Pb on fusion barrier distributions will enable its use as a tool, to study the behaviour of lighter nuclei in fusing with ^{208}Pb. Because of its high Z, and the relatively weak coupling to its excited states (as compared with statically deformed nuclei), it could be considered as a "magnifier" of the couplings, and may be most useful in investigations of fusion reactions with radioactive beams.

IV INFLUENCE OF TRANSFER

The effect on fusion of a single positive Q-value transfer was seen for the ^{17}O + ^{144}Sm reaction discussed previously. Reactions with larger Coulomb barriers (resulting in greater overlap of the matter densities of the two interacting nuclei at the fusion barrier energy) have shown much larger effects of transfer on the barrier distribution. This was shown clearly in Ref. [18], where

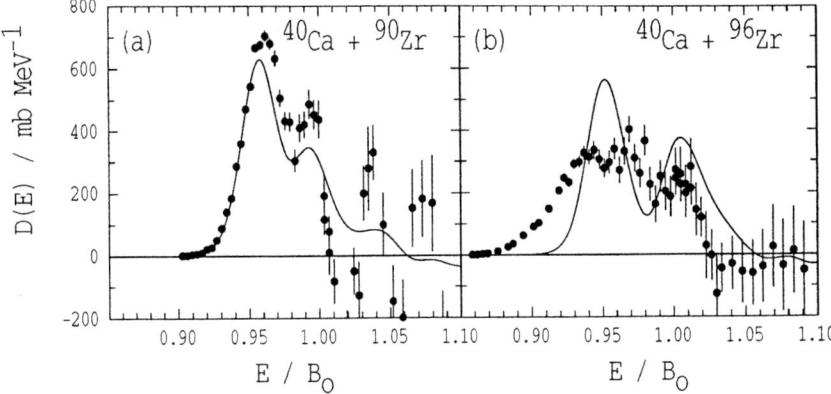

FIGURE 5. Fusion barrier distributions for the ^{40}Ca + ^{90}Zr, ^{96}Zr reactions, showing the change in the experimental barrier distribution as the neutron excess changes. The calculations only take into account coupling to multi-phonon states in the Zr nuclei.

the reactions ^{40}Ca + ^{90,96}Zr were studied. The fusion barrier distributions are shown in Fig. 5 for the two reactions.

The former can be reasonably explained by coupling to phonon states in ^{90}Zr (solid curve), whilst the same procedure applied to ^{96}Zr fails completely to reproduce the measured low–energy structure. The authors of Ref. [18] point out that in the former reaction no neutron transfers have a positive Q-value, whilst for the latter, transfers of up to 8 neutrons have a positive Q-value. Thus a correlation between the low–energy fusion, and multi–neutron transfer is suggested. This may in some ways be analagous to the macroscopic picture of neck–formation, which is has been predicted to result in a lowered fusion barrier.

The observation of yields of very heavy nuclei (fusion products) following the reactions of Ni and heavier nuclei with Pb and Bi, at energies close to the Bass barrier, may also be a result of such a macroscopic mechanism, acting counter to the extra–extra–push energy expected to be required to initiate fusion for systems with such a large Coulomb repulsion.

V APPLICATIONS

The improved quantitative understanding of the fusion process as a result of the new measurements allows the use of fusion as a tool to aid the study of other processes.

A Fusion angular momentum distributions

It has been shown that if centrifugal and Coulomb energies can be considered as equivalent, the fusion transmission coefficient for each ℓ-wave can be mapped from the barrier distribution [19]. Equivalently, a model which reproduces the barrier distribution and fusion cross-sections is expected to predict accurately the angular momentum distribution of the fused nuclei.

A reliable knowledege of angular momentum distributions is useful to explain and predict cross-sections for populating super-deformed bands, for nuclear spectroscopy studies. It is also valuable in the interpretation of measured fission probabilities in heavy–ion reactions, since the fission width varies strongly with angular momentum. It is vital to know the angular momentum distribution in order to understand and obtain information from fission angular distributions.

B Fission Angular Anisotropies

The fission angular distribution can be characterised by the anisotropy A, defined as the ratio of yields at 180° or 0° to that at 90° to the beam axis. The anisotropy is approximately given by the expression based on the transition state model [23]:

$$A = 1 + \frac{<\ell^2>\hbar^2}{4I_{eff}T}. \qquad (2)$$

Here I_{eff} is the effective moment of inertia at the saddle–point and T is the temperature, whilst $<\ell^2>$ is closely related to the mean square angular momentum following fusion. The equation shows that A is very sensitive to $<\ell^2>$, as well as to properties of the fission process. Using this equation resulted in under–prediction of measured fission anisotropies [15] at beam energies near the fusion barrier. This was particularly puzzling for the reaction ^{16}O + ^{208}Pb, where the TSM was expected to be applicable. Precise remeasurements of the anisotropies and fusion cross-sections (the latter giving more reliable ℓ distributions) together with more detailed TSM calculations, resulted in agreement [16], indicating that the fusion and TSM models are applicable.

There remained a reported large anomaly for the ^{19}F + ^{208}Pb reaction [20]. Since for projectiles of A $\geq$ 24 incident on ^{208}Pb, quasi–fission was established as the cause of large anisotropies at all energies [21], A detailed measurement was recently performed o investigate whether quasi–fission could be the cause of the anomaly for the ^{19}F + ^{208}Pb reaction. These measurements had the further aim of investigating whether information on fission dynamics can be extracted from the anisotropy data at above–barrier energies.

The experimental results are shown in the left panels of Fig. 6. A very good reproduction of the fusion cross–sections and the barrier distribution could be obtained, taking into account coupling to phonon states in both projectile and target, and transfer channels. This gives confidence in the extracted fusion angular momentum distributions, which were used as input into a statistical model calculation of the fission anisotropies (see lower panel). The measured anisotropies in the fusion barrier region were lower than previously reported, and together with inclusion of a fission transient time, and assuming breakdown of the TSM at angular momenta greater than that at which the fission barrier has fallen to 1 MeV, excellent reproduction of the new data was obtained.

This result gives added support to the view that neither failure of the fusion models nor breakdown of the TSM itself are responsible for the very large anisotropies which had been observed [15,22] for reactions with actinide targets at sub–barrier energies. The most detailed such measurements, for the ^{16}O + ^{238}U reaction [24,25], showed a striking correlation of the anisotropies with the fusion barrier distribution, the low fusion barriers appearing to be correlated with higher anisotropies, and the transition from higher to lower anisotropies occuring at an energy close to the average barrier. The measured barrier distribution was very similar in shape to that for ^{16}O + ^{154}Sm, showing experimentally that to a good approximation, the ^{238}U behaves as a statically deformed, prolate nucleus in the reaction. This conclusion was vital for the subsequent interpretation of the fission anisotropies, and the fission mass-distribution data [25], implying a relationship between the anisotropy and the relative orientations of the colliding nuclei.

Considering the collision classically, the higher fusion barriers correspond to contact of the projectile with the flattened side of the prolate target, resulting initially in a compact dinuclear system. Conversely, the lower barriers correspond to contact with the tip, giving an elongated dinuclear system. Intuitively, it seems reasonable that the former configuration would be more likely to result in fusion-fission, and the latter in quasi–fission (fission–like fragments from a system which never formed a compact compound nucleus inside the unconditional fission barrier, resulting in a large anisotropy). A simple geometrical model [25] was able to reproduce the energy dependence of the anisotropy data.

In contrast, recently reported experimental results for ^{11}B, ^{12}C, ^{16}O, ^{19}F + ^{232}Th [26] show a dramatic fall in anisotropies at the lowest beam energies, almost down to TSM–predicted values in some cases. However, this feature is not seen in other measurements [27,28] for ^{12}C + ^{232}Th. To investigate this disagreement for another reaction, and to investigate the origin of the feature, measurements were recently performed for the ^{19}F + ^{232}Th reaction, down to below 75% of the mean fusion barrier energy. The results are presented in the right panels of Fig. 6. The upper panel shows that below 80 MeV, fission following transfer reactions is dominant. The lower panel in the figure shows

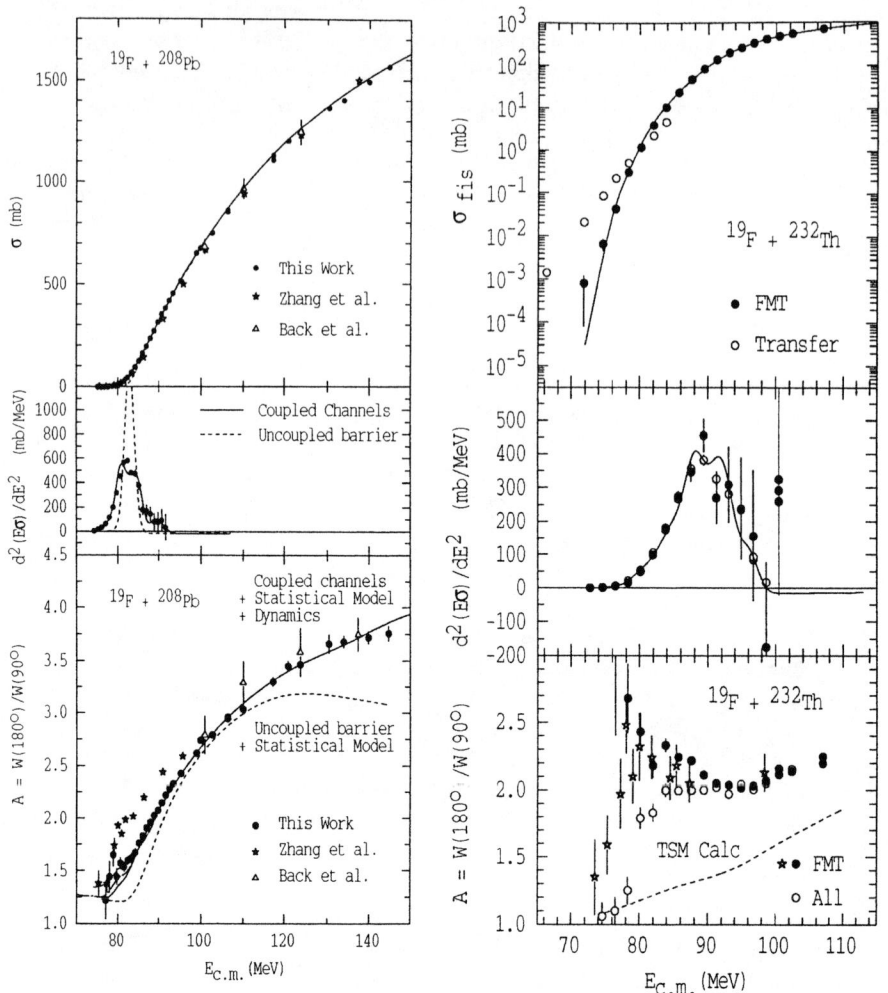

FIGURE 6. The left panels show the fusion excitation function, the fusion barrier distribution and the measured fission anisotropies for the ^{19}F + ^{208}Pb reaction. The data can be explained within the framework of the TSM, taking into account some dynamical effects (see text). The right panels show preliminary results for the ^{19}F + ^{232}Th reaction. Despite the good fit to the fusion cross-sections and barrier distribution, the predicted anisotropy using the TSM is far below the data for fusion–fission (FMT fission). The anisotropies for FMT fission continue to rise as the beam energy decreases.

that in this measurement, the anisotropy (A) falls dramatically when transfer-induced fission events are included (all fission), but remains high when they are rejected, as indicated by the full points, for full momentum transfer fission (FMT). These results suggest that it is possible that the fall in anisotropy at low energies measured for this reaction in Ref. [26] (star points in the figure) may be due to inadequate rejection of transfer fission events.

Further experimental studies, particularly of evaporation residue survival probabilities, should be made to obtain a complete picture of the processes occurring in sub-barrier reactions of deformed nuclei.

It should be possible to learn more of nuclear reaction dynamics in general through the measurement of fusion barrier distributions. In particular, a deeper understanding of the extra–push and extra–extra–push energies should be forthcoming, together with the possibly compensating feature of neck formation, or multi-nucleon transfer. When it becomes possible to measure fusion barrier distributions for reactions involving unstable nuclei, novel fusion mechanisms may present themselves, which could be a new path to superheavy element formation.

VI SUMMARY

The experimental measurement of fusion barrier distributions represents a new stage in the study of the fusion of heavy nuclei. A more quantitative understanding is emerging, based on the high precision measurements of fusion excitation functions, from which experimental fusion barrier distributions have been extracted. It may be asked "why take the second derivative?" There are three answers: the first is that it removes a trivial geometrical factor from the data, giving on a linear scale what was previously dispayed over many orders of magnitude, the second is that it shows us clearly whether the experimental data are adequate to give us the information we ask, and thirdly, it is a function which in certain limits actually represents the fusion barrier distribution. Although it may be argued from a strict theoretical point of view there is no true fusion barrier distribution, the experimental data show that the actual situation appears generally to be within the limits where a barrier distribution is meaningful. Because of the above arguments in its favour, we believe that the "barrier distribution" representation of fusion data is certainly here to stay, and will play a valuable role in gaining a full understanding of fusion of atomic nuclei.

REFERENCES

1. C.Y. Wong, Phys. Rev. Lett. **31** (1973) 766.
2. L.C. Vaz and J.M. Alexander, Phys. Rev. C **10** (1974) 464.

3. R.G. Stokstad, Y. Eisen, S. Kaplanis, D. Pelte, U. Smilansky, and I. Tserruya, Phys. Rev. Lett. **41** (1978) 465; Phys. Rev. C **21** (1980) 2427.
4. C.H. Dasso, S. Landowne and A. Winther, Nucl. Phys. A **405** (1982) 381; **407** (1983) 221.
5. S.G. Steadman and M.J. Rhoades-Brown, Ann. Rev. Nucl. Part. Sci. **36** (1986) 649 and references therein.
6. M. Beckerman, Rep. Prog. Phys. **51** (1988) 1047 and references therein.
7. J.R. Leigh, M. Dasgupta, D.J. Hinde, J.C. Mein, C.R. Morton, R.C. Lemmon, J.P. Lestone, J.O. Newton, H. Timmers, J.X. Wei, and N. Rowley, Phys. Rev. C **52** (1995) 3151.
8. N. Rowley, G.R. Satchler, and P.H. Stelson, Phys. Lett. B **254** (1991) 25.
9. J.X. Wei, J.R. Leigh, D.J. Hinde, J.O. Newton, R.C. Lemmon, S. Elfström, J.X. Chen, and N. Rowley, Phys. Rev. Lett. **67** (1991) 3368.
10. R.C. Lemmon, J.R. Leigh, J.X. Wei, C.R. Morton, D.J. Hinde, J.O. Newton, J.C. Mein, and M. Dasgupta, Phys. Lett. B **316** (1993) 32.
11. C.R. Morton, M. Dasgupta, D.J. Hinde, J.R. Leigh, R.C. Lemmon, J.P. Lestone, J.C. Mein, J.O. Newton, H. Timmers, N. Rowley, and A.T. Kruppa, Phys. Rev. Lett. **72** (1994) 4074.
12. A.M. Stefanini, D. Ackermann, L. Corradi, J.H. He, G. Montagnoli, S. Beghini, F. Scarlassara, and G.F. Segato, Phys. Rev. C **52** (1995) R1727.
13. R. Vandenbosch, A.A. Sonzogni, and J.D. Bierman, to appear in J. Phys. G (1997).
14. A.M. Stefanini, D. Ackermann, L. Corradi, D.R. Napoli, C. Petrache, P. Spolaore, P. Bednarczyk, H.Q. Zhang, S. Beghini, G. Montagnoli, L. Mueller, F. Scarlassara, G.F. Segato, F. Soramel, and N. Rowley, Phys. Rev. Lett. **74** (1995) 864.
15. T. Murakami, C.-C. Sahm, R. Vandenbosch, D.D. Leach, A. Ray, and M.J. Murphy, Phys. Rev. C **34** (1986) 1353.
16. C.R. Morton, D.J. Hinde, J.R. Leigh, J.P. Lestone, M. Dasgupta, J.C. Mein, J.O. Newton, and H. Timmers, Phys. Rev. C **52** (1995) 243.
17. D.J. Hinde, C.R. Morton, M. Dasgupta, J.R. Leigh, J.C. Mein and H. Timmers, Nucl. Phys. **A592** (1995) 271.
18. H. Timmers, L. Corradi, A.M. Stefanini, D. Ackermann, J.H. He, S. Beghini, G. Montagnoli, F. Scarlassara, G.F. Segato and N. Rowley, Phys. Lett. B **399** (1997) 35.
19. N. Rowley, J.R. Leigh, J.X. Wei and R. Lindsay, Phys. Lett. B **314** (1993) 179.
20. H. Zhang, Z. Liu, J. Xu, K. Xu, J. Lu and M. Ruan, Nucl. Phys. **A512** (1990) 531.
21. B.B. Back et al., Phys. Rev. C **32** (1985) 195.
22. H. Zhang, Z. Liu, J. Xu, X. Qian, Y. Qiao, C. Lin, and K. Xu, Phys. Rev. C **49** (1994) 926, and references therein.
23. R. Vandenbosch and J.R. Huizenga, *Nuclear Fission*, (Academic Press NY) 1973.
24. D.J. Hinde, M. Dasgupta, J.R. Leigh, J.P. Lestone, J.C. Mein, C.R. Morton, J.O. Newton, and H. Timmers, Phys. Rev. Lett. **74** (1995) 1295.

25. D.J. Hinde, M. Dasgupta, J.R. Leigh, J.C. Mein, C.R. Morton, J.O. Newton, and H. Timmers, Phys. Rev. C **53**, (1996) 1290.
26. N. Majumdar et al., Phys. Rev. Lett. **77**, 5027 (1996).
27. J.P. Lestone et al., Phys. Rev. C**55**, R16 (1997).
28. J.C. Mein et al., Phys. Rev. C**55**, R995 (1997).

Fusion reactions of deformed nuclei near Coulomb barriers

H. Ikezoe, T. Ikuta, S. Mitsuoka, T. Kuzumaki, Lu Jun,
Y. Nagame, I. Nishinaka and K. Tsukada

Japan Atomic Energy Research Institute, Tokai-mura, Naka-gun, Ibaraki-ken, 319-11 Japan

Abstract. The fusion evaporation residues were measured in the ^{232}Th + ^{30}Si reaction near the Coulomb barrier. The measured xn cross sections were compared with the statistical model calculation by taking into account the target deformation. It was found that the calculated $3n$ and $4n$ cross sections were considerably overestimated and the predicted enhancement of fusion due to the target deformation does not contribute to the formation of an equilibrated compound nucleus near the Coulomb barrier. This fact is also supported by the statistical model analysis for the reported xn cross sections in the reactions with ^{232}Th and ^{238}U targets.

INTRODUCTION

Recently several authors have measured anomalously large anisotropies of the fission fragment angular distributions in the reaction systems of light projectiles ^{12}C, ^{16}O and ^{19}F on deformed targets ^{232}Th and ^{238}U [1]. These anisotropies $W(0°)/W(90°)$, which are 1.5 ~ 2 times larger than the predictions of the transient state model assuming the formation of an equilibrated compound nucleus, become evident near and below Coulomb barriers. Hinde et al. [2] have pointed out that the quasi-fission is the main reaction process below Coulomb barriers and the nuclear orientation of the deformed target with respect to the projectile plays important role in the formation of the compound nucleus, that is, the collisions with the tips of the deformed nuclei lead to quasi-fission, while the collisions with the sides lead to the complete fusion process.

In order to investigate this phenomena, we measured the fusion-evaporation residues produced in the ^{232}Th + ^{30}Si reactions at the bombarding energy near the Coulomb barrier, because the fusion-evaporation residue give us the direct evidence that the projectile really fuses with the deformed target. Especially, the magnitudes of the $3n$ and $4n$ cross sections in the fusion reactions with actinide as a target are sensitive to the absolute value of the fusion cross section near the Coulomb barrier.

TABLE 1. Measured xn cross sections in the ^{232}Th + ^{30}Si reactions.

E_{lab} MeV	dose $\times 10^{16}$	Cross sections (nb) 3n	4n	5n
150.2	5.28	< 0.55	< 0.39	< 0.55
152.2	2.03	< 1.4	< 1.0	< 1.4
153.8	5.26	< 0.55	< 0.39	0.55 ±0.60
155.4	1.85	< 1.6	< 1.1	3.2 ±2.5
157.4	3.77	< 0.77	0.54 ±0.60	3.9 ±2.3

EXPERIMENTAL RESULTS

A target of ^{232}Th (the thickness of 400 $\mu g/cm^2$) was bombarded by ^{30}Si beams of 150.2, 152.2, 153.8, 155.4 and 157.4 MeV from the tandem accelerator of the Japan Atomic Energy Research Institute. The evaporation residues (ER's) emitted from the target at the beam direction were separated in flight from the primary beams by the recoil mass separator (JAERI-RMS) [3]. The ER's were implanted into a double-sided position sensitive strip detector (PSSD) placed at the focal plane, which provided two-dimensional positions and energies associated with the implantation of the ER and subsequent decays. The PSSD was surrounded by other four single-sided strip detectors to detect particles (α particles and fission fragments) escaped from the PSSD. Beam intensity was typically 70-90 pnA and the event rate of the coming particles through the JAERI-RMS was about 2 Hz. The details of the detection method of the ER's and the subsequent decays from the ER's are shown in [4].

The energy spectrum of decay particles in coincidence with the ER within the time interval of 50 s at the same position in the PSSD within the uncertainties of 0.6 mm for the horizontal position and 0.8 mm for the vertical position are shown in Fig. 1. The α decay events of ^{257}Rf were observed at the bombarding energies of 153.8 MeV (1 event), 155.4 MeV (2 events) and 157.4 MeV (5 events). The observed energies of the decay events and the observed time intervals between the ER events and the subsequent decay events were consistent with the reported values [5]. One fission event with the energy of about 103 MeV was observed at the bombarding energy of 157.4 MeV. The time interval between the fission event and the ER event was 102 ms. This fission event was identified to be the spontaneous fission decay of ^{258}Rf whose half-life was reported to be 12 ms [5]. Since no decay event was observed in the bombarding energy of 150.2 MeV and 152.2 MeV, the upper limits for the cross sections of $3n$, $4n$ and $5n$ channels were obtained as listed in Table 1. Here the transmission efficiency 8% of the JAERI-RMS was used to obtain the absolute values of the cross sections.

The fusion cross sections in the reactions ^{232}Th + ^{28}Si and ^{238}U + ^{28}Si were also measured by detecting the fission fragments.

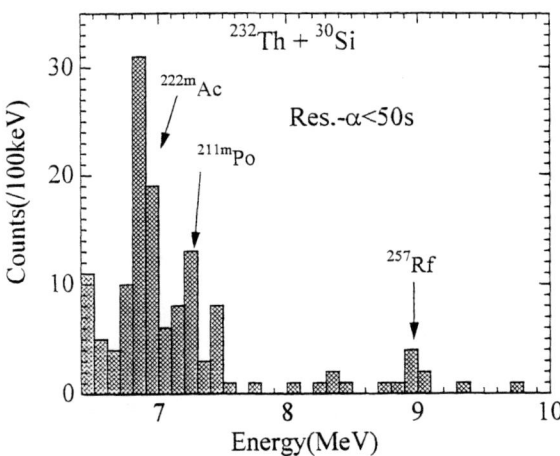

FIGURE 1. Energy spectrum of particles decaying from implanted evaporation residues within the time interval of 50 s in the ^{232}Th + ^{30}Si reaction. The data obtained at all bombarding energies were summed.

DISCUSSION

The measured xn cross sections were compared with the calculated results of the statistical model using the code HIVAP [6]. The total fusion cross section near the Coulomb barrier, including the cross sections of fusion-fission, fusion-evaporation and quasi-fission processes, is the input quantity to the HIVAP code. It was calculated by the code CCDEF [7] taking into account the target deformation and the coupling of inelastic excitations of the projectile and the target to the fusion process. The Christensen-Winther potential [8] was used for the nuclear potential without any modification.

In order to check how good the predicted fusion cross sections agree with the measured fusion cross sections, the calculated results of the CCDEF were compared with the observed data for the reactions ^{232}Th + ^{28}Si and ^{238}U + ^{28}Si, where the static deformations $\beta_2^s = 0.22$ and $\beta_4^s = 0.09$ for ^{232}Th and $\beta_2^s = 0.25$ and $\beta_4^s = 0.052$ for ^{238}U were taking into account together with the coupling of the inelastic excitations for ^{28}Si (see Table 2). The octupole excitations β_3 of 0.09 and 0.086 were also taken into account for ^{232}Th and ^{238}U, respectively. Since the calculated results agreed with the data very well, the fusion cross sections calculated by the CCDEF code were used for all the reactions considered here. The calculated fusion cross sections for the reactions ^{238}U + ^{16}O and ^{238}U + ^{22}Ne are shown in Fig. 3 and Fig. 5 together with the experimental data [15–17].

The calculated results of the statistical model strongly depend on the level density parameters a_n and a_f at the ground state deformation and the saddle point deformation, respectively, and its ratio a_n/a_f and also the fission barrier height

TABLE 2. Deformation parameters and excitation energies of 2^+ and 3^- states of various projectiles.

Nuclei	2^+ states		3^- states	
	β_{2^+}	E_x (MeV)	β_{3^-}	E_x (MeV)
^{16}O	0.36	-6.917	0.70	-6.129
^{18}O	0.355	-1.982	0.35	-5.098
^{22}Ne	0.562	-1.274	0.33	-5.909
^{28}Si	0.408	-1.779	0.283	-6.879
^{30}Si	0.316	-2.236	0.275	-5.488

$B_f = C \times B_f^L - E_{shell}$, where B_f^L is the liquid drop value of the fission barrier [9] and C is a fitting parameter and E_{shell} is the shell correction energy at the ground state deformation. In the present calculation, the level density parameters were varied with excitation energies according to the prescription of [10] with the shell damping constant of 18.5 MeV, where the macroscopic values $\tilde{a}_n$ and $\tilde{a}_f$ [6] were calculated by $\tilde{a}_n = \tilde{a}_f = A/10$. Here A is the mass number. Since the $6n$ channel cross sections are located well above the Coulomb barrier in the reactions with ^{232}Th and ^{238}U as shown in Fig. 3-5, they are insensitive to the fusion enhancement due to the target deformations. Therefore the parameter C was adjusted so as to reproduce the measured $6n$ cross sections in these reactions and also the ^{232}Th + ^{26}Mg reaction [11]. The parameter C was determined to be 1.2 for all the reactions considered here except the ^{238}U + ^{18}O reaction, where C was determined to be 1.1.

The calculated results for the ^{232}Th + ^{30}Si are shown in Fig. 2 together with the observed $4n$ and $5n$ cross sections. The figure 2a represents the calculated cross sections without taking into account the deformation of ^{232}Th as well as the coupling of the inelastic excitation of ^{232}Th and ^{30}Si. The observed $5n$ cross sections are larger than those calculated at the center-of-mass energy E_{cm} less than 138 MeV. The calculated $4n$ cross section is extremely small but not inconsistent with the data for which the upper limit was measured at E_{cm} <138 MeV.

The middle figure 2b shows the calculated results by taking into account the coupling of the inelastic excitations in the fusion calculation. The calculated $4n$ and $5n$ cross sections are not inconsistent with the data.

The figure 2c shows the full calculations where the deformation of ^{232}Th and the coupling of the inelastic excitations of ^{232}Th and ^{30}Si are taken into account. If the target deformation is taken into account, the fusion barrier is lowered by the amount of more than 10 MeV due to the collision of the projectile with the tip of the deformed nucleus and then the statistical model predicts the $3n$ and $4n$ cross sections comparable with the $5n$ cross section even below the spherical Coulomb barrier shown by the dotted line. The present experimental result shows no $3n$ and $4n$ cross sections in the region of E_{cm} <138 MeV. This means that the calculated results assuming the complete fusion of the projectile and the deformed target below the spherical Coulomb barrier disagree with the present data.

We also analyzed the xn cross section data for the reactions ^{238}U + ^{16}O [12,13],

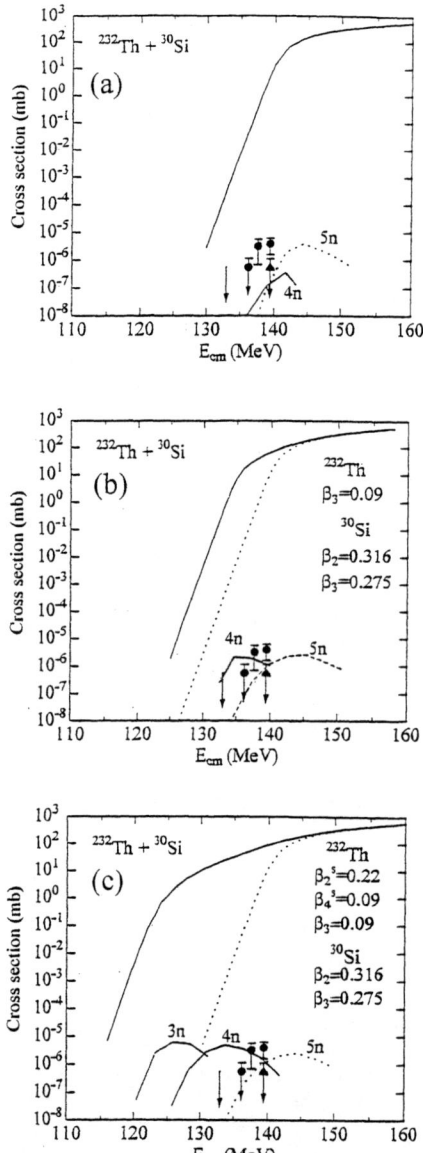

FIGURE 2. Measured 4n (the solid triangles) and 5n (the solid dots) cross sections in the ^{232}Th + ^{30}Si reaction. The CCDEF calculation without the target deformation and the coupling of inelastic excitation is shown as the solid line in (a), the dashed lines in (b) and (c). The CCDEF calculation with the target deformation and the coupling of the inelastic excitations is shown as the solid line in (c). The calculated result taking into account only the coupling of the inelastic excitations is shown as the solid line in (b). Calculated 3n, 4n and 5n cross sections are shown by the dashed-dotted line, the solid line and the dashed line, respectively.

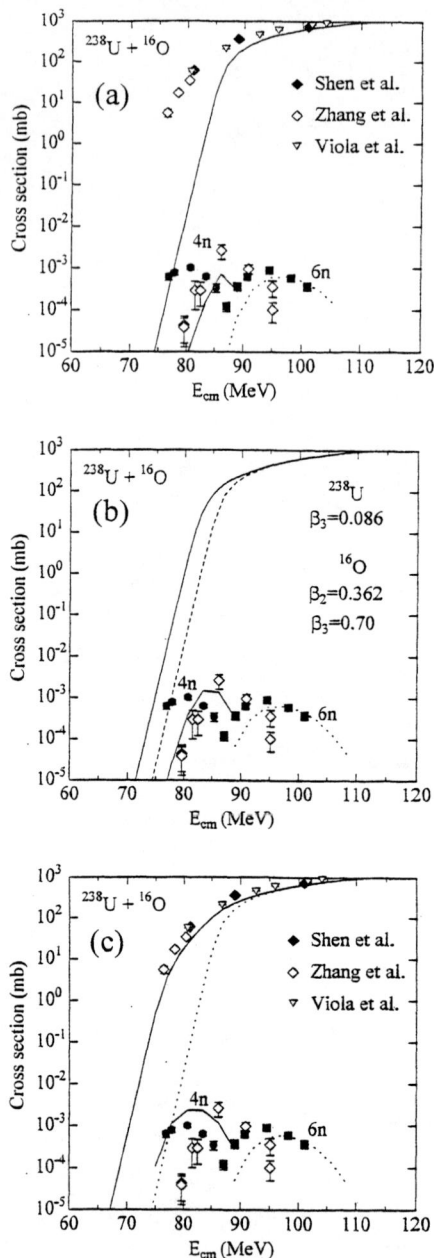

FIGURE 3. Same as Fig. 2, but for the ^{238}U + ^{16}O reaction. The xn cross section data are taken from [12] (the solid squares and the solid dots) and [13] (the open diamonds). The fusion-fission data [15-17] are also shown.

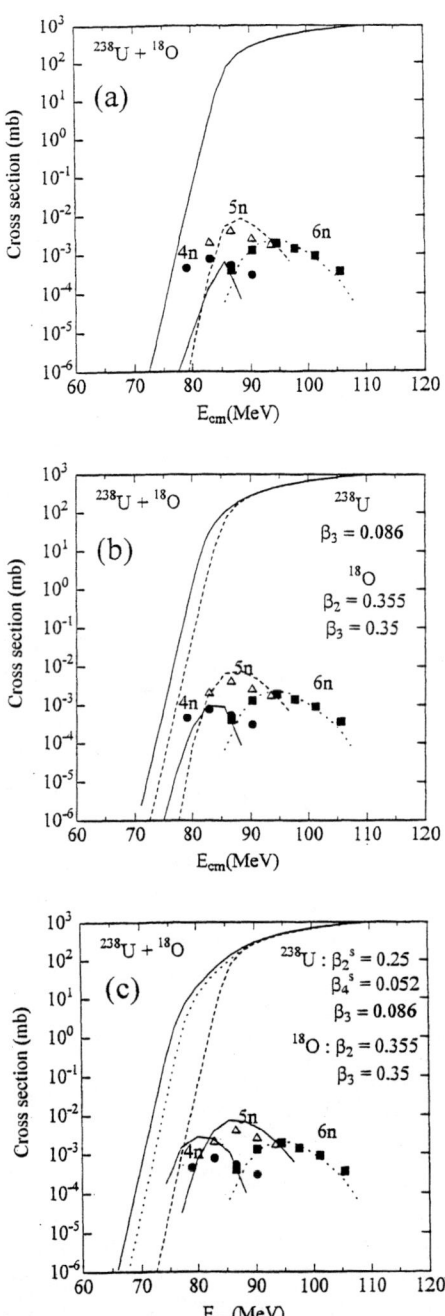

FIGURE 4. Same as Fig. 2, but for the ^{238}U + ^{18}O reaction. The xn cross section data are taken from [14].

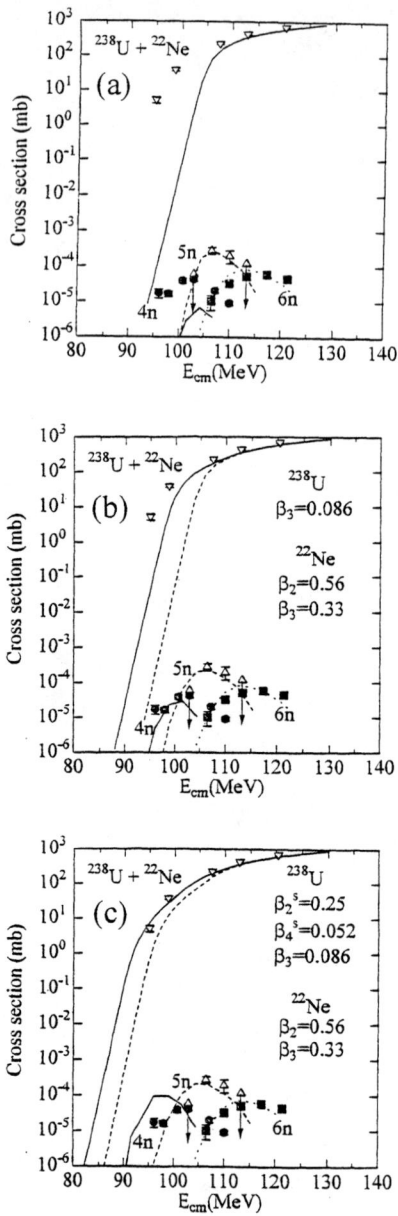

FIGURE 5. Same as Fig. 2, but for the ^{238}U + ^{22}Ne reaction. The xn cross section data are taken from [14]. The fusion-fission data of the ^{238}U + ^{20}Ne reaction [17] are shown as the open triangles. The CCDEF calculation with the couplings of the 2^+ state ($\beta_2 = 0.728$) and the 3^- state ($\beta_3 = 0.43$) of ^{20}Ne for the ^{238}U + ^{20}Ne reaction shows almost the same result as the solid line in (c) for the ^{238}U + ^{22}Ne reaction at $E_{cm} > 90$ MeV.

^{238}U + ^{18}O [14] and ^{238}U + ^{22}Ne [14]. The parameters used in the CCDEF calculations are listed in Table 2. The results of the analysis are shown in Fig. 3-5. In the reactions ^{238}U + ^{16}O, two data sets for the $4n$ cross section are rather inconsistent with each other. The statistical model calculations assuming the fusion enhancement due to the deformation of ^{238}U disagree with the $4n$ cross section data (the open diamonds) of [13], suggesting that the collisions of ^{16}O with the tips of deformed nucleus ^{238}U do not contribute to the formation of the fully equilibrated compound nucleus at the bombarding energies below the spherical Coulomb barrier. On the other hand, the same statistical model calculations are not inconsistent with the data (the solid points) of [12]. Therefore, it is difficult to get any conclusion about the complete fusion possibility near the Coulomb barrier in the ^{238}U + ^{16}O reaction.

In the cases of the reactions ^{238}U + ^{18}O and ^{238}U + ^{22}Ne, the calculated results by taking into account only the coupling of the inelastic excitations of both the targets and the projectiles agree well with the $4n$ and $5n$ cross section data as shown in Fig. 4b and Fig. 5b. The calculated results without the coupling of the inelastic excitations underestimate considerably the $4n$ cross section as shown in Fig. 4a and Fig. 5a. On the other hand, the calculated results by taking into account the target deformations as well as the coupling of the inelastic excitations overestimate the $4n$ cross sections in both reaction systems as shown in Fig. 4c and Fig. 5c. This fact suggests that the fusion enhancement due to the target deformation does not contribute to the formation of the equilibrated compound nucleus below the Coulomb barrier. On the other hand, the fusion enhancement due to the coupling of the inelastic excitations of 2^+ and 3^- of the targets and the projectiles may contribute to the formation of the equilibrated compound nucleus.

CONCLUSION

The measured xn cross section data in the ^{232}Th + ^{30}Si reaction and the statistical model analysis for the reactions of projectiles ^{18}O and ^{22}Ne on deformed target ^{238}U as well as the present reaction suggest that the fusion enhancement due to the target deformation does not result in the complete fusion-fission and fusion evaporation processes, while the coupling of the inelastic excitations (2^+ and 3^- states) of targets and projectiles may contributes to the formation of the fully equilibrated compound nucleus at the bombarding energies below the spherical Coulomb barrier. This result is consistent with the conclusion obtained by Hinde et al. from the analysis of the anomalous anisotropy of fission fragments.

REFERENCES

1. Samant, A.M. and Kailas, S, *Z. Phys.* **A356**, 309-312 (1996) and see references therein.
2. Hinde, D.J., Dasgupta, M., Leigh, J.R., Lestone, J.P., Mein, J.C., Morton, C.R., Newton, J.O. and Timmers, H., *Phys. Rev. Lett.* **74**, 1295-1298 (1995).

3. Ikezoe, H., Nagame, Y., Ikuta, T., Hamada, S., Nishinaka,I., and Ohtsuki, T., *Nucl. Instr. and Meth.* **A376**, 420-427 (1996).
4. Ikezoe, H., Ikuta, T., Hamada, S., Nagame, Y., Nishinaka, I., Tsukada, K., Oura Y. and Ohtsuki, T. *Phys. Rev.* **C54**, 2043-2046 (1996).
5. Somerville, L.P., Nurmia, M.J. Nitschke, J.M. Ghiorso, A., Hulet, E.K. and Lougheed, R.W. *Phys. Rev.* **C31**, 1801-1815 (1985).
6. Reisdorf, W. and Schädel M., *Z. Phys.* **A343**, 47-57 (1992).
7. Fernandez, J., Dasso, C.H. and Landowne, S., *Comp. Phys. Comm.* **54**, 409-412 (1989).
8. Christensen, P.R. and Winther, A., *Phys. Lett.* **65B**, 19-22 (1976).
9. Cohen S., Plasil,F. and Swiatecki,W.J., *Ann. Phys.* **82**, 557-596 (1974).
10. Ignatyuk,A.V., Smirenkin, G.N.., Tishin A.S., *Yad. Phys.* **21**, 485 (1975).
11. Andreyev, A.N., Bogdanov,D.D.,Chepigin,V.I.,Kabachenko, A.P., Malyshev, O.N.,Sagaidak, R.N., Salamatin,L.I.,Ter-Akopian, G.M. and Yeremin, A.V., *Z. Phys.* **A345**, 389-394 (1993).
12. Akap'ev, G.N., Demin, A.G., Druin, V.A., Imaev, É.G., Kolesov, I.V., Lobanov, Yu.V. and Pashchenko, L.P., *Atomnaya Énergiya* **21**, 243-246 (1966).
13. Shinohara, N., Usuda, S., Ichikawa, S., Suzuki, T., Magara, M., Okashita, H., Yoshikawa, H., Horiguchi, T., Iwata, Y., Shibata, S. and Fujiwara, I., *Phys. Rev.* **C 34**, 909-913 (1986).
14. Donets, D., Shchegolev,V.A. and Ermakov, V.A., *Sov. Jour. Nucl. Phys.* **2**, 723-729 (1966).
15. Zhang, H.Q, Liu, Z., Xu, J., Qian, X., Qiao, Y., Lin, C and Xu.,K., *Phys. Rev.* C **49**, 926-931(1994).
16. Shen, W.Q., Albinski, J., Gobbi, A., Gralla, S., Hildenbrand, K.D., Herrmann, N., Kuzminski, J., Müller, W.F., Stelzer, H., Tōke, J., Back, B.B., Bjørnholm, S. and Sørensen, S.P., *Phys. Rev.* **C36**, 115-142 (1987).
17. Viola, V.E. and Sikkeland, T., *Phys. Rev.* **128**, 767-774 (1962).

Anharmonic phonon excitations in subbarrier fusion reactions

K. Hagino*, N. Takigawa*, and S. Kuyucak[†]

*Department of Physics, Tohoku University, Sendai 980-77, Japan
[†]Department of Theoretical Physics, Research School of Physical Sciences,
Australian National University, Canberra, ACT 0200, Australia

Abstract. Recently measured high precision data of fusion excitation function have enabled a detailed study on the effects of nuclear collective excitations on fusion reactions. Using such highly accurate data of the ^{16}O + ^{144,148}Sm reactions, we discuss the anharmonic properties of collective phonon excitations in ^{144,148}Sm nuclei. It is shown that subbarrier fusion reactions are strongly affected by the anharmonic effects and thus offer an alternative method to extract the static quadrupole moments of phonon states in a spherical nucleus.

INTRODUCTION

It has been well recognized that cross sections of heavy-ion fusion reactions at energies near and below the Coulomb barrier are strongly influenced by couplings of the relative motion of the colliding nuclei to several nuclear intrinsic motions [1]. In the eigen-channel approach, such couplings give rise to a distribution of potential barriers [2,3]. Based on this idea, a method was proposed to extract barrier distributions directly from fusion excitation functions using the second derivative of the product of the fusion cross section and the center of mass energy $E\sigma$ as a function of energy E [4]. Based on coupled-channels calculations, it was shown that the fusion barrier distribution, i.e. $d^2(E\sigma)/dE^2$, is very sensitive to the details of the couplings. In order to deduce meaningful barrier distributions, excitation functions of fusion cross sections have to be measured with high precision at small energy intervals. Thanks to the recent developments in experimental techniques [5], such data are now available for several systems, and they have clearly demonstrated that the barrier distribution is indeed a sensitive quantity to channels couplings [6].

In this contribution, we analyse the recently measured accurate data on the ^{16}O + ^{144}Sm fusion reaction to discuss effects of nuclear surface vibrations on heavy-ion fusion reactions [7]. The barrier distribution analysis of the recent high precision data on the ^{58}Ni + ^{60}Ni fusion reaction has shown clear evidence for coupling of multi-phonon states in ^{58}Ni and ^{60}Ni [8], while no evidence for double phonon couplings is seen in the ^{16}O + ^{144}Sm reaction [6,9]. We show that anharmonicities in nuclear vibrations play an important role in the latter reaction. We estimate the magnitude as well as the sign of the quadrupole moments of the quadrupole and octupole single-phonon states of ^{144}Sm from the experimental fusion barrier distribution. A similar analysis is performed also for the ^{16}O + ^{148}Sm reaction.

ANHARMONICITIES IN NUCLEAR VIBRATIONS

Collective phonon excitations are common phenomena in fermionic many-body systems. In nuclei, low-lying surface oscillations with various multipolarities are typical examples. The harmonic vibrator provides a zeroth order description for these surface oscillations, dictating simple relations among the level energies and the electromagnetic transitions between them. For example, all the levels in a phonon multiplet are degenerate and the energy spacing between neighboring multiplets is a constant. In realistic nuclei, however, there are residual interactions which cause deviations from the harmonic limit, e.g., they split levels within a multiplet, change the energy spacings, and also modify the ratios between various electromagnetic transition strengths. There are many examples of two-phonon triplets ($0^+, 2^+, 4^+$) of quadrupole surface vibrations in even-even nuclei near closed shells. Though the center of mass of their excitation energies are approximately twice the energy of the first 2^+ state, they usually exhibit appreciable splitting within the multiplet. A theoretical analysis of the anharmonicities for the quadrupole vibrations was first performed by Brink et al. [10], where they related the excitation energies of three-phonon states to those of double-phonon triplets. For a long time, however, the sparse experimental data on three-phonon states had caused debates on the existence of multi-phonon states. The experimental situation has improved rapidly in recent years, and data on multi-phonon states are now available for several nuclei. As a consequence, study of multi-phonon states, and especially their anharmonic properties, is attracting much interest [11]. It is worthwhile to mention that anharmonic effects are not restricted to low-lying vibrations but have also been observed in multi-phonon excitations of giant resonances in heavy-ion collisions [12].

In many even-even nuclei near closed shells, a low-lying 3^- excitation is observed at a relatively low excitation energy, which competes with the quadrupole mode of excitation [13]. These excitations have been frequently interpreted as collective octupole vibrations arising from a coherent sum of one-particle one-hole excitations between single particle orbitals differing by three units of orbital angular momentum. This picture is supported by large

E3 transition probabilities from the first 3^- state to the ground state, and suggests the possibility of multi-octupole-phonon excitations. In contrast to the quadrupole vibrations, however, so far there is little experimental evidence for double-octupole-phonon states. One reason for this is that E3 transitions from two-phonon states to a single-phonon state compete against E1 transitions. This makes it difficult to unambiguously identify the two-phonon quartet states $(0^+, 2^+, 4^+, 6^+)$. Only in recent years, convincing evidences have been reported for double-octupole-phonon states in some nuclei, including ^{208}Pb [14] and ^{144}Sm [15].

EFFECTS OF PHONON EXCITATIONS ON FUSION

Let us now discuss the effects of nuclear surface vibrations on heavy-ion fusion reactions. In this section we use the linear coupling approximation to describe the coupling between the relative motion of the colliding nuclei and the surface vibrations. This simple model enables us to understand easily the effects of anharmonicity. Extension of the model so as to include the couplings to all orders and comparisons with the experimental data is given in the next section.

Harmonic limit

The effects of nuclear surface vibrations on heavy-ion fusion reactions at energies below and near the Coulomb barrier has been investigated by many groups (see Ref. [1] for a recent review). These studies were later extended to include the effect of multi-phonon states within the harmonic oscillator approximation [16,17]. Using the no-Coriolis approximation [16] and the linear coupling approximation, the coupling Hamiltonian, which describes the coupling between the relative motion and the quadrupole surface oscillations, is assumed to be

$$V_{coup}(r,\xi) = \frac{\beta}{\sqrt{4\pi}} f(r)(a_{20}^\dagger + a_{20}), \tag{1}$$

where $a_{20}^\dagger$ and a_{20} are the creation and the annhilation operators for the quadrupole phonon, respectively, and β is the quadrupole deformation parameter. The coupling form factor $f(r)$ consists of the nuclear and Coulomb parts and reads

$$f(r) = -R_T \frac{dV_N}{dr} + \frac{3}{5} Z_P Z_T e^2 \frac{R_T^2}{r^3}. \tag{2}$$

Here R_T is the radius of the target nucleus and V_N is the nuclear potential.

For the quadrupole surface vibrations, the two phonon state has three levels $(0^+, 2^+, 4^+)$. In the harmonic limit, this two-phonon triplet is degenerate in the excitation energy. One can then introduce the two-phonon channel by taking particular linear combinations of the wave functions of the two-phonon triplet [16,17]. The wave function of the two-phonon channel then reads

$$|2> = \sum_{I=0,2,4} <2020|I0> |I0> = \frac{1}{\sqrt{2!}}(a_{20}^\dagger)^2|0>. \qquad (3)$$

In the same way, one can introduce the n-phonon channel as

$$|n> = \frac{1}{\sqrt{n!}}(a_{20}^\dagger)^n|0>. \qquad (4)$$

The dimension of the coupled-channels equations is reduced drastically with the introduction of the n-phonon channels. If we truncate to the two phonon states, the corresponding coupling matrix is given by

$$V_{coup} = \begin{pmatrix} 0 & F(r) & 0 \\ F(r) & \hbar\omega & \sqrt{2}F(r) \\ 0 & \sqrt{2}F(r) & 2\hbar\omega \end{pmatrix}. \qquad (5)$$

Here, $F(r)$ is defined as $\frac{\beta}{\sqrt{4\pi}}f(r)$.

Anharmonic vibrator

The sd-interacting boson model (IBM) in the vibrational limit provides a convenient calculational framework to discuss the effects of anharmonicity in the surface vibrations [18]. The vibrational limit of the IBM and the anharmonic vibrator (AHV) in the geometrical model are very similar, the only difference coming from the finite number of bosons in the former [19]. A model for subbarrier fusion reactions, which uses the IBM to describe effects of channel couplings, has been developed in Ref. [20]. Following Ref. [20], we assume that the coupling Hamiltonian is given as

$$V_{coup}(r,\xi) = \frac{\beta}{\sqrt{4\pi N}}f(r)Q_{20}. \qquad (6)$$

Here, N is the boson number and we have introduced the scaling of the coupling strength with $\sqrt{N}$ to ensure the equivalence of the IBM and the geometric model results in the large N limit [20]. Q_{20} is the quadrupole operator in the IBM, which we take as

$$Q_{20} = s^\dagger d_0 + sd_0^\dagger + \chi_2(d^\dagger \tilde{d})_0^{(2)}, \qquad (7)$$

where tilde is defined as $\tilde{b}_{l\mu} = (-)^{l+\mu} b_{l-\mu}$.

As in the harmonic limit, one can introduce the multi-phonon channel if one assumes that the multi-phonon multiplets are degenerate in the excitation energy. The wave function of the n-phonon channel in the framework of the IBM then reads

$$|n> = \frac{1}{\sqrt{n!(N-n)!}} (s^\dagger)^{N-n} (d_0^\dagger)^n |0>. \qquad (8)$$

The corresponding coupling matrix, truncated to the two-phonon states, is given by

$$V_{coup} = \begin{pmatrix} 0 & F(r) & 0 \\ F(r) & \hbar\omega - \frac{2}{\sqrt{14N}} \chi F(r) & \sqrt{2(1-1/N)} F(r) \\ 0 & \sqrt{2(1-1/N)} F(r) & 2\hbar\omega + \delta - \frac{4}{\sqrt{14N}} \chi F(r) \end{pmatrix}. \qquad (9)$$

The parameter δ is introduced to represent deviation of the energy spectrum from the harmonic limit. When the χ parameter in the quadrupole operator is zero, quadrupole moments of all states vanish, and one obtains the harmonic limit in the large N limit. Non-zero values of χ generate quadrupole moments and, together with finite boson number, they are responsible for the anharmonicities in electric transitions.

It has been shown that anharmonicities in level energies have only a marginal effect on the fusion excitation function and the barrier distribution [7]. In fact, our studies show that the fusion barrier distribution does not depend so much on the excitation energies of the multi-phonon states once the energy of the single-phonon state is fixed. We therefore set δ to be zero in the following discussion. As we will see later, the main effects of anharmonicity on fusion barrier distributions come from the deviation of the transition probabilities from the harmonic limit as well as the reorientation effects.

COMPARISON WITH EXPERIMENTAL DATA

The ^{16}O + ^{144}Sm reaction

We now discuss the effects of anharmonicities on the ^{16}O + ^{144}Sm fusion reaction, whose excitation function has recently been measured with high accuracy [6]. It has been reported that inclusion of the double-phonon excitations of ^{144}Sm in coupled-channels calculations in the harmonic limit destroys the good agreement between the experimental fusion barrier distribution and the theoretical predictions obtained when only the single-phonon excitations are taken into account [21]. On the other hand, there are experimental [15,22] as well as theoretical [23] support for the existence of the double-octupole-phonon states in ^{144}Sm. Reconciliation of these apparently contradictory facts may

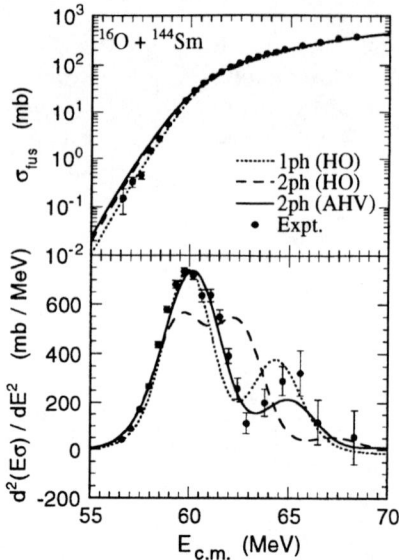

FIGURE 1. Comparison of the experimental fusion cross section (the upper panel) and fusion barrier distribution (the lower panel) with the coupled-channels calculations for ^{16}O + ^{144}Sm reaction. The experimental data are taken from Ref. [6].

be possible if one includes the anharmonic effects, which are inherent in most multi-phonon spectra.

In order to address these questions, it is necessary to extend the models which were discussed in the previous section so that they include the octupole mode as well as the couplings to all orders. The full order treatment is crucial in order to quantitatively, as well as qualitatively, describe heavy-ion sub-barrier fusion reactions [20,24]. We therefore assume the following coupling Hamiltonian based on the sdf-IBM.

$$V_{coup}(r,\xi) = V_C(r,\xi) + V_N(r,\xi), \tag{10}$$

$$V_C(r,\xi) = \frac{Z_P Z_T e^2}{r}\left(1 + \frac{3}{5}\frac{R_T^2}{r^2}\frac{\beta_2 \hat{Q}_{20}}{\sqrt{4\pi N}} + \frac{3}{7}\frac{R_T^3}{r^3}\frac{\beta_3 \hat{Q}_{30}}{\sqrt{4\pi N}}\right),$$

$$V_N(r,\xi) = -V_0\left[1 + \exp\left(\frac{1}{a}\left(r - R_0 - R_T(\beta_2\hat{Q}_{20} + \beta_3\hat{Q}_{30})/\sqrt{4\pi N}\right)\right)\right]^{-1}. \tag{11}$$

The quadrupole and the octupole operators are defined as

$$\hat{Q}_2 = s^\dagger \tilde{d} + sd^\dagger + \chi_2(d^\dagger \tilde{d})^{(2)} + \chi_{2f}(f^\dagger \tilde{f})^{(2)},$$
$$\hat{Q}_3 = sf^\dagger + \chi_3(\tilde{d}f^\dagger)^{(3)} + h.c., \tag{12}$$

respectively.

The results of the coupled-channels calculations are compared with the experimental data in Fig. 1. The upper and the lower panels in Fig. 1 show the excitation function of the fusion cross section and the fusion barrier distributions, respectively. The experimental data are taken from Ref. [6]. The dotted line is the result in the harmonic limit, where couplings to the quadrupole and octupole vibrations in ^{144}Sm are truncated at the single-phonon levels and all the χ parameters in Eq. (12) are set to zero. The deformation parameters in Eq. (11) are estimated to be β_2=0.11 and β_3=0.21 from the electric transition probabilities. The dotted line reproduces the experimental data of both the fusion cross section and the fusion barrier distribution reasonably well, though the peak position of the fusion barrier distribution around $E_{cm} = 65$ MeV is slightly shifted. As was shown in Ref. [21], the shape of the fusion barrier distribution becomes inconsistent with the experimental data when the double-phonon channels are included in the harmonic limit (the dashed line). To see whether this discrepancy is due to neglecting of anharmonic effects, we have repeated the calculations including the χ parameters in the fits and using $N = 2$ in the IBM. The χ^2 fit to the fusion cross sections resulted in the set of parameters, $\chi_2 = -3.30 \pm 2.30$, $\chi_{2f} = -2.48 \pm 0.07$, and $\chi_3 = 2.87 \pm 0.16$, regardless of the starting values. The resulting fusion cross sections and barrier distributions are shown in Fig. 1 by the solid line. They agree with the experimental data much better than those obtained in the harmonic limit. Thus, inclusion of the anharmonic effects in vibrational motion appear to be essential for a proper description of barrier distributions in the reaction ^{16}O + ^{144}Sm.

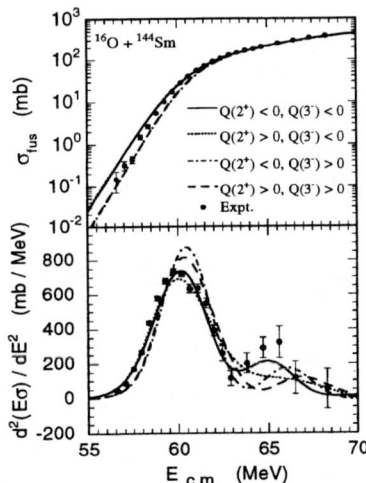

FIGURE 2. Dependence of the fusion cross section and barrier distribution on the sign of the quadrupole moment of the excited states in ^{144}Sm.

One of the pronounced features of an anharmonic vibrator is that the excited states have non-zero quadrupole moments [13]. Using the χ parameters extracted from the analysis of fusion data in the E2 operator, $T(E2) = e_B \hat{Q}_2$, we can estimate the static quadrupole moments of various states in ^{144}Sm. Here, e_B is the effective charge, which is determined from the experimental $B(E2; 0 \to 2_1^+)$ value as $e_B = 0.16$ eb. For the quadrupole moment of the first 2^+ and 3^- states, we obtain -0.89 ± 0.63 b and -0.70 ± 0.02 b, respectively. Fig. 2 shows the influence of the sign of the quadrupole moment of the excited states on the fusion cross section and the fusion barrier distribution. The solid line is the same as in Fig. 1 and corresponds to the optimal choice for the signs of the quadrupole moments of the first 2^+ and 3^- states. The dotted and dashed lines are obtained by changing the sign of the χ_2 and χ_{2f} parameters in Eq. (12), respectively, while the dot-dashed line is the result where the sign of both χ_2 and χ_{2f} parameters are inverted. The change of sign of χ_2 and χ_{2f} is equivalent to taking the opposite sign for the quadrupole moment of the excited states. Fig. 2 demonstrates that subbarrier fusion reactions are sensitive to the sign of the quadrupole moment of excited states. The experimental data are reproduced only when the correct sign of the quadrupole moment are used in the coupled-channels calculations.

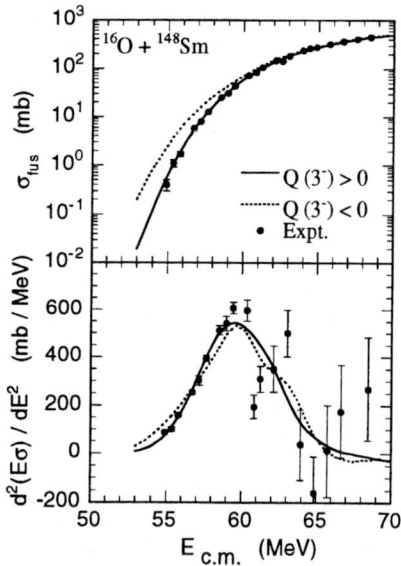

FIGURE 3. Dependence of the fusion cross section and barrier distribution for the ^{16}O + ^{148}Sm reaction on the sign of the quadrupole moment of the first 3^- state in ^{148}Sm. The experimental data are taken from Ref. [6].

The ^{16}O + ^{148}Sm reaction

A similar analysis was performed also for the ^{16}O + ^{148}Sm reaction [25]. The best fit to the experimental fusion cross section [6] was obtained with the quadrupole moments of $-1.00 \pm 0.25b$ and $+1.52 \pm 0.14b$ for the first 2^+ and 3^- states, respectively. The total boson number N was assumed to be 4 and the deformation parameters were estimated from the electric transition probabilities. Note that the value of the quadrupole moment of the first 2^+ state which we obtained from the fusion analysis is very close to that measured from the Coulomb excitation technique, i.e. $-0.97 \pm 0.27b$ [26]. Fig. 3 shows the sensitivity of the fusion cross section and the fusion barrier distribution to the sign of the quadrupole moment of the first 3^- state. The experimental data are taken from Ref. [6]. The solid line corresponds to the optimal choice for the sign of the first 3^- state, while the dotted line was obtained by inverting it. We again observe that the use of the incorrect sign of the quadrupole moment destroys the good fit to the experimental data. This strongly suggests that subbarrier fusion can provide an alternative method to determine the sign as well as the magnitude of the quadrupole moments in spherical nuclei.

SUMMARY

We discussed the effects of multi-phonon excitations on subbarrier fusion reactions. We especially focused on the anharmonic properties of the phonon excitations. The experimental fusion excitation functions for the ^{16}O + ^{144,148}Sm reactions were analyzed with a model which explicitly takes into account the effects of anharmonicity of the vibrational modes of excitation in ^{144,148}Sm. We found that the best fit to the experimental data requires negative quadrupole moments for the first 2^+ and the first 3^- states of ^{144}Sm. For the ^{148}Sm nucleus, we obtained a negative quadrupole moment for the first 2^+ state and a positive one for the first 3^- state. As a general conclusion, we find that heavy-ion subbarrier fusion reactions, and in particular, barrier distributions extracted from the fusion data, are very sensitive to the sign of the quadrupole moments of phonon states in the target nucleus and thus offer an alternative method to determine them.

ACKNOWLEDGMENTS

The authors thank J.R. Leigh, M. Dasgupta, D.J. Hinde, and J.R. Bennett for useful discussions. The work of K.H. was supported by the Japan Society for the Promotion of Science for Young Scientists. This work was also supported by the Grant-in-Aid for General Scientific Research, Contract No.08640380, Monbusho International Scientific Research Program: Joint Research, Contract No. 09044051, from the Japanese Ministry of Education,

Science and Culture, and a bilateral program of JSPS between Japan and Australia.

REFERENCES

1. A.B. Balantekin and N. Takigawa, Rev. Mod. Phys., in press.
2. K. Hagino, N. Takigawa, and A.B. Balantekin, Phys. Rev. C **56**, 2104(1997).
3. C.H. Dasso, S. Landowne, and A. Winther, Nucl. Phys. **A405**, 381 (1983); *ibid* **A407**, 221 (1983).
4. N. Rowley, G.R. Satchler, and P.H. Stelson, Phys. Lett. **B254**, 25 (1991).
5. J.X. Wei *et al.*, Phys. Rev. Lett. **67**, 3368 (1991).
6. J.R. Leigh, M. Dasgupta, D.J. Hinde, J.C. Mein, C.R. Morton, R.C. Lemmom, J.P. Lestone, J.O. Newton, H. Timmers, J.X. Wei, and N. Rowley, Phys. Rev. C **52**, 3151 (1995).
7. K. Hagino, N. Takigawa, and S. Kuyucak, Phys. Rev. Lett. **79**, in press (1997).
8. A.M. Stefanini *et al.*, Phys. Rev. Lett. **74**, 864 (1995).
9. C.R. Morton *et al.*, Phys. Rev. Lett. **72**, 4074 (1994).
10. D.M. Brink, A.F.R. de Toledo Piza, and A.K. Kerman, Phys. Lett. **B19**, 413 (1965).
11. R.F. Casten and N.V. Zamfir, Phys. Rep. **264**, 81 (1996), and references therein.
12. E.G. Lanza *et al.*, Nucl. Phys. **A613**, 445 (1997).
13. A. Bohr and B. Mottelson, *Nuclear Structure* (Benjamin, New York, 1975), vol. 2.
14. M. Yeh *et al.*, Phys. Rev. Lett. **76**, 1208 (1996).
15. R.A. Gatenby *et al.*, Phys. Rev. C **41**, R414 (1990); Nucl. Phys. **A560**, 633 (1993).
16. N. Takigawa and K. Ikeda, in *Proceedings of the Symposium on Many Facets of Heavy Ion Fusion Reactions*, edited by W. Henning *et al.*(Argonne National Laboratory Report No. ANL-PHY-87-1), 1986, p.613.
17. A.T. Kruppa, P. Romain, M.A. Nagarajan, and N. Rowley, Nucl. Phys. **A560**, 845 (1993).
18. F. Iachello and A. Arima, *The Interacting Boson Model* (Cambridge University Press, Cambridge, England, 1987).
19. R.F. Casten and D.D. Warner, Rev. Mod. Phys. **60**, 389 (1988).
20. A.B. Balantekin, J.R. Bennett, and S. Kuyucak, Phys. Rev. C **48**, 1269 (1993); C **49**, 1079 (1994); C **49**, 1294 (1994).
21. C.R. Morton, Ph.D. thesis, the Australian National University, 1995.
22. M. Wilhelm *et al.*, Phys. Rev. C **54**, R449 (1996).
23. M. Grinberg and Ch. Stoyanov, Nucl. Phys. **A573**, 231 (1994).
24. K. Hagino, N. Takigawa, M. Dasgupta, D.J. Hinde, and J.R. Leigh, Phys. Rev. C **55**, 276 (1997).
25. K. Hagino, N. Takigawa, and S. Kuyucak, in preparation.
26. L.K. Peker, Nucl. Data Sheets **59**, 393 (1990).

PHYSICS WITH EXOTIC NUCLEI

Mass and Nuclear Moment Measurement with High and Low Energy RIB's

W. Mittig

GANIL, B.P. 5027, 14076 Caen Cedex 5, France

Introduction

I want to present here an introduction to the subject of the title for non-specialists of the domain. It is related to the ground state properties of nuclei. If we describe the nucleus by a Schrodinger equation

$$E\psi = H\psi$$

the ground state corresponds to the solution with the lowest energy E, or the highest binding energy E_B. The value of E_B is related to the nuclear mass by the relation

$$m_A c^2 = N\, m_n c^2 + Z\, m_p c^2 + E_B$$

The wave function $\psi(r)$ determines the density distributions

$$|\psi(r)|^2 = \rho\ \mathbf{matter}\ (r)$$

$$|\psi_{proton}(r)|^2 = \rho\ \mathbf{proton}\ (r)$$

$$|\psi_{neutron}(r)|^2 = \rho\ \mathbf{neutron}\ (r)$$

These density distributions may be explored by total reaction cross-section measurements or elastic scattering of electrons, protons, neutrons or heavier particles. This subject will not be treated here, it may be found in other contribututions to the present workshop. In a odd-even or odd-add nucleus having a spin $J \neq 0$, the electric and magnetic moments may interact with external fields (hyperfine interaction) with the characteristic frequencies

$$\hbar\,\omega_B = \mu\cdot B \qquad \text{and}$$

$$\hbar\,\omega_Q = \frac{eQ\, V_{zz}}{4J(J+1)}$$

where B and V_{zz} is the external magnetic field and the external electric field gradient respectively; μ is the magnetic dipole and Q is the electric quadrupole moment.

For a single particle, coupled to a spherical core, these two quantities are given by

$$\vec{\mu} = (g_\ell \vec{l} + g_s \vec{S}) \mu_N$$

$$Q = \langle J\, M=J \,|Q_{operator}|\, J, M=J \rangle$$

and hence

$$\mu = J\left\{g_\ell \pm (g_s - g_\ell)\frac{1}{2\ell+1}\right\} \mu_N$$

with $J = \ell \pm \frac{1}{2}$

and

$g_\ell = \{{0 \atop 1}}$ $g_s = \{{5.58 \atop -3.82}}$ for neutron / for proton

$$Q = \frac{2J-1}{2J+1} \int r^4 R_{n\ell j}\, dr \quad \text{for proton with s.p. wavefunction } R_{n\ell j}$$

For a nucleus with a strong intrinsic deformation β one has

$$Q = -\frac{I}{2I+3} Q_{intr} \quad \text{with} \quad Q_{intr} = \frac{3}{\sqrt{5\pi}} \beta\, A\, R^2$$

where Q_{intr} is the intinsic Q-moment. So the magnetic moment is essentially related to the single particle structure of the ground state and the coupling of neutrons and protons, the Q-moment is mainly related to the collectivity of the g.s. wave function. These relations may be found in standard textbooks, such as ref.1.

II. Mass-measurements far from stability

1. Experiments

This subject was treated in a recent review article (ref 2), to which an interested reader may refer.

From experimental point of view, one may distinguish essentially two methods. One is the determination of a reaction

A(a,b) B with

$$E_{bB}^{cin} = E_{aA}^{cin} + Q \quad \text{in the center of mass.}$$

If 3 masses are known in this equation, a fourth one may be deduced. Reactions may be β-decay, transfer reactions or invariant mass spectroscopy.

The other method is a direct mass measurement with

$$mc^2 = N m_n c^2 + Z\, m_p c^2 + E_B$$

This method implies high precision, typically in the domain of 10^{-6}. This may be achieved by mainly high resolution spectrometers using one or both of the basic relations

$$B\rho = m\,v/(e\,q) \quad\quad \text{magnetic rigidity}$$

$$E\rho = 2E/q \quad\quad \text{electric rigidity}$$

or the measurement of the cyclotron frequency ω_{cycl}

$$B/\omega_{cycl} = m/(qe)$$

2. Some examples

I will illustrate the experimental result by figure 1, concerning a recent measurement of the mass of ^{100}Sn (ref 3). With respect to smooth extrapolations by Audi and Wapstra, one has significant deviations. This may be related to the question of the conservation of the Wigner-term along the N=Z line, for heavier nuclei (ref 4).

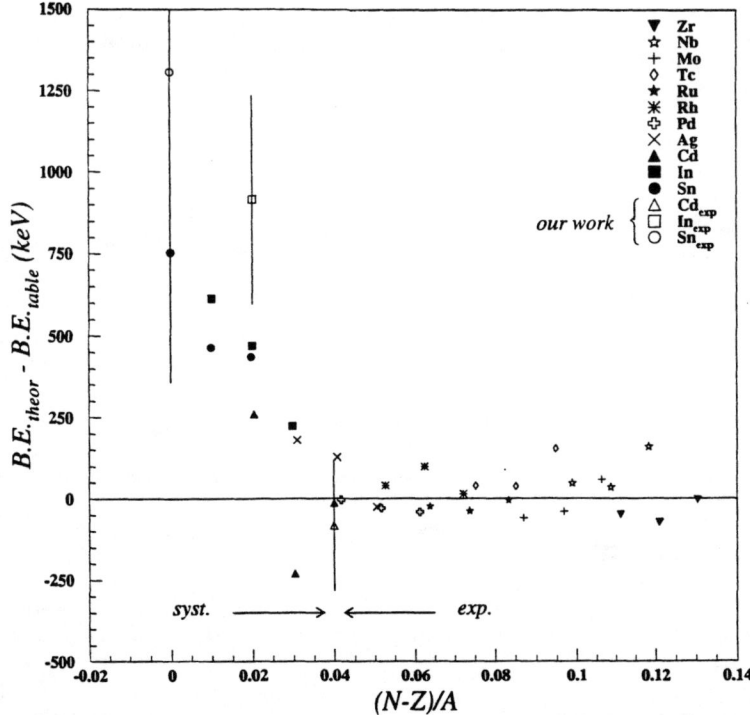

Figure 1: *Comparison between the shell model calculations of Johnstone and Skouras* ($B.E._{theor}$) *and the mass table of Audi and Wapstra* ($B.E._{table}$) *for the binding energies of isotones N = 50, 51 and 52 from ^{90}Zr to ^{102}Sn. The binding energies of the mass table come either from experimental values (right side of the dashed vertical line) or from systematic trends extrapolations (left side of the line). Our experimental results are represented by black stars.*

In a more general way, one may say at the moment nuclear models have still strongly diverging predictions of nuclear masses, as one goes far from stability. One of the questions is the evolution of the spin-orbit term far from stability, and eventual changes of magic numbers. A quenching of this strength gives a better agreement of stellar abundances of nuclear synthesis calculations of the r-process (ref 5).

Nuclear Moment Measurements

1. Experimental techniques

A big variety of methods are used for measuring nuclear moments by hyperfine interaction. I will restrict the discussion here to the measurement of g.s. nuclear moments, measured in β-decay.

As is well known, a polarised source emits the β-s with a parity violating up-down asymmetry, and this was the first experiment showing parity violation. So the measurement of an asymmetry in the β-emission is a measure of a polarisation, and a change of the polarisation by an external field will therefore give a change of the β-decay asymmetry.

Live-times of nuclei far from stability are too short to achieve a low-temperature polarisation. Nuclear reactions may produce products that are polarised. This was shown to be done for fragmentation products, and these polarised products were implanted in crystals, and magnetic moments and Q-pole moments were measured using NMR (nuclear magnetic resonance) techniques. A review of this technique can be found in reference 6.

This method can not be applied at an angle of zero degres : due to symmetry properties, there is no polarisation possible at the beam axis. However, there may be strong alignment of the reaction products. A classical example are the fusion evaporation products, that show very strong alignment. Alignment does not produce any asymmetry in allowed l = 0 β-decay. So alignment must be transformed in polarisation. This is the LMR (level mixing resonance technique) as first proposed and used for fusion-evaporation products in ref 7. This technique is illustrated on fig. 2.

The nucleus is implanted in a crystal having an electric field gradient V_{zz}. Applying an external magnetic field, one sees that there will be a crossing of magnetic sublevels, here of m = 0 and m = +1. If there is at this crossing point a small interaction mixing these two sublevels, the alignment will be transformed in polarisation. This mixing is obtained by a small tilt angle between the magnetic field and the symmetry axis of the electric field gradient V_{zz}.

At this cossing point the β-decay asymmetry will show a resonant like structure. It was proposed to use this method for exotic nuclei produced in fragmentation at zero degress by G. Neyens et al. (ref 8).

As a first application, the LMR was measured for ^{18}N. A sufficiently strong alignment at the center of the momentum distribution was found to do such a measurement, of the order of 10 to 20 %.
As in the case of polarisation in fragmentation products, the value of alignment was found to be very small for reaction products having a very strong mass-difference with respect to the beam ; as for ^{12}B produced by a ^{22}Ne beam the alignment was found to be less than 1 %. The result for ^{18}N is shown on figure 3 (ref 9).

Fig 2: Schematic illustration of the LMR (Level Mixing Resonance) technique for a spin 1 nucleus. In the upper part the hyperfine energy is shown. For B=0, the magnetic sublevels of m=+-1 are degenerate. As a function of the magnetic field B, the magnetic sublevels m=0 and m=-1 will cross, and for misalignment between the magnetic field and the electric field they will mix at this crossing point. Due to the mixing, alignment is transformed in polarisation, and the β-decay angular distribtion will show an asymmetry, that appears as a resonance as a function of B (lower part of the figure).

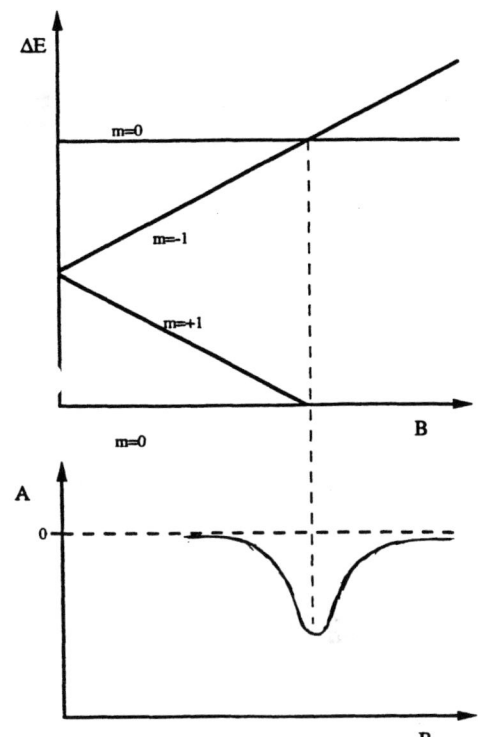

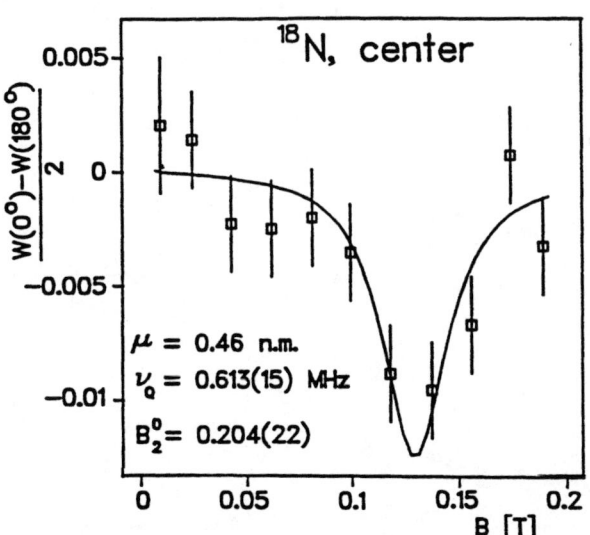

Fig 3: LMR curve for 18N in the center of the longitudinal momentum distribution, this means for the most probable velocity of the fragments at zero degrees. The Mg-crystal was mounted on a Copper cold finger cooled to near liquid He temperature, in order to have long enough relaxation times.

Unfortunately, the LMR technique provides only one equation for two unknowns, the magnetic and the electric moment. Only the ratio of g/Q is obtained. Amoung other possibilities to obtain separately g and Q, a mixed NMR-LMR method was first tested at Louvain la Neuve for 12B. The principle of the method is indicated on figure 4.

Preliminary on line results have been obtained for ^{18}N in july of this year. The actual values obtained show a very low g-value, and a very strong quadrupole moment. This very strong Q-moment, if confirmed in the final analysis, is in good agreement with deformation parameters as predicted in RMF-calculations by S. Patra et al., ref 11. The figure 5 shows that 18N is situated in a transition region of deformation and thus the results of this experiment will provide a stringent test of theories in this region.

A quantitative analytic description of the mixed NMR-LMR method is under the way. An interesting question is if the contribution of LMR and NMR is coherent or nncoherent. Some resonance narrowing seems to be apparent from the data, thus indicating a coherent contribution. In this case, eventually the relative sign of the magnetic and the electric moment may be obtained.

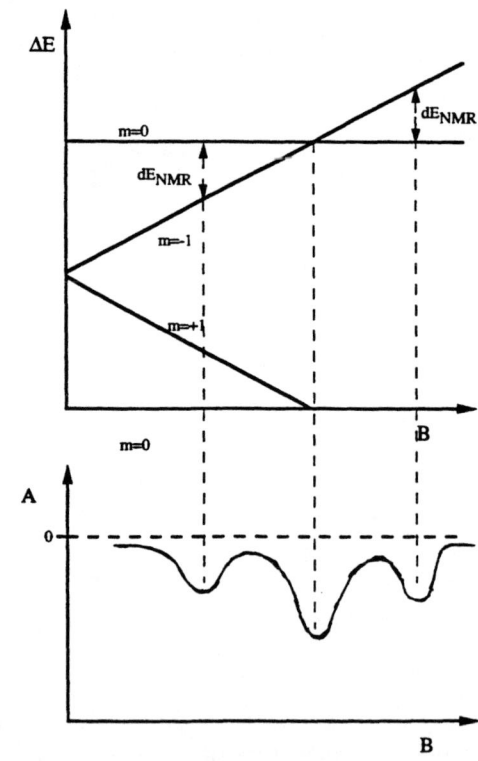

Fig 4: Same as figure 2, except that now a time dependant magnetic field is superimposed. This will introduce a transition-energy $dE_{NMR}= h\nu$, where ν is the frequency. This will introduce two other magnetic fields for which there will be a a level-mixing and thus 2 more resonances will show up in the b asymmetry, as indicated int the lower part of the figure.

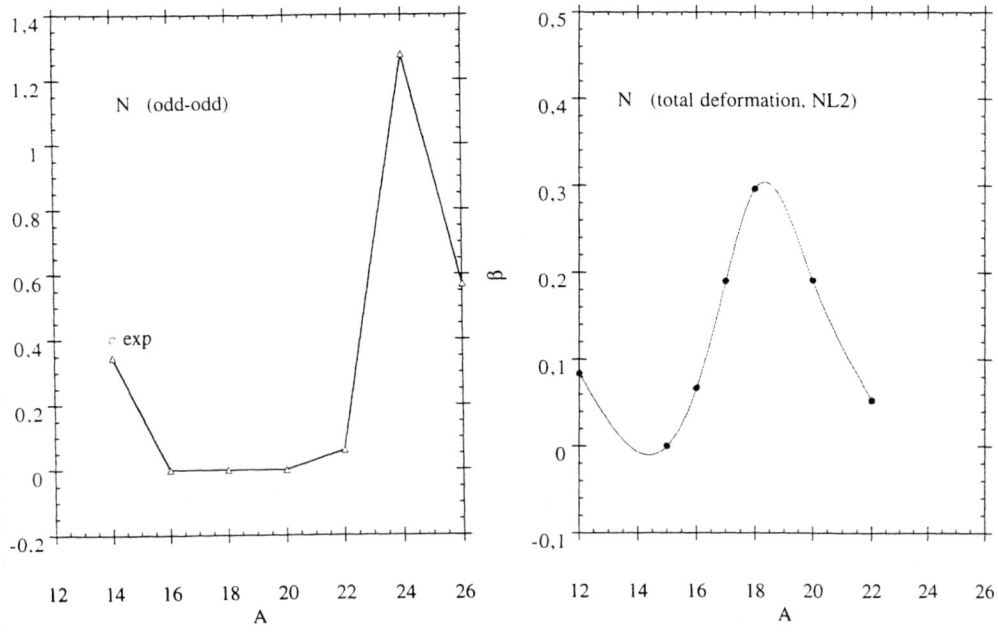

Fig 5: predicted magnetic moments for odd N-isotopes, as obtained in the shell model calculation of N.A.F. Popellier etal, ref 10, left part, and deformations predicted as a function of A for different nuclei in a relativistic mean field approximation including pairing, from S.K.Patra et al, ref 11, right part

Conclusion

Mass and nuclear moments may be determined for nuclei far from stability, with rapidly evolving experimental techniques. They represent a very powerful way to check model predictions far from stability. Nuclear masses and Q-moments are related essentially to global properties of the ground-state, whereas the magnetic moment is essentially related to single particle properties. The measure of these properties along long isotopic chains will provide a good test for nuclear models, and provide fundamental information on nuclear forces far from stability.

REFERENCES

1) A. Bohr and B. Mottelson, Nuclear Structure, W. A. Benjamin, New York, Amsterdam, 1969
2) W.Mittig, A.Lepine and N.Orr, Annual Review of Nuclear Science, in Press
3) M. Cartier et, Phys. Rev. Lett. 77 (1996) 2400
   W.Mittig et al., Nucl. Phys. A 616 (1997) 329c
4) P. van Isacker et al, Phys.Rev. Lett. 74 (1995) 4607
5) B.Pfeiffer etal, Z.Phys. A in print
6) K. Asahi et al, Phys Rev. C 43 (1991) 456
7) G. Scheveneels et al, Hyp. Int 23(1989) 179 and 257 , and ref. cited
8) G. Neyens et al, NIM A 555 (1994) 555
9) G. Neyens et al, Phys. Lett. 393B (1997) 36
10) N.A.F.M. Poppelier et al, Zeitschrift Phys. A 346 (1993) 11
11) S.K Patra et al, Nucl. Phys. A 559 (1993)173

Gamma Spectroscopy with Low and High Energy Radioactive Beams.

W. Gelletly

School of Physical Sciences, University of Surrey, Guildford GU2 5XH, U.K.

Abstract

Gamma ray spectroscopy has been one of our main tools for the study of nuclear structure, and will continue to be so with the newly available beams of radioactive ions from fragmentation and ISOL-based facilities. Examples of γ-ray studies at the end of the LISE3 spectrometer at GANIL are described. The likely programme of γ-ray spectroscopy at facilities such as SPIRAL is discussed.

1. INTRODUCTION

Our understanding of physical systems relies generally on varying key parameters which describe the system and observing how its properties change. A straightforward example is that if we could not change the temperature of water then we would never know that it can become a vapour or a solid. The same is true of the atomic nucleus. Figure 1, taken from Richter [1], shows schematically the landscape created by some of the key parameters for the nucleus namely energy E (or temperature T), the angular momentum I and the ratio of neutrons/protons N/Z.

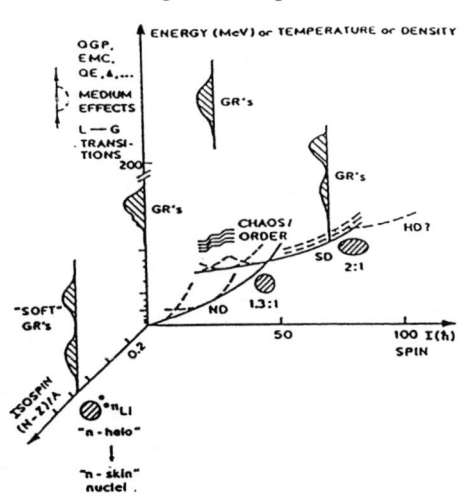

Fig 1: Parameters of importance in studying nuclear properties

We have been and are exploring how nuclear properties vary with all of these parameters. For example there has been considerable advance in recent years in our knowledge of how nuclei behave under extreme rotational stress. Much of this advance comes from the study of the long cascades of gamma rays emitted in the de-excitation of rotational bands populated in fusion-evaporation reactions induced by heavy ions. Such studies have also provided much of our information about neutron deficient nuclei far from stability.

All of our knowledge of nuclear properties is derived from studies of radioactive decay and nuclear reactions. The flexibility of nuclear reactions, the ease with which we can vary the key parameters, means that this is the main route to studying and understanding the properties of atomic nuclei. So far our efforts to explore the N/Z dimension have been very limited, because it is constrained by the fact that, in general, we have been able to use only stable beams and stable target nuclei. Given that the resulting compound nuclei are stable or neutron-deficient, it is not surprising that we have much more information from gamma ray spectroscopy about stable and proton-rich nuclei than about neutron-rich nuclei. At the same time even the information we do have on these nuclei is limited because of the nature of the reactions used.

All of this is being transformed by the rapid development of beams of radioactive nuclei, produced both by the fragmentation [2] and ISOL based techniques [3]. Gamma-ray spectroscopy will play a major role in exploiting these new beams to learn about nuclear structure. In this article we will be concerned with the beginnings of gamma-ray spectroscopy with beams of radioactive nuclei.

2. GAMMA-RAY STUDIES WITH HIGH ENERGY RNBs

Beams of radioactive nuclei at high energy, typically >25 MeV/u, can be produced in the fragmentation [2] and fission [4] of very high energy particles. The beams are weak and poorly defined but they offer the possibility of studying gamma rays from the fragments or from secondary reactions induced by them.

The nuclei produced in fragmentation and projectile fission have a wide range of A and Z. To do anything with them it is essential to be able to identify them and tag them with A and Z. There are four major laboratories which have exploited this technique, namely GSI, GANIL, MSU and RIKEN. In each case a spectrometer has been developed which does this tagging job. All of these spectrometers have much in common. Here as an example I will use the A1200 spectrometer at MSU. It is shown schematically in Fig. 2. Energetic heavy ions, typically with energies 50-100 MeV/u or greater, are focused on a thick production target. Fragments produced in peripheral collisions, with a lineat momentum close to that of the beam particles are

Fig 2: The MSU A1200 spectrometer

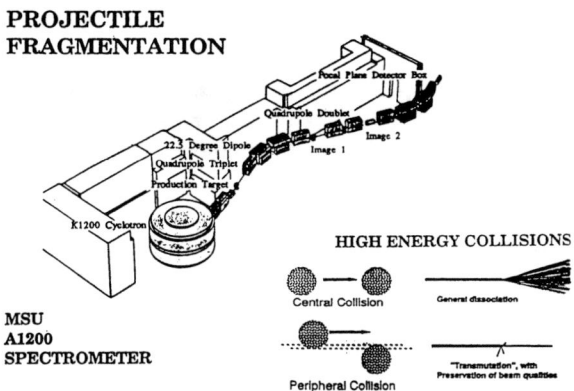

focused at various points in the spectrometer. Measurements of the fields in the magnets combined with the position of an ion at the dispersive focal point marked Image 2, measured with a position-sensitive, parallel plate avalanche counter, allow the determination of magnetic rigidity, $B\rho$. At the final achromatic focus a silicon detector telescope is used to give two energy loss signals and a total energy signal. These measurements, combined with the measured time-of-flight over a distance of 14m, from Image 1 to the first element of the detector telescope, give a unique measure of Z and A for each ion.

The LISE3 spectrometer [5] at GANIL performs exactly the same function. It has been used to observe and identify a large number of both neutron-deficient and neutron-rich nuclei. The final element in the LISE3 spectrometer is a stack of Si detectors. The first two Si detectors give energy loss signals providing redundant information on the atomic number of each of the transmitted fragments. The final Si detector was a strip detector of square shape. Each of the 12 strips is 2 mm wide. This strip detector allows one to record and use the spatial correlation between the implanted ions and the following radioactive decay.

This method has been further developed [6] to allow the detection, identification and study of short-lived, isomeric states in fragmentation products. In essence the standard method of selection and identification of the fragments is used. The stack of Si detectors can be surrounded by high efficiency, Ge detectors and one can search for gamma decays in the period after the implantation of ions. In this method the electronic signals from the implanted ion and the delayed photons are recorded in the same event in the data acquisition. One can thus select in software a gamma spectrum coincident with ions selected by A and Z.

Figure 3a shows the particle identification spectrum recorded at LISE3 [7] in the fragmentation of a ^{92}Mo beam on natural Ni targets with thickness between 50 and 100 mm. The primary beam intensity was typically 100 enA. Each 'blob' in this ΔE-ToF spectrum represents a particular nuclear species. The $T_Z=-\frac{1}{2}$ nuclei ^{77}Y, ^{79}Zr and ^{83}Mo were observed for the first time in this experiment. This despite the fact that the lighter odd - Z, $T_Z = -\frac{1}{2}$ systems ^{69}Br [8] and ^{73}Rb [9] are known to be particle unstable. One possible explanation [7] could lie in the shape polarising effect of the deformed shell gaps in this region.

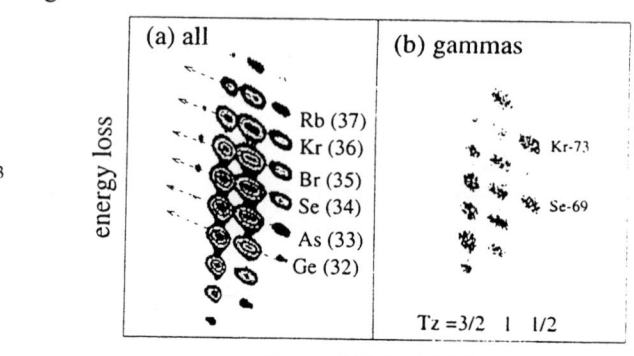

Fig 3

Figure 3b shows the same spectrum in coincidence with gamma rays. It is immediately clear that there are µs isomers in a number of the nuclei. One isomer in ^{74}Kr is of particular interest. Figure 4 shows the time and γ-ray spectra within 150 µs of the ^{74}Kr ion stopping in the Si stack. The measured meanlife is 42(8) ns. The 456 keV, $2^+ - 0^+$ γ-ray is seen clearly, but the $4^+ - 2^+$ transition is not seen. The lifetime of the 2_1^+ state is known to be ~ 25 ps and the flight time through the spectrometer is 480 ns. The only explanation for this anomaly is that the isomeric level has $J^\pi=0^+$ and normally decays mainly by E0 conversion to the ground state. However, if the ion is fully stripped when in passage through the spectrometer then both the E0 transition to the ground state and the internal conversion of the $0_2^+ - 2_1^+$ transition cannot occur. It will then survive until it stops in the Si stack, where the ion is clothed quickly and decays normally.

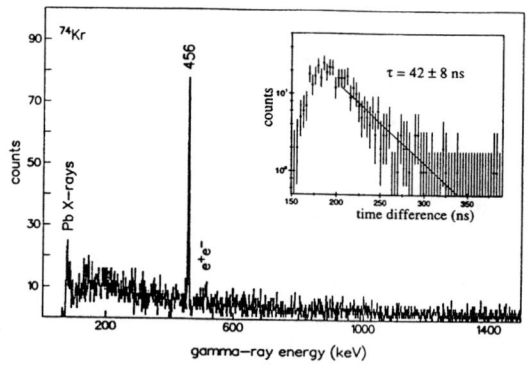

Gamma-ray and time spectra for ^{74}Kr showing the decay from the proposed 0$^+$ isomer.

Fig 4

Model calculations [10] suggest the presence of a 0_2^+ state at ~ 600 keV excitation energy in ^{74}Kr with oblate shape in contrast with a prolate ground state. Analysis of the observed transition rate is consistent with $\beta_2=-0.32$ for this oblate state. It seems probable that this is the first example of a well deformed oblate band close to the N=Z line in this region.

A number of similar experiments [11] had been performed earlier at GANIL to study isomers in both n-rich and n-deficient nuclei.

3. GAMMA RAY SPECTROSCOPY AT SPIRAL

Radioactive beams produced by ISOL facilities [3] will have properties better suited to traditional style gamma-ray studies than beams produced in fragmentation. Although the intensities will be weaker than those of stable beams they will be stronger than from fragmentation facilities. The beams will have superior emittance and the energy will be more readily varied and more precisely defined.

3.1 Coulomb Excitation

In principle Coulomb excitation should be amongst the simplest and most straightforward techniques to use with radioactive beams. The cross-section for Coulomb excitation is generally large, and even with weak beams it should be possible to measure gamma ray transition probabilities in nuclei far from stability. The experiments will be carried out in inverse reactions, that is in projectile excitation.

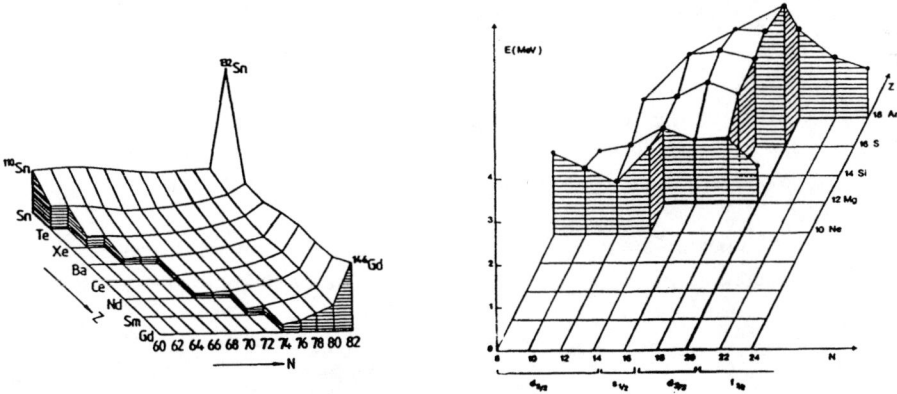

Fig 5: Plot of E(2^+) vs Z and N for two regions of the Periodic Table

At an early stage there are many obvious problems to be tackled. Figure 5 shows energies E(2^+) of the 2_1^+ states in even-even nuclei in two regions of the Periodic Table. In general the B(E2;2_1^+-0_1^+) values for these states have not been measured. In many cases the deformation of the nucleus has been deduced from empirical relationships. Direct measurements of the B(E2) in projectile Coulomb excitation would put all of this on a much sounder footing.

The methods used will depend critically on the lifetime of the state involved. If the lifetime is of the order of ps all of the gamma rays will be emitted in or near the target. In this case one can extract the B(E2) by measuring the total yield of gamma rays from the target. If, however, the lifetime is of the order of ns then the excited projectile nuclei will recoil out of the target and it will be possible to measure the lifetime directly by measuring the number of gamma rays emitted as a function of distance along the path.

The quality of the results will be dictated not only by the low counting rates and potentially high backgrounds but by the purity of the beam. This will be, of course, a key feature of beams from ISOL facilities and will dictate not only the quality but feasibility of some experiments.

3.2 Transfer Reactions

Another classical technique, which will be widely exploited at ISOL facilities, is the study of transfer reactions. Here again the experiments will be carried out in inverse kinematics.

Traditionally studies of transfer reactions have been used to measure excitation energies of states, and from angular distributions of the emitted particles, the angular momentum transfer (ℓ) and hence sometimes J^π. This means that it can provide information on the location and spectroscopic strength of single particle or hole states. Comparison of the results from a range of transfer reactions will provide matrix elements of two-body residual interactions. For example if one compares n- and 2n- pickup one will obtain information about pairing correlations.

With radioactive beams the use of transfer reactions will allow us to test the Shell Model and determine the location of single particle (hole) states far from stability. This information is a vital ingredient for Shell Model calculations. For example studies of the d (^{132}Sn,p)^{133}Sn reaction and of transfer reactions on the nuclei which neighbour the doubly-magic ^{132}Sn, are of vital importance as input to the Shell Model. Again ^{19}C is currently under study as a potential one-neutron halo nucleus. Interpretation of the experiments is made difficult because the spin and parity of the ground state of ^{19}C is not known. If the p(^{19}C,^{18}C)d reaction can be studied it would solve this part of the problem.

Winfield et al [12] have analysed the factors which will affect the study of transfer with radioactive beams. They conclude that it will be difficult to achieve the angular and energy resolutions needed to extract the required structure information. The quality of the results will also depend on whether the light or heavy reaction product is detected or whether they are detected in coincidence. However Rehm [13] has studied the d(^{56}Ni, p)^{55}Ni reaction, with a radioactive beam of ~ 10^4 pps. He reports a resolution of ~ 80keV. In cases where the particle energy resolution is inadequate the detection of gamma rays from transfer reactions will be important in terms of characterising the states and, in some cases, resolving peaks seen in the particle spectra.

Studies of transfer reactions will undoubtedly be one of the tools widely used at an early stage with ISOL-produced radioactive beams.

3.3 Fusion-Evaporation Reactions

As stated earlier gamma rays from fusion-evaporation reactions have been widely used to study how nuclear properties vary with angular momentum. It is not immediately obvious that they will be equally useful with beams of radioactive nuclei.

To this end we set out to test whether one could successfully use the ^{19}Ne beam available at Louvain-La-Neuve to study gamma rays from the ^{19}Ne + ^{40}Ca reaction. The details are given in [14]. A beam of 10^9pps of 70 MeV ^{19}Ne particles was incident on a 4 mg cm^{-2}, self-supporting ^{40}Ca metal target. Charged particles were detected in the LEDA array, an array of 128 silicon strip segments arranged in an octagonal shape and placed in the forward direction. It covered a solid angle of ~ 10%. The gamma rays were detected in seven, Compton-suppressed Ge detectors of TESSA [37] geometry. Each of these detectors was placed at 25 cm from the target in the backward direction and at 1332 keV. The data were recorded in event-by-event mode and the signals included the timing pulse from the CYCLONE accelerator. One awkward feature of the experiment was the fact that the properties of the CYCLONE beam meant that ~ 5% of the beam was stopped on the final beam collimators, 150 mm before the target. A 20 mm thickness of Pb shielding was fitted around each of the BGO suppression shields and around the beam collimators. The beam stop was 2m beyond the target and well shielded from the gamma ray detectors with lead bricks.

The singles and coincidence spectra recorded in the experiment were dominated, as feared, by the radioactive background. However the elimination of non-reaction counts from the γ-ray spectra was achieved using the measurements of γ-ray

Fig 6: γ-rays from a) ^{19}Ne + ^{40}Ca, b)^{19}F + ^{40}Ca

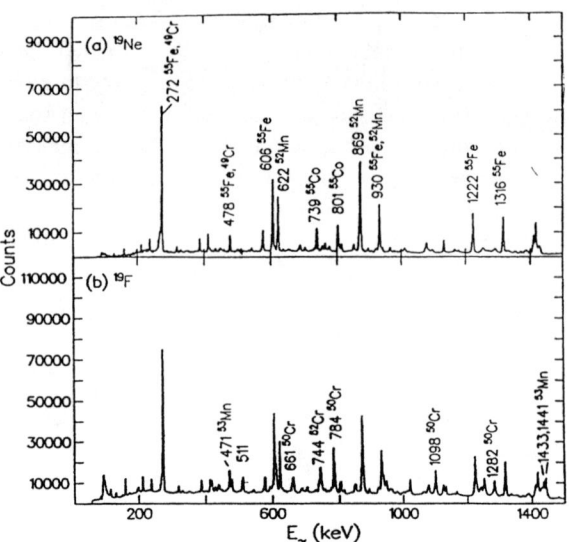

timing relative to the beam pulses. The spectra obtained by subtracting the spectrum recorded between beam pulses from that recorded in the beam pulses is shown in the upper half of fig 6 and compared with the spectrum of γ-rays from the ^{19}F + ^{40}Ca reaction.

The main conclusion is that active Compton suppression with BGO shields gave data of comparable quality to that of stable beams. To achieve this software gating on the beam pulse with background subtraction was essential. This was also essential to obtain clean coincidence spectra. The most important source of background was beam radioactivity which had been stopped on the collimators. The conclusion here must be that good emittance is an essential requirement for the post-accelerator of a radioactive beam facility.

3.4 N=Z nuclei

Having established that gamma rays from fusion-evaporation reactions induced by energetic radioactive ions can be studied, the question is what does it allow us to study? The N=Z nuclei will provide an example.

Figure 7 shows a schematic version of the Segré chart. The banana shaped curve represents the stable nuclei. The drip-lines are shown as smooth lines. The N=Z line is shown. It diverges from the line of stability above ^{40}Ca and it crosses the proton drip-line just above ^{100}Sn.

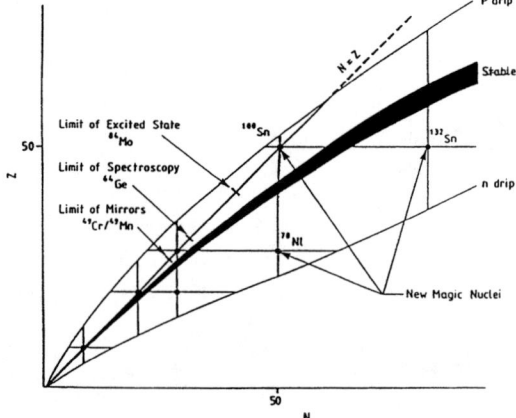

Fig 7: Where have we reached in studying N=Z nuclei?

The N=Z nuclei are of particular interest for a number of reasons. Firstly up to ^{100}Sn neutrons and protons fill the same shells. As a result effects associated with a particular value of Z or N are reinforced. Hence deformation depends strongly on Z, N and J. It turns out that we expect the co-existence of states with prolate, oblate and spherical shapes at the same excitation energy. It is obvious that N=Z nuclei will also reveal symmetries and their breakdown since nuclear forces are known to be charge independent. As A increases we expect to see the growing effects of isospin mixing. It can be seen in beta decay or in the strength of isospin-forbidden E1 transitions in even-even nuclei. Finally the astrophysical rp-process passes close to the N=Z nuclei so that information on nuclei near the N=Z line is important for the understanding of this process.

Figure 7 attempts to summarise where we have reached in studying N=Z nuclei [15]. Most of the experiments have been undertaken with the prompt gamma rays tagged by A and Z using a recoil mass separator. In general the 2n reaction channel was used. With beam energies close to the Coulomb barrier this meant that the nucleus of interest had the highest Z, which eased Z identification. It also meant that the number of channels open was minimised. However the cross-section of interest is small relative to the total cross-section. For example the heaviest, even-even nucleus from which a prompt gamma ray has been seen is ^{84}Mo. Here the measured cross-section is 7(3) µb out of a total of the order of 1b. The expectation is that the use of radioactive ions to create such nuclei will allow one to use reaction channels with much larger absolute cross-sections, which also take a much larger fraction of the total reaction cross-section. In the case of ^{64}Ge the ^{30}S + ^{40}Ca reaction is expected to have a cross-section some 50-60 times larger than the measured cross-section in the earlier studies. Given a gamma ray array with ε_{ph} ~ 10% and a recoil separator such as the FMA at Argonne one will obtain data with comparable statistics to the earlier stable beam study in the same running time, with a beam of ~10^6 þþs. At the same time the grazing angular momentum will be a factor of two larger and the γ-ray multiplicity will be higher which will enhance the sensitivity of the experiment.

Figure 7 also shows the limit where we have been able to study mirror nuclei at high spin. Here ^{49}Cr/^{49}Mn is the heaviest pair studied to date. Similarly ^{64}Ge is the heaviest, even-even nucleus where we have been able to do 'real' spectroscopy.

The even-even nuclei are also expected to be the place where we might expose the effect of n-p pairing. The level scheme [19] for the odd-odd nucleus ^{74}Rb resembles the 0-2-4 sequence in the T_z=1 isobar ^{74}Kr. The ground state of ^{74}Rb is thought to be T=1, J=0 from its observed fast Fermi decay to ^{74}Kr. The resemblance of the level structure suggests that the lowest levels in ^{74}Rb are rotational levels below the pairing gap built on the ground state and are hence due to n-p pairing. This interpretation is consistent with a cranking model calculation with a residual pn-interaction. In this interpretation it is assumed that the observed gamma ray cascade is built on the 0^+ state observed to beta decay to ^{74}Kr. In odd-odd nuclei the odd-particles in the ground state often couple to form two states of quite different spin. If the two states lie close in energy then we might expect the fusion-evaporation reaction to populate the higher spin state preferentially. So far experiments to search for such an isomeric state have not been carried out.

3.5 K Isomers

Isomers, nuclear states with significantly longer lifetimes than their neighbours, can occur in a variety of ways. Amongst them are the so-called K isomers, where a particular state, which carries a large amount of angular momentum in non-collective motion, lies lower in energy than collective states of lower seniority. The decay of such isomers requires a transition to a state of very different shape.

Such states are only accessible in the region of Z ~ 74, N ~ 104 where the Fermi surface lies high in the shell and the $i_{13/2}$ neutrons and $h_{9/2}$ protons lie near the Fermi surface. They have been intensively studied at ANU, Canberra and ref 16 provides a good example. Figure 8 shows part of the level structure of ^{178}W. Here we see a series of K isomers with rotational bands built on them. The states at 3237, 5316 and 6576 keV are interpreted as 4, 6 and 8 qp states. From the rotational bands one can derive a value for the moment-of-inertia. With increasing seniority the moment-of-inertia increases but remains well below the value for a rigid body.

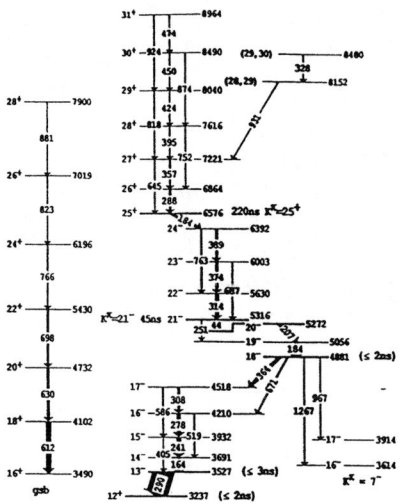

Fig 8: Partial level scheme for ^{178}W

These isomers provide an excellent way of studying the effects of rotation since one can effectively "decouple" one pair of particles at a time. Unfortunately many of the predicted K isomers lie to the right of stability and can only be reached in fusion reactions induced by radioactive ions. Sletten [17] has made an estimate of the beam current needed to populate K isomers in neutron-high Hf, Ta and W isotopes in the ^{48}Ti + ^{142}Xe reaction by comparison with the beams of ^{136}Xe used by Pedersen et al [18] to study the K isomers in ^{182}Os. In essence, with a beam of 6×10^6 pps one can obtain publishable spectra in about 15 hours running.

4 SUMMARY

Gamma ray spectroscopy will continue to be one of our main tools for the study of nuclear structure with radioactive as well as stable ions. It is already in use with the beams produced in fragmentation but is likely to be more important when intense beams of radioactive ions with energies around the Coulomb barrier become available. A wide range of reactions will be studied including Coulomb excitation, transfer reactions and fusion-evaporation. Studies of beta decay, which have not been mentioned here, will acquire greater significance because of the rapid increase in Q-value as we move away from stability. Gamma spectroscopy will play a significant role. The instrumentation needed for such experiments will differ from the arrays aimed at fusion reactions with stable beams where there is a very high γ-ray multiplicity. New arrays must have high efficiency, since the beams will be weak, and a high degree of segmentation in order to deal with Doppler corrections. This is the basis of the EXOGAM and MINIBALL arrays planned for SPIRAL and REX-ISOLDE facilities. At the same time attention must be paid to the purity of radioactive beams, to the emittance of the beams and to experimental design of collimators, shielding and beam stops.

References:

1. A.Richter, Nucl. Phys. A553, 417c (1993)
2. B.M.Sherrill, Proc. 2nd Int. Conf. on Radioactive Nuclear Beams, Louvain-La-Neuve, Belgium, p3 (1991)
3. H.Ravn, ibid. P85
4. M.Bernas et al., Phys. Letters, B331, 19 (1994)
5. A.C.Mueller and R.Anne, N.I.M. B56/57, 559 (1991)
6. R.Grzywacz et al., Phys Letters 355B, 439 (1995)
7. P.H.Regan et al., Acta Physica Polonica B28, 431 (1997)
8. B.Blank et al., Phys. Rev. Letters 74, 4611 (1991)
9. J.M.D'Auria et al., Phys. Letters B66, 233 (1977)
10. A.Petrovici et al., Nucl. Phys. A605, 290 (1996)
    W.Nazarewicz et al., Nucl. Phys. A435, 397 (1985)
11. K Rykaczewski et al.,Phys. Rev. C52, R2310 (1995)
12. J.Winfield, N.Orr and W.N.Catford, NIMB, in press
13. E.Rehm, private communication
14. W.N.Catford et al., NIM A371, 449 (1996)
15. W.Gelletly, Acta Physica Polonica B26, 323 (1995)
16. C.S.Purry et al., Phys Rev. Letters 75 406 (1995)
17. G.Sletten, private communication
18. G.Pedersen et al., Z. Physik A321, 567 (1985)

Study of Deformations in Light Neutron-Rich Nuclei

D. Guillemaud-Mueller

Institut de Physique Nucléaire, CNRS-IN2P3, 91406 Orsay Cedex, France

Abstract. The influence of shell effects far from stability is one of the most fascinating challenge of the study of exotic nuclei. For the neutron magic number N=20, a new region of deformation was found in the very neutron-rich Na and Mg isotopes. The study of this region has been extended through the search of the stability of the doubly-magic nucleus ^{28}O. Evidence for the unbound character of ^{28}O was obtained. In the same experiment, the half-lives of the very neutron-rich ^{27,29}F and ^{30}Ne were measured and those for ^{28,29}Ne and ^{30,31}Na reexamined. The results are compared to shell-model predictions and conclusions drawn regarding the extent of the region of deformation around N=20. The study of half-lives of P to Ar isotopes led to the evidence of a new region of deformation around N=28. The closure of the N=14 and N=40 subshells was also studied through Coulomb excitation experiments.

AROUND N=20

Search for the stability of the doubly-magic isotope ^{28}O

The study of the extremely neutron-rich isotopes of light elements is of considerable interest both for locating the neutron drip-line and for testing models describing the properties of exotic nuclei. One of the up-to-date questions is the problem of the stability of the neutron-rich isotopes of oxygen.

The ground state properties of the neutron-rich oxygen isotopes were studied recently in several theoretical papers. In particular, a large basis shell model (1), a Skyrme-Hartree-Fock (SHF) approach (2), and a relativistic mean-field (RMF) theory (3) have been used to calculate the properties, including a possible neutron halo, of ^{28}O. Quadrupole excitation in ^{28}O was also examined within the framework of the random phase approximation (4,5,6). The beta-decay of ^{28}O was studied theoretically by Poves et al. (7) and found to exhibit beta-delayed deuteron emission characteristics of a neutron halo. Most of the mass models predict a bound ^{26}O while the doubly-magic ^{28}O is predicted unbound against two-neutron emission (8). Some models, however (Möller and Nix (8), Möller et al. (8), Liran and Zeldes (9)) suggest a positive S_{2n} for ^{28}O. Recent shell-model calculations dedicated to this region of nuclei predict both ^{26}O and ^{28}O to be unstable (10). On the other hand, Poves et al. (7) find an $S_{2n}(^{28}O)$ of 1.3 MeV. Similar positive values were found in the SHF (2) and RMF (3) calculations. From a theoretical point of view it is clear that the question of the particle stability of ^{28}O is still open. Perhaps the most interesting aspect of the N=20 region is the transition from spherical to deformed shapes, resulting in the so-called island of inversion (11-15). A large B(E2; $0^+ \to 2^+$) recently measured by Motobayashi et al. for ^{32}Mg (16) confirmed the deformation as suggested by earlier decay studies (17). This phenomenon has been explained within the framework of the shell-model by inclusion of fp-shell intruder states (18,19).

The nucleus ^{24}O is the heaviest experimentally known oxygen isotope. As it has been shown experimentally at GANIL with the LISE spectrometer, the heaviest known oxygen isotope is ^{24}O while ^{26}O has been found unbound using the reaction 44 MeV/u ^{48}Ca +Ta (20). This unstability of ^{26}O has been confirmed in an experiment at MSU (10) using a ^{40}Ar beam at 90 MeV/u. Interestingly, however, a new isotope in the vicinity of ^{28}O, namely ^{31}Ne, was observed in a recent experiment with a 50Ti beam at RIKEN (21) and confirmed in an experiment made at GANIL with a ^{36}S beam.

Recently, we have made an experiment whose aim was to search for the existence of a bound ^{28}O nucleus and study the beta-decay of nuclei in the N=20 region. The fragmentation of an intense beam of ^{36}S - a neutron-rich isotope - was chosen as the tool to produce these extremely neutron-rich nuclei. The production yield of the neutron-rich isotopes with N=20 near the drip-line has been increased using the high energy and intensity beam of the rare isotope ^{36}S and a careful optimisation of the production target. As a result the production yield of ^{29}F, the lightest known N=20 nucleus, was increased with respect to our previous experiment (20) by almost two orders of magnitude. Another advantage was obtained by the use of an increased magnetic rigidity (up to 4.2 Tm) available after the recent upgrade of the first dipole of the LISE3 spectrometer (22). The Si identification and implantation telescope was surrounded by ^{3}He neutron counters and a 70% HPGe detector for the measurement of beta-n and beta-gamma coincidences and to search for microsecond isomeric states. The beta-decay time spectra were obtained on an event-by-event basis using the time difference measured between the ion of interest and the first decay electron subsequently detected in the same detector. The unambiguous identification was helped by the absence of the unstable nuclei ^{21}C, ^{25}O and ^{28}F; the A/Q=3 series of nuclei is clearly visible for ^{24}O, ^{27}F and ^{30}Ne (Fig.1). The identification was also confirmed using the observation of the characteristic gamma-lines of known isomeric states in ^{16}N, ^{26}Na, and ^{32}Al. Two other settings of the spectrometer were used in the experiment which were optimized for the production and half-life measurements of ^{27}F and ^{30}Ne, ^{31}Na. The results of a 53 hour measurement with an average beam intensity of 800 enA are shown in Fig. 1. A dashed line is drawn through nuclei with N=20. The heaviest known isotope of fluorine, ^{29}F, is clearly visible, for which 519 events were detected. In contrast, no events corresponding to ^{26}O and ^{28}O were observed.

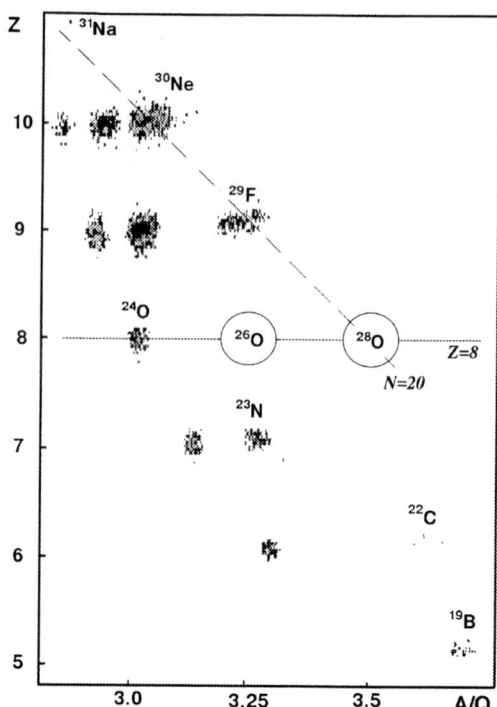

FIGURE 1. Two-dimensional identification plot A/Q versus Z. No counts were observed corresponding to ^{28}O after 53 hours of irradiation of the Ta target with a 78 AMeV ^{36}S beam at 800 enA The ^{29}F isotope is clearly visible but no counts of ^{26}O and ^{28}O are observed (see text).

Fig. 2 shows the experimentally measured yields for the N=20 isotones. The solid line represents calculations performed using the LISE program (23).This code accounts for the transmission through the spectrometer and uses a modified EPAX parametrization of Sümmerer et al. (24) for the production cross sections. The beam "memory effect" coefficients (eq.13 in (24)) were modified in order to better describe yields for the extremely neutron-rich nuclei observed in the fragmentation of a neutron-rich beam. This was done in a manner similar to that of Ref. (25). A more detailed analysis of the parametrization modification will be published elsewhere (26). According to this estimation one would expect about 11 events in 53 hours corresponding to ^{28}O. The symbol with the arrow indicates the counting rate corresponding to the observation of one event. The results of the present experiment indicate that ^{28}O is most probably particle unstable and confirm the earlier results concerning the instability of ^{26}O. Upper limits for the cross sections for the formation of the oxygen isotopes are estimated to be 0.7 pb and 0.2 pb for ^{26}O and ^{28}O, respectively.

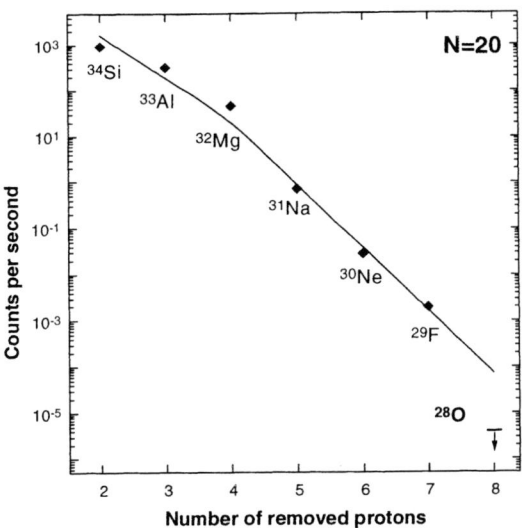

FIGURE 2. Yields of the N=20 nuclei measured in the present experiment. The solid line represents yields calculated with the modified EPAX parametrization (see text). The point with arrow for ^{28}O corresponds to the upper limit of one event.

Half-lives measurements around N=20

The N=20 shell closure was also studied through β decay half-life and β delayed neutron emission measurements. The beta-decay time spectra for ^{27,29}F and ^{30}Ne, shown (Fig. 3) provide the first measurement of their half-lives ($T_{1/2}$). Additionally, ^{28,29}Ne and ^{30,31}Na have been remeasured. The experimental $T_{1/2}$ values are summarized in Table 1. The measured half-lives for ^{28}Ne and ^{30,31}Na agree within the uncertainties with previous experiments (27, 11). The only important discrepancy is observed for ^{29}Ne; 15±3ms in this work and 200±100ms reported by Tengblad et al. (27). The ISOLDE experiment suffered from a high contamination of ^{87}Kr^{++} ions in the spectrum of the beta-decay of mass 29 which prevented an unambiguous half-life measurement for ^{29}Ne.

TABLE 1. Experimental and shell-model half-lives in milliseconds and predicted (present work) ground state spin and parity for nuclei in vicinity of N=20

Nucleus	Half-life (ms)	Other Experiments	Shell Model (28)	Shell Model (this work)	J^π
^{27}F	5.3 ± 0.9	-	7.8	4.7	$5/2^+$
^{29}F	2.4±0.8	-	2.7	1.4	$5/2^+$
^{28}Ne	21±5	17± 4 [27]	16.9	12.9	0^+
^{29}Ne	15±3	200±100 [27]	7.4	28.2	$7/2^-$
^{30}Ne	7±2	-	3.7	17.8	0^+
^{30}Na	48±4	50±3 [11]	24.7	56	2^+
^{31}Na	18±2	17±0.4 [11]	11.8	16.4	$3/2^+$
^{31}Mg	-	230±20 [17]	27	308	$7/2^-$
^{32}Mg	-	120±20 [17]	11	195	0^+

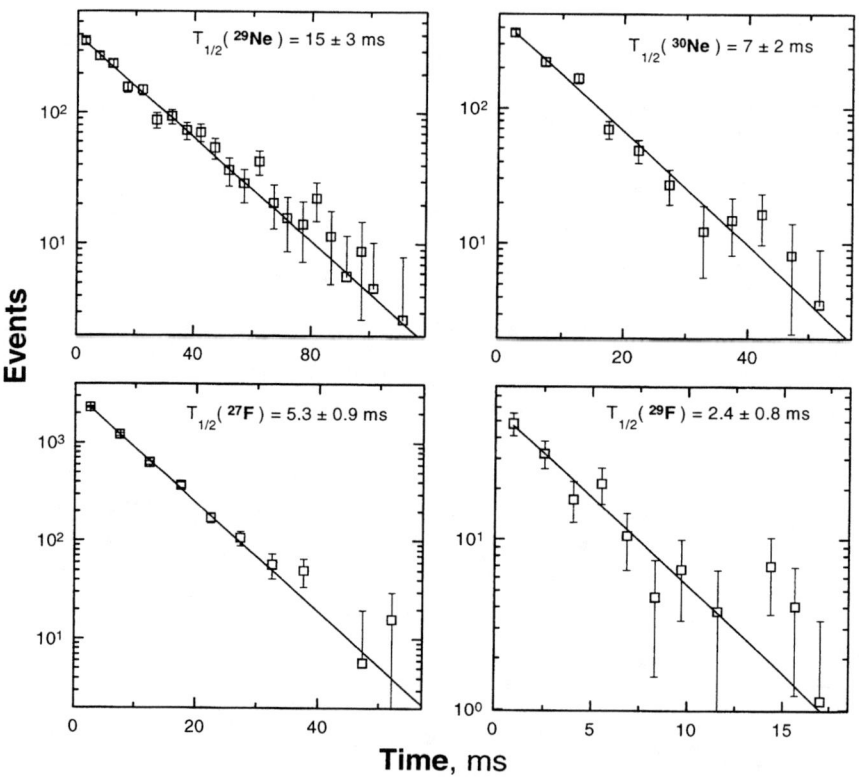

FIGURE 3. The beta-decay time spectra and corresponding half-lives measured for ^{27,29}F and ^{29,30}Ne.

The present work used an event-by-event correlation between the identified implanted ion and the subsequent coincident beta-decay, thus providing a first reliable value for the half-life of ^{29}Ne. To compare the experimental results with theoretical predictions, we have performed shell-model calculation for several isotopes in the N=20 region with the Strasbourg-Madrid shell-model code ANTOINE. Instead of including explicitly $n\bar{h}\omega$ mixing we determined the lowest state for fixed 0-particle 0-hole (0p0h), 1p1h and 2p2h excitations in the full sdfp space. The interaction of Retamosa et al. (29) which gives very satisfying results for $0\bar{h}\omega$ states around N=28 was used. In particular, use of this interaction correctly reproduces the deformation occurring around ^{40}S and ^{42}S (30). Gamow-Teller decays were computed from the lowest state predicted by the calculations. The results and comparison with experimental values and the calculations of Wildenthal et al. (28) are given in Table 1. A relatively good overall agreement exists between our calculations and the experimental values. The results for ^{31}Mg and ^{32}Mg have been added to demonstrate the reliability of the calculations. The values for the spherical nuclei ^{27,29}F, ^{28}Ne and ^{30}Na are slightly different from those of Wildenthal et al. (28). This is essentially because of the values of the masses used, which were extracted from the recent compilation of Audi and Wapstra (31). Importantly, in the calculations of ref. 28 the ground states (g.s.) are pure sd spherical states, while our deformed g.s. are sometimes 2 or 3 MeV lower than the spherical ones. For example, in the case of ^{31}Na, the half-life in the sd calculations was computed for the transition from the $3/2^+$ state instead of the deformed $5/2^+$ (g.s.) which is 1.5 MeV more bound. The calculation performed here suggests that the region of inversion where the fp-shell intruder configurations dominate the g.s. starts at Z=10 and N=19. Consequently the oxygen isotopes may be described by pure sd configurations in agreement with the previous studies on the the subject (10) predicting the unstability of ^{28}O. This conclusion disagrees with the results of Poves et al. (19), where the intruder limits were predicted at Z=9 and N=19. The disagreement may be explained by the larger model space and improved interaction used in the present calculations.

A more detailed analysis of the experimental results including the measured probabilities for the emission of delayed neutrons and observed beta-gamma decays are in progress.

In addition, an unexpected isomeric state was observed in the N=19 nucleus ^{32m}Al (32) through the detection of a γ radiation in a slow (up to 100 μs) correlation with the identified implanted nucleus. Two γ lines at energies of 221.6 ± 0.3 and 734.1 ± 0.3 keV have been observed in cascade. A half-life for ^{32m}Al of 200 ± 20 ns was measured and an excitation energy of 956 keV deduced from the observed cascade. These observations have been compared to various model predictions. Shell model calculations provide a candidate for the isomer having $J^\pi=4^+$. However, this analysis reveals a rare inversion of low-energy levels by Wildenthal's "universal" sd-shell USD interaction (12), extending to the N=19, Z=13 range of USD anomalies (previously identified for N=20, Z=11,12) probably linked with the vanishing of the N=20 shell gap far from stability

AROUND N=28

Determinations of half-lives and of β-delayed neutron emission probabilities have been performed from ^{40}P to ^{47}Ar produced using the fragmentation of a ^{48}Ca beam at about 60MeV/u in order to clarify the situation concerning the overabundance of ^{48}Ca as compared to ^{46}Ca in the solar system. Their decay was studied by a β-n time correlation measurement using a 4π ^{3}He-proportional counters imbedded in a polyethylene moderator matrix surrounding the implantation detector. The major results (33) arise from the systematically shorter values of nuclei with a neutron number N = 28 (0.12 s compared to 0.31 s for ^{44}S, and 0.4 compared to 1.7 s for ^{45}Cl). This vanishing of the N=28 shell strength changes the predicted nucleosynthesis path (34). With these experimental half-lives and P_n-values, nuclear deformations can be derived by QRPA parameter studies (see for instance figure 4 for the case of ^{44}S (35), where a prolate deformation of at least e_2=0.30 has to be invoked to fit the experimental values of $T_{1/2}$ and P_n). Taking into account these deformations, QP-levels below and close to the neutron separation energy can be determined. These measurements around N=28 reveal as well as for N=20 the vanishing of the shell closure effect for very exotic nuclei. Recent theoretical calculations based on completely different approaches (shell model (29), HFB with the Gogny force (36), relativistic mean field theory (37)) seem to reproduce this result.

AROUND THE N=14 AND N=40 SUBSHELLS

An alternative method to study the structure of exotic nuclei near shell or subshell closures is Coulomb excitation of radioactive beams. Low intensity radioactive beams as low as few tens per seconds offer the possibility of studying collective modes of even-even nuclei through the energy of the 2^+ state and its excitation probability B(E2). This represents the first generation experiments to obtain relevant information on the strength of shell gaps when large neutron or proton excess is reached and to avaluate core-vibration or rotation coupling with weakly bound neutrons and their rôle in halo systems.

The closure of the subshell $1d_{5/2}$ at N=14 leads to spherical or strongly oblate energy gap depending on the associated proton number. This gives either spherical or deformed shapes for isotones. For example, through HFB calculations (36), oxygen or carbon N=14 isotones exhibit different shapes. ^{22}O is predicted to be spherical with a steep potential-energy surface whereas ^{20}C is predicted strongly oblate deformed. Such an oblate shape should be associated with a large collectivity and with a high 2^+ energy, the ratio of the 4⁺ and 2⁺ energies being that of a typical rotor as in ^{28}Si. On fig. 4 is reported the experimental 2⁺ energies in this mass region (38). One sees an increase for Mg and Ne isotopes. Nothing is known concerning the lighter isotones particularly when a larger excess of neutrons is encountered. This was the aim of the study of the isotopes of oxygen to check whether the expected magicity would remain. These nuclei cannot be easily deformed and will not minimize their total energy by adopting strong deformations as done by the carbon isotopes. They will rapidly become unbound when increasing the neutron number further from stability. This seems to be in agreement with the unstability of ^{26}O and ^{28}O observed at GANIL.

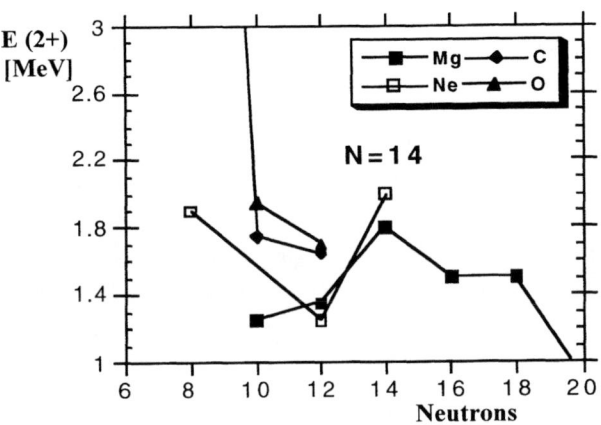

FIGURE 4. The experimental 2⁺ energies in the mass region of C to Mg

Concerning the Z=40 subshell, its closure is clearly established by the well known excitation energy of the first 2⁺ state at 2.19 MeV and the B(E2)value of ^{90}Zr$_{50}$ (38). In ^{68}Ni$_{40}$, only the the excitation energy of the first 2⁺ state at 2.03 MeV is known (38). These two nuclei are "valence mirror" nuclei having both a 5⁻ isomeric state corresponding respectively to neutron and proton particle-hole (p1/2⊗g9/2) excitation. In addition, they have both a short-lived 0$_2^+$ isomeric state lying at 1.76 in ^{90}Zr and at 1.77 MeV in ^{68}Ni. The structure of these states are not really understood and self-consistent Hartree-Fock calculations using effective interactions (Gogny or Skyrme type) predict that a shape isomeric state with a large quadrupole deformation of β_2=0.5 is expected at low energy (39, 40).

Two experiments of Coulomb excitation have been recently performed at GANIL using the LISE3 spectrometer to study the behaviour of the N=14 and of the N=40 subshells (41). Secondary beams of ^{22}O and ^{26}Ne with a ^{36}S primary beam and ^{76}Ge, ^{72}Zn and ^{68}Ni with a ^{86}Kr primary beam have been produced. The secondary beams are identified in two large-area silicon detectors mounted at around 50 cm from the reaction target. The deflection angle of the fragments can be determined by positive sensitive gas detectors located before and after the interaction target. The scattered fragments can be detected up to an angle of 3° in the laboratory frame where Coulomb contribution dominates the total cross-section. The target was surrounded by the 70 BaF2 detectors of the "Château de Cristal" in the 4π geometry of the first-generation TAPS detectors to detect in flight the gammas of deexcitation. The Doppler corrected energy spectra obtained for ^{76}Ge, ^{72}Zn are shown on fig. 5. For ^{72}Zn, the deduced 2^+ energy of 650 ± 30 keV is in good agreement with the known value 652.5 keV. The 2^+ energy for ^{76}Ge (563 ± 45 keV) is obtained. ^{76}Ge is a well-known case of deformed nucleus exhibiting a low-lying 2^+ state (562.9 keV) and a very strong excitation probability $B(E2)\uparrow=2680$ e^2fm^4 which serves at a test case for the determination of the value of excitation energy. The deduced value for ^{72}Zn is 1700 ± 400 e^2fm^4. In the case of ^{68}Ni, the energy of the 2^+ state at 2.03 MeV is not observed but only a peak at around 500 keV is visible. This peak corresponds clearly to a radiation emitted in flight and may possibly correspond to the $2_2^+ \rightarrow 0_2^+$ state of a super deformed nucleus predicted by the theory. A deduced upper limit for ^{68}Ni of the $B(E2) < 680$ e^2fm^4 is obtained. From the completed systematics of B(E2), it seems that for Ge isotopes N=40 correspond to the middle of a shell with a maximum collectivity; for Zn this behaviour is less pronounced but for Ni the behaviour is opposite and B(E2) seem to be minimum corresponding to a shell effect with a large energy gap. These results are still under analysis and need further confirmation.

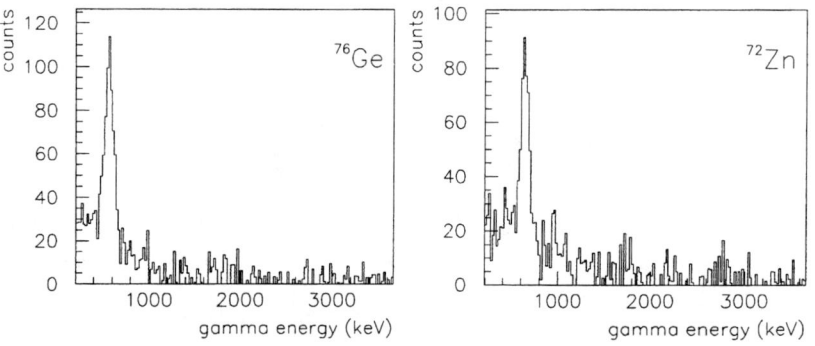

FIGURE 5. Doppler corrected, background substracted γ-ray spectra of ^{76}Ge and ^{72}Zn

In summary, β decay studies and Coulomb excitation experiments using 4π high-efficiency γ-ray devices offer complementary, unique and promising perspectives to study nuclear structure far from stability and of isomeric states.

REFERENCES

1. Otsuka, T. et al., *Phys. Rep.* **264**, 297 (1966).
2. Shen, Y.-S. et al., *Z. Phys.* **A356**, 133 (1996).
3. Ren, Z. et al., *Phys. Rev.* **C52**, R20 (1995).
4. Yokoyama, M. et al., *Phys. Rev.* **C52**, 1122 (1995) .
5. Hamamoto, I. et al., *Phys. Rev.* **C54**, 2369 (1996).
6. Ghielmetti, F. et al., *Phys. Rev.* **C54**, R2143 (1996).
7. Poves, A. et al., *Z. Phys.* **A347**, 227 (1994).
8. Haustein, P.E. (special editor), *ADNT* **39**,185 (1988) and references therein.
9. Liran, S. et al., *ANDT* **17**, 431 (1976).
10. Fauerbach, M. et al., *Phys. Rev.* **C53**, 647 (1996) and references therein.
11. Guillemaud-Mueller, D. et al., *Nucl. Phys.* **A426**, 37 (1984) and references therein.
12. Warburton, E.K. et al., *Phys. Rev.* **C41** 1147 (1990).
13. Lyutostansky,Yu. et al., *Proc. of the 5th Int. Conf. on Nucl. Far from Stab.*, Rosseau-Lake, Ontario, Canada, 1987, Conf. Proc. 164, edited by Ian S. Towner (AIP, New York, 1987), pp.727.
14. Orr, N. et al., *Phys. Lett.* **B258**, 29 (1991).
15. Zhou, X.G. et al., *Phys. Lett.* **B 260**, 285 (1991).
16. Motobayashi, T. et al., *Phys. Lett.* **B346**, 9 (1995).
17. Langevin, M. et al., *Nucl. Phys.* **A414**, 151 (1984) and references therein.
18. Klotz, G. et al., *Phys. Rev.* **C47**, 2502 (1993).
19. Poves, A. et al., *Nucl. Phys.* **A571**, 221 (1994).
20. Guillemaud-Mueller, D. et al., *Phys. Rev.* **C41**, 937 (1990).
21. Sakurai, H. et al., *Phys. Rev.* **C54**, R2802 (1996).
22. Anne, R. et al., *NIM.* **A257**, 215 (1987).
23. Bazin, D. et al., *Phys. Rev.* **E50**, 4017 (1994), and computer code LISE, to be published and Web site: http://www.nscl.msu.edu/~bazin/LISE.html.
24. Sümmerer, K. et al., *Phys. Rev.* **C42** ZZZ (1990).
25. Pfaff, R. et al., *Phys. Rev.* **C53**, 1753 (1996).
26. Tarasov, O., PhD thesis, Dubna, 1997.
27. Tengblad, O. et al., *Z.Phys.* **A342**, 303 (1992).
28. Wildenthal, B.H. et al., *Phys. Rev.* **C28**, 1343 (1983).
29. Retamosa, J. et al., *Phys. Rev.* **C55**, 1266 (1997).
30. Scheit, H. et al., *Phys. Rev. Lett.* **77**, 3967 (1996).
31. Audi, G. et al., *Nucl. Phys.* **A595**, 409 (1995).
32. Robinson, M. et al., *Phys. Rev.* **C53**, R1465 (1996).
33. Sorlin, O. et al., *Phys. Rev.* **C47**, 2941 (1993).
34. Kratz, K.-L. et al., in *Proceedings of the 1st European Biennal Workshop on Nuclear Physics*, ed. D.Guinet and J.P. Pizzi (World Scientific, Singapore, 1991).
35. Böhmer, W., Diploma Thesis Univ. Mainz (1996), and to be published.
36. Girod, M., private communication.
37. Werner, T.R. et al., *Nucl.Phys.***A597**, 327 (1996).
38. Table of isotopes, eight edition, Firestone, R.B. editor, John Wiley and Sons.
39. Girod, M. et al., *Phys. Rev.* **C37**, 2600 (1988).
40. Bonche, P. et al., *Nucl. Phys.* **A500**, 308 (1989).
41. Sorlin, O. et al., in *Proceeding of the XXXV International Winter Meeting on Nuclear Physics* Bormio, Italy, I. Iori ed., Ricerca Scientifica ed Educacione Permanente Supplemento no 110, University of Milan (1997), pp. 566.

The September 1997 status of SPIRAL project and future developments

Nicolas Alamanos

CEA/DSM/DAPNIA/SPhN Saclay, 91191 Gif-sur-Yvette Cedex, France

Abstract. In this report I will give some information concerning the construction of SPIRAL. I will enumerate the new equipments suited for experiments with the low energy beams provided by SPIRAL, and finally I will give indications concerning the phase II of the project.

PROGRESS REPORT CONCERNING THE CONSTRUCTION OF SPIRAL

The status of the SPIRAL project at GANIL is based on the ISOL technique for production of Radioactive Ion Beams (RIB's). The construction of SPIRAL can be devised into three major operations, see ref. (1),(2).

i) Increase of the intensity of the primary beam up to 6kW for ions up to ^{40}Ar of 96MeV/u.
ii) Construction of the RIBs production system (target plus ionization source)
iii) Construction of a new cyclotron.
In the following, I will give some information concerning the present status of these operations.

i) *Increase of the intensity of the primary beam or THI (transfert de haute intensité) project.*

The aim of this project is to increase the beam intensity of the accelerator by a factor 15 at least for the light ions. For this it is essential to improve the transport of the high intensity beam through the second cyclotron up to the SPIRAL target, (verification of the majority of power supplies of the accelerator, reduction of the phase width at the entrance of the cyclotron CSS2 using a new rebuncher, fast shut down of the beam if an abnormal intensity loss is detected,..). A test was performed recently to test the equipment of the THI project using a 2kW, 95MeV/u ^{36}Ar beam, (3). The results of this test are rather positive. The beam intensity (pps) in the source and the injector cyclotron is in agreement with the expectations. The transmission of the first cyclotron was of the order of 93.3%. The transmission of the second cyclotron CSS2 was 80% due to a failure of two magnetic extraction elements. Beam losses have been detected in the beam line L1 (between the injector and the first cyclotron) with 60% transmission. Beam losses have been detected also in the beam line L2 (between the first and second cyclotron) due essentially to the rotating stripper. The new rebuncher was also tested. The results are satisfactory although its effect on the transmission of the beam through the CSS2 could not be observed due to the mentioned problems. The beam loss detection system was also tested during this run, it seems to be extremely sensitive to unavoidable sparks generated by DC or RF voltages and to have a very short response time (less than 1 ms). Most of the problems encounter during this test run can be easily fixed. It is expected that new test runs, in the forthcoming months, would successfully validate the THI operation.

ii) Construction of the RIB's production system (target plus ionization source)

The primary beam accelerated by the GANIL cyclotrons will impinge on a production target, which will be at high temperature ~2300K. The radioactive species produced by nuclear reactions and released from the target will pass through a transfer tube into the ionization source. The radioactive atoms will be ionized to q/m ratio ranging from 0.09 to 0.40. After extraction from the source they will be injected into the new (K=265) compact cyclotron CIME (cyclotron de moyenne énergie). The production system will be located inside a well shielded cave.

The equipment of the source cave is now completed. The technology of the carbon target is under control. The target was tested successfully at Louvain-la-Neuve, in Belgium using a 6kW proton beam of 30 MeV. The characteristics of the Bragg peak (energy losses inside the target) for this beam are very similar to those of a ^{20}Ne beam at 100 MeV. The target withstood this power for long periods of time (two consecutive weekends). The first NANOGAN2 source was installed. Finally the remote-controlled pick-up and place-in robot will be installed very soon.

iii) Construction of the new cyclotron.

The CIME cyclotron has been installed at GANIL. The magnetic field mapping was completed and appeared to be in agreement with the theoretical predictions.

Acceleration of stable beams is expected in the forthcoming months, hopefully in November. The first test beam is ^{18}O at 10.6 MeV/u. This corresponds to the acceleration of 4+ charge state using the 3rd harmonics of 14.4 MHz RF.

NEW EQUIPMENTS FOR THE EXPERIMENTS WITH THE SPIRAL BEAMS

The SPIRAL international scientific committee has organized an ensemble of workshops in order to define the best suited experimental apparatus for the experiments with the SPIRAL beams. From the scientific discussions during these meetings it became clear that two new major experimental equipments are necessary in this domain. A high acceptance spectrometer for identifying the reaction products induced by the SPIRAL beams and a high efficiency γ–ray spectrometer.

i) VAMOS, a *Variable Mode* high acceptance *Spectrometer* for identifying products of reactions induced by SPIRAL beams

The technical requirements for VAMOS were formulated in three working meetings, each with around one hundred participants. Starting with the experimental needs expressed in the workshops, the following spectrometer properties have been defined in four meetings with around 15 persons.

  a) A very large geometrical acceptance, of the order of 100msr, which is equivalent to an angular acceptance of 200mr.
  b) A nominal dispersion which should not exceed 2cm/% over the focal length.
  c) A momentum acceptance of the order of +-5%
  d) A velocity filter function
  e) The possibility to rotate the spectrometer around the target point by an angle close to 90°.

These requirements are met by an optical structure consisting of a :

Doublet of quadrupoles, a Wien filter and a variable angle dipole. A description of the project is given in ref. (4).

ii) EXOGAM, A γ-ray spectrometer for nuclear reaction studies at SPIRAL.

The design of EXOGAM is also based on the demands of the physics community. The beam intensity, at least at the start up of the facility is expected to be much lower than with stable beams. EXOGAM must therefore be designed to maximize the total photo-peak efficiency at both low and intermediate γ-ray multiplicities. The radioactive nature of the beams is also a concern and shielding of the detectors becomes an important design criterion. The new γ-spectrometer has to be adapted to a variety of reaction mechanisms including compound nucleus formation, deep inelastic or transfer reactions and Coulomb excitation.

The design of EXOGAM comprises an array of 16 segmented germanium detectors, (5). Each segmented Clover detector consists of four co-axial n-type Ge crystals. Each individual crystal in the segmented Clover detector is electronically segmented into four regions. The main advantage of this segmentation is the reduction of the opening angle of the Ge crystals and consequently a reduction of the Doppler broadening.

The expected performance of this design for a (typical) γ-ray energy of 1.3 MeV are :
a) An intrinsic resolution $\Delta E=2.3$ keV (FWHM)
b) Photo-peak efficiency ~20%
c) Peak-to-total ratio P/T=47%

Both Projects are based on broad European collaborations. A Memorandum of Understanding is prepared for each equipment specifying also the financial contribution of the partners. The final decision is expected to be taken at the December meeting of the GANIL Board of Directors.

THE LETTERS OF INTENT

The international scientific committee of the SPIRAL Project has launched a call for letters of intent for the first experiments with SPIRAL. It had received 60 letters coming from about 350 scientists of 60 laboratories in 18 countries interested by the project. As a general conclusion, the committee noted the high scientific interest of many proposed programs. Many experiments can be made from the beginning, with noble gas beams of realistic intensities. From the analysis of the experimental requirements expressed by the letters of intent, the need of a highly efficient gamma ball and a charged-particle spectrometer is clear.

THE PHASE II OF SPIRAL

Discussions have already started about a possible phase II of SPIRAL. It is based :
i) On the replacement of the actual NANOGAN2 source by a two steps solution (1+/n+). The primary ion source for the production of 1+ ions will be

located in the production cave of SPIRAL. An ECRIS source (Electron Cyclotron Resonance Ion Source) outside from the production cave can be used for charge multiplication. The advantage of this solution is an increased reliability in the production of radioactive species and a low cost. The first results obtained at Grenoble for the production of Rb(9+) and Ar(8+) are encouraging, (6).

ii) The use of primary deuteron beams coupled with a thick U target for the production of fission fragments. Tests at the Orsay Tandem have been very successful. Secondary neutron-rich isotopes of the noble gazes Xe and Kr have been rapidly released and collected at a detection station from a 20g Uranium target bombarded by 10 MeV neutrons. As a next step, IPN Orsay (collaboration leaded by A. Mueller) will connect an ion-source to the target and implant a small-scale experimental separator at the Tandem in order to investigate the efficiency and diffusion constats, with the goal to make very thick but fast targets. A fission target for SPIRAL will increase the possible radioactive beams.

The construction of SPIRAL is progressing towards its end. The experimental community around the accelerator is well identified and the first experiments are emerging.

ACKNOWLEDGMENTS

I am indebted to Dr. F. Auger for a careful reading of the manuscript

REFERENCES

1) A.C.C. Villary and the SPIRAL group, GANIL S 96 04
2) Minutes of the meetings of the International Scientific committee of the SPIRAL project
3) Eric Baron and the GANIL operation team, Nouvelles du GANIL APRIL 1997
4) W. Mittig, Nouvelles du GANIL APRIL 1997
5) F. Azaiez et al., Nouvelles du GANIL JULY 1997
6) A.C.C. Villari et al., Nouvelles du GANIL JULY 1997

Inelastic Proton Scattering of Unstable Nuclei

F. Maréchal[1], A. Azhari[2,3], D. Bazin[2], Y. Blumenfeld[1], J.A. Brown[2], P.D. Cottle[4], M. Fauerbach[2,3], T. Glasmacher[2,3], S. Hirzebruch[1,2], J.K. Jewell[4], J.H. Kelley[1], K.W. Kemper[4], P.F. Mantica[2,5], D.J. Morissey[2,5], S. Ottini[6], L.A. Riley[4], J.A. Scarpaci[1], M. Steiner[2] and T. Suomijärvi[1,2]

[1] *Institut de Physique Nucléaire, IN$_2$P$_3$-CNRS, 91406 Orsay, France*
[2] *NSCL/MSU, East Lansing, MI 48824, USA*
[3] *Depart. of Physics and Astronomy, MSU, East Lansing, MI 48824, USA*
[4] *Depart. of Physics, FSU, Tallahassee, FL 32306, USA*
[5] *Depart. of Chemistry, MSU, MI 48824, USA*
[6] *SPhN, DAPNIA, CEA Saclay, 91191 Gif sur Yvette, France*

Abstract. The availability of radioactive beams with relatively high intensities and good optical qualities makes possible the study of direct reactions induced by unstable nuclei. The problems of nuclear matter distributions, deformation and the modification of shell structure far from stability can be addressed through inverse kinematics reactions on light targets. The experimental method is presented and two examples in the case of the ^{38}S (p,p') and ^{40}S (p,p') reactions measured at the NSCL-MSU are given. Finally, the newly built silicon strip detector system, MUST, dedicated to such studies, is presented.

INTRODUCTION

Direct reactions can provide a wealth of information on nuclear structure and interaction potentials which are fundamental problems in nuclear physics. Such reactions have been performed extensively with stable nuclei and form the foundation of our current understanding of nuclear systems. However, the valley of β-stability is only a very limited area of the nuclear chart, and several theoretical calculations predict drastic modifications in the nuclear structure outside the valley of stability. In particular, the magic number shell closures are expected to vanish for nuclei approaching the drip-line yielding new deformation regions [1,2]. The recent availability of radioactive beams with sizeable intensities at facilities such as GANIL, GSI, NSCL-MSU and

RIKEN has broadened the scope of such nuclear structure studies to a wide range of unstable nuclei.

Elastic scattering yields information on the nuclear matter distributions and the effective nucleon-nucleon potentials. Folding model analyses, using well tested effective potentials such as the JLM potential [3], do not have any free parameters apart from the matter density of the nucleus studied, and should be sensitive to new manifestations of nuclear structure such as neutron halos or skins.

Inelastic scattering towards low lying collective states gives access to transition probabilities and nuclear deformations, and is a well suited tool to scan new regions of deformation and the modification of shell closures far from stability. Comparison of inelastic transition probabilities measured through Coulomb and hadronic excitation should give access to the isoscalar or isovector character of the states through the ratio of multipole transition matrix elements M_n/M_p [4]. Moreover, inelastic scattering should lead to the observation of new types of nuclear excitations such as the soft modes predicted for halo nuclei.

We have recently undertaken a study of neutron rich sulphur isotopes through elastic and inelastic scattering of protons in inverse kinematics. We performed two experiments on ^{38}S and ^{40}S by using secondary fragmentation beams of ^{38,40}S delivered by the National Superconducting Cyclotron Laboratory at Michigan State University. Excitation energy spectra and elastic and inelastic angular distributions were obtained through the measurement of the angle and energy of recoiling protons. In the following, the experimental method, experimental results and the extracted β_2 value and M_n/M_p ratio for both ^{38,40}S nuclei will be discussed.

In order to pursue similar studies at GANIL, using unstable beams produced by projectile fragmentation with the SISSI superconducting solenoid, or the future beams accelerated by the SPIRAL facility, the Institut de Physique Nucléaire at Orsay, the DPTA/SPN CEA Bruyères le Châtel and the DAPNIA/SPhN CEA Saclay have engaged in a collaboration to design and build *MUST*, a new modular double sided silicon-strip array [5]. The characteristics of this detector will be presented as well as the results obtained in a test run performed at GANIL in November 1996 with an ^{40}Ar beam.

EXPERIMENTAL METHOD

The most straightforward approach is to induce direct reactions induced by light particles which present the advantage of not having any excited states, and for which the interaction potentials are better determined than for heavy ions. Inverse kinematics must then be used, the unstable projectile impinging on the light target, which entails specific experimental requirements. It is possible to determine the two-body kinematics of the reaction, either by detecting the outgoing heavy nucleus, for example with a spectrometer, or by measuring the energy and angle of the recoiling light particle. In the first case, however, as soon as the projectile is somewhat heavy (A $\geq$ 20), due to the kinematic focusing the angular resolution is no longer sufficient to obtain meaningful angular distributions. Moreover, a simple measurement of the outgoing quasi-projectile does not allow one to access unbound states which will particle-decay in flight. Therefore the recoiling proton method seems to be of more general use, and is in addition more cost-effective. Some experiments with unstable beams have already been performed using this method [6,7]. In the following we will concentrate on the ^{38,40}S(p,p') reactions investigated as part of a IPN-Orsay, Michigan State University and Florida State University collaboration using the beams from the NSCL-MSU.

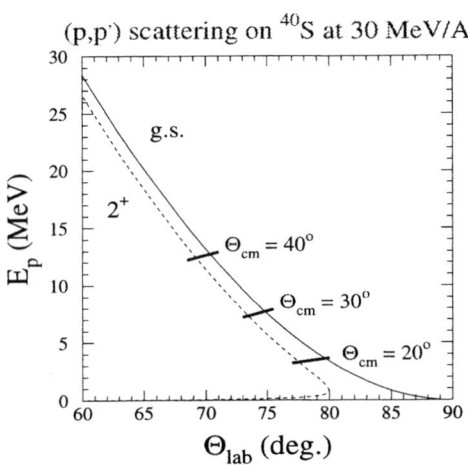

FIGURE 1. Kinematics of recoiling protons for elastic scattering (solid curve) and inelastic scattering to the first 2^+ state at 0.9 MeV (dashed curve) in ^{40}S(p,p') at 30 MeV/u.

In order to more precisely pinpoint the experimental requirements, an example of kinematics of recoiling protons for elastic (p,p) and inelastic (p,p') scattering to the first 2^+ state of ^{40}S is shown for an incident energy of 30 MeV/n in Fig. 1. To separate the elastic and inelastic scattering, an angular resolution of better than $1°$ is needed, along with a reasonable energy resolution. Due to energy and angular straggling effects, such performances require the use of a thin target, at most a few mg/cm^2 in the case of CH$_2$. A system of telescopes using strip detector technology will then provide the necessary resolutions, along with particle identification through energy-loss, energy and time of flight measurements. In the majority of cases, due to the poor emittance of secondary fragmentation beams, such a system must be supplemented with beam-tracking detectors, allowing an event by event determination of the incident beam angle and beam position on the target.

SCATTERING OF ^{38,40}S ON PROTONS

Experimental Setup

The MSU silicon strip array [8] consists of telescopes made up of a 5 X 5 cm^2, 300 μm thick Si-strip detector with 16 vertical strips 3mm wide, followed by a 300μm or 500μm thick PIN diode and a 1cm thick CsI stopping detector read out by 4 photodiodes. These telescopes were placed at 30 cm from the target, yielding a recoil angle measurement with an accuracy of 0.6° and a total angular range of 9.8° for each telescope. This setup allowed us to measure protons from 1 MeV up to 70 MeV.

The secondary ^{38}S (resp. ^{40}S) beam was produced by fragmenting an 85 MeV/A ^{40}Ar beam (resp. 60 MeV/A ^{48}Ca beam), provided by the K1200 cyclotron at the National Superconducting Cyclotron Laboratory, in a ^{9}Be target. The fragments were identified using the A1200 fragment separator, and the beams were purified with an aluminiun wedge placed in the second dispersive image of the A1200. The final energy of the secondary ^{38}S beam was 39 MeV/nucleon. The intensity of the beam was 3×10^5 particles per second, and the purity was higher than 99%. In order to avoid using beam tracking detectors, the angular emittance and momentum spread of the beam were limited using collimators. The final beam intensity on the reaction target was 20000 particles per second. For the ^{40}S beam, the final intensity was 2000 particles per second but the purity was not higher than 15%, the beam being mainly contaminated by ^{43}Ar. However, the incident particles were identified event by event. In the both experiments, the sulfur projectiles were scattered on a very thin 2 mg/cm^2 CH$_2$ target which allowed to obtain accurate data also for the low energy protons. The data in the silicon strip telescopes were taken in coincidence with a zero degree plastic detector, which identified the outgoing fragments and allowed us to eliminate fusion or fragmentation reactions, thus very effectively reducing the background. This plastic detector also yields a start signal for the proton time-of-flight measurements.

Experimental Results

In the case of ^{38}S, the first 2^+_1 state is observed at 1.2 MeV, compatible with the adopted value of 1.29 MeV. Figure 2 displays a scatterplot of laboratory energy vs. angle for recoiling protons from the ^{38}S scattering. The elastic scattering and inelastic scattering to the first 2^+ state are clearly separated. Indications for the presence of higher lying states are also observed.

Figure 3 shows the angular distributions for the ground state and the first 2^+ excited state of ^{38}S. Each angular distribution is obtained by projecting

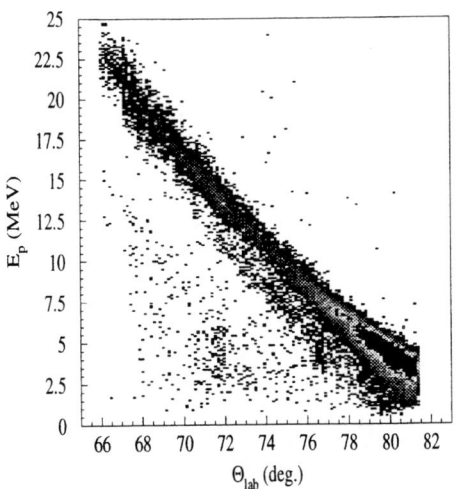

FIGURE 2. ^{38}S scattering on protons at 39 MeV/nucleon : energy vs. laboratory angle scatterplot for recoiling protons.

the contents of the corresponding contour in the excitation energy vs. θ_{CM} plane.

Coupled-channels predictions using the ECIS code [9] are shown in comparison with the data. Note that no arbitrary normalization is involved here. A calculation based on the Becchetti-Greenlees parameterization [10], which was developed for (p,p) scattering on A$\geq$40 nuclei, is shown by the solid line in Fig. 3. A second calculation, shown as the dotted line in Fig. 3, uses optical model parameters for ^{40}Ar(p,p) [11] and gives slightly better agreement with the measured ground state distribution, in particular at small angles. The shape of the experimental inelastic angular distribution is in full agreement with the calculation. Using the Becchetti-Greenlees parameterization (dashed line) a β_2=0.35±0.04 value is obtained while the ^{40}Ar optical parameters (solid line) give β_2=0.36±0.04. In the following we will adopt the value β_2=0.35±0.04 [8]. This value is clearly larger that the electromagnetic value β_2=0.25±0.02 obtained from the COULEX experiment [12].

For ^{40}S, the statistics are too low to get a complete inelastic angular distribution like in the case of ^{38}S. But we can obtain the β_2 value by considering the excitation energy spectrum. Figure 4 displays the excitation energy spectrum for ^{40}S, first integrated over the total center of mass angular range (left) and then integrated over an angular bin where the elastic contribution is minimum (right). We clearly see the first 2$^+$ excited state located at 0.9 MeV which is in a good agreement with the value obtained by coulomb excitation [12]. Fitting the two contributions with gaussians and comparing the ratio of the

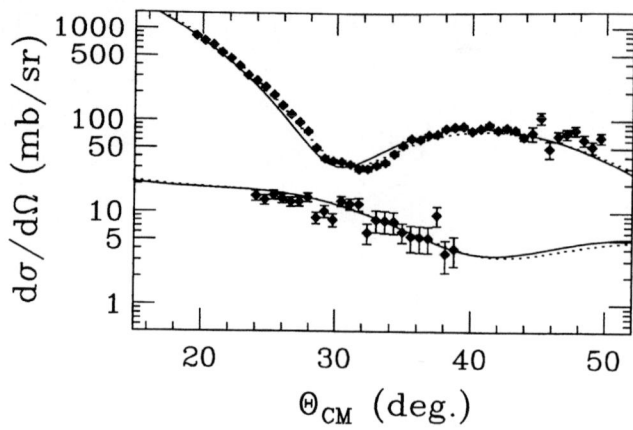

FIGURE 3. Angular distributions for the ground state and the 2_1^+ state in the ^{38}S(p,p') reaction at 39 MeV/nucleon, obtained by projecting the contents of contours (see text). The calculations are coded as follows ; solid line: coupled-channel calculation with Becchetti-Greenlees potential; dotted line: coupled-channel calculation with ^{40}Ar(p,p') potential.

integrals with a calculation will allow us to extract the β_2 value for ^{40}S. This analysis is currently in progress.

Combined measurements of inelastic hadron scattering and electromagnetic excitation such as COULEX are interesting since they can provide information to disentangle proton and neutron contributions to nuclear deformation. In the case of inelastic proton scattering at a few tens of MeV incident energy the neutron interaction strength is 3 times larger than the proton one [13] and the scattering probes mainly neutrons in the nucleus while COULEX yields proton deformation parameters. The measurement of these two different deformation parameters allows to determine the ratio between the neutron and proton multipole transition matrix elements M_n/M_p. This ratio is derived in ref. [4]:

$$\frac{M_n}{M_p} = \frac{b_p}{b_n}\Big(\frac{\delta}{\delta_{e.m.}}\Big(1 + \frac{b_n}{b_p}\frac{N}{Z}\Big) - 1\Big). \tag{1}$$

where b_p and b_n are the interaction strengths of protons with protons and neutrons respectively, δ is the deformation length from (p,p') and $\delta_{e.m.}$ is the electromagnetic deformation length ($\delta = \beta_2 r_0 A^{1/3}$).

Low lying 2^+ and 3^- states are generally well described by an isoscalar collective model with equal neutron and proton deformation, yielding a ratio $M_n/M_p = N/Z$ [4]. However, deviations from this isoscalar picture have been observed typically for single closed shell nuclei such as ^{116}Sn-^{124}Sn where

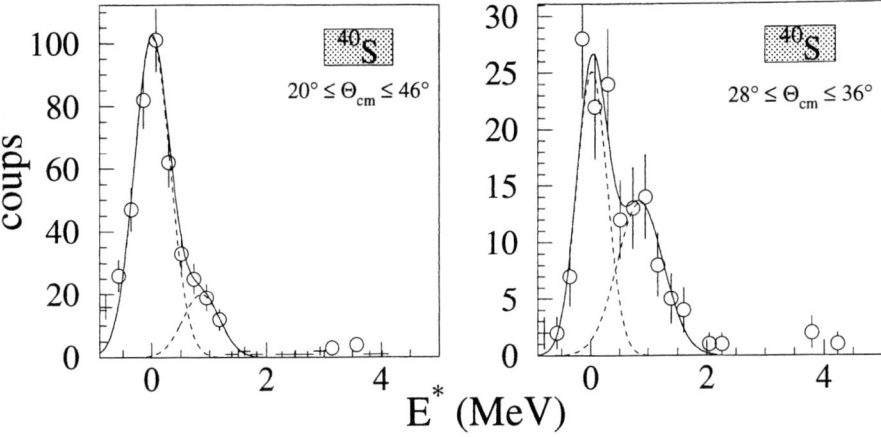

FIGURE 4. Excitation energy spectra for ^{40}S scattering on protons at 30 MeV/nucleon integrated over the total cm angular range (left) and over an angular bin where the elastic contribution is minimum (right).

the valence neutrons above the inert proton core dominate in the 2^+ and 3^- excitations yielding an isovector component in the excitation [13].

We have tentatively extracted the M_n/M_p ratio for the 2^+_1 state in ^{38}S by using the formula given above with β_2 values from the COULEX experiment [12] and from the proton scattering experiment. An r_0 value of 1.17 fm corresponding to the optical parameters of the Bechetti-Greenlees systematics was used for (p,p') scattering, while $r_0 = 1.20$ fm was taken for electromagnetic excitation. The b_p and b_n values were taken as 0.3 and 0.7, respectively [13]. This yields $M_n/M_p = (1.5 \pm 0.3)N/Z$. This value is incompatible with the value of N/Z expected for a pure isoscalar mode and thus indicates an isovector component in the excitation. The method used here, follows the prescription of ref. [13]. However, it would be important to compare this value with the result of a full microscopic analysis.

It is interesting to observe the trend of β_2 and M_n/M_p values for the 2^+_1 state as a function of neutron number in the sulphur isotopes (see table 1).
One should first note the very low β_2 values and high excitation energy of the 2^+_1 state in ^{36}S, as well as the M_n/M_p value compatible with N/Z. Therefore ^{36}S exhibits features akin to those of a well closed nucleus. When moving away from ^{36}S, the measured β_2 values increase and a large difference in M_n/M_p values is observed between ^{36}S and ^{38}S, showing a rapid change of the structure as a function of neutron number. The large M_n/M_p value for ^{38}S can be qualitatively understood by considering the ^{38}S nucleus as a ^{36}S core plus two valence neutrons. In this case, the two neutrons drive the oscillation and

TABLE 1. Compilation of 2_1^+ states for sulphur isotopes. Energies and $\beta_2(em)$ values are from ref. [15]. $\beta_2(p,p')$ values are from refs. [16] (^{32}S),[17] (^{34}S),[18] (^{36}S) and from this work (^{38}S).

	E (MeV)	$\beta_2(p,p')$	$\beta_2(em)$	$(M_n/M_p)/(N/Z)$
^{32}S	2.23	0.28	0.31	0.84
^{34}S	2.12	0.28	0.25	1.12
^{36}S	3.29	0.18	0.16	1.12
^{38}S	1.29	0.35	0.25	1.5

the core polarization is not sufficient to restore the isoscalar character of the excitation.

Preliminary results of the ^{40}S(p,p') experiment analysis seem to confirm this isovector character of the 2^+ oscillation in the very neutron rich sulfur isotopes.

THE *MUST* DETECTOR SYSTEM

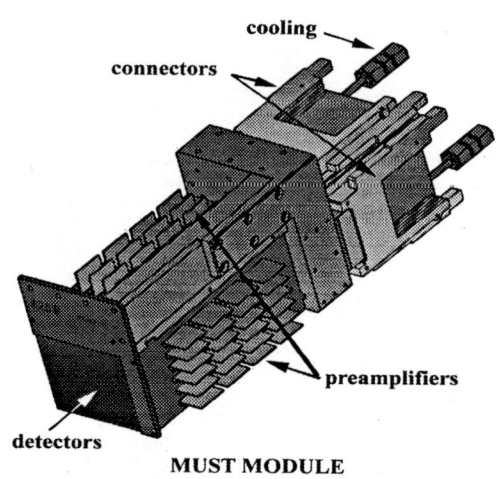

MUST MODULE

FIGURE 5. Artistic view of one of the 8 telescopes.

This new detection system built by an IPN-Orsay, DAPNIA/SPhN CEA Saclay and DPTA/SPN CEA Bruères le Châtel, and devoted to the measurement of light particles in reactions induced by radioactive beams, consists of 8 telescopes with a 6X6 cm² 300μm thick double sided silicon strip detector with 60 horizontal and vertical strips 1mm wide, backed by a 3mm thick

Si(Li) detector covering the same surface. A third CsI stage, read out by a photodiode, may be added for experiments where protons of energies above 25 MeV must be detected. Particle identification is obtained by a ΔE-E measurement for particles traversing the strip detector, and by E-time of flight for the slower particles. The system is modular to adapt to diverse experimental requirements. A picture of one telescope is presented on Fig. 5.

Each strip is connected to a preamplifier located under vacuum directly behind the detector. Compact custom-made electronics in VXI standard were designed. All the necessary functions (discriminators, amplifiers, energy and time digitalization) for one telescope are housed in one VXI module, which means that all the electronics for the system, which consist of over 1000 logic and analogic channels, hold in one VXI crate. These electronics are read out by a dedicated data acquisition system based on a SUN workstation.

Tests of one telescope with its associated electronics and data acquisition were performed recently at GANIL. The energy resolution measured was 50 keV whereas the time resolution was 900 ps at 2.0 MeV proton energy. ΔE, E and time of flight measurements allowed identification of light charged particles between 300 KeV and 25 MeV proton energy (see figure 6), with the detector located at 15 cm from the CH_2 target.

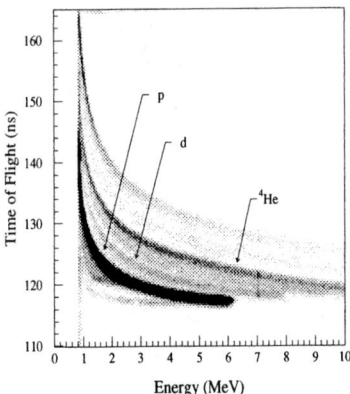
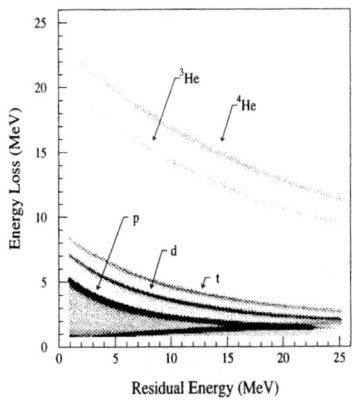

FIGURE 6. E-Time of Flight identification for particles which stop in the silicon strip detector (left) and ΔE-E identification for more energetic particles (right).

Using the $^{40}Ar(p,p')$ reaction at 77 MeV/A in inverse kinematics, an excellent separation of elastic scattering from inelastic scattering to the first 2^+ and 3^- excited states at 1.4 and 3.6 MeV was observed. Angular distributions for the ground state and the first 2^+ and 3^- states were measured. of the elastic and inelastic distributions calculated with the code ECIS [9] are in excellent agreement with the data. The relative normalizations yield a β_2 value of 0.22

± 0.01 and a β_3 value of 0.24 ± 0.02 which are in remarkable agreement with the values obtained in previous experiments performed in direct kinematics [18], demontrating the excellent performances of the set-up.

CONCLUSIONS

The measurement of recoiling light charged particles is shown to be a powerful and straightforward method to measure excitation energy spectra and angular distributions for unstable nuclei with good resolution and low background.

The comparison of the β_2 values obtained by Coulomb excitation and (p,p') scattering to low lying states yields information on the nature of the vibration and on the relative importance of the neutron and proton contributions to this vibration. This kind of nuclear structure analysis is of great importance to constrain the theoretical models which aim to describe the nuclei far from the valley of stability.

The *MUST* detector system, a modular array of telescopes based on silicon strip technology will allow us to undertake an extensive program to pursue such studies, using the unstable beams delivered by the GANIL facility.

REFERENCES

1. A. Poves and J. Retamosa, *Nucl. Phys.* **A571**, 221 (1994).
2. T. R. Werner et al., *Nucl. Phys.* **A597**, 327 (1996).
3. J.P. Jeukenne, A. Lejeune, and C. Mahaux, *Phys. Rev.* **C16**, 80 (1977).
4. A.M. Bernstein et al., *Comm. Nucl. Part. Phys.* **11**, 203 (1983).
5. Y. Blumenfeld et al., Internal Report IPNO-DRE-96-12, Institut de Physique Nucléaire, Orsay, France (1996).
6. G. Kraus et al., *Phys. Rev.* **C73**, 1773 (1994).
7. A.A. Korshenenikov et al., *Phys. Rev. Lett.* **78**, 2317 (1997).
8. J.H. Kelley et al., *Phys. Rev.* **C56**, R1206 (1997).
9. J. Raynal, *Phys. Rev.* **C23**, 2571 (1981).
10. F.D. Becchetti Jr., and G.W. Greenlees, *Phys. Rev.* **182**, 1190 (1969).
11. E. Fabrici et al., *Phys. Rev.* **C21**, 830 (1980).
12. H. Scheit et al., *Phys. Rev. Lett.* **77**, 3967 (1996).
13. M.A. Kennedy, P.D. Cottle and K.W. Kemper, *Phys. Rev.* **C46**, 1811 (1992).
14. S. Raman et al., *Atomic Data and Nuclear Data Tables* **36**, 1 (1987).
15. R. de Leo et al., *Nuovo Cimento* **59A**, 101 (1980).
16. R. Alarcon et al., *Phys. Rev.* **C31**, 697 (1985).
17. A. Hogenbirk et al., *Nucl. Phys.* **A516**, 205 (1990).
18. R. De Leo et al., *Phys. Rev.* **C31**, 362 (1985).

Double Giant Resonance States

Haruki Kurasawa* and Toshio Suzuki[†‡]

*Department of Physics, Faculty of Science, Chiba University, Chiba 263, Japan
†Department of Applied Physics, Fukui University, Fukui 910, Japan
‡RIKEN, 2-1 Hirosawa, Wako-shi, Saitama 351-01, Japan

Abstract. The strength, excitation energy and width of the double giant dipole and quadrupole resonances are studied. The model-independent sum rules for the strength are derived. It is shown that the folding model satisfies the sum rule, but can not reproduce experimental data. The energy and width of the double giant dipole resonance can be expressed rather model-independently in terms of those of the single giant dipole resonance. Their relationship is in good agreement with experiment.

INTRODUCTION

Pioneering work on double giant resonance states was carried out at GANIL. The double giant quadrupole resonance(DGQR) was observed with nuclear interaction using heavy ion beams at intermediate energy [1]. Recent experiments of (π^+, π^-) and relativistic heavy ion reactions have revealed also the existence of the double giant dipole states (DGDR) in various nuclei [1]. The giant resonance states are characterized by their excitation energy, width and excitation strength. Emling et al. [1] have summarized the data on the excitation energy and width of DGDR in terms of those of the single giant dipole resonance states(GDR),

$$\omega_2 = (1.91 \pm 0.02)\omega_1, \qquad (1)$$

$$\Gamma_2 = (1.60 \pm 0.03)\Gamma_1, \qquad (2)$$

where $\omega_2(\omega_1)$ stands for the excitation energy and $\Gamma_2(\Gamma_1)$ the width of DGDR(GDR). They have also compared the strength of DGDR with various theoretical results, and found that all the theoretical values obtained so far underestimate the data by more than factor 2,

$$S_2 \geq 2 S_{\text{th}}. \qquad (3)$$

The recent data for ^{208}Pb by LAND collaboration [2] have provided $S_2 = (1.33 \pm 0.16)S_{\text{th}}$, where S_{th} is calculated by the folding model. In contrary to

DGDR, the strength of GDR is well known to be understood by the Thomas-Reiche-Kuhn sum rule. The same difficulty on the strength is found for DGQR [1].

The purpose of the present paper is to study the above relationship between the excitation energies and widths of DGDR, and to explore the strength of DGDR and DGQR model-independently [3]. In the following section, we will derive the model-independent sum rule for the strength of DGDR and DGQR which correspond to TRK sum rule for GDR. In Section III, we will show that the relationship on the energy and width is explained for DGDR rather model-independently according to Tomonaga theory [4]. The final section will be devoted to a brief conclusion.

EXCITATION STRENGTH

Model-Independent Sum Rules

First let us review the TRK sum rule for the strength of GDR [5]. The strength function of GDR is defined by

$$S_1(E) = \sum_n |\langle n|D|0\rangle|^2 \delta(E_n - E), \tag{4}$$

where the dipole operator is given by

$$D = \frac{Z}{A}\sum_{n=1}^{N} x_n - \frac{N}{A}\sum_{p=1}^{Z} x_p. \tag{5}$$

The integration of the strength function over the excitation energy gives the non energy-weighted sum,

$$S_1^{(0)} = \sum_n |\langle n|D|0\rangle|^2 = \langle 0|D^2|0\rangle^2. \tag{6}$$

In the last step the completeness of the intermediate states has been used. The energy-weighted sum of the strength is given by

$$S_1^{(1)} = \sum (E_n - E)|\langle n|D|0\rangle|^2. \tag{7}$$

This is expressed by the double commutator between D and the hamiltonian, H,

$$S_1^{(1)} = \frac{1}{2}\langle 0|[D,[H,D]]|0\rangle, \tag{8}$$

and calculated model-independently as

$$S_1^{(1)} = \frac{1}{2m}\frac{NZ}{A}, \qquad (9)$$

for the nuclear interaction, V, satisfying $[D,[V,D]] = 0$. The last equation is known as the TRK sum rule. We note that eq.(9) is obtained, since the double commutator between D and the kinetic part of the hamiltonian, T, is a constant

$$\frac{1}{2}[D,[T,D]] = \frac{1}{2m}\frac{NZ}{A}. \qquad (10)$$

Thus, TRK sum rule holds not only as to the ground state, but also as to any state.

With use of the non energy-weighted and energy-weighted sum we can define the mean energy of the resonance,

$$\omega_1 = S_1^{(1)}/S_1^{(0)}, \qquad (11)$$

which is a measure of the peak energy given in eq.(1).

Next we derive a model-independent sum rule for DGDR in the same way as for GDR. We define the strength function of DGDR,

$$S_2(E) = \sum_f |A_f|^2 \delta(E - E_f), \qquad (12)$$

where A_f is given by

$$A_f = \sum_n \langle f|D|n\rangle\langle n|D|0\rangle = \langle f|D^2|0\rangle. \qquad (13)$$

For this function the non energy-weighted and energy-weighted sum are given as

$$S_2^{(0)} = \sum_f |\langle f|D^2|0\rangle|^2, \qquad (14)$$

$$S_2^{(1)} = \sum_f (E_f - E_0)|\langle f|D^2|0\rangle|^2. \qquad (15)$$

The non energy-weighted sum on $|f\rangle$ should not include the ground state, $|0\rangle$, but the energy-weighted sum can include $|0\rangle$ as $|f\rangle$. Because of this fact the energy-weighted sum can be expressed in terms of the double commutator,

$$S_2^{(1)} = \frac{1}{2}\langle 0|[D^2,[H,D^2]]|0\rangle. \qquad (16)$$

The commutator with D^2 is rewritten with D,

$$S_2^{(1)} = \frac{1}{2}\langle 0|D^2[D,[H,D]] + 2D[D,[H,D]]D + [D,[H,D]]D^2|0\rangle. \qquad (17)$$

The double commutator is given by eq.(10), when $[D, [V, D]] = 0$. Therefore we obtain the model-independent sum rule for DGDR [3],

$$S_2^{(1)} = 4\frac{1}{2m}\frac{NZ}{A}\langle 0|D^2|0\rangle = 4S_1^{(0)}S_1^{(1)}. \tag{18}$$

If we define the mean energy of DGDR by

$$\omega_2 = S_2^{(1)}/S_2^{(0)}, \tag{19}$$

the above sum rule gives the relationship between mean energies of DGDR and GDR in terms of their non energy-weighted sum,

$$\frac{\omega_2}{\omega_1} = 4\frac{\{S_1^{(0)}\}^2}{S_2^{(0)}}. \tag{20}$$

Thus the ratio of the mean energies is determined model-independently by that of the non energy-weighted sum.

Third, the sum rule for DGQR is derived in the same way as for DGDR. The energy-weighted sum is given by

$$S_2^{(1)} = \sum_f (E_f - E_0)|\langle f|Q^2|0\rangle|^2$$
$$= \frac{8}{m}\{2\langle 0|Q^2 M|0\rangle + \langle 0|Q^3|0\rangle\}. \tag{21}$$

where Q and M are defined as

$$Q = \sum_{i=1}^{A}(2z_i^2 - x_i^2 - y_i^2), \quad M = \sum_{i=1}^{A} r_i^2. \tag{22}$$

In contrast to the case of DGDR, the above sum rule is not expressed in terms of those for GDR,

$$S_1^{(1)} = \sum_n (E_n - E_0)|\langle n|Q|0\rangle|^2, \quad S_1^{(0)} = \sum_n |\langle n|Q|0\rangle|^2. \tag{23}$$

However, if assume that the nucleus is spherical and that

$$\langle 0|Q^2|n\rangle\langle n|M|0\rangle = 0 \quad (|n\rangle \neq |0\rangle), \tag{24}$$

then we obtain

$$S_2^{(1)} = 4S_1^{(0)}S_1^{(1)}, \tag{25}$$

as in eq.(18) for DGDR. Therefore, eq.(20) also holds for DGQR in the same assumption.

Results of the Sum Rules

The sum rules provide us with interesting results in some limit. First, when the mean energy of DGDR is twice that of GDR, $\omega_2 = 2\omega_1$, eq.(20) gives

$$S_2^{(0)} = 2\{S_1^{(0)}\}^2. \tag{26}$$

As mentioned in INTRODUCTION, the observed excitation strength of DGDR can not be reproduced theoretically so far. Experiments are frequently analyzed by the folding model [1] which assumes independent phonon excitations. The strength function of the folding model is given by

$$S_2(E) = \sum_{n,n'} |\sum_{n''} \langle n, n'|D|n''\rangle \langle n''|D|0\rangle|^2 \delta(E - E_0 - E_{n'}), \tag{27}$$

which can be rewritten as

$$S_2(E) = 2\sum_{n,n'} |\langle n|D|0\rangle|^2 |\langle n'|D|0\rangle|^2 \delta(E - E_n - E_{n'}). \tag{28}$$

This immediately gives

$$\omega_2 = 2\omega_1, \quad S_2^{(0)} = 2\{S_1^{(0)}\}^2. \tag{29}$$

Thus the folding model satisfies the sum rule. Furthermore, when we define the variance of the strength function of GDR, instead of FWHM,

$$\sigma_1 = \sum_n (E_n - \omega_1)^2 |\langle n|D|0\rangle|^2 / \sum_n |\langle n|D|0\rangle|^2, \tag{30}$$

and in the same way for DGDR, it is easily shown that

$$\sqrt{\sigma_2} = \sqrt{2\sigma_1}, \tag{31}$$

which explains well the experimental result, eq.(2). The folding model, however, has been shown to reproduce only a half of the experimental value of the strength [1].

If the strength of DGDR is twice that of the folding model, $S_2^{(2)} = 4\{S_1^{(0)}\}^2$, the sum rule, eq.(20), gives $\omega_2 = \omega_1$, which is unrealistic.

We note that the above arguments are valid also for DGQR.

Second, let assume that there exists a single collective state, $|\omega_1\rangle$, which exhausts TRK sum rule for GDR, and $|\omega_2\rangle$ for DGDR. Since, as mentioned before, TRK sum rule holds not only with respect to the ground state, but also to the excited state, we have [6]

$$|\langle \omega_2|D|\omega_1\rangle|^2 / |\langle \omega_1|D|0\rangle|^2 = 2\omega_1/(\omega_2 - \omega_1). \tag{32}$$

the new sum rule, eq.(20), on the other hand, provides

$$\omega_2/\omega_1 = 4|\langle\omega_1|D|0\rangle|^2/|\langle\omega_2|D|\omega_1\rangle|^2. \tag{33}$$

The above two equations yield

$$\omega_2 = 2\omega_1, \qquad S_2^{(0)} = 2\{S_1^{(0)}\}^2, \tag{34}$$

as in the folding model. We note that eqs.(33) and (34) give the Bose factor, 2,

$$|\langle\omega_2|D|\omega_1\rangle|^2/|\langle\omega_1|D|0\rangle|^2 = 2. \tag{35}$$

RELATIONSHIP BETWEEN MEAN ENERGIES AND VARIANCES

The relationship of the mean energy and variance of DGDR to those of GDR is studied rather model-independently according to the Tomonaga theory [4]. We need not specify the explicit form of the nuclear hamiltonian in order to derive the relationship.

We define the collective variables,

$$\xi = \sqrt{\frac{A}{NZ}}D, \qquad \pi = \sqrt{\frac{A}{NZ}}\left(\frac{Z}{A}\sum_i^N p_{xi} - \frac{N}{A}\sum_i^Z p_{xi}\right), \tag{36}$$

which satisfy the canonical relationship,

$$[\xi, \pi] = i. \tag{37}$$

Using the canonical relationship, we can expand the hamiltonian in terms of ξ and π [4],

$$H = H_0 + H_1^\xi \xi + H_1^\pi \pi + \frac{1}{2}(H_2^\xi \xi^2 + H_2^\pi \pi^2 + H_2^{\xi\pi} \xi\pi + H_2^{\pi\xi} \pi\xi) + \cdots, \tag{38}$$

where the coefficients of ξ and π satisfy

$$[H_i, \xi] = [H_i, \pi] = 0, \tag{39}$$

and are given for the second order approximation by

$$H_2^\xi = -[\pi, [\pi, H]], \tag{40}$$

$$H_2^\pi = -[\xi, [\xi, H]], \qquad H_2^{\xi\pi} + H_2^{\pi\xi} = 2[\xi, [\pi, H]], \tag{41}$$

$$H_1^\xi = i[\pi, H] + \xi[\pi, [\pi, H]] - \pi[\xi, [\pi, H]], \tag{42}$$

$$H_1^\pi = -i[\xi, H] + \pi[\xi, [\xi, H]] - \xi[\pi, [\xi, H]]. \tag{43}$$

When we have
$$H_1^\xi = H_1^\pi = H_2^{\xi\pi} = H_2^{\pi\xi} = 0, \tag{44}$$
$$H_2^\xi = \text{const.}, \quad H_2^\pi = \text{const.}, \tag{45}$$

the collective dipole state has no width and its excitation energy is given by
$$\omega = H_2^\xi \cdot H_2^\pi. \tag{46}$$

Generally, however, eqs.(44) and (45) do not hold, and the collective state has the damping width.

Now let us introduce A and $A^\dagger = (A)^\dagger$,
$$A = \sqrt{\frac{B\omega}{2}}\xi + \frac{i}{\sqrt{2B\omega}}\pi, \tag{47}$$

which satisfies $[A, A^\dagger] = 1$. We rewrite eq.(38) in terms of A and $A^\dagger$,

$$\begin{aligned} H =\ & H_0 + H_1^\xi \frac{1}{\sqrt{2B\omega}}(A + A^\dagger) - H_1^\pi i\sqrt{\frac{B\omega}{2}}(A - A^\dagger) \\ & + \frac{1}{4}\{\langle 0|H_2^\xi|0\rangle\frac{1}{B\omega} + \langle 0|H_2^\pi|0\rangle B\omega\}(A^\dagger A + AA^\dagger) \\ & + \frac{1}{4}\{(H_2^\xi - \langle 0|H_2^\xi|0\rangle)\frac{1}{B\omega} + (H_2^\pi - \langle 0|H_2^\pi|0\rangle B\omega\}(A^\dagger A + AA^\dagger) \\ & + \frac{1}{4}(H_2^\xi\frac{1}{B\omega} - H_2^\pi B\omega)(AA + A^\dagger A^\dagger) \\ & - \frac{i}{4}(H_2^{\xi\pi} + H_2^{\pi\xi})(AA - A^\dagger A^\dagger) + \frac{i}{4}(H_2^{\xi\pi} - H_2^{\pi\xi}). \end{aligned} \tag{48}$$

When we take into account the first two lines and define
$$B^{-1} = \langle 0|H_2^\pi|0\rangle, \quad B\omega^2 = \langle 0|H_2^\xi|0\rangle, \tag{49}$$

the hamiltonian is given by
$$H = H_0 + H_1^\xi\frac{1}{\sqrt{2B\omega}}(A + A^\dagger) - H_1^\pi i\sqrt{\frac{B\omega}{2}}(A - A^\dagger) + \frac{\omega}{2}(A^\dagger A + AA^\dagger). \tag{50}$$

The above hamiltonian may be diagonalized in the space of $|\omega\rangle = A^\dagger|0\rangle$ and states, $|m\rangle$, which are orthogonal to $|\omega\rangle$, to have the dipole states,
$$|n\rangle = \alpha_n|\omega\rangle + \sum_m \beta_{nm}|m\rangle. \tag{51}$$

With use of the above equation, we can easily shown that the mean energy and variance of GDR are given by [3]

$$\omega_1 = \omega \tag{52}$$

$$\sigma_1 = \sum_m |\langle m|H|\omega\rangle|^2 = \sum_m |\langle m|\frac{1}{\sqrt{2B\omega}}H_1^\xi - i\sqrt{\frac{B\omega}{2}}H_1^\pi|0\rangle|^2. \tag{53}$$

We note that the variance includes both Landau and collision damping.

For DGDR we diagonalize the hamiltonian, eq.(50), in the space, $|2\omega\rangle = (1/\sqrt{2})A^\dagger A^\dagger|0\rangle$ and states, $|i\rangle$, which are orthogonal to $|2\omega\rangle$. As a result we may have for DGDR,

$$|f\rangle = c_f|2\omega\rangle + \sum_i d_{fi}|i\rangle. \tag{54}$$

This equation gives the mean energy and variance of DGDR [3],

$$\omega_2 = 2\omega \tag{55}$$

$$\sigma_2 = \sum_i |\langle i|H|2\omega\rangle|^2 = 2\sigma_1. \tag{56}$$

Thus the mean energy of DGDR is twice that of GDR, while the width of DGDR, which is $\sqrt{\sigma_2}$, is given by $\sqrt{2}$ times GDR width. These results are in good agreement with experiment mentioned in INTRODUCTION. The small difference may be due to higher order terms of the hamiltonian, eq.(48).

In order to apply the above arguments to GQR and DGQR, we need further assumptions, but do not discuss those in this paper.

CONCLUSION

The relationship of the mean energy and width of the double giant dipole resonance states(DGDR) to those of the single dipole states(GDR), which is recently observed in (π^+,π^-) and relativistic heavy ion reactions [1], is explained rather model-independently. The mean energy of DGDR is twice that of GDR, while the width of DGDR $\sqrt{2}$ times GDR one. The observed excitation strength of DGDR, however, can not be reproduced in the present investigation. We have the same problem for the double giant quadrupole resonance states(DGQR). It has shown that there is a model-independent sum rule of DGDR and DGQR strength as Thomas-Reiche -Kuhn sum rule for GDR. According to the ref. [1], the folding model, which is shown to have the sum rule strength, reproduces only a half of the observed strength. The disagreement between the present sum rule and experiment is a serious problem for nuclear physics.

REFERENCES

1. Emling H., *Prog. Part. Nucl. Phys.* **33**, 729 (1994);
   Chomaz Ph., and Frascaria N., *Phys. Reports* **252** 275 (1995).

2. Boretzky K., et al., it Phys. Lett. **B384** 30 (1996).
3. Kurasawa H., and Suzuki T., *Nucl. Phys.* **A597** 374 (1996).
4. Tomonaga S. I., *Prog. Theor. Phys.* **13** 467, 482 (1955); Suzuki T., *J. de Phys.* **45** C4-251 (1984).
5. Suzuki T., *Ann. de Phys.* **9** 535 (1984).
6. Bertulani C. A., and Zelevinsky V., *Nucl. Phys.* **568** 931 (1994).

THERMONUCLEAR REACTIONS I

Reactions with radioactive beams : direct measurements

W.Galster

Université Catholique de Louvain, Institut de Physique Nucléaire, Chemin du Cyclotron 2,
1348 Louvain-la-Neuve, Belgium

Abstract. The radioactive beam facility at Louvain-la-Neuve is dedicated to a large extend to nuclear reaction studies of astrophysical interest. Since becoming fully operational in 1991, several key reactions in the hot CNO cycles were investigated experimentally : $^{13}N(p,\gamma)^{14}O$ the onset of the hot CNO cycles, $^{19}Ne(p,\gamma)^{20}Na$ the escape to the rp process and others. These reactions have to be studied close to the astrophysically relevant energies (temperatures) to allow a meaningful extrapolation into the energy region of the Gamow window. The talk will summarize the results obtained with radioactive beams so far. A few experimental details are discussed and an outlook into future experiments is given. Finally, implications of the results for astrophysical scenarios are presented.

MOTIVATION

Nuclear reaction chains determine the time scales, production of energy and the nucleosynthesis of the stellar burning phases. The quiescent phases at low stellar temperature evolve to complex reaction networks at temperatures above 10^8 °K. In this explosive burning phase of a nova, an X-ray burster or a supernova, the effective timescale becomes sufficiently short so that unstable nuclei with short halflives play a vital role in the explosive scenario.

The stellar site prior to explosion contains heavy elements A > 12 produced in earlier generation stars, which were ejected into stellar space and accumulated subsequently as seeds in newer generation stars. An essential feature of nucleosynthesis is that heavy elements are being produced and ejected from explosive stellar events continuously. Realistic stellar models therefore require a good knowledge of the initial elemental composition, the nuclear reaction rates and the dynamics of the explosive event.

Let us recall that an operating reaction cycle is closed in the quiescent burning phase. E.g. the pp or the CNO cycle convert 4 protons into one alpha particle under the release of energy, respectively, and apart from the conversion of hydrogen into helium, the initial stellar composition remains unchanged. To what extent can matter escape (or be transferred to another cycle) from a hot cycle in the explosive burning phase ?

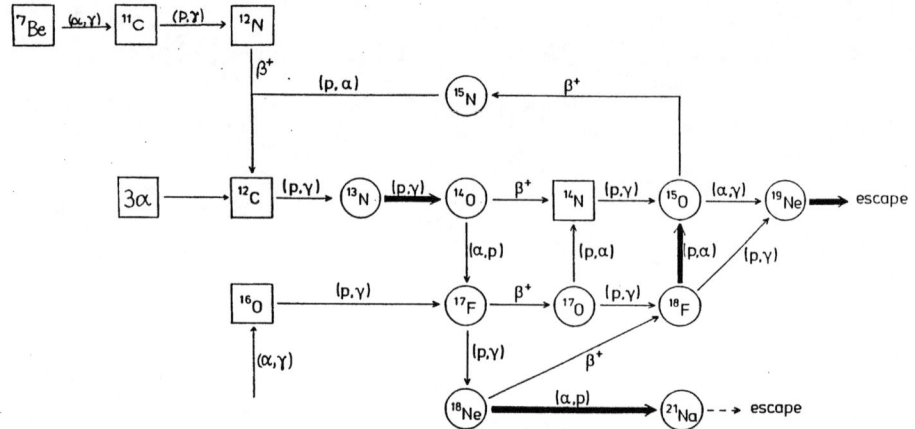

FIGURE 1. Simplified reaction network for the hot CNO cycles. The reactions, which were studied in Louvain-la-Neuve, are indicated by thick arrows. The most abundant seed nuclei in novae are ^{12}C and ^{16}O. ^{13}N(p,γ)^{14}O constitutes the onset of the hot CNO cycle. The bottleneck in the reaction path is ^{15}O(α,γ)^{19}Ne due to the high coulomb barrier for α-capture. The most likely escape path is through ^{19}Ne(p,γ)^{20}Na.

Fig.1 illustrates a part of the reaction network in the hot CNO cycles. In the explosion of a nova, a sizeable amount of matter may escape from the hot CNO cycles synthesizing A > 19, thus feeding the rp-process of rapid proton capture on beta-unstable nuclei. If the rp-process indeed operates out of novae effectively, our understanding of the nucleosynthesis up to ^{56}Ni (and beyond) will be profoundly affected. In a standard scenario, the elements A < 56 are synthesized in a sequence of He-, C-, O-, Si-burning in massive stars. Note that He-burning alone cannot synthesize heavy elements beyond oxygen efficiently; although, at the effective stellar temperature, the Gamow window embraces a resonance in ^{20}Ne ; this resonance has unnatural parity and could only be populated through parity violation. Therefore heavy elements are synthesized at much higher temperature at later stages of burning.

The nuclear reaction rates of interest (mostly proton and alpha capture) can in principle be obtained directly from accelerator based experiments. These reactions have to be studied in (or close to) the astrophysically relevant energy region, the Gamow window. The experiments are difficult to perform due to the low reaction cross sections and the need for intense radioactive beams; as so often, the most important reaction rates are the most difficult ones to obtain. Clearly, a hot reaction cycle is determined primarily by certain types of reactions, such as the onset (or first reaction initiating the hot cycle), a bottleneck (the lowest cross section in the reaction path), a waiting point (for β-decay, where further capture reactions cannot take place, e.g. at the limit of p-stability) and the escape (from the cycle, contributing to heavy element nucleosynthesis). At Louvain-la-Neuve, we have been concentrating our efforts on investigating these vital reactions in the hot CNO cycles.

EXPERIMENTS

The investigation of explosive stellar burning requires the development of accelerated radioactive beams with intensities in the 10^7 - 10^9 pps (particles per second) range, accelerated to energies in the region of astrophysical interest (0.2-1 MeV/u). Fully operational since 1991, the radioactive beam facility at Louvain-la-Neuve provides beams of the apropriate intensities and energies in the mass range A = 6-35 by using two cyclotrons coupled via a high intensity positive plasma ion source (ECR). The radioactive element is produced through (p,n), (p,2n), (p,αn), (p,2p) etc reactions on thick targets such as ^{13}C, LiF using a high intensity (> 150 µA) 30 MeV proton beam from CYCLONE 30, which is stopped in the target. The latter reaches its release temperature (2000 °K) by balancing beam heating and water cooling of the production target. The radioactive species is transported to the ion source in gaseous form, where it is subsequently ionized. The ions are then extracted and axially injected into a second cyclotron (K = 110), where they are accelerated (in a high harmonic mode) to the required (low) energy of usually < 1 MeV/u.

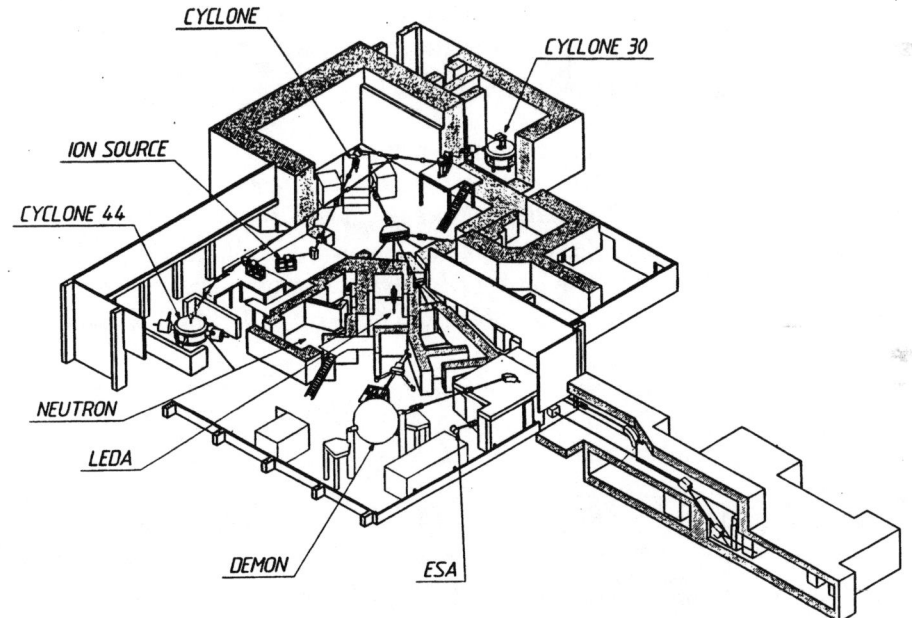

FIGURE 2. The radioactive beam facility at Louvain-la-Neuve. The location of the three cyclotrons is indicated : 30 MeV p production site - CYCLONE 30 - for radioactive elements, K=110 accelerator CYCLONE, K = 44 accelerator CYCLONE 44. Three external positive ion sources (ECR) feed the K = 110 (2) and K = 44 (1) accelerators, respectively. A few features are shown : the NEUTRON beam line, the large Si-strip detector array LEDA, the neutron detector array DEMON and the irradition facility for the European Space Agency (ESA). A recoil mass separator ARES is currently being installed.

The facility provides intense radioactive ion beams of ^{6}He, ^{13}N, ^{15}O, ^{18}F, ^{18}Ne and ^{19}Ne of a few (< 15) MeV/u for nuclear physics and nuclear astrophysics experiments. New beams are under development. A third K = 44 cyclotron, dedicated to nuclear astrophysics is expected to deliver its first beam by the end of 1997. A view of the first generation facility for radioactive beams at Louvain-la-Neuve is displayed in Fig. 2.

With the exception of ^{6}He, all radioactive beams available are on the proton-rich side of the periodic table. This limitation arises from the use of CYCLONE 30 (30 MeV p) for production, a high intensity but low energy proton accelerator. The Louvain-la-Neuve facility is therefore particularly suited to study reactions in the hot CNO cycles and in the rp-process. While some information relevant to astrophysics has been obtained in the past by indirect methods using stable beams, e.g. spectroscopic data from (^{3}He,t) and (^{3}He,n) reactions, it is preferable to measure the quantities of interest directly, whenever possible.

Accelerated radioactive beams of similar properties may soon become available at Rex-ISOLDE (Cern), SPIRAL (Ganil, France), ISAC (Triumf, Canada), CNS-Tokyo (Japan), Oak Ridge (USA), GSI (Germany), RAL (UK) and elsewhere. Preliminary results have been obtained at Oak Ridge. The ISOLDE and ISAC facilities have provided radioactive ions in the 10-200 keV range for many years. The experience gained in our facility will certainly enter into the "inital composition" of next generation facilities.

In experiments, very efficient detection systems have to be deployed due to the low cross sections and the (relatively) low intensities of accelerated radioactive beams. Note that an intense radioactive beam of 6×10^9 pps amounts to only 1 pnA (particle nano Amp.). At the same time, the detection system should be highly selective for optimal background suppression. As most radioactive elements are beta emitters, the large detector arrays should also be able to cope with the hostile radiation environment.

Onset of the hot CNO cycles

The standard or cold CNO cycle operates predominantly in the quiescent phase of burning in massive stars. Hydrogen is converted to helium through the well-known sequence of reactions: ^{12}C(p,γ)^{13}N(β^+)^{13}C(p,γ)^{14}N(p,γ)^{15}O(β^+)^{15}N(p,α)^{12}C. At the end of the sequence, ^{12}C is restored to the initial abundance; ^{12}C acts as a catalyzer in the conversion process. As the temperature and/or density increases, p-capture on ^{13}N competes with the beta-decay of ^{13}N ($T_{1/2}$ = 10 min), thus bypassing ^{13}C. The capture reaction ^{13}N(p,γ)^{14}O characterizes the transition from the cold to the hot CNO cycle. It is therefore considered to be the onset of the hot CNO cycle. At even higher temperatures/densities, unstable nuclei of shorter halflives (than ^{13}N) contribute to an increasingly complex network of reactions in the hot CNO cycles. The hot CNO reaction chains are important once the burning process becomes fast that is in the explosive stellar environment. Novae and X-ray bursters are most likely ignited by the hot CNO cycles operating on hydrogen accreting at their surfaces, respectively.

The first radioactive beam developed at Louvain-la-Neuve was ^{13}N (1,2). The radioactive species is obtained through the ^{13}C(p,n)^{13}N reaction. ^{13}N must still be considered to be a "difficult" beam even by todays standards; unlike inert gases such as the noble gases He, Ne,...., nitrogen is reactive at the high temperatures needed for an efficient release of the activity from the production target. A release efficiency of about 30 % of the total ^{13}N activity produced is achieved routinely with our production target (3). Final ^{13}N beam intensities on target are 3×10^8 pps. The purity of the ^{13}N beam is routinely better than 99 % after stripping.

After a thorough investigation of various detection and target techniques (4), we decided on using thin polyethelene foils $(CH_2)_n$ of 180 µg/cm^2 mounted on a rotating motor-driven annulus. For the final experiment (2), a high resolution Ge detector of 90 % relative efficiency was chosen over a setup using six 3"x3" NaI(Tl) detectors. In test experiments with the ^{13}C(p,γ)^{14}N reaction, it was realized (4) that, due to the high inherent γ-background in the radioactive beam experiment, high resolution was more important than high efficiency, - in this particular case of ^{13}N(p,γ)^{14}O. In these early experiments (4) it became also clear that beam contaminants of A = 13, essentially ^{13}C and ^{12}CH, had to be effectively suppressed, in order to carry out a successful experiment. Finally, the problem of absolute normalization was solved by measuring the resonant scattering of the proton recoils in Si detectors simultaneously with the capture γ-rays. The resonant part of the scattering amplitude can be deduced from the proton yield by a Breit-Wigner type formula incorporating coulomb, hardsphere and resonant scattering amplitutes. This method was verified (4) by measuring the resonant scattering ^{13}C(p,p)^{13}C in the regime of the $\ell = 0$ resonance at 512 keV in the ^{13}C+p center-of-mass frame.

In the region of astrophysical interest, the ^{13}N(p,γ)^{14}O rate was expected to be dominated by the tail of a $\ell = 0$ resonance at 545 keV in the ^{13}N+p center-of-mass frame, corresponding to the 1$^-$ level in ^{14}O at 5.17 MeV excitation energy. Our experiments revealed (2,6) that the resonance is in fact situated at 526 ± 1 keV. Its total width is 37 ± 1.1 keV. The correct excitation energy of the first excited 1$^-$ level in ^{14}O is 5.15 MeV. The γ-width of this level was measured to be $\Gamma_\gamma = 3.3 \pm 0.7 \pm 0.6$ eV; the first number is the statistical and the second one the systematic error of the measurement. Subsequently, we carried out experiments to determine the component of direct capture (as opposed to resonant) through the ^{13}N(d,n)^{14}O reaction (5). Not unexpectedly, the direct capture component turned out to be quite small even in the tail region of the resonance. However, due to constructive (destructive) interference below (above) the resonance energy, the effect of direct capture on the reaction rate in the tail region is important and cannot be neglected, as can be seen in (2,5,6).

In conclusion, the direct measurements of the $^{13}N(p,\gamma)^{14}O$ and $^{13}N(d,n)^{14}O$ rates made it possible to delineate the temperature/density regions where the cold and hot CNO cycles dominate. Under nova burning conditions, $^{13}N(p,\gamma)^{14}O$ bypasses the beta-decay to ^{13}C. This may have consequences for the s-process, which relies on $^{13}C(\alpha,n)^{16}O$ as a source of neutrons. Further astrophysical consequences have been discussed in (7).

Escape from the hot CNO cycles

The most likely escape from the hot CNO cycles at temperatures and densities prevalent in nova explosions is through the reaction sequence $^{15}O(\alpha,\gamma)^{19}Ne(p,\gamma)^{20}Na$. This leakage may remove heavy nuclides from the CNO cycles and feed the rp-process. In the explosion of a nova, matter is always ejected into the interstellar medium. The situation is different for X-ray bursters, which operate at much higher temperature and density through a variety of possible reaction chains. However, matter ejected from the massive X-ray burster is pulled back in by gravition, and consequently they do not contribute to the ongoing nucleosynthesis.

^{19}Ne beams were produced from a LiF target contained in a graphite box through the $^{19}F(p,n)^{19}Ne$ primary reaction. The LiF production target becomes liquid under irradation with the intense 30 MeV p beam (> 150 µA). As ^{19}Ne is an inert gas, extraction from the target is easy and highly efficient. The halflife of ^{19}Ne is $T_{1/2}$ = 17.2 s, which is sufficiently long so that all activity diffuses out of the liquid LiF target. Final ^{19}Ne beam intensities on target are 2×10^9 pps. The ^{19}F impurity in the beam is < 1%.

At nova temperatures of $2-5 \times 10^8$ °K, the reaction rate $^{19}Ne(p,\gamma)^{20}Na$ is expected to be dominated by a resonance at E_{cm} = 450 keV above the $^{19}Ne+p$ threshold. A comparison with isobaric nuclei has led to a tentative assignment of its spin, parity and strength (8). Four independent experiments (9,10) were carried out to deduce an upper limit for the reaction strength of this resonance. ^{20}Na ($T_{1/2}$ = 0.45 s) decays through β^+ emission (E_β < 11.2 MeV) to ^{20}Ne; a 20 % branch populates unbound levels in ^{20}Ne, followed by α-decay to ^{16}O (E_α = 2.15, 16 % and 4.43 MeV, 4 %). In two of the experiments we detected α particles in (i) 2 thin double-sided Si (X-Y) strip detectors of 48 x 16 strips each (30 and 60 mm, respectively), and in (ii) solid state nuclear track detectors; passive polycarbonate foils, which were etched chemically after irradiation thus revealing possible a-tracks. Both types of detectors were essentially insensitive to the high flux of minimum ionizing particles of 10^9 β^+ per sec. The targets used were thin polyethylene foils. In the other two experiments, β^+ particles were detected in a stack of scintillators at the end of a solenoid of longitudinal magnetic field. Thin polyethylene foils were used as targets in one case and a H-gas target (H depleted in deuterium) in the other. The total efficiency of the experimental setup was roughly 1%, in each of the four cases, inspite of considerable efforts. A fifth experiment served to determine the resonance parameters of two higher lying levels at E_{cm} =797 and 887 keV above the ^{19}Ne+p threshold in ^{20}Na (11).

The results of the measurements are summarized as follows: (i) an upper limit of the resonance strength $\omega\gamma$ < 18 meV was obtained for the 450 keV level (9); (ii) resonance strengths for several energy intervals above 600 keV were determined (10), resonances at higher energy may dominate the ^{19}Ne(p,γ)^{20}Na rate in the high temperature/density regime; (iii) resonance energy, total width, spin and parity of two strong resonances in ^{20}Na were deduced (11). As in the previous case, the direct capture component was determined by means of the ^{19}Ne(d,n)^{20}Na reaction (9,10). Furthermore, the upper limit could only be obtained after subtracting the contribution of the (d,n) reaction on the (natural abundance) deuterium in the polyethylene targets.

The astrophysical implications of these results are : (i) The upper limit obtained for ^{19}Ne(p,γ)^{20}Na is low ; given that the estimated strength $\omega\gamma$ = 10 μeV for ^{15}O(α,γ)^{19}Ne is roughly correct, this reaction impetes the flow out of the hot CNO cycles at standard nova temperatures and densities, it constitutes the bottleneck as only an insufficient number of ^{19}Ne nuclei is produced to feed the rp-process; (ii) at higher temperature and density, a strong additional feeding of ^{15}O, starting from ^{16}O seed nuclei via the sequence ^{16}O(p,γ)^{17}F(β^+)^{17}O(p,γ)^{18}F(p,α)^{15}O, may arise, however, and "push enough flux" through the bottleneck to produce a sufficient number of ^{19}Ne nuclei to operate the rp-process.

The ^{18}F(p,α)^{15}O reaction

The standard sequence of the hot CNO cycle creates ^{15}O nuclei through the chain ^{12}C(p,γ)^{13}N(p,γ)^{14}O(β^+)^{14}N(p,γ)^{15}O, awaiting beta decay of ^{14}O ($T_{1/2}$ = 70.6 s). In the high temperature explosive environment, two additional chains can produce ^{15}O nuclei originating from ^{12}C and ^{16}O seeds (Fig.1) : ^{16}O(p,γ)^{17}F(β^+)^{17}O(p,γ)^{18}F(p,α)^{15}O and ^{12}C(p,γ)^{13}N(p,γ)^{14}O(α,p)^{17}F(β^+)^{17}O(p,γ)^{18}F(p,α)^{15}O.

A ^{18}F beam was produced by irradiating ^{18}O-enriched water with a 12 μA proton beam of 15 MeV. Fluor from the primary reaction ^{18}O(p,n)^{18}F was then treated chemically "offline" substituting iodine in CH_3I to form $CH_3^{18}F$ molecules. These molecules were transported to the ECR ion source, cracked in the electron plasma and ionized to the q = 2$^+$ charge state. Due to the long lifetime of ^{18}F ($T_{1/2}$ = 110 min), the activity > 1 Curie could be produced and fed to the ion source in batch mode in 2 hour intervals. Final ^{18}F intensity on target during the experiments was 10^6 pps averaged over 2 hour irradiation periods.

Two energy regions in ^{19}Ne above ^{18}F+p threshold were investigated, E_{cm} = 265-535 and 550-740 keV. The targets were thin polyethelene foils of 275 and 200 µg/cm^2, respectively. Charged particles were detected in the Louvain-Edinburgh Detector Array (LEDA), a large annular Si strip detector array of 100 mm inner and 260 mm outer diameter, arranged in 8 segments of 16 strips each. The angular range covered in these particular experiments was 12°-28° in the laboratory frame. A clear separation of ^{18}F, ^{12}C-recoils, p-recoils and α particles was achieved by using energy and time-of-flight information.

In the 550-740 keV energy range, a broad resonance dominates the yield (12). The level energy and total width were deduced from the alpha spectra to be E_{res} = 638 ± 15 keV and Γ_{tot} = 37 ± 5 keV. Angular momentum, spin and parity of the level and its partial width Γ_p were obtained from a Breit-Wigner one-level-formula fit to the proton spectra, ℓ =0, J^π=3/2$^+$ and Γ_p/Γ_{tot} = 0.5 ± 0.1. The resonance strength of the level was determined from the (p,α) reaction yield to be ωγ = 5.6 ± 0.6 keV. The strength of the resonance can also be obtained independently from a complete determination of the resonance parameters. With the statistical factor ω = 4/6 and Γ_p/Γ_{tot} = 0.5 ± 0.1 we obtain ωγ = 5.9 ± 0.9 keV in agreement with the value above.

A recent experiment at Argonne (13) quotes Γ_{tot} = 13.6 ± 4 keV and ωγ = 2.1 ± 0.7 keV, in disagreement with our results. However, in the Argonne experiment only the ^{15}O recoil yield was measured, as compared to the measurement of both the (p,α) yield and the (p,p) resonance parameters in our case.

The experimental techniques employed at lower energy are as above. The 265-535 keV energy range is dominated by the tail of the broad 638 keV resonance and by a narrow resonance at about 324 keV. Additional levels are expected (14) to occur at 287 ± 7 and at 450 ± 7 keV. No evidence for these levels has been found, possibly due their weak strengths. The reaction rate for ^{18}F(p,α)^{15}O is the highest among all rates in the competing sequences and processes the initial ^{12}C and ^{16}O abundances efficiently into ^{15}O. These experimental results and astrophysical implications are discussed in detail in a comprehensive paper (15).

The high ^{18}F(p,α)^{15}O reaction rate implies that ^{16}O is rapidly converted into ^{15}O under prelevant nova conditions (15). While ^{15}O may mostly β–decay to ^{15}N and remain in the CNO cycles, a high initial abundance of ^{16}O in the pre-nova nevertheless enhances the possibility for escape through the ^{15}O(α,γ)^{19}Ne bottleneck.

Outlook

Recently, the ^{18}Ne(α,p)^{21}Na reaction was investigated at Louvain-la-Neuve. This possible escape route may be active in a very high temperature/density environment.

An array of Si strip detectors was placed inside the He target chamber in the forward hemisphere behind a thin stopper foil. The analysis of the data is in progress.

Some remaining uncertainties can be eliminated over the next years. First, more precise data for $^{18}F(p,\alpha)^{15}O$ in the low energy range can be expected with the higher intensity ^{18}F beams from the new K = 44 cyclotron. In addition, it is planned to study the capture reaction $^{18}F(p,\gamma)^{19}Ne$ with the recoil mass separator ARES, currently being installed. We could also lower the present upper limit on the $^{19}Ne(p,\gamma)^{20}Na$ rate (9) by an order of magnitude in a new experiment using the K = 44 cyclotron and the recoil separator.

The most important remaining problem is the determination of the $^{15}O(\alpha,\gamma)^{19}Ne$ rate. This capture reaction constitutes the bottleneck for escape from the hot CNO cycles. However, the experimental problems are such that a direct measurement, as carried out in the above cases, is extremely difficult. A succesfull attempt would require typically 10^{11} pps ^{15}O beam, 10^{18} He target atoms and close to 100 % detection efficiency. We have therefore adopted a different strategy, deploying an indirect method but using a radioactive beam (16). The idea is to populate excited states in ^{19}Ne through a transfer reaction in inverse kinematics and investigate the subsequent α–decay. Using a nuclear reaction rather than the (α,γ) capture of electromagnetic origin, several orders of magnitude are gained in the cross section. The reaction chosen for this purpose is $^{18}Ne(d,p)^{19}Ne \rightarrow {}^{15}O+\alpha$. This reaction exhibits some advantages :
(i) inverse kinematics allows to detect all three particles of interest with high efficiency, as they are focussed forwards $(\alpha, {}^{15}O)$ and backwards (proton), respectively;
(ii) by detecting all three particles with high energy and angular resolution, the system is kinematically overdetermined and accidental background events are reduced;
(iii) using triple coincidences in energy and time-of-flight, a further strong rejection of unrelated events can be achieved. A first experiment is scheduled for the end of '97.

The $^{15}O+\alpha$ yields from the transfer reaction, at different excitation energies in ^{19}Ne, have to be related to the $^{15}O(\alpha,\gamma)$ capture rate. For this purpose, the "isobaric" $^{18}F(d,p)^{19}F \rightarrow {}^{15}N+\alpha$ reaction will be investigated in parallel. The $^{15}N(\alpha,\gamma)^{19}F$ capture yield is a wellknown quantity (17) in a wide energy range, so that in this case direct measurement and indirect method can be related to one another.

Once an intense ^{7}Be beam becomes available at the K = 44 cyclotron, one could also investigate alternative routes to the triple alpha process for producing ^{12}C, such as $^{7}Be(\alpha,\gamma)^{11}C(p,\gamma)^{12}N(\beta^{+})^{12}C$. Other reaction chains may also be important in bypassing the mass = 8 gap initiated by $^{7}Be(\alpha,p)^{10}B$ ...... and $^{7}Be(^{3}He,\gamma)^{10}C$ .... This merits further investigation.

CONCLUSIONS

Over the last 7 years, we have investigated the hot CNO cycles at the Louvain-la-Neuve facility through direct measurements using radioactive beams. Valuable results have been obtained and are summarized in this paper. Nuclear astrophysics research with radioactive beams is continuing. Such measurements of the key reaction rates in the explosive stellar environment will further our understanding of the ongoing stellar nucleosynthesis.

ACKNOWLEDGEMENTS

The experimental work is a collaboration of the two Catholic Universities of Louvain (UCL Louvain-la-Neuve, KUL Leuven), the Free University of Brussels (ULB), the University of Edinburgh (UK), the University of Notre Dame (USA) and the Technical University of Budapest (Hungary) and is supported by the Belgian Government, the UK ESPSRC and by NATO Grant CRG.940213.

REFERENCES

1. Darquennes, D. et al., *Phys. Rev. C* **42**, R804 (1990)
2. Decrock, P. et al., *Phys. Rev. Lett.* **67**, 808 (1991)
3. Delbar, Th. et al., *Nucl. Instr. Meth B* **84**, 512 (1994)
4. Galster, W. et al., *Phys. Rev. C* **44**, 2776 (1991)
5. Decrock, P. et al., *Phys. Rev. C* **48**, 2057 (1993)
6. Delbar, Th. et al.., *Phys. Rev. C* **48**, 3088 (1993)
   Decrock, P. et al., *Phys. Lett. B* **304**, 50 (1993)
7. Arnould, M. et al.., *Astr. Astrophys.* **254**, L9 (1992)
8. Kubono, S. et al., *Z.Phys. A* **331**, 359 (1988)
   Lamm, L. et al., *Nucl. Phys. A* **510**, 503 (1990)
   Brown, B. et al., *Phys. Rev. C* **48**, 1456 (1993)
9. Page, R. et al., *Phys. Rev. Lett.* **73**, 3066 (1994)
10. Michotte, C. et al., *Phys. Lett. B* **381**, 402 (1996)
    Vancraeynest, G. et al., *Nucl. Phys. A* **616**, 107c (1997)
11. Coszach, R. et al., *Phys. Rev. C* **50**, 1695 (1994)
12. Coszach, R. et al., *Phys. Lett. B* **353**, 184 (1995)
13. Rehm, E. et al., *Phys. Rev. C* **53**, 1950 (1996)
14. Utku, S. et al., submitted to *Phys. Rev. C*
15. Graulich, J. et al., submitted to *Nucl. Phys. A*
16. Louvain-la-Neuve, Catania, Edinburgh Collaboration
17. Magnus, P. et al., *Nucl. Phys. A* **470**, 206 (1987)

A plan for ^{4}He(^{12}C,^{16}O)γ experiment

Kenshi Sagara

Department of Physics, Kyushu University, Hakozaki, Fukuoka, 812-81 Japan

Abstract. An experiment on ^{4}He(^{12}C,^{16}O)γ reaction cross section for an astrophysical interest has been planned at the Kyushu University tandem accelerator laboratory (KUTL) and equipments for the experiment are under development. The present status of the preparation is described.

General

The cross section of ^{4}He-^{12}C capture reaction at low energy is very small because of the Coulomb barrier. The total cross section is of the order of 10^{-18} barn at E_{cm} = 0.3 MeV, the energy of astrophysically importance, and roughly 10^{-12} barn even at E_{cm} = 0.7 MeV. Therefore, the measurement has to be made with a high production rate and a high detection efficiency, and in low-background circumstance. Hence, we need a high intensity beam of well-defined energy, a target of enough thickness containing no other materials, a high-resolution separator of the reaction products, a high-efficiency detector, and effective methods to suppress backgrounds.

To increase the detection efficiency, we use the inverse kinematics. A ^{12}C beam is incident on a ^{4}He target. The reaction products of ^{16}O are emitted in a forward-angle cone of about ±2° as seen in Fig. 1. By using a recoil mass separator, we completely separate the ^{12}C beam away, collect and focus the ^{16}O products on a charged particle detector, and detect the ^{16}O products with a 100% efficiency. The layout of the experimental equipments is shown in Fig. 2.

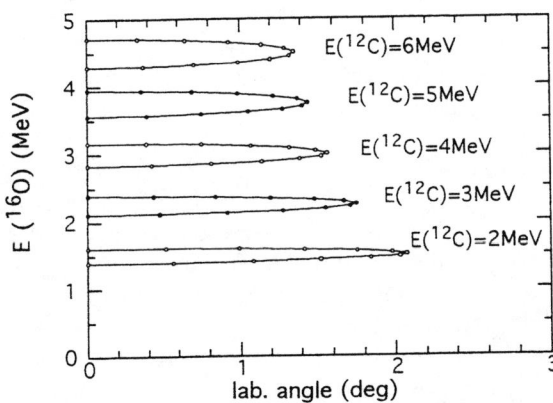

FIGURE 1. Kinematics of ^{4}He(^{12}C,^{16}O)γ reaction. At E(^{12}C) = 4 MeV, E_{cm} is 1 MeV.

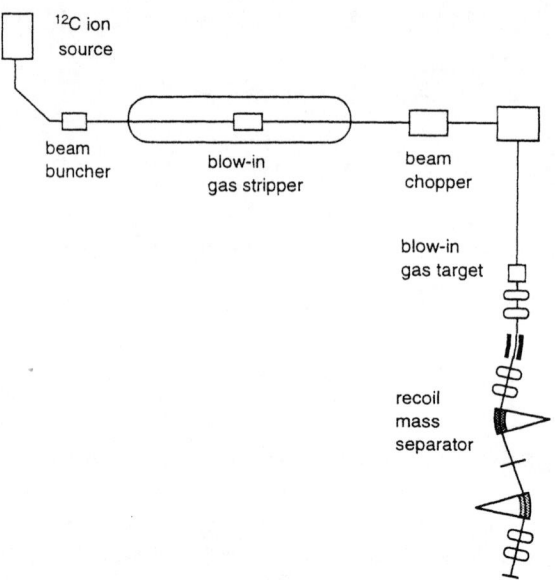

FIGURE 2. Layout of the experimental setup for ^{4}He(^{12}C,^{16}O)γ reaction cross section.

Ion source

A negative ^{12}C ion beam is extracted from a sputter ion source, SNICSII. The ion source has a capability to produce a ^{12}C beam of intensity more than 100 μA according to the specification document. The negative ^{12}C ion beam is pre-accelerated to 100 keV and is injected into the tandem accelerator. The ion source has already been installed.

Accelerator

The Kyushu University tandem accelerator is a home-designed accelerator having home-made pellet chains to carry the electric charge. The maximum terminal voltage is 10 MV, and the total length of the accelerator is about 10 m. To obtain the ^{12}C beam of energy of 3 - 6 MeV for the ^{4}He(^{12}C,^{16}O)γ experiment, the terminal voltage is set below 2 MV. The low voltage operation brings forth two problems. One is that the beam focusing of the accelerator becomes weak and some improvement to push up the focusing force may be necessary. We have ideas to effectively shorten the acceleration length and to restore the focusing force, though the practical design work has to be made hereafter. The other problem is that the charge stripping carbon foil at the terminal may be broken in a very short time by the low-energy and high-intensity ^{12}C beam. The foil stripper has to be replaced by a gas stripper. We are developing a large-acceptance gas stripper based on a new idea. The gas stripper is described later together with our gas target which has already been developed on the same idea.

Beam buncher and chopper

For the complete separation of the reaction product ^{16}O from the ^{12}C beam and other backgrounds, we use the time of flight of the particles from the target to the detector. To determine the time when the reaction takes place, we use the bunched beam. According to our design, the time interval between the beam pulses is about 150 ns and the time width of the beam pulse is about 1 ns. The use of the pulsed beam decreases the continuum backgrounds at least to 1/150 level.

Since it is difficult to supply a high voltage of perfectly saw-teeth shape to bunch the beam, the bunching procedure is divided into three stages. First by supplying low voltage of saw-teeth like shape onto the extraction voltage in the ion source, about 90 % of the beam is roughly pre-bunched. The time width of the pre-bunched beam is then shorten by the beam buncher in front of the tandem accelerator. When the beam comes out of the accelerator, the width of the beam may be about 1 ns, that is 90 % of the beam is gathered in a beam pulse of about 1 ns width. The remaining 10 % component of the beam may be homogeneously distributed in all over the time. The continuum beam component may produce harmful backgrounds in the time spectrum of the true events which is very small in number. Hence we cut the continuum component by a beam chopper placed after the accelerator. The chopper allows only the beam pulses of 1 ns in width to pass through and stops all the other beam components.

The pre-buncher in the ion source and the chopper are now being designed. A prototype buncher was installed in front of the tandem accelerator. It was found in the first test experiment that a 60 % of the beam was bunched within the time width of about 6 ns. There is no fatal technical problem foreseen, and the pulsed ^{12}C beam will be obtained after the budget problem is settled.

Gas target and gas stripper

When ^4He target gas is sealed in a cell having thin window foils, and the low-energy and high-intensity ^{12}C beam enters and exits through the foils, the foils may be severely damaged in a short time by the beam. Hence the ^4He target has to be a windowless one.

Although the gas jet target is one possible way to make a windowless target, we took another way because the gas jet target needs an enormous gas flow to maintain the jet and consequently needs a huge pumping system. The conventional windowless gas target of the differentially pumping method needs a little gas flow, however, the target thickness is not enough. We have, therefore, developed a new windowless gas target, called a blow-in type gas target (BIGT)[1]. The gas is blown into the central target region symmetrically from the walls of the upstream and the downstream holes which are the beam entrance and the exit, respectively. Figure 3 shows the pressure distributions of

BIGT and of the conventional blow-out target in which the gas is injected directly into the central target region. The distributions were measured by using the proton scattering by the target gas. In BIGT the target pressure is high. Moreover, since the target boundary is sharp the target region is well defined.

The ^{4}He gas target thickness necessary for the ^{4}He(^{12}C,^{16}O)γ experiment is about 0.1 atm•cm. The target thickness indicated in Fig. 3 is about half of this value. Cooling of the target gas to decrease the gas diffusion and to increase the gas density, or enlarging the pumping speed to increase gas flow may be necessary.

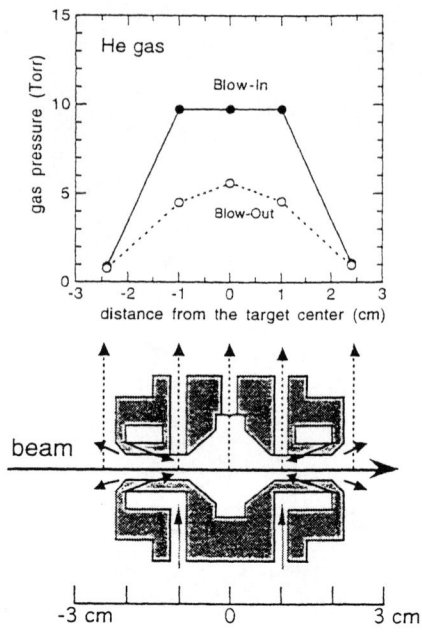

FIGURE 3. Pressure distributions of the blow-in and the conventional blow-out gas targets.

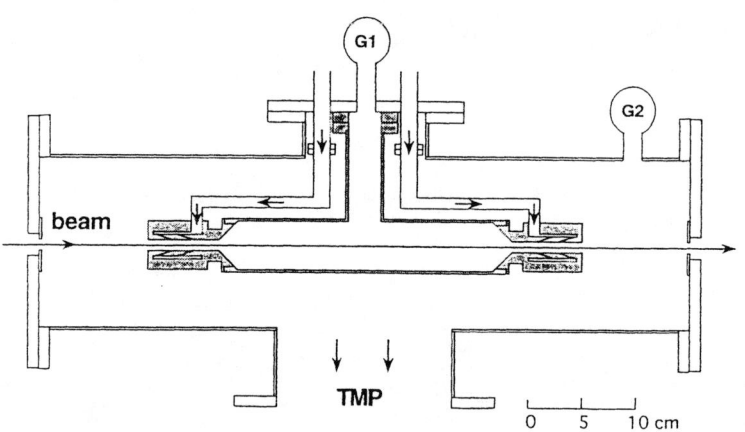

FIGURE 4. The first model of the gas stripper. G1 and G2 are the pressure gauges.

We are now examining whether the same blow-in method is applicable also for the gas stripper of the tandem accelerator. The most different point is in the fact that thickness of the stripper gas (~μg/cm^2) is two or three orders of amplitude thinner than that of the gas target. When the gas pressure decreases and the gas flow becomes a molecular flow, the confinement force of the blow-in gas flow may disappear. Hence the confinement action was first investigated using the first model in Fig. 4.

The first model had a stripper length of about 40 cm. The number of the blow-in gaps was varied from 1 to 3 (2 in Fig. 2). The beam entrance and the exit holes of the stripper had the diameter of 8 mm. A vacuum pump had the pumping speed of 150 l/s.

For the 3-gap blow-in stripper, the pressures P1 and P2 were measured in wide pressure range and the confinement factor (P1/P2) was evaluated for several different gases. When the stripper gas pressure (P1) was in the range from 10^{-3} Torr to 10^{-2} Torr, the confinement factors were found to take the maximum values. The maximum values were 60, 50 and 10 for argon, nitrogen and hydrogen gases, respectively. When the stripper gas pressure became below 10^{-3} Torr, the confinement factor was found to decrease gradually with the pressure.

It is considered that the decrease of the confinement factor at higher pressure may be due to the decrease of the pumping speed of the molecular pump, and that the factor may be constant if the pumping speed is constant. On the other hand, the decrease of the factor in the low pressure region may be attributed to the decrease of the molecular collisions in the blow-in process.

Nitrogen gas of 10^{-2} Torr times 40 cm has thickness of 0.66 μg/cm^2, which may be an accessible value for the gas stripper. We have therefore an optimistic anticipation for the blow-in gas stripper although some improvement may be necessary to enlarge the diameter of the beam holes.

Recoil mass separator

The recoil ^{16}O are emitted within a forward cone of a half angle of less than 2°. To separate them from the ^{12}C beam and focus them on the detector plane, we have designed a recoil mass separator in Fig. 5. The electric deflector has a rather small deflection angle of 12° with the radius of curvature of 2.5 m, so that the ^{12}C beam does not hit the deflector electrode. The ^{12}C beam is completely stopped by shutters placed between the two magnetic dipoles. The recoil ^{16}O's having the same charge are focused at the same position on the detector plane.

For ^{16}O of energy of a few MeV, 3$^+$ charge state may have the maximum fraction of 0.3 - 0.4 in the charge state distribution. Since our recoil mass separator can accept particles emitted within the angle of ± 2°, and since charged particles can be detected with a 100 % efficiency, then 30-40 % of all the ^{16}O particles produced by ^{4}He(^{12}C,^{16}O)γ reaction are detected by a detector. We measure the energy and the time of flight of the focused particles, and plot the data in a two-dimensional (energy-time) plane to reject still remaining backgrounds.

All the focusing elements have already been fabricated and the bench test of the electric deflector is going to start. Installation of all the elements will be carried out in Spring 1998.

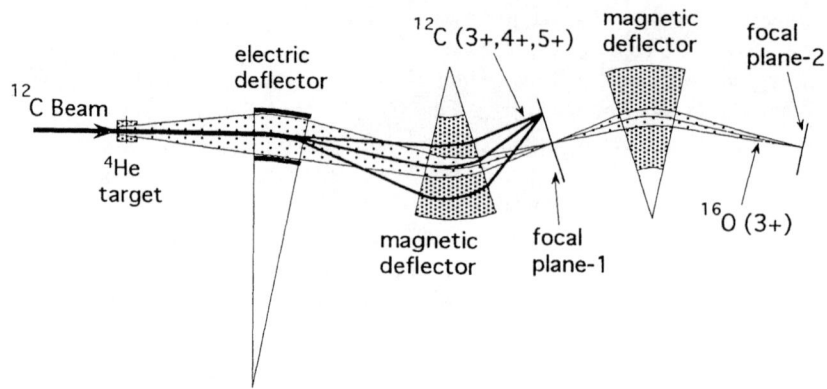

FIGURE 5. Recoil mass separator. On the first and the final focal planes, particles are focused with the velocity dispersion and the mass (per charge) dispersion, respectively.

Summary

The direct measurement of the ^{4}He-^{12}C capture reaction cross section has been planned by the inverse kinematics ^{4}He(^{12}C,^{16}O)γ experiment. Development of the windowless gas target as well as design and fabrication of the recoil mass separator have already been made. The beam buncher, the beam chopper, and the gas stripper are now under development. The first test experiment is expected to be done at the end of 1998.

Our goal is to measure the ^{4}He(^{12}C,^{16}O)γ reaction cross section within an accuracy of about 10 % in the energy range from E_{cm} = 3 MeV down to E_{cm} = 0.7 McV.

ACKNOWLEDGMENTS

The preparation of the experimental equipments has been made by the collaboration with S. Morinobu, T. Nakashima, N. Ikeda, S. Kojima, A. Motoshima, K. Tsuruta, M. Koga, K. Ogata, M. Nakamura, and H. Urabe.

The present work is supported in part by Japan Ministry for Education, Science and Culture under Grant-in-Aid No. 08404017.

REFERENCES

1. K. Sagara et al, Nucl. Instr. and Methods **A378** (1996) pp. 392-398.

Time Scale for Non-Resonant Breakup of 7Li Over the Gamow Energy Region

H. Utsunomiya,* Y. Tokimoto,* K. Osada,* T, Yamagata,* M. Ohta,*
Y. Aoki,#1 K. Hirota,#1 K. Ieki,#2 Y. Iwata,#2 K. Katori,#3
S. Hamada,#4 Y.-W. Lui#5 and R.P. Schmitt#5

*Department of Physics, Konan University, 8-9-1 Okamoto, Higasinada, Kobe 658, Japan
#1 Institute of Physics, University of Tsukuba, Tsukuba, Ibaragi 305, Japan
#2 Department of physics, Rikkyo University, Nishi-Ikebukuro, Tokyo 171, Japan
#3 Department of Physics, Osaka University, Toyonaka, Osaka 560, Japan
#4 Advanced Science Research Center, JAERI, Tokai, Ibaragi, 319-11, Japan
#5 Cyclotron Institute, Texas A&M University, College Station, Texas 77843, USA

Abstract. Cross sections for α-t coincidences were measured at energies of astrophysical relevance in non-resonant breakup of 42 MeV-7Li by various targets. The data have revealed delayed nature of non-resonant breakup over the Gamow energy region arising from the tunneling through the Coulomb barrier between the two constituent clusters. Quantitative discussions on meanlives of continuum states and post-Coulomb acceleration are presented.

INTRODUCTION

Coulomb dissociation is considered to be a useful tool of determining radiative capture cross sections for primordial and stellar nucleosyntheses.[1] One of fundamental questions in the astrophysical context of Coulomb dissociation is related to the time scale for the process called *direct breakup*. One might assume *a priori* a prompt nature for direct breakup that breakup of the system instantaneously follows the excitation to an intermediate continuum state. In the prompt breakup of c into a + b, both a and b may suffer from post-acceleration in a Coulomb field of a target nucleus. When the two constituents (a and b) have different charge-to-mass ratios, relative energies between a and b at the time of breakup of c can differ from those at the time of detecting a and b in the asymptotic region. This post-Coulomb acceleration is considered to be a complication for Coulomb dissociation.[2-5]

In this paper, it is shown that breakup of 7Li through continuum states over the Gamow energy region is characteristic of delayed nature arising from the tunneling effect which strongly suppresses post-Coulomb acceleration.

EXPERIMENTAL METHOD

A 42 MeV-^{7}Li beam was provided by the Tandem Accelerator Center at the University of Tsukuba. Self-supporting foils of ^{197}Au (10.3 mg/cm2), ^{144}Sm (3.3 mg/cm2), ^{120}Sn (6.4 mg/cm2), ^{90}Zr (5.2 mg/cm2), ^{58}Ni (4.8 mg/cm2), and ^{27}Al (2.2 mg/cm2) were irradiated. A 2.74 msr collimator was used to select collinear α and triton breakup pairs. The detector system was the same as that used in previous experiments,[2] but was mounted in a slightly different configuration in the focal plane of an Enge split-pole spectrograph, which had a momentum acceptance, $p_{max}/p_{min} \approx 2.8$. This configuration enabled us to simultaneously measure two types of collinear emissions with $v_\alpha \geq v_t$ and $v_\alpha \leq v_t$ in the energy region of $E_{\alpha t} = 0 - 500$ keV. The present dynamic range fully covered the Gamow energy window at $E_o \approx 55 - 256$ keV with widths $\approx 54 - 369$ keV for primordial synthesis at $T \approx 0.01 - 0.1$ MeV.[6] The experimental set-up is depicted in Fig. 1. Measurements were made inside the classical grazing angles. A decrease of the breakup cross sections in the forward direction was observed for ^{197}Au. The energy calibration of the focal plane detectors was performed with the reactions ^{12}C(α, α')^{12}C (0_1^+, 2_1^+, 0_2^+, 3_1^-) for the α counter and with the reactions ^{27}Al(α, t)^{28}Si(0_1^+, 2_1^+, 4_1^+, 0_2^+, 3_1^-) for the triton counter at 29 MeV.

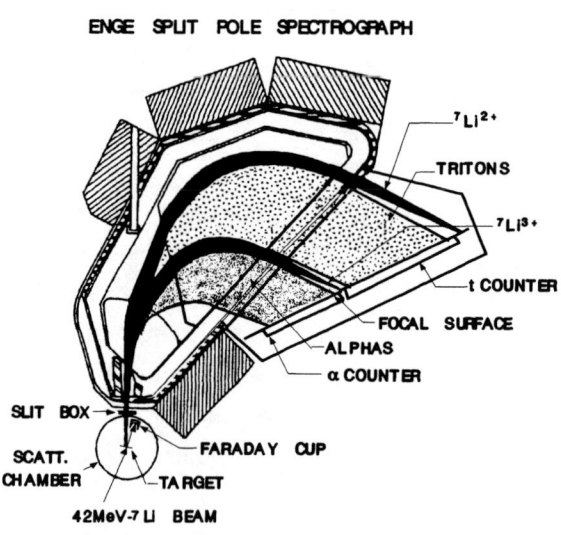

FIGURE 1. Spectrograph method of particle-particle coincidences.

RESULTS AND ANALYSES

The α - t coincidence events were dominated by elastic breakup, which was unambiguously identified for all targets except for ^{197}Au. For ^{197}Au, the experimental separation of low-lying states in the target at 77, 269, and 279 keV was not possible with the present energy resolution. The kinetic energies of the α's and the triton were corrected for the energy lost in the target foils using the stopping power table of Northcliffe and Schilling,[7] assuming that the reaction occurred at the midpoint of the target foil. The corrected energies were kinematically converted into c.m. energies with respect to the center-of-mass of the ^{7}Li + target system.

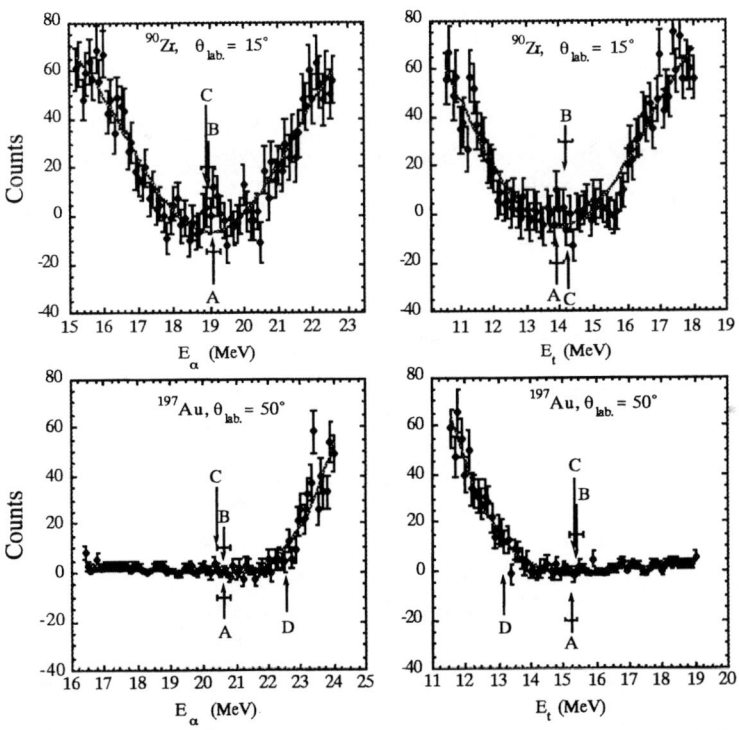

FIGURE 2. The c.m. energy distributions of α-particles and tritons measured with ^{90}Zr and ^{197}Au.

Delayed Non-Resonant Breakup

Figure 2 shows c.m. energy distributions of α's and tritons for ^{90}Zr and ^{197}Au. Highly asymmetrical distributions were observed at large angles for heavy targets, favoring the $v_\alpha \gtrsim v_t$ component. The distributions show that the depletion of coincidence events occurs within the present dynamic range. The depletion ought to correspond to breakup at $E_{\alpha t} = 0$ MeV. We searched for the location of the yield minima using two methods: (A) fits to the spectra with a fourth order polynomial function and (B) event-by-event analyses of energies leading to zero relative energy. The results of the two methods agree with each other within the estimated uncertainties (± 200 keV). In the figure, the results of the methods are indicated by arrows (A, B) along with a third arrow, C, which gives the energy expected for asymptotic breakup. The c.m. energies for the breakup at the distance of closest approach in the Rutherford orbit were calculated with kinematics treating the α and the t as free particles in the target Coulomb field without interacting each other. The results are indicated by the arrow D. Clearly the location of the depletion are consistent with that expected for the asymptotic breakup without significant post-acceleration for all targets. This holds for all of the six targets measured. The target independence suggests that the strongly suppressed post-Coulomb acceleration is associated with the projectile ^{7}Li.

Tunneling Effect

The continuum states of astrophysical relevance in ^{7}Li decay into α + t by tunneling through the Coulomb barrier. We have examined this for S-states because of their dominant role in radiative capture. The transition rate (ω), i.e., the number of decays per second,[8] can be evaluated by

$$\omega = \omega_o P \qquad (1)$$

Here ω_o is the transition rate without potential barrier and P represents the transmission probability through the Coulomb barrier. The quantity ω_o is of the same order as the *frequency of the S-wave vibration* in the potential; thus $\omega_o \approx v/2R$, where v is the velocity of the α - t relative motion and R is the nuclear radius. In the WKB approximation, the transmission probability P is expressed[8-10] by

$$P = \exp\left\{-4\eta\left[\frac{\pi}{2} - \arcsin\left(\frac{E_{\alpha t}}{B}\right)^{1/2} - \left(\frac{E_{\alpha t}}{B}\right)^{1/2}\left(1 - \frac{E_{\alpha t}}{B}\right)^{1/2}\right]\right\}, \qquad (2)$$

where $\eta = Z_t Z_\alpha e^2 / \hbar v$ is the Sommerfeld parameter and $B = Z_t Z_\alpha e^2 / R$ is the height of the Coulomb potential, and Z is the atomic number. Note that the Gamow factor which approximates P at $E_{\alpha t} \ll B$ for S-wave capture appears as the first term in the bracket in Eq. (2).

The quantity P depends solely on B, or equivalently R, within the formula for a sharp cut-off Coulomb potential. We allowed the value of R to vary in a range 4.75 - 6.0 fm. The former and latter values are the radii at which a microscopic intercluster nuclear plus Coulomb potential crosses zero and reaches a maximum, respectively (see the MHN result of Fig. 7 in Ref. 12). It was found that P is not very sensitive to the choice of R.

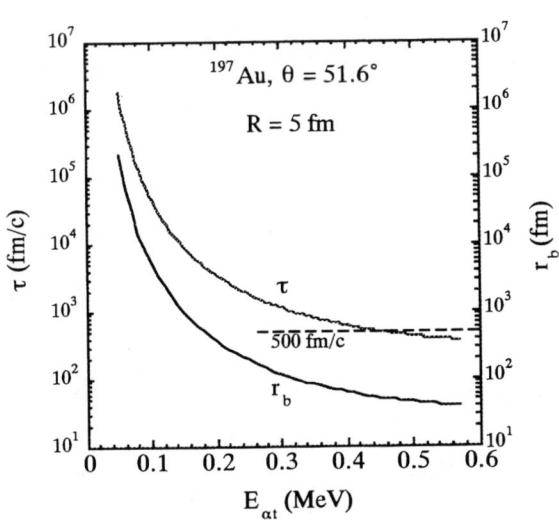

FIGURE 3. The meanlife (τ) and the location of breakup (r_b) as a function of $E_{\alpha t}$.

The resultant meanlives (the reciprocal of ω) are shown in Fig. 3 as a function of $E_{\alpha t}$. The meanlives over the Gamow energy region are larger than the nuclear transit times (60 fm/c for ^{27}Al and 120 fm/c for ^{197}Au). They are even larger than the meanlife of the

7/2- state at 4.63 MeV ($\Gamma = 93 \pm 8$ keV; $\tau \approx 2130$ fm/c) at $E_{\alpha t} \leq 200$ keV. Thus, reacceleration of the breakup fragments in the target Coulomb field would be expected to be vanishingly small compared to that for prompt breakup. Of course, the post-acceleration would become important at relative energies closer to or larger than the barrier height. It should be noted that lifetimes of $L \neq 0$ states are prolonged by the centrifugal potential.

Post-Coulomb Acceleration

In view of the lifetime of the continuum state over the Gamow energy region, post acceleration of the breakup fragments in the target Coulomb field is expected to be strongly suppressed. If we assume that the excitation of the projectile to the continuum states takes place at the distance of closest approach in a Rutherford orbit, it is straightforward to estimate the post-Coulomb acceleration. In the focal system of the hyperbolic orbit,[13] the distance between projectile and target at which breakup takes place (r_b) is given by

$$r_b = a(\varepsilon \cosh s + 1) . \tag{3}$$

Here $a = Z_P Z_T e^2 / 2E_{cm}$ is half the distance of closest approach in a head-on collision, where E_{cm} is the incident energy. The eccentricity parameter (ε) is related to the scattering angle (θ) by $\varepsilon = 1/\sin(\theta/2)$. The parameter s is related to the meanlife of the continuum state by

$$\tau = \frac{a}{v_o}(\varepsilon \sinh s + s), \tag{4}$$

where v_o is the incident velocity.

The distance thus determined for ^{197}Au, r_b, is shown in Fig. 3. The Coulomb energy which the particle i ($i = \alpha$ or t) gains after breakup is expressed by

$$E_i^\infty - E_i = \frac{A_T}{A_P + A_T} \frac{Z_i Z_T e^2}{r_b} , \tag{5}$$

where A is the mass number. The asymptotic kinetic energy E_i^∞ is the experimental observable and E_i is the kinetic energy at the breakup point. Using E_i determined from Eq. (5), relative energies E_α at the breakup point that are free from Coulomb distortion were deduced. The difference (ΔE) between $E_{\alpha t}^\infty$ and $E_{\alpha t}$ are shown in Fig. 4. For the

$v_\alpha \geq v_t$ ($v_\alpha \leq v_t$) emission in the Coulomb field of a ^{197}Au target, $E_{\alpha t}$ is shifted by 0.47 keV (- 0.80 keV) at $E_{\alpha t}$ = 100 keV, by 7.9 keV (- 17 keV) at 200 keV and by 26 keV (- 67 keV) at 300 keV.

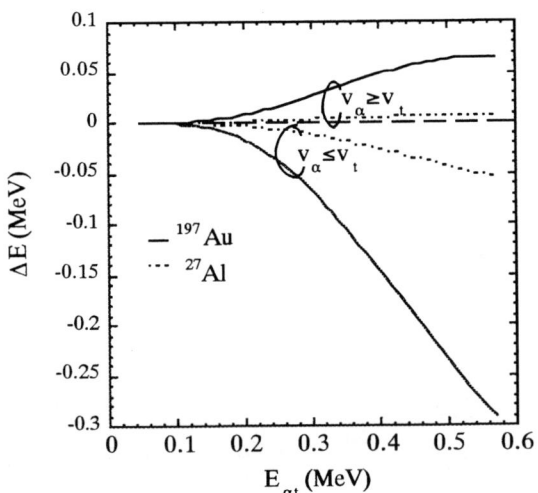

FIGURE 4. The effect of post-Coulomb acceleration on $E_{\alpha t}$.

To see the validity of the assumption that the excitation itself takes place at the distance of closest approach, let us study the property of the $df_{E1}(\theta, \xi)/d\Omega$ function.[13] For this purpose, we define the quantity $h(\theta, \xi, t)$ as

$$df_{E1}(\theta, \xi) / d\Omega = \lim_{t \to \infty} h_{E1}(\theta, \xi, t), \qquad (6)$$

where $h_{E1}(\theta, \xi, t)$ is

$$h_{E1}(\theta, \xi, t) = \frac{4\pi^2 a^2 v_o^2}{27} \sin^{-4}\frac{\theta}{2} \sum_\mu \left|Y_{1\mu}(\tfrac{\pi}{2}, 0)\right|^2 \left|\int_{-t}^{+t} \frac{(x+iy)^\mu}{r^{\mu+2}} e^{i\omega t} dt\right|^2. \qquad (7)$$

The (x, y) are cartesian coordinates and r is the radial coordinate in the focal system. The $Y_{1\mu}(\tfrac{\pi}{2}, 0)$ is the spherical harmonics. The $\xi = a E_\gamma / \hbar v_o$ is the adiabaticity parameter defined at $\theta = \pi$. Figure 5 shows the time dependence of $h_{E1}(\theta, \xi, t)$ for ^{197}Au. It is seen that the excitation primarily takes place during the time interval ± 500 fm/c. This time interval is nearly ± 4 times the nuclear transit time (t_{tr} = 120 fm/c) around t = 0, where t_{tr} was estimated by $t_{tr} \approx (R_P + R_T)/v_D$ using the velocity v_D at the distance of closest approach (t =0). This time interval is well concentrated around t = 0 in view of the meanlives of the continuum states. The time interval is indicated in Fig. 3.

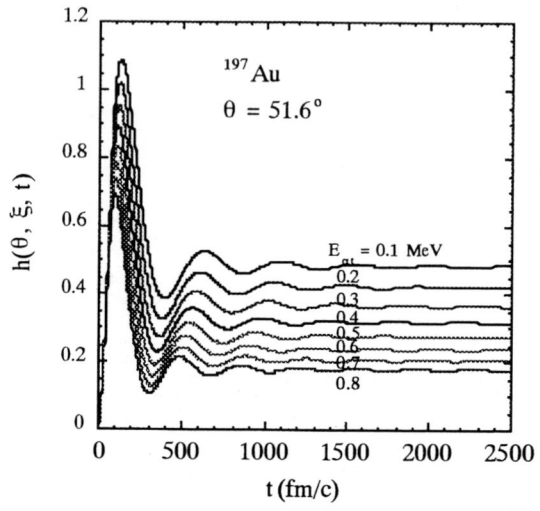

FIGURE 5. The time dependence of the h(θ, ξ, t) function.

SUMMARY

Breakup of 42 MeV-^{7}Li through continuum states at energies of astrophysical relevance was studied by measuring α-t coincidences in the collinear configuration. The data have revealed the delayed nature of non-resonant breakup over the Gamow energy region; the lifetime of the continuum state is an intrinsic quantity of the projectile determined by tunneling though the Coulomb barrier between the two constituent clusters. The estimated lifetimes vary from 5.1×10^4 fm/c at $E_{\alpha t}$ = 100 keV to 1.2×10^3 fm/c at 300 keV which are much larger than the nuclear transit time (60 –120 fm/c). Post-Coulomb acceleration is determined by the flight distance of the projectile from the target during the lifetime, which is therefore expected to become unimportant with increasing bombarding energy.

The effect of post-Coulomb acceleration on $E_{\alpha t}$ in the Gamow energy region was numerically evaluated for the collinear detection geometry based on the classical theory of Coulomb excitation. The post-acceleration effect was found to be very small for the $v_\alpha \geq v_t$ collinear branch compared to that for the $v_\alpha \leq v_t$ branch.

ACKNOWLEDGMENTS

This work was supported by the Japan Ministry of Education, Science, Sport, and Culture, by the Japan Private School Promotion Foundation, and by the US Department of Energy under grant DE-FG03-93ER40773.

REFERENCES

1. Baur, G., Bertulani, C.A., and Rebel, H., *Nucl. Phys.* **A458** 188-204 (1986).
2. Utsunomiya, H. *et al.*, *Phys. Rev. Lett.* **65** 847-850 (1990).
3. Baur, G., and Weber, M., *Nucl. Phys.* **A504** 352-366 (1989).
4. Mason, J. B. *et al.*, *Phys. Rev.* **C45** 2870-2878 (1992); reference therein.
5. Baur, G., and Rebel, H., *J. Phys. G: Nucl. Part. Phys.* **20** 1-33 (1994).
6. Smith, M.S., Kawano, L.H., and Malaney, R.A., *Astrophys. J. Suppl. Ser.* **85** 219-247 (1993).
7. Northcliffe, L. C., and Schilling, R. F., *Nucl. Data. Tables.* **A7** 233 (1970).
8. Bethe, H. A., *Rev. Mod. Phys.* **9** 69-244 (1937).
9. Rolfs, C.E., and Rodney, W.S., *Cauldrons in the Cosmos*, Chicago and London: The University of Chicago Press, 1988, ch. 4, pp. 150-189.
10. Clayton, D.D., *Principles of Stellar Evolution and Nucleosynthesis*, Chicago and London: The University of Chicago Press, 1983, ch. 4, pp. 281-361.
11. Kajino, T., *Nucl. Phys.* **A460** 559-580 (1986); references therein.
12. Alder, K.*et al.*, *Rev. Mod. Phys.* **28** 432-542 (1956).

THERMONUCLEAR REACTIONS II

New Experiments for the Breakout off the Hot-CNO Cycle

S. Kubono, M. Kurokawa[a], S.H. Park[b], K.I. Hahn[a], Y. Fuchi[c],
S.C. Jeong[c], S. Kato[d], H. Kawashima[c], J.C. Kim[b], C.H. Lee[b],
C.S. Lee[e], J.H. Lee[e], X. Liu, T. Minemura[f], T. Miyachi,
T. Motobayashi[f], P.D. Parker[g], T. Shimoda[h], M. Smith[i],
P. Strasser[a], M.H. Tanaka[c], H. Utsunomiya[j], and M. Yasue[k]

Center for Nuclear Study (CNS), University of Tokyo, 3-2-1 Midori-cho, Tanashi, Tokyo, 188 Japan
[a] *RIKEN, Wako, Saitama, 351-01 Japan*
[b] *Department of Physics, Seoul National University, Seoul, Korea*
[c] *IPNS, KEK-Tanashi, Tanashi, Tokyo, 188 Japan*
[d] *Department of Physics, Yamagata University, Yamagata, 990 Japan*
[e] *Department of Physics, Chun Ann University, Seoul, Korea*
[f] *Department of Physics, Rikkyo University, Toshima, Tokyo, 171 Japan*
[g] *Department of Physics, Yale University, New Haven, CT06511, USA*
[h] *Department of Physics, Osaka University, Toyonaka, Osaka, 560 Japan*
[i] *Physics Division, P.O. Box 2008, Oak Ridge, TN37831, U.S.A.*
[j] *Department of Physics, Konan University, Kobe, 658 Japan*
[k] *Miyagi University of Education, Sendai, 980 Japan*

Abstract. Experimental investigation of critical nuclear reactions for the breakout off the Hot-CNO cycle are in progress. One is the $^{15}O(\alpha,\gamma)^{19}Ne$ reaction, which could be the limiting reaction of the breakout at relatively low temperature ($T < 5 \times 10^8$ K), and the others are $^{14}O(\alpha,p)^{17}F(p,\gamma)^{18}Ne$ at higher temperatures. Some new results on these experiments and the astrophysical implication are discussed.

I. EXPLOSIVE HYDROGEN BURNING

Nuclear reactions on possible breakout processes off the Hot-CNO (HCNO) cycle are of great interest in nuclear astrophysics. This subject is probably directly related to the problem of nucleosynthesis in novae and X-ray bursts, for instance. An interesting question here is if CNO material is really transmuted into heavier elements in explosive phenomena such as novae. This is

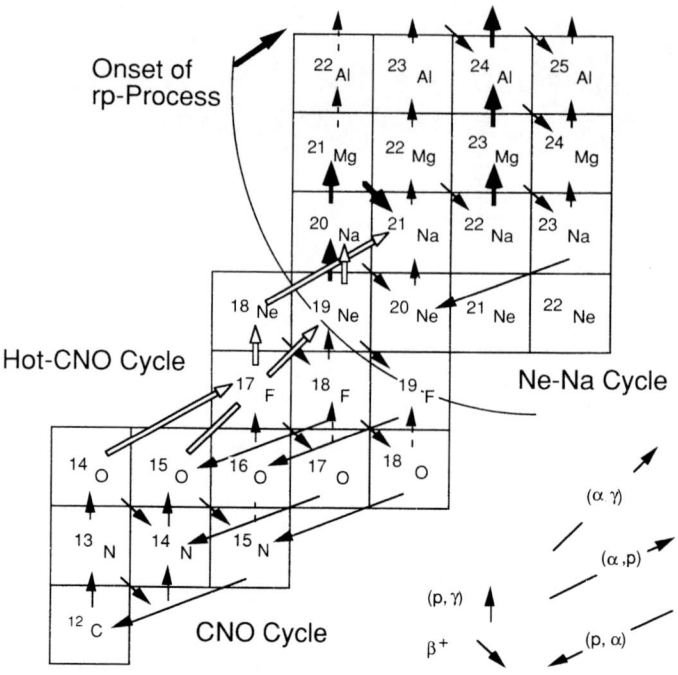

FIGURE 1. The nucleosynthesis flow diagram of the HCNO cycle and the rp-process. The white arrows indicate possible breakout processes off the HCNO cycle and the thick arrows the nuclear processes studied in our project before.

also an important question for energy generation. However, crucial reactions for this problem are not well known yet.

The breakout process off the HCNO cycle, which leads to explosive hydrogen burning (rp-process), was first pointed out by Wallace and Woosely [1]. The reaction sequence of $^{15}O(\alpha,\gamma)^{19}Ne(p,\gamma)^{20}Na$ is considered to be one of them [2]. Because of lack of experiment, the $^{19}Ne(p,\gamma)^{20}Na$ reaction was considered to be the limiting reaction for ignition of the rp-process at the time.

We performed before a series of experiments [7] to clarify the onset mechanism at the SF cyclotron of the Center for Nuclear Study (CNS), University of Tokyo. Many new resonances were discovered above the proton thresholds, including the one at 2.64 MeV in ^{20}Na [8], which is much lower in energy than the level known before at 2.9 MeV, where the proton threshold energy is 2.20 MeV. This new resonance reduces the ignition temperature nearly by a factor of two. The new estimates, obtained based on the experimental results, for the ignition of the onset and early stage of the rp-process is summarized in

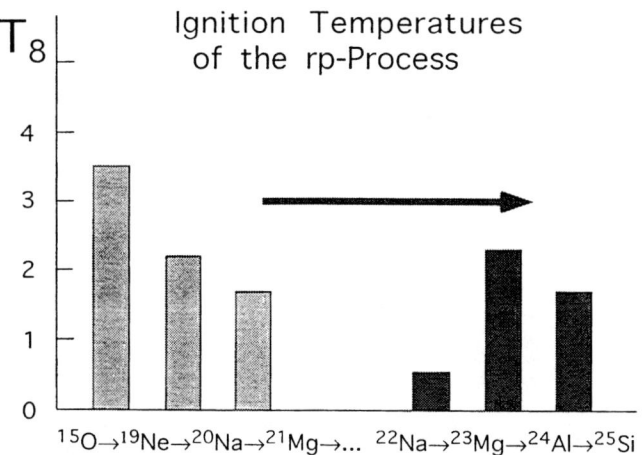

FIGURE 2. Estimates of ignition temperatures of the onset and the early stage of the rp-process. See text for details.

Fig. 2. Here, the estimate for the $^{15}O(\alpha,\gamma)^{19}Ne$ reaction was taken from ref. [9]. Thus, the $^{15}O(\alpha,\gamma)^{19}Ne$ reaction is likely the limiting reaction of the onset of the rp-process. However, note that none of the rates of the reactions in Fig. 2 is determined experimentally at the temperature region of interest.

The second possible reaction sequence for the breakout off the HCNO cycle [1], more likely at higher temperature, is $^{14}O(\alpha,p)^{17}F(p,\gamma)^{18}Ne(\alpha,p)^{21}Na$. Here, a new type reaction (α,p) sets in for the explosive burning phase in the low mass region of the nuclear chart. The reaction rate of $^{14}O(\alpha,p)^{17}F$ was investigated with updated data [3] and with a theoretical model [4]. However, there still is a large uncertainty in the rate, although many experimental efforts were made [5,6]. The other reactions of this sequence also are not known well yet.

Therefore, we have started to investigate these breakout reactions experimentally. Since the experiments and the analysis are in progress, I will just discuss the problems and some preliminary results.

II. THE RP-PROCESS IN NOVAE

As was discussed in sec. I, the limiting reaction for the onset of the rp-process in novae is considered to be $^{15}O(\alpha,\gamma)^{19}Ne$ from the estimates shown in Fig. 2 [7]. The relevant levels in ^{19}Ne were studied before by the $^{20}Ne(^{3}He,\alpha)^{19}Ne$ reactions, the $^{21}Ne(p,t)^{19}Ne$ reaction, etc. [11]. The total widths are not known for the states in this energy region in ^{19}Ne. Since the mirror nucleus ^{19}F is well studied, the analog relation is established up to

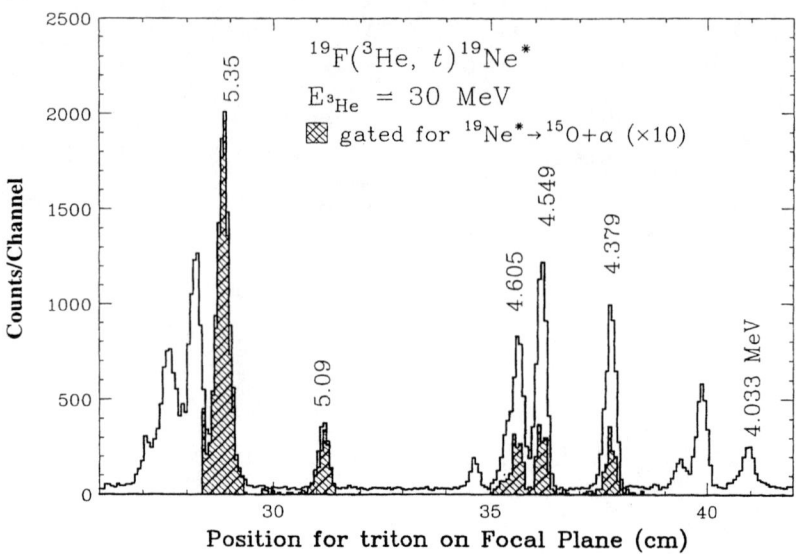

FIGURE 3. Triton spectra of ^{19}F(^{3}He,t)^{19}Ne of singles (white part) and of coincidence with decay particles (shaded part), measured at $0°$. The coincidence spectrum was multiplied by a factor of ten.

around the alpha threshold energy, 3.528 MeV. However, it is not so clear above the threshold. Nevertheless, the resonance strengths for the important resonances seem to be predominantly determined by the alpha decay widths.

The most crucial level predicted for burning under nova conditions (T= 2 ∼ 5 x 10^8 K, $\rho = 10^{3\sim4}$ g/cm^3) is the 4.033 MeV 3/2$^+$ state, which is about 505 keV above the α threshold in ^{19}Ne. This state was very strongly excited by the (p,t) reaction, but not so by the (^{3}He,α) reaction [11]. Thus, it is considered to have a main shell-model component of 5p-2h configuration, but the resonance strength of this state is not known. The decay widths calculated by Langanke et al. [2] are Γ_α = 7.2 μeV and Γ_γ = 73 meV. Recently, the alpha width was estimated from α-transfer data of ^{15}N(^{6}Li,d)^{19}F and ^{16}O(^{6}Li,d)^{20}Ne, giving Γ_α = 9.9 ± 1.5 μeV [10].

An experimental investigation of the astrophysical ^{15}O(α,γ)^{19}Ne reaction is underway at CNS using the ^{19}F(^{3}He,t)^{19}Ne*(α) reaction at 30 MeV. Since the α-decay width is the crucial parameter, we are trying to measure the branching ratios of the α-decays first. The tritons from the reaction were measured at $0°$ with a magnetic spectrograph, and the decay α particles were measured in coincidence using strip silicon detectors placed in the scattering chamber. Alpha decays were observed from some states above the α threshold in ^{19}Ne.

Figure 3 shows two triton spectra measured on the focal plane of the spectrograph. The shaded bars indicate the triton spectrum measured in coincidence with the particle detection in the scattering chamber, whereas the other part (open bars) is the singles spectrum of tritons. The yields of the coincidence spectrum are multiplied by a factor of ten, since the solid angle of the decay particle detector is roughly 10 %.

The 5.35 MeV state shows a large α-decay branching ratio, whereas the levels at lower energies near the α-threshold have less yields. Random coincident events are also included in the coincident events shown in the figure; for instance, roughly a half of the yields is the background for the 4.379 MeV state. This poor signal to noise ratio comes from the setup condition and partly from the electronic noises. These will be improved for the final run.

III. HIGH TEMPERATURE RP-PROCESS

The second possible breakout process, $^{14}O(\alpha,p)^{17}F(p,\gamma)^{18}Ne$ . . ., was also investigated here by the $^{20}Ne(p,t)^{18}Ne$ reaction. The nuclear structure of ^{18}Ne is directly related to the first two reactions in the sequence. Specifically, the properties of the levels just above the proton and α thresholds are the problems in ^{18}Ne. However, even spin parities are not determined for them.

Figure 4 shows the level schemes of ^{18}Ne and the mirror nucleus ^{18}O. The levels at 5.11 and 5.15 MeV states were assigned to have 3^- and 2^+, respectively, in the previous predictions [3,4]. Hahn et al. [6], however, discussed inverted assignments for them from Thomas-Ehrman shift calculations.

Similar discussions were made for the states at 6.29 and 6.35 MeV states. The state at 6.15 MeV, which was observed only by the reactions of $^{16}O(^3He,n)$ and $^{12}C(^{12}C,^6He)$ [6], was not seen at any angles measured in the present (p,t) spectra. A possibility is that this state has many-particle many-hole configuration. There should be analog states of the 6.20 MeV 1^- state and the 6.40 MeV 3^- state of ^{18}O in this energy region. These correspondences will be clarified by identifying the spin parity from measurements of the (p,t) angular distributions. These states could have major contributions to the reaction rate of $^{14}O(\alpha,p)^{17}F$ [3].

A special effort was placed on identifying a possible 3^+ state around 4.5 MeV. A 3^+ state is known at 5.38 MeV in ^{18}O. A Thomas-Ehrman shift calculation [12] predicted the 3^+ state to be around 4.33 MeV in ^{18}Ne, suggesting that this state will enhance considerably the reaction rate of $^{17}F(p,\gamma)^{18}Ne$, although there was no clear experimental evidence seen so far for the presence of the 3^+ state. Only a possibility was suggested at 4.56 MeV experimentally by the $^{16}O(^3He,n)^{18}Ne$ reaction [5]. Since 4.56 MeV is just in the middle of the known doublet of 4.52 and 4.59 MeV, we made a careful search with high resolution measurement using a Ne-implanted carbon target of about 50 $\mu g/cm^3$ (about 7 $\mu g/cm^3$ of Ne). Several spectra were taken for the $^{20}Ne(p,t)^{18}Ne$

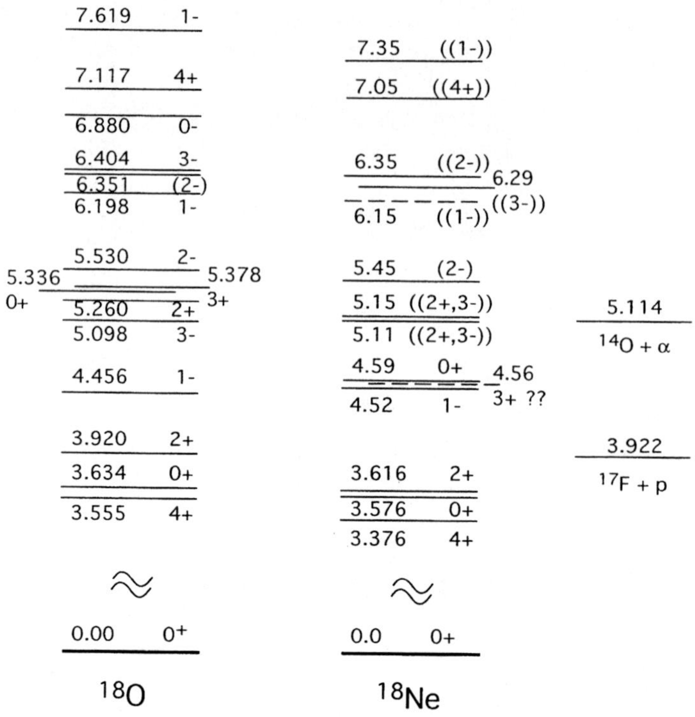

^{18}O

7.619	1-	
7.117	4+	
6.880	0-	
6.404	3-	
6.351	(2-)	
6.198	1-	
5.530	2-	5.378
5.336 0+		
5.260	2+ 3+	
5.098	3-	
4.456	1-	
3.920	2+	
3.634	0+	
3.555	4+	
0.00	0+	

^{18}Ne

7.35	((1-))
7.05	((4+))
6.35	((2-)) 6.29
6.15	((1-)) ((3-))
5.45	(2-)
5.15	((2+,3-))
5.11	((2+,3-))
4.59	0+ 4.56
4.52	1- 3+ ??
3.616	2+
3.576	0+
3.376	4+
0.0	0+

5.114 ^{14}O + α

3.922 ^{17}F + p

FIGURE 4. Level schemes of A = 18.

reaction using a magnetic spectrograph. The overall energy resolution was about 12 keV, which is sufficient enough to see the peak suggested if the energy suggested is correct and the intensity is of the same order as in the (^{3}He,n) reaction. However, there was no evidence seen in the spectra for a state between the doublet states. The background level is also better than a previous spectrum [6]. We observed small contamination peaks which were not seen before. It should be noted that the (p,t) reaction is not the best reaction for this search because the residual state has an unnatural parity.

Since the energy resolution is good, the excitation energies and the natural widths of some states crucial for the (p,γ) and (α,p) reactions will be determined precisely. Eventually, the resonance strengths of the (α,p) reaction are needed to be determined. However, the direct simulation study of the ^{14}O(α,p)^{17}F reaction using an ^{14}O beam is really awaited for since the direct reaction contribution would not be negligible, but it will be determined only by this method.

REFERENCES

1. Wallace, R.K., and Woosley, S.E., *Astrophys. J. Suppl.* **45**, 389 (1981).
2. Langanke, K., Wiescher, M., Fowler, W.A., and Görres, J., *Astrophys. J.* **301**, 629 (1986).
3. Wiescher, M., Harms, V., Görres, J., Thielemann, F.-K., and Rybarcyk, L.J., *Astrophys. J.* **316**, 162 (1987).
4. Funk, C., and Langanke, K., *Nucl. Phys.* **A480**, 188 (1988).
5. Garcia, A., et al., *Phys. Rev.* **C43**, 2012 (1991).
6. Hahn, K.I., et al., *Phys. Rev.* **C54**, 1999 (1996).
7. Kubono, S., *Prog. Theor. Phys.* **96**, 275 (1996).
8. Kubono, S., Orihara, H., Kato, S., and Kajino, T., *Astrophys. J.* **344**, 460 (1989).
9. Magnus, P.V., et al., *Nucl. Phys.* **A506**, 332 (1990).
10. Mao, Z.Q., Fortune, H.T., and Lacaze, A.G., *Phys. Rev. Lett.* **74**, 3760 (1995).
11. Ajzenberg-Selove, F., *Nucl. Phys.* **A475**, 1 (1987).
12. Wiescher, M., Görres, J., Thielemann, F.-K., *Astrophys. J.* **326**, 384 (1988).

Nuclear astrophysics with intermediate-energy RI beams

Tohru Motobayashi*

*Department of Physics, Rikkyo University, Toshima, Tokyo 171, Japan

Abstract. Recent development of RI beam production opens a new possibility of research in nuclear astrophysics. At intermediate-energies (several tens to several hundreds MeV/u), thermonuclear reactions can be studied indirectly by the Coulomb dissociation and Coulomb excitation reactions, and possibly also by the transfer and charge-exchange reactions with inverse kinematics. A brief review and perspectives for studies using these reactions are given.

INTRODUCTION

Various nuclear reactions at low energies play important roles in thermonuclear burning processes in stars or in early universe. In so called explosive burning situation, where both temperature and matter density are high, nuclear reactions on unstable nuclei can be faster than their competing β decays. To study directly the reactions in such explosive nuclear burning scenarios, radioactive targets should be used. Therefore measurements are very difficult or sometimes not possible when short-lived isotopes are involved.

Recent development on radioactive (or RI) beam production allows for new approaches to those difficult-to-measure reactions. One possibility is to realize the reaction system with the unstable nucleus as the beam at a low energy. This approach becomes possible only if a strong RI beam at several hundreds keV/u is available. The first example is the $^{13}N(p,\gamma)^{14}O$ reaction studied at Louvain-la-Neuve [1] using a ^{13}N beam of 3×10^8 s^{-1} intensity at 630 keV/u incident energy.

Another possibility is to use intermediate-energy RI beams, which are now available at many laboratories with variety of ions. Low-energy cross sections can be extracted indirectly from the data of Coulomb dissociation, Coulomb excitation, charge-exchange reaction, particle transfer reaction, and so on at several hundred MeV/u. An advantage is availability of thick targets. Targets with several tens or several hundred mg/cm^2 thickness are commonly used in actual experiments. This is larger by an order of 3 or 4 than the typical thickness of targets in low-energy direct measurements, where the thickness is limited by the energy loss of the incident beam.

Among the reactions quoted above, the most studied one is the Coulomb dissociation. The Coulomb excitation has been studied for the purpose of nuclear spectroscopy. However it might be applicable to some cases of astrophysical interest. The charge-exchange and transfer reactions are not yet seriously considered. In this report, we discuss about what has been achieved and what can be studied in future by using intermediate-energy radioactive nuclear beams from the nuclear-astrophysics point-of-view.

COULOMB DISSOCIATION

In the Coulomb dissociation method, the residual nucleus B of the capture reaction $A(x,\gamma)B$ bombards a high-Z target and is Coulomb excited to an unbound state that decays to the A+x channel. Since the process is regarded as absorption of a virtual photon, i.e. $B(\gamma,x)A$, the radiative capture (the inverse of the photoabsorption) cross section can be extracted from the dissociation yield. This idea was first proposed by Baur [2] based on the virtual photon theory. Topical reviews were given by Baur and Rebel [3]. In addition to the advantage discussed before, the Coulomb dissociation method enhances the original capture cross section by a large factor. This is due to the large virtual-photon number and the phase space factor. The two factors can be in the order of 100 or 1000 in some actual cases like the ^{14}O and ^{8}B dissociation, which will be discussed later in this report.

Pioneering studies of the Coulomb dissociation were made for the stable Li isotopes, ^{6}Li$\to \alpha$+d and ^{7}Li$\to \alpha$+t, at around E_{in}=10 MeV/u [4-7]. Several questions were raised in view the validity of the method. Large asymmetries observed in the fragment's energy spectra [6,7] may be due to the post acceleration effect or higher order excitation. Importance of the nuclear force was pointed out with a quantum mechanical calculation [8]. Recently Utsunomiya et al. made a new experiment of the ^{7}Li breakup, and found that the symmetry is recovered at a certain kinematical condition where successful comparison of the breakup data and capture data are possible [9].

In the ^{6}Li dissociation experiment at a higher energy of 26 MeV/u [10], the asymmetry observed at 10 MeV/u [6] looks to vanish. However, Hirabayashi and Sakuragi [11] argue possible strong contribution of nuclear breakup based on a microscopic cluster model applied to the wave function of ^{6}Li-d+α.

The first Coulomb dissociation experiments with radioactive beams were made for the ^{208}Pb(^{14}O,^{13}N p)^{208}Pb reaction at yet higher incident energies of 87.5 MeV/u [12] and 70 MeV/u [13]. The results demonstrates the usefulness of the method, and stimulated further studies of the Coulomb dissociation such as ^{12}N$\to^{11}$C+p [14] and ^{8}B$\to^7$Be+p [15-18].

Coulomb Dissociation of ^{14}O and Hot CNO-Cycle

In the core of super massive stars, novae [19] or at the surface of neutron stars [20], the ^{13}N(p,γ)^{14}O reaction becomes faster than the β^+ decay of ^{13}N, and the hot CNO cycle starts instead of the regular CNO cycle. This key reaction ^{13}N(p,γ)^{14}O was difficult to be studied because the radioactive nucleus ^{13}N (10 min. half life) is involved. An alternative method is to measure the E1 electromagnetic-width Γ_γ of the first 1^- state in ^{14}O, because this resonant state dominates in the hot CNO burning. This resonance was populated by reactions like ^{12}C(^{3}He,n), and the branching ratio of γ decay was measured [21–24]. However such "indirect measurements" could only provide data with rather large experimental uncertainties due to the small ($\approx 10^{-4}$) γ-branching ratio. We performed an experiment of Coulomb dissociation [12] hoping that the experimental accuracy for Γ_γ might be improved.

An ^{14}O beam was obtained by the projectile-fragmentation scheme with the RIPS separator at RIKEN [25]. It bombarded a thick (350 mg/cm^2) ^{208}Pb target. Outgoing particles were detected in coincidence with a large efficiency of approximately 40%. In addition to the large Coulomb dissociation cross section, this large efficiency together with the thick target enable collection of about 10,000 events in one day and a half with 3×10^4 s^{-1} ^{14}O beam intensity.

The Coulomb dissociation yield was analyzed with a coupled-channel (essentially DWBA in this case) calculation after correcting for the detection efficiency. Assuming pure E1 Coulomb excitation mechanism, a radiative width Γ_γ=3.1±0.6 eV was extracted for the 1^- state.

For possible nuclear contribution, Hirabayashi and Sakuragi constructed a transition form-factor microscopically with a one-body potential for the p+^{13}N system [26]. Their DWBA cross section for the nuclear excitation is only about 1/100 of the Coulomb one. This is a general feature of E1 transitions of intermediate-energy Coulomb excitation [27]. Possible effects of the multistep excitation, $0^+\to 1^-\to 0^+_2$ were discussed by Delbar [28]. However, a theoretical estimate of higher order excitation was given by Typel and Baur [29], and it turned out that the effects can be safely neglected.

The ^{13}N$\to^{12}$C+p dissociation data were also taken in the same experiment. The E1 transition from the ground $1/2^-$ state to the first excited $1/2^+$ state in ^{13}N was measured. The resultant width, 0.59±0.16 eV, agrees with a tabulated value of 0.50±0.04 eV [30]. This demonstrates the validity of the Coulomb dissociation method applied to E1 transitions to resonant states.

Almost at the same time as our Coulomb dissociation result was reported, the direct measurement mentioned earlier gave the first result using a post-accelerated ^{13}N beam [1]. Their result Γ_γ=3.3±0.7(stat)±0.6(syst) eV [28] agrees well with that extracted from our Coulomb dissociation. Another Coulomb dissociation result from GANIL [13] is also consistent with these values. Recent compilation takes an average using all available data and leads to Γ_γ=3.0±0.4 eV [24] or Γ_γ=3.1±0.4 eV [31]. Significant contribution to the compiled value is from our result due to its good accuracy.

Coulomb Dissociation of ^{8}B and Solar Neutrino Problem

Because of its connection to solar neutrino production, the ^{7}Be(p,γ)^{8}B reaction has attracted much attention. So called "solar neutrino puzzle" is based on neutrino fluxes observed on the earth, which are much lower than the ones predicted in standard solar models [32,33]. Most neutrinos detected in the experiments of Homestake [34] and Kamiokande [35] originate from β^+ decay of ^{8}B with its high end-point energy of 14 MeV. Therefore the predicted neutrino flux is directly connected to the ^{7}Be(p,γ)^{8}B cross section.

In order to provide new data of the ^{7}Be(p,γ)^{8}B cross section in addition to existing data from direct measurements, a series of experiments has been and is being made at RIKEN with radioactive ^{8}B beams of about 50 MeV/u and ^{208}Pb target. So far three experiments have been performed. The first one is described in two articles [15], and a part of the second result is reported in Ref. [16]. The data obtained in the third experiment is now being analyzed.

Beams of ^{8}B were produced by the ^{12}C+^{9}Be interaction at 92 MeV/u for the first two experiments and 135 MeV for the third one. The ^{8}B energies in the center of the target, 50 mg/cm^2 ^{208}Pb, were approximately 50 MeV/u. We employed the time-of-flight (TOF) technique to determine the energies of the fragments. A plastic scintillator hodoscope of 1×1 m^2 active area was set 3-5 m from the target. The outgoing particles of the Coulomb dissociation, ^{7}Be and p, are detected in coincidence, and their energies and scattering angles are determined by their TOF's and the positions of their hits, respectively. A p-^{7}Be relative energy spectrum is constructed from the measurement and it is converted to the ^{7}Be(p,γ)^{8}B cross section with the help of a Monte-Carlo simulation calculation on detection efficiency and theoretical calculation for the Coulomb dissociation mechanism.

In the second and third experiments, a NaI(Tl) scintillator array DALI [36] was also installed at the target position. It consisted of fifty-four identical crystals with $6\times6\times12$ cm^3 volume. The DALI setup measured the deexcitation γ rays from the first excited state of ^{7}Be at 429 keV populated in the dissociation process. The correction due to this excited state population was about 5%.

The resultant S_{17}-factors of the first and second experiments are shown in Fig. 1 together with those of direct measurements, which are renormalized by Filippone [37]. The Coulomb dissociation results shown in the figure were obtained by assuming pure E1 transition. They are slightly different from each other, but consistent with the (p,γ) data by Vaughn et al. [38] and Filippone et al. [39], lower values among the available direct capture results.

The angular distributions obtained in the second experiment are displayed in Fig. 2. The predictions with E2-mixture calculated by two theoretical models [29,40] are shown by dashed and dotted curves in the first two graphs. Possible effects of the E2-mixture was discussed [41,42] on our first result. The original E2 component in the (p,γ) yield can be enhanced in the Coulomb dissociation. However, as seen in Fig. 2, no E2 mixture is favored at relative energies lower than 1.75 MeV. Note that $\ell=2$ nuclear breakup is in the same order of magnitude as the E2 Coulomb dissociation.

In the third experiment, most of the helium gas region is replaced by vacuum to reduce background events caused by parasitic reactions and hence to lower the limit of relative energy. The data are now being analyzed. Almost background-free spectrum was obtained above 200 keV.

An example of the energy dependence of the E1-, E2- and M1-Coulomb

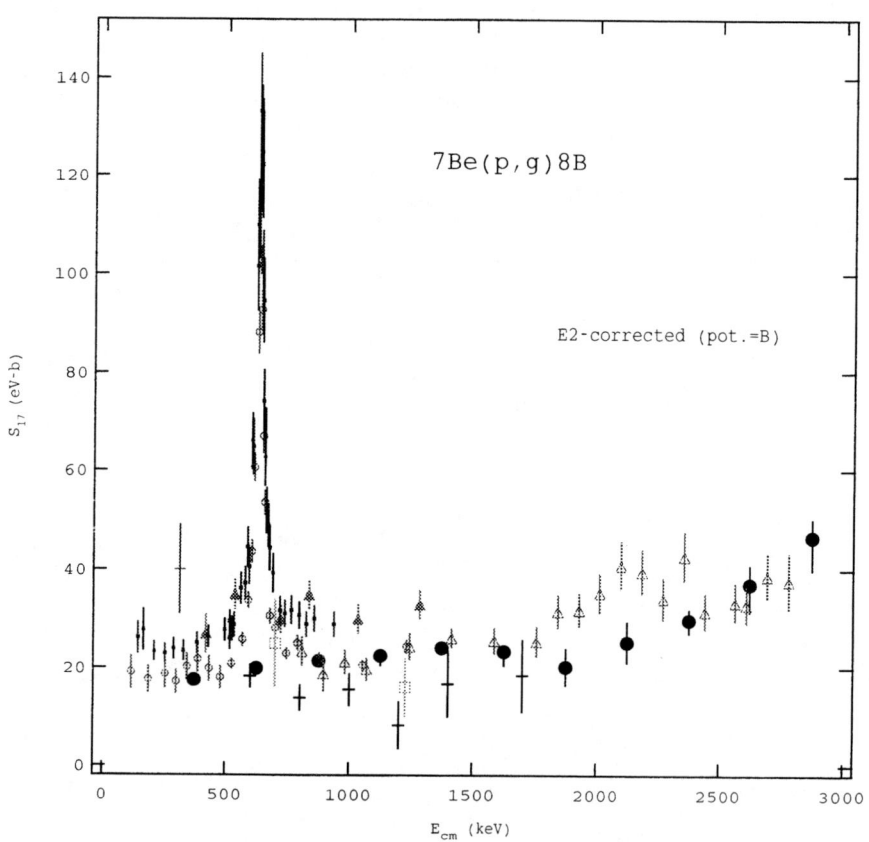

FIGURE 1. The astrophysical S_{17}-factors extracted from the first (thick crosses) and second (large solid dots) experiments together with existing (p,γ) data.

dissociation cross sections is shown in Fig. 3. It is seen that the E2/E1 ratio is larger at lower energies, and the M1 transition becomes important at higher energies.

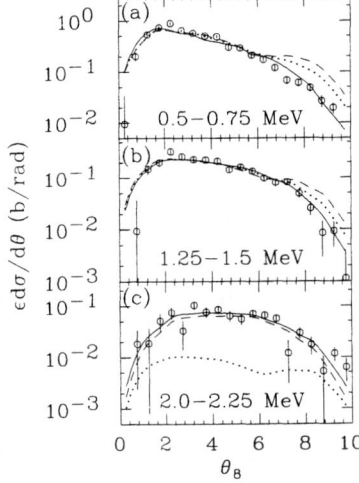

FIGURE 2. Angular distribution of the ^{8}B Coulomb dissociation reaction.

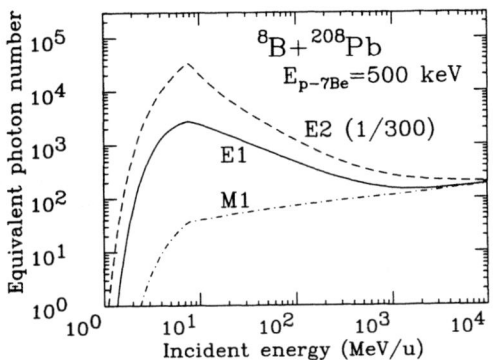

FIGURE 3. Virtual photon numbers plotted as a function of the incident ^{8}B energy for M1, E1 and E2 transitions.

Notre Dame group [18] measured exclusive ^{8}B dissociation yield at a much lower incident energy of 3.25 MeV/u to see the enhanced E2 component. Their result suggests also that the E2 components are very small. A measurement at a higher energy of 254 MeV/u has been performed at GSI [17], and its preliminary results will be presented in this symposium by N. Iwasa. At higher energies, multistep excitation effects or post acceleration effects are expected to be smaller [43].

Reaction mechanism of the Coulomb dissociation

Since first-order pure Coulomb excitation mechanism is assumed in the analysis, mixture of different mechanisms should be considered. As mentioned before, nuclear contribution is negligibly small for the E1 Coulomb dissociation of ^{14}O at 87.5 MeV/u [26]. This is the case also for the ^{8}B dissociation in its E1 component [46,47]. However, for E2 transitions, nuclear $\ell=2$ component can be considerably large as in the case of the ^{8}B dissociation. This is a major difficulty for the Coulomb dissociation of ^{16}O [44,45], the reaction simulating the key reaction in helium burning ^{12}C$(\alpha,\gamma)^{16}$O, where the E2 transition is more important than in the case of ^{8}B breakup due to the isospin selection rule. Possible strong nuclear contribution in the ^{6}Li breakup discussed in the beginning of this section may be caused by the same reason.

The post acceleration is one of the effects of higher order processes. According to many theoretical investigations performed so far [29,46,48–51] the effect is small in the ^{14}O and ^{8}B dissociation at 90 and 50 MeV/u incident energies, respectively.

For transitions to continuum as the ^{8}B dissociation, mixture of E1, E2 and M1 components should be considered. The angular distribution measurement performed in our second ^{8}B dissociation experiment is one of the methods of its determination. Another possibility is to measure the angular correlation or the angular distribution of the fragments in their rest frame [52]. Recently, a parallel momentum distribution was measured at NSCL/MSU for ^{7}Be from the ^{8}B dissociation [53]. Their observed asymmetric distribution affects the interference in the angular correlation. More detailed information should be obtained from our third ^{8}B dissociation experiment, where the angular correlation data might be extracted with fixed relative energies.

COULOMB EXCITATION

In certain cases, states just below the particle threshold affects the behavior of the capture cross section. The well known example is the ^{12}C$(\alpha,\gamma)^{16}$O reaction where the tail of the subthreshold 1^{-} state contributes to possible increase of the S-factor close to $E_{cm}=0$. Therefore, Coulomb excitation to such subthreshold states might be useful, because their electromagnetic transition-strength or electromagnetic width (= total width) can be determined, and the influence of the states to unbound region can be evaluated together with their reduced particle width determined by independent measurements.

Study of the Coulomb excitation using intermediate-energy RI beams has been just started. For example, the 0^{+}-2^{+} transition of the very neutron-rich ^{32}Mg was measured [54], though the motivation of the experiment was purely in nuclear physics. A thick (350 mg/cm^2) ^{208}Pb target was bombarded by ^{32}Mg with only 300 s^{-1} intensity, and γ rays from the 2^{+} state in ^{32}Mg were measured in coincidence with the ejectiles. A fairy accurate transition probability B(E2) was extracted demonstrating the usefulness of the method.

TRANSFER AND CHARGE-EXCHANGE REACTIONS WITH REVERSED KINEMATICS

Nucleon or cluster transfer reactions have been used to populate resonant states important for thermonuclear burning processes. They provide spectroscopic information of the states, such as the level energy, the angular momentum transfer which helps to assign the spin and parity, and the spectroscopic factor. Another important quantity is the branching ratio of the resonance,

$\Gamma_\gamma/\Gamma_{tot}$ or Γ_p/Γ_{tot}, which can be determined by measuring γ- or particle-decay. If the total width Γ_{tot} is known, the branching ratio leads to the capture cross section with the help of the single-level Breit-Wigner formula. An example is the $^{12}C(^3He,n)^{14}O$ measurement mentioned earlier.

If such direct reactions involve unstable targets, reversed kinematics can be employed with radioactive beams and a target of a light nuclide as hydrogen. Recently, such type of study was made for the $^1H(^{11}Li,^{11}Be)n$ reaction. Measurements of decay particles from excited ^{11}Be could provide detailed information on the structure of the isobalic analog state of ^{11}Li [55].

SUMMARY

Since the idea to investigate astrophysical thermo-nuclear reactions was proposed in 1950s, much progress has been achieved in both experimental and theoretical studies. Recent development of producing radioactive nuclear beams allows now for access to unstable nuclei important in explosive nuclear burning. So called indirect method using projectiles of unstable nuclei at intermediate-energies can also be applied to "hard-to-access" cases. Especially the Coulomb dissociation method has been studied for several important reactions. The ^{14}O and ^8B dissociation provide typical examples of resonant and non-resonant breakups. Their experimental results encourage new measurements of astrophysical reactions There remain questions on the reaction mechanism including those discussed in this report. Further experimental investigation to control the possible ambiguities as well as theoretical works are desirable. Other direct reactions with radioactive nuclear beams can also provide various spectroscopic information useful for determination of astrophysical reaction rates.

The works present in this report were performed in collaboration with T. Kikuchi, T. Takei, N. Iwasa, Y. Ando, K. Ieki, M. Kurokawa, S. Moriya, H. Murakami, T. Nishio, J. Ruan (Gen), S. Shirato, S. Shimoura, T. Uchibori, Y. Yanagisawa (*Rikkyo University*), A. Goto, T. Ichihara, N. Inabe, T. Kubo, H. Sakurai, T. Teranishi, Y. Watanabe, M. Ishihara (*RIKEN*), M. Hirai, T. Nakamura (*University of Tokyo*), S. Kubono (*INS*), Y. Furutaka(*TIT*), Y. Futami(*Tsukuba University*), S. Kox, C. Perrin, F. Merchez, D. Rebreyend (*ISN Grenoble*), M. Gai, R. France III, K.I. Hahn, Z. Zhao (*Yale University*), Th. Delbar, P. Lipnik and C. Michotte (*Louvain la Neuve*).

REFERENCES

1. P. Decrock et al., Phys. Rev. Lett. 67 (1991) 808; Th. Delbar et al., Phys. Rev. C 48 (1993) 3088.

2. G. Baur, C.A. Bertulani, and H. Rebel, Nucl. Phys. A 458, 188 (1986); C.A. Bertulani and G. Baur, Phys. Rep. 163 (1988) 299.
3. G. Baur and H. Rebel, J. Phys. G 20 (1994) 1; Ann. Rev. Nucl. and Part. Sci., 46 (1996) 321.
4. A.C. Shotter, V. Rapp, T. Davinson, D. Brandord, N.E. Sanderson and N.A. Nagarajan, Phys. Rev. Lett. 53 (1984) 1539; A.C. Shotter, V. Rapp, T. Davinson and D. Brandord, J. Phys. G: Nucl. Phys. 14 (1988) L169
5. H. Utsunomiya et al., Phys. Lett. B 211 (1988) 24; Nucl. Phys. A 511 (1990) 379; Phys. Rev. Lett 65 (1990) 847.
6. J. Hesselbarth, S. Khan, Th. Kim and K.T. Knöpfle, Z. Phys. A 331 (1988) 365; J. Hesselbarth and K.T. Knöpfle, Phys. Rev. Lett. 67 (1991) 2773.
7. S.B. Gazes, J.E. Mason, R.B. Roberts and S.G. Teichmann, Phys. Rev. Lett. 68 (1992) 150; J.E. Mason, S.B. Gazes, R.B. Roberts and S.G. Teichmann. Phys. Rev. C 45 (1992) 2870.
8. R. Shyam, G. Baur and P. Banerjee, Phys. Rev. C 44 (1991) 915.
9. H. Utsunomiya et al., private communication
10. J. Kiener, H.J. Gils, H. Rebel and G. Baur, Z. Phys. A 332 (1989) 359; J. Kiener et al., Phys. Rev. C44 (1991) 2195.
11. Y. Hirabayashi and Y. Sakuragi, Phys. Rev. Lett. 69 (1992) 1892.
12. T. Motobayashi et al., Phys. Lett. B 264 (1991) 259.
13. J. Kiener et al., Nucl. Phys. A 552 (1993) 66.
14. A. Lefebvre et al., Nucl. Phys. A 592 (1995) 69.
15. T. Motobayashi et al., Phys. Rev. Lett. 73 (1994) 2680; N. Iwasa et al., J. Phys. Soc. Jpn. 65 (1996) 1256.
16. T. Kikuchi et al., Phys. Lett. B 391 (1997) 261.
17. K. Sümmerer et al., GSI proposal (1994); N. Iwasa, Int. Workshop Physics of Unstable Nuclear Beams, São Paulo, Aug 1996; N. Iwasa, this symposium.
18. J. Schwarzenberg et al., Phys. Rev. C 53 (1996) R2598.
19. R.K. Wallace and S.E. Woosley, Astrophys. J., Supplement 45 (1981) 289
20. M.Y. Fujimoto, T. Hanawa and S. Miyaji, Astrophys. J., 247 (1981) 267.
21. T.F. Wang, Thesis, Yale 1986.
22. P. Aguer et al., Porc. Int. Symp. Heavy Ion Physics and Nuclear Astrophysical Problems, 1989, p.107.
23. P.B. Fernandez, E.G. Adelberger and A. Garcia, Phys. Rev. C 40 (1989) 1887.
24. M. Smith et al., Phys. Rev. C 47 (1993) 2740.
25. T. Kubo et al., Nucl. Instr. Meth. B 70 (1992) 309.
26. Y. Hirabayashi and Y. Sakuragi, private communication.
27. T. Motobayashi, proc. 3rd IN2P3-RIKEN Symp. Heavy Ion Collisions, Shinrin-Koen, Saitama, Oct. 1994, ed. T. Motobayashi, N. Frascaria, M. Ishihara (World Scientific, Singapore, 1995) p. 47.
28. Th. Delbar, Phys. Rev. C 47 (1993) R14.
29. S. Typel and G. Baur, Phys. Rev. C 49 (1994) 379.
30. R.B. Firestone, *Table of Isotopes, 8th edition*, ed. V.S. Shirley (John Wiley & Sons, New York, 1996).
31. A.E. Champagne and M. Wiescher, Ann. Rev. Nucl. Part. Sci. 42 (1992) 39.

32. J.N. Bahcall and M.H. Pinsonneault, Rev. Mod. Phys. 60 (1992) 885.
33. S. Turck-Chièze et al., Phys. Rep. 230 (1993) 57.
34. R. Davis, Progr. Part. Nucl. Phys. 32 (1994) 13.
35. Y. Fukuda, et al., Phys. Rev. Lett. 77 (1996) 1683.
36. T. Nishio et al., in preparation; T. Motobayashi et al., Phys. Lett. B 346 (1995) 9.
37. B. Filippone, Ann. Rev. Nucl. Part. Sci. 36 (1986) 717.
38. F.J. Vaughn, R.A. Chalmers, D. Kohler, and L.F. Chase, Jr., Phys. Rev. C 2 (1970) 1657.
39. B. Filippone, S.J. Elwyn, C.N. Davids, and D.D. Koetke, Phys. Rev. Lett. 50 (1983) 412; Phys. Rev. C 28 (1983) 2222.
40. K.H. Kim, M.H. Park, and B.T. Kim, Phys. Rev. C 35 (1987) 363.
41. K. Langanke and T.D. Shoppa, Phys. Rev. C 49 (1994) R1771; Erratum, Phys. Rev. C 51 (1995) 2844; Phys. Rev. C 52 (1995) 1709.
42. M. Gai and C.A. Bertulani, Phys. Rev. C 52 (1995) 1709.
43. C.A. Bertulani, Nucl. Phys. A587 (1995) 318.
44. V. Tatischeff, J. Kiener, P. Aguer and A. Lefebvre, Phys. Rev. C51 (1995) 2789.
45. J. Kiener et al., Proc. 4th Intern. Conf. Nuclei in the Cosmos, June 1996, Notre-Dame, USA, Nucl. Phys. in press.
46. C.A. Bertulani, Phys. Rev. C49 (1994) 2688.
47. R. Shyam, I.J. Thompson and A.K. Dutt-Mazumder, Phys. Lett. B 371 (1996) 1.
48. S. Typel and G. Baur, Phys. Rev. C 50 (1995) 2104.
49. H. Esbensen, G.G. Bertsch, C.A. Bertulani, Nucl. Phys. A 581 (1995) 107.
50. T. Kido, K. Yabana and Y. Suzuki, Phys. Rev. C 50 (1994) R1276.
51. C.A. Bertulani, L.F. Canto and M.S. Hussein, Phys. Lett. B 353 (1995) 413.
52. H. Esbensen and G.F. Bertsch, Phys. Lett. B 359 (1995) 13; Nucl. Phys. A 600 (1996) 37.
53. J.H. Kelley et al., Phys. Rev. Lett. 77 (1996) 5020.
54. T. Motobayashi et al., Phys. Lett. B 346 (1995) 9.
55. T. Teranishi et al., Phys. Lett. B 407 (1997) 110.

Information About the $^{12}C(\alpha,\gamma)^{16}O$ Reaction from the β-delayed Proton Decay of ^{17}Ne

J.D. King,[1] J.C. Chow,[1] A.C. Morton,[1] R.E. Azuma,[1] N. Bateman,[1,2,4] R.N. Boyd,[3] L. Buchmann,[2] J.M. D'Auria,[4] T. Davinson,[5] M. Dombsky,[2] W. Galster,[6] E. Gete,[2] U. Giesen,[2,4] C. Iliadis,[1,2,*] K.P. Jackson,[2] G. Roy,[7] T. Shoppa[2] and A. Shotter[5]

[1] *Physics Department, University of Toronto, Toronto, Ontario, Canada M5S 1A7*
[2] *TRIUMF, 4004 Wesbrook Mall, Vancouver, British Columbia, Canada V6T 2A3*
[3] *Departments of Physics and Astronomy, Ohio State University, Columbus, OH 43210, USA*
[4] *Department of Chemistry, Simon Fraser University, Burnaby, British Columbia, Canada V5A 1S6*
[5] *Department of Physics and Astronomy, University of Edinburgh, Edinburgh, United Kingdom EH9 3JZ*
[6] *Département de Physique, Université Catholique de Louvain, Louvain-la-Neuve, Belgium 1348*
[7] *Department of Physics, University of Alberta, Edmonton, Alberta, Canada T6G 2J1*

Abstract. We are studying the β-delayed proton decay of ^{17}Ne with the goal of determining the $E2$ part of the $^{12}C(\alpha,\gamma)^{16}O$ cross section at energies relevant to helium burning in stars. We have determined branching ratios for proton and α-decay for states in ^{17}F from 8.08 to 11.193 MeV. In addition, we have observed the break-up of the isobaric analogue state (IAS) at 11.193 MeV into three particles via two channels: proton decay to the 9.59 MeV state in ^{16}O which breaks up into an α-particle plus ^{12}C; and α-decay to the 2.365 MeV state in ^{13}N which breaks up into a proton plus ^{12}C. This is the first reported observation of the decay of the IAS to the 1^- state in ^{16}O at 9.59 MeV.

---

*) Present addresss: Department of Physics and Astronomy, The University of North Carolina at Chapel Hill, Chapel Hill, N.C. 27599-3255

INTRODUCTION

In the helium-burning phase of stellar evolution, ^{12}C is converted into ^{16}O via the ^{12}C$(\alpha,\gamma)^{16}$O reaction. The ^{12}C/^{16}O abundance ratio at the end of helium burning is important in the subsequent evolution of massive stars, both for determining the abundances of heavier elements produced during later stages of stellar evolution, and for determining the nature of the remnant after the subsequent supernova explosion [1,2]. In particular, a small ^{12}C/^{16}O ratio leads to the formation of a more massive iron core during later burning stages, and thus increases the probability of leaving a remnant black hole rather than a neutron star [2]. The ^{12}C$(\alpha,\gamma)^{16}$O rate at a center-of-mass energy of $E = 300$ keV (the "most effective energy") depends mainly on the properties of two ^{16}O states (which produce sub-threshold resonances) that lie below the ^{12}C $+ \alpha$ threshold: a $J^\pi = 1^-$ state at an energy of -45 keV ($E_x = 7.117$ MeV); and a 2^+ state at -245 keV ($E_x = 6.917$ MeV) to which feasible radiative capture measurements are only weakly sensitive. Measurements are, instead, dominated by electric dipole ($E1$) contributions from a broad 1^- state at 2.42 MeV ($E_x = 9.59$ MeV) and by direct electric quadrupole ($E2$) radiative capture.

We have recently used the β-delayed α-particle spectrum from ^{16}N ($t_{\frac{1}{2}} = 7.13$ s) to constrain the $E1$ cross section at low energies [3,4]. From simultaneous fits to the ^{16}N β-delayed α-spectrum, to the four sets of ^{12}C$(\alpha,\gamma)^{16}$O cross section data, and to the ^{12}C$(\alpha,\alpha)^{12}$C phase shifts, we were able to determine the α-width of the sub-threshold 1^- state at 7.117 MeV, and thereby much reduce the uncertainty in the ^{12}C$(\alpha,\gamma)^{16}$O $E1$ cross section at 300 keV [3,4]. Recently, Buchmann et al. [5] have included all available angular distribution data in the fit and confirm the result of ref. [4] for the $E1$ contribution. The $E2$ cross section at 300 keV is dominated by the tail of the sub-threshold 2^+ state at 6.917 MeV. However, 2^+ states are not populated in the decay of ^{16}N and our knowledge of the $E2$ component was not improved by that experiment.

The energy available for the β^+-decay of ^{17}Ne is 13.51 MeV [6]. Since ^{17}F is bound by only 0.6005 MeV against proton decay to ^{16}O, most states populated in the β-decay of ^{17}Ne will decay by proton emission [6–8]. States in ^{17}F with energy greater than 7.762 MeV may decay into α-unbound states in ^{16}O, including the tails of the sub-threshold 6.917 and 7.117 MeV states. Therefore, the possibility exists that an experiment similar in principle, but different in detail, to the β-delayed α-decay of ^{16}N could be carried out that would reveal the contribution of the tail of the sub-threshold 2^+ state at 6.917 MeV to the ^{12}C$(\alpha,\gamma)^{16}$O cross section. The partial level scheme for ^{17}Ne decay [6,8–10] shown in Fig. 1 indicates that all states in ^{17}F above 5.819 MeV may also decay into ^{13}N plus an α-particle.

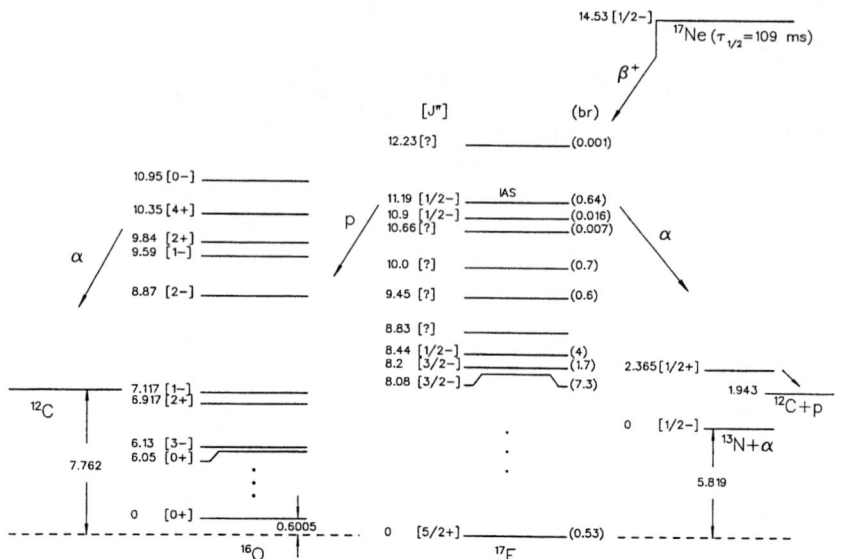

FIGURE 1. Partial decay scheme of ^{17}Ne [6,8–10]. Reprinted from Nucl. Phys. **A621** pp. 169–72c, 1997 with kind permission from Elsevier Science - NL, Sara Burgerhartstraat 25, 1055 KV Amsterdam, The Netherlands.

EXPERIMENT

The TISOL facility [11] at the TRIUMF laboratory has been used to investigate the β-delayed proton decay of ^{17}Ne to α-unbound states of ^{16}O and to look for the break-up into $\alpha + {}^{12}$C as was observed with ^{16}N [3,4]. ^{17}Ne was produced by bombarding a MgO target with 500 MeV protons and extracting a mass 17 beam from an on-line ECR source. Beam intensities of up to 2×10^5 s^{-1} were produced.

Branching Ratios

The branching ratios for the decay to excited states in ^{16}O were determined initially by measuring proton–γ-ray coincidences between a 300 μm annular silicon detector and an HPGe detector. Transitions were observed to the 2$^+$ state at 6.917 MeV in ^{16}O from states in ^{17}F at 11.193, 10.0, 9.45, 8.83 and 8.44 MeV [9,10]. Many new transitions to the 1$^-$ state at 7.117 MeV and to the 3$^-$ state at 6.130 MeV were also observed. In addition, a new transition from the isobaric analogue state (IAS) at 11.193 MeV in ^{17}F to the 8.872 MeV 2$^-$ state in ^{16}O was seen.

From the branching ratio studies, the 9.45 MeV state in ^{17}F has been identified as a possible initial state for investigating α-decay through the tail of

TABLE 1. Branching ratios for the decay of the IAS of ^{17}F.

Particle	E_x (MeV)	$J^\pi; T$	B.R. (%) [7]	B.R. (%) [8]	Present
	(^{16}O)				
p	0	$0^+; 0$	10 ± 2	10.7 ± 0.6	8.5 ± 0.5[a]
	6.049	0^+	< 3	11 ± 3	9.6 ± 0.8
	6.130	3^-	22 ± 2	25 ± 2	17.6 ± 0.6
	6.917	2^+	24 ± 6	< 4	0.48 ± 0.05
	7.117	1^-	44 ± 4	18 ± 3	13.4 ± 0.3
	8.872	2^-			8.7 ± 1.2
	9.59	1^-			22.7 ± 11.3
	(^{13}N)				
α	0	$\frac{1}{2}^-; \frac{1}{2}$	–	1.1 ± 0.5	0.7 ± 0.2
	2.365	$\frac{1}{2}^+$	–	29 ± 9	14.6 ± 1.1
	(^{17}F)				
γ	0.495	$\frac{1}{2}^+; \frac{1}{2}$			3.4 ± 1.5[a]

[a] Branching ratios for protons to the ground state of ^{16}O relative to the total observed particle decay and for γ-decay taken as given in ref. [8].

the 6.917 MeV state in ^{16}O, since the proton decay from the 9.45 MeV state to the 6.917 MeV state is more than an order of magnitude greater than the decay to either the 7.117 or 6.130 MeV states in ^{16}O [9,10].

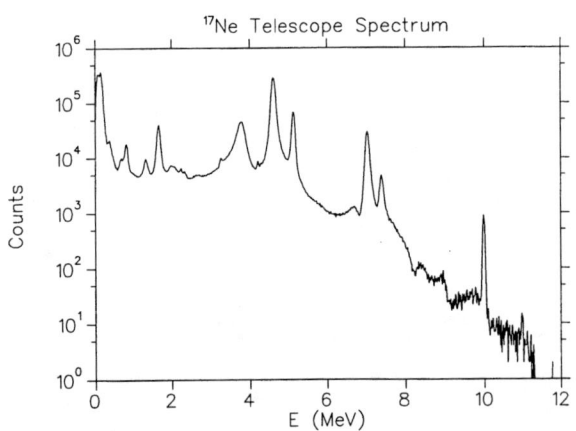

FIGURE 2. Particle spectrum from ^{17}F following the β-decay of ^{17}Ne.

In the triple coincidence experiment described below, yet another proton decay branch of the IAS was revealed – to the broad 9.59 MeV 1^- state. Its branching ratio was determined by comparison with the α-particle decay to the 2.365 MeV state in ^{13}N, which was also observed (see below). This state cannot be observed in the particle spectrum of Fig. 2 because of its large width of 420 ± 20 keV [6]. The branching ratios for the decay of the IAS of

^{17}F are given in Table 1.

Spectrum

In a separate experiment, $\Delta E - E$ telescopes were used to record the complete particle decay spectrum. Fig. 2 shows the reconstructed telescope spectrum. Weak narrow peaks at 3.3 and 4.2 MeV are due to proton transitions from the IAS to the 7.117 and 6.130 MeV states in ^{16}O, respectively.

Triple Coincidences

To collect a low-background α-particle spectrum with its main component from the tail of the sub-threshold state in ^{16}O at 6.917 MeV we shall need to detect a triple coincidence between the α-particle, the recoiling ^{12}C nucleus and the proton emitted from the parent ^{17}F state. The IAS in ^{17}F has a width of only 0.18 keV [6] and is populated in $\approx 0.7\%$ of the ^{17}Ne decays. Before this work, it was known to decay by proton emission to all states in ^{16}O up to the 8.872 MeV state [10] and by α-emission to the ground and first excited states of ^{13}N. Thus, the IAS seemed a good candidate state in ^{17}F for the initial observation of p–α–^{12}C triple coincidences.

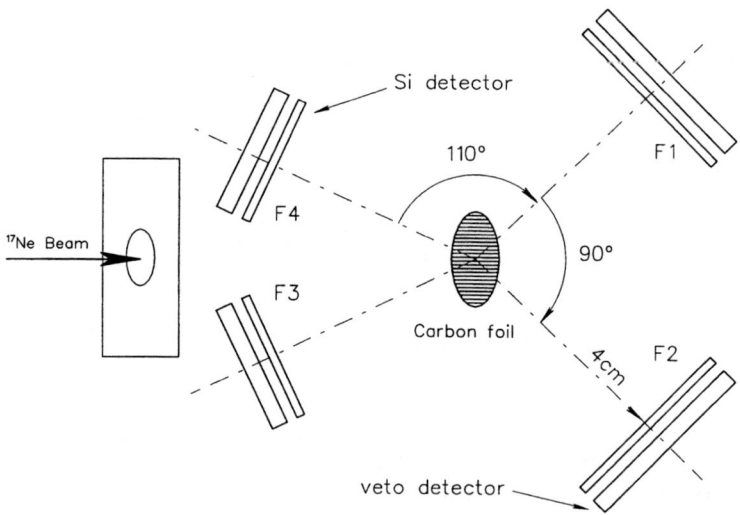

FIGURE 3. Detector arrangement for the detection of p–α–^{12}C triple coincidences from the decay of the IAS in ^{17}F.

Experimental Set-up

A Monte Carlo simulation of the proton decay from the IAS of ^{17}F to the 1^- state at 7.117 MeV in ^{16}O (and to the broad 1^- state at 9.59 MeV if it were also populated) was used to define an optimum arrangement in the laboratory reference frame for detecting triple coincidences. The experimental arrangement (see Fig. 3) consisted of two 900 mm^2 ion-implanted Si detectors at right angles, and two 450 mm^2 ion-implanted Si detectors placed at 110° to one of the larger detectors. Each of these was backed up by a second detector to record particles passing through the front detector. The detector plane was inclined at 45° to the incoming ^{17}Ne beam in order to provide access for the beam to a 10 μg cm^{-2} carbon collector foil at the center of the array.

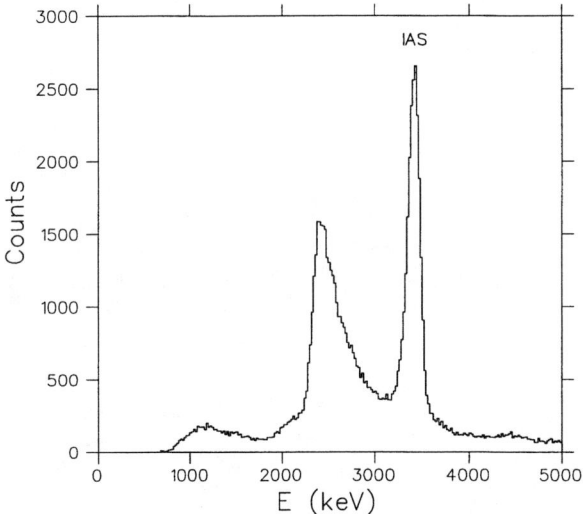

FIGURE 4. Triple coincidence energy-sum spectrum for the β-delayed particle decay of ^{17}Ne. See text for details.

Results

Triple coincidence data for the detector combinations F1F2F3 and F1F2F4 were combined into the triple-energy-sum spectrum of Fig. 4. The break-up of the recoiling ^{16}O and ^{13}N nuclei produces a back-to-back α–^{12}C or proton–^{12}C pair, respectively, in the center-of-mass system of each pair. The kinematic constraint imposed on the relative energies of the three particles observed in the laboratory frame of reference by invoking conservation of momentum was used to reduce the background from the spectrum due to two-body decays. Since the decay of excited states in ^{17}F is predominantly by proton emission

[7–10], and each proton decay is accompanied by an ^{16}O recoil, an accidental event in the third counter (at 90°) will provide a triple coincidence.

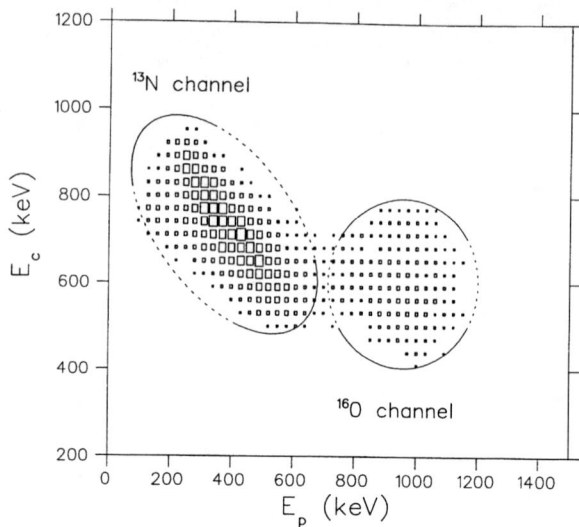

FIGURE 5. Two-dimensional plot of protons versus ^{12}C recoil nuclei emitted during the break-up of ^{17}F in its IAS. The ellipses show the gates used to obtain the particle spectra of Fig. 6.

The energy difference between the IAS in ^{17}F and the threshold for break-up into $\alpha + {}^{12}$C in ^{16}O is 3.43 MeV [6]. This energy is well-defined since the IAS has a width of only 0.18 keV. A narrow peak at about 3.4 MeV is apparent in Fig. 4. The peak near 2.4 MeV is due primarily to the strong α-decay of the 8.08 MeV state to the ground state of ^{13}N, which has not been removed completely by the kinematic cut. The width of the peak associated with the IAS decay is compounded from the inherent resolution of the system and the inaccuracy of energy calibration of lower-energy α-particles and ^{12}C recoils, plus a possible contribution from misidentified low-energy particles.

Spectra were obtained with a gate set on the 3.4 MeV peak of Fig. 4. Fig. 5 shows a plot of ^{12}C recoil energy versus proton energy summed over the two detector configurations. Protons from the IAS to the 9.59 MeV state in ^{16}O should peak at about 1 MeV in the laboratory frame of reference, while α-particles from the break-up of the 9.59 MeV state should appear at about 1.8 MeV, with the ^{12}C recoils at about 0.6 MeV.

If the IAS decays by emission of an α-particle to the excited state in ^{13}N at 2.365 MeV and the α-particle is detected in F2(F1), then there is a high probability that the proton from the break-up of ^{13}N will be detected in F1(F2) and the ^{12}C recoil in F3(F4), or vice-versa. For this decay mode the proton, α and ^{12}C energies are near 0.4, 2.3 and 0.7 MeV, respectively, in the laboratory.

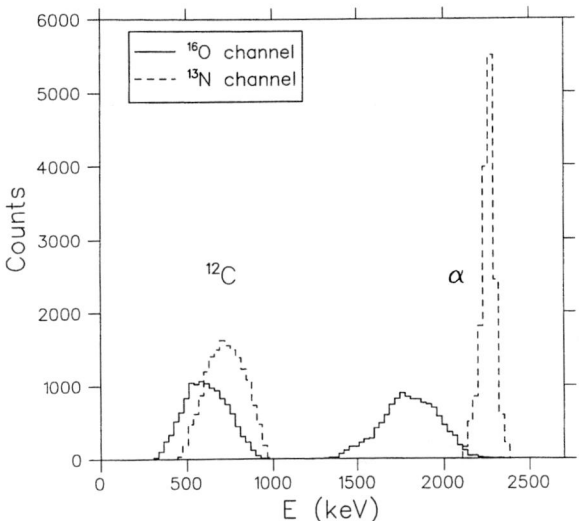

FIGURE 6. Spectrum of α-particles and ^{12}C recoil nuclei in triple coincidence with protons resulting from the break-up of the IAS of ^{17}F. The full curve is the spectrum for proton decay to the 9.59 MeV state of ^{16}O followed by break-up into α + ^{12}C. The dashed curve is the spectrum for α-decay to the 2.365 MeV state in ^{13}N followed by break-up into p + ^{12}C. The spectra were obtained by setting conditions on proton–^{12}C coincidences as indicated in Fig. 5.

The two coincidence peaks outlined in Fig. 5 show that these two decay modes are fairly well separated. Fig. 6 shows the α and ^{12}C spectra obtained by setting gates on the two ellipsoidal areas outlined in Fig. 5. This observation of βp-delayed α-particles from ^{17}Ne through the IAS of ^{17}F to the 9.59 MeV state of ^{16}O adds yet another branch to the proton decay of the IAS reported in ref. [10].

The background under the peak at 3.4 MeV in Fig. 4 prevents the low-energy tail of the α-peak from the break-up of the 9.59 MeV state from being observed as it was, for example, in the β-delayed decay of ^{16}N [4].

Detector Improvements

In June and July of 1997 we repeated the triple coincidence measurement with two separate modifications to the detector arrangement of Fig. 3. In the first experiment, we replaced detectors F3 and F4 with a 4400 mm^2 microchannel plate (MCP) in the chevron configuration [12] and a 3850 mm^2 microsphere plate (MSP) [13], respectively. Because of the large solid angle subtended by these detectors, the F1F3 and F2F4 pairs were well-placed (at

160°) to detect protons (α's) and ^{16}O (^{13}N) recoils in coincidence, as well as to record triple coincidences. These data are currently under analysis.

In the second experiment, detectors F3 and F4 were replaced by two identical double-sided silicon strip detectors of active area 5 cm x 5 cm with 16 strips per side, with adjacent strips connected together to give 64 pixels per detector. These data are currently being analysed. However, on-line analysis indicates that there should be sufficient data to produce triple coincidence spectra of high angular resolution (with considerably lower accidental background) for each pixel of the strip detectors.

SUMMARY

We have undertaken a detailed study of the β-delayed particle decay of ^{17}Ne with the goal of determining the α-width of the 2^+ state in ^{16}O at 6.917 MeV so as to better constrain the $E2$ component of the ^{12}C$(\alpha,\gamma)^{16}$O reaction. We have identified a state at 9.45 MeV in ^{17}F whose proton decay is predominantly to the 6.917 MeV state, which makes it a good candidate for initiating proton-delayed break-up of ^{16}O into $\alpha + {}^{12}$C through the tail of the 6.917 MeV state. In our first attempt to observe triple coincidences, we have recorded two decay modes of the IAS of ^{17}F resulting in the emission of a proton, α-particle and ^{12}C recoil nucleus. Because of the large solid angle of the detectors, background events prevented observation of the tails of the α-particle peak associated with the break-up of the 9.59 MeV state in ^{16}O. A recent measurement using strip detectors will result in much cleaner spectra because the substantial improvement in angular resolution will considerably reduce the background.

ACKNOWLEDGEMENTS

We wish to thank H. Biegenzein, D. Jones, P. Machule and H. Sprenger for help with the technical aspects of this experiment and M. Trinczek for assistance with the data collection. The work was supported in part by the Natural Sciences and Engineering Research Council of Canada, the National Science Foundation, the United Kingdom Science and Engineering Research Council, and TRIUMF.

REFERENCES

1. Fowler W.A., *Rev. Mod. Phys.* **56**, 149 (1984).
2. Weaver T.A., and Woosley S.E., *Phys. Reports* **227**, 65 (1993).
3. Buchmann L., et al., *Phys. Rev. Lett.* **70**, 726 (1993).
4. Azuma R.E., et al., *Phys. Rev. C* **50**, 1194 (1994).

5. Buchmann L., Azuma R.E., Barnes C.A., Humblet J., and Langanke K., *Phys. Rev. C* **54**, 393 (1996).
6. Tilley D.R., Weller H.R., and Ceves C.M., *Nucl. Phys.* **A564**, 1 (1993).
7. Hardy J.C., Esterl J.E., Sextro R.G., and Cerny J., *Phys. Rev. C* **3**, 700 (1971).
8. Borge M.J.G., et al., *Nucl. Phys.* **A490**, 287 (1988).
9. King J.D., et al., *Proceedings of the Fourth International Symposium on Nuclear Astrophysics: Nuclei in the Cosmos, Notre Dame, USA, 1996*, *Nucl. Phys.* **A**, in press.
10. King J.D., et al., *Proceedings of the International Workshop on the Physics of Unstable Nuclear Beams, Serra Negra, São Paulo, Brazil, 1996*, edited by C.A. Bertulani, L. Felipe Canto, and M.S. Hussein (World Scientific, New Jersey, in press).
11. D'Auria J.M., Buchmann L., Dombsky M., McNeely P., Roy G., Sprenger H., and Vincent J., *Nucl. Instrum. Methods Phys. Res.* **B70**, 75 (1992).
12. Wiza, J.L., *Nucl. Instrum. Methods* **162**, 587 (1979).
13. Tremsin, A.S., Pearson, J.F., Lees, J.E., and Fraser, G.W., *Nucl. Instrum. Methods Phys. Res.* **A368**, 719 (1996).

Coulomb Dissociation of ^{8}B at 254 MeV/u

N. Iwasa,[1] K. Sümmerer, F. Boue, G. Surowka, F. Uhlig,[a]
J. Speer,[b] P. Senger, T. Baumann, H. Geissel, M. Hellström,
P. Koczon, F. Laue, A. Ozawa, E. Schwab, W. Schwab,
A. Surowiec, E. Grosse,[2] A. Förster,[a] H. Oeschler,[a] C. Sturm,[a]
A. Wagner,[a] B. Kohlmeyer,[b] B. Blank,[c] C. Marchand,[c]
M. S. Pravikoff,[c] S. Czajkowsky,[c] R. Kulessa,[d] W. Walus,[d]
T. Motobayashi,[e] T. Teranishi,[f] M. Gai[g]

Gesellschaft für Schwerionenforschung m.b.H., Planckstr. 1, D-64291 Darmstadt, Germany.
[a] *Institut für Kernphysik, Technische Hochschule Darmstadt, D-6100 Darmstadt, Germany*
[b] *Fachbereich Physik, Philipps Universität, D-3550 Marburg, Germany*
[c] *Centre d'Etudes Nucléaires de Bordeaux-Gradignan, F-33175 Gradignan Cedex, France*
[d] *Institute of Physics, Jagiellonian University, PL-30-059 Krakow, Poland.*
[e] *Department of Physics, Rikkyo University, Toshima, Tokyo 171, Japan.*
[f] *RIKEN (Institute of Physical and Chemical Research), Hirosawa, Wako, Saitama 351-01, Japan.*
[g] *Department of Physics, University of Connecticut, Storrs, CT 06269-3046, U.S.A.*

Abstract. We have measured the Coulomb-dissociation cross section for ^{8}B→^{7}Be + p in the field of ^{208}Pb for a wide range of 0.15-2.95 MeV at E_{in}=254 MeV/u. The preliminary results agree with Monte Carlo simulations that used the ^{7}Be(p,γ)^{8}B cross sections measured by Filippone et al., and Vaughn et al. and assuming E1 (+M1) Coulomb dissociation process. Further analysis is in progress.

The Coulomb dissociation of ^{8}B into ^{7}Be+p is an alternative method for extracting the ^{7}Be(p,γ)^{8}B cross section [1]. The astrophysical S factor of the ^{7}Be(p,γ)^{8}B is of crucial importance for predicting the high-energy solar-

---

[1]) Present Address: *RIKEN (Institute of Physical and Chemical Research), Hirosawa, Wako, Saitama, 351-01, Japan.*
[2]) Present Address: *Institut für Kern- und Hadronenphysik, Forschungszentrum Rossendorf, Postfach 510119, D-01314 Dresden, Germany.*

neutrino flux which is relevant to the "solar neutrino problem" defined by a discrepancy by a factor of two to three between measured and predicted neutrino fluxes from the sun [2]. Among four existing neutrino observation, the Homestake [3] and Kamiokande [4] experiments measured the high-energy solar-neutrino mainly or solely from ^{8}B produced predominantly by the reaction.

Six direct measurements were reported [5-10]. Since the target nuclei ^{7}Be decay to ^{7}Li with $T_{1/2} = 53$ days, the normalization relies on the ^{7}Li(d,p)^{8}Li reaction which was used for determining effective target thickness. The difficulty reflects the fact that the four high-precision results [6-8,10] group into two distinct pairs which agree on their energy dependences, but disagree on their absolute values. The extracted absolute values of the S-factors for these two groups disagree by approximately 30%. Since the discrepancy is larger than the denoted error of the adopted S factor in the standard solar model, experimental studies with different methods were highly desirable for improving the reliability of the input to the standard solar model. [2]

Recently, Motobayashi et al. succeeded to extract the ^{7}Be(p,γ)^{8}B cross section at 0.6-1.7 MeV with the Coulomb-dissociation method at $E(^8{\rm B}) = 46.5$ MeV/u [1]. The method is free from the uncertainty of determining effective target thickness, because the dissociation process of ^{8}B into ^{7}Be and proton in the Coulomb field of a ^{208}Pb target was measured, instead of measuring the ^{7}Be(p,γ)^{8}B cross section. The process is regarded as absorption of a virtual photon [11]. The ^{7}Be(p,γ)^{8}B cross section is extracted from the Coulomb-dissociation cross section, because the photo-absorption reaction is directly related to the radiative-capture reaction. Another advantage is to be able to use a thicker target. High detection efficiency is expected because of high-energy charged-particle measurement. Moreover the (p,γ) cross section of 1-1000 nb is enhanced to the Coulomb-dissociation cross section of 0.1-10 mb for 10 keV-bin.

The extracted (p,γ) cross section [1] is consistent with the results of the lower group [8,10] within their errors. The cross section at lower energies than 500 keV could not be deduced due to large background from the dissociation in helium gas filled between the target and detector. The M1 peak at 633 keV is suppressed, because of relatively small flux of the M1 virtual photon. Although the E2 amplitude can be neglected in the (p,γ) reaction, it is expected to be enhanced in the Coulomb-dissociation due to large flux of the E2 virtual photon. For precise extraction of the ^{7}Be(p,γ)^{8}B cross section from the Coulomb-dissociation cross section, information of the M1 and E2 admixtures to the dominant E1 component are desired. Several experimental [12-14] and theoretical [15] studies on the E2 component of the Coulomb-dissociation were reported.

In this article, we report an experiment of the Coulomb dissociation at higher energy at 254 MeV/u. Since the M1 component is enhanced compared with the one at 50 MeV/u incident energy, the M1 cross section is expected to be extracted by comparing the results at 50 MeV/u [1,13] and 254 MeV/u. Another advantage is that effects of the post-dissociation Coulomb acceleration or higher-order effects in the Coulomb dissociation are reduced [16,17]. Influence of straggling on the experimental resolution is reduced. Forward focusing allows us to use a magnetic spectrometer for a kinematical complete measurement with high detection efficiency over a wider range of the p-^{7}Be relative energy.

The experiment was performed at the radioactive-beam facility at GSI [19]. A ^{8}B beam at 254.5 MeV/u was produced by a ^{12}C beam at 350 MeV/u accelerated by SIS impinging on a beryllium target with a thickness of 8.01 g/cm^2. The secondary beam bombarded on an enriched ^{208}Pb target with a thickness of 199.7 mg/cm^2. The mean beam energy at the middle of the target is 253.9 MeV/u and a typical intensity of the beam is 10^4/sec. For reconstructing relative energy, the momenta and the scattering angle of the reaction products, ^{7}Be and proton, were measured as shown in Fig.1.

The scattering angles of the products were measured by four silicon microstrip detectors with a thickness of 300 μm and an active area of 56 × 56 mm^2 placed at 14.0 cm, 14.8 cm, 30.2 cm, and 31.0 cm downstream of the target. They are position sensitive in the horizontal, vertical, horizontal, and vertical directions, respectively, with a pitch of 100 μm. Individual readout for each strip was performed by incorporating with GASSIPLEX chips [20] for analog multiplexing of the energy-deposit signals and CRAMS modules [21] for the digital readout of analog-multiplexed signals. The two-dimensional position information allow us to reconstruct the position of dissociation events with an

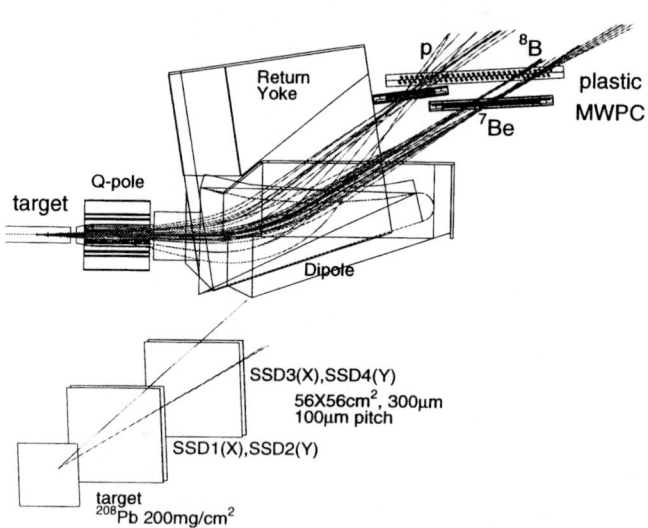

FIGURE 1. Schematic view of the experimental setup.

uncertainty of 5.5 mm × 0.4 mm × 0.4 mm (one sigma) for beam-, horizontal- and vertical-directions, respectively. Breakup events at the target were well identified event-by-event.

Momenta of the reaction products were analyzed by KaoS [22]. In order to reduce angular and energy straggling and background reaction, KaoS was filled with helium gas of one atmosphere. The trajectories were reconstructed by the micro-strip detectors and a two-dimensional multwire proportional chambers (MWPC) [22] placed at around the focal plane of KaoS. Thirty plastic scintillators were placed just behind of the MWPC for producing trigger conditions to data acquisition system.

KaoS has larger momentum and angular acceptance compared with the experimental momentum range for both proton and ^{7}Be and the grazing angle, respectively. Averaged coincidence efficiency is calculated to be about 90%. The angular and energy resolutions of KaoS were estimated to be 0.7 mrad and 0.04 %, respectively. They are considerably smaller than the angular straggling in the target material and strip detector evaluated [18] to be 5.5 mrad. Similarly the energy straggling in the target material, strip detectors, vacuum windows, and helium gas were all estimated to be 0.3 % for proton at around 254 MeV/u. The influence of the detector resolution on the experimental resolution is negligibly small.

The measured complete kinematics of the breakup products allowed us to determine the p-^{7}Be relative energy, and scattering angle of the excited ^{8}B.

The p-^{7}Be coincidence yields for 0.15-2.95 MeV were plotted in Fig. 2 as a function of p-^{7}Be relative energy. The loss of the coincidence events due to nuclear reactions in the layers of matter downstream of the targets were corrected by using recent experimental results for the total reaction cross

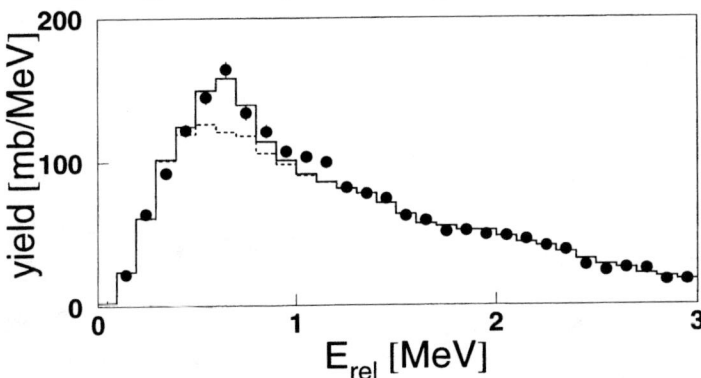

FIGURE 2. A preliminary relative energy spectrum The solid and dashed histograms denote results of Monte Carlo simulations assuming the ^{7}Be(p,γ)^{8}B cross sections measured by Filippone et al. [10] and Vaughn et al. [8] and E1+M1 and E1 Coulomb dissociation process.

section by Blank [23] with an empirical formula for the total reaction cross section reported by Kox et al. [24]. The correction of the dissociation process to the excited state of ^{7}Be, ^{8}B→^{7}Be*+p→^{7}Be+p+γ, was not performed as yet. The relative-energy resolution is 0.09 MeV at E_{rel} = 0.6 MeV and 0.15 MeV at E_{rel} = 1.7 MeV, which was calculated with Monte-Carlo simulations.

The Monte-Carlo simulations were performed by using the code GEANT [25] It evaluates the response of the detection system to the products of the ^{8}B → ^{7}Be+p dissociation reaction. Events were generated with probabilities proportional to the Coulomb-dissociation cross section calculated with a semi-classical formula proposed by Baur, Bertulani and Rebel [11]. The incident ^{8}B beam spread in energy, angle, and position at the target, were also included in the calculation. The influence of the angular and energy straggling as well as energy loss in the layers of the matter was taken into account in the simulations.

The solid (dashed) histogram in the figure are the E1+M1 (E1) yields calculated by the simulations assuming the Coulomb dissociation cross section is dominated by the E1 excitations (plus an M1 resonance), with the ^{7}Be(p,γ)^{8}B capture cross section measured by Filippone et al. [10] and Vaughn et al. [8] normalized by the resonant cross section of the ^{7}Li(d,p)^{8}Li to be 157 ± 10 mb. The shape and magnitude of the experimental energy dependence are

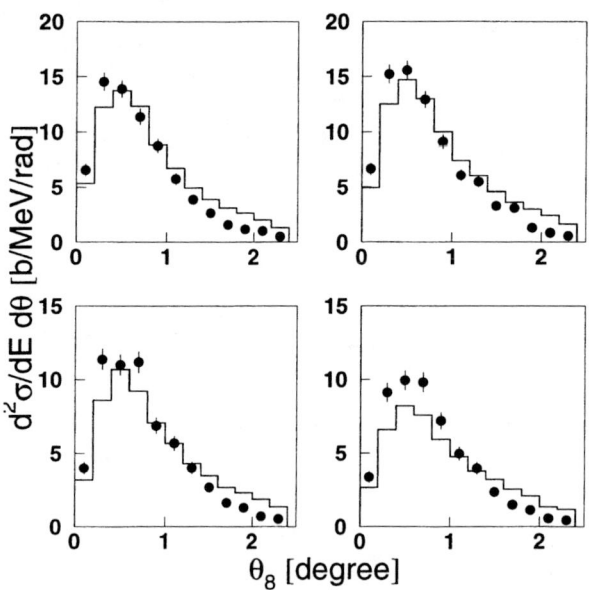

FIGURE 3. A preliminary angular distribution of the excited ^{8}B (center of mass of the p-^{7}Be system) for four relative energy bins, 0.4-0.6 (upper left), 0.6-0.8 (upper right), 0.8-1.0 (lower left), and 1.0-1.2 (lower right) MeV The histograms show the simulation results.

fairly well described by the simulated results. This agreement denotes that the partial γ width of the M1 resonance is consistent with that measured by Filippone et al. [10].

Figure 3 shows the experimental yield plotted against the scattering angle of the excited ^{8}B for four relative energy bins, 0.4-0.6, 0.6-0.8, 0.8-1.0, and 1.0-1.2 MeV together with results of the Monte Carlo simulations assuming E1 and M1 excitation mechanism. The angular resolution was estimated by the simulation to be 0.35°. The experimental distributions are well reproduced by the simulated results. The agreement between the experimental and theoretical dependences both on the relative energy and the scattering angle suggests the E1 dominant nature of the first-order Coulomb dissociation.

Further studies are necessary to elucidate the role of E2 and M1 amplitudes, for estimating the dissociation yield to the first excited state of ^{7}Be, and other issues. We plan to extract the zero-energy astrophysical S factor in the near future.

We would like to thank C. A. Bertulani, S.Typel, and G. Baur for their help in theoretical aspects.

REFERENCES

1. Motobayashi,T., Iwasa,N., Ando,Y., Kurokawa,M., Murakami,H., Ruan,J., Shimoura,S., Shirato,S., Inabe,N., Ishihara,M., Kubo,T., Watanabe,Y., Gai,M., France III,R.H., Hahn,K.I., Zhao,Z., Nakamura,T., Teranishi,T., Futami,Y., Furutaka,K., and Delbar,Th., *Phys. Rev. Lett.* **70**, 2680 (1994); Iwasa.N. et al., *J. Phys. Soc. Japan* **65**, 1256 (1996).
2. Bahcall,J.N., and Pinsonneault,M.H., *Rev. Mod. Phys.* **64**, 885 (1992).
3. Davis,R., *Prog. Part. Nucl. Phys.* **32**, 13 (1994).
4. Fukuda,Y. et al., *Phys. Rev. Lett.* **77**, 1683 (1996).
5. Kavanagh,R.W., *Nucl. Phys.* **15**, 411 (1960).
6. Parker,P.D., *Astrophys. J.* **153**, L85 (1968).
7. Kavanagh,R.W., Tombrello,T.A., Mosher,J.M., and Goosman,D.R., *Bull. Am. Phys. Soc.* **14**, 1209 (1969).
8. Vaughn,F.J., Chalmers,R.A., Kohler,D., and Chase,L.F., *Phys. Rev.* **C2**, 1657 (1970).
9. Wiezorek,C., Kräwinkel,H., Santo,R., and Wallek,L., *Z. Phys.* **A282**, 121 (1977).
10. Filippone,B.W., Elwyn,A.J., Davids,C.N., and Koetke,D.D., *Phys. Rev.* **C28**, 2222 (1983).
11. Baur,G., Bertulani,C.A., and Rebel,H., *Nucl. Phys.* **A 458**, 188 (1986); Bertulani,C.A., and Baur,G., *Phys. Rep.* **163**, 300 (1988).
12. Schwarzenberg,J.v., Kolata,J.J., Peterson,D., Santi,P., Belbot,M., and Hinnefeld,J.D., *Phys. Rev.* **C53**, R2598 (1996).

13. Kikuchi,T., Motobayashi,T., Iwasa,N., Ando,Y., Kurokawa,M., Moriya,S., Murakami,H., Nishio,T., Ruan,J., Shirato,S., Shimoura,S., Uchibori,T., Yanagisawa,Y., Kubo,T., Watanabe,Y., Ishihara,M., Hirai,M., Nakamura,T., Sakurai,H., Teranishi,T., Kubono,S., Gai,M., France III,R., Hahn,K.I., Delbar,Th., Lipnik,P., and Michotte,C., *Phys. Lett.* **B 391**, 261 (1997).
14. Kelley,J.H., Austin,S.M., Azhari,A., Bazin,D., Brown,J.A., Esbensen,H., Fauerbach,M., Hellström,M., Hirzebruch,S.E., Kryger,R.A., Morrissey,D.J., Pfaff,R., Powell,C.F., Ramakrishnan,E., Sherrill,B.M., Steiner,M., Suomijarvi,T., and Thoennessen,M., *Phys. Rev. Lett.* **77**, 5020 (1996).
15. Esbensen,H., and Bertsch,G.F., *Nucl. Phys.* **A600**, 37 (1997).
16. Bertulani,C.A., *Phys. Rev.* **C49**, 2688 (1994).
17. Typel,S., and Baur,G., *Phys. Rev.* **C50**, 2104 (1994); Typel,S., Wolter,H.H., and Baur,G., *Nucl. Phys.* **A613**, 147 (1997).
18. A calculation code for energy loss, energy-loss and angular straggling, and so on by GSI, Darmstadt, Germany.
19. Geissel,H. *et al.*, *Nucl. Instr. Meth.* **B 70**, 286 (1992).
20. Santiard,J.C., private communication.
21. Module V-550 manufactured by C.A.E.N., Viareggio, Italy.
22. Senger,P. *et al.*, *Nucl. Instr. Meth.* **A 327**, 393 (1993).
23. Blank,B., Marchand,C., Pravikoff,M.S., Baumann,T., Boué,F., Geissel,H., Hellström,M., Iwasa,N., Schwab,W., Sümmerer,K., and Gai,M., *Nucl. Phys.*, in print.
24. Kox,S., Gamp,A., Perrin,C., Arvieux,J., Bertholet,R., Bruandet,J.F., Buenerd,M., Cherkaoui,R., Cole,A.J., El-Masri,Y., Longequeue,N., Menet,J., Merchez,F., and Viano,J.B., *Phys. Rev* **C35**, 1678 (1987).
25. Detector description and simulation tool by CERN, Geneva, Switzerland.

Astronuclear Physics with Coulomb Dissociation

Stefan Typel

Sektion Physik, Universität München, Am Coulombwall 1, D-85748 Garching, Germany

Abstract. The method of Coulomb dissociation can be applied to the investigation of nuclei and reactions relevant to nuclear astrophysics. General aspects of the theoretical description are discussed. Some experimental results from the electromagnetic breakup of stable and radioactive nuclei are surveyed. Suggestions for future applications are mentioned.

INTRODUCTION

Nuclear physics constitutes a significant basis of many astrophysical models, especially in the description of nucleosynthesis under various, partly extreme, conditions[1]. Information on the structure and reactions of stable and radioactive nuclei is needed but the direct processes are often not accessible in experiments.

Nowadays, exotic nuclei are available as secondary beams and can be studied in scattering experiments. Nuclear collisions where the colliding nuclei interact only by the theoretically well understood electromagnetic forces are of particular interest. There can be a clean interpretation of the experimental data.

A projectile nucleus a passing the Coulomb field of a highly charged target X can break up into fragments b and c. Their momentum distribution in this Coulomb dissociation process provides us with information on the nuclear structure of the projectile and the inverse radiative capture reaction $b(c,\gamma)a$[2-5]. To reduce the nuclear interaction one can either use bombarding energies below the Coulomb barrier or choose very forward scattering angles in high energy collisions.

In general, electromagnetic excitation is not limited to a first order process. Higher order effects can lead to the excitation of new nuclear states, like the double phonon giant dipole resonance[6]. They have to be taken into account as a correction when cross sections for first orders processes, like the radiative capture reaction, are to be extracted[7-9].

THEORY OF ELECTROMAGNETIC EXCITATION

The theory of Coulomb excitation is well developed. Assuming a first order process, the cross section can be written as

$$\frac{d\sigma}{d\omega} = \frac{dn}{d\omega}\,\sigma_\gamma(\omega) \qquad (1)$$

where $\sigma_\gamma(\omega)$ denotes the corresponding cross section for the photo-induced process. The quantity $\frac{dn}{d\omega}$ is called the equivalent photon number. Besides the excitation energy $\hbar\omega$, it depends on the beam velocity v, the impact parameter b and the multipolarity (E1, E2, M1, ...). Usually, $\frac{dn}{d\omega}$ is calculated in the semiclassical approximation where the motion of the projectile is treated classically assuming a straight line trajectory. Refined expressions, which take corrections due to Coulomb-deflection as well as relativistic effects into account, are available in the literature[3,8,10].

The characteristic parameter of electromagnetic excitation is the adiabaticity parameter

$$\xi = \frac{\omega b}{\gamma v} \qquad (2)$$

For $\xi \ll 1$ the equivalent photon numbers become largest, whereas for $\xi \gg 1$ we only have an adiabatic excitation. This parameter determines suitable beam energies and scattering angles for given excitation energies.

The contributions of different multipolarity also depend on these quantities. In the limit $\xi \to 0$ we have

$$\frac{n_{E2}}{n_{E1}} \approx \frac{4}{(k_\gamma b)^2} \qquad \frac{n_{M1}}{n_{E1}} \approx \frac{v^2}{c^2}. \qquad (3)$$

In principle, these dependences can be used to separate the various contributions in the cross section.

The projectile motion can be treated quantally in Glauber theory or a Distorted Wave Born Approximation[3,9,11–13]. In this kind of calculations, nuclear effects can also be added[12,14]. The main effects due to the strong absorption at impact parameters less the sum of the two nuclear radii are characteristic diffraction effects. Additionally, there is the nuclear excitation. This effect depends largely on the used model of nuclear interaction, but it is generally small and its characteristic angular dependence can be used to disentangle nuclear and electromagnetic excitation[15].

When the number of equivalent photons becomes large, there is not only the first order excitation, but a final state can be reached via multiple excitation through intermediate states. These higher order effects can be treated in a coupled channels approach or in higher order perturbation theory. This involves a sum over all intermediate states considered to be relevant. But the treatment of intermediate continuum states is quite complicated. Another approach is to integrate the time-dependent Schrödinger equation directly for a given model Hamiltonian[16].

For a sudden collision, i.e. $\xi = 0$, the time ordering in the usual perturbation approach can be neglected and the interaction can be summed up to infinite order. No intermediate states appear explicitly in the calculation of the excitation cross section. The summation is implicitly included in the corresponding operator between initial and final states in the calculation of the relevant matrix-elements.

For small values of ξ, a related approach was developed in Refs. 7–9 for the calculation of second and third order corrections in perturbation theory. Analytical results were obtained in a simple zero-range model for the neutron-core interaction, a prototype for a loosely bound system[7]. In such a model, the strength of the higher order effects essentially depends on the parameter

$$\chi = \frac{2Z_X Z^{(1)}_{eff} e^2}{vb\hbar k} \quad (4)$$

with the effective (dipole) charge

$$Z^{(1)}_{eff} = -\frac{Z_c m_n}{m_c + m_n} \quad (a = n + c) \quad (5)$$

and the excitation energy

$$E_f - E_i = \frac{\hbar^2 k^2 (m_n + m_c)}{2 m_n m_c}. \quad (6)$$

From Eq. 4 it is seen that higher order effects are largest for high charge number of the projectile, large scattering angle, small beam and excitation energies, as is intuitively obvious.

DISCUSSION OF EXPERIMENTS

Coulomb excitation has become a useful spectroscopic tool for the investigation of exotic nuclei with beams of intermediate energy. The following examples show the possibilities for an application of the Coulomb excitation, even if they are not directly relevant to astrophysics. The experiments mainly concern ground-state properties (e.g. halo structure), the excitation of special states, or the determination of electromagnetic matrix-elements.

In the breakup of the two-neutron halo nucleus ^{11}Li the correlation of the outgoing neutrons and the dipole strength in the continuum was studied[17–23]. Similar dissociation experiments with ^{11}Be also aimed at the fragment momentum distribution and low-lying continuum E1 strength[24–26]. In both cases post-acceleration of the fragments was studied.

The electromagnetic excitation of the first excited state in ^{11}Be has been investigated experimentally[27-29]. A comparison with the known B(E1)-value of the corresponding ground-state transition showed that higher order effects are small as expected from theoretical estimations[30,31]. By measuring the excitation energy and the corresponding B(E2)-value of the first 2^+-state in ^{32}Mg, a strong deformation of the nucleus was inferred[32]. This kind of experiment can also be applied to the study of nuclear structure for other neutron-rich nuclei[33].

Coulomb breakup experiments with ^{8}B probed the existence of a possible proton-halo in the ground state[34]. The momentum distribution of the fragments can be explained without assuming a special halo state. For the interpretation of the momentum distribution it is neccessary to consider the excitation mechanism correctly.

Coulomb dissociation is a useful indirect method to measure radiative capture reactions of the type $b + c \rightarrow a + \gamma$ relevant to nuclear astrophysics. Instead of the direct process, the time-reversed reaction $a + \gamma \rightarrow b + c$ is studied, where the photons are provided by the equivalent photons in a fast peripheral collision. Of course, the method is limited to ground-state transitions. In the following, we concentrate on a few examples. More detailed information can be found in recent reviews[4].

A test case of the Coulomb dissociation method was the breakup of ^{6}Li into $\alpha + d$[4]. The abundance of primordial ^{6}Li, determined by the reaction $\alpha + d \rightarrow {}^6\text{Li} + \gamma$, can be a test for big bang nucleosynthesis[35]. Predicted by theoretical calculations[36,37], the astrophysical S-factor shows a significant energy dependence, where at energies less than ≈ 100 keV the E1 component dominates over the E2 contribution. Coulomb breakup experiments tested both resonant and continuum breakup, in good agreement with theoretical results[38,39]. Unfortunately, the E1 contribution is suppressed as compared to E2 in the Coulomb dissociation. For a precise determination of the E1 component, an analysis of E1-E2 interferences in angular distributions is required.

The role of E1 and E2 is reversed in the cases of $^7\text{Li} \rightarrow \alpha + {}^3\text{H}$ or $^7\text{Be} \rightarrow \alpha + {}^3\text{He}$ Coulomb dissociation[40-44]. Here, the E1 contribution completely dominates the corresponding capture reaction[45]. The enhanced E2 component in the breakup could be responsible for the large asymmetry observed in the fragment angular distribution of the experiment[43].

From the resonant E1 breakup of ^{14}O the gamma-width of the 1^--resonance in the continuum was extracted in agreement with other methods[46,47]. Higher order effects are expected to be small from theoretical considerations[48]. The corresponding $^{13}\text{N}(p,\gamma)^{14}\text{O}$ reaction is important in the hot CNO-cycle, as well as the capture reactions $^{12}\text{C}(p,\gamma)^{14}\text{N}$ and $^{11}\text{C}(p,\gamma)^{12}\text{N}$, where Coulomb dissociation experiments have been carried out, too[47,49]. A particular challenge is the Coulomb breakup of ^{16}O to determine the low-energy S-factor of the $^{12}\text{C}(\alpha,\gamma)^{16}\text{O}$ reaction (mainly E2 component). The high excitation energy requires high projectile velocities and nuclear breakup can severely interfere with electromagnetic breakup.

The Coulomb dissociation of ^{8}B → ^{7}Be + p, studied already at intermediate[50–54] and rather low energies[55], can give information on the low-energy ^{7}Be(p,γ)^{8}B radiative capture reaction, which is relevant to the solar neutrino problem. The rate of produced ^{8}B in this reaction determines the solar high-energy neutrino flux. A model-independent separation of E1 and E2 contributions should be possible by a careful study of angular distributions[52,56,57]. At high projectile energies, the resonant M1 breakup can also be observed[54]. Nuclear and higher order electromagnetic effects are expected to be negligibly small at high enough beam energies[9]. From studies under different experimental conditions a reliable astrophysical S-factor should emerge, consistent with direct measurements[58].

Recently, it was proposed to use electromagnetic dissociation of very neutron-rich nuclei to obtain information on (n,γ)-reactions[59]. Some of these nuclei far off the valley of stability, which are relevant for the r-process[1,60], can be produced with the new radioactive beam facilities. The low neutron-thresholds are favourable for the Coulomb dissociation method. The simple zero-range model for the neutron-core interaction[7] can be useful to estimate the importance of higher order effects and the minimum beam energy to extract reliable results from the experiment.

In the case of two-nucleon capture reactions, which cannot be studied in a direct way in the laboratory, the Coulomb dissociation method can be useful if the relevant information cannot be obtained by other means. As discussed in Ref. 61, a sequence of two-neutron capture reactions, ^{4}He(2n,γ)^{6}He(2n,γ)^{8}He, can bridge the A=5 and 8 gaps at high neutron abundances. A two-step resonant process through the broad 2^+-state could dominate the synthesis of ^{6}He and ^{8}He. Another key process is the ^{4}He(αn,γ)^{9}Be reaction[61]. Despite the direct study of the photo-dissociation of ^{9}Be, a ^{9}Be Coulomb breakup experiment could be rewarding[62]. Two-proton capture reactions can bridge the waiting points in the rp-process[63–65] at conditions typical for X-ray bursts on neutron stars. A Coulomb dissociation experiment, e.g. ^{40}Ti → p + p + ^{38}Ca, should give more insight to the 2p-capture process. The emission of the two protons during the breakup is expected to be sequential rather than correlated.

CONCLUSIONS

The dissociation of stable or radioactive nuclei in the electromagnetic field of a highly charged target during a peripheral, medium or high energy collision is a useful tool of nuclear physics research. The method is well developed, both in theory and experiment. Higher order electromagnetic effects and contributions from nuclear excitation can be kept under control by choosing suitable kinematic conditions. A variation of these parameters influences the contributions of different multipolarity. Already, the method has successfully been applied to several reactions. There are still many cases relevant to nuclear astrophysics. Coulomb dissociation is a favourable method for the investigation of nuclei near the drip lines because of the small binding energies and the existence of few excited states. Some reactions cannot be studied in a direct way, but the indirect method of electromagnetic breakup can supply us with useful information.

ACKNOWLEDGMENTS

The author is grateful to G. Baur, M. Gai, N. Iwasa, T. Motobayashi, K. Sümmerer, H. H. Wolter, and H. Utsunomiya for interesting discussions. This work was supported by GSI, Darmstadt.

REFERENCES

1. Rolfs, C. E., and Rodney, W. S., *Cauldrons in the Cosmos*, Chicago and London: The University of Chicago Press, 1988.

2. Baur, G., Bertulani, C. A., and Rebel, H., *Nucl. Phys.* **A458**, 188 (1988).

3. Bertulani, C. A., and Baur, G., *Phys. Rep.* **163**, 299 (1988).

4. Baur, G., and Rebel, H., *J. Phys. G: Nucl. Part. Phys.* **20**, 1 (1994); *Ann. Rev. Nucl. Part. Sci.* **46**, 321 (1996).

5. Vervier, J., *Prog. Part. Nucl. Phys.* **37**, 435 (1996).

6. Baur, G., and Bertulani, C. A., *Phys. Lett.* **B174**, 23 (1986).

7. Typel, S., and Baur, G., *Nucl. Phys.* **A573**, 486 (1994).

8. Typel, S., and Baur, G., *Phys. Rev.* **C50**, 2104 (1994).

9. Typel, S., Wolter, H. H., and Baur, G., *Nucl. Phys.* **A613**, 147 (1997).

10. Winther, A., and Alder, K, *Nucl. Phys.* **A319**, 518 (1979).

11. Bertulani, C. A., and Nathan, A. M., *Nucl. Phys.* **A554**, 158 (1993).

12. Muendel, A., and Baur, G., *Nucl. Phys.* **A609**, 254 (1996).

13. Baur, G. and Bertulani, C. A., *Phys. Rev.* **C56**, 581 (1997).

14. Hencken, K., Bertsch, G., and Esbensen, H., *Phys. Rev.* **C54**, 3043 (1996).

15. Shyam, R., Baur, G., and Banerjee, P., *Phys. Rev.* **C44**, 915 (1991).

16. Esbensen, H., Bertsch, G. F., and Bertulani, C. A., *Nucl. Phys.* **A581**, 107 (1995).

17. Ieki, K., Galonski, A., et al., *Phys. Rev.* **C54**, 1589 (1996).

18. Kobayashi, T., et al., *Phys. Lett.* **B232**, 51 (1989).

19. Orr, N. A., et al., *Phys. Rev.* Lett. **74**, 2050 (1992).

20. Orr, N. A., et al., *Phys. Rev.* **C51**, 3116 (1995).

21. Ieki, K., et al., *Phys. Rev.* Lett. **70**, 730 (1993).

22. Sackett, D., et al., *Phys. Rev.* **C48**, 118 (1993).

23. Kobayashi, T., et al., *Phys. Lett.* **B232**, 51 (1989).

24. Nakamura, T., et al., *Phys. Lett.* **B331**, 296 (1994).

25. Kelley, J. H., et al., *Phys. Rev.* Lett. **74**, 30 (1995).

26. Nakamura, T., et al., *Phys. Lett.* **B331**, 296 (1994).

27. Anne, R., et al., *Z. Phys.* **A352**, 397 (1995).

28. Nakamura, T., et al., *Phys. Lett.* **B394**, 11 (1997).

29. Fauerbach, M., et al., *Phys. Rev.* **C56**, R1 (1997).

30. Bertulani, C. A., Canto, L. F., and Hussein, M. S., *Phys. Lett.* **B353**, 413 (1995).

31. Typel, S., and Baur, G., *Phys. Lett.* **B356**, 186 (1995).

32. Motobayashi, T., et al., *Phys. Lett.* **B346**, 9 (1995).

33. Scheit, H., et al., *Phys. Rev.* Lett. **77**, 3967 (1996).

34. Schwab, W., et al., *Z. Phys.* **A350**, 285 (1995).

35. Nollett, K. M., et al., *Phys. Rev.* **C56**, 1144 (1997).

36. Mukhamedzhanov, A. M., et al., *Phys. Rev.* **C52**, 3483 (1995).

37. Typel, S., Blüge, G., and Langanke, K., *Z. Phys.* **A339**, 335 (1991).

38. Kiener, J., et al., *Phys. Rev.* **C44**, 2195 (1991).

39. Hesselbarth, J., and Knöpfle, K. T., *Phys. Rev.* Lett. **67**, 2773 (1991).

40. Utsunomiya, H., et al., *Phys. Lett.* **B211**, 24 (1988).

41. Utsunomiya, H., et al., *Nucl. Phys.* **A511**, 379 (1990).

42. Utsunomiya, H., et al., *Phys. Rev.* Lett. **65**, 847 (1990); *Phys. Rev.* Lett. **69**, 863 (1992).

43. Utsunomiya, H., contribution, this symposium.

44. Mason, J. E., et al., *Phys. Rev.* **C45**, 2870 (1992).

45. Igamov, S. B., et al., Physics of Atomic Nuclei **60**, 1126 (1997).

46. Kiener, J., et al., *Nucl. Phys.* **A552**, 66 (1993).

47. Motobayashi, T., et al., *Phys. Lett.* **B264**, 259 (1991).

48. Typel, S., and Baur, G., *Phys. Rev.* **C49**, 379 (1994).

49. Lefebvre, A., et al., *Nucl. Phys.* **A592**, 69 (1995).

50. Motobayashi, T., et al., *Phys. Rev. Lett.* **73**, 2680 (1994).

51. Motobayashi, T., contribution, this symposium.

52. Kelley, J. H., et al., *Phys. Rev. Lett.* **77**, 5020 (1996).

53. Sümmerer, K., et al., GSI Darmstadt, experimental proposal and private communication

54. Iwasa, N., contribution, this symposium.

55. von Schwarzenberg, J., et al., *Phys. Rev.* **C53**, R2598 (1996).

56. Gai, M., and Bertulani, C. A., *Phys. Rev.* **C52**, 1706 (1995).

57. Motobayashi, T., et al., *Phys. Lett.* **B391**, 261 (1997).

58. Bogoert, G., et al., presented at the SOLAR Neutrino Conference IV, Heidelberg, April, 1997.

59. Gai, M., presented at the ISOL workshop, Columbus/Ohio, July 30 – August 1, 1997.

60. Cowan, J. J., Thielemann, F.-K., and Truran, J. W., *Phys. Rep.* **208**, 267 (1991).

61. Görres, J., Herndl, H., Thompson, I. J., and Wiescher, M., *Phys. Rev.* **C52**, 2231 (1995).

62. Kalassa, D. M., and Baur, G., *J. Phys. G: Nucl. Part. Phys.*, 115 **22** (1996).

63. Baraffe, I., et al., Thielemann, F.-K., (convener), NuPECC Report, Nuclear and Particle Astrophysics, July 16, 1997.

64. Görres, J., Wiescher, M., and Thielemann, F.-K., *Phys. Rev.* **C51**, 392 (1995).

65. Schatz, H., et al., *Phys. Rep.* (1997), in press.

THERMONUCLEAR REACTIONS III

Pulsed keV Neutrons for Nuclear Astrophysics and Recent Results of (n,γ) Reaction of Light Nuclei

Y.Nagai[*], T.Shima[*], T.Kii[*], T.Baba[*], K.Takaoka[*], S.Naito[*], A.Tomyo[*], T.Takahashi[*], Y.Nobuhara[*], M.Kinoshita[*], M.Igashira[+], and T.Ohsaki[+]

[*]*Department of Applied Physics and* [+]*Research Laboratory for Nuclear Reactors, Tokyo Institute of Technology, O-okayama, Meguro, Tokyo 152, Japan*

Abstract. Pulsed keV neutrons play an important role in studying problems related to the nuclear astrophysics. In the present study we have measured the neutron capture cross sections on D, ^{7}Li and ^{18}O samples in the neutron energies between 10 and 550 keV. They are important not only for the primordial nucleosynthesis theory, but also for the stellar nucleosynthesis theory of less evolved stars and for the solar neutrino problem. While in order to extend the present study of the keV neutron induced reaction of nuclei further we have been working on the improvement of the performance of the NaI(Tl) spectrometer and testing the property of the spallation neutron source at KEK.

INTRODUCTION

Pulsed keV neutrons have been playing important roles in studying problems related to the nuclear astrophysics.[1,2,3] The keV neutron capture reaction by a nucleus is important in the problems relevant to the primordial-nucleosynthesis[3] and stellar-nucleosynthesis.[1,2] In standard big-bang models of cosmology, the observed abundances of the light elements, such as D, ^{3}He, ^{4}He and ^{7}Li, are nicely explained with the baryonic density of about 4% of the critical density.[4] While recently several groups have reported their observed primordial D abundance in Lyman-α absorption line at high redshift. Their values are quite different from each other, by an order of magnitude; namely, one of the reported D abundance relative to hydrogen is 2.0×10^{-4} and another one is 2.3×10^{-5}.[5,6] It has been discussed that although one can attain concordance for the observed primordial abundances of ^{4}He and ^{7}Li with a single parameter of the baryon-to-photon ratio with the low value of D/H, one can not attain concordance with the high value of D/H.[7] Although the origin of the difference has not yet been clarified, one would get a stringent value of the primordial D abundance in near future. Then one can determine the baryonic density and thereby to derive the amount of the dark matter, since D abundance is the best of the nuclear species among the primordial light elements to determine the baryonic matter density of the universe.[8] In order to derive the the baryonic density by comparing the observed primordial D abundance with the calculated one, it is necessary to know the cross section of the nuclear reactions which creates or destroys D in the primordial nucleosynthesis and/or during chemical evolution. D was created via the p(n,γ)D reaction, and the cross section was measured recently at astrophysically relevant energies between 10 and 80 keV,[9] and 550 keV.[10] While D was destroyed by several reactions, such as $D(D,p)^3H$, $D(D,n)^3He$, $D(p,\gamma)^3He$ and $D(n,\gamma)^3H$ reactions. Among these reactions, the cross section of the $D(n,\gamma)^3H$ reaction, σ_{nD}, has not ever been measured directly at the astrophysically relevant energy and thus, the reaction rate of the $D(n,\gamma)^3H$ reaction was evaluated by correcting for the Coulomb effect[11] for the measured cross section $D(p,\gamma)^3He$, σ_{pD}.[12] Recently, however, the astrophysical S-factor of the $D(p,\gamma)^3He$ reaction was reported to be about 37% smaller than the previous one,[13] which affects the reaction rate the $D(n,\gamma)^3H$ reaction.

Whereas a non-standard inhomogeneous big-bang model predicts a significant amount of intermediate-heavy nuclei during the primordial nucleosynthesis in the neutron rich-region mainly via neutron capture reactions,[14] such as, $D(n,\gamma)^3H(d,n)^4He(^3H,\gamma)^7Li(n,\gamma)^8Li$ $(\alpha,n)^{11}B(n,\gamma)^{12}B(\beta^-,\nu)^{12}C(n,\gamma)^{13}C(n,\gamma)^{14}C$ etc.. The $D(n,\gamma)^3H$ reaction is the important one to produce ^4He in the model, and thus, it is necessary to measure the cross section to estimate the production yields of ^4He and other intermediate nuclei. Here it should be mentioned about the recent work by N.Bahcall et al. on the mass to light ratio of galaxy system; they suggest a total mass density of $\Omega \sim 0.2$-0.3,[15] which is compared with the critical density $\Omega=1$. While it is interesting to note that the inhomogeneous model can attain concordance for the observed primordial light elements with the above baryonic density of $\Omega \sim 0.2$,[16] although the standard model gives $0.01 < \Omega < 0.1$.

Thus, it is necessary to directly measure the cross section σ_{nD} at the astrophysically relevant energy to estimate the production yields of the primordial elements in both the standard and inhomogeneous models, respectively. It should be added that the measurement would provide useful information to solve the discrepancy between the measurements of the $D(p,\gamma)^3He$ reaction.

In the stellar nucleosynthesis the neutron capture reaction by light nuclei is important, especially for the slow process nucleosynthesis in metal deficient stars[17] where abundant light nuclei, such as ^{12}C, ^{16}O, ^{18}O and ^{20}Ne, are considered to play an important role as a possible neutron poison. Namely, if the capture cross sections of these nuclei would be large, the yields of heavier s-process isotopes decrease and similarly, the yields of p-process nuclei would decrease.[18] Thus, the neutron capture cross section of ^{18}O is necessary to know for the construction of the models of the s-process isotope production.

Concerning the nucleosynthesis in the sun, deficit of the observed solar neutrino flux compared with the calculated value based on the standard solar model is one of the current topics. In all the reactions relevant to the solar neutrino production, the cross section of the $^7Be(p,\gamma)^8B$ reaction has been claimed to have a large uncertainty.[19] So far the cross section of the $^7Be(p,\gamma)^8B$ reaction has been measured by two different methods;

a direct (p,γ) experiment using the ^{7}Be target[20] and an inverse Coulomb break up reaction using the radioactive ^{8}B beam.[21] In the former case, the experiment was carried out using the proton beam with the energy as low as 100 keV, and the astrophysical S factor, S(E=0), has been obtained by extrapolating the experimental data with use of a theoretical model as 22.4(21) eV barn by Johnson et al.[22] and 17(3) eV barn by Barker.[23] Barker has evaluated the cross section assuming that the non-resonant contribution of the ^{7}Be(p,γ)^{8}B reaction can be well described by a direct capture model; the optical potential parameters and spectroscopic factors of ^{8}B are same as those of ^{7}Li+n system. Here it should be noted that the ^{7}Li(n,γ)^{8}Li reaction at astrophysically relevant energy is quite important to extract the neutrino flux from the ^{7}Be(p,γ)^{8}B reaction by assuming the charge symmetry of the reactions. So far, the cross section of the ^{7}Li(n,γ)^{8}Li reaction has been measured by several groups. Quite recently Blackmon et al. measured the partial cross section from the capture state to the ground state for En=10-1500 eV,[24] and Nagai et al. measured the partial capture cross section at En=30 keV.[25] The experimental accuracy, however, is not so good, and also in the previuos experiments only the partial capture cross section from a captured state to the ground state have been measured, but not the one from a capture state to the first excited state. The partial cross section is quite important to understand the neutron capture reaction mechansim. Therefore, in this study we aimed at measuring both the partial cross cross section from the capture state to the ground and first excited states precisely.

EXPERIMENTAL PROCEDURE AND RESULTS

The neutron capture cross section by a nucleus has been measured with use of a prompt γ-ray detection method using pulsed keV neutrons, produced by the ^{7}Li(p,n)^{7}Be reaction. The pulsed proton beams were provided from the 3.2 MV Pelletron Accelerator of the Research Laboratory for Nuclear Reactors at the Tokyo Institute of Technology. The neutron energy spectrum was measured with a time-of-flight (TOF) method by a ^{6}Li-glass scintillation detector. Samples of D$_2$O, enriched ^{7}Li$_2$O and H$_2^{18}$O, and a gold (Au) were used, where Au was for normalization of the cross section, since the capture cross section of Au has been well known within uncertainty of 3-5 %.[26]

Prompt γ-rays from the capture state of the low energy 10-80 keV neutron by samples to the low-lying states were detected by four NaI(Tl) spectrometers,[27] while the prompt γ-rays from the capture state produced by the 550 keV neutron capture by D to the low-lying states was detected by an anti-Compton NaI(Tl) spectrometer.[28] In order to determine the γ-ray branching ratio from a capture state to low lying states in ^{8}Li we used the high purity Ge spectrometer.[29] These detectors were situated at angles of 125° with respect to the proton beam direction, where the second Legendre polynomial is zero, and thus the γ-ray intensity measured at this angle gives an angle integrated one for a dipole transition. The captured events were stored in a hard disk drive in a list mode.[30]

The TOF spectrum was measured by either the central NaI(Tl) or HpGe detectors in order to obtain the keV neutron capture events by sample free from the background ones.

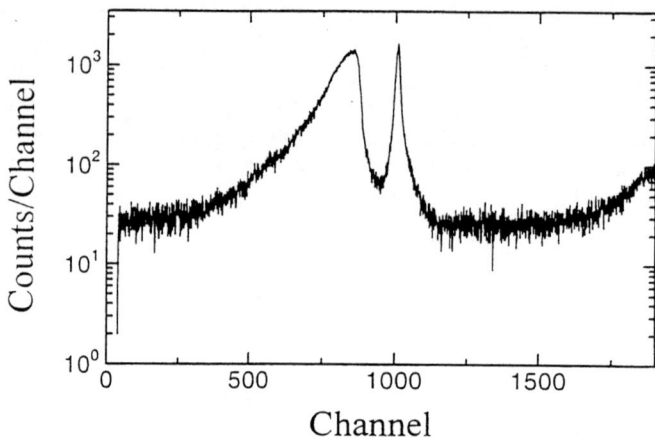

FIGURE 1. A time-of-flight spectrum measured for the 10-80 keV neutron capture reaction by Au. Sharp and broad peaks are due to the γ-rays from the ^{7}Li(p,γ)^{8}Be reaction and the keV neutron capture by Au, respectively.

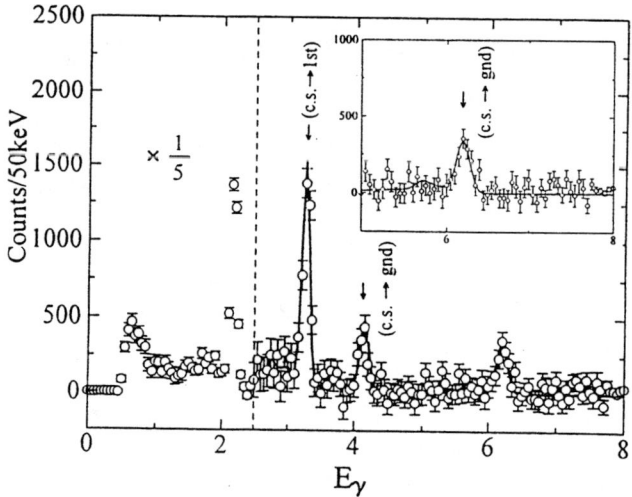

FIGURE 2. A background-subtracted γ-ray spectrum from the D(n,γ)^{3}H reaction. Two peaks around 4 MeV are from the ^{16}O(n,γ)^{17}O reaction.

A typical TOF spectrum measured for the 10-80 keV neutron induced reaction on Au is shown in fig.1. The background subtracted (net) γ-ray spectrum for the $D(n,\gamma)^3H$ reaction is shown in fig.2, where one sees two γ-rays of 3.3 and 4.2 MeV. These are from the capture state of ^{17}O, populated by the $^{16}O(n,\gamma)^{17}O$ reaction in the D_2O sample, to the first ($1/2^+$) and ground ($5/2^+$) states, respectively. While the 6.3 MeV γ-ray is from the neutron capture state of ^3H to the ground state.

As for the $^7Li(n,\gamma)^8Li$ reaction, the net γ-ray spectrum taken in the neutron energy 20-70 keV by the HpGe spectrometer is shown in fig.3, where one sees clearly both γ-rays from a capture state of ^8Li to the ground one and from the first excited state to the ground one. Here it should be noted that since the capture state has width of the incident neutron of 50 keV the width of the former γ-ray is much larger than that of the latter one.

Concerning the $^{18}O(n,\gamma)^{19}O$ reaction it was measured recently at several energies above 25 keV by employing an activation technique, and large capture cross sections were obtained compared to the extrapolated ones using a thermal neutron capture assuming a 1/v law.[31] Therefore, in this study we aimed at obtaining the temperature dependence of the partial capture cross section by detecting the discrete γ-rays from the capture state to low-lying ones.

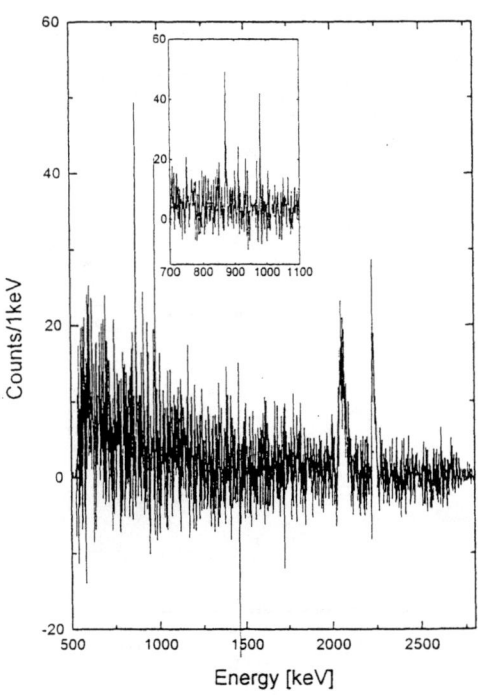

FIGURE 3. A background subtracted γ-ray spectrum from the $^7Li(n,\gamma)^8Li$ reaction measured by the Ge detector in the neutron energy between 20-70 keV.

The γ-rays, shown in the net spectrum for each reaction, were analyzed by a stripping method with use of the response function of the NaI(Tl) detector. We thus obtained the absolute capture cross section, $\sigma_\gamma(D)$ for D sample, as

$$\sigma_\gamma(D) = C \frac{(\phi)_{Au}}{(\phi)_D} \cdot \frac{(r^2 n)_{Au}}{(r^2 n)_D} \cdot \frac{Y_\gamma(D)}{Y_\gamma(Au)} \cdot \sigma_\gamma(Au) \quad (1)$$

Where C is the factor correcting for the multiple-scattering effect, the shielding of the incident neutrons in the sample, and the γ-ray absorption by the sample, and the finite size of the sample, respectively. The first two factors were obtained using a Monte-Calro code, TIME-MULTI.[32] Here, ϕ(Au) and Y_γ(Au) are the yields of the neutron and the γ-ray for Au, respectively; σ_γ(Au) is the absolute capture cross section of Au, r and n are the radius and thickness (atoms/ barn) of the sample, respectively.

Finally, the cross section σ_{nD} was obtained at the average neutron energies of 30.5, 54.2, and 550 keV as 2.12(35), 2.04(33), and 3.76(41) µbarns, respectively. Being combined the present results with the latest measured cross section of 0.508(15) mb for thermal neutrons,[32] the cross section σ(E) is given as

$$\sigma_{nD}(E) = 0.357 E^{-0.425} + 3.75 E^{0.5} \quad (1)$$

Hence, the reaction rate <σv> is obtained as

$$<\sigma v> = 3.05 \times 10^2 \times T_9^{0.075} + 4.96 \times 10^2 \times T_9 \quad (cm^3/sec/mole) \quad (2)$$

and it is shown in Fig.4 together with the previously reported one. The present value is smaller (larger) at high (low) temperature by about a factor 2 compared to the old one. In the standard models the present reaction rate, however, changes little the abundances of the primordial light elements.

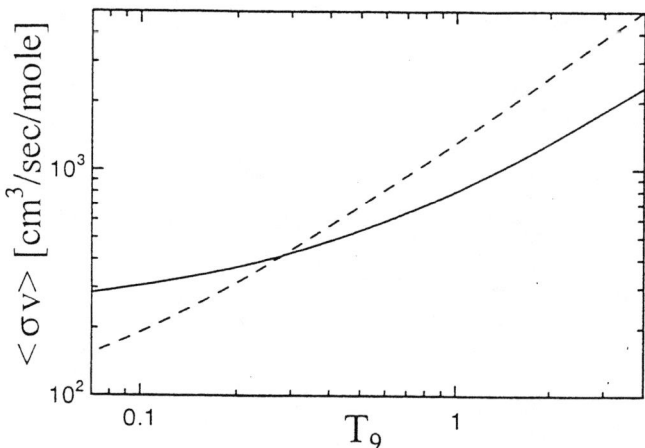

FIGURE 4. Reaction rate of the $D(n,\gamma)^3H$ reaction, where the prsent result and the previous one are shown by solid and dotted lines, respectively.

The partial and total cross sections of the ^{7}Li(n,γ)^{8}Li reaction were obtained; the ratio of the partial cross section from the capture state to the ground and first excited states agrees with that for thermal neutrons, and the total cross section was derived as 41(2), 32(3), and 25(2) μb at the average neutron energies of 60, 40 and 20 keV, respectively. They are in good agreement with the ones derived from a measured thermal neutron capture cross section of ^{7}Li by assuming a 1/v law. Thus, the present results of the partial capture cross section and total one indicate that the s-wave neutron capture by ^{7}Li dominates in the keV region.

Concerning the capture reaction by ^{18}O, the partial γ-ray transition strengths from the capture state to low-lying states were obtained for the firts time in the keV neutrons, and they were quite different from those for thermal neutrons, indicating that the keV neutron capture proceeds via a p-wave capture. The total capture cross section, 14.5(14) μb at En=48 keV, agrees with the previous result.[31]

MEASUREMENT OF THE KENS NEUTRON FLUX

The pulsed keV neutrons have been produced either by the ^{7}Li(p,n)^{7}Be reaction or the spallation reaction; in the latter case one can obtain the white neutron spectrum from thermal energy to 1 MeV. The neutrons produced by the spallation reaction have been rarely used for studying problems related to the nuclear astrophysics. In this study we have measured the neutron flux from the pulsed spallation neutron source (KENS) at KEK in the energy range between thermal neutrons and 70 keV using the B-doped plastic scintillation counter. The spallation neutrons are produced by bombarding the 500 MeV protons with the beam current of 5 μA on W target. The produced neutrons were moderated by moderators down to the thermal and cold neutron energy. The neutron flux has been measured with a time-of-flight (TOF) method; the signals from the detector and the kicker pulse from the proton synchrotron were used to obtain the TOF spectrum. Both spectra of TOF and PH were stored on the hard disk drive in a list mode. In order to obtain the neutron flux we made the correction for the prompt γ-ray from the spallation reaction by comparing the TOF spectrum measured by the B-doped detector with that by the non-B-doped one. The latter detector was only sensitive to the γ-ray. Thus, the neutron flux was obtained, and it is compared with the estimated flux using the measured flux at LANSCE[33] in fig.5. Although the flux is weaker than our expected flux (we are investigating the property of the neutron more in detail), the low-energy neutron (En<1 keV) can be used for studying the problem related to the astrophysics.

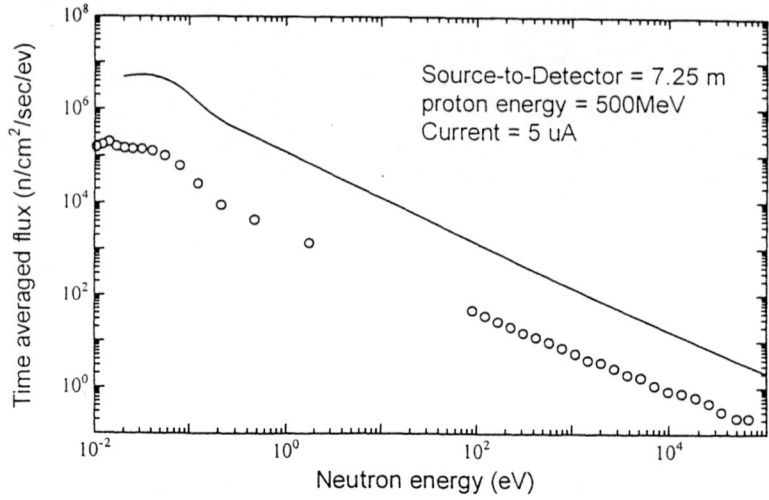

FIGURE 5. Measured neutron flux at KENS using the B-doped plastic scintillation counter. Circle and solid line are the present measured flux and the expected one, respectively.

DISCUSSION AND SUMMARY

We have successfully measured the small neutron capture cross sections of deuteron, ^{7}Li, and ^{18}O in the neutron energies between 10 and 80 keV, and at 550 keV. The cross section for the D(n,γ)^{3}H reaction is larger than the value so far used, but the present value affects little for the light elements abundance in the standard models. It would be interesting to see the effect of the present result on the abundance in the inhomogeneous models, and also to derive the astrophysical S factor of the D(p,γ)^{3}He reaction by using the present result. Here it is worthwhile to note that the D(p,γ)^{3}He reaction is important for the nucleosynthesis in the presolar system, where the temperature is quite low. Through the precise measurement of the ^{7}Li(n,γ)^{8}Li reaction cross section together with the new determination one can be sure that the reaction proceeds via the direct s-wave capture. We are studying the effect of the present result on the ^{8}B neutrino flux from the ^{7}Be(p,γ)^{8}B reaction. The large capture cross section of the ^{18}O(n,γ)^{19}O reaction would have important implications for nucleosynthesis theories of s- and p-processes, especially in metal deficient massive stars as an important neutron poison.

ACKNOWLEDGMENT

Finally, we would like to thank to M.Hashimoto, T.Kajino, T.Otsuka, A.Mengoni, T.Sato, and H.Ohtsubo for useful discussions. We are grateful to Profs. S. Ikeda, H.Ikeda, Y.Masuda, and Drs. K.Shibata, T.Otomo for their useful discussions and The part of the present work was supported by Grant-in-Aid for Specially Promoted Research (No. 08102004) of the Japan Ministry of Education, Science, Sports and Culture.

REFERENCES

1. Macklin, R.L. & Gibbons, J.H., Rev.Mod.Phys., 37 (1965) 166
2. Kappeler,F., Beer,H, & Wisshak, K., Rep.Prog.Phys. 52 (1989) 945
3. Nagai, Y., Igashira, M. et.al. ApJ, 372 (1991) 683
4. Schramm, D.N., & Wagoner, R.V., Ann.Rev.Nucl.Part.Sci., 27 (1977) 37
5. Songaila, A., Cowie,L.L., Hogan, C., & Rugers, M., Nature, 368 (1994) 599
   Carswell, R.F., Mon. Not.R.Astr.Soc. 268 (1994) L1
6. Tytler, D., Fan, X.-M., & Burles, S., Nature 381 (1996) 207
7. Hata, N. et al., Phys.Rev.Lett., 75 (1995) 3977
8. Epstein, R.I., Lattimer, J.M., & Schramm, D.N., Nature, 263 (1976) 198
9. Suzuki, T.S., Nagai, Y., Shima, T., Igashira, M. et al, ApJ, 439 (1995)L59
10. Nagai, Y., Suzuki, T.S., Kikuchi, T., Shima, T., Kii, T., and Igashira, M., Phys. Rev. C in press
11. Fowler, W.A., Caughlan, G.R., & Zimmerman, B.A., ARA & A, 5 (1967) 525
12. Griffiths,G.M., Lal,M., and Scarfe,C.D., Can.J.Physics, 41(1963)724
13. Schmid, G.J. et al., Phys.Rev.Lett. 76(1996)3088
14. Applegate, J.H., Hogan, C.J., & Scherrer, R.J., Phys.Rev.D35 (1987) 1151, T.Kajino, G.J.Mathews. & G.M.Fuller, ApJ,364 (1990) 7, Malaney, R.A., and Fowler, W.A., ApJ, 333 (1988) 14
15. Bahcall,N., Lubin,L.M., and Dorman,V., ApJ, 447 (1995) L81
16. Mathews,G.J., Kajino,T., and Orito,M., ApJ, 456 (1996) 98
17. Prantzos, N., Hashimoto, M., & Nomoto, K., Astron. & Astrophys. 234 (1990) 211
18. Lambert, D., Astronomy & Astrophysics Rev.3 (1992) 201
19. Bahcall, J.N. & Ulrich, R.K., Rev.Mod.Phys. 60 (1988) 297
20. Kavanagh, R.W., Nucl.Phys. 15 (1960) 411, Parker,P.D., ApJ, 153 (1968) L85, Filippone, B.W., et al., Phys.Rev.Lett. 50 (1983) 412
21. Motobayashi, T., et al., Phys.Rev.Lett. 73 (1994) 2680
22. Johnson,C.W.,Kolbe,E.,Koonin,S.E.,and Langanke,K.,ApJ. 392(1992)320
23. Barker,F.C., Nucl.Phys. A588(1995)693
24. Blackmon,J.C. et al., Phys.Rev.C 54(1996)383
25. Nagai, Y., Igashira, M., et al., 381 (1991) 444
26. Macklin, R.L. & Gibbons, H., Phys. Rev., 159 (1967) 1007, ENDF/B-V data file for 197Au (MAT=1379). 1979, evaluated by S.F.Mughabghab
27. Ohsaki,T.,Nagai,Y.,Shima,T.,Igashira,M., et al., private communication
28. Igashira, M.,Tanaka,K. and Masuda,K., in Proc. 8th. Int. Symp. on Capture Gamma-ray spectroscopy and Related Topics, Switzerland, ed. J.Kern (World Scientific, Singapore, 1994) 992
29. Igashira, M. et al., private communication
30. Ohsaki, T., Nagai, Y., Igashira,M. and Shima,T., private communication
31. Meissner,J. et al., Phys.Rev.C53 (1996) 459
32. Senoo,K., Nagai,Y., Shima,T.,Ohsaki,T., & Igashira,M., Nucl. Instr. Meth. A339 (1994) 556
33. Koehler, P.E., Nucl. Instr. Meth. A292(1990)541

Constraints for s–Process Scenarios: Neutron Capture Studies in the Lanthanide Region

F. Käppeler and K. Wisshak

Forschungszentrum Karlsruhe, IK-III, Postfach 3640
D-76021 Karlsruhe, Germany

Abstract. The standard abundance distribution of the elements in the universe is particularly well defined in the mass region of the lanthanides thanks to the chemical similarity of these elements. Therefore, the respective isotope patterns are reliably linked and provide an important basis for the investigation of several s–process branchings. New (n,γ) cross sections on Pr, Nd, Dy, and Er isotopes are presented which are based on measurements with the Karlsruhe $4\pi \text{BaF}_2$ detector and with the activation method. The implications of these data for s–process analyses are discussed.

I ORIGIN OF THE HEAVY ELEMENTS

Stellar nucleosynthesis of the heavy elements can only proceed via neutron capture reactions and subsequent beta decays since in realistic astrophysical scenarios the Coulomb barriers are prohibitive for charged particle reactions in the mass range A>60. The characteristic features of the observed abundance distribution between iron and the actinides [1] require essentially three processes: The dominating mechanisms are the slow and the rapid neutron capture processes (s and r process), each accounting for approximately 50% of the heavy element abundances, while the p process accounts for the rare proton rich nuclei.

The s process is associated with stellar helium burning where relatively low neutron densities imply neutron capture times of the order of several months, much longer than typical β–decay half–lives. Therefore, the s–process path follows the stability valley. The resulting s abundances are determined by the respective (n,γ) cross sections averaged over the stellar neutron spectrum in such a way that isotopes with small cross sections are built up to large abundances. In contrast, the r and p processes are related to explosive scenarios, presumably to supernovae. In these cases, the reaction paths are driven off

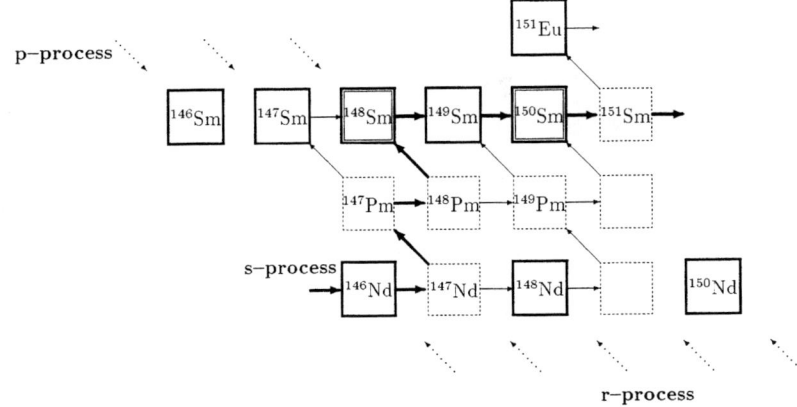

FIGURE 1. Reaction flow and abundance contributions in the Nd–Pm–Sm region.

the stability valley where experimental data are extremely difficult to obtain. Correspondingly, the synthesized material consists of short–lived nuclei, which decay back to the stability valley after the explosion.

The various reaction paths are sketched in Fig. 1, where the s–process flow is directly shown by arrows connecting the nuclei in the stability valley while the contributions from explosive events are schematically indicated. Obviously most isotopes received abundance contributions from both main mechanisms, the s and the r process. But there are stable isotopes – e.g. ^{150}Nd in Fig. 1 – that are not reached by the s process because of their short–lived neighbors. Consequently, this species is of pure r–process origin. On the other hand, there is an ensemble of s–only nuclei which are shielded by stable isobars against the r–process β–decays, e.g. ^{148}Sm and ^{150}Sm. The existence of these two groups is of vital importance for nucleosynthesis, since the credibility of any model depends on how well the abundances of these particular set of nuclei can be reproduced. Since the respective p–process abundances represent only a few % of the s and r contributions, the corresponding effects are much less pronounced and often difficult to quantify.

A second observation from Fig. 1 has significant impact for the characterization of the astrophysical s–process sites: At some points, the neutron capture chain encounters isotopes with β–half–lives that are comparable to the neutron capture times which are between a few months and a few years. The competition between β–decay and neutron capture causes the s–process path to split, e.g. at A=147/148. The resulting abundance pattern of these branchings carry information on the physical conditions in the stellar plasma, in particular on the mean neutron density and temperature.

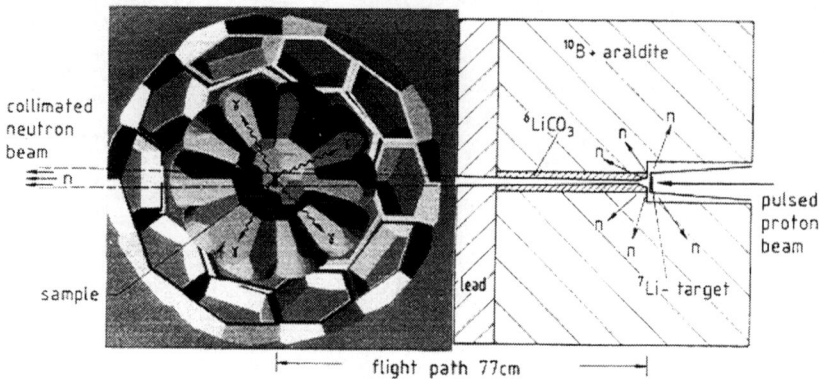

FIGURE 2. The Karlsruhe $4\pi\text{BaF}_2$ detector.

The following discussion is restricted to the s process, which is more easily accessible to laboratory experiments and which can be studied by stellar models and astronomical observations as well [2]. A particularly interesting group of elements along the reaction path of the s process are the rare earth elements (REE). Their relative abundances are known to better than 2% since these elements are chemically almost identical. This feature facilitates the interpretation of several important branchings in this region provided that the stellar (n,γ) cross sections are reliably known.

Therefore, a program was initiated for measuring the (n,γ) cross sections of all stable REE isotopes using the time–of–flight (TOF) method in conjunction with the Karlsruhe 4π BaF_2 detector [3]. These measurements were complemented by additional studies based on the activation technique [4,5].

II STELLAR CAPTURE CROSS SECTIONS

A The Karlsruhe 4π BaF_2 Detector

The concept of the Karlsruhe 4π BaF_2 detector and the way it is operated is sketched in Fig. 2, showing the setup at the accelerator. The detector is indicated schematically by a computer simulation, which presents one half of the spherical BaF_2 shell (15 cm thickness and 20 cm inner diameter) consisting of 42 individual BaF_2 crystals, and the supporting structure. All other details are omitted for the sake of clarity.

Neutrons are produced by means of a pulsed Van de Graaff accelerator (repetition rate 250 kHz, average beam current 2 μA, pulse width 0.7 ns) via the $^7\text{Li}(p,n)^7\text{Be}$ reaction. The collimated neutron beam hits the sample in the center of the detector at a flight path of 77 cm. Up to 6 samples can be investigated simultaneously in the same experiment. The samples are mounted on a vertical ladder which carries also a gold sample for determination of the neutron flux. Two more positions are used for a scattering sample and an empty sample canning, both serving for background determination. The various samples are cycled into the measuring position in intervals of about 10 min which are defined by integrating the beam current on target. Additional neutron monitors are used to check for equal neutron exposure per sample and to get a rough information on the respective total cross sections.

The essential features of the detector are a resolution in γ–ray energy ranging from 14% at 662 keV to 6% at 6.13 MeV, a time resolution of 500 ps, and an overall efficiency for capture events of better than 98% (for a detailed description and typical measurements see [3,6]).

With this detector, (n,γ) cross sections can be determined with an accuracy of 1 to 2%. Recently, it was used for investigating the stable neodymium isotopes [7] in the neutron energy range from 3 to 225 keV. The efficiency for capture events was evaluated by means of the experimental pulse height spectra shown for the examples of ^{142}Nd and ^{143}Nd (Fig. 3). After correction for backgrounds due to isotopic impurities and scattered neutrons, most of the events are found in a line at the neutron separation energy. The remaining tail towards lower energies corresponds to events where not all γ–rays of the cascade could be detected, mainly due to losses through the openings for the neutron beam and the sample changer.

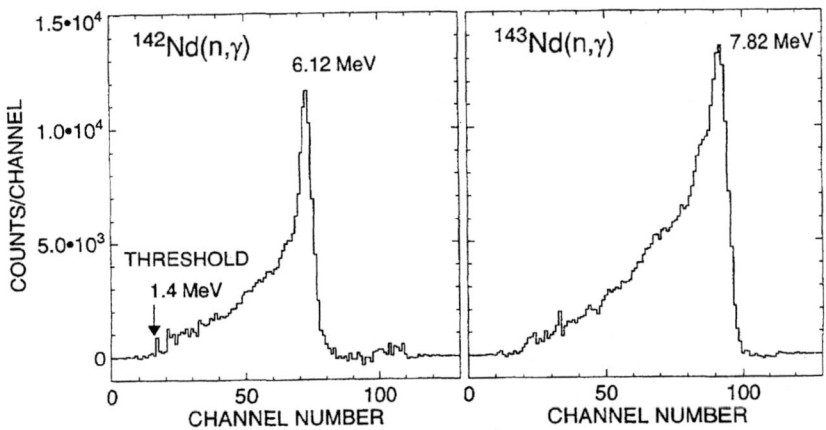

FIGURE 3. The γ–spectra for capture events in ^{142}Nd and ^{143}Nd measured with the 4πBaF$_2$ detector.

The stellar cross sections were obtained by folding the differential data, $\sigma(E_n)$, with the stellar neutron spectra for various temperatures. The Maxwellian averages for the "standard" thermal energy of $kT=30$ keV could be determined with uncertainties of only 1.5% to 2%. This accuracy is particularly important for the role of the s–only isotope ^{142}Nd as a normalization point of the overall s–process abundance pattern.

TABLE 1. Some recent results for Maxwellian averaged (n,γ) cross sections in the REE region compared to previous compilations

Target	Stellar (n,γ) Cross Sections at $kT=30$ keV (mb)			
	Ref. [8]	Ref. [9]	Recent Results	
^{141}Pr	119±	119±15	110±3[1]	[10]
^{142}Nd	46±4	46±4	35.0±0.7	[11]
^{143}Nd	242±10	242±10	244.6±3.7	[7]
^{144}Nd	110±6	108±6	81.3±1.5	[11]
^{145}Nd	485±10	485±10	424.8±5.9	[7]
^{146}Nd	157±40	157±40	91.2±1.3	[7]
^{148}Nd	192±40	192±40	146.6±2.3	[7]
^{160}Dy	738±35	772±39	917±28[1]	[10]
^{161}Dy	2007±72	2006±60	2000±60[1]	[10]
^{162}Dy	473±50	473±48	454±14[1]	[10]
^{163}Dy	1142±44	1140±38	1128±34[1]	[10]
^{164}Dy	268±27	268±27	215±6[1]	[10]
^{164}Er	714±61	714±61	1084±51	[12]
^{170}Er	223±33	223±33	170±7	[12]

[1] Preliminary values

Other recent measurements were performed on ^{141}Pr and on a series of Dy isotopes. For the near future, it is planned to continue with measurements on Yb nuclei and to complete the REE project eventually with the stable Er isotopes. All recent results, that have not yet been published, are summarized in Table 1.

B The Activation Technique

The activation technique represents an alternative method for the determination of stellar (n,γ) rates. Compared to the techniques using the detection of the prompt capture γ–rays, this method offers the advantages of higher sensitivity (which means that much smaller samples can be measured reliably – an important aspect for the investigation of radioactive isotopes on the s–process path), of selectivity (which means that isotope mixtures can be studied via the characteristic γ–ray energies of the respective decay products), and of a comparably simple experimental setup.

This method is based on the fact that stellar neutron spectra can be simulated in the laboratory. With the ^{7}Li(p,n)^{7}Be reaction, a continuous energy spectrum with a high–energy cutoff at $E_n = 106$ keV and a maximum emission angle of 60° can be obtained that corresponds closely to a Maxwell–Boltzmann distribution for $kT=25$ keV [4,5]. Hence, the reaction rate measured in such a spectrum yields immediately the proper stellar cross section.

The scheme of the activation technique is shown in the left part of Fig. 4. The sample is sandwiched between thin gold foils and is directly attached to the neutron target. The induced activity of the gold foils provides the normalization to this common cross section standard. The neutron yield is continuously monitored during the irradiations by means of a ^{6}Li glass detector, an option that allows to consider the decay of the induced activity during the irradiation properly.

The close geometry and the fact that the irradiations can be carried out with the accelerator operating in DC mode provide approximately 6 orders of magnitude better sensitivity compared to the TOF technique. Accordingly, much smaller samples can be used, even in measurements of μb cross sections. On the other hand this technique is restricted to reactions leading to unstable nuclei, and is also affected by uncertainties of typically 3 to 5%. Within these uncertainties, the stellar cross sections for the isotopes ^{146}Nd and ^{148}Nd determined via the activation technique [13] were found in perfect agreement with the more recent results obtained with the 4πBaF$_2$ detector [7].

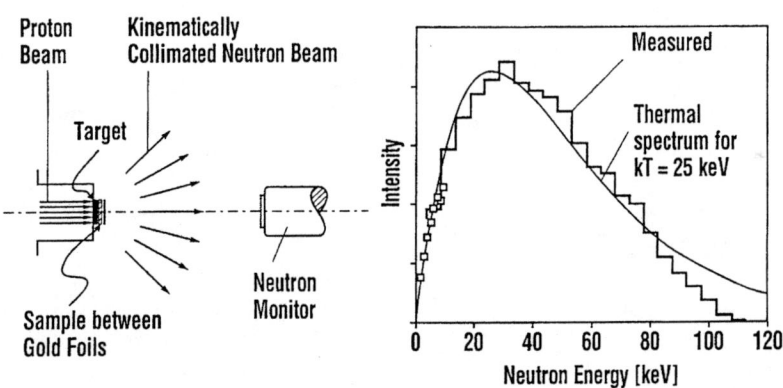

FIGURE 4. The activation technique: irradiation of a sample sandwhich at the accelerator(left) and the angle-integrated neutron spectrum (right).

An actual example in the REE region was an activation measurement on natural erbium, which includes the astrophysically important s isotope ^{164}Er. Fig. 5 shows the γ-spectrum of an activated sample. While the cross section of ^{170}Er could be derived by several clear γ-lines, a careful analysis of the X–

ray multiplet (see inset) was required to extract the ^{164}Er cross section from the pure holmium component. The stellar cross section of ^{164}Er of $\langle\sigma v\rangle/v_T = 1084 \pm 51$ mb was found to be significantly larger than reported earlier [14], allowing for an improved interpretation of the s–process branching at ^{163}Dy.

Finally, it should be emphasized that the superior sensitivity of the activation technique can be used for measuring the stellar cross sections for a number of unstable nuclei [15]. In connection with the REE project, the recent determination of the cross section for the branch point nucleus ^{155}Eu ($t_{1/2}$=4.96 yr) was an important success [16].

III ASTROPHYSICAL IMPLICATIONS

A The $\langle\sigma\rangle N_s$ Systematics in the REE Region

Together with the here reported results, stellar (n,γ) cross sections have been remeasured for 32 of the 55 stable isotopes in the REE region over the last 4 years, 23 with the 4πBaF$_2$ detector. These investigations included the s–only nuclei ^{142}Nd, ^{150}Sm, and ^{160}Dy which are important normalization points of the s–process abundance distribution. The general shape of this distribution versus mass number yields information on the required seed nuclei as well as on the corresponding neutron exposure [2].

Fig. 6 shows the product of stellar cross section times abundance, $\langle\sigma\rangle N_s(A)$, that is characteristic of the classical approach, an often used phenomenological s–process model [2]. The symbols in Fig. 6 denote the empirical products for the s–only isotopes. In the REE region, the new values carry uncertainties smaller than the size of the symbols. Obviously, the differences between the classical approach (solid line) and the empirical values for the expected normalization points ^{142}Nd and ^{160}Dy are considerably larger than these uncertainties. The value of ^{142}Nd lying *below* the calculated curve means that this isotope is *over*produced by the model, even if a possible p contribution is neglected. This discrepancy is inherent to the assumption of an exponential distribution of neutron exposures, on which the classical approach is based. Correspondingly, this result reveals a severe deficiency of that model, which should, therefore, be used with some reservation in the future.

In contrast to ^{142}Nd, the value for ^{160}Dy is 11% higher than the curve. Principally, such a difference could be explained by a corresponding p–process component. In this case, however, current p–process models [17–19] predict this component to be comparable to that of the p–only neighbors ^{156}Dy and ^{158}Dy, which would explain only 30% of that difference. Very likely, this apparent excess of ^{160}Dy is the first experimental evidence for a thermal effect: At s–process temperatures the excited state in ^{160}Dy at 87 keV becomes significantly populated. Though this state may have a larger cross section than the

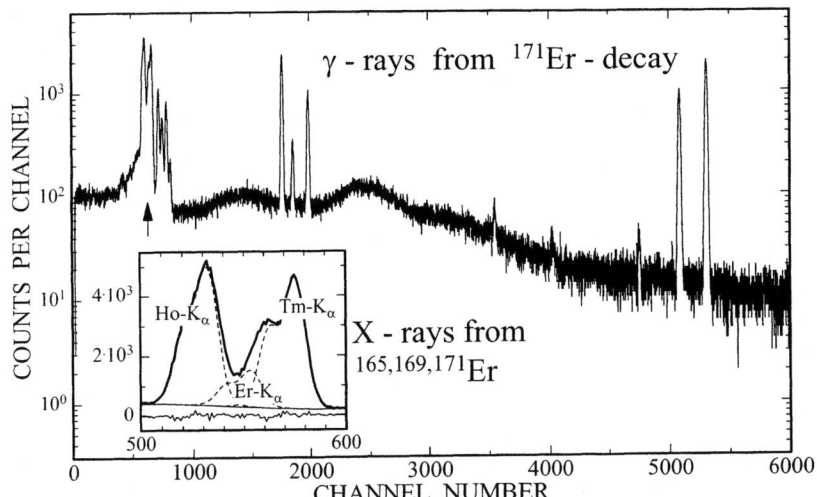

FIGURE 5. The γ-ray spectrum of an Er sample after irradiation in a quasi–stellar neutron spectrum. The inset shows the decomposition of the X-ray multiplet into the Ho component from the EC decay of ^{165}Er and into the Tm part due to the β^--decays of ^{169,171}Er, respectively.

ground state due to its larger spin, the total (n,γ) cross section can be reduced because of the competing reaction channel for inelastic scattering from the excited state down to the ground state. This process is sometimes denoted as *superelastic* scattering. In contrast to this experimental result, previous statistical model calculations predicted a 10% enhancement for the stellar ^{160}Dy cross section [20,21]. In view of this discrepancy, further investigations of this effect appear rewarding.

B   *s*–Process Branchings

All other *s* nuclei which fall below the calculated curve in Fig. 6 do not experience the full reaction flow due to branchings in the *s*–process path (Fig. 1). Accordingly, these nuclei define the effective branching strengths for interpreting the resulting abundance patterns in terms of neutron density, temperature, and mass density. Since these quantities are critical conditions for the He burning plasma, the *s*–process branchings represent important tests for the various *s*–process models.

The information obtained from recent analyses of the REE branchings which are defined by the partially bypassed *s*–only isotopes ^{148}Sm, ^{154}Gd, ^{164}Er, and ^{170}Yb is summarized in Table 2. These data were obtained quasi model–free by

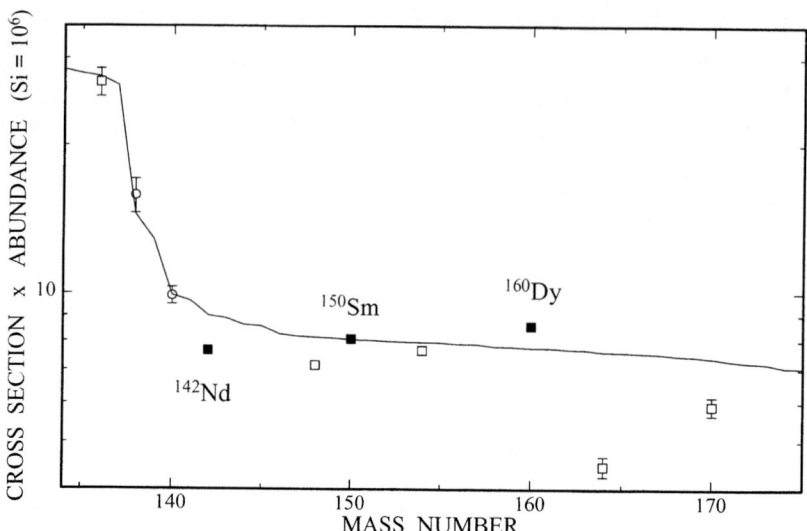

FIGURE 6. The $\langle\sigma\rangle N_s(A)$-curve in the mass region of the REE. The differences between the classical s-process approach (solid line) and the empirical values for the expected normalization points ^{142}Nd and ^{160}Dy (black squares) are discussed in the text. Note that in some cases the uncertainties have been reduced to the symbol size.

using the classical approach and characterize the situation during He burning in low mass stars. More realistic descriptions of this scenario are based on stellar models for the He burning stage of low mass stars (e.g. [24]).

Furthermore, the fairly quantitative determination of the s-process abundances can be used for separating the respective r components independently of the hitherto uncertain r-process models by subtracting the s-process yields from the observed distribution [2,7,13,25].

REFERENCES

1. Anders, E. & Grevesse, N., *Geochim. Cosmochim. Acta* **53**, 197 (1989).
2. Käppeler, F., Beer, H., & Wisshak, K., *Rep. Prog. Phys.* **52**, 945 (1989).
3. Wisshak, K., Guber, K., Käppeler, F., Krisch, J., Müller, H., Rupp, G., & Voss, F., *Nucl. Instr. Meth. A* **292**, 595 (1990).
4. Beer, H. & Käppeler, F., *Phys. Rev. C* **21**, 534 (1980).
5. Ratynski, W. & Käppeler, F., *Phys. Rev. C* **37**, 595 (1988).
6. Wisshak, K., Voss, F., Käppeler, F., Guber, K., Kazakov, L., Kornilov, N., Uhl, M., & Reffo, G., *Phys. Rev. C* **52**, 2762 (1995).

TABLE 2. Results from various branching analyses in the REE region characterizing the neutron density, temperature, and mass density during the s process

Branch point isotope	s-Process parameter	Reference
^{147}Nd/^{147}Pm/^{148}Pm	$n_n = (4.1 \pm 0.6) \cdot 10^8$ cm^{-3}	[22]
^{151}Sm/^{154}Eu	$T_8 = 3.5 \pm 0.4$	[6]
^{163}Dy/^{163}Ho	$\rho_s = (6.5 \pm 3.5) \cdot 10^3$ gcm^{-3}	[12]
^{176}Lu	$T_8 = 3.1 \pm 0.6$	[23]

7. Wisshak, K., Voss, F., Käppeler, F., Kazakov, L., & Reffo, G., Report FZKA-5967, Forschungszentrum Karlsruhe, 1997.
8. Bao, Z.Y. & Käppeler, F., *Atomic Data Nucl. Data Tables* **36**, 411 (1987).
9. Beer, H., Voss, F. & Winters, R.R., *Ap. J. Suppl.* **80**, 403 (1992).
10. Wisshak, K., Voss, F., Käppeler, F., Kazakov, L., & Reffo, G., in preparation
11. Wisshak, K., Voss, F., & Käppeler, F., Report FZKA-5968, Technical report, Forschungszentrum Karlsruhe, 1997.
12. Best, J., Report FZKA-5824, Forschungszentrum Karlsruhe, 1996.
13. Toukan, K.A., Debus, K., Käppeler, F., & Reffo, G., *Phys. Rev. C* **51**, 1540 (1995).
14. Beer, H., Walter, G., & Macklin, R.L., in *Capture Gamma–Ray Spectroscopy and Related Topics–1984*, ed. S. Raman, New York: AIP, 1985, pp. 778-781.
15. Käppeler, F., Wiescher, M., & Koehler, P.E., in *Production and Use of Intense Radioactive Beams at the Isospin Laboratory*, ed. J. D. Garrett, Oak Ridge: Joint Institute for Heavy Ion Research, 1992, pp. 163-166.
16. Jaag, S. & Käppeler, F., *Phys. Rev. C* **51**, 3465 (1995).
17. Rayet, M., Prantzos, N., & Arnould, M., *Astron. Astrophys.* **227**, 271 (1990).
18. Howard, W.M., Meyer, B.S., & Woosley, S.E., *Ap. J.* **373**, L5 (1991).
19. Rayet, M., Arnould, M., Hashimoto, M., Prantzos, N., & Nomoto, K., *Astron. Astrophys.* **298**, 517 (1995).
20. Holmes, J.A., Woosley, S.E., Fowler, W.A., & Zimmerman, B.A., *Atomic Data Nucl. Data Tables* **18**, 305 (1976).
21. Harris, M.J., *Ap. Space Sci.* **77**, 357 (1981).
22. Wisshak, K., Guber, K., Voss, F., Käppeler, F., & Reffo, G., *Phys. Rev. C* **48**, 1401 (1993).
23. Klay, N., Käppeler, F., Beer, H., & Schatz, G., *Phys. Rev. C* **44**, 2839 (1991).
24. Gallino, R., Arlandini, C., Busso, M., Lugaro, M., Travaglio, C., Straniero, O., Chieffi, A. & Limongi, M., *Ap. J.*, in print.
25. Käppeler, F., Toukan, K.A., Schumann, M., & Mengoni, A., *Phys. Rev. C* **53**, 1397 (1996).

Recent progress in theoretical nuclear astrophysics

P. Descouvemont

Physique Nucléaire Théorique et Physique Mathématique, CP229
Université Libre de Bruxelles, B1050 Bruxelles - Belgium

Abstract. We review and compare some theoretical models used in nuclear astrophysics. Applications are presented for the ^{3}He(^{3}He,2p)α and ^{12}C(p,γ)^{13}N reactions (R-matrix), for the ^{3}He(α,γ)^{7}Be reaction (potential model), and for the ^{8}Li(α,n)^{11}B reaction (microscopic model). Screening effects in the ^{3}He(^{3}He,2p)α reaction are briefly discussed.

I INTRODUCTION

Nuclear astrophysics plays a crucial role in the understanding of nucleosynthesis in the universe [1]. Nuclear reactions determine stellar evolution and are the major energy source in stars. Current modelisations of star evolution require a very large amount of data, which remains a challenge for nuclear physicists. Experimental investigations meet two main limitations. (i) For charged-particle reactions, relevant energies are much lower than the Coulomb barrier, which makes the cross sections too small to be measured. Recent technological developments allow measurements at stellar energies for some reactions with low charges, such as ^{3}He(^{3}He,2p)α [2], but for most important reactions the experimental lower limit is far above the Gamow energy. (ii) Investigations of explosive burning are fairly sensitive to reactions involving unstable nuclei. Development of radioactive ion beams [3] in many laboratories provides useful information on such reactions, but much work remains to be carried out, more especially for the rp-process which runs through the proton-rich unstable region.

In view of these limitations, theoretical models of nuclear reactions appear to be a necessary complement to experimental studies. Theoretical calculations can be done at any energy, and are not restricted by instability of the nuclei. However, they face many other problems. Transfer and radiative-capture reactions involved in stellar nucleosynthesis are often difficult to treat, and no systematics can be applied. Estimating the reliability of a theoretical model is very important, but is often a delicate problem.

Theoretical models can be roughly classified in three categories:

(i) Models involving adjustable parameters, such as the R-matrix [4] or the K-matrix [5] methods; parameters are fitted to the available experimental data and the cross sections are extrapolated down to astrophysical energies. These fitting procedures of course require the knowledge of data, which are sometimes too scarce for a reliable extrapolation.

(ii) "Ab initio" models, where the cross section is determined from the wave functions of the system. The potential model [6], the Distorted Wave Born Approximation (DWBA) [7], and microscopic models [8] are, in principle, independent of experimental data. More realistically, these models depend on some physical parameters, such as a nucleus-nucleus or a nucleon-nucleon interaction which can be reasonably determined from experiment only. The microscopic Generator Coordinate Method (GCM) provides a "basic" description of a A-nucleon system, since the whole information is obtained from a nucleon-nucleon interaction. Since this interaction is nearly the same for all light nuclei, the predictive power of the GCM is important. Many applications have been considered these recent years (see references in ref. [8]).

(iii) Models (i) and (ii) can be used for low level-density nuclei only. This condition is fulfilled in most of the reactions involving light nuclei ($A \leq 20$). However when the level density near the threshold is large (i.e. more than a few levels per MeV), statistical models are better adapted. Statistical models, such as the Hauser-Feschbach theory [9] can be used in a systematic way, but are not expected to provide more than a rough estimate of the reaction rates.

In this paper, we review different theoretical approaches, and stress their advantage and drawbacks. With some physical examples, we try to derive some general trends for their use in astrophysical applications.

II OVERVIEW OF SOME THEORETICAL MODELS

In table 1, we give some general properties of different models, and we specify if they can be applied to radiative capture (RC) or transfer (Tr) reactions.

The R-matrix method: assumes two regions of the configuration space. The cross section is determined from energy independent parameters. These parameters are related to energies and widths of resonances but the link is not simple in a general case. The K-matrix formalism [5] solves this problem, but presents other drawbacks (see ref. [10] for a comparison of both methods). Simplification of the R-matrix formalism to the very low-energy regime yields the extra-nuclear capture model [11], where the electromagnetic matrix elements are calculated with the asymptotic parts of the wave functions.

TABLE 1. Comparison of different theoretical models

Model	Ref	RC	Tr	Comments
R-matrix	[4]	Yes	Yes	Requires many data for a good reliability. The link between parameters and experimental information is indirect.
Potential model	[6]	Yes	No	The potential is not unambiguously defined. Requires spectroscopic factors.
DWBA	[7]	No	Yes	Needs potentials in entrance and exit channels. The validity is not well established at low energies.
Microscopic models	[8]	Yes	Yes	Start from a nucleon-nucleon force, with antisymmetrized wave functions. Calculations are in general very long.

The potential model: the initial and final wave functions are obtained from a numerical integration of the Schrödinger equation involving a nucleus-nucleus potential. Some external constraints must be available to determine this potential. The choice of the potential depth remains however an open problem [6].

The DWBA method: assumes that the transfer cross section is small respect to elastic scattering. The transfer reaction is considered as

$$A + (B + x) \to (A + x) + B \quad (1)$$

where x is the exchanged particle. Interaction between particle B and x is usually assumed to have a zero range [7].

Microscopic models: the information is derived from a A-body Hamiltonian

$$H = \sum_{i=1}^{A} T_i + \sum_{i<j}^{A} V_{ij} \quad (2)$$

where T_i is the kinetic energy of nucleon i, and V_{ij} a nucleon-nucleon interaction. The wave functions are defined from the cluster wave functions ϕ_1 and ϕ_2 of the colliding nuclei; the total wave function reads, in a schematic notation

$$\Psi = \mathcal{A}\, \phi_1\, \phi_2\, g(\boldsymbol{\rho}) \quad (3)$$

where $g(\boldsymbol{\rho})$ is the relative function depending on the relative coordinate $\boldsymbol{\rho}$; it is determined from the Schrödinger equation. In (3), $\mathcal{A}$ is the antisymmetrization operator which ensures the Pauli principle to be satisfied. Projection over good quantun numbers is performed exactly. The reliability on the model mostly depends on the accuracy of the internal wave functions ϕ_1 and

ϕ_2. Many efforts have been done to go beyond the simple shell model approximation: multicluster description [12], monopole distortion [13] and extended shell model developments [14] aim at improving the anzatz (3). Microscopic models are however uneasy to handle and require much computer time.

III THE R-MATRIX METHOD: APPLICATION TO ^{3}He(^{3}He,2p)α AND ^{12}C(p,γ)^{13}N

In the R-matrix theory [4,15] the physics of the problem is assumed to be determined from some selected "poles" with energies E_λ, and reduced widths γ_i^λ in channel i. Radiative capture reactions are treated in the first-order perturbation theory, and each pole is characterized by a γ width Γ_γ^λ. The poles are associated to bound-states or resonances, but the link between "calculated" values and "observed" values is not straightforward in a general case, and is one of the main drawbacks of the R-matrix. In a multichannel problem involving N poles, the R-matrix is given by

$$R_{ij}(E) = \sum_{\lambda=1}^{N} \frac{\gamma_i^\lambda \gamma_j^\lambda}{E_\lambda - E} \tag{4}$$

where γ_i^λ and E_λ are energy-independent parameters which must be fitted to experiment. Some simplifications appear for elastic scattering, where $R(E)$ is a single function, or in non-resonant calculations, where $E \ll E_1$ in (4), and hence the R-matrix is constant.

Let us consider the ^{3}He(^{3}He,2p)α reaction, which is important for the solar-neutrino problem, and has been investigated by many authors. Since this reaction is well known to be non-resonant, a constant R-matrix is adopted, with R_{12} given by $R_{12}^2 = R_{11}R_{22}$. For a two-channel problem, elements of the collision matrix U read [15]

$$U_{11} = \frac{I_1}{O_1} \frac{1 - R_{11}L_1^* - R_{22}L_2}{1 - R_{11}L_1 - R_{22}L_2}, \quad U_{22} = \frac{I_2}{O_2} \frac{1 - R_{11}L_1 - R_{22}L_2^*}{1 - R_{11}L_1 - R_{22}L_2}, \tag{5}$$

$$U_{12} = U_{21} = \frac{2ia\sqrt{k_1 k_2} R_{12}}{O_1 O_2 (1 - R_{11}L_1 - R_{22}L_2)} \tag{6}$$

where k_i is the wave number in channel i, a is the channel radius, and

$$L_i = k_i a \frac{O_i'}{O_i} \tag{7}$$

O_i being the outgoing Coulomb function. The collision matrix is easily shown to be symmetric and unitary. From (5) and (6), one deduces the elastic phase shifts δ_i, and the transfer cross section $\sigma_{1\to 2}$:

$$\delta_i = \frac{1}{2}\arg(U_{ii}) \tag{8}$$

$$\sigma_{1\to 2} = \frac{\pi}{k_1^2}\frac{(2J+1)(\delta_{12}+1)}{(2I_1+1)(2I_2+1)}|U_{12}|^2$$

where I_1 and I_2 are the spins of the colliding nuclei, and δ_{12} is 1 or 0 according to whether the system is symmetric or not.

We consider the s and p partial waves and, for reducing the number of parameters, we assume identical R-matrices. The channel radius is $a = 4$ fm. Parameter R_{11} is fitted on the ^{3}He + ^{3}He elastic phase shifts yielding $R_{11} = 0.200$; the remaining parameter R_{22} is fitted on the ^{3}He(^{3}He,2p)α cross section up to $E_{cm} \approx 1$ MeV, yielding $R_{22} = 0.065$. The resulting fits are displayed in Fig. 1.

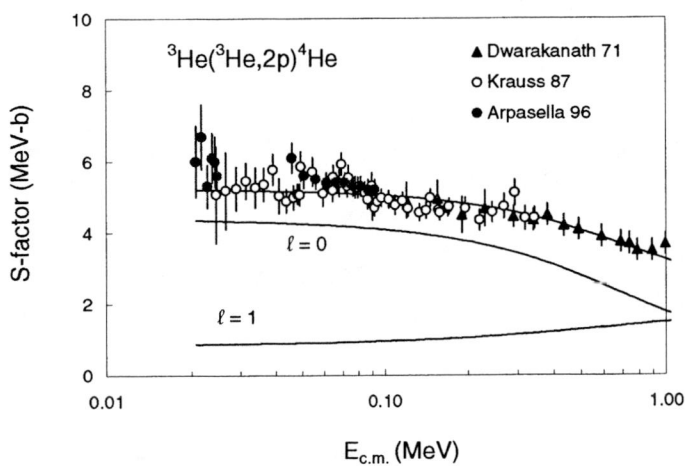

FIGURE 1. ^{3}He(^{3}He,2p)α S-factor in the R-matrix theory. The data are from refs. [16,17,2]

Using the very-low energy data $\sigma_{\exp}$ and the present extrapolation σ_{th}, an electron screening potential U_e can be determined using the formula [18]

$$\frac{\sigma_{\exp}}{\sigma_{\text{th}}} = \exp(\pi\eta\frac{U_e}{E}) \tag{9}$$

From the recent data of Arpasella et al. [2], we find

$$U_e = 226 \pm 185 \text{ eV} \tag{10}$$

whose mean value is close to that quoted in ref. [2] (290±40 eV) who used the extrapolation of a second order polynomial fit above 100 keV as bare nucleus cross section. However our error bar is much larger than in ref. [2]; it is still underestimated since the uncertainty on the theoretical fit is not taken into account. Using the data of Krauss et al. [17] yields

$$U_e = 30^{+180}_{-30} \text{eV} \qquad (11)$$

a value compatible with zero. It is obvious that an accurate determination of screening potentials would require very precise measurements and theoretical extrapolations.

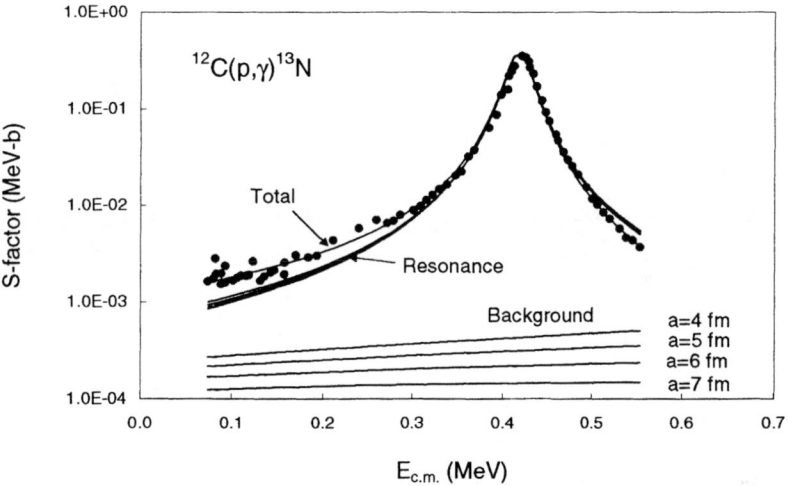

FIGURE 2. ^{12}C(p,γ)^{13}N S-factor in the R-matrix theory. The total S-factor is decomposed in a Breit-Wigner term, and a non-resonant contribution for different channel radii. The experimental data are from ref. [19]

Let us now come to the ^{12}C(p,γ)^{13}N reaction which exemplifies a resonant capture reaction. The R-matrix (4) reduces to a single function parametrized as

$$R(E) \approx R_0 + \frac{\gamma_1^2}{E_1 - E} \qquad (12)$$

where pole 1 refers to the $\frac{1}{2}^+$ ($E_{cm} = 0.42$ MeV) resonance. This pole and the background pole have a γ width which enters the calculation of the capture cross section [4]. For the sake of simplicity, external contribution to the E1 matrix elements is neglected. Fig. 2 shows the fit obtained for a values equal to 4, 5, 6 and 7 fm; as expected, this total fit does not depend on a. We have calculated the contribution of pole 1 and background alone for different a values. Such a situation, involving a single isolated s resonance, often occurs in proton capture reactions. In most cases, properties of the resonance are known, but a correction, sometimes called "direct-capture contribution" is introduced to go beyond the Breit-Wigner approximation which corresponds to an R-matrix approach with a single pole. Fig. 2 shows that the background term is very sensitive to the condition of the calculation. A given background contribution is associated to a precise a-value whereas the total S-factor does not depend on a. This simple example shows that the mixing of resonant and non-resonant information should be treated very carefully.

IV THE POTENTIAL MODEL: APPLICATION TO ^{3}He$(\alpha,\gamma)^7$Be

There are many works devoted to the ^{3}He$(\alpha,\gamma)^7$Be reaction. Here, we do not aim at giving the most accurate extrapolation, but we use this well known reaction to illustrate different approaches of the potential model. In this model, the colliding nuclei are assumed to be structureless, with a radial interaction between them. Antisymmetrization effects are simulated by an appropriate choice of the potential [6]. The ^{3}He + α potential is assumed here to have a gaussian shape

$$V(\rho) = V_0 \exp(-(\rho/a)^2) \tag{13}$$

with a point-sphere Coulomb term ($R_C = 3.25$ fm). The ^{7}Be ground state is defined with $a = 2.35$ fm, $V_0 = -95.0$ MeV, which reproduce the binding energy and the quadrupole moment of ^{7}Li (-3.4 ± 0.6 e.fm^2 – see ref. [20]). For the first excited state, the spin-orbit splitting yields $V_0 = -93.2$ MeV. Many data on ^{3}He + α are available (spectroscopy of ^{7}Be, elastic phase shifts, capture cross sections), and we use the different constraints for determining the initial potential. These approaches are typical of reactions where only a few data are available.

- Method 1: The initial potential is equal to the final potential.

- Method 2: The initial potential is fitted on the ^{3}He + α phase shifts; with $a = 2.35$ fm, we get $V_0 = -75.0$ MeV.

- Method 3: The initial potential is fitted on the mirror ^{3}H$(\alpha,\gamma)^7$Li reaction, yielding $V_0 = -81.0$ MeV.

The resulting S-factors, calculated with a spectroscopic factor equal to unity, are given in Fig. 3. Method 1, which is the simplest possible, overestimates the data, but gives a fair approximation with spectroscopic information only. Method 2 underestimates the data but the sensitivity of the phase shift to the potential is rather weak. The best constraint seems to be provided by the mirror $^3\text{H}(\alpha,\gamma)^7\text{Li}$ cross section which gives an $^3\text{He}(\alpha,\gamma)^7\text{Be}$ S-factor in very nice agreement with experiment.

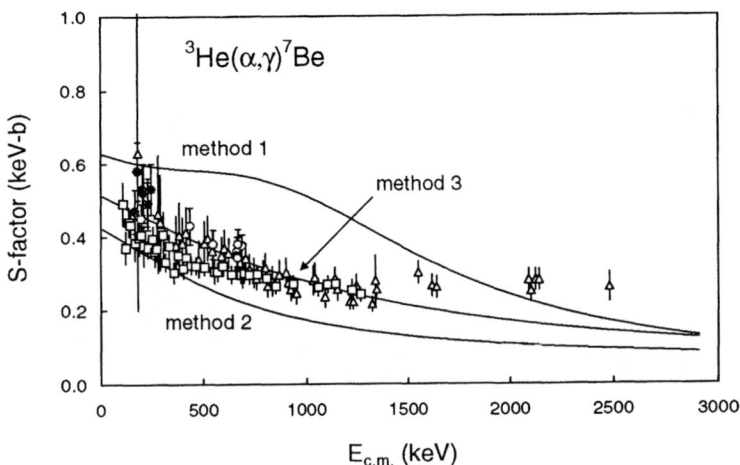

FIGURE 3. $^3\text{He}(\alpha,\gamma)^7\text{Be}$ S factors with the 3 methods. The data are from ref.[21]($\triangle$), [22]($\bullet$), [23]($\square$), and [24]($\circ$)

V THE MICROSCOPIC METHOD: APPLICATION TO $^8\text{Li}(\alpha,n)^{11}\text{B}$

The $^8\text{Li}(\alpha,n)^{11}\text{B}$ reaction is expected to play an important role in the primordial nucleosynthesis. A first experiment, using the reverse reaction has been done by Paradellis et al. [25], but was limited to the $^{11}\text{B}_{gs}$ contribution. More recently, new data have been obtained with a ^8Li radioactive beam [26,27], and show that some n + $^{11}\text{B}^*$ excited channels are even more important that the n + $^{11}\text{B}_{gs}$ channel.

The microscopic study of the $^8\text{Li}(\alpha,n)^{11}\text{B}$ cross section has been done using an extended cluster model [14] where the ^8Li and ^{11}B internal wave functions are defined in the shell model involving all p shell configurations. This yields 28 states for ^8Li and 62 states for ^{11}B. The ground state of these nuclei is therefore described by a flexible mixing of shell-model states, and several excited states can be accurately reproduced. Details on the conditions of calculation are given in [14]. The astrophysical S-factor is given in Fig. 4 and compared with experimental data [25,27]. The data of ref. [25] correspond to the ^{11}B $3/2^-$ ground state only. The GCM total S-factor is in good agreement with the data of Gu et al. [27], and the partial wave decomposition shows that the ground-state contribution is lower than one half of the total. The S-factor is dominated, at low energies, by a broad 2^+ resonance which determines the reaction rate at temperatures below 10^9 K.

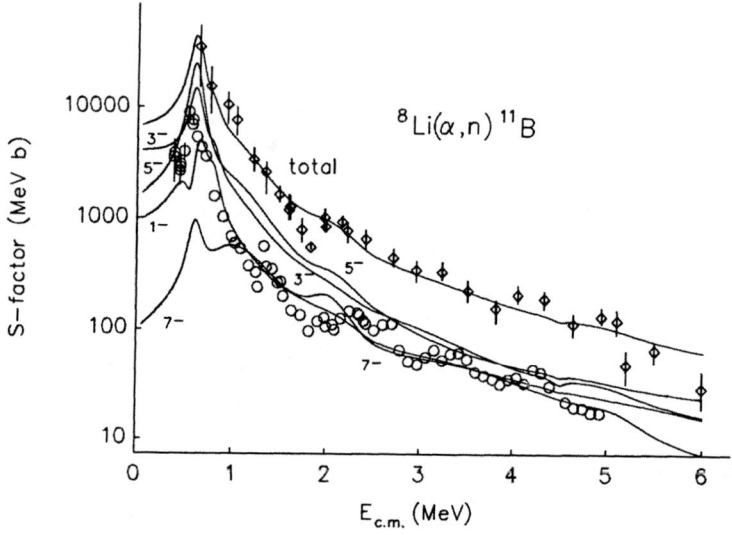

FIGURE 4. ^{8}Li$(\alpha,n)^{11}$B S-factor, with the individual contribution of each ^{11}B state (labeled by 2 times the spin). The experimental data are from ref. [25] (circles - ^{11}B ground state only), and ref.[27] (diamonds - sum over all ^{11}B states).

VI CONCLUSION

The present review shows that each model has advantages and drawbacks. The choice mainly depends on the amount of available experimental data. When many experimental data exist, the R-matrix formalism is a convenient way to extrapolate the cross section down to low energies. The potential model and the DWBA method have some flexibility but, in practice, are limited to rather simple systems. Microscopic models can be, in principle, applied on any reaction. In practice, the length of calculation and the uncertainties on the nucleon-nucleon forces are important limitations. However, owing to their predictive power, microscopic models are the first candidates for investigations of badly known reaction rates.

REFERENCES

1. Rolfs, C. and Rodney, W.S. *Cauldrons in the Cosmos*, University of Chicago Press, 1988.
2. Arpasella, C. et al., Phys. Lett. **B389**, 452 (1996).
3. Vervier, J., Nucl. Phys. **A616**, 97c (1997).
4. F. C. Barker and N. Ferdous, Aust. J. Phys. **33**, 691 (1980).
5. Humblet, J., Nucl. Phys. **A187**, 65 (1972).
6. Baye, D. and Descouvemont, P., Ann. Phys. **165**, 115 (1985).
7. Oberhummer, H. and Staudt, G., in "Nuclei in the Cosmos", ed. H. Oberhummer et al., Springer, Berlin, 1991, p.29.
8. Descouvemont, P. , J. Phys. **G19**, S141 (1993).
9. Thielemann, F.-K., Arnould, M. and Truran, J.W., in "Advances in Nuclear Astrophysics", eds. Vangioni-Flam et al., 525 (1987).
10. Barker, F.C., Nucl. Phys. **A575**, 361 (1994).
11. Christy, R.F. and Duck, I., Nucl. Phys. **24**, 89 (1961).
12. Dufour, M. and Descouvemont, P., Nucl. Phys. **A605**, 160 (1996).
13. Baye, D. and Kruglanski, M., Phys. Rev. **C45**, 1321 (1992).
14. Descouvemont, P., Nucl. Phys. **A596**, 285 (1996).
15. Lane, A.M. and Thomas, R.G., Rev. Mod. Phys. **30**, 257 (1958).
16. Dwarakanath, M.R. and Winkler, H., Phys. Rev. **C4**, 1532 (1971).
17. Krauss, A., Becker, H.W., Trautvetter, H.P., and Rolfs, C., Nucl. Phys. **467**, 273 (1987).
18. Langanke, K., in "Nuclei in the Cosmos", ed. H. Oberhummer et al., Springer-Verlag Berlin, 1991, p.61.
19. Vogl, J.L., Ph.D. Thesis, California Institute of Technology, 1963.
20. Ajzenberg-Selove, F., Nucl. Phys. **A490**, 1 (1988).
21. Parker, P.D. and Kavanagh, R.W., Phys. Rev. **131**, 2578 (1963).
22. Nagatani, K., Dwarakanath, M.R. and Ashery, D., Nucl. Phys. **A128**, 325 (1969).
23. Kräwinkel, H. et al., Z. Phys. **A304**, 307 (1982).
24. Hilgemeier, M., Becker, H.W., Rolfs, C., Trautvetter, H.P. and Hammer, J.W., Z. Phys. **A329**, 243 (1988).
25. Paradellis, T. et al., Z. Phys. **A337**, 211 (1990).
26. Boyd, R.N. et al., Phys. Rev. Lett. **68**, 1283 (1992).
27. Gu, X. et al., Phys. Lett. **B343**, 31 (1995).

The Cross Section of the Neutron Capture Reaction $^{13}C(n,\gamma)^{14}C$

Harald Herndl, Rainer Hofinger and Heinz Oberhummer

Institut für Kernphysik, TU Wien,
Wiedner Hauptstr. 8-10, A-1040 Vienna, Austria

Abstract. We calculate the cross section and reaction rate of the neutron capture reaction $^{13}C(n,\gamma)^{14}C$ which plays an important role in helium burning zones of red giant stars as well as in inhomogeneous big bang models. We consider both the direct capture to bound states and resonant capture to the 143 keV resonance. Interferences between the direct and resonant contributions are taken into account. The resulting total cross section is compared with experimental cross sections.

INTRODUCTION

The cross section of the neutron capture reaction $^{13}C(n,\gamma)^{14}C$ is of importance for nucleosynthesis theories of the s–process as well as in inhomogeneous big bang models. In stellar helium burning it can act as neutron poison for the slow neutron capture process (s–process). As one of the main products of the CNO–cycle ^{13}C is also abundant in helium burning zones. In low mass asymptotic giant branch (AGB) stars the reaction $^{13}C(\alpha,n)^{16}O$ is considered to be the main source of neutron production for the main component of the s–process. In this case the capture $^{13}C(n,\gamma)^{14}C$ can reduce not only the neutron but also the ^{13}C abundance [1].

Furthermore the reaction is also important in the nucleosynthesis of inhomogeneous Big Bang models. In neutron–rich zones intermediate–mass nuclei might be produced via a reaction sequence passing through the following reaction sequence $^{12}C(n,\gamma)^{13}C(n,\gamma)^{14}C\ldots$ [2].

In the next section we will present the methods used to determine the reaction cross section. Then we will discuss the experimental and theoretical input parameters for our calculations. In the following section we give the results for the cross section of $^{13}C(n,\gamma)^{14}C$. Finally we will summarize and discuss our results.

CALCULATION OF THE CROSS SECTION

We calculate the direct capture (DC) cross section in a potential model described in [3-5]. The total nonresonant cross section σ_{nr} is determined by the direct capture transitions σ_i^{DC} to all bound states with the single particle spectroscopic factors $C^2 S_i$ in the final nucleus:

$$\sigma_{tot}^{DC} = \sum_i (C^2 S)_i \sigma_i^{DC} \quad . \tag{1}$$

The DC cross sections σ_i^{DC} are determined by the overlap of the scattering wave function in the entrance channel, the bound-state wave function in the exit channel and the multipole transition-operator. In most cases only E1-transition need to be taken into account.

For the calculation of the scattering and bound-state wave function we use real folding potentials which are given by [4,6]

$$V(R) = \lambda V_F(R) = \lambda \int \int \rho_a(\mathbf{r}_1) \rho_A(\mathbf{r}_2) v_{\text{eff}}(E, \rho_a, \rho_A, s) \, d\mathbf{r}_1 d\mathbf{r}_2 \quad , \tag{2}$$

with λ being a potential strength parameter close to unity, and $s = |\mathbf{R} + \mathbf{r}_2 - \mathbf{r}_1|$, where R is the separation of the centers of mass of the projectile and the target nucleus. The density can been derived from measured charge distributions [7] and the effective nucleon-nucleon interaction v_{eff} has been taken in the DDM3Y parameterization [6]. The imaginary part of the potential is very small because of the small flux into other reaction channels and can be neglected in most cases involving neutron capture by neutron-rich target nuclei.

The cross section of a neutron capture to a single isolated resonance can be described by the resonant Breit-Wigner formula [8,9]

$$\sigma_r(E) = \frac{\pi \hbar^2}{2\mu E} \frac{(2J+1)}{2(2J_t+1)} \frac{\Gamma_n(E)\Gamma_\gamma(E)}{(E_r - E)^2 + \left(\frac{\Gamma_{tot}(E)}{2}\right)^2} \quad , \tag{3}$$

where J and J_t are the spin of the resonance level and the target nucleus, respectively. The resonance energy is given by E_r and the partial widths in the entrance and exit channel by Γ_n and Γ_γ, respectively. The energy dependence of a particle width with angular momentum l is given by [9]

$$\Gamma_l(E) = \Gamma_l(E_r)\sqrt{\frac{E}{E_r}}\frac{P_l(E)}{P_l(E_r)} \qquad (4)$$

where $P_l(E)$ is the penetrability.

The energy dependence of the γ width is given by

$$\Gamma_\gamma(E) = \Gamma_\gamma(E_r)\frac{(E - E_r + E_\gamma)^{2l+1}}{(E_\gamma)^{2l+1}} \qquad (5)$$

where E_γ denotes the energy and l the multipolarity of the electromagnetic transition.

The total width Γ_{tot} is the sum of the partial widths. In most cases $\Gamma_n \gg \Gamma_\gamma$ so that $\Gamma_{tot} \approx \Gamma_n$. In that case the neutron width can be obtained directly from the half live of the resonant state. On the other hand, the neutron width is related to the single-particle width by

$$\Gamma_n = C^2 S\ \Gamma_{s.p.} \qquad (6)$$

where C is the isospin Clebsch–Gordan coefficient and S the spectroscopic factor. The single-particle width $\Gamma_{s.p.}$ can be calculated from the scattering phase shift of a scattering potential with the potential depth determined by matching the resonance energy.

The gamma width Γ_γ can, if it is not known experimentally, be calculated from the reduced transition probabilities obtained from shell-model wave functions.

If the resonance width is broad, interferences between direct and resonant capture can be important. The total cross section is then obtained by [18]

$$\sigma(E) = \sigma_{nr}(E) + \sigma_r(E) + 2\left[\sigma_{nr}(E)\sigma_r(E)\right]^{1/2}\cos[\delta_r(E)] \ . \qquad (7)$$

In this equation $\delta_r(E)$ is the resonance phase shift which is given by $\delta_r(E) = \arctan[\Gamma(E)/2(E - E_r)]$.

DETERMINATION OF THE NUCLEAR INPUT PARAMETERS

The input parameters for direct capture transitions of the reaction $^{13}C(n,\gamma)^{14}C$ are listed in Table 1. The spectroscopic factors C^2S are important input parameters. They can be obtained experimentally from single-particle transfer reaction studies. In our case we have used the spectroscopic factors obtained from the reaction ^{13}C(d,p)^{14}C [11] for all transitions except the ground state transition. The spectroscopic factor for the transition to the ground state of ^{14}C is taken from a shell–model calculation [12].

TABLE 1. Transitions for the direct contribution to the reaction ^{13}C(n,γ)^{14}C.

J^π	E_x (MeV)	Q-value (MeV)	Transition	C^2S
0^+	0.000	8.176	s→1p$_{1/2}$	1.734
			d→1p$_{1/2}$	1.734
1^-	6.094	2.082	p→2s$_{1/2}$	0.75
0^+	6.589	1.587	s→2p$_{1/2}$	0.14
3^-	6.728	1.448	p→1d$_{5/2}$	0.65
2^+	7.012	1.164	s→2p$_{3/2}$	0.065
2^-	7.341	0.835	p→1d$_{5/2}$	0.72

The bound state wave functions are obtained from the folding potential with the parameter λ being adjusted to the known binding energies. In the scattering state we determine λ from the scattering lengths. For neutron scattering on ^{13}C the free scattering lengths are given by $a_c = 5.76$ fm and $a_i = -0.48$ fm, where the subscript c and i denote the coherent and incoherent scattering length, respectively. Since the incoherent scattering is not negligible in this case we determine two different potentials for the channel spins in the entrance channel $J_a = 0$ and $J_a = 1$, which are $\lambda_0 = 1.1396$ and $\lambda_1 = 1.2465$, respectively.

In the energy region of astrophysical interest there is only one resonance at a resonance energy of 143 keV. The resonance parameters of this 2^+ state are known experimentally. In Table 2 we list the experimentally known parameters. We also calculated the widths. The single particle width is determined from the phase shifts of a folding potential where the parameter λ is adjusted to the resonance energy. This single particle width is multiplied by the spectroscopic factor of 0.065 known from the ^{13}C(d,p)^{14}C reaction [11]. The resulting width of 3.15 keV is in excellent agreement with the experimental width of (3.4 ± 0.6) eV [11]. The γ-width calculated from shell model wave function is 0.396 eV which is a bit larger than the width of 0.215 eV determined by Raman et al. [14].

TABLE 2. Resonance parameters of the reaction ^{13}C(n,γ)^{14}C.

J^π	E_x (MeV)	E_r (MeV)	Γ_n(exp.) (keV) [a]	Γ_γ(exp.) (eV) [b]	$\omega\gamma$ (eV)
2^+	8.320	0.143	3.4 ± 0.6	0.215	0.269

[a] From Ref. [11]
[b] From Ref. [14]

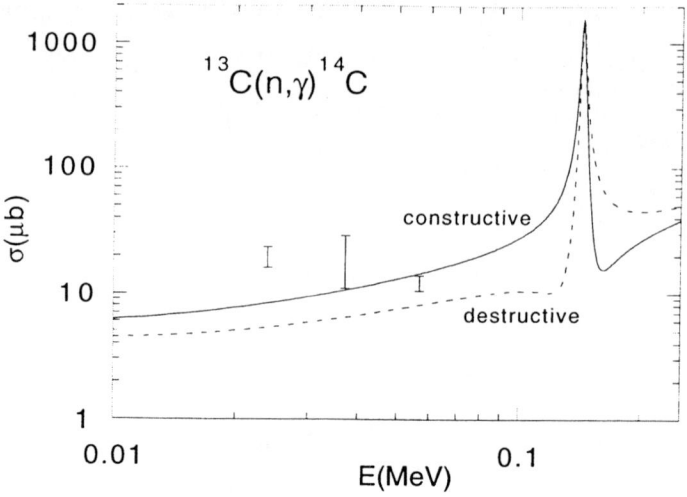

FIGURE 1. The calculated total cross section with constructive (solid line) and destructive (dashed line) interferences below the resonance energy compared with experimental data.

RESULTS

Using the input parameters discussed in the previous chapter we can now calculate the total cross section. We calculate the p– and d–wave contributions in our potential model and multiply the cross sections with the spectroscopic factors of Table 1. The s–wave is an extrapolation of the experimentally known thermal cross section [13].

The s–wave cross section can also be calculated in the potential model. Since the s–wave capture mainly takes place in the nuclear interior the wave functions couple to the giant dipole resonance which causes a severe hindrance of the s–wave capture cross section. If we take this effect into account the thermal cross section can be reproduced nicely (cf. Ref. [15] for a detailed discussion).

The resonance is calculated using the Breit–Wigner formula given in Eq. 3 with the input parameters from Table 2. We take into account the energy dependence of the widths. Using Eq. 4 and $P_1 = \rho^2/(\rho^2 + 1)$ for the penetrability of a p–wave neutron ($\rho = kr$) we can evaluate the energy dependence of the neutron width as

$$\Gamma_n(E) = 70.4577 \frac{E^{3/2}}{0.84328E + 1} \quad \text{keV} \tag{8}$$

with the energy E given in MeV. The electromagnetic decay of the 8.318 MeV state

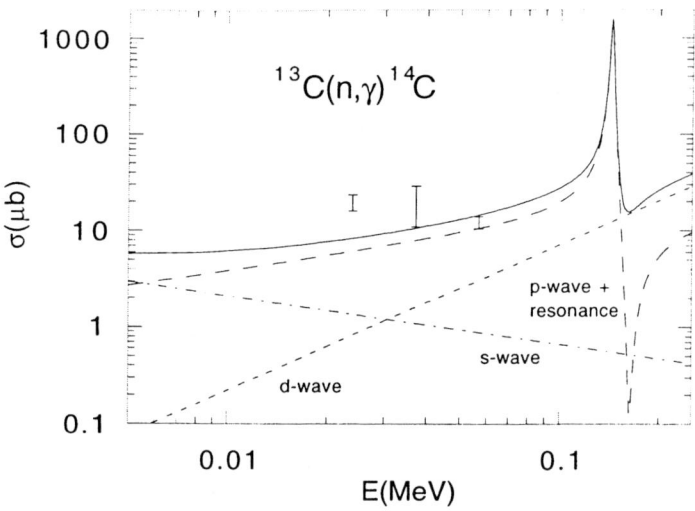

FIGURE 2. The total cross section of ^{13}C(n,γ)^{14}C decomposed in the different contributions.

is dominated by an E1-transition to the 1^- state at 6.094 MeV [14]. Therefore we adopt an energy dependence (see Eq. 5)

$$\Gamma_\gamma(E) = 0.0195(E + 2.082)^3 \quad \text{eV} \tag{9}$$

with the energy E in MeV.

Finally we calculate the interferences between the p-wave resonance and p-wave capture. In order to determine the sign of the interference term in Eq. 7 we proceed as follows. We determined a potential strength λ which reproduces a potential resonance at an energy of 143 keV. The p-wave cross section calculated with this potential shows an interference pattern where the resonance and the direct p-wave add constructively for energies below the resonance energy and destructively for energies higher than the resonance energy. With the small spectroscopic factor of 0.065 the 143 keV resonance cannot be approximated through a potential resonance. However, this fact only affects the width of the resonance but not the principal interference pattern. Furthermore it is observed that the agreement of the calculated cross section with the experimental data is much better in case of constructive interference before the resonance (see below). The same interference pattern was also observed in some previous investigations for low-energy resonances of nucleon capture by p-nuclei [16-19].

The thermonuclear cross section of the reaction ^{13}C(n,γ)^{14}C was measured by

Raman et al. [14] and Shima et al. [20]. Raman et at. reported a value of $(20\pm9)\,\mu$b at an average neutron energy of $\bar{E}_n = 40\,\text{keV}$. Shima et al. measured $(19.8\pm3.7)\,\mu$b at $\bar{E}_n = 25.7\,\text{keV}$ and $(12.2 \pm 1.7)\,\mu$b at $\bar{E}_n = 61.1\,\text{keV}$.

The total cross section is compared with the experimental data for both kind of interference patterns in Fig. 1. If we assume constructive interference before the resonance the agreement with the experimental data is quite good. Only the first data point differs from our calculation by a factor of 2.

In Fig. 2 the various components of the total cross section are displayed. The s–wave dominates up to about 5 keV. In the thermonuclear energy range the p–wave becomes the most important component. For energies above the resonance energy the d–wave is dominating due to the strong d–wave transition to the ground state of ^{14}C.

SUMMARY AND DISCUSSION

We have presented an detailed analysis of the low–energy cross section of the reaction ^{13}C(n,γ)^{14}C. All important direct capture transitions to bound states as well as the 143 keV p–wave resonance are taken into account. We also include interference effects between resonant capture and direct capture. We find that the interferences are constructive for neutron energies below the resonance energy and destructive above the resonance energy. The total cross section agrees well with the available experimental data.

ACKNOWLEDGMENTS

This work was supported by Fonds zur Förderung der wissenschaftlichen Forschung (FWF project S7307-AST).

REFERENCES

1. I. Iben Jr., Astrophys. J. **395**, 202 (1976).
2. T. Kajino, G.J. Mathews and G.M. Fuller, Astrophys. J. **364**, 7 (1990).
3. K. H. Kim, M. H. Park, and B. T. Kim, Phys. Rev. C **35**, 363 (1987).
4. H. Oberhummer and G. Staudt, in *Nuclei in the Cosmos*, edited by H. Oberhummer, (Springer-Verlag, Berlin, New York, 1991) p. 29.
5. P. Mohr, H. Abele, R. Zwiebel, G. Staudt, H. Krauss, H. Oberhummer, A. Denker, J. W. Hammer, and G. Wolf, Phys. Rev. C **48**, 1420 (1993).

6. A. M. Kobos, B. A. Brown, R. Lindsay, and G. R. Satchler, Nucl. Phys. A **425**, 205 (1984).
7. H. de Vries, C. W. de Jager, and C. de Vries, At. Data Nucl. Data Tables **36**, 495 (1987).
8. R. G. Breit and E. P. Wigner, Phys. Rev. **49**, 519 (1936)
9. J. M. Blatt and V. F. Weisskopf, *Theoretical Nuclear Physics*, (Wiley & Sons, New York, 1962).
10. C. Rolfs and R. E. Azuma, Nucl. Phys. A **227**, 291 (1974).
11. F. Ajzenberg–Selove, Nucl. Phys. A **449**, 1 (1986).
12. S. Cohen and D. Kurath, Nucl. Phys. A **101**, 1 (1967).
13. V. F. Sears, Neutron News **3**, 26 (1992).
14. S. Raman, M. Igashira, Y. Dozono, H. Kitazawa, M. Mizumoto and J.E. Lynn, Phys. Rev. C **41**, 458 (1990).
15. H. Herndl and H. Oberhummer, Proceedings of the Conference "Advances in Nuclear Physics and Related Areas", Thessaloniki, July 8–12, 1997, in press.
16. R. G. Thomas, Phys. Rev. **88**, 1109 (1952).
17. C. Mahaux, Nucl. Phys. **71**, 241 (1965).
18. C. Rolfs and R. E. Azuma, Nucl. Phys. A **227**, 291 (1974).
19. P. Decrock, M. Gaelens, M. Huyse, G. Reusen, G. Vancraeynest, P. Van Duppen, J. Wauters, Th. Delbar, W. Galster, P. Leleux, I. Licot, E. Lienard, P. Lipnik, C. Michotte, J. Vervier and H. Oberhummer, Phys. Rev. C **48**, 2057 (1993).
20. T. Shima, F. Okazaki, T. Kikuchi, T. Kobayashi, T. Kii, T. Baba, Y. Nagai and M. Igashira, Nucl. Phys. A (Suppl.), in press.

Radiative neutron captures by exotic nuclei

S. Goriely[1]

Institut d'Astronomie et d'Astrophysique
Université Libre de Bruxelles
Campus de la Plaine, CP 226
1050 Brussels – Belgium

Abstract. The radiative direct capture of neutrons is calculated within the potential model. The direct transitions to all the excited states predicted by a combinatorial model of nuclear level densities are calculated with the introduction of an average spectroscopic factor. The uncertainties related to the prediction of the excitation spectrum in the residual nucleus, and in particular of the spectroscopic factor, are discussed. The resulting direct contribution to the neutron capture rate is estimated for all the neutron-rich nuclei potentially involved in the r-process nucleosynthesis and compared with the statistical compound nucleus contribution. The total radiative capture cross section is approximated taking into account a possible overestimate of the compound nucleus contribution for exotic nuclei, and used in parametric r-process calculations to study the impact of the direct capture mechanism on the r-process nucleosynthesis.

I INTRODUCTION

The radiative neutron captures by exotic neutron-rich nuclei are known to be of fundamental importance for our understanding of the rapid neutron-capture process (or r-process) of nucleosynthesis. The r-process is one of the major mechanisms able to explain the origin of the stable nuclides heavier than iron observed in nature. It is believed to take place in environments characterized by high neutron densities ($N_n \gtrsim 10^{20}$ cm^{-3}), so that successive neutron captures can proceed into neutron-rich regions well off the β-stability valley. The abundance distribution of the elements produced by the r-process is closely bound to the nuclear, electromagnetic and weak interactions taking place in such conditions. For this reason, our understanding of the r-process nucleosynthesis strongly depends on the reliability of our predictions concern-

---

[1] S.G. is FNRS senior research assistant

ing the different nuclear quantities involved, and in particular the radiative neutron capture rates by the very neutron-rich nuclei.

So far, the r-process calculations made use of neutron capture rates only evaluated within the statistical model of Hauser-Feshbach. Such a model makes the fundamental assumption that the capture process takes place with the intermediary formation of a compound nucleus (CN) in thermodynamic equilibrium. The formation of a CN is usually justified by assuming that the level density in the CN at the projectile incident energy is large enough to ensure an average statistical continuum superposition of available resonances. However, when the number of available states in the compound system is "relatively" small, the validity of the Hauser-Feshbach predictions has to be questioned, the neutron capture process being possibly dominated by direct electromagnetic transitions to a bound final state rather than through a CN intermediary. Direct captures (DCs) are of first importance for light or closed shell systems for which no resonant states are available [1], and are expected to play a non-negligible role for heavy nuclei close to the neutron drip line [2]. The DC by exotic nuclei (such as those produced by the r-process) has unfortunately never been estimated and is the subject of the present contribution, which aims to report and complement a previous study [3]. The DC rates are calculated within the potential model, the main features of which are described in Sect. 2. Emphasis is put on the uncertainties affecting the determination of the spectroscopic factor (SF) in the residual nucleus. Two prescriptions of the SF are used to estimate the possible DC contribution to the neutron capture rates. In Sect. 3, the total (n,γ) rate is approximated taking into account the possible overestimate of the CN contribution for neutron-rich nuclei. Finally, the influence of the newly-derived (n,γ) rates on the r-process abundance distribution is discussed.

II THE DC RATES

Many studies on neutron capture reactions have been devoted to the description of the DC mechanism, in which the incoming neutron is scattered directly into a final bound state without forming a CN. In a general way, the total DC cross section of a nucleus (Z, N) can be expressed as

$$\sigma^{DC}(E) = \sum_f C_f^2 \, S_f \, \sigma_f^{DC}(E) , \qquad (1)$$

where E is the energy of the incident neutron and the sum runs over all the available final states f of the residual nucleus. C_f is the isospin Clebsh-Gordan coefficient and S_f the SF describing the overlap between the antisymmetrized wave function of the initial system $(Z, N)+n$ and the final state f in $(Z, N+1)$. Equation 1 emphasizes the importance of the single-particle configuration of the final state (for which $S_f \simeq 1$) in the direct transition (other configurations being characterized by a small or even negligible S_f). The DC cross section σ_f^{DC} is calculated within the potential model [4] in which the wave functions of the initial and final systems are determined by solving the respective Schrödinger equations. For both channels, the same analytical form of the potential is used, although the depth of the potential is scaled in order to reproduce, in the final system the exact binding energy of the state f, and in the initial system the so-called volume integral per nucleon. The widely-used (especially in CN models) optical potential of Jeukenne et al. [5] is considered in the present work as it has the advantage of predicting fairly well the experimental volume integrals in a wide range of the periodic table. More details on the influence of the potential on the DC rates can be found in [3].

A crucial aspect in the calculation of the DC cross-section concerns the determination of the excited spectrum of the final system, especially when dealing with experimentally unknown nuclei. Most of the previous DC studies have tried to predict the 1 neutron particle-hole configuration derived from a single-particle level spectrum [6]. Unfortunately, the remarkable sensitivity of the DC cross section to the exact determination of the very few ph states inevitably gives rise to considerable scattered predictions when use is made of different nuclear input. To avoid such difficulties, we consider here all the possible nucleon excitations predicted by a model of nuclear level density. In this case, the full single-particle strength is fragmented among the different excited states, so that to each of them should be attributed its corresponding SF, or in a first approximation an average SF. The total DC cross section is then expressed as

$$\sigma^{DC}(E) = \sum_{f=0}^{x} C_f^2 \, S_f \, \sigma_f^{DC}(E)$$
$$+ \int_{E_x}^{S_n} \sum_{J_f, \pi_f} \langle C^2 S \rangle \rho(E_f, J_f, \pi_f) \, \sigma_f^{DC}(E) \, dE_f \,, \qquad (2)$$

where x corresponds to the last experimentally known level of excitation energy E_x (smaller than the neutron separation energy S_n). Above E_x, the summation is replaced by a continuous integration over the spin (J)- and parity (π)-dependent level density ρ, and the SF and isospin Clebsh-Gordan coefficient by an average quantity $\langle C^2 S \rangle$. Different models of nuclear level densities are now available, although most of them are based on the statistical approach which does not enable a detailed description, especially at low energies, of the spin and parity distributions. Because of the high sensitivity of the DC cross section to these quantities, only the microscopic combinatorial approach seems appropriate. In particular, the combinatorial Monte Carlo technique [7] has been shown to be a reliable tool for calculating spin- and parity-dependent level densities with a sufficient accuracy and a short computation time which enable an exact treatment of the BCS pairing and its

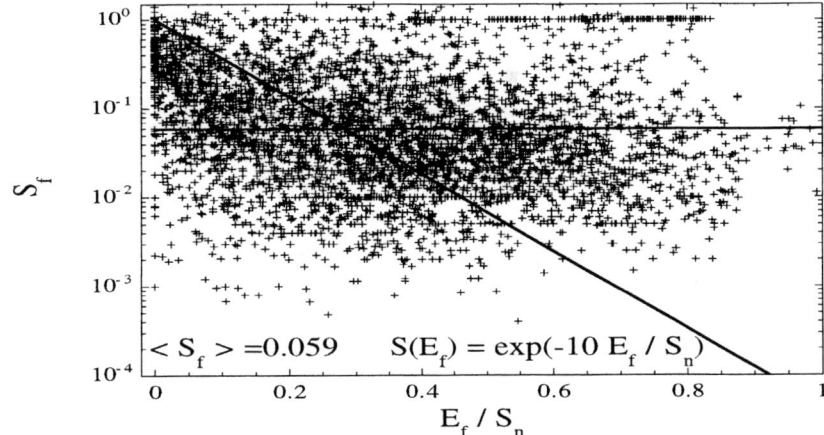

FIGURE 1. Experimental SFs for 4270 nuclear levels in 146 different nuclei extracted from (d,p) reactions (Nuclear Data Sheets) as a function of E_f/S_n. The two lines correspond to approximations SFa and SFb used in the calculations.

application to large-scale calculations.

So far, the determination of the SF remains an open problem. Closed shell nuclei often have low-lying levels with a high-purity single-particle configuration, but away from magic numbers, or at an increasing excitation energy, residual interactions, as well as couplings of the single-particle motion to other degrees of freedom, distribute the spectroscopic strength of a single-particle state among several nuclear levels. Experimental (d,p) reactions clearly show the spreading and fragmentation of the single-particle states which give rise to a continuous spectroscopic strength if the fragmentation width is larger than the spacing of the single-particle states [8]. The SF remains a quantity which is difficult to estimate theoretically, although some attempts have been made in the quasi-particle [9] or shell model [10], but such calculations are outside the scope of the present work, because of the large number of transitions studied and nuclei involved. For this reason, a more phenomenological approach is adopted by considering either

a) an energy-independent SF, $\langle C^2 S \rangle = 0.06$, corresponding to the geometrical average of the 4270 experimental values shown in Fig. 1; this approximation is referred to as SFa, or

b) an energy-dependent SF, $\langle C^2 S \rangle = \exp(-10 E_f/S_n)$, which empirically simulates a decreasing SF at increasing energies, as inferred from experimental analyses of the reduced width [11]; this approximation is referred to as SFb. Finally, let us note that all the calculations assume a spherical symmetry. Due to the large uncertainties already existing in the spherical approximation of the DC model, as well as the nuclear level density model, it seems presumptuous to

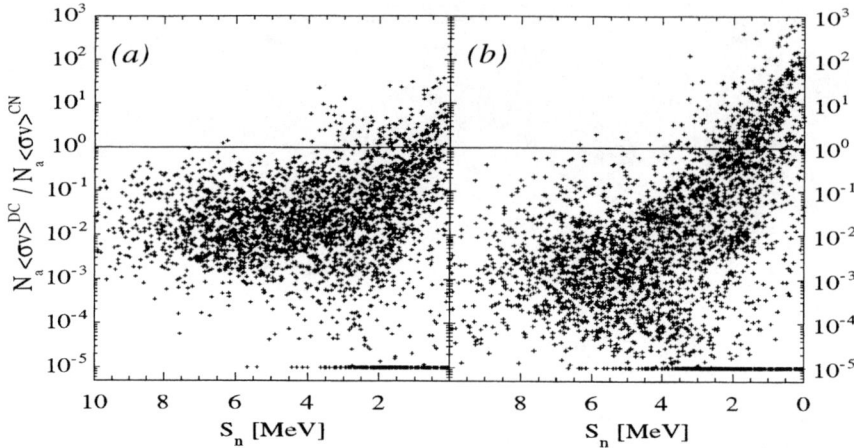

FIGURE 2. a) Ratio of the Maxwellian-averaged DC rate, $N_a\langle\sigma v\rangle$ (where N_a is the Avogadro number and v the relative velocity between target and projectile), to the CN rate for 3100 neutron-rich nuclei at $T = 1.5\ 10^9$K. The DC is calculated with approximation SFa. A lower limit of 10^{-5} is imposed. b) same as a) when the DC is calculated with the approximation SFb.

treat deformation in detail at the moment. In the same spirit, we only consider $E1$- and $E2$-transitions. $M1$-transitions could contribute significantly if $E1$-transitions are forbidden, but in this case, the resulting neutron capture rate is expected to be small in comparison with the β-decay rate, at least for very neutron-rich nuclei.

The DC rates are calculated by Eq.(2) for all the 3100 nuclei potentially involved in the r-process nucleosynthesis, i.e lying in the nuclear chart (with $20 \leq Z \leq 92$) between the valley of β-stability and the neutron drip line. Figure 2 compares the calculated Maxwellian-averaged CN rates [12] at $T = 1.5\ 10^9$K with the DC rates obtained with our two prescriptions of the SF. In contrast to previous calculations [2], the DC rates decrease with decreasing S_n, as does the CN contribution. This is especially true in our approximation SFa, as also illustrated in Fig. 3. This result is to be expected since the number of levels with excitation energy smaller than S_n decreases with decreasing S_n, and with it, the DC cross section. The high-lying levels have a relatively small contribution to σ^{DC}–and even much smaller in approximation SFb–, but the exponentially rising number of levels with increasing excitation energy compensates for this effect to some extent, at least if the SF does not vanish completely. In our approximation SFb, the low-energy transitions are suppressed and the DC rate is reduced in comparison with SFa, at least when such low-energy transitions are dominant. Figure 2 also shows that for many nuclei with low S_n the DC rates can become negligible, the selection rule for-

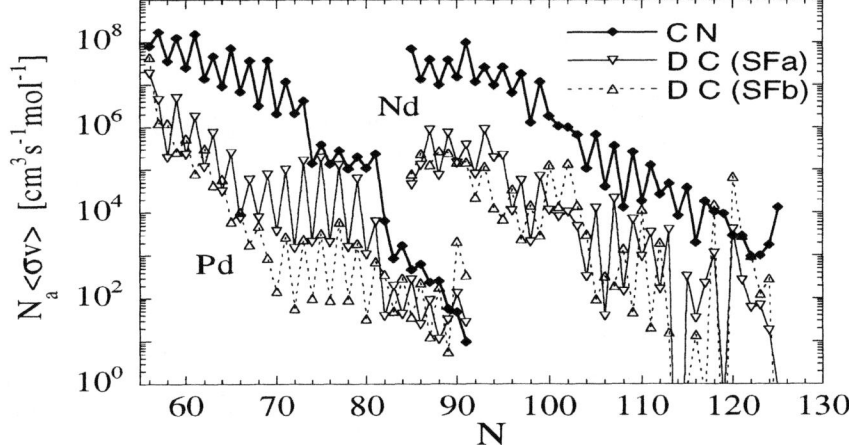

FIGURE 3. Comparison of CN and DC (with approximation SFa and SFb) rates at $T = 1.5\ 10^9$K for the isotopic chains of Pd and Nd.

bidding any $E1$- or $E2$-transition to excited states in the residual nucleus. This feature is independent from our adopted prescriptions for the average SF and is more clearly illustrated in Fig. 4 where in the (N, Z) plane the half-lives against neutron capture, $\tau_n = 1/N_n \langle \sigma v \rangle^{DC}$, are estimated at $T = 1.5\ 10^9$K for a typical neutron density of $N_n = 10^{24}$cm^{-3}. At such a density, many nuclides close to the neutron drip line have a neutron capture half-life longer than 10 ms or even 1 s. In particular, many neutron-rich isotopes of the same element (e.g. Kr, Ba, Yb) show the same hindrance against DC. It mainly originates from the small number of available states below S_n and the resulting small number of allowed transitions. In many cases, no allowed transitions are found at all, and the direct channel is consequently inhibited. However, it should be noted that level crossings resulting from deformation effects could modify the spin and parity assignment of the low-lying states, and thus transform forbidden into allowed transitions. Inversely, some of the allowed transitions in the spherical approximaiton could become forbidden with the onset of deformation.

III NEUTRON CAPTURES AND THE R-PROCESS

Direct and compound nuclear reactions are known not to be mutually exclusive. Both mechanisms may contribute to the radiative capture of a neutron. For this reason, the total capture rate is often taken as the simple sum of both contributions, neglecting all possible interferences between them. However, the Hauser-Feshbach prediction of the capture rate is valid only if the num-

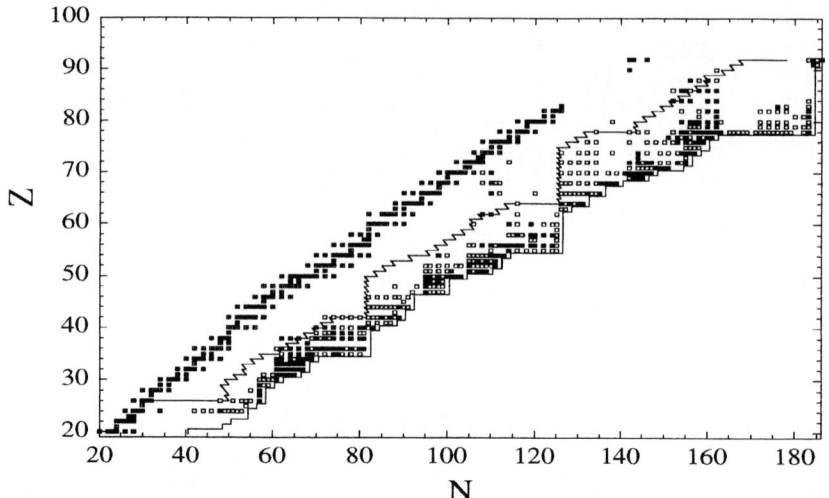

FIGURE 4. Representation in the (N, Z) plane of the nuclides with a half-life against neutron DC larger than 1s (full squares) or ranging between 10ms and 1s (open squares) for $N_n = 10^{24} \text{cm}^{-3}$ and $T = 1.5 \; 10^9 \text{K}$. The upper left curve corresponds to a canonical r-process path in the same thermodynamic conditions. The right line represents the neutron drip line, and the upper left squares the stable isotopes. The DC rates are calculated with the SFa prescription.

ber of levels available to the incident nucleon in the CN is high enough and their energy and width randomly distributed within the contributing energy interval to ensure a continuum superposition of resonances. These conditions might not be fulfilled for very neutron-rich nuclei, so that special care should be taken when extrapolating the statistical predictions to nuclei close to the neutron drip line. One way to account for such an overestimate of the neutron capture rate is to damp artificially the statistical contribution as

$$\langle \sigma v \rangle = \langle \sigma v \rangle^{DC} + \langle \sigma v \rangle^{CN} \frac{1}{1 + (N_{sp}^*/N_{sp}(S_n))^a}. \quad (3)$$

where $N_{sp}(S_n)$ is the number of levels available to s- and p-neutrons in an energy interval of $2kT$ around $U = S_n + kT$ in the CN, and a is a damping parameter characterizing the disappearance of the CN contribution for nuclei with $N_{sp} < N_{sp}^*$. N_{sp} is calculated for each nucleus with the nuclear level density formula used to estimate the CN rate [12] according to the spin and parity of the target nucleus, and the damping parameters are taken subjectively as $N_{sp}^* = 5$ and $a = 5$. Approximation (3) is obviously extremely schematic and much work (for example in a R-matrix or Breit-Wigner approach) remains to be done to estimate the resonant neutron capture rate of exotic nuclei when only a few CN states are available.

Approximation (3) is used to study the competition between neutron captures and β-decays taking place during the r-process. For this purpose, we consider the simple non-equilibrium canonical model in which a full reaction network is solved for a given set of astrophysical conditions $(T, N_n, \tau_{irr})^2$. For low-T, low-N_n (i.e such that $T[10^9\text{K}] < 4.4 - 0.12 \log N_n[\text{cm}^{-3}]$) r-processes, no $(n,\gamma)-(\gamma,n)$ equilibrium is reached and the newly-derived rates inevitably influence the r-process path [13]. Even when an $(n,\gamma)-(\gamma,n)$ equilibrium is believed to be achieved, non-negligible effects are found in the vicinity of the shell closures. For example, in the $(T = 1.5\ 10^9\text{K}, N_n = 10^{24}\ \text{cm}^{-3})$ conditions, the isotopic chains $Z = 41 - 42$ are out of equilibrium and differences in the predictions of the neutron capture rates lead to variations in the final r-abundance distribution (Fig. 5a). Many exotic nuclei close to the $N \lesssim 82$ shell present a deficiency in sp-resonances $(N_{sp}(S_n) < N_{sp}^*)$, so that the total n,γ rate is given by the DC component only. In turn, the DC rate is found to be very small for many of those nuclei, the number of available states below S_n in the residual nucleus being limited. The low rates build up bottle-necks for the r-process flow, and an increase in the production of the $A \sim 120$ r-elements. For the same reason, the high-N_n r-processes proceeding close to the neutron drip line are also affected by the low DC and CN rates found in the $N \lesssim 82$ (Fig. 5b) and $N \lesssim 126$ region. Obviously, the impact of the DC on the r-process depends on the approximation used for the SF, although it does not play a crucial role.

IV CONCLUSIONS

The neutron capture reactions play a pivotal role in the r-process nucleosynthesis. We have tried to improve the description of the neutron captures by exotic nuclei by adding to the traditional statistical estimate two major effects, namely the direct component and the overestimate of the CN predictions for resonance-deficient nuclei. The DC rates are calculated within the potential model for which the excited level spectrum in the residual nucleus is predicted by a combinatorial model of nuclear level densities, and the resulting single-particle configuration is assumed to be characterized by an average SF. In this approach, the DC mechanism is often not negligible compared with the CN process for nuclei close to the valley of β-stability. Neutron-rich nuclei present DC rates which show large variations according to the existence or absence of allowed transitions between the initial and final states. Large uncertainties still affect the DC predictions, but our results emphasize the possibility of a negligible DC rate for many neutron-rich nuclei, independently of the uncertainties related to the SF. This new feature might have an important impact on the r-process nucleosynthesis.

---

[2]) where τ_{irr} is the time during which the neutron irradiation takes place [13].

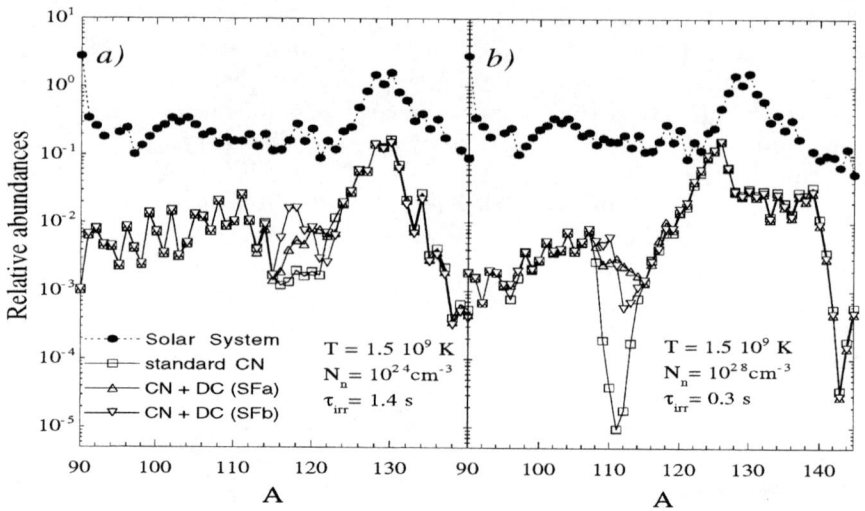

FIGURE 5. a) r-abundance distributions obtained for $T = 1.5\ 10^9$K, $N_n = 10^{24}$ cm^{-3} and $\tau_{irr} = 1.4$ s with 3 estimates of the neutron capture rates, namely the standard CN rates and the damped CN plus DC contribution with approximation SFa or SFb. The top curve corresponds to the solar r-abundances arbitrarily normalized. b) same as a) for $T = 1.5\ 10^9$K, $N_n = 10^{28}$ cm^{-3} and $\tau_{irr} = 0.3$ s.

REFERENCES

1. Mengoni, A., Otsuka, T., Ishihara, M., *Phys. Rev.* **C52**, R2334 (1995); Krausmann, E., Balogh, W., Oberhummer, H., et al., *Phys. Rev.* **C53**, 469 (1996).
2. Mathews, G.J., Mengoni, A., Thielemann, F.-K., et al., *Astrophys. J.* **270**, 740 (1983).
3. Goriely, S., *Astr. Astrophys.* in press (1997).
4. Christy, R.F., and Duck, I., *Nucl. Phys.* **A24**, 89 (1961); Baye, D., and Descouvemont, P., *Ann. Phys.* **165**, 115 (1985).
5. Jeukenne, L.-P., Lejeune, A., Mahaux, C., *Phys. Rev.* **C16**, 80 (1977).
6. Oberhummer, H., Balogh, W., Bieber, R., et al., *Conference on Exotic Nuclei and Atomic Masses*, Gif-sur-Yvette: Editions Frontières, p. 649 (1995).
7. Cerf, N., *Phys. Rev.* **C50**, 836 (1994).
8. Back, B.B., Bang, J, Björnholm, S., et al., *Nucl. Phys.* **A222**, 377 (1974).
9. Yoshida, S., *Phys. Rev.* **123**, 2122 (1961).
10. Harvey, T.F., Clement, D.M., *Nucl. Phys.* **A176**, 592 (1971).
11. Macfarlane, M.H., French, J.B., *Rev. Mod. Phys.* **32**, 567 (1960).
12. Goriely, S., *International Conference on Nuclear Data for Science and Technology*, Trieste (Italy), (1997); Goriely, S., *International Conference on Nuclear Nuclear Structure and Related Topics*, Dubna (Russia), (1997).
13. Goriely, S., Arnould, M., *Astr. Astrophys.* **312**, 327 (1996).

NON-THERMAL REACTIONS AND NEUTRINO ASTROPHYSICS

Spallation Reactions in Extraterrestrial Matter

Rolf Michel

Zentrum für Strahlenschutz und Radioökologie, Universität Hannover
Am Kleinen Felde 30, D-30167 Hannover, Germany

Abstract. This paper describes the cosmic-ray-induced production of stable and radioactive residual nuclides, the so-called cosmogenic nuclides. In extraterrestrial solar-system matter, i.e. planetary surfaces, meteorites, cosmic dust and the heavy component of the galactic cosmic radiation, these nuclides are experimentally observable as positive anomalies of isotopic abundances. They preserve a record of cosmic ray exposure which can be interpreted with respect to the collision and exposure history of the irradiated objects as well as to intensities and spectral distributions of cosmic ray particles in the past. To decipher the cosmic ray record in extraterrestrial matter and to obtain information which cannot be obtained by any other means reliable models are needed for the calculation of the production rates of cosmogenic nuclides. On the basis of thin-target and thick-target accelerator experiments such a model has been developed which is applied here exemplarily to interprete cosmogenic nuclide abundances in stony meteorites and lunar surface materials.

COSMIC RADIATION AND COSMOGENIC NUCLIDES

The term spallation reactions is used here in an historical context independently of it's exact meaning in nuclear physics as synonym for nuclear reactions induced by interactions of primary particles with energies above 100 MeV/A. Such particles occur naturally in space as galactic cosmic ray particles and in the vicinities of stars as energetic stellar wind particles. Interactions of cosmic ray particles with matter are of importance for many fields of sciences such as astro- and cosmophysics and -chemistry, planetology, geophysics and -chemistry, climatology, glaciology, hydrology, space and aviation technology as well as radiation protection in atmospheric and space flights.

Solar system matter is exposed to solar (SCR) and galactic (GCR) cosmic radiation. SCR particles are emitted during solar flares from the sun and consist on the average of 98 % protons and 2 % α-particles (1). Their spectral distributions and intensities vary from flare to flare, typical energies going up to a few hundred MeV/A. GCR consists out of 87 % protons, 12 % α-particles and 1 % heavier ions which show similar energy spectra if looked at as function of energy per nucleon (2). Mean GCR energies are a few GeV/A. GCR spectra are modulated by interaction of GCR particles with the solar magnetic field and thus depend on the solar activity. SCR and GCR proton spectra at 1 A.U. are shown in Figure 1.

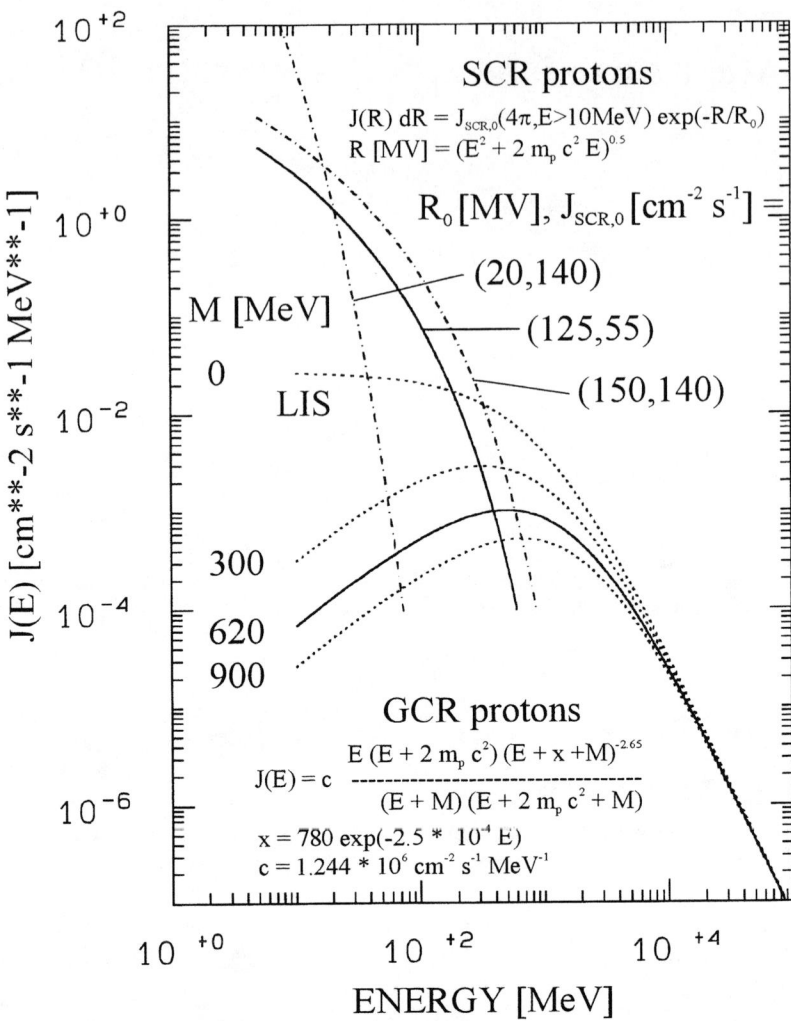

FIGURE 1. Spectra of solar and galactic cosmic ray protons at 1 A.U.. GCR spectra (broken lines) can be unambiguously described (3) by a modulation parameter M [MeV] representing the energy loss of a particle penetrating the solar magnetic field to 1 A.U.. Modulation parameters between 300 MeV (quiet sun) and 900 MeV (active sun) have been observed during the last three solar cycles (3). A modulation parameter of 0 MeV gives the local interstellar GCR spectrum (LIS). SCR spectra (dashed-dotted lines) can be described for individual flares as well as for long-term averaged spectra by exponentially falling rigidity spectra with characteristic rigidities R_0 and 4π integral flux densities $J_{SCR,0}$ of SCR particles with energies above 10 MeV (4). Characteristic rigidities between 20 MV and 150 MV have been observed during the last three solar cycles (1). Such spectra are shown here arbitrarily normalized to $J_{SCR,0} = 140$ cm^{-2} s^{-1}. The solid lines are long-term averaged SCR and GCR spectra for the last 10 Ma derived from the analysis of cosmogenic nuclides in meteorites (5).

A wide variety of stable and radioactive cosmogenic nuclides is produced by SCR and GCR interactions with extraterrestrial matter such as cosmic dust, meteoroids and lunar and planetary surface materials. In these materials, stable cosmogenic nuclides can be observed as positive isotopic anomalies for rare gas isotopes where the small amounts of cosmogenic origin are not mimicked by high natural abundances. Cosmogenic radionuclides can either be observed by their decay, e.g. ^{37}Ar, ^{56}Co, ^{22}Na, ^{55}Fe, ^{60}Co, ^{3}H, ^{44}Ti, and ^{26}Al, or in case of long-lived radionuclides such as ^{10}Be, ^{14}C, ^{26}Al, ^{36}Cl, ^{41}Ca, ^{53}Mn, ^{59}Ni, and ^{129}I by accelerator mass spectrometry (AMS), by radiochemical neutron activation analysis (^{53}Mn) or by conventional rare gas mass spectrometry as for ^{39}Ar and ^{81}Kr. The longest-lived cosmogenic radionuclide, ^{40}K, can only be observed in special matrices with low potassium contents such as iron meteorites by mass spectrometry.

PRODUCTION OF COSMOGENIC NUCLIDES

For the interpretation of the observed abundances of cosmogenic nuclides in extraterrestrial matter reliable model calculations of production rates are a basic requirement. The production rate $P_i(d,R,c_s,c_b)$ [s^{-1} g^{-1}] of a cosmogenic nuclide i in a sample with chemical composition c_s at the depth d inside a meteoroid with shape parameter (radius) R and bulk chemical composition c_b is given by

$$P_i(d,R,c_s,c_b) = N_L \cdot \sum_j c_{s,j} \cdot A_j^{-1} \cdot \sum_k \int \sigma_{i,j,k}(E_k) \cdot J_k(E_k,d,R,c_b) dE_k \qquad (1)$$

with N_L being Avogadro´s number, $c_{s,j}$ being the chemical abundance of target element j with atomic mass A_j, $\sigma_{i,j,k}$ being the cross sections for the production of nuclide i from target element j by particle type k and J_k the differential flux densities of particle type k. The particles to be considered are primary SCR and GCR protons and α-particles as well as secondary GCR protons and neutrons. Other types of GCR secondaries as well as SCR secondaries, in general, can be neglected for most cosmogenic nuclides produced. According to their different energies, SCR and GCR interactions with matter exhibit strong differences with respect to interaction lengths and types of nuclear reactions. Due to their relatively low energies, SCR interactions are restricted to the outmost surface (depth < 15 g cm^{-2}) of the irradiated material. For GCR particles, energies up to 10 GeV/A have to be taken into account. Therefore, secondary particles, in particular neutrons, become important and the depth scale on which GCR interactions occur extends to several hundreds of g · cm^{-2}.

The specific activity A_i of a cosmogenic radionuclide i with a decay constant λ_i in a sample with a production rate P_i depends of the exposure time t_{exp} as well as of its terrestrial residence time t_{terr}

$$A_i(t_{exp}, t_{terr}) = P_i \cdot \left[1 - \exp(-\lambda_i \cdot t_{exp})\right] \cdot \exp(-\lambda_i \cdot t_{terr}) \qquad (2)$$

while the concentration of a stable cosmogenic nuclide C_i just is increasing with the exposure time

$$C_i(t_{\exp}) = P_i \cdot t_{\exp} \qquad (3)$$

Thus, by combining the observed abundances of stable and radioactive nuclides one can determine the exposure as well as the terrestrial ages provided that the respective production rates are known. Historically, different approaches have been used to derive cosmogenic nuclide production rates; see (5,6) for reviews. In this work, a purely physical model (5,7) is described for the calculation of production rates on the basis of spectra of primary and secondary cosmic ray particles and of the cross sections of the underlying nuclear reactions.

PARTICLE SPECTRA AND CROSS SECTIONS

While the spectra of primary SCR particles as function of depth in an irradiated object can be calculated with good accuracy by simply taking into account electronic stopping and nuclear attenuation (8), the spectra of primary and secondary GCR particles in dependence of depth, size and bulk chemical composition can only be calculated by Monte Carlo techniques describing the intra- and internuclear cascades giving a complete picture of attenuation, production and transport of primary and secondary GCR particles. In our model, the spectra are calculated using the HET-code (9) within the HERMES code system (10). Figure 2 gives an example of GCR spectra in the center of a stony meteoroid with a radius of 25 cm irradiated by a GCR proton spectrum with a modulation parameter of 650 MeV. The secondary neutrons show roughly an E^{-1} energy dependence. The spectrum of total (= primary plus secondary) protons, in contrast, exhibits a shape similar to that of the free space spectrum for energies above 1 GeV and fades away below a few hundred MeV because of the removal of particles from the spectrum by electronic stopping. Since the calculational methods by which SCR and GCR spectra are derived are validated state of the art, the quality of the theoretical production rates depends mainly on that of the cross sections used.

In order to provide the necessary reaction data systematic measurements of cross sections for residual nuclide production by proton-induced reactions from thresholds up to 2.6 GeV were performed using accelerators at CERN/Geneva, IPN/Orsay, KFA Jülich, LANL/Los Alamos, LNS/Saclay, PSI/Villigen, TSL/Uppsala, UCL/Louvain La Neuve. The target elements C, N, O, Mg, Al, Si, Ca, Ti, Mn, Fe, Co, Ni, Cu, Sr, Y, Zr, Nb, Te, Ba and Au were investigated. Stable and radioactive residual nuclides were measured by X- and γ-spectrometry as well as by accelerator and conventional rare gas mass spectrometry; (11) and references therein. The entire consistent dataset now covers excitation functions from thresholds up to 2.6 GeV for ca. 550 target-product combinations with about 15,000 cross sections.

Though nuclide production by α-particles is of minor importance due to their relatively low abundances in the cosmic radiation, we also investigated α-induced reactions on cosmophysically relevant target elements for energies up to 180 MeV and at 2.4

GeV. Residual radioactive nuclides were measured by γ-spectrometry and accelerator mass spectrometry; (12) and references therein. The resulting data for 485 reactions with 6368 cross sections provide a basis for an adequate consideration of α-particles when modeling cosmogenic nuclide production rates.

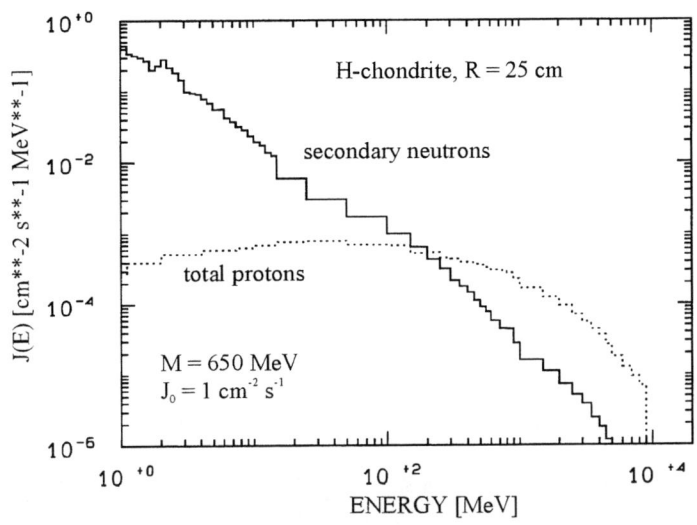

FIGURE 2. Spectra of total (= primary and secondary) GCR protons and of secondary neutrons in the center of a stony meteoroid with a radius of 25 cm an H-chondritic composition.

In spite of the fact that secondary neutrons are dominating the production of cosmogenic nuclides, there is an extreme lack of experimental neutron cross sections. Most available data are for energies equal to or below 14.7 MeV, just a minority of investigations went up to 30 MeV. Therefore, the only first hand sources of information on neutron cross sections at medium energies are by nuclear reaction models. This causes uncertainties of *a priori* GCR model calculations and calls for validations of the calculational methods and/or for other means to obtain information about the required neutron cross sections. Such means are for instance thick-target experiments by which the exposure of meteoroids to GCR particles is simulated as close as possible under completely controlled conditions.

THICK TARGET SIMULATION OF GCR INTERACTIONS

During recent years, we have performed six such experiments to simulate the exposure of meteoroids to galactic protons, (13-15) and references therein. In these experiments artificial stony and iron meteorites with radii between 5 cm and 25 cm were irradiated isotropically with 600 MeV and 2.6 GeV protons at CERN/Geneva and LNS/Saclay, respectively. The artificial meteoroids contained large numbers of individual small targets of up to 27 elements in which the depth-dependent production of radioactive and stable nuclides was measured by γ-spectrometry and accelerator and conventional mass spectrometry. In each experiment, the depth-dependent production of residual nuclides was measured for more than 500 target-product combinations. Thus, the elemental production rates of most cosmogenic nuclides from all relevant target elements were determined as function of depth and meteoroid size.

A theoretical analysis of the measured production rates was performed using depth- and size-dependent spectra of primary and secondary particles calculated by the HERMES code system (10) and experimental and theoretical thin-target cross sections of the underlying nuclear reactions. The results of all thick-target experiments were used to extract information about the excitation functions of the neutron-induced reactions by least-squares adjustment. Thereby, a consistent set of excitation functions of the underlying neutron-induced reactions was established (16). With the new neutron cross sections it is possible to describe simultaneously all data from the simulation experiments with an accuracy better than 10 % (16).

Though the spectra of secondary particles closely match those calculated for a real GCR exposure no monoenergetic accelerator experiment can completely simulate the GCR exposure in space with an accuracy required for the analysis of cosmogenic nuclides. The accuracy of cosmogenic nuclide measurements in meteorites is about 5 % for radionuclides, 3 % for absolute amounts and 1 % - 2 % for isotope ratios of rare gases. To obtain an adequate accuracy of the model calculations it is necessary to account also for the differences between the monoenergetic exposure in the simulation experiments and the reality in space. This can be done by performing the Monte Carlo calculations with realistic GCR spectra as in Fig. 2 and then calculating production rates with the experimental proton cross sections and with the neutron cross sections extracted from the thick-target accelerator experiments.

INTERPRETATION OF COSMOGENIC NUCLIDES

Cosmogenic nuclide production rates in meteoroids depend on size and bulk chemical composition of the meteoroid, on the shielding depth and the chemical composition of a sample in it, on spectral distribution, composition and intensity of the solar and galactic cosmic radiation and on the - possibly complex - exposure histories. Except for bulk and sample chemical compositions, all parameters are unknown and must be reconstructed. Thus, the goals of interpreting cosmogenic nuclide abundances in meteorites are to reconstruct the preatmospheric shapes of the meteoroids and the locations of the samples in them, to describe their exposure histories, and to draw conclusions about long-term spectral distributions and intensities of the cosmic radiation. All these goals can be met by our model.

Extending and improving earlier model calculations (7,14,17,18) of cosmogenic nuclide production in extraterrestrial matter, we now have established a physical model without free parameters for a consistent calculation of GCR production rates (5). The model takes explicitly into account proton- and neutron-induced reactions. Galactic ^{4}He-particles are only approximately considered. In addition, calculations of SCR production rates are done straight forward by folding the depth-dependent spectra of SCR primary particles with the experimental cross sections. The new model calculations cover the cosmogenic nuclides ^{10}Be, ^{14}C, ^{26}Al, ^{36}Cl, ^{41}Ca, ^{53}Mn, ^{129}I as well as He, Ne, Ar, Kr and Xe isotopes in stony and iron meteoroids as well as in lunar surface materials. They have an accuracy of better than 10 % for production rates and better than 3 % for production rate ratios. Exemplary, GCR production rate depth profiles of ^{10}Be and ^{26}Al in L-chondrites, which are a particular class of stony meteorites, are presented in Figure 3. Note that in meteoritics radionuclide production rates are given in disintegrations per min and kg [dpm/kg].

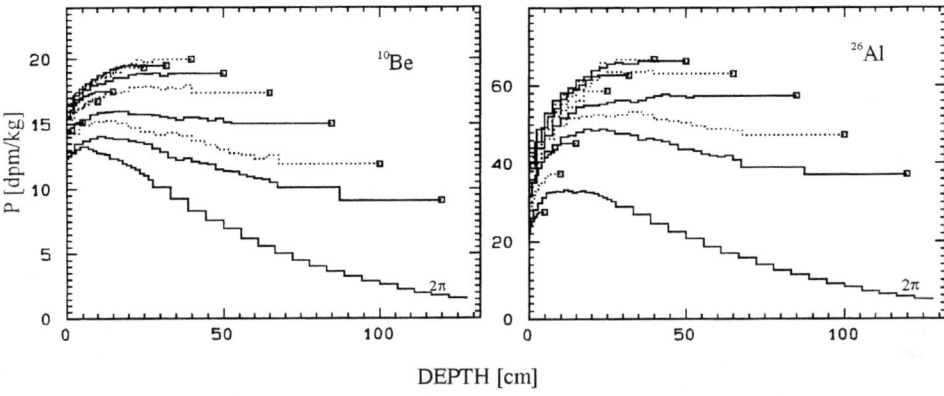

FIGURE 3. Production rates of ^{10}Be and ^{26}Al in L-chondrites as function of depth and size.

Since the multiplicities for the production of secondary particles in medium-energy reactions depend on the energies of the primary particles, the secondary particle field inside a meteoroid depends on the solar modulation of GCR spectra and consequently the production rate of a cosmogenic nuclides depends on the GCR modulation parameter M. For example, the production rates of ^{26}Al in the center of an L-chondrite with a radius of 40 cm vary by a factor of about 2.5 for modulation parameters between 300 MeV and 900 MeV. This dependence is more pronounced for „low-energy" products such as ^{26}Al than for „high-energy" ones such as ^{10}Be. For the latter, production rates differ just by a factor of about 1.5 under the same conditions. This dependence on solar modulation of production rates provides a tool to determine from observed cosmogenic radionuclide abundances the spectral GCR parameters averaged over typical time scales of three half-lives of the nuclide. As described elsewhere in detail (5), we deduced the average GCR spectrum during the last 10 Ma to have a modulation parameter M = 620 MeV (Figure 1). Since model calculations with this spectrum reproduce all the depth profiles of ^{10}Be ($T_{1/2}$ = 1.6 Ma), ^{26}Al ($T_{1/2}$ = 716 ka), ^{36}Cl ($T_{1/2}$ = 300 ka), and ^{53}Mn ($T_{1/2}$ = 3.7 Ma) measured in meteorites and lunar drill cores we conclude that the GCR spectrum was constant within 10 % during the last 10 Ma. Figure 4 shows exemplarily experimental and theoretical production depth profiles of ^{10}Be and ^{26}Al in the Apollo 15 drill core. An extension of time scales can be made by modeling the production of ^{129}I ($T_{1/2}$ = 15.7 Ma) and ^{40}K ($T_{1/2}$ = 1.28 Ga). For ^{129}I such work is underway (19).

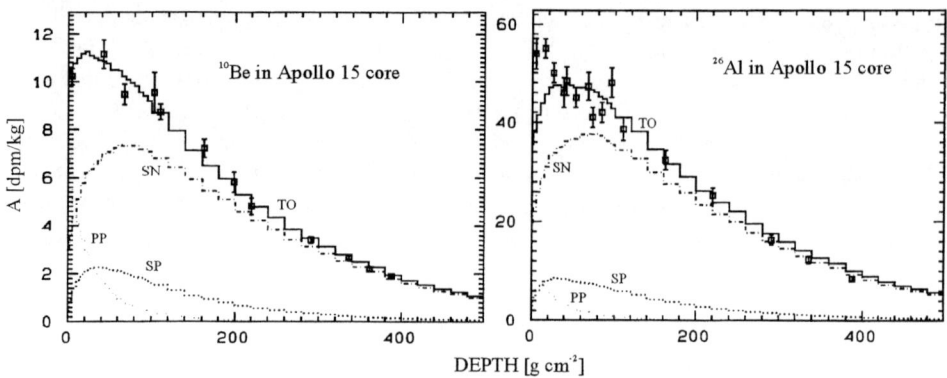

FIGURE 4. Production rate depth profiles of ^{26}Al and ^{10}Be in the Apollo 15 drill core. The experimental data are from Nishiizumi et al. (20,21). For the model calculations the total production (TO) as well as the production by primary protons (PP), by secondary protons (SP) and by secondary neutrons (SN) are distinguished.

The differences between theory and experiment seen for ^{26}Al in Figure 4 at shallow depths are due to SCR contributions in these samples which are not considered in the calculations. Generally, SCR production of cosmogenic nuclides is superimposed to GCR production. SCR production rates all show steep monotonous decreases from surfaces to larger depths. In order to derive long-term spectral parameters of SCR particles from cosmogenic nuclide depth profiles of lunar rocks or meteorites, the GCR production which is not accessible by direct measurements but only on by GCR model calculations has to be subtracted. Then, a comparison with the deduced „experimental" SCR production can be compared with SCR model calculations and thereby the long-term SCR spectral parameters can be determined. The quality of calculated SCR production rates depends exclusively on the quality of the experimental cross sections since the depth dependent SCR spectra can be calculated with much higher accuracy. In the past, lack of high-quality cross sections led to a number of widely disagreeing determinations of long-term SCR parameters (22-26). Moreover, uncertainties about erosion rates of extraterrestrial matter in space added to ambiguities of such determinations.

From recent investigations of ^{3}He, ^{21}Ne, ^{22}Ne, and ^{38}Ar in lunar rocks 68815 and 61016 Rao et al. (27) derived long-term SCR parameters of $J_{SCR,0}$ = 66 cm^{-2} s^{-1} and R_0 = 85 MeV, while an analysis of ^{26}Al and ^{53}Mn depth profiles in various lunar rocks by yielded $J_{SCR,0}$ = 55 cm^{-2} s^{-1} and R_0 = 125 MeV (5). A close-up of this discrepancy is given in Figure 5 where ^{10}Be and ^{26}Al activities of the small meteorite Salem are analyzed for both spectra. Salem had a preatmospheric radius of only 2.5 cm. While both SCR spectra under discussion yield total ^{10}Be production rates compatible with the experimental data within their uncertainties, the ^{26}Al data can only be explained by a long-term spectrum with $J_{SCR,0}$ = 55 cm^{-2} s^{-1} and R_0 = 125 MeV (5). The most likely explanation of the discrepancy to results from the rare gas analyses is that there are deficits in the mostly very old cross section measurements of these nuclides.

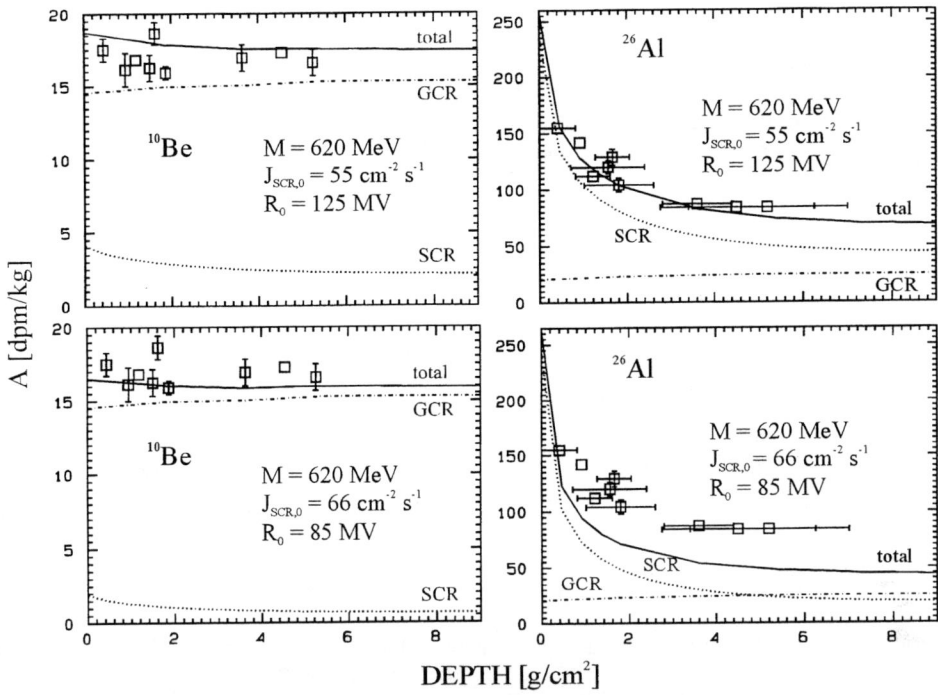

FIGURE 5. Experimental and calculated production rate depth profiles of ^{26}Al and ^{10}Be in the L-chondrite Salem. The experimental data are from Nishiizumi et al. (26).

In conclusion, systematic measurements of thin-target cross sections for the production of cosmogenic nuclides from their relevant target elements and of thick-target production rates obtained by accelerator experiments have provided a basis for a physical model of SCR and GCR production of cosmogenic nuclides which describes all aspects of spallation reactions in extraterrestrial matter and by which the cosmic ray record in meteorites and lunar surface materials can be deciphered.

Acknowledgment

This work was partially supported by the Deutsche Forschungsgemeinschaft/Bonn.

References

1. Goswami, J.N., McGuire, R.E., Reedy, R.C., Lal, D., Jha, R., *J. Geophys. Res.* **A93**, 7195-7205 (1988).

2. Alsmiller jr., R.G., Santoro, R.T., Barish, J., and Claiborne, H.C., ORNL-RSIC-35 (1972).
3. Castagnoli, G.C., and Lal, D., *Radiocarbon*, **22**, 133-159 (1980).
4. R.E. McGuire, R.E., and von Rosenvinge, T.T., *Adv. Space Res.* **4**, 117-125 (1984).
5. Michel, R., Borges, L., and Leya, I., *Nucl. Instr. Meth. Phys. Res.* **B113**, 434-444 (1996).
6. Vogt, S., Herzog, G.F., and Reedy, R.C., *Rev. Geophys.*, **28**, 253-275 (1990).
7. Michel, R., Dragovitsch, P., Cloth, P., Dagge, G., and Filges, D., *Meteoritics*, **26**, 221-242 (1991).
8. Michel, R., Brinkmann, G., and Stück, R., *Earth Planet. Sci. Letters*, **59**, 33-48 (1982), Erratum: *ibid*, **64**, 174 (1983).
9. Armstrong, T.W., and Chandler, K.C., *Nucl. Sci. Eng.*, **49**, 110 (1972).
10. Cloth, P., Filges, D., Neef, R.D., Sterzenbach, G., Reul, Ch., Armstrong, T.W., Colborn, B.L., Anders, B., and Brueckmann, H., Juel-2203 (1988).
11. Michel, R., Bodemann, R., Busemann, H., Daunke, R., Gloris, M., Klug, B., Krins, A., Lange, H.-J., Leya, I., Lüpke, M., Neumann, S., Reinhardt, H., Herpers, U., Schiekel, Th., Sudbrock, F., Holmqvist, B., Condé, H., Malmborg, P., Suter, M., Dittrich-Hannen, B., Kubik, P.W., and Synal, H.-A., *Nucl. Instr. Methods in Phys. Res.* **B129**, 153-193 (1997).
12. Lange, H.-J., Hahn, T., Michel, R., Schiekel, Th., Rösel, R., Herpers, U., Hofmann, H.-J., Dittrich-Hannen, B., Suter, M., and Wölfli, W., *Appl. Rad. Isotop.* **46**, 93-112 (1994).
13. Michel, R., Peiffer, F., Theis, S., Begemann, F., Weber, H., Signer, P., Wieler, R., Cloth, P., Dragovitsch, P., Filges, D., and Englert, P., *Nucl. Instr. Meth. Phys. Res.*, **B42**, 76-100 (1989).
14. Michel, R., Lange, H.-J., Lüpke, M., Herpers, U., Rösel, R., Suter, M., Dittrich-Hannen, B., Kubik, P.W., Filges, D., and Cloth, P., *Planetary and Space Science* **43**, 557-572 (1995).
15. Michel, R., Lüpke, M., Herpers, U., Filges, D., Dragovitsch, P., Wölfli, W., Dittrich, B., Hofmann, H.-J., *J. Radioanal. Nuclear Chem.* **169**, 13-25 (1993).
16. Leya, I., and Michel, R., „Determination of neutron cross sections for nuclide production at intermediate energies by deconvolution of thick-target production rates", in *Proc. of the Int. Conf. Nuclear Data for Science and Technology, May 19-24, 1997, Trieste, Italy, G. Reffo (ed.)*, 1997, in press.
17. Bhandari, N., Mathew, K.J., Rao, M.N., Herpers, U., Bremer, K., Vogt, S., Wölfli, W., Hofmann, H.-J., Michel, R., Bodemann, R., and Lange, H.-J., *Geochim. Cosmochim. Acta* **57**, 2361-2375 (1993).
18. Herpers, U., Vogt, S., Bremer, K., Hofmann, H.-J., Suter, M., Wölfli, W., Wieler, R., Lange, H.-J., and Michel, R., *Planetary and Space Science* **43**, 545-556 (1995).
19. Schnabel, Ch., Gartenmann, P., Lopez-Guitierrez, P.M., Dittrich-Hannen, B., Suter, M., Synal, H.-A., Leya, I., Gloris, M., Michel, R., Sudbrock, F., and Herpers, U., „Determination of proton-induced production cross sections and production rates of ^{129}I and ^{41}Ca", in *Proc. of the Int. Conf. Nuclear Data for Science and Technology, May 19-24, 1997, Trieste, Italy, G. Reffo (ed.)*, 1997, in press.
20. Nishiizumi, K., Regnier, S., and Marti, K., *Earth Planet. Sci.* **70**, 164-168 (1984).
21. Nishiizumi, K., Klein, J., Middleton, R., and Arnold, J.R., *Earth Planet. Sci.* **70**, 157-163 (1984).
22. Reedy, R.C., and Marti, K., „Solar-cosmic-ray fluxes during the last ten million years", in *The Sun in Time* (Sonett, C.P., Giampapa, M.S., Matthews, M.S., eds), The University of Arizona Press, Tucson, 1991, 260-287.
23. Finkel, R.C., Arnold, J.R., Imamura, M., Reedy, R.C., Fruchter, J.S., Loosli, H.H., Evans, J.C., and Delany, A.C., „Depth variation of cosmogenic nuclides in a lunar rock and lunar soil", in *Proc. 2nd Lunar Science Conference*, 1971, pp. 1773-1789.
24. Kohl, C.P., Murrell, M.T., Russ III, G.P., and Arnold, J.R., „Evidence for the constancy of the solar cosmic ray flux over the past ten million years: Mn-53 and Al-26 measurements", in *Proc. 9th Lunar Science Conference*, 1978, pp. 2299-2310.
25. Bhandari, N., Bhattacharya, S.K., and Padia, J.T., „Solar proton fluxes during the last million years", *Proc. 7th Lunar Science Conference*, 1976, pp. 513-523.
26. Nishiizumi, K., Nagai, N., Imamura, M., Honda, M., Kobayashi, K., Kubik, P.W., Sharma, P., Wieler, R., Signer, P., Goswami, J.N., Reedy, R.C., and Arnold, J.R., *Meteoritics* **25**, 392-393 (1990).
27. Rao, M.N., Garrison, D.H., Bogard, D.D., and Reedy, R.C., *Geochim. Cosmochim. Acta* **58**, 4231-4245 (1994).

NUCLEAR DATA FOR GAMMA RAY ASTRONOMY AND THE COSMOCHEMISTRY OF ISOTOPIC ANOMALIES

Alain Coc and Marie–Geneviève Porquet

Centre de Spectrométrie Nucléaire et de Spectrométrie de Masse,
IN2P3-CNRS, 91405 Orsay Campus, France

Abstract. The ^{26}Al nucleus is of prime interest both for gamma ray astronomy (observation of its 1.809 MeV line) and for the cosmochemistry of isotopic anomalies (anomalous ^{26}Mg abundances in some meteorites). It is thus important to improve the nuclear data involved in ^{26}Al nucleosynthesis. For instance, we discuss here the modes of ^{26}Al destruction and in particular its beta–decay through off-equilibrium thermal population of the lowest lying levels. Available nuclear data are used to improve the estimates for this mode of destruction.

INTRODUCTION

The discovery and the subsequent mapping of ^{26}Al in the interstellar medium through the detection of its 1809 keV γ-ray line by satellites (HEAO-3, COMPTEL-CGRO, and for the future: INTEGRAL) has increased the interest for ^{26}Al nucleosynthesis (see [1] for a review). In addition, the study of interstellar grains extracted from primitive meteorites has provided new data on correlated abundances, that may improve our understanding of ^{26}Al formation. (The ^{26}Mg overabundances are interpreted as the result of ^{26gs}Al decay.) The sites for ^{26}Al production include supernovae (SNII), Wolf-Rayet, AGB stars and novae. Consequently, the nuclear physics involved in ^{26}Al nucleosynthesis is of renewed importance. Thermonuclear reaction rates involved in ^{26}Al formation have been discussed recently in the context of explosive [2] and non-explosive [3] hydrogen burning and new data have just been published [4] on ^{26}Al destruction by neutron capture. Here we discuss its off-equilibrium destruction rate through the thermal population of excited levels. This was first studied by Ward and Fowler [5] who set the rules used in ^{26}Al nucleosynthesis calculations, stating that ^{26gs}Al and ^{26m}Al have to be considered either as separate isotopes below $T_9 \approx 0.4$ or as a single one above. This temperature is uncomfortably close to the peak temperature attained in novae outburst ($T_9 \lesssim 0.35$) or in AGB thermal pulses ($T_9 \lesssim 0.3$) for instance. Moreover, some of the nuclear data used by Ward and Fowler are estimates that are known to be valid only to within many orders of magnitude. Hence, we felt that it was time to reconsider the *onset* of thermal equilibrium in ^{26}Al. We study the available experimental data for the neighboring nuclei to obtain better estimates, or set limits on the branching ratios, and calculate the transition to equilibrium with these new values. We consider the four first levels as separate nuclides and all possible internal transitions between them together with their beta decays. The set of equations is solved numerically and the results are compared to the previous ones.

NUCLEAR DATA

The ^{26}Al ground state ($J^\pi=5^+$) has a mean life of 1.07 My and decays predominantly (97.3%) to an excited level at 1.809 MeV in ^{26}Mg. Its subsequent deexcitation is the source of the observed gammas. With its long lifetime, ^{26gs}Al is also the source for the above mentioned meteoritic anomalies. At $E_{cm}=0.228$ MeV above, lies an isomer ($J^\pi=0^+$, $\tau=9.15$ s) which decays to the ground state of ^{26}Mg, i.e. without emitting any gamma. At stellar temperature, the effective mean life of ^{26}Al is reduced through the population of excited levels, in particular ^{26m}Al as described by [5] (see Figure 1.L). Since equilibration between ^{26m}Al and ^{26gs}Al proceeds through $E_x \lesssim 1$ MeV [5] levels, the next two levels (labeled a and b) are also displayed in this figure. In Figure 1.R (R or L refers to right or left panel) is displayed the effective ^{26}Al lifetimes obtained by Ward and Fowler [5], (or Fuller, Fowler, and Newman [6], for $\log(\rho/\mu_e) \lesssim 5$) and Vogelaar [7]. They present large discrepancies in the $0.1 \leq T_9 \leq 0.4$ interval i.e. at the onset of equilibrium; this comes from different assumptions on to which level is responsible for the leak through beta decay. Outside this interval, all calculations agree. At the lowest temperatures, only the ground state is populated and hence the mean-life does not differ from its laboratory value. At higher temperatures, levels are in thermal equilibrium and, $\tau_{\beta^+}^{-1}(^{26}\text{Al}) = 9.94 \times 10^{-3} \exp(-2.646/T_9)$ in the case where ^{26m}Al is the only significantly populated level. At even higher temperatures, decays from other levels become important. For high electrons densities ($\log(\rho/\mu_e) \gtrsim 6$), their Fermi energy becomes large enough to open up more favored electron capture channels [6,8].

Here we concentrate on the onset of equilibrium at moderate densities. The time scale to reach equilibrium is long in ^{26}Al because of the large spin difference between the two first levels so that equilibrium is reached through intermediate states located above the isomer [5]. The gamma ray transitions that link these levels are represented in Figure 1.L. The thick or thin arrows correspond to those for which the transition probabilities (λ_{ij}) are known experimentally or, respectively, have to be estimated. They are labeled by their electric (EL) or magnetic (ML) multipolarity (L). No arrow links ^{26m}Al and ^{26gs}Al since a transition with such a high multipolarity ($M5$) is strongly inhibited ($\tau(m \to o) \sim 3 \times 10^7$ years), even though this rate is slightly enhanced by the photon bath, as described in [5]. These two levels are only connected indirectly through transitions via the higher lying levels and in particular the E_{cm}=0.417 and 1.058 MeV.

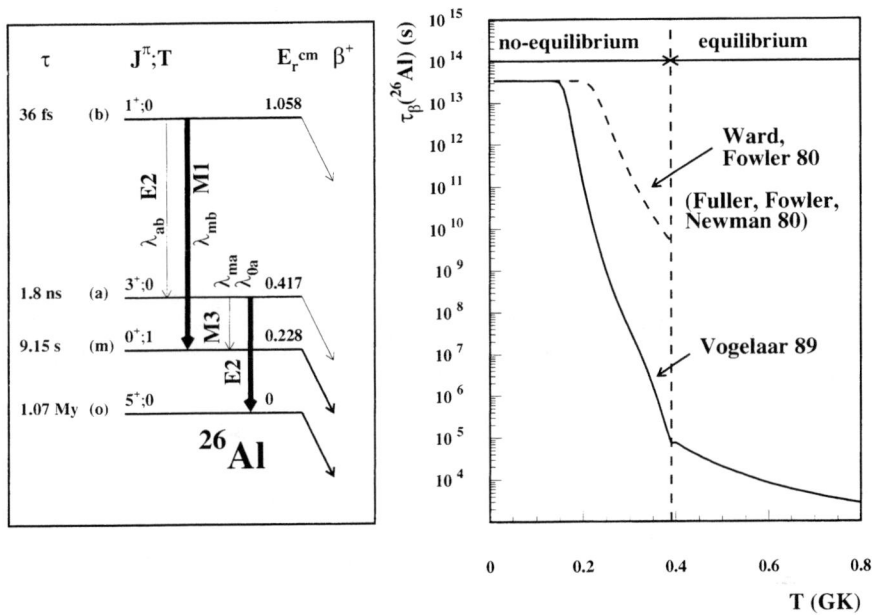

FIGURE 1. Level scheme for ^{26}Al and effective lifetime. The four first levels of ^{26}Al with their characteristics are represented in the left panel, together with the gamma ray transitions that link them. The right panel shows the various rates [5,7,6] for $\tau_{\beta+}(^{26}$Al) as a function of the temperature.

Due to the huge dynamics in gamma widths, or transition probabilities, it is customary in nuclear physics [9] to express their values in Weisskopf *units* (W.u. for EL) or Moszkowsi *units* (M.u. for ML). These *units* are single particle estimates, functions of L and E_γ. The experimental values, when measured in these units, are scattered within many ($\lesssim 6$) rather than much more orders of magnitude. To estimate the branching ratios $M3(a \to m)/E2(a \to o)$ and $E2(b \to a)/M1(b \to m)$ that govern the time scale for equilibration, Ward and

Fowler have assumed 1 M.u. (EL) or 1 W.u. (ML) for the four transitions shown in Figure 1.L. These branching ratios are too small to be measured, but one can gain information from the neighboring nuclei. Such level scheme, with a spin isomer (Figure 1.L), can be easily understood as it is a known feature of some odd–odd nuclei to possess low lying levels with large spin difference. The ^{26}Al odd–A neighbors, ^{25}Al and ^{25}Mg, have both a $J^\pi = \frac{5}{2}^+$ ground state and a $J^\pi = \frac{1}{2}^+$ first excited level at $E_x \approx 500$ keV. This can be readily interpreted in the shell model framework by the localization of the unpaired nucleon in the $d\frac{5}{2}$ or $s\frac{1}{2}$ shell. In ^{26}Al, the coupling of the two unpaired nucleons can account for the first levels. The ground state (o) and isomer (m) correspond to $\frac{5}{2}^+ \otimes \frac{5}{2}^+ \rightarrow J^\pi = 5^+, 0^+$ while $\frac{1}{2}^+ \otimes \frac{1}{2}^+$ leads to the 1^+ level (b) located at $E_x = 1.058$ MeV, as expected from ^{25}Al and ^{25}Mg spectra. The $E_x = 0.417$ MeV, $J^\pi = 3^+$ (a) can be interpreted as a $\frac{1}{2}^+ \otimes \frac{5}{2}^+$ coupling. Hence, all levels of interest can be simply interpreted in terms of two particle couplings. As shown in Figure 1.L, the main transitions are an $E2$, and a $M1$, linking the 3^+ level with the ground state and, the 1^+ level with the isomer, respectively. However, the possibility remains of a $M3$ transition between the 3^+ and isomeric levels and of an $E2$ between the 1^+ and ground state levels. Even though the branching ratios $M3/E2$ and $E2/M1$ are expected to be very small, they provide the link between the ground state and isomer but delay the onset of equilibrium. Following the above discussion, the $b \rightarrow a$ transition corresponds to $\frac{1}{2}^+ \otimes \frac{1}{2}^+ \rightarrow \frac{1}{2}^+ \otimes \frac{5}{2}^+$ i.e. a $\Delta J = 2$ single nucleon $E2$ transition from the $s\frac{1}{2}$ to the $d\frac{5}{2}$ shell. The other $E2$ transition ($a \rightarrow o$) can be viewed as $\frac{1}{2}^+ \otimes \frac{5}{2}^+ \rightarrow \frac{5}{2}^+ \otimes \frac{5}{2}^+$ i.e. the same single nucleon transition than the former. Hence, one can safely assume that the two $E2$ transitions have the same reduced transition probabilities and use the measured one ($a \rightarrow o$, 7.7 W.u. [10]). The $a \rightarrow m$, $M3$ transition corresponds to $\frac{1}{2}^+ \otimes \frac{5}{2}^+ \rightarrow \frac{5}{2}^+ \otimes \frac{5}{2}^+$ i.e. to a $\Delta J = 3$ nucleon transition between shells $s\frac{1}{2}$ and $d\frac{5}{2}$. No similar $M3$ transition is experimentally known in ^{26}Al or its neighbors. For instance, a $M3$ transition is found in the odd–odd nucleus ^{24}Na between its 4^+ ground state and 1^+ isomer. The ground states of ^{23}Ne and ^{25}Mg are interpreted as a neutron in the $d\frac{5}{2}$ shell (i.e. $\nu\frac{5}{2}^+$) while the ^{23}Na, ^{25}Na and ^{25}Na first excited state by the proton configurations $\pi\frac{3}{2}^+$, $\pi\frac{5}{2}^+$ and $\pi\frac{3}{2}^+$ respectively. Accordingly, the $J^\pi = 1^+$ and 4^+ ^{24}Na states are accounted for by different couplings of the same configuration: $\pi\frac{3}{2}^+ \otimes \nu\frac{5}{2}^+$. Such $M3$ transitions are also known in other odd–odd nuclei: ^{24}Al ($\pi\frac{5}{2}^+ \otimes \nu\frac{3}{2}^+$), ^{34}Cl ($\pi\frac{3}{2}^+ \otimes \nu\frac{3}{2}^+$) and ^{38}Cl ($\pi\frac{3}{2}^+ \otimes \nu\frac{7}{2}^-$). In the absence of data for $M3$ transitions between configurations similar to the ^{26}Al case, we estimated the possible range for the $a \rightarrow m$ gamma width from the whole statistics of experimental values. Figure 2.L represents an histogram of $M3$ experimental transition probabilities compiled by Roussière et al. [11]. The value used by Ward and Fowler [5] lies well outside the range spanned (7 orders of magnitude) by experimental values. (They have used for the $a \to m / a \to o$ branching ratio, 1 M.u./1 W.u.. Since experimentally $a \to o$ is found to correspond to 7.7 W.u., this is equivalent to assume 7.7 M.u. for the $a \to m$ transition.) On the basis of this histogram we adopt instead the range 10^{-3}–1. M.u. for the $M3$ reduced transition probability as shown on the figure. (The few points well outside this range concern heavy (A≈180), nuclei, or possibly mixed transitions $M3/E2$)

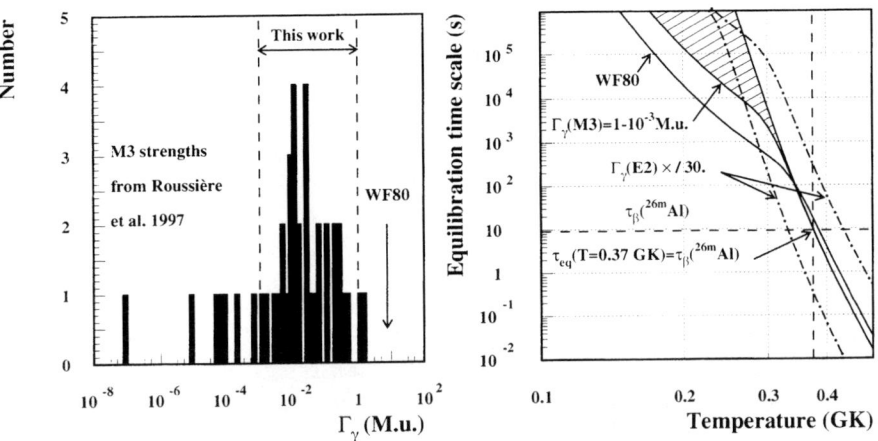

FIGURE 2. Range of variation of M3 transition probabilities (left) and time scale for equilibration (right). The left panel display the histogram of M3 gamma widths measured in Moszkowski's units (adapted from [11]). The right panel shows the calculated time scales for equilibration obtained with different gamma widths, compared to the ^{26m}Al mean life.

INTERNAL EQUILIBRATION OF ^{26}Al

Within stellar environment, the gamma transition probabilities are modified by the thermal photon gas and we use the method exposed in ref. [5]. Let i and j two levels such that $E_j > E_i$ and where the only mode of deexcitation for level j is a gamma transition to i. The evolution of the populations (N_j, N_i) of these levels is governed by the set of coupled equations [12]: $\frac{dN_j}{dt} = -\lambda_{ij}N_j + (-\lambda_{ij}N_j + \lambda_{ji}N_i)u(T) = -\frac{dN_i}{dt}$. The first term of the second member represents the spontaneous decay ($j \to i$), the following the stimulated ($j \to i$) and induced ($i \to j$) transitions and $u(T)$ is the photon density: $\left(\exp\left(\frac{E_j - E_i}{kT}\right) - 1\right)^{-1}$. The λ coefficients are readily obtained by considering the limits i) $T = 0$ where only spontaneous decays occurs, i.e. $\hbar\lambda_{ij} = \Gamma_{\gamma;j}$

and ii) thermal equilibrium ($\frac{dN_i}{dt} = 0$), that is : $\frac{\lambda_{ji}}{\lambda_{ij}} = \frac{2J_j+1}{2J_i+1} \exp\left(\frac{E_j - E_i}{kT}\right)$. To be more general, the evolution of the population of the various level ($i, j \in \{o, m, a, b, ..\}$) linked by all possible internal gamma transition can be written in matrix form as: $d\mathbf{N}/dt = -\mathbf{\Lambda N}$. The inverse of the non-zero eigenvalues of $\mathbf{\Lambda}$ give the time scales to achieve internal equilibrium. The longest of these time scales $\tau_{\text{eq}}(T)$ has to be compared with the time scale for beta decay. Then according to [5], ^{26gs}Al and ^{26m}Al can be considered at equilibrium when $\tau_{\text{eq}}(T) < \tau_\beta(^{26m}\text{Al})$. On the contrary, when $\tau_{\text{eq}}(T) > \tau_\beta(^{26m}\text{Al})$, ^{26gs}Al and ^{26m}Al have to be considered as separate species. To calculate the limit in temperature between these two processes, it is sufficient (see [5]) to consider the first four levels of Figure 1.L. Then as $\det(\mathbf{\Lambda}) = 0$ the eigenvalues are obtained numerically from the zeros of a third order polynomial. As a test, we first performed the calculation with the numerical values given in [5] and obtained τ_{eq} as a function of T as depicted by the solid line in Figure 2.R.. It intercepts the line $\tau_\beta(^{26m}\text{Al})$ at $T_9 \approx 0.39$ as found by Ward and Fowler. When the updated nuclear data of the previous section is used instead, the equilibration time remains uncertain (hatched area) below $T_9 = 0.3$ as it can be seen in Figure 2.R due to the $M3$ transition. On the contrary, in the region of interest where the curve reach the $\tau_\beta(^{26m}\text{Al})$ line at $T_9 \approx 0.37$, there is negligible uncertainty. Even if we allow the $E2$ ($b \rightarrow a$) transition probability to change by a factor of 30 (much larger than the estimated uncertainty) the temperature where equilibrium is set changes very little as depicted in Figure 2.R by the dashed curves.

EFFECTIVE ^{26}Al LIFETIME

The limit shown in Figure 2.R between the equilibrium and non–equilibrium regions can now be regarded as a precise one. It remains to see the effect of the updated nuclear data on the effective lifetime. The difference between the Ward and Fowler, and Vogelaar *off–equilibrium* formulas are explained by Vogelaar [7]. The formers assume that at the relevant temperature ^{26}Al decays to ^{26}Mg predominantly through the beta decay of the 3^+ (a) level while the latter favor the decay from the 0^+ (m) level. To clarify this situation, we calculate the effective ^{26}Al lifetime by numerical integration of the set of differential equations. The matrix elements for the beta decay to ^{26}Mg of the four levels considered (see Figure 1.L) are supplemented to the internal transition matrix $\mathbf{\Lambda}$ discussed above. The lifetime of the a and b levels with respect to beta decay are obtained from the $\log(ft)$ values calculated by Kajino et al. [8]. The numerical integration is performed with an implicit code for nucleosynthesis calculation where the *isotopes* are the four ^{26}Al levels plus ^{26}Mg and the *reactions* are the $\mathbf{\Lambda}$ matrix elements and beta decay rates. In these conditions, no assumption is made, at any stage, on the degree of equilibrium nor on the preferred beta decay channel. Initially, the ^{26gs}Al abundance is set

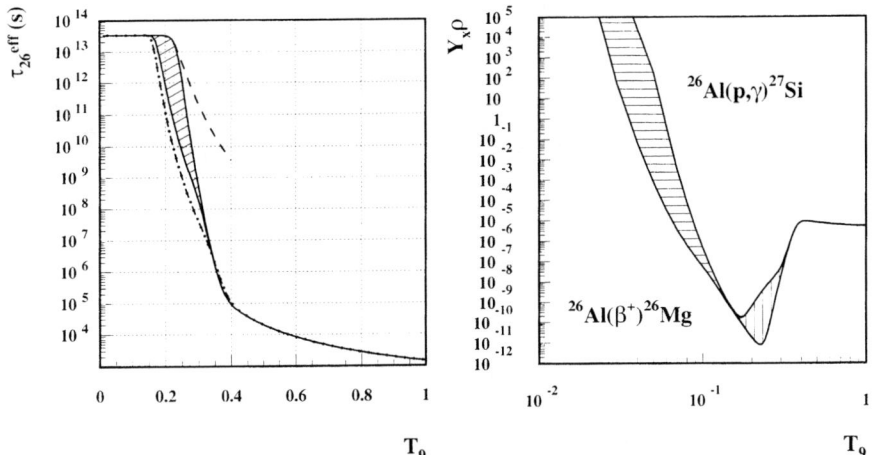

FIGURE 3. Effective lifetime and modes of destructions. The dashed area (left panel) represents the uncertainty on ^{26}Al effective life time compared with other calculations (see Figure 1). The right panel shows the limit between the ^{26}Al destruction through ^{26}Al$(p,\gamma)^{27}$Si or ^{26}Al$(\beta^+)^{26}$Mg. The dashed areas, horizontal and vertical, represent the corresponding rate uncertainties.

to 1, all the others to 0 and the calculation is stopped when the ^{26}Al abundance reach $1/e$. Again as a test we start with the input data of ref. [5,7] and obtain the (dotted) curve displayed in Figure 1.R but barely distinguishable from the (dashed) Vogelaar's [7] one (save for $T_9 \approx 0.4$). Above, $T_9 \approx 0.4$, as expected, the curve obtained by numerical integration merges with the one given by the formula for equilibrated ^{26}Al. The study of the flows confirm that, in these conditions, ^{26}Al decays via the 0^+ level. Using now the updated nuclear data, the effective life time is depicted in Figure 3.L where the dashed area represent the uncertainty induced by the unknown $M3$ gamma width. It reach up to several orders of magnitude between $T_9 = 0.2$ and 0.3. Only in the extreme case, 10^{-3} M.u., does the curve reach the Ward and Fowler one around $T_9 = 0.2$ and beta decay from the 3^+ level becomes significant as the transition to the 0^+ level is strongly inhibited.

CONCLUSIONS

One can conclude that the temperature which limits the domain where ^{26gs}Al and ^{26m}Al have to be considered separately is insensitive to the remaining nuclear uncertainties. We showed that even when equilibrium is not set, decay via ^{26m}Al is important above $T_9 \approx 0.2$ where there remains a rather large uncertainty which is unlikely to be improved experimentally due to the very small branching ratio $M3/E2$. Even though the discussed uncertainty lies well within the temperature range of nova nucleosynthesis, it has no consequence as destruction via proton capture remains much faster. In Figure 3.R is represented the uncertainties (hatched areas) on the limit between the $^{26}\text{Al}(\beta^+)^{26}\text{Mg}$ (this work) and $^{26}\text{Al}(p,\gamma)^{27}\text{Si}$ (ref. [2]) destruction modes in the $T \times Y_p \rho$ plane. In the absence of protons, destruction of by $^{26}\text{Al}(n,p)^{26}\text{Mg}$ and $^{26}\text{Al}(n,\alpha)^{23}\text{Na}$ may become important (see new experimental results in ref. [4]). Consequences of the uncertainties on $\tau_{\beta^+}(^{26gs}\text{Al})$ and on other rates in various astrophysical sites, potential sources of ^{26}Al, are however beyond the scope of this study.

ACKNOWLEDGMENTS

We are grateful to H.T. Duong for clarifying discussions and to B. Roussière for communicating us her compilation prior to publication.

REFERENCES

1. Prantzos, N., and Diehl, R. *Phys. Rep.* **267**, 1 (1996).
2. Coc, A., Mochkovitch, R., Oberto, Y., Thibaud, J.-P., and Vangioni-Flam, E., *Astron. Astrophys.* **299**, 479 (1995).
3. Arnould, M., Mowlavi, N., and Champagne, A., "Non-Explosive Hydrogen Burning: Where Do We Stand?", 32nd Liège Int. Astroph. Coll., Liège, Belgium July 3–5, 1995.
4. Koehler, P.E., Kavanagh, R.W., Vogelaar, R.B., Gledenov, Yu.M., and Popov, Yu.P., *Phys. Rev. C* **56**, 1138 (1997).
5. Ward, R.A., and Fowler, W.A., *Astrophys. J.* **238**, 266 (1980).
6. Fuller, G.M., Fowler, W.A., and Newman, M.J., *Astrophys. J. Sup.* **42**, 447 (1980).
7. Vogelaar, R.B., Ph. D. Thesis, California Institute of Technology (1989).
8. Kajino, T., Shiino, E., Toki, H., Brown, B.A., Wildenthal, B.H., *Nucl. Phys.* **A480**, 175 (1988).
9. e.g.: Lederer et al., *Table of the Isotopes*, John Wiley & Sons, 1978.
10. Endt, P.M., *Nucl. Phys.* **A521**, 1 (1990).
11. Roussière, B., et al., *private communication* and IPNO-DRE 97-05 preprint, Orsay, (1997).
12. e.g.: Cohen-Tannoudji, C., Dupont-Roc, J, and Grynberg, G, *Processus d'interaction entre photons et atomes*, Paris, InterEditions/Editions du CNRS, 1988, ch 4, p. 249,
    or: Kastler, A., *Optique*, Paris, Masson & Cie, 1965, ch 36, p. 949.

The Gamma-Ray Line Emission of Orion

Michel Cassé [1,2], Elisabeth Vangioni-Flam[1], and Sean T. Scully[1]

[1] Institut d'Astrophysique de Paris, 98bis Boulevard Arago, 75014 Paris, France

[2] Service d'Astrophysique, DSM, DAPNIA, CEA, France

Abstract. Observations of gamma-ray emission lines of C and O in the Orion star forming region indicate that shock waves produced by stellar winds and exploding stars may be accelerating these nuclei. This mechanism may have a significant impact on the production of Li, Be, and B through spallation reactions. The nature of the acceleration makes strong predictions which may be tested by the upcoming INTEGRAL mission.

INTRODUCTION

Recently, observations of the Orion complex by COMPTEL on board of the Gamma Ray observatory [1,2] have caused a revival of non-thermal gamma-ray line astronomy [3,4]. It has been further suggested that the gamma-ray line emission of this active star forming region is powered by stellar winds and supernovae. Indeed, the most massive stars explode close to their birthplace due to their short lifetimes. The progenitor stars, which likely have lost a substantial fraction of their initial mass through stellar winds [5], should explode as WC stars producing shock waves leading to a turbulent environment in which efficient acceleration of nuclei at moderate energy can take place [6,7].

ORION

The COMPTEL data show no sign of significant emission above 6.1 MeV in the Orion region, imposing stringent constraints on the maximum energy of the oxygen nuclei [8]. In the low energy regime considered most of the photons come from the de-excitation of nuclei close to the nuclear excitation threshold. In this case, fast nuclei, including for example ^{16}O and ^{12}C, are predominantly excited by collisions with the surrounding H and He (5 and

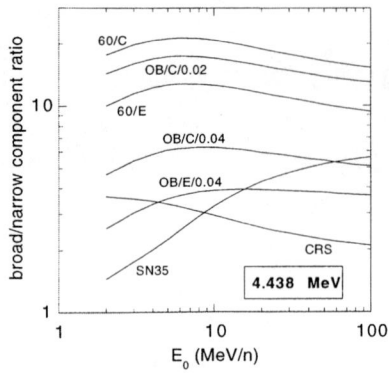

FIGURE 1. Broad/narrow component ratio of the carbon line for differing compositions of the beam(SN35: 35 $M_\odot$ SN, CRS:cosmic-ray source,OB:OB association with differing metallicities,60/C:single star of 60 $M_\odot$ with enhanced mass loss,60/E: same as 60/C but with normal mass loss).

7 MeV/n for C and O respectively). Orion is the prototype of an active star forming region, where indeed many supernovae have already exploded. Observations from radio lines, X-rays [1], and gamma-ray lines are clues that there have been stellar explosions within or in the close vicinity of the Orion clouds. In addition, the large scale structure in the interstellar medium around Orion bears traces of past supernova (SN) explosions [9,10] as witnessed by expanding motions and soft X-ray emission (the so called "Barnard's loop" and "Orion Cloath"). There is also clear observational evidence of ongoing interactions of the stars of Orion OB1 and the surrounding region. The Orion giant molecular cloud (GMC) itself lies in a superbubble evacuated by winds and explosions of massive stars. The physical conditions prevailing in these agitated low density media seem propitious to the generation of a non thermal population of particles. C and O nuclei injected at 2000 km s^{-1} could be accelerated to moderate energies (E < 100 meV) in the hot rarefied SN bubble [7,11].

The main sources of kinetic energy are stellar winds of massive stars and SNII explosions. Orion harbours 56 stars between B0 and O6, grouped in 4 subgroups (Ia,b,c,d). The ages and total energy output of the subgroups since its birth have been estimated recently by [12]. The more massive objects dominate the energy output. Thus the energy budget is affected by large statistical errors. The energy injection by SN and stellar winds (SW) in the past 10^7 yrs lies therefore between $5 - 17.5 \times 10^{51}$ ergs, which correspond to an average of $2 - 6 \times 10^{37}$ erg s^{-1}, with strong variations about the average due to WC winds (lasting a few 10^5 yrs) and SN explosions.

The gamma-ray luminosity however amounts to 3×10^{39} ph cm^{-2} s^{-1}, or $6L_\odot$ in gamma-rays which seems small but the energetic cost of production of

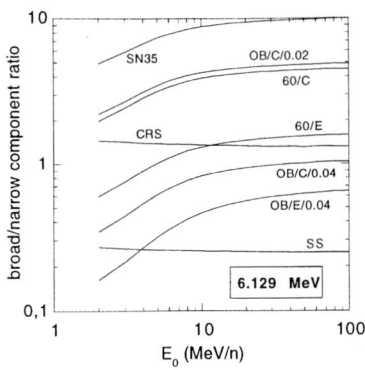

FIGURE 2. Same as figure 1 but for Oxygen. Additionally shown is SS:solar system composition.

gamma-ray lines is very high. Ten erg are required to produce a gamma-ray photon in the 3-7 MeV range in the most favourable case [13], [14]. Thus the minimum energy injection in the form of energetic nuclei (EN) is about 10^6 $L_\odot$ which is comparable to the whole IR emission measured by the COBE satellite [15]. The energy required to sustain the gamma-ray line emission (a few 10^{38} erg s^{-1}) is high compared to the total mechanical luminosity of the present winds (10^{37} erg s^{-1}). However the energy stocked in the superbubble is high enough to sustain the gamma-ray line emission for a period of more than 10^5 yrs [7].

Nuclei injected mainly by WC stars, accelerated and confined in the superbubble by shocks and turbulence, in turn interact with nuclei within the clouds. They suffer ionization losses and nuclear interactions producing gamma-ray lines and non thermal nucleosynthesis (see our accompanying review). The excitation of C and O in flight and the subsequent de-excitation followed by the emission of gamma-ray photons produces broad lines [14,16] due to Doppler broadening. Sharp lines are produced by the reverse reaction where C and O at rest in the ISM are excited by collisions with fast protons and α's.

The thin/broad line ratios are shown in figure 1 and figure 2 for an average wind composition and a spectrum of the form $n(E)dE = k\, E^{-1.5}e^{(-E/Eo)}$ [17]. The contribution of the various excitation reactions to the carbon and oxygen lines as a function of E_0 for a given composition (OB or solar metallicity) is shown in figures 3 and 4. Synthetic spectra have been made that would allow, by detailed comparison with data [16], the determination of the beam and spectrum composition. These predictions may be directly confirmed by the upcoming INTEGRAL European satellite, to be launched early next century ([18] and references therein).

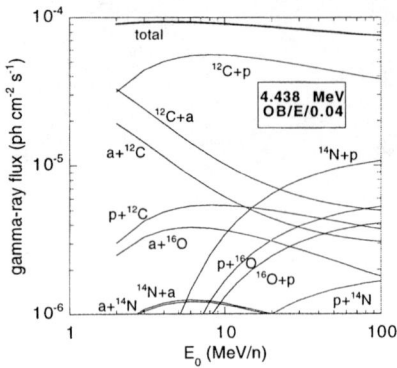

FIGURE 3. Contribution of the various reactions to the carbon line as a function of E_0.

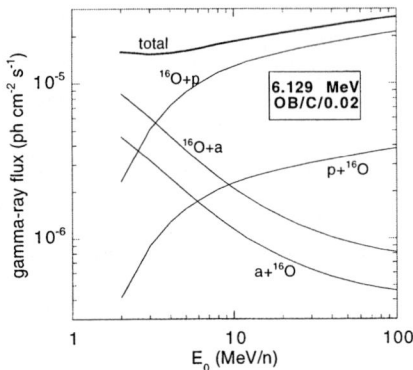

FIGURE 4. Contribution of the various reactions to the oxygen line as a function of E_0.

WC STARS IN ORION-ERIDANUS AND OTHER SUPERBUBBLES

A systematic study of gamma-ray lines produced by WC stars has been worked out in the framework of a model in which nuclei injected by WR stars and accelerated in the superbubble surrounding them, interact with the cloud from which they originate [17]. Differing initial masses, mass loss rates, $^{12}C(\alpha,\gamma)^{16}O$ reaction rates and convection/mixing criteria have been considered to appreciate the sensitivity of the model to the input parameters. Results have been compared to the COMPTEL observations of Orion to determine the main characteristics of the possible WC implied in the gamma-ray line emission of this region and to predict the intensity and the width of the accompanying lines. The main conclusions are the following: i) The 1-3/3-7 MeV band ratio (sensitive to the He/C+O abundance ratio) is consistent with the COMPTEL observation in all the cases analyzed. ii) The flux of the

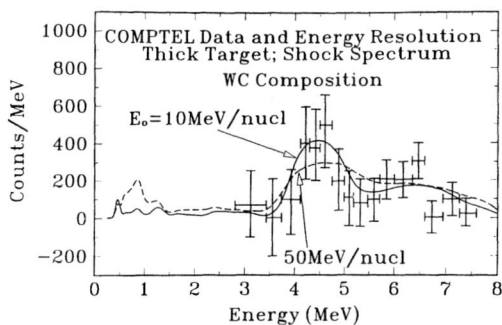

FIGURE 5. Observed[1] and calculated gamma-ray spectra for Orion for 2 values of the characteristic energy E_0 (from [14])

Li-Be feature due to the $\alpha + \alpha$ reaction, close to 500 keV coming from Orion is potentially observable by INTEGRAL. iii) The C/O line ratio is considerably enhanced compared to previous work and in better agreement with the COMPTEL observations (see figure 5).

Various source compositions have been invoked to explain the COMPTEL observation of an intense gamma-ray C and O line emission in Orion [13,14,19,20]. In all cases, the composition of the fast nuclei producing these lines was found to be C and O rich to explain the line intensity at 4.1 and 6.3 MeV, devoid of protons and helium poor to avoid undesirable line production in the 1-3 MeV range, and to minimize the energy budget [13,21]. On this basis, three compositions have been suggested by [16]: carbon rich Wolf-Rayet stars (WC), supernovae with massive progenitors (~ 60 $M_\odot$) and interstellar grains. The WC case has been analyzed in greater detail since these stars offer generous sources of kinetic energy. They also provide an injection site of fresh products of nucleosynthesis (namely helium burning) along with an acceleration mechanism through the shock wave induced by their supersonic wind. Injecting their own material in the shock region, hydrogen-less WC stars would be the sources of energetic nuclei strongly enriched in C, O and ^{22}Ne. Thus they represent a privileged way to get a beam strongly enriched into C and O and devoid of protons. On the basis of this argument WC stars have been invoked as the source of the ^{22}Ne-rich component of the Galactic cosmic radiation (GCR) [22–24]. Thus a link should exist between the Orion phenomenon and the richness in ^{22}Ne of the GCR.

WC stars could accelerate nuclei individually or they could participate to the collective energization of nuclei in a superbubble around OB associations. Indeed such a superbubble exists around Orion and one can assume that most of the low energy particles involved in the production of gamma-ray lines and LiBeB in Orion are confined in it. If this is the case, the severe power problem presented by the gamma-ray line emission of Orion (5×10^{38} ergs s^{-1} injected under the form of fast nuclei) is essentially solved. At present no WR stars are observed in Orion. They have either all exploded or faded to insignificance. It is however likely that WR stars have been active quite recently (i.e. within 10^4 to 10^5 yrs ago), together with supernova with massive progenitors ($> 40 M_\odot$) they should have filled the cavity with C and O rich fast nuclei. The adopted spectra should be truncated below 100 MeV/n to avoid overproduction of high energy gamma rays arising from π^0 disintegration and excessive broadening of the C and O lines. This truncation finds a natural explanation in the framework of the superbubble model due to the predominance of weak reflected shocks [7,11]. Neglecting selection effects associated with acceleration related to different mass/charge ratios of the ions [24], the composition of the shock accelerated energetic particles should reflect that of the stellar wind. Thus the source composition should be taken as the wind composition averaged from the ZAMS to the explosion. The mean wind composition depends quite sensitively on the mass of the WC progenitor and on still uncertain parameters such as the mass loss rates in the pre-WC and WC phases, on the prescription of convection and internal mixing, and finally, on the $^{12}C(\alpha,\gamma)^{16}O$ reaction rate.

Using the various models kindly made available by the Geneva group [5], we have evaluated the various line ratios produced by a single WC star. The particles are injected in the cloud medium with the prescribed source composition and a shock spectrum (truncated power law of $N(E)dE = k\ E^{-1.5} e^{(-E/Eo)}$. The interactions are calculated, as previously [13,14,19] in the thick target steady-state approximation. For calculating line ratios this approximation is sufficient. For other aspects of the problem a time dependent formalism is required [25]. The line intensities are obtained after normalization to the Orion gamma-ray line emission observed in the 1-3 MeV band. In general, high concentration of C and O go in the wind, but also of He. The "WC" composition of [13,14] taken as the surface abundances of the last model of [26] is considerably different from the average wind composition derived from [5]. The low helium abundances (He/O = 0.6) of [16] avoids undesirable production of lines by alpha excitation of heavy nuclei in the ISM (Ne-Fe). High C/O ratios result in all the WC models except those of [14]. This should influence the line profile in the 3-7 MeV range and could explain why the observed gamma-ray spectrum is C dominated.

WR OF SOLAR AND SUPER-SOLAR METALLICITIES AND THE EMISSION OF THE GALACTIC DISK

Nuclei propelled to MeV/n by winds and explosions of massive stars are further accelerated up tens MeV/n by ambient shocks and/or turbulence in the superbubble maintained by the high concentration of active objects grouped in OB associations. They propagate in the parent molecular cloud inside or at the border of this superbubble. In this dense medium they suffer dramatic energy losses, and occasionally fertile nuclear collisions with H and He nuclei, producing LiBeB and gamma-ray lines. At the low energies considered, ionization losses dominate other losses (escape of the system and nuclear destruction). Details of the model can be found in [17]. WR models at solar and super-solar metallicities ($Z = 0.02$ and 0.04, [5]) are of interest to study possible variation of the disk emission (C, O lines and, Li-Be feature) with Galactic longitude due to the increase of the metallicity toward the center of the Galaxy. Low metallicities ($Z < 10^{-3}$), on the other hand, are of interest for computing the early production of LiBeB in the Galaxy. The mean wind composition of stars with initial mass of $40 M_\odot$ and metallicity $Z = Z_\odot$ and $Z = 2 Z_\odot$ can be found in table 1 of [17]. The corresponding gamma ray line emissions are quite contrasted (figures 1 and 2), leading to interesting effects at the galactic scale. In the inner Galaxy and especially in the "4 kpc ring" where active star formation is taking place, Z is higher. The wind of WR is more intense with its mean composition enriched with helium and more $\alpha + \alpha$ reactions are induced, leading to enhanced Be and B production in their excited states. The $\alpha + {}^{16}O$ reaction, producing a thin O line at 6.1 MeV is also increased. Thus, we predict a modification of the gamma-ray line spectrum in the central radian of the Galaxy. This is of interest for the future INTEGRAL mission.

CONCLUSION

Gamma-ray lines arising from nuclear excitation and prompt de-excitation should serve as a diagnostic of the collective effects of stellar winds and supernovae in hot and dilute superbubble, whereas radio synchrotron emission and high energy gamma ray emission produced by π^0 decay would be more appropriate to study isolated SN interacting with the normal ISM. Besides its general interest to nucleosynthesis and galactic radioactivity, the INTEGRAL satellite, devoted to gamma-ray line astronomy, could open a new era in the understanding of the origin of fast nuclei and their interaction with galactic matter.

ACKNOWLEDGMENTS

We wish to thank E. Parizot for his collaboration. We are grateful to Roland Lehoucq for his numerical expertise. This work was supported by PICS n° 319, "Gamma-ray line astrophysics and associated nucleosynthesis", CNRS.

REFERENCES

1. Bloemen, H., et al. , 1994, A&A 281, L5.
2. Bloemen, H., et al. , 1997, ApJ 475, L25.
3. Meneguzzi, M., & Reeves, H., 1975, A&A. 40, 99,
4. Ramaty, R., Kozlovsky, B., & Lingenfelter, R.E., 1975, Space Sci. Rev. 18, 341.
5. Meynet, G., et al. , 1994, AAS, 103, 97.
6. Bykov, A., & Bloemen, H., 1994, A& A, 283, L1.
7. Parizot, E.M.G., 1997, A& A, in press.
8. Tatischeff, V., Cassé, M., Kiener, J., Thibaud, J.-P., & Vangioni-Flam, E., 1996, ApJ 472, 205.
9. Reynolds, S.,& Ogden, 1979, ApJ, 229, 942.
10. Cowie, L.L., Songaila, A., & York, D.G., 1979, ApJ 230, 469.
11. Bykov, A.M., 1995, Space Sci. Rev. 74, 397.
12. Brown, A.G.A., de Geus, E.J., & de Zeeuw, P.T., 1994, A&A 289, 101.
13. Ramaty, R., Kozlovsky, B., & Lingenfelter, R.E., 1995, ApJ. 438, L21.
14. Ramaty, R., Kozlovsky, B., & Lingenfelter, R.E., 1996, ApJ 456, 525.
15. Wall, W.F., et al. , 1996, ApJ, 456, 566.
16. Ramaty, R., Kozlovsky, B., & Lingenfelter, R.E., 1997, in "The Transparent Universe" ed. C. Winkler, T.J.-L. Courvoisier and Ph. Durouchoux, ESA Publications Division, p. 75.
17. Parizot, E.M.G., et al. , 1997, A& A, in press.
18. Winckler, C., 1997, in " The Transparent Universe, 2nd INTEGRAL Workshop"ed. C. Winkler, T.J.-L. Courvoisier and Ph. Durouchoux, ESA Publications Division, p. 573.
19. Cassé, M., Lehoucq R., & Vangioni–Flam, E. 1995, Nature, 373, 318.
20. Vangioni-Flam, E., Cassé, M., Olive, K., Fields, B., 1996, ApJ, 468, 199.
21. Ramaty, R., Kozlovsky, B., & Lingenfelter, R.E., 1995, in "Proceedings of the 17th Texas Symposium", New York Acad. Sci., p. 392.
22. Cassé, M., & Paul, J., 1982, ApJ, 258, 860.
23. Meyer, J.P., 1987, in "Origin and Evolution of the Elements," ed. G. Mathews, World Scientific, p. 310.
24. Ellison, D.C., Drury, L.O'C. & Meyer, J.-P., 1997, ApJ, in press.
25. Parizot, E.M.G., et al. , 1997, in " The Transparent Universe, 2nd INTEGRAL Workshop"ed. C. Winkler, T.J.-L. Courvoisier and Ph. Durouchoux, ESA Publications Division, p. 93.
26. Maeder, A.,& Meynet, G., 1987, A& A. 182, 243.

OTHER BASIC NUCLEAR DATA FOR ASTROPHYSICS

The ETFSI Mass Formula - Recent Developments

J. M. Pearson [a], R. C. Nayak [a],
F. Tondeur [b,c], A. Mamdouh [b], M. Rayet [b] and I. N. Borzov [b,d]

[a] *Département de Physique, Université de Montréal, Montréal (Québec), H3C 3J7 Canada*
[b] *Institut d'Astronomie et d'Astrophysique, Université Libre de Bruxelles*
[c] *Institut Supérieur Industriel de Bruxelles*
[d] *State Scientific Centre - Institute of Physics and Power Engineering, 249020 Obninsk, Russia*

Abstract

a) Using the force SkSC4, on which the ETFSI-1 mass formula is based, we have calculated the fission barriers of nearly 1400 nuclei, both in the known region of the nuclear chart and in the highly neutron-rich region relevant to the r-process. b) A new version of the ETFSI mass formula has been developed, the staggered-pairing feature of which leads to an improved fit to neutron-separation and beta-decay energies. The underlying force SkSC13 has the further advantages over SkSC4 of: (i) fitting well the position of the Gamow-Teller resonance, a necessary condition for the calculation of the beta-strength function involved in the r-process; (ii) avoiding the collapse of neutron matter, thereby making it suitable for calculating the equation of state in neutron stars and the cores of collapsing stars.

1. Introduction

We are involved in a program to develop a microscopic theory of nuclear systems applicable to the wide variety of phenomena encountered at subnuclear and nuclear densities during and after stellar collapse, and in particular to describe all these phenomena, as far as possible, in terms of a single, universal, effective interaction. The main achievement so far has been the development, for the first time, of a mass formula based entirely on microscopic forces, the ETFSI-1 mass formula [1, 2, 3, 4, 5]. The astrophysical interest of such a mass formula lies in the fact that the r-process of nucleosynthesis depends crucially on, among other properties, the masses, or more precisely, on the neutron-separation energies S_n, and the beta-decay energies Q_β, of nuclei that are so neutron-rich that there is no hope of being able to measure them in the laboratory. It is thus of the greatest importance to be able to make reliable extrapolations of masses away from the known region, relatively close to the stability line, out towards the neutron-drip line.

A full elucidation of the r-process also requires a knowledge of the fission barriers of a large number of highly neutron-rich nuclei. It is fission, in the form of either neutron-induced fission or beta-delayed fission, that sets an upper limit on the mass number of nuclei that can be synthesized by the r-process. Beta-delayed fission can also intervene in the return to the stability line during freeze-out. A nucleus is energetically capable of undergoing neutron-induced fission provided the height B of its fission barrier satisfies

$$B < S_n \quad , \tag{1.1}$$

neglecting the thermal energy of the captured neutron. Likewise, the necessary, but not sufficient, energy condition for beta-delayed fission is

$$B < Q_\beta \quad , \tag{1.2}$$

where Q_β refers to the beta-decay energy of the precursor of the nucleus in question. Most modern mass formulas, including the ETFSI method, can be extended to the calculation of fission barriers.

The ETFSI method is essentially a high-speed approximation to the Hartree-Fock (HF) method, with a macroscopic part given by the extended Thomas-Fermi (ETF) method, and shell corrections calculated by the so-called Strutinsky-integral (SI) method [1, 5]. Pairing is handled in the BCS approximation with a δ-function force. Although this is really a microscopic-macroscopic mass formula, there is a much greater coherence between the two parts than is the case with mass formulas based on the drop(let) model, since the same Skyrme force underlies both parts. In fact, it has been shown [1, 2] that the ETFSI method is equivalent to the HF method in the sense that when the two methods fit the same *form* of Skyrme force to the mass data they give essentially the same extrapolation out to the neutron-drip line. This equivalence to the HF method presumably accounts for the fact that with just 8 parameters the underlying force of the ETFSI-1 mass formula, SkSC4, fits 1492 mass data for $A \geq 36$ with an rms error of only 0.736 MeV [4].

However, neither this empirical success nor the conformity to the Skyrme-HF method are sufficient to guarantee a reliable extrapolation out to the neutron-drip line. In particular, the fact that the two-body spin-orbit term of the conventional Skyrme force (see Eq. (3.1) below) has a single parameter, W_0, means that there is no degree of freedom available to represent the isospin dependence of the spin-orbit term in the single-particle field. This raises the question as to how reliable our extrapolations out to the neutron-drip line can be, particularly with regard to the behaviour of the gaps associated with magic numbers. However, we have introduced into the ETFSI mass formula two different generalizations of the Skyrme spin-orbit force, both of which contain more than one parameter, and in neither case was it possible to obtain an improved fit to the mass data. Moreover, equivalent fits with different parameter sets led to virtually identical extrapolations out to the neutron-drip line. These results suggest that the conventional single-parameter spin-orbit force might have, rather fortuitously, the correct isospin dependence built into it, at least up to the rather modest neutron excesses corresponding to the drip line. This was confirmed by showing that this force can reproduce satisfactorily the isospin dependence of the spin-orbit field generated in relativistic Hartree calculations on semi-infinite nuclear matter with different neutron excesses [6]. This conclusion enhances considerably our confidence in the reliability of the extrapolations made by the ETFSI mass formulas out to the neutron-drip line.

In Section 2 we extend the ETFSI-1 mass formula to the large-scale computation of fission barriers, calculating the barriers of nearly 1400 nuclei, both in the known region of the nuclear chart and in the highly neutron-rich region relevant to the r-process. In Section 3 we present a new version of the ETFSI mass formula, which responds to a number of limitations of the ETFSI-1 model.

2. Fission Barriers with SkSC4

The generalization of the ETFSI method to the calculation of fission barriers is straightforward, although computationally laborious. We impose the constraint of axial symmetry, but optimize the fission path with respect to the three parameters c, h and α, as defined, for example, in Ref. [7], the last referring to the left-right asymmetry. In this paper we give results only for force SkSC4, the underlying force of the ETFSI-1 mass formula; we stress that the parameters of this force have been determined entirely by the mass fit, and that no further adjustment has been made to fit measured barriers.

Our results for the nuclei whose barriers have been measured are shown in Table 1 (we take all data from the compilation of Smirenkin [8]. When there is more than one barrier, as is usually the case for $Z > = 88$, we quote only the highest. For the actinides ($Z > 88$) the agreement with experiment is quite satisfactory, but for lighter nuclei some serious discrepancies will be noticed. However, these problems arise only with barriers that are very high, and thus are less likely to be of relevance to the r-process, on account of the energy constraints of Eqns. (1.1-2). Nevertheless, it is desirable to have some understanding of the discrepancies, and we note that since the high barriers in question are invariably associated with very large deformation there is a suggestion that the dimensionality of our oscillator basis might be too low.

As for the barriers that we have computed in the neutron-rich region, full details will be published elsewhere, and we confine ourselves here to a few general remarks. Fig. 1 shows our results for the full range of uranium isotopes on the neutron-rich side of the stability line. Here it will be seen that whereas the highest barrier is the inner one for relatively low neutron number, it is the outer barrier that is

Table 1. Fission barriers (highest only) for force SkSC4; comparison with experiment.

Z	A	B_{cal}	B_{exp}	ϵ	Z	A	B_{cal}	B_{exp}	ϵ
81	200	25.1	22.2	2.9	92	236	5.7	5.6	0.1
	201	26.9	22.8	4.1		237	6.2	6.2	0.0
82	204	27.4	23.3	4.1		238	6.0	6.0	0.0
	205	28.6	24.4	4.2		239	6.4	6.3	0.1
	206	30.2	25.3	4.9		240	6.3	6.1	0.2
	207	31.6	26.9	4.7	93	233	4.9	5.0	-0.1
	208	33.0	27.4	5.6		234	5.4	5.5	-0.1
83	206	25.1	22.1	3.0		235	5.6	5.5	0.1
	207	26.4	22.4	4.0		236	6.2	5.8	0.4
	208	27.9	23.8	4.1		237	6.0	5.7	0.3
	209	29.2	24.1	5.1		238	6.6	6.0	0.6
	210	28.9	23.4	5.5		239	6.4	5.8	0.6
84	207	21.7	19.2	2.5	94	235	5.3	5.7	-0.4
	208	23.1	20.3	2.8		236	5.4	5.7	-0.3
	209	24.5	21.1	3.4		237	6.0	5.6	0.4
	210	25.8	21.3	4.5		238	5.9	5.9	0.0
	211	25.2	20.4	4.8		239	6.6	6.2	0.4
	212	24.9	19.5	5.4		240	6.3	5.8	0.5
85	212	21.7	19.6	2.1		241	6.7	6.2	0.5
	213	21.0	17.4	3.6		242	6.6	5.7	0.9
86	216	16.9	13.7	3.2		243	7.0	5.9	1.1
88	225	8.9	7.7	1.2		244	6.6	5.5	1.1
	226	9.0	8.5	0.5		245	6.9	5.5	1.4
	227	8.9	8.3	0.6		246	6.4	5.4	1.0
	228	8.8	7.8	1.0	95	239	6.2	6.3	-0.1
89	218	8.4	7.4	1.0		240	6.7	6.4	0.3
	219	7.3	7.4	-0.1		241	6.5	6.2	0.3
	220	6.7	7.4	-0.7		242	7.0	6.4	0.3
	221	5.7	7.3	-1.6		243	6.9	6.1	0.8
	222	5.3	7.3	-2.0		244	7.3	6.2	1.1
	226	6.9	8.3	-1.4		245	6.9	6.1	0.8
	227	7.1	7.9	-0.8		246	7.1	5.8	1.3
	228	7.1	7.1	0.0		247	6.7	5.7	1.0
90	220	6.7	6.8	-0.1	96	241	6.5	6.4	0.1
	221	5.4	6.7	-1.3		242	6.5	6.0	0.4
	222	5.1	6.7	-1.6		243	7.0	6.5	0.5
	223	4.5	6.9	-2.4		244	6.8	6.1	0.7

continued on next page

90	227	4.8	6.6	-1.8	96	245	7.3	6.3	1.0
	228	5.0	6.5	-1.5		246	6.8	6.0	0.8
	229	5.2	6.3	-1.1		247	6.9	6.1	0.8
	230	5.5	6.1	-0.6		248	6.4	5.9	0.5
	231	5.4	6.1	-0.7		249	6.3	5.7	0.6
	232	5.6	6.2	-0.6		250	5.9	5.4	0.5
	233	5.4	6.3	-0.9	97	244	7.4	6.6	0.8
	234	5.6	6.3	-0.7		245	7.3	6.4	0.9
91	230	4.6	5.4	-0.8		246	7.7	6.5	1.1
	231	4.6	5.7	-1.1		247	7.2	6.5	0.7
	232	5.0	6.1	-1.1		248	7.3	6.3	1.0
	233	5.1	6.0	-0.9		249	6.8	6.1	0.7
92	231	4.6	5.2	-0.6		250	6.7	6.1	0.6
	232	4.6	5.3	-0.7	98	250	6.2	5.6	0.6
	233	5.1	5.7	-0.6		251	6.2	6.2	0.0
	234	5.3	5.9	-0.6		252	5.9	5.3	0.6
	235	5.8	6.0	-0.2		253	5.9	5.4	0.5

Fig. 1

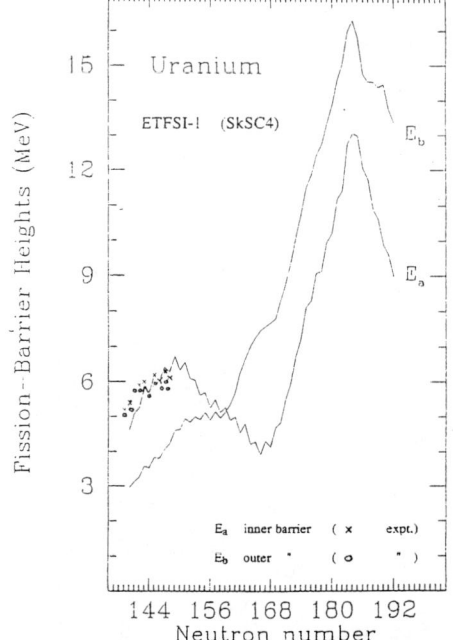

the highest for nuclei closer to the drip line. Experimental results are also shown in this figure for both barriers, and while the agreement for the inner barrier, which is always the highest for those nuclei for which data are available, is excellent, our computed outer barriers for the uranium isotopes are always about 2 MeV low compared to experiment. This is puzzling, since in the case of the $Z \leq 88$ nuclei large deformations are associated rather with too high values for the calculated barriers. Deformation is clearly not the only factor that determines the accuracy of our calculations, and it seems that we can either underestimate or overestimate an outer barrier, depending on whether or not there is a substantial inner barrier. We do not pursue these considerations here, beyond making the point that the measurement of second barriers, i.e., barriers that are not the highest, is subject to considerably more experimental uncertainty. Nevertheless, we are left with the possibility that the computed highest barriers for very neutron-rich uranium isotopes (E_b in Fig. 1) might be significantly in error, and that we cannot even be sure of the sign of this error.

Despite these uncertainties, it seems clear that with force SkSC4 very high barriers are inevitable for very neutron-rich uranium isotopes. This is the result of a strong magic neutron number at $N = 184$. However, the chances of being able to synthesize superheavy nuclei in the r-process are not good, since the lightest plausible candidate has mass number $A = 298$, which lies considerably beyond the neutron-drip line for uranium. Thus to form superheavies the r-process path would have to take us to significantly higher values of Z, and it turns out that the $N = 184$ barrier becomes much lower as Z increases: already by $Z = 96$ we find that this barrier height has fallen to 9 MeV. (This lowering of the $N = 184$ barrier with increasing Z simply reflects the fact that the gap at $N = 184$ is strongly Z-dependent, and becomes much smaller as the stability line is reached [9].)

But even if superheavies are not formed, the high barriers in very neutron-rich uranium must still have a considerable impact on r-process nucleosynthesis, in particular by reducing the role of beta-delayed fissions, which were generally expected (see e.g. [10]) to hinder the formation of the very heavy beta-stable nuclei that are the progenitors of the naturally-occurring actinides. However, there is some question as to whether these predictions of ETFSI are correct. The ETFSI method has been shown to be a very good approximation to the HF method when pairing is treated in the BCS approximation, but full HF-Bogolyubov calculations [11] show a strong quenching of the $N = 184$ gap far from stability. If these latter calculations are correct then the barriers in very neutron-rich uranium will be much lower than predicted by ETFSI.

Clearly, even though with the ETFSI method we have put the study of the r-process on a much more microscopic basis than was previously the case, we might still have not gone far enough.

3. The ETFSI-2 Mass Formula

A new mass table, ETFSI-2, has been constructed on the basis of a modified force, SkSC13, that in three ways constitutes an improvement on the force SkSC4 of ETFSI-1. i) The new force, SkSC13, fits the g_0' parameter of Landau, thereby ensuring a reliable calculation of the Gamow-Teller resonance in the beta-strength function, a quantity of vital importance in the r-process. ii) The unphysical collapse of neutron matter that was found with force SkSC4 is avoided. iii) The pairing properties are significantly improved.

To explain the changes we note first the form of the Skyrme force used:

$$v_{ij} = t_0(1 + x_0 P_\sigma)\delta(\mathbf{r}_{ij})$$
$$+ t_1(1 + x_1 P_\sigma)\frac{1}{2\hbar^2}\{p_{ij}^2 \delta(\mathbf{r}_{ij}) + h.c.\}$$
$$+ t_2(1 + x_2 P_\sigma)\frac{1}{\hbar^2}\mathbf{p}_{ij}\cdot\delta(\mathbf{r}_{ij})\mathbf{p}_{ij}$$
$$+ \frac{1}{6}t_3(1 + x_3 P_\sigma)\rho^\gamma \delta(\mathbf{r}_{ij})$$
$$+ \frac{i}{\hbar^2}W_0(\sigma_i + \sigma_j)\cdot\mathbf{p}_{ij} \times \delta(\mathbf{r}_{ij})\mathbf{p}_{ij} \quad , \tag{3.1}$$

where P_σ is the two-body spin-exchange operator. Also the (t_1, x_1) and (t_2, x_2) parameters are constrained through the relations

$$t_2 = -\frac{1}{3}t_1(5 + 4x_1) \quad , \qquad x_2 = -\frac{4 + 5x_1}{5 + 4x_1} \tag{3.2}$$

in order for the effective nucleon mass M^* to be equal to the real nucleon mass M, a condition that has been shown to improve the mass fit and the description of fission barriers (see [3]), besides simplifying enormously the ETF formalism. Fitting the Landau parameter g_0' simultaneously with the masses is possible only because the mass fit is rather insensitive to the x_2 parameter (see, however, the remark below). The value of g_0' to which we fitted was taken as 0.45.

As for the pairing force, we retain the δ-function form

$$v_{pair}(\mathbf{r}_{ij}) = V_\pi \delta(\mathbf{r}_{ij}) \quad , \tag{3.3}$$

handled in the BCS approximation (see Section 4 of Ref. [3] for a discussion of why this seems to work better than the Lipkin-Nogami method). In ETFSI-1 (force SkSC4) the pairing strength V_π takes the same value for both neutrons and protons, and in all nuclei. However, as discussed at length in Section 4 of Ref. [3], this simplified choice of pairing parameters results in a tendency to overestimate even-odd mass differences, with serious implications for both the S_n and Q_β, especially in the case of heavy nuclei. We could, of course, have reduced the errors in the even-odd mass differences simply by taking a weaker pairing force, and since these errors also contribute to the overall error of the mass fit it might be expected that the latter would improve at the same time. This is not so, since the pairing force not only generates even-odd fluctuations, but also contributes a much smoother (though shell-dependent) background term to the total energy. Thus optimizing the overall mass fit and optimizing the fit to the even-odd differences are in conflict if we take a unique strength for the pairing parameter. Our solution to this problem is to allow V_π to be slightly stronger for an odd number of neutrons ($V_{\pi n}^-$) than for an even number ($V_{\pi n}^+$), i.e., the pairing force between neutrons depends on whether N is even or odd; the proton pairing strength, on the other hand, remains unchanged from SkSC4.

We fit the force to the 1995 data compilation of Audi and Wapstra (AW) [12]. As with the ETFSI-1 fit [5], we limit ourselves to nuclei with $A \geq 36$, because of problems with the model for lighter nuclei. In the new compilation used here [12] there are 1772 such nuclei for which experimental masses are given.

Table 2. Errors in the data fit of ETFSI-1 (SkSC4) and ETFSI-2 (SkSC13). The first force is fitted to 1492 masses from the 1988 data of Ref. [13], and the second to 1772 masses from the 1995 data of Ref. [12]. All quantities are in MeV.

	SkSC4	SkSC13
$\sigma(M)$	0.736	0.741
$\epsilon(M)$	-0.0610	-0.0206
$\sigma(S_n)$	0.524	0.463
$\epsilon(S_n)$	0.0120	0.0137
$\sigma(Q_\beta)$	0.683	0.620
$\epsilon(Q_\beta)$	0.0465	0.0410

The parameters of our final fit SkSC13 are: $t_0 = -1789.48$, $t_2 = -105.828$, $t_3 = 12781.0$, $x_0 = 0.573361$, $x_2 = 0.826389$, $x_3 = 0.821481$, $W_0 = 140.310$, $\gamma = 0.333333$, $V_{\pi n}^+ = V_{\pi p}^+ = V_{\pi p}^- = -220$, and $V_{\pi n}^- = -224$. Table 2 gives the associated rms errors σ in the absolute masses, the S_n and the Q_β, respectively. Also shown are the corresponding mean errors ϵ. For comparison Table 2 also gives the fitting errors for ETFSI-1 (SkSC4). With the new force there is actually a slight deterioration in the quality of fit to the absolute masses, but this is because SkSC4 was fitted to a more restricted data set [13]; a preliminary version of the new force was fitted to the old data set, and showed a distinct improvement. In any case, we see that even with the larger data set the fit to the S_n and Q_β has improved significantly with the new force. This improvement is entirely due to the "staggered-pairing" feature described above, and in fact we have found that imposing the fit to $g'_0 = 0.45$ has had a slightly deleterious effect on the quality of the mass fit. However, this is a small price to pay for the excellent agreement that we find with the measured positions of the Gamow-Teller resonances (Table 3); this agreement confirms that our value of g'_0 was well chosen.

It should be remarked that the mass fits are insensitive to the x_2 parameter only if we retain in the energy density the terms that are quadratic in the spin-density current. We find that in the usual approximation of dropping these terms the mass fits become quite sensitive to the value of x_2, and in fact are unacceptably bad if this parameter is fitted to g'_0. The retention of the terms that are quadratic in the spin-density current thus becomes an essential feature of our parametrization.

Table 3. Q_β values and excitation energies of Gamow-Teller resonances calculated for SkSC13.

Nucleus	Q_β (MeV) calc.	Q_β (MeV) exp.	E_x (MeV) calc.	E_x (MeV) exp.
^{112}Sn	7.30	7.06	8.8	8.9
^{114}Sn	6.07	5.88	9.3	9.4
^{116}Sn	4.51	4.71	10.1	10.0
^{117}Sn	1.73	1.76	12.6	12.9
^{118}Sn	3.49	3.66	10.6	10.6
^{119}Sn	0.73	0.59	13.5	13.7
^{120}Sn	2.54	2.68	10.9	11.5
^{124}Sn	1.03	0.62	13.2	13.3
^{128}Te	1.35	1.25	13.3	13.1
^{130}Te	0.71	0.42	14.3	13.6
^{208}Pb	3.08	2.88	15.9	15.6

Imposing the constraint $g'_0 = 0.45$ on the mass fit has also the effect of shifting the optimal value of the nuclear-matter symmetry coefficient J from 27 to 28 MeV. As a result, it is found that neutron matter no longer collapses: see Fig. 2, in which the curve FP corresponds to the realistic nucleon-nucleon force v_{14} [14]. That is, by taking care of the problem of g'_0 we have automatically solved the problem of the unphysical collapse of neutron matter.

To summarize, the new force SkSC13 not only gives an improved fit to the S_n and Q_β, the mass-related quantities of crucial importance in the r-process, but also permits a more unified treatment of the r-process, because it can be used to calculate reliably the beta-strength function. We have not yet computed the fission barriers for this new force, but there is no reason to believe that they will be significantly different than those reported above for the old force SkSC4. Once these barriers are available, we will use them in a new calculation of the r-process, based entirely on the force SkSC13. Moreover, the use of this force will not be confined to the r-process, since its physically acceptable behaviour in neutron matter means that it can be used to determine the equation of state in neutron stars (up to nuclear densities) and in the cores of collapsing stars.

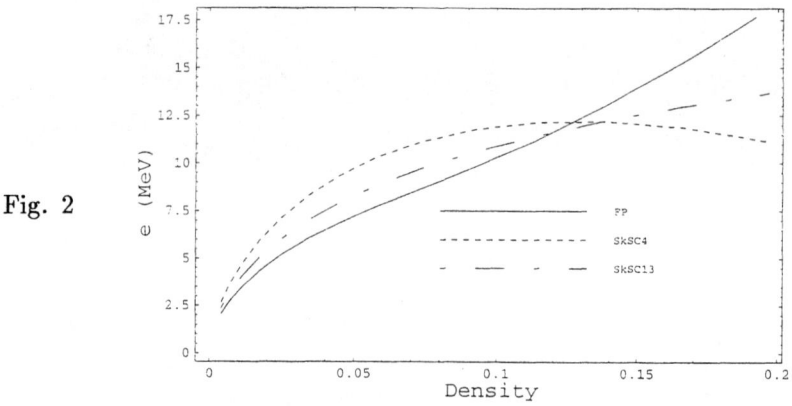

Fig. 2

References

[1] A. K. Dutta *et al*, Nucl. Phys. **A458** (1986) 77

[2] F. Tondeur *et al*, Nucl. Phys. **A470** (1987) 93

[3] J. M. Pearson *et al*, Nucl. Phys. **A528** (1991) 1

[4] Y. Aboussir *et al*, Nucl. Phys. **A549** (1992) 155

[5] Y. Aboussir *et al*, At. Data Nucl. Data Tables **61** (1995) 127

[6] M. Onsi *et al*, Phys. Rev. C **55** (1997) 3166

[7] M. Brack *et al*, Rev. Mod. Phys. **44** (1972) 320

[8] G. N. Smirenkin, Report 359, International Nuclear Data Committee, IAEA (Vienna 1993)

[9] J. M. Pearson *et al*, Phys. Lett. **B387** (1996) 455

[10] A. Staudt and H. V. Klapdor-Kleingrothaus, Nucl. Phys. **A549** (1992) 254

[11] J. Dobaczewski *et al*, Phys. Scr. **T56** (1995) 15; J. Dobaczewski *et al*, Phys. Rev. C **53** (1996) 1

[12] G. Audi and A. H. Wapstra, Nucl. Phys. **A595** (1995) 409

[13] A. H. Wapstra *et al*, At. Data Nucl. Data Tables **39** (1988) 281

[14] B. Friedman and V. R. Pandharipande, Nucl. Phys. **A361** (1981) 502

Beta–decay rates: towards a self-consistent approach

I.N.Borzov*, S.Goriely, J.M.Pearson[†]

*Institut d'Astronomie et d'Astrophysique, Université Libre de Bruxelles
CP 226 Bvd. du Triomphe 1050 Brussels – Belgium*

Abstract. An approximation to a self-consistent model of the ground state properties and spin-isospin excitations of neutron-rich nuclides is outlined. The structure of the Gamow-Teller strength functions in stable nuclei and short-lived nuclides undergoing high-energy β–decay is discussed. The results of large-scale calculations of the β–decay rates for spherical and slightly deformed nuclides of relevance to the r-process are analysed and compared with the results of existing global calculations.

1.INTRODUCTION

The β–decay rates of short-lived nuclides are at the present time among the most uncertain nuclear quantities in the modeling of the r-process nucleosynthesis. For some neutron-rich nuclides the β–decay rates have been measured with RIB facilities and high-flux reactors providing often the only valuable source of information on nuclear structure far from stability. However, given the lack of experimental data for the vast majority of the neutron-rich nuclei produced by the r-process, theoretical predictions of nuclear properties are of fundamental importance.

Since most of the nuclear models are fitted to experimental data along the valley of β-stability, sometimes only crude extrapolations to remote regions of the nuclear chart are available. Thus, among the different nuclear models used for practical purposes, those having the most reliable predictive power far from stability are to be preferred for the r-process modeling. In addition, for a reliable calculation of the r-process nucleosynthesis, it is well known that the different nuclear quantities, such as masses, deformations, β-decay rates, etc. should be estimated coherently within the same model.

[1)] * On leave from Institute of Physics & Power Engineering, 249020 Obninsk, Russia
[2)] † Laboratoire de Physique Nucléaire, Département de Physique, Université de Montréal, Montréal (Québec) H3C 3J7 Canada

The main characteristic of nuclear β–decay processes is the β–strength function. For the short-lived nuclides produced by the r- or rp-processes ($T_{1/2} <$ 1s), the approximation of allowed Gamow-Teller (GT) transitions (Δ L=0, Δ S=1, $\Delta\pi$ =no) is usually accurate enough. Experimental measurements of β–strength function show a typical resonance structure which is quite pronounced in spherical nuclides and more spread-out in deformed ones. The resonance shape is essentially due to single-particle transition effects. This gives grounds to the "gross theory" [1] (a statistical model corrected for a single-particle and pairing effects). The latest versions of the "gross theory" are widely used in astrophysical applications.

At the same time, realistic microscopic models like the proton-neutron quasi-particle random phase approximation (QRPA), show that not only the single-particle properties, but also the pairing and spin-isospin components of the effective NN-interactions play an important role. Various microscopic predictions of the β–decay properties [2,3] have been performed. In particular, the recent work [3] contains the ground state and β–decay characteristics obtained within the Finite-Range Droplet Model (FRDM) and simplified QRPA approach based on an empirical one body single-particle potentials, a simple pairing interaction and a separable A-dependent effective NN-interaction.

Despite these important efforts, the lack of consistent theoretical models still makes the predictions of the β-decay properties quite unreliable. The half-lives calculated from empirical models [1–3], show deviations among each other, but also with new experimental data near neutron-rich closed-shell nuclei (see NUBASE [4]). A common problem to all the non-self-consistent models is related to the empirical choice of a suitable mean field, for which the extrapolation to exotic nuclei has to be questioned.

A fully self-consistent Hartree-Fock-Bogoliubov (HFB) plus QRPA approach applicable to large-scale calculation of the ground state and β-decay properties far from stability is not feasible so far. However, as a practical step in this direction, it is possible to develop the QRPA on the ground state description given by self-consistent mean-field models [5] and especially by those used for large-scale calculations of nuclear masses [6,7]. The predictive power of the latter approaches can be significantly improved if the form and parameters of the corresponding density functional or effective force are fitted not only to the nuclear ground state properties near the line of stability, but also to the ones away from stability.

For astrophysical application, the simultaneous evaluation of many different nuclear properties for a great number of nuclides is required, hence we are forced to consider microscopicaly founded approximations to self-consistent methods. Much effort has been devoted to build up such a unified microscopic model based on the Extended Thomas-Fermi plus Strutinsky Integral (ETFSI) approximation to the HF method [6]. The present work is devoted to ETFSI based approximation for description of the spin-isospin excitations in very neutron-rich nuclei.

2. THE ETFSI+CQRPA MODEL OF β–DECAY PROPERTIES

For large-scale calculations of β-decay rates, existing fully self-consistent HF+RPA schemes based on the Skyrme effective interaction [8–10] are of limited practical use. For this reason we consider here the continuum QRPA (cQRPA) based on the ETFSI ground state description [6]. The main features of the resulting ETFSI+cQRPA method are:

1. The nuclear mean field is obtained by folding the Skyrme force over the smooth (fourth order) ETF nuclear density. The density-dependent two-body option of the Skyrme force with an effective mass $m^*/M=1$ [6] is used.

2. The quasi-particle basis is constructed by including a δ-pairing [12] on top of the HF solution (in practice, the pairing strength is taken to be 2% stronger for an odd neutron number than for an even one).

3. In order to control the description of the spin properties of the Skyrme force, its parameters can be constrained by the values of the dimensionless Landau–Migdal spin-isospin constant g'_0 or spin-spin constant g_0 [11].

4. Based on 1-3, the Skyrme force parameters are determined by optimizing the fit to known nuclear masses. The resulting force corresponds to $g'_0=0.45$, (normalized to $N_0^{-1} = 150.5$MeV fm^3), the highest value of the spin-isospin constant that can be used to constrain the Skyrme parameters without violating the spin stability of nuclear matter ($g_0 >-1$). (A higher value of g'_0 also makes the mass fit worse). At the same time, the nuclear-matter symmetry energy of the Skyrme force is chosen to be equal to J=28 MeV in order to avoid the unphysical collapse of neutron matter at lower J values. The resulting force (SkSC13) predicts nuclear masses with a root mean square deviation of 732 keV from the 1722 $A > 35$ masses of the 1995 Audi-Wapstra compilation (practically the same deviation as for the SkSC4 force [6], though a better description of the nuclear mass differences S_n and Q_β [13]).

5. The strength function of the charge-exchange excitations and the resulting β–decay rate is calculated within the spherical cQRPA with the exact account for the single-particle continuum in the particle-hole (ph) channel.

7. Only allowed transitions are included in the global calculations of the β–decay half-lives of short-lived nuclides.

8. The effective spin-isospin NN-interaction strength used in the cQRPA calculations corresponds to a renormalized Landau–Migdal constant g'_0 with respect to the one constrained by the ETFSI Skyrme parameters. We adopt for the spin-isospin interaction in the ph channel a renormalized one-pion exchange term F_π plus a contact δ–term with a Landau-Migdal constant g'_0=1.94 obtained by a fit to the experimental position of the GT resonance in ^{208}Pb. (Note, that in the present approach with the constraint on the $m^*/M=1$ it is impossible to obtain a spin-stable nuclear matter and to keep a good quality fit to the masses using the empirical value of the $g'=1.6$-1.8 ($F_\pi=0$) [11]). In

the *pp* channel a form similar to that of the pairing is used. More details can be found in [14,5] and refs. therein.

3. GAMOW-TELLER STRENGTH IN STABLE NUCLEI

Making use of the SkSC13 force to describe the ground state properties, we performed continuum QRPA calculations to determine some experimentally known energies of the giant GT resonances in order to check the derived value of g'_0. First, let us discuss the single-particle energies which are the basic ingredients in the calculation of the GT strength function. Experimental [19] singe-particle energies in ^{132}Sn and ^{208}Pb are compared in Table 1 with ETFSI predictions obtained with force SkSC4 ($g'_0 = 0.21$, J=27 MeV), SkSC13 ($g'_0 = 0.45$, J=28 MeV) and SkSC17 ($g'_0 = 0.21$, J=28 MeV).

For ^{208}Pb, all the theoretical energies are close to each other and to experimental ones (except for the $3s_{1/2}$ and $3p_{1/2}$ levels). For the unstable doubly-magic ^{132}Sn, larger deviations are observed between SkSC4 and two other forces indicating their different behaviour with increasing neutron excess which is mainly due to the difference in J. Only insignificant changes are produced by the additional constraint on the g'_0=0.45, as it can be seen from the comparison of the SkSC13 and SkSC17 columns of Table 1. It is important that in the specific case of the GT decay, the single-particle transition energies turn out to be quite close for all forces. The energies of main GT resonance component $\nu 1g_{9/2} \pi 1g_{7/2}$ are -6.73 and -5.74 for SkSC4 and SkSC13, respectively. The same conclusion is found for the high-energy GT β-decay component $\nu 1g_{7/2} \pi 1g_{9/2}$. Note that no special adjustment to the experimental single-particle spectra has been performed.

The GT resonance energies calculated for the chain of tin isotopes in a comparison with available experimental data from (^{3}He,t) reactions [15] can be found in Table 3 of Ref. [13] in which these energies have been calculated with a renormalized value of g'_0=1.94 (F_π included) and not with 0.45 as claimed by the authors. The overall good agreement with experiment show that the estimated value of g'_0 can be used for large scale calculations of the β-decay half-lives. Note that the underestimation of about 5% on the value of g'_0 cannot be rule out. The RPA approaches are known to overestimate the experimental energy of the giant resonances by about 1 MeV. Only the extended RPA approach including higher order (damping) effects predicts a spread of the GT resonance and the shift of its maximum by about 1 MeV downwards [18]. Thus, deriving the strength parameter g'_0 from the direct reproduction of the experimental position of the giant GT resonance within the 1st order QRPA might lead to some underestimate of its value.

Table 1: Single-particle levels (in MeV) in ^{132}Sn and ^{208}Pb calculated for Skyrme forces SkSC4, SkSC13 and SkSC17. Experimental data are taken from Ref. [19]

^{132}Sn

Level	ϵ_{exp}	ϵ_{SkSC4}	ϵ_{SkSC13}	ϵ_{SkSC17}	Level	ϵ_{exp}	ϵ_{SkSC4}	ϵ_{SkSC13}	ϵ_{SkSC17}
$\nu 1g_{9/2}$		-15.67	-15.12	-15.19	$\pi 2p_{3/2}$		-16.82	-17.24	-17.32
$\nu 2d_{5/2}$	-9.041(38)	-11.21	-10.68	-10.64	$\pi 2p_{1/2}$	-16.131(52)	-15.39	-15.82	-15.91
$\nu 1g_{7/2}$	-9.820(38)	-9.96	-9.54	-9.63	$\pi 1g_{9/2}$	-15.778(44)	-14.39	-14.76	-14.81
$\nu 3s_{1/2}$	-7.718(38)	-9.29	-8.81	-8.80			***		
$\nu 1h_{11/2}$	-7.628(38)	-7.87	-7.46	-7.43	$\pi 1g_{7/2}$	-9.653(43)	-8.94	-9.38	-9.28
$\nu 2d_{3/2}$	-7.386(38)	-8.52	-8.06	-8.42	$\pi 2d_{5/2}$	-8.691(43)	-9.18	-9.62	-9.66
		***			$\pi 1h_{11/2}$	-6.860(43)	-6.85	-7.22	-7.24
$\nu 2f_{7/2}$	-2.445(49)	-3.32	-2.97	-2.88	$\pi 2d_{3/2}$	-6.945(43)	-6.80	-7.23	-7.34
$\nu 3p_{3/2}$	-1.5913	-1.56	-1.31	-1.24	$\pi 3s_{1/2}$		-6.87	-7.32	-7.38
$\nu 1h_{9/2}$	-0.8841	0.11	0.33	-0.21	$\pi 2f_{7/2}$		-1.02	-1.47	-1.48
$\nu 3p_{1/2}$	(-0.75)	-0.87	-0.61	-0.56					
$\nu 2f_{5/2}$	-0.4404	-0.27	-0.04	0.08					

^{208}Pb

Level	ϵ_{exp}	ϵ_{SkSC4}	ϵ_{SkSC13}	ϵ_{SkSC17}	Level	ϵ_{exp}	ϵ_{SkSC4}	ϵ_{SkSC13}	ϵ_{SkSC17}
$\nu 1h_{9/2}$	-10.85	-11.03	-11.03	-11.09	$\pi 1g_{7/2}$	-11.25	-11.51	-11.79	-11.69
$\nu 2f_{7/2}$	-9.72	-11.26	-11.23	-11.16	$\pi 2d_{5/2}$	-9.89	-9.70	-9.82	-9.90
$\nu 1i_{13/2}$	-9.01	-9.29	-9.13	-9.31	$\pi 1h_{11/2}$	-9.37	-9.03	-8.94	-8.99
$\nu 3p_{3/2}$	-8.27	-8.99	-8.98	-8.92	$\pi 2d_{3/2}$	-8.38	-8.07	-8.05	-8.10
$\nu 1f_{5/2}$	-7.95	-8.33	-8.32	-8.39	$\pi 3s_{1/2}$	-8.03	-7.66	-7.00	-7.70
$\nu 3p_{1/2}$	-7.38	-7.96	-7.96	-8.92			***		
		***			$\pi 1h_{9/2}$	-3.80	-4.11	-4.07	-3.96
$\nu 2g_{9/2}$	-3.94	-4.17	-4.21	-4.10	$\pi 2f_{7/2}$	-2.91	-3.23	-3.19	-3.24
$\nu 1_{11/2}$	-3.16	-2.30	-2.37	-2.46	$\pi 1i_{13/2}$	-2.20	-2.42	-2.35	-2.38

4. β–DECAY RATES FOR SPHERICAL NUCLIDES NEAR THE CLOSED SHELLS

The major uncertainties in the QRPA prediction of β–decay half-lives arise for nuclides characterized by a low Q_β-value and/or a low β-decay transition energy ω. In the first case, the uncertainty is essentially related to the typical errors (0.5-1 MeV) in the prediction of the Q_β- and ω-values, as well as to the contribution of the first forbidden transitions. In the case of low-energy GT β-decay the high-order effects beyond QRPA cannot be neglected.

Experimental β-decay data on short-lived nuclides ($T_{1/2} < 1$s) are still very scarce. However, some experimental information, in particular in the vicinity of the doubly magic ^{132}Sn nucleus, are of particular interest for microscopic studies. Recent experimental studies provided us with single-particle energies at the "magic cross" at ^{132}Sn [19] and β-decay half-lives in this region [4].

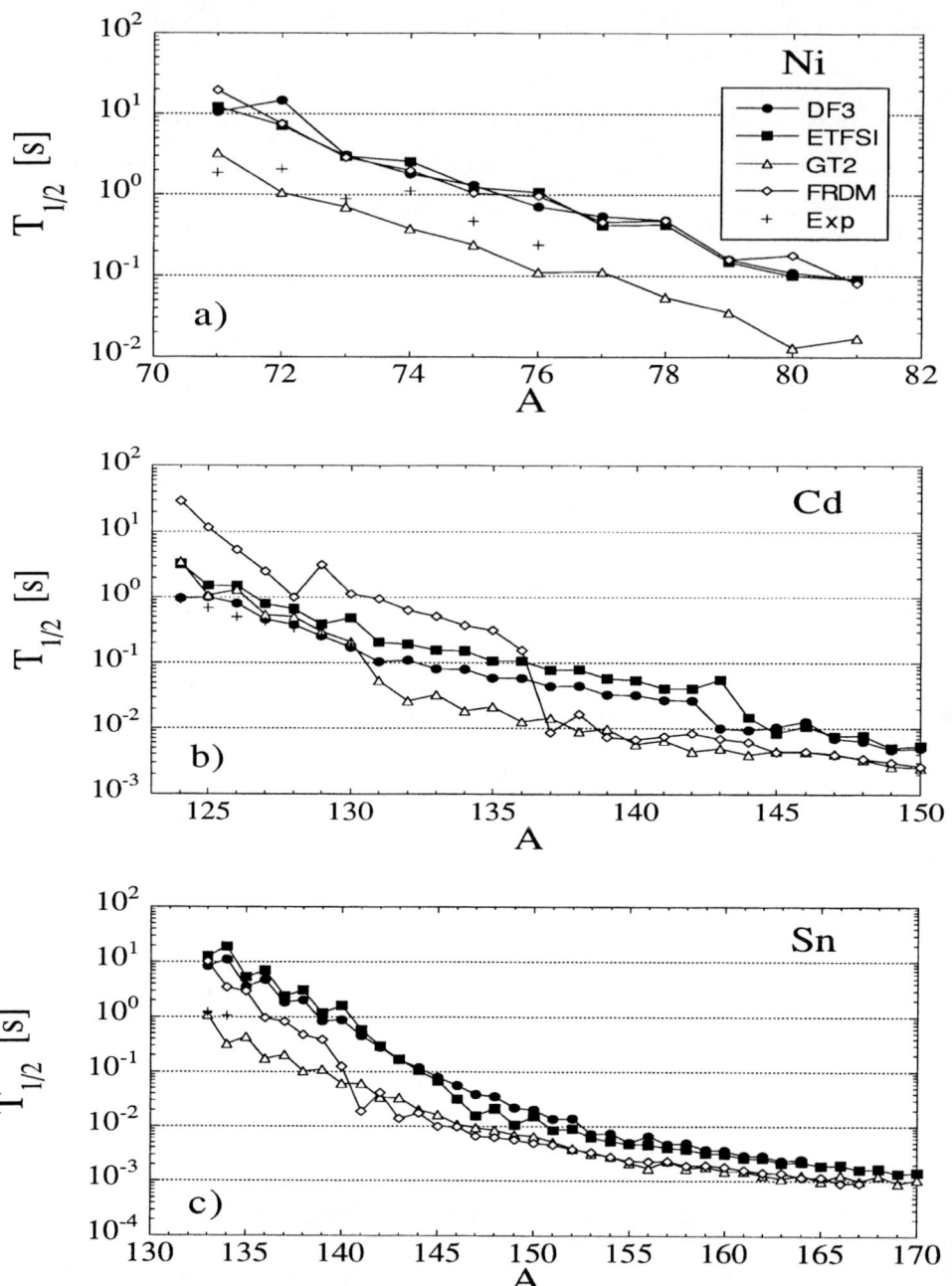

FIGURE 1. Comparison for Ni, Cd and Sn isotopic chains of the experimental half-lives with the ETFSI+cQRPA, GT2 [1], FRDM+QRPA [3] and DF3 [5] predictions.

Most of these nuclei have a spherical shape in the ground state, hence their β-decay half-lifes are sensitive to nuclear structure effects. Moreover, the nuclei with $Z<50$ in the ^{132}Sn region are predicted to undergo a high-energy Gamow-Teller β-decay ($\omega \approx Q_\beta$). In these conditions, the QRPA should give a reliable description of the β-decay half-lives. Therefore, a detailed study in this region of the nuclear chart provides a clear and unambiguous pattern both for application of the self-consistent QRPA and for a comparison of "exact" and approximate methods.

For spherical nuclides with a high-energy GT β–decay, our ETFSI+cQRPA calculations [20,21] shows a fairly good agreement with experimental β-decay half-lives as illustrated in Figs.1a,b which compares the half-lives for the Ni and Cd isotopic chains with other predictions [1,3] and available experimental data [4]. Of course, within the spherical approximation only nuclei with $\beta_2 \lesssim 0.1$ can be treated (all the Ni isotopes, $^{125-141}$Cd and for $^{133-145}$Sn in Fig.1). The half-lives calculated with self-consistent single-particle potentials [5,21] reveals a quite regular A-dependence which can be understood in terms of the monotonic A-dependence of the energies of those quasiparticle responsible for the GT-transitions.

Note that in the vicinity of the Z=28 closed shell, calculations predict the existence of a high-energy GT β-decay transition both for isotopes below and above the proton shell closure. On the contrary, for the Z>50 nuclides in the region of the double magic ^{132}Sn a switch to low-energy GT β–decay regime is observed by all QRPA approaches. This naturally results in an overestimate of the experimental half-lives as seen in Fig.1c. It can be explained by the fact that for low-energy GT β-decay, even a weak additional strength located at higher transition energies should have an impact on the total half-life because of the strong energy dependence of the phase space factor. An additional strength may come from possible contribution of forbidden transitions or effects beyond the QRPA (such as the spreading of the 1p1h strength over more dense neighbouring np-nh configurations or/and the appearance of an additional GT strength due to np-nh correlations in the ground state). For ^{133}Sn, the 1st forbidden non-unique transitions have been found to give only 6.2% of the total half-life, the contribution of the unique forbidden transitions estimated by [22] to be of 11.4%. To account for the spreading width ($\Gamma^\downarrow$), a semi-microscopic scheme [17] has been used including the Green functions averaging procedure and some elements of the optical model. The energy dependent quasi-particle damping is described in terms of a complex self-energy $\Sigma_{ph}(E_x)$ depending on the excitation energy E_x, spin S and isospin T transfered by the (ph) pair [18]. We have used a parametrization for the spreading width which is numericaly close to [18]. The line shape of the individual GT excitation has been described by a Lorentz function with $\Gamma^\downarrow = \alpha(Q_\beta - \omega)^2$, where $\alpha \approx 0.02$ roughly corresponds to the experimental width of the GT resonance in medium and heavy nuclei. Such a simple recipe allows us to improve the half-lives description significantly, especially for the nuclides with a low-

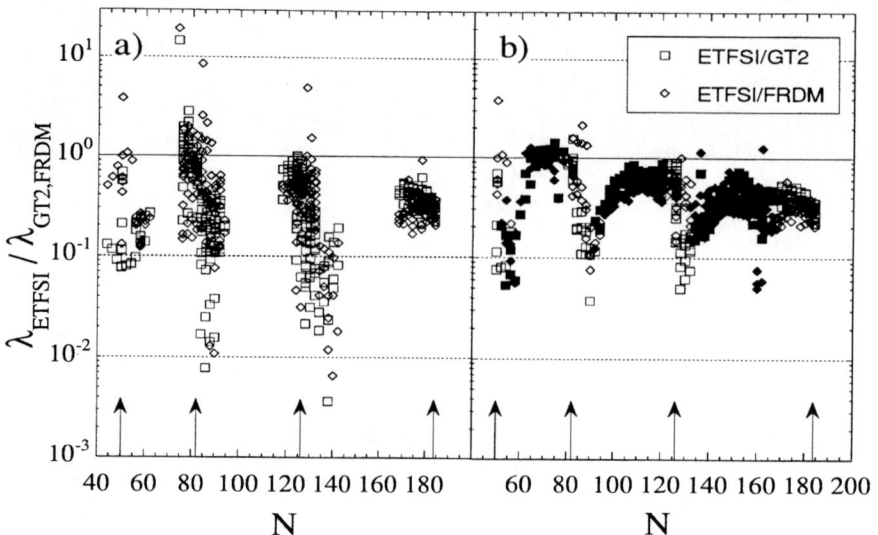

FIGURE 2. a) Ratios of the ETFSI+cQRPA to GT2 and to FRDM+QRPA β-rates for even-N spherical nuclei. b) same as a) for even-nuclei with $2 < S_n[\text{MeV}] < 3$. Full symbols correspond to deformed ($\beta_2 > 0.1$) nuclei.

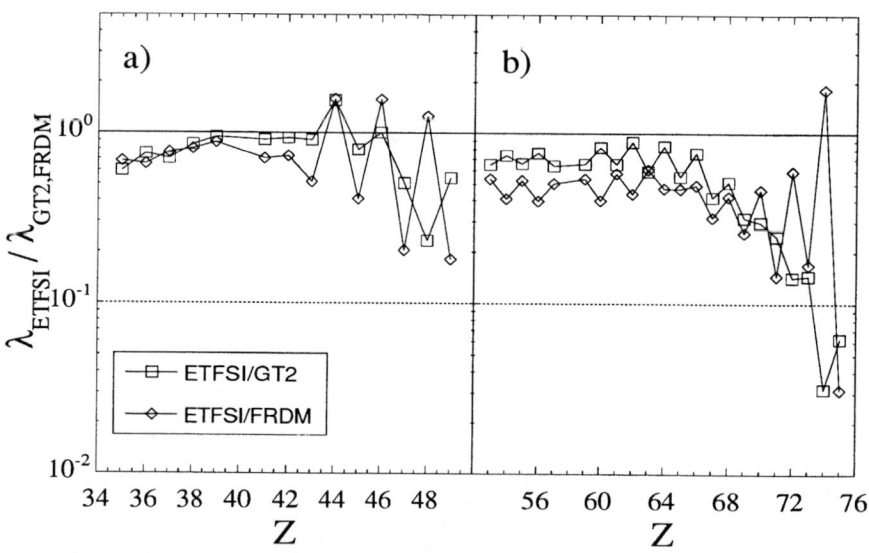

FIGURE 3. a) Ratios of the ETFSI+cQRPA to GT2 and to FRDM+QRPA β-rates for the neutron-rich $N=82$ isotones. b) same as a) for the $N = 126$ isotones.

energy GT β–decay in the Z=50-54 region near ^{132}Sn. It remains of high interest to elucidate the influence of higher-order effects on the β–strength function on more microscopic level.

Large-scale calculations of the β–decay rates of spherical and slightly deformed (β_2 <0.1, β_4 <0.02) short-lived nuclides ranging from Fe to U have been performed within ETFSI+cQRPA. Our predictions are compared in Fig. 2 with the phenomenological "gross-theory" (GT2) [1] and empirical FRDM+QRPA [3] (for clarity only even-N nuclei are included). Although the diffrent models give quite similar predictions of the very high-energy β–decays around N=184, large deviations between them can be observed near the other neutron shell closures (N=28, 50 and 126), especially between the ETFSI+cQRPA and the FRDM+QRPA models. These deviations are partially due to differences in the ground state description. The β–rates of nuclei of interest for the r-process nucleosynthesis, i.e even-N nuclei with 2 MeV< S_n <3 MeV, are compared in Fig. 2b. For completeness, the ETFSI β-decay rates (obtained in the spherical cQRPA) of deformed nuclei are also shown. The inclusion of deformation effects will reduce these half-lives.

For r-process calculations it is of high importance to predict correctly the β–decay rates of nuclides along the specific neutron closed shells. Figure 3 compares theoretical predictions of the β–decay rates for the N=82 and 126 neutron shells. The observed deviations with [1,3] are expected to lead to different predictions of the r-process timescales, and consequently of r-abundance distributions.

4. CONCLUSIONS

The calculated set of β-decay half-lives of spherical neutron-rich nuclei near exotic shell closures can be regarded as more coherent and reliable than the one derived by previous large-scale calculations [1–3]. It is based on a self-consistent

description of the ground state properties which reproduces masses and radii of experimentally known nuclei remarkably well, but also gives single-particle energies of relevance to high-energy GT decay in good agreement with available experimental data. A consistent account for deformation effects will allow us to complete a fully coherent table of β-decay half-lives, which would help to improve our understanding of the explosive nucleosynthesis.

Acknowledgments: The work was partially supported by CGRI, Belgium. S.G. is FNRS senior research assistant.

REFERENCES

1. K.Takahashi, M.Yamada, T.Kondoh ADNDT., **12** (1975) 101
   T.Tachibana,M.Yamada,N.Yoshida, Prog. Theor. Phys., **84** (1990) 641.

2. M. Hirsch, A. Staudt, H-V. Klapdor-Kleingrothaus., ADNDT **51** (1992) 244.
3. P.Möller,J.R.Nix,K.-L.Kratz., ADNDT **66**(1997) 131.
4. Audi G., Wapstra A.H., 1997, Nuclear Physics A., (1997) (to be published).
5. I.N.Borzov,S.A.Fayans,E.Kromer,D.Zawischa, Zeit.Phys., **A335** (1996) 127.
6. Y.Aboussir,J.M.Pearson,A.K.Dutta,F.Tondeur., ADNDT **61** (1995) 127.
7. J.Dobaczewski, W.Nazarewicz, T.R.Werner Phys. Scr., **T56** (1995) 15.
8. S.Krewald,, V.Klempt, J.Speth, A.Faessler, Nucl.Phys., **A281** (1977) 166.
9. Nguen Van Giai, H.Sagawa, Phys.Lett., **106B** (1981) 379.
10. K-F.Liu, H-D.Luo, Z.Ma, M.Feng, Nucl.Phys., **A534** (1991) 48.
11. A.B.Migdal. "Theory of finite Fermi systems and atomic nuclei properties" Interscience, N.-Y.,1983 [Russ.original 2nd ed., Nauka, Moscow, 1983]
12. R.C.Nayak, J.M. Pearson, Phys. Rev., **C52** (1995) 2254.
13. J.M. Pearson, R.C.Nayak, A. Mamdouh, F.Tondeur, M.Rayet, I.Borzov., (this conference).
14. I.N.Borzov, S.A.Fayans, E.L.Trykov., Nucl.Phys.,**A584**, 335 (1995).
15. K.Pham, J. Janeke, D.Roberts et.al., Phys.Rev. **C51** ,(1995) 526.
16. P.J.Horen, C.D.Goodman, C.C.Foster et al., Phys.Rev.Lett. **B95** (1980) 27.
17. S.Adachi, N.Auerbach., Phys.Lett. **131B** (1983) 11.
18. S.Drozd, S.Nishizaki, S.Speth, J.Wambach., Phys.Repts. **197**,(1990) 1.
19. K.A. Mezilev, Y.N. Novikov, V.A. Popov, B. Fogelberg, L. Spanier., Phys. Scripta **T56** (1995) 227; ISOLDE COLLABORATION (1996) in press.
20. I.N.Borzov, S.Goriely, J.M.Pearson., Nucl.Phys. **A621** (1997) 307c.
21. I.N.Borzov, S.Goriely, J.M.Pearson, M.Arnould., to appear in Proc. Int.Conf."Nucl.Data in Science and Technology", (1997) Trieste,Italy.
22. M.Hirsh, A.Staudt, H.V.Klapdor-Kleingrothaus., Phys.Rev. **C54** (1996) 2972.

β-decay Rates
—— The semi-gross theory ——

Takahiro Tachibana, Hidehiko Nakata* and Masami Yamada

Advanced Research Institute for Science and Engineering, Waseda University
3-4-1 Okubo, Shinjuku-ku, Tokyo 169, Japan
*Optical Access Technology Laboratory Network Systems Laboratories, Fujitsu Laboratories Ltd.,
10-1, Morinosato-Wakamiya, Atugi 243-01, Japan

Abstract. The β-decay rate of a nucleus is one of the basic data not only in nuclear physics but also in nuclear astrophysics through the study of nucleosynthesis. This rate is obtained from the β-decay strength function, and the semi-gross theory is one of the theories to enable one to estimate the strength function. This theory has been obtained by refining the conventional gross theory to take into account some shell effects of the parent nucleus. In the semi-gross theory, the one-particle energy-levels are assumed to be discrete and non-uniform, and the one-particle strength function depends on the orbital and total angular momenta of the decaying nucleon. Some results, such as half-lives, which are proportional to the inverse of the β-decay rates, and the delayed neutron emission probabilities are given and compared with those from the conventional gross theory and those from a microscopic theory.

INTRODUCTION

In the usual V-A β-decay theory, the decay rate of the Ω–type β-transition, λ_Ω, is obtained from the following nuclear matrix elements. For example, in the case of the Gamow-Teller transition, it is written as,

$$\lambda_{GT} = \frac{m_e^5 c^4}{2\pi^3 \hbar^7} |g_A|^2 \sum_l \left| \langle \Psi_l, \sum_k \tau^\pm \sigma_k \Psi_0 \rangle \right|^2 f(E_0 - E_l) , \qquad (1)$$

where g_A is the coupling constant, the subscripts 0 and l respectively indicate the initial and the l-th final states, Ψ_0 and Ψ_l (E_0 and E_l) are their wave functions (energies), and f is the integrated Fermi function. In our gross treatment, the β-decay strength function of the Ω–type transition, $|M_\Omega(E)|^2$, is defined as

$$|M_\Omega(E)|^2 = \overline{|(\Psi_l, \Omega\Psi_0)|^2} \rho(E) \ . \tag{2}$$

Then Eq. (1) becomes

$$\lambda_{GT} \approx \frac{m_e^5 c^4}{2\pi^3 \hbar^7} |g_A|^2 3 \int_{-Q}^{0} |M_{GT}(E)|^2 f(-E) dE \ . \tag{3}$$

Here, Q represents the ground-state Q-value, and E the transition energy measured from the parent state. If we take the allowed transitions and first-forbidden transitions, the total decay rate is given as

$$\lambda = \lambda_F + \lambda_{GT} + \lambda_1^{(0)} + \lambda_1^{(1)} + \lambda_1^{(2)} \ , \tag{4}$$

where the subscripts and superscripts on the right hand side indicate Fermi, Gamow-Teller, and the first-forbidden transitions with rank $L=0$, 1, and 2 respectively. The β-decay half-life is obtained from the total decay rate as

$$T_{1/2} = \frac{\ln 2}{\lambda} \ . \tag{5}$$

In the region far from the β-stability line, the β-decay Q-value becomes larger than the neutron separation energy of the daughter nucleus, and it has a possibility of delayed neutron emission. The emission rate of the delayed (one-) neutron is given by

$$\lambda_n = \int_{-Q+S_n}^{0} S_\beta(E) f(-E) \frac{\Gamma_n}{\Gamma_n + \Gamma_\gamma} dE \ , \tag{6}$$

where Γ_n and Γ_γ represent the neutron and radiation widths, respectively, and the total β-strength function S_β is the sum of allowed-equivalent strengths of all transitions $|M_\Omega(E)|^2$. From Eqs. (4) and (6), the delayed neutron emission probability P_n is obtained as

$$P_n = \frac{\lambda_n}{\lambda} \ . \tag{7}$$

The estimation of $T_{1/2}$ and P_n for very neutron rich nuclei, whose experimental data do not exist, is useful for the study of nuclear structure and necessary for the calculation of r-process nucleosynthesis. We explain the semi-gross theory briefly in the next section, and give some results related to $T_{1/2}$ and P_n in the last section.

SEMI-GROSS THEORY

In the semi-gross theory (1), as in the gross theory (2,3), the allowed transitions and the first-forbidden transitions are considered, and it is assumed that the strength function $|M_\Omega(E)|^2$ is given as

$$|M_\Omega(E)|^2 = \int_{\varepsilon_{min}}^{\varepsilon_{max}} D_\Omega(E,\varepsilon) W(E,\varepsilon) \frac{dn_1}{d\varepsilon} d\varepsilon , \qquad (8)$$

where $D_\Omega(E,\varepsilon)$ is the one-particle strength function, $dn_1/d\varepsilon$ the one-particle energy distribution of decaying nucleon, and $W(E,\varepsilon)$ a weight function to take into account the Pauli exclusion principle. We give a schematic illustration of $D_\Omega(E,\varepsilon)$ and $dn_1/d\varepsilon$ for β^- – decay of odd-odd nucleus in Fig. 1.

In the 2nd generation of the gross theory (2, 3) [referred to as GT2 hereafter], $dn_1/d\varepsilon$ is a continuous function calculated by the Fermi gas model with an effective mass except for the nucleons lying near the Fermi surface where pairing gaps expressed

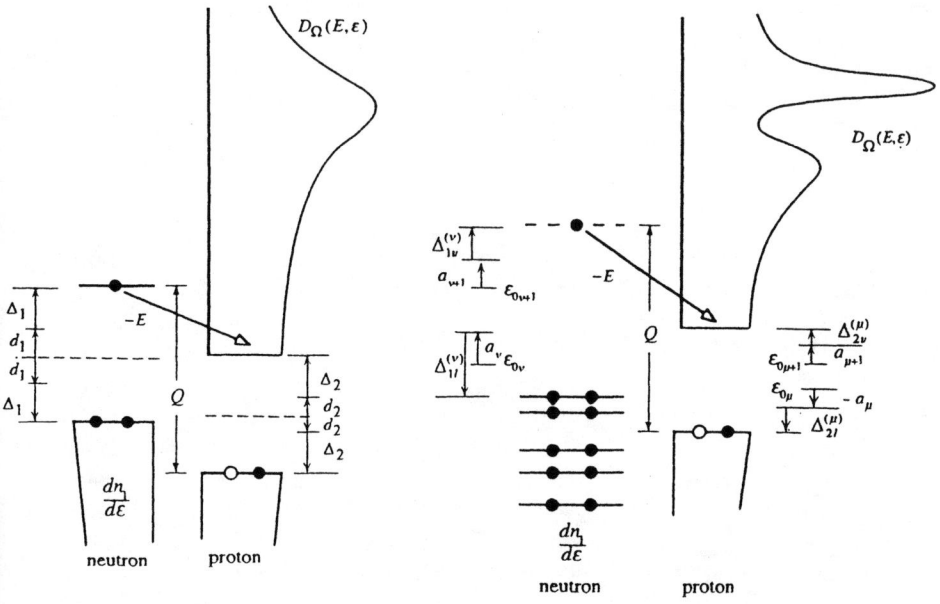

FIGURE 1. Schematic illustration of the one-particle energy distributions $dn_1/d\varepsilon$ and one-particle strength functions $D_\Omega(E,\varepsilon)$ for GT2 (the left hand side) and SGT (the right hand side). β^- – decay of odd-odd nucleus is taken as an example. The pairing gaps, which are sums of d_i and Δ_i (i=1,2), for GT2, and $\Delta_{1u}^{(v)}$ etc. for SGT are shown. The energy shifts a_v etc. due to the shell effect, and the standard energy levels ε_{0v} etc. are also shown for SGT. See Refs. (1-3) for more details.

by sums of d_i and Δ_i [i=1,2] (see Fig. 1). The function $D_\Omega(E,\varepsilon)$ is a superposition of two parts. For example, in the case of Gamow-Teller transition, one is a large peak near the isobaric analogue state [IAS] corresponding to the giant resonance of (p, n) reactions, and the other is a distribution spreading widely with long tails. The occupation and vacancy probabilities of the paired nucleon (the UV factors of the BCS paring theory) are introduced on an average. This GT2 is a certain average theory and almost independent of the characteristics of individual nuclei. The shell effect is taken into account only through the Q-values which are the input data of this model.

In the semi-gross theory [SGT] (1), however, $dn_1/d\varepsilon$ and $D_\Omega(E,\varepsilon)$ have been modified from those of the gross theory to take into account the shell effect of the parent nucleus. As for the daughter nucleus, the treatment is almost the same as that in the GT2, which is the reason for naming of the new theory.

To employ the extreme spherical single-particle model will be one of the easiest ways to obtain the one-particle energy distribution $dn_1/d\varepsilon$. However, this method is known to be inadequate for treating β– decay. For nonspherical nuclei, deformed potentials may be useful, but it will be laborious to find appropriate deformed potentials in the whole nuclidic region. For these reasons, we derive approximate one-particle energy distributions from experimental mass data (4). For the nucleus without mass measurement, we use TUYY mass formula (5). The practical prescription is explained in detail in Ref. (1). The calculated energy shifts a_v etc. due to the shell effect, and the pairing gaps $\Delta_{1u}^{(v)}$ etc. are schematically shown in Fig. 1. Discrete, non-uniform, and doubly degenerated energy levels are thus obtained. For each energy level, we assign a set (n, l, j) or a mixture of sets (n, l, j)'s with the use of a modified Woods-Saxon potential (6), where n, l, j are the principal quantum number, the orbital and total angular momenta of the nucleon, respectively. This mixing of quantum numbers is obtained from a rough consideration of the configuration mixing.

According to Refs. (1) and (3), the smooth function $F_\Omega(E)$ represents the main E-dependence of the one-particle strength function $D_\Omega(E,\varepsilon)$. In SGT, it is assumed that this function $F_\Omega(E)$ depends on the quantum numbers of the initial state of the decaying nucleon (n, l, j). If we take the Gamow-Teller transition as an example, the function $F_{GT}(E)$ is a superposition of two functions reflecting the effect of the change of the spin-orbit interaction caused by the spin flip. In the case of the transition from the state with $j = l + 1/2$ [$j = l - 1/2$], one function corresponds to the transition $j = l + 1/2 \to j = l + 1/2$ [$j = l - 1/2 \to j = l - 1/2$] with a peak at IAS, and the other function corresponds to $j = l + 1/2 \to j = l - 1/2$ [$j = l - 1/2 \to j = l + 1/2$] with a peak locating at a somewhat higher [lower] excitation energy than IAS. (See Fig. 1) Hence, the function $F_{GT}(E)$ is expressed as a sum of non-flip and flip terms

$$F_{GT}(E) = C_{GT}^{nfl} F_{GT}^{nfl}(E) + C_{GT}^{fl} F_{GT}^{fl}(E) \ , \qquad (9)$$

where, C_{GT}^{nfl} and C_{GT}^{fl} are the relative weights of these two terms.

For the first-forbidden transitions, the functions $F_\Omega(E)$'s also depend on (n, l, j) of the decaying nucleon. Their constitutions, however, are more complicated than that expressed by Eq. (9). This is because there exist vector and axial-vector type transitions, and each of them is assumed to be a superposition of several functions reflecting the effect of spin flip as well as of change of oscillator quanta $\pm\hbar\omega$. As for the Fermi transition, we use the same type $F_F(E)$ function as used in GT2, which has a single sharp peak at IAS and is independent of the quantum numbers of the decaying nucleons.

Finally, we fixed the parameter values included in SGT model; the widths and the peak positions of the functions $F_\Omega(E)$'s, the spin-orbit splitting, etc. Their values are determined with due consideration of the experimental strength functions, half-lives, giant resonance in (p, n) reactions, etc.

RESULTS AND DISCUSSIONS

In this section we show some results related to β-decay strength functions, delayed neutron emission probabilities and β-decay half-lives. In the following, necessary nuclear data for the calculations are taken mainly from Refs. (4) and (7).

Strength Function

It is difficult to make a detailed examination of the strength functions in a whole nuclidic region because of the existence of too many nuclei and poor experimental data particularly in the highly excited energy regions of daughter nuclei. From Eqs. (1)-(5), we notice that the inverse of half-life is proportional to the (appropriately weighted) strength function. Since GT2 gives an average property of β-decay, $1/T_{1/2}$ obtained from GT2 can be regarded as a standard to measure the strengths. In Fig. 2, we show the half-life ratios between GT2 and SGT, and GT2 and experimental data as a function of neutron number. The ratio GT2/EXP decreases around the

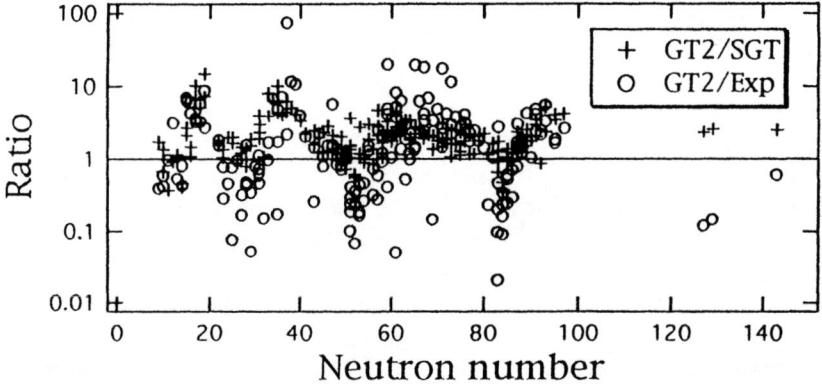

FIGURE 2. β^- – decay half-life ratios between GT2 and SGT (+), and between GT2 and experiments (o). These values serve as ratios of appropriately weighted strengths. Only the nuclei with $5 \leq Q \leq 10$ (MeV) are considered.

neutron magic numbers $N=28$, 50 and 82. This means GT2 does not take into account the shell effect correctly. The qualitatively tendency is similar in both ratios GT2/EXP and GT2/SGT although the amplitude of the oscillation of GT2/SGT is smaller than that of GT2/EXP. As a result, the shell effect seems to be included in SGT fairly well except for its magnitude. In Ref. (1), we examined similar ratios for all nuclei with experimental half-lives in a whole nuclidic region, and found that averages of these ratios over the whole available nuclei are,

$$\langle T_{1/2}(\text{GT2})/T_{1/2}(\text{EXP})\rangle = \begin{cases} 3.01 & \text{for } \beta^- - \text{decay} \\ 2.94 & \text{for } \beta^+ - \text{decay and electron capture,} \end{cases} \quad (10)$$

$$\langle T_{1/2}(\text{GT2})/T_{1/2}(\text{SGT})\rangle = \begin{cases} 2.61 & \text{for } \beta^- - \text{decay} \\ 2.66 & \text{for } \beta^+ - \text{decay and electron capture.} \end{cases} \quad (11)$$

For the strength functions in the giant resonance region, we give calculated Gamow-Teller strengths for ^{208}Pb (p, n) ^{208}Bi and ^{90}Zr (p, n) ^{90}Nb reactions in Ref. (1). They are in fair agreement with experimental data if we consider the ambiguity in the subtraction of the experimental background.

Delayed Neutron Emission Probability

The delayed neutron emission probabilities P_n, besides the half-lives given in the next subsection, are very important data in the study of r-process nucleosynthesis. In Fig. 3, we show the relative abundance of r-process elements calculated with and without the P_n-values obtained by using GT2 model. The result without P_n-values shows excessive even-odd staggering. In this calculation, the stellar model is a simple one (8), but even in a more realistic stellar model, such as a hot bubble model, the P_n-values are still indispensable.

The P_n-values calculated with the use of SGT are shown in Fig. 4 compared with the experimental data (9). In this calculation, we put the competition factor $\Gamma_n/(\Gamma_n + \Gamma_\gamma)$ in Eq. (6) unity for simplicity. The P_n-values of the microscopic calculation [QRPA] by Hirsch et al. (10), in which only allowed transitions are considered, are also shown in Fig. 4. In QRPA calculation, a simple but energy-dependent competition factor is used. We should note that the experimental P_n-values in Ref. (9) are limited to those in the fission product region ($28 < Z < 40$ with $45 < N < 64$, and $46 < Z < 58$ with $73 < N < 94$). As seen in Fig. 4, many P_n-values obtained from SGT underestimate the experimental values. Consequently, the strength functions estimated by SGT seem to be smaller than the actual strengths in the delayed neutron window. In contrast, many P_n-values calculated by QRPA overestimate the experimental data. These theoretical P_n-values calculated by either theory converge as the Q-value becomes large.

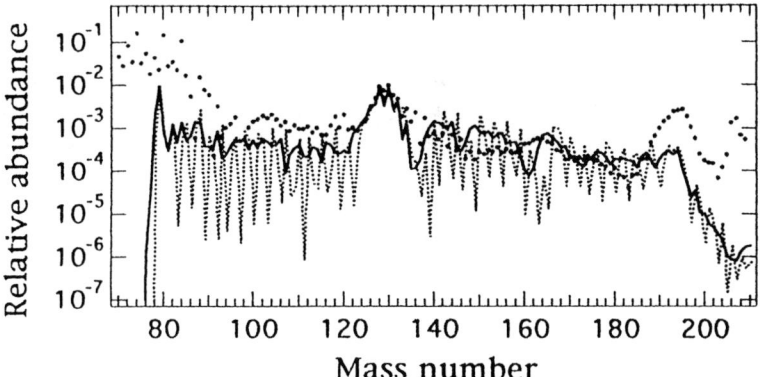

FIGURE 3. Relative r-process abundance calculated with the use of the canonical model. The dots represent the solar system r-process abundances, and the solid line (the dotted line) the estimated abundances with (without) P_n-values. GT2 is used to obtain the β^-- decay half-lives and P_n-values, and TUYY mass formula (5) is also adopted to estimate Q-values and neutron separation energies S_n (see Eqs. (6) and (7)). The stellar conditions are given in Ref. (8).

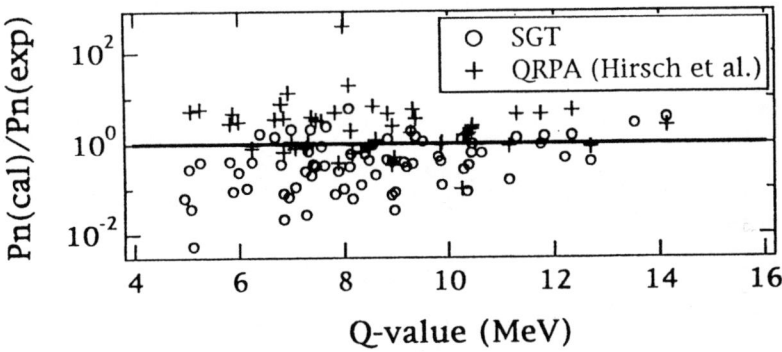

FIGURE 4. The ratio of calculated and experimental P_n-values. The experimental data and the QRPA results are obtained from Refs. (9) and (10), respectively.

In Fig. 5, we examine the strength in the region of delayed neutron window. Here, we consider a value I defined as

$$I = \frac{\lambda_n}{\ln 2} = \frac{1}{\ln 2} \int_{-Q+S_n}^{0} S_\beta(E) f(-E) \frac{\Gamma_n}{\Gamma_n + \Gamma_\gamma} dE . \tag{12}$$

According to the previous subsection, $\ln 2 \times I$ is considered to be the (appropriately weighted) strength function within the delayed neutron window. In Fig. 5, the notation I_{GT2} (I_{SGT}) indicates the value calculated by Eq. (12) with the use of GT2 (SGT) model; note that, in this calculation, the competition factor $\Gamma_n /(\Gamma_n + \Gamma_\gamma)$ is unity. A similar value corresponding to the experiments, I_{exp}, may be written as

$$I_{exp} = \frac{P_n^{exp}}{T_{1/2}^{exp}} , \qquad (13)$$

where P_n^{exp} and $T_{1/2}^{exp}$ are the experimental P_n and $T_{1/2}$, respectively.

Similarly to Fig. 2, the ratio I_{SGT}/I_{GT2} decreases around the neutron magic numbers $N=50$, and 82. This suggests that SGT includes some shell effects in the right direction. Although the tendency of the ratio I_{exp}/I_{GT2} is somewhat vague, it may be said that the ratios I_{SGT}/I_{GT2} and I_{exp}/I_{GT2} have a similar tendency for the nuclei with $N \geq 60$. We can not examine further because the number of the experimental P_n -values are limited.

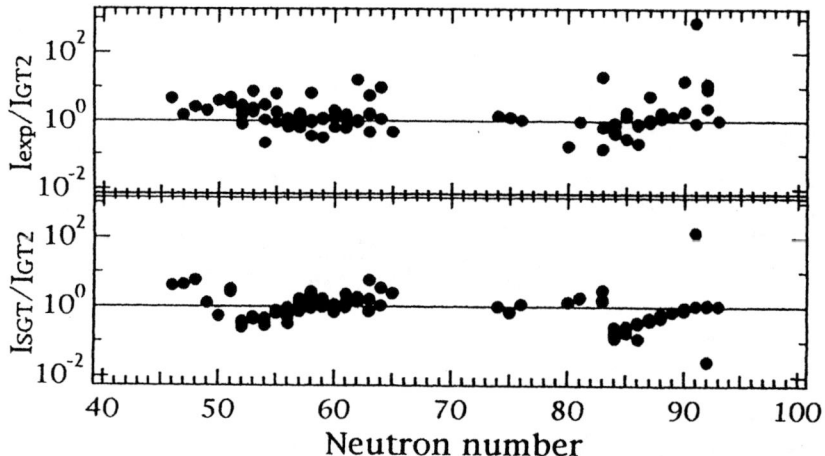

FIGURE 5. The ratios of the appropriately weighted strength function I, Eq. (12).

Half-life

In Fig. 6, we show ratios of calculated to experimental half-lives on a logarithmic scale. The QRPA calculation was performed for both β^- and β^+-decay of nuclei with the experimental half-lives up to $A=150$ (see Ref. 11). We also select the same nuclei in the SGT calculation to make a comparison. Although the allowed and the first forbidden transitions are taken into account in SGT calculation, only allowed transitions are taken in QRPA. A special prescription referred to as "bottom raising" (1) is applied in SGT to consider highly forbidden transitions to low-lying levels in the daughter nucleus. The accuracy of SGT is almost same as that of QRPA, and the estimated half-lives converge to the experimental values as the Q-value becomes large. In Fig. 7, ratios between QRPA and SGT half-lives are shown. The systematic deviation around the neutron magic numbers is found, which is partly due to the

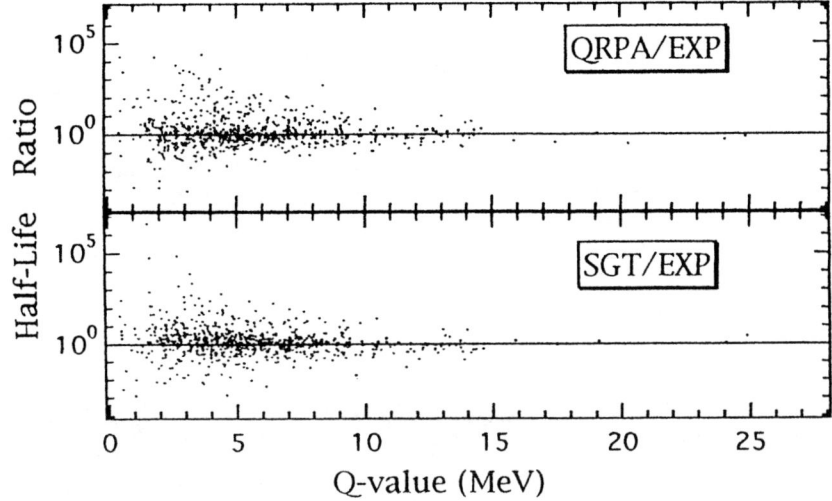

FIGURE 6. Ratios of half-lives between the theory and the experiment.

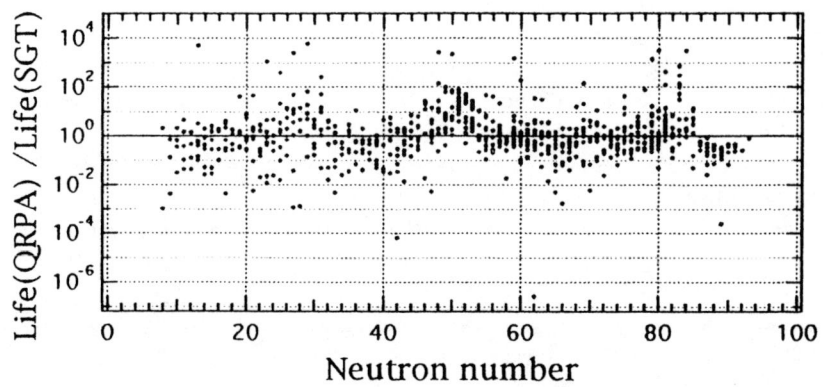

FIGURE 7. Ratios of half-lives between QRPA and SGT estimations.

insufficient inclusion of the shell effects in SGT, and partly due to the inaccurate estimation by QRPA near the magic numbers.

CONCLUSION

We have developed the semi-gross theory to include the shell effect of the parent nucleus in the gross theory. A non-uniform and discrete one-particle energy distribution of decaying nucleons has been introduced, and one-particle strength functions have been modified. Their effects are clearly seen in Figs. 2 and 5. In Ref. (1), we have checked, for the half-lives, the mean value and the deviation of the ratios between the semi-gross theory and the experimental data, and found that the accuracy and the predictive power of the semi-gross theory are comparable to the microscopic calculations QRPA. Although some parts of the shell effects are correctly considered in the semi-gross theory, there still remain deviations from the experimental data, such as the insufficient inclusion of the shell effects shown in Fig. 2, and the underestimation of P_n-values in Fig. 4. Hence refinement of the semi-gross theory, particularly inclusion of the shell effects of daughter nucleus, will become a future problem.

ACKNOWLEDGMENTS

The authors thank Dr. H. Homma, Dr. K. Muto and Professor T. Oda for providing the half-life values calculated by using their QRPA model. This work was financially supported, in part, by the Grant for Special Research Projects, Waseda University.

REFERENCES

1. Nakata, H., Tachibana, T., Yamada, M., to be published in *Nuclear Physics* **A**, (1997).

2. Kondoh, T., Tachibana, T., and Yamada, M., *Progress of Theoretical Physics*, **74**, 708-724 (1985).

3. Tachibana, T., Yamada, M., and Yoshida, Y., *Progress of Theoretical Physics*, **84**, 641-657 (1990).

4. Audi, G., and Wapstra, A. H., *Nuclear. Physics.* **A565**, 1-66, (1993) and **A595**, 409-480 (1995).

5. Tachibana, T., Uno, M., Yamada, M., and Yamada, S., *Atomic Data and Nuclear Data Tables* **39**, 251-258 (1988).

6. Koura, H., private communication, (1997)

7. *Evaluated Nuclear Structure Data Files (ENSDF)*, 1995 November version, communicated through Nuclear Data Center, Japan Atomic Energy Research Institute.

8. Tachibana, T., Arnould, M., *Nuclear Physics* **A588**, 333c-338c (1995).

9. Rudstam, G., Aleklett, K., and Sihver, L., *Atomic Data and Nuclear Data Tables* **53**, 1-22 (1993).

10. Hirsch, M., Staudt, A., Klapdor-Kleingrothaus, H. V., *Atomic Data and Nuclear Data Tables* **51**, 243-271 (1992).

11. Homma, H., Bender, E., Hirsch, M., Muto, K., Klapdor-Kleingrothaus, H. V., and Oda, T., *Physical Review* **C54**, 2972-2985, (1996).

STELLAR EVOLUTION AND NUCLEOSYNTHESIS

NUCLEOSYNTHESIS IN LOW- AND INTERMEDIATE-MASS STARS: AN OVERVIEW

Nami Mowlavi

Geneva Observatory, CH-1290 Sauverny, Switzerland

Abstract. An overview of the main phases of the evolution of low- and intermediate-mass stars is presented, and the different types of nucleosynthesis operating from the pre-main sequence up to and including the asymptotic giant branch phase described. The surface abundance modifications brought by each nucleosynthesis process is also briefly discussed.

I INTRODUCTION

Low- and intermediate mass (LIM) stars are defined as those who end their life without proceeding through the carbon and heavier elements burning phases. Part of them experience only the core H-burning phase, ending their life as He white dwarfs (WDs). This is the case for stars with initial masses between ~ 0.08 and $\sim 0.5\,M_\odot$ ($M_\odot$ being the solar mass). Stars with masses between ~ 0.5 and $6-8\,M_\odot$, on the other hand, proceed further to the core He-burning phase, and end as C-O WDs. Stars less massive than $\sim 0.08\,M_\odot$ never reach central temperatures high enough to ignite H, and end as brown dwarfs. Stars more massive than $\sim 10\,M_\odot$ (called massive stars), on the other hand, continue their evolution through central C and heavier element burning, eventually ending their life in a supernova explosion. The mass limits defining these categories are obtained through evolutionary model calculations (e.g. [19]). They are still subject to some uncertainties mostly due to the shortcomings in the mixing prescriptions and mass loss rates. As for stars with initial masses between $6-8$, their fate is still uncertain. We refer to [19] for a discussion (see also [16,15]).

In this paper, we consider the evolution of *single* stars with masses between ~ 0.5 and $6-8\,M_\odot$ (the case of massive stars is reviewed by G. Meynet in this volume). The nucleosynthesis operating in these LIM stars involve mainly the light elements up to Al, and the *s*-process elements. Because of the high mass loss rates characterizing the late stages of the evolution of these stars,

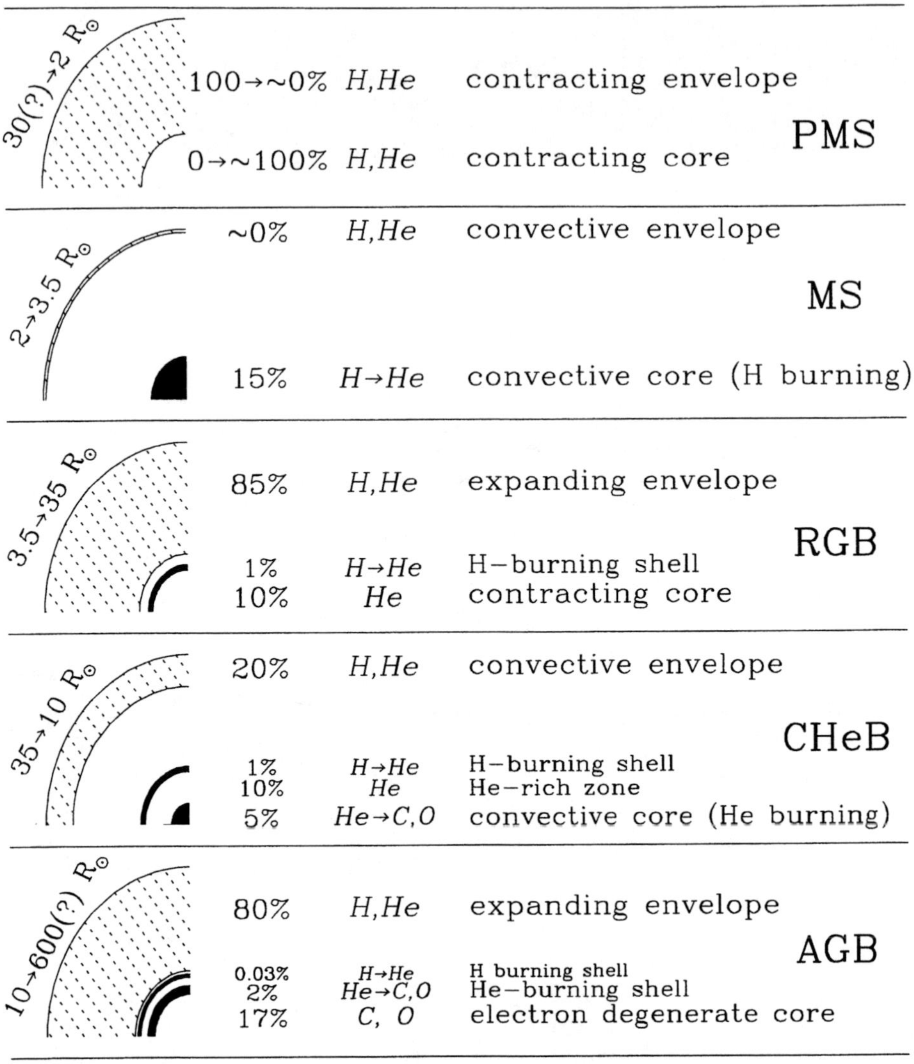

$30(?) \to 2\,R_\odot$	100→~0%	H,He	contracting envelope	
	0→~100%	H,He	contracting core	**PMS**
$2 \to 3.5\,R_\odot$	~0%	H,He	convective envelope	**MS**
	15%	H→He	convective core (H burning)	
$3.5 \to 35\,R_\odot$	85%	H,He	expanding envelope	
	1%	H→He	H-burning shell	**RGB**
	10%	He	contracting core	
$35 \to 10\,R_\odot$	20%	H,He	convective envelope	
	1%	H→He	H-burning shell	**CHeB**
	10%	He	He-rich zone	
	5%	He→C,O	convective core (He burning)	
$10 \to 600(?)\,R_\odot$	80%	H,He	expanding envelope	
	0.03%	H→He	H burning shell	**AGB**
	2%	He→C,O	He-burning shell	
	17%	C, O	electron degenerate core	

FIGURE 1. Schematic representations of the structure of a star at different phases of its evolution. The numbers on the upper left of each diagram indicate the variation of the surface radius during the considered phase of evolution. The columns on the right of each diagram indicate, from left to right, the percent of the total mass of the star contained in the given region, its main chemical composition and a description of it. hatched areas in the diagrams indicate convective regions, while filled areas denote regions where nuclear energy is produced. The quantities refer to a $3\,M_\odot$ Z=0.02 star. Adapted from [21].

the nuclides synthesized in their deep interior, and dredged-up to the surface, efficiently contribute to the chemical enrichment of the interstellar medium. Such is the case for He, ^{7}Li, C, N, ^{19}F, ^{26}Al, and the s-process elements.

This paper aims at presenting a general overview of the nucleosynthesis occurring in LIM stars. A more detailed review of the subject is presented in [22]. The main phases of the structural evolution of these stars are presented in Sect. II, while Sect. III analyses the nucleosynthesis operating in the different phases of their evolution. Some concluding remarks on surface abundance predictions in LIM stars are discussed in Sect. IV.

II STRUCTURAL EVOLUTION

The structure of a $3\,M_\odot$ Z=0.02 model star, with Z being the mass fraction of all elements heavier than He (called the "metallicity"), is displayed in Fig. 1 at different phases of its evolution. Five phases are distinguished:

1) The *pre-main sequence (PMS) phase*, which constitutes the overall contraction phase prior to H ignition in the center. It is characterized by an accretion phase during which mass from a circumstellar disk is accreted on the forming star ([5] and references therein), and an increasing central temperature until H ignites in the core at $12 \lesssim T_6 \lesssim 25$ (where T_6 is the temperature expressed in units of 10^6 K). The chemical composition is homogeneous throughout the star, except for some of the light nuclides (up to C) which already burn at temperatures of a few 10^6 K.

2) The *main sequence (MS) phase*, characterized by the transformation of H to He in the core. It represents the star's longest duration phase. The core is convective for stars with $M \gtrsim 1.2\,M_\odot$, while H burns radiatively in lower mass stars.

3) The *red giant branch (RGB) phase*, which follows the MS phase. As a result of core contraction, the radius of the star increases and its surface temperature decreases. Hydrogen burns now in a shell surrounding a H-depleted core. *Low-mass stars* ($M \lesssim 2\,M_\odot$), defined as those containing an electron-degenerate core at this phase of their evolution, may pass as much as 20% of their life as red giants. For *intermediate-mass stars* ($M \gtrsim 2\,M_\odot$), on the other hand, this phase is short, generally less than 7% of the MS lifetime. This phase is also characterized by the convective envelope penetrating into the deep layers, the material of which has been affected by H-burning. The ashes of H-burning are thus mixed to the surface. This is called the *"first dredge-up"*.

4) The *core He-burning (CHeB) phase*, characterized by the transformation of He to C and O in a convective core surrounded by a thin H-burning shell. This is the second longest lived phase in the life of the star. In low-mass stars, He ignition proceeds in a degenerate core, which leads to a thermal runaway called "core helium flash".

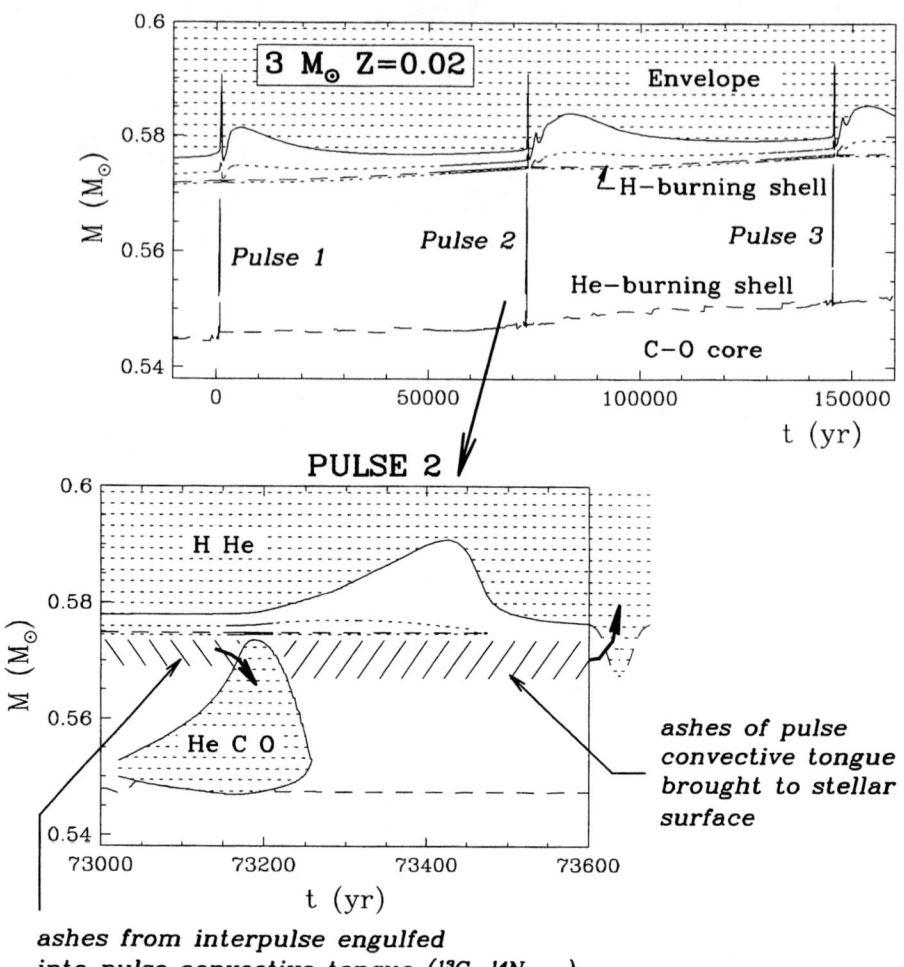

FIGURE 2. *Upper panel:* Structural evolution of a 3 M$_\odot$ Z=0.02 AGB star, during the first three thermal pulses in the He-burning shell. Filled areas denote convective regions. The long-dashed lines locate the maximum energy production in the H-burning (top) and He-burning (bottom) layers. Short-dashed lines locate the extensions of the H-burning shell, defined by the region where the energy production exceeds 1 erg g^{-1} sec^{-1}. The three vertical lines in the He-burning shell denote the occurrence of the thermal instabilities. *Lower panel:* Enlargement of the structural evolution during the second pulse of the 3 M$_\odot$ Z=0.02 star. A third dredge-up is simulate on the outer right-hand side of the panel. The hatched region to the left of the pulse indicates the layers containing the ashes left behind by the H-burning shell during the interpulse phase, and which are injected into the pulse. The layers containing the ashes of the pulse and which are mixed into the envelope are delimited by the hatched region on the right of the pulse. Adapted from [21].

5) The *asymptotic giant branch (AGB) phase*, following the CHeB phase. Helium now burns in a shell and the star becomes a red giant for the second time. The envelope penetrates in the deep layers and, in stars more massive than about $4\,M_\odot$, the products of H burning are transported for the second time to the surface (*"second dredge-up"*). The star is now characterized by an electron degenerate C-O core of mass between 0.5 and $1.2\,M_\odot$, one thin He-burning shell capped by a thin H-burning shell, and a deep convective envelope. The nucleosynthesis and energy production are confined to a region comprising less than 3% of the total mass of the star. A distinctive feature of the AGB phase is the fact that the He-burning shell becomes thermally unstable (see Fig. 2) and liberates, periodically and on a short time-scale (several tens of years), 10^2 to 10^6 times the energy provided by the H-burning shell. These energy bursts are called "pulses". They lead to the development of a convective zone in the He-burning shell. The quiescent evolutionary phase, or "interpulse" period, lasts several 10^4 years.

The lower panel of Fig. 2 gives an enlargement of the convective regions during a thermal instability. The thermal pulses have two important consequences. From a chemical point of view, the material synthesized in the He-burning shell is convectively mixed with layers close to the H-burning shell. The layers left behind by the H-burning shell (hatched region on the left of the pulse in Fig. 2), containing the ashes of that combustion phase, are engulfed by the pulse and contribute to a rich nucleosynthesis therein.

From a structural point of view, an important envelope response is predicted to occur after the pulse extinction. The convective envelope penetrates into the H-burning shell, and can even reach the H-depleted regions. Eventually, it can sink into the carbon-rich layers. The material processed by the pulse could then be transported to the surface. This scenario is called the *"third dredge-up"*.

III NUCLEOSYNTHESIS

A During the PMS phase

Nuclear reactions involving light elements from D to C begin to occur in PMS stars at temperatures varying from $T_6 = 1$ to 12. The case of deuterium is of particular interest in relation with Big Bang Nucleosynthesis. This element burns at $T_6 \sim 1 - 1.5$, mainly through $D(p,\gamma)^3He$. As a result, the total mass of $D + {}^3He$ is conserved. This property, together with the fact that D is not produced in Galactic environments, has been, and still often is, used to derive an information on primordial D abundance (e.g. [32] and references therein).

Let us also mention the case of ^6Li and ^7Li, which burn in the deep layers by p-capture at $T_6 \sim 3$. To what degree the surface Li is affected depends on the initial stellar mass and on various physical conditions such as mixing or rotation. We refer to [20,11] and references therein for a discussion on this issue.

B From the MS up to the AGB phase

As far as stellar surface abundances and interstellar chemical enrichment are concerned, H burning is the only nucleosynthesis process of interest in LIM stars from the MS up to the AGB phase. Indeed, the nuclides produced by He burning during the CHeB phase remain trapped into the white dwarfs.

Four non-explosive H-burning modes have been identified to date (e.g. [28]): the pp-chains, the "cold" CNO cycles, and the NeNa and MgAl chains. The pp-chains and CNO cycles are the main contributors to the energy production. The pp-chains are dominant at $T_6 \lesssim 20$ (i.e. in the cores of MS stars with $M \lesssim 1.2\,M_\odot$), and the CNO cycles at higher temperatures (i.e. in the cores of MS stars with $M \gtrsim 1.2\,M_\odot$, and in the H-burning shells). When the latter cycles are active in the central regions of MS stars, convection develops in the core due to the steep temperature dependence of the energetics of the CNO reactions. All the four modes, however, are of importance from a nucleosynthetic point of view.

The pp-chains are described at length in the literature (see [10,2] for a general description, and [4,12] for a discussion related to the "solar neutrino problem"). We discuss here only the implications of this mode of H burning in relation with ^{3}He. In addition to its production through D burning during the PMS phase (see Sect. III A), ^{3}He is produced in low-mass stars[1] by $p(p,e^+\nu)\,D\,(p,\gamma)\,^3$He. This occurs essentially in stellar regions where $T_6 \lesssim 15$ (in deeper layers where the temperatures are higher, ^{3}He is destroyed by ^{3}He$(^3$He$,2p)\,^4$He or ^{3}He$(\alpha,\gamma)\,^7$Be), and which are engulfed by the first dredge-up. Low-mass stars would thus be important contributors to the galactic ^{3}He enrichment. According to current galactic chemical evolution models, however, this leads to a present interstellar ^{3}He abundance prediction much higher than what is observed (e.g. [33,24,26]). We refer to [22] for a more extensive discussion on this subject.

Hydrogen burning by the CNO cycles operates mainly through the transformation of ^{12}C and ^{16}O to ^{14}N ([10], see also G. Meynet in this volume). We refer to [3] for a description of the cycles and their yields, as well as for a discussion of the uncertainties still affecting the involved reactions rates. Two nuclides, ^{13}C and ^{17}O, play an important role in confronting stellar model predictions with observations. The nuclear transformation of ^{12}C to ^{14}N proceeds first through the production of ^{13}C by ^{12}C$(p,\gamma)\,^{13}$N$(\beta^+)\,^{13}$C,

---

[1]) In stars more massive than $3-4\,M_\odot$, the MS lifetime is shorter than the characteristic ^{3}He production time-scale, and no ^{3}He is produced. In these stars, ^{3}He is essentially destroyed in the deep layers

and then through its destruction by $^{13}C(p,\gamma)^{14}N$ (see Fig. 1 of [3]). As a result, ^{13}C is overproduced in intermediate layers of LIM stars, and destroyed in deeper regions ([22]). The same is true for ^{17}O, which is produced by $^{16}O(p,\gamma)^{17}F(\beta^+)^{17}O$ and destroyed by $^{17}O(p,\alpha)^{14}N$. The surface abundances of both ^{13}C and ^{17}O are then increased by the first dredge-up scenario. Indeed, a severe decrease in the $^{12}C/^{13}C$ and $^{16}O/^{17}O$ ratios is observed at the surface of red giant stars (e.g. [13]).

Arnould, Mowlavi and Champagne [3] have also reviewed the NeNa and MgAl chains. These become mainly active at $T_6 \gtrsim 40$. The first of these chains contributes to the production of ^{23}Na by proton capture on ^{22}Ne. The second one is of importance, prior to the AGB phase, in the production of ^{27}Al from ^{25}Mg and, at $T_6 \gtrsim 65$, from ^{24}Mg. Overabundances of both ^{23}Na and ^{27}Al are also confirmed at the surface of red giants (e.g. [31,8,25]).

C During the AGB phase

The nucleosynthesis occurring in AGB stars is summarized in Fig. 3. Four sites can be distinguished:

1) The first site is obviously the H-burning shell during the interpulse period. The ashes of that combustion are brought to the surface through the third dredge-up scenario. The temperature in the H-burning shell increases with time, and reaches values above T_6. In those conditions, the MgAl chain leads, in particular, to an efficient production of ^{26}Al ([14]). The interest in this nuclide resides both in relation to γ-ray line astronomy (e.g. [27]) and to cosmochemistry (e.g. [1,18]).

2) The He-burning shell constitutes the second obvious nucleosynthesis site. It contributes mainly to the production of ^{12}C through the 3-α process. The carbon is brought by the convective tongue of the thermal pulses to regions close to the H-burning shell. Successive occurrences of the third dredge-up scenario are then responsible for a gradual surface carbon enrichment, turning eventually a M star (whose surface C/O ratio is less than 1) into a C star (with C/O>1).

Another important contribution from the He-burning shell is provided by the nucleosynthesis resulting from the injection into the pulse of the H-burning shell ashes. In particular, the injection of ^{13}C leads to the production of neutrons through the $^{13}C(\alpha,n)^{16}O$ reaction. These neutrons can then be used to produce elements heavier than iron via the s-process nucleosynthesis, as well as ^{19}F (the nucleosynthesis path of which is described in [23]). Unfortunately, the amount of ^{13}C left behind by the H-burning shell turns out to be too low to result in any significant s-process nucleosynthesis, as well as to produce the fluorine in the amounts required by the observations ([23]).

3) An independent source of ^{13}C is expected to result from a partial mixing of

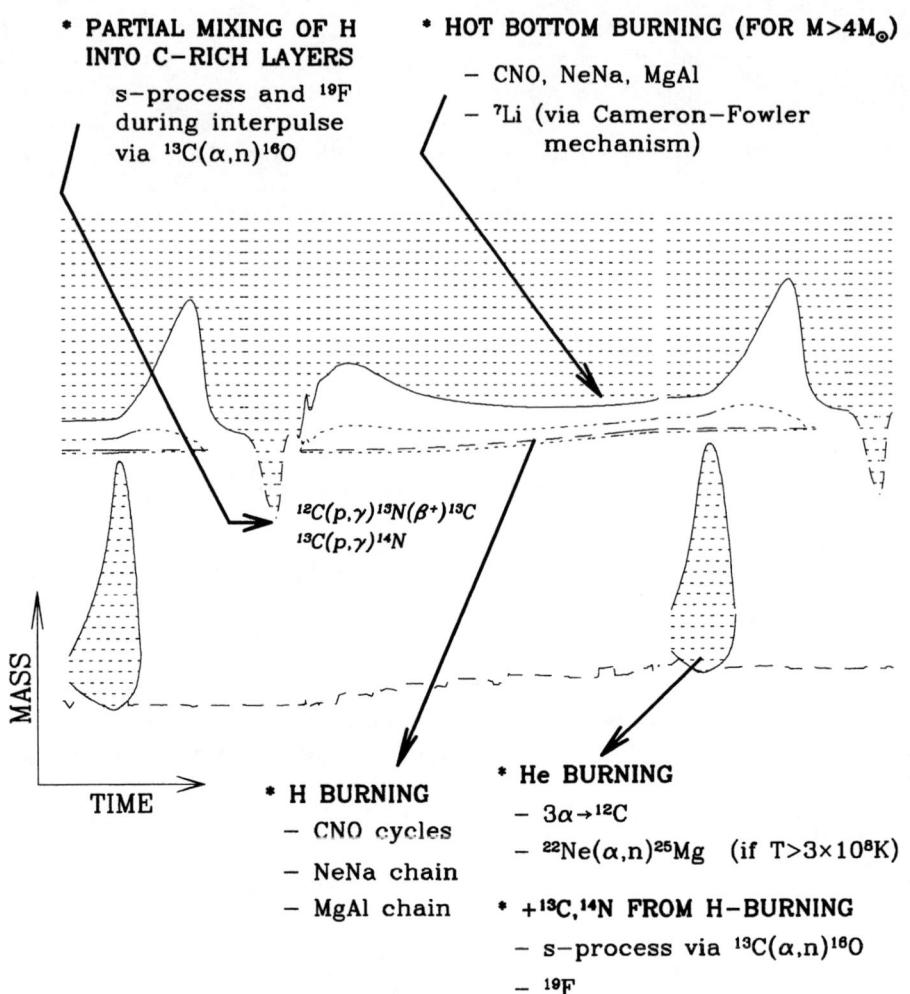

FIGURE 3. Overview of the nucleosynthesis occurring in AGB stars. The figure displays a schematic representation of two successive pulses. The interpulse phase is drawn on a much reduced time-scale, and lasts in reality *much* longer than each pulse. Hatched areas and dashed lines have the same meanings as in Fig. 2. Dredge-ups following each pulse (identified by the dashed envelope border) are also displayed. Adapted from [21].

protons into the carbon-rich region ([17]). This occurs below the convective envelope during a dredge-up scenario (see Fig. 3). When the temperature increases during the interpulse period, the protons are captured by ^{12}C and form ^{13}C. The low proton density prevents *all* of this ^{13}C to be transformed into ^{14}N. The surviving ^{13}C is most probably burned radiatively before the occurrence of the next pulse, leading to the production of ^{19}F and the s-process elements during the interpulse phase.

4) Finally, in stars more massive than about $4\,M_\odot$, the temperature at the bottom of the envelope increases above $T_6 \sim 50$, activating the H-burning modes *within* the envelope. This phenomenon, known as hot bottom burning (HBB), modifies the surface composition without the need to invoke a dredge-up scenario. The signatures of HBB are quite characteristic. In particular, it can explain the low ^{12}C/^{13}C at the surface of some stars. It destroys ^{12}C and ^{18}O through the CNO cycles ([6]), and ^{19}F by ^{19}F$(p,\alpha)^{16}$O ([23]). It provides also an efficient site for the production of ^{26}Al through the MgAl chain. Moreover, HBB can lead to the synthesis of ^{7}Li through the Cameron and Fowler ([7]) mechanism, which combines the production of ^{7}Be at the bottom of the convective envelope by ^{3}He$(\alpha,\gamma)^7$Be and its convective transport to the surface where it is transformed to ^{7}Li by electron capture ([29]).

IV   FINAL REMARKS

The first and second dredge-ups mix the ashes of hydrogen burning in the envelope. In contrast, the third dredge-up has the distinctive characteristic of bringing to the surface, in addition to the ashes from the H-burning shell, products resulting from helium burning. Nucleosynthesis occurring *in* the envelope, on the other hand, affects surface abundances without the need to invoke a dredge-up scenario. This concerns the light elements during the PMS phase, mainly D, ^{3}He and Li, or the elements involved in the CNO cycles and NeNa and MgAl chains in the case of HBB during the AGB phase.

Quantitative confrontation of surface abundance predictions from stellar models with various observations, however, reveal several disagreements which attest of our still poor knowledge of some physical processes occurring in the stars. The main shortcomings concern the mixing processes. Some examples in relation to LIM stars can be found in [34,9,30]. These are reviewed in more details in [22].

Finally, let us emphasize that the transition from the AGB to the post-AGB phase, during which the envelope of the AGB star is ejected and a planetary nebula eventually forms, is still poorly known. In particular, the history and rates of mass loss, whether the envelope is ejected intermittently or through a single "super wind" episode, and when this (these) strong wind(s) occur during the AGB phase, all limit the predictive capabilities of current AGB models concerning the final yields of LIM stars.

REFERENCES

1. Anders E., Zinner E., *Meteoritics* **28**, 490 (1993)
2. Arnould M., Mowlavi N.: 1993, in *Inside the Stars*, eds. Weiss W.W., Baglin A., ASP Conf. Ser. Vol. 40, pp 310-323 (1993)
3. Arnould M., Mowlavi N., Champagne A.: 1995, in *32nd Liège Int. Astroph. Coll.*, eds. Noels A., Fraipont-Caro D., Gabriel M., Grevesse N., Demarque P., pp 17-29 (1995)[2]
4. Bahcall J.N., in *"Neutrino Astrophysics"*, Cambridge University Press, Cambridge (1989)
5. Bernasconi P.A., *A&AS* **120**, 57 (1996)
6. Boothroyd A.I., Sackmann I.-J., Ahern S., *AJ* **418**, 457 (1993)
7. Cameron A.G.W., Fowler W.A., *ApJ* **164**, 111 (1971)
8. Cavallo R.M., Sweigart A.V., Bell R.A., *ApJ* **464**, L79 (1996)
9. Charbonnel C. *ApJ* **453**, L41 (1995)
10. Clayton D.D., in *"Principles of stellar evolution and nucleosynthesis"*, McGraw-Hill, New-York (1968)
11. D'Antona F., Mazziteeli I., *ApJS* **90**, 467 (1994)
12. Dzitko H., Turck-Chièze S., Delbourgo-Salvador P., Lagrange C., *ApJ* **447**, 428 (1995)
13. El Eid M.F., *A&A* **285**, 915 (1994)
14. Forestini M., Paulus G., Arnould M., *A&A* **252**, 597 (1991)
15. Garcia-Berro E., Ritossa C., Iben I. *ApJ* **485**, 765 (1997)
16. Hashimoto M., Iwamoto K., Nomoto K. *ApJ* **414**, L105 (1993)
17. Herwig F., Bloecker T., Schoenberner D., El Eid M., *A&A* **324**, L81 (1997)
18. MacPherson G.J., Davis A.M., Zinner E.K., *Meteoritics* **30**, 365 (1995)
19. Maeder A., Meynet G., *A&A* **210**, 155 (1989)
20. Martin E.L., Claret A., *A&A* **306**, 408 (1996)
21. Mowlavi N., Ph.D. Thesis (unpublished)[2]
22. Mowlavi N., in *Cosmic Chemical Evolution*, IAU Symp. 187, in press[2]
23. Mowlavi N., Jorissen A., Arnould M., *A&A* **311**, 803 (1996)[2]
24. Olive K.A., Schramm D.N., Scully S.T., Truran J.W., *ApJ* **479**, 752 (1997)
25. Pilachowski C.A., Sneden C., Kraft R.P., Langer G.E., *AJ* **112**, 545 (1996)
26. Prantzos N., *A&A* **310**, 106 (1996)
27. Prantzos N., *A&A Sup.Ser.* **120**, 303 (1996)
28. Rolfs C.E., Rodney W.S., in *Cauldrons in the Cosmos*, The University of Chicago Press, Chicago (1988)
29. Sackmann I.-J., Boothroyd A.I., *ApJ* **392**, L71 (1992)
30. Shetrone M.D., *AJ* **112**, 2639 (1996)
31. Takeda Y., Takada-Hidai M., *PASJ* **46**, 395 (1994)
32. Turner M.S., Truran J.W., Schramm D.N., Copi C.J., *ApJ* **466**, L59 (1996)
33. Vangioni-Flam E., Olive K.A., Prantzos N., *ApJ* **427**, 618 (1994)
34. Wasswerburg G.J., Boothroyd A.I., Sackmann I.-J., *ApJ* **442**, L21 (1995)

---

[2] postcript files available by anonymous ftp on 'obsftp.unige.ch' in the directory 'pub/mowlavi'

Supernova Nucleosynthesis

M. Hashimoto[1], S. Nagataki[2], K. Sato[2,3], and S. Yamada[2,4]

[1] Department of Physics, Faculty of Science, Kyushu University, Fukuoka 810, Japan
[2] Department of Physics, School of Science, the University of Tokyo, Tokyo 113, Japan
[3] Research Center for the Early Universe, School of Science, the University of Tokyo, Tokyo 113, Japan
[4] Max-Planck-Institute für Physik und Astrophysik Karl-Schwarzschild Strasse 1, D-8046, Garching, Germany

Abstract. Nucleosynthesis in supernovae is reviewed considering the recent development. There is a consensus that supernovae have two distinct origins. One is the explosion of white dwarfs which we call SNIa. The other is the explosion of massive stars of $M_{ms} \gtrsim 10 M_\odot$ which we call SNII (M_{ms} is the stellar mass in the main-sequence stage). However, the mechanism of the explosion for both supernovae is now in debate. SNIa is the result of the mass accretion onto a white dwarf from a companion star. The deflagration model which has been believed as the standard model of SNIa may be modified because many different types of SNIa have been observed recently. Furthermore, multi-dimensional hydrodynamical calculations suggest a slow deflagration model rather than a fast deflagration model like W7. On the other hand, to explain new types of SNIa, delayed or late detonation model has been proposed. For SNII, there does not still exist the reliable calculations from the core collapse to the explosion after a bounce; the present delayed explosion calculation cannot explain the observed explosion energy of SN1987A. Though these calculations are limited to the spherical cases, from the observation and multi-dimensional calculations, asymmetric explosion is suggested. Therefore, we must explore possible explosion models of supernovae for both SNIa and SNII. Recently, developed are two-dimensional hydrodynamical calculations to simulate an asymmetric explosive nucleosynthesis; we have found significant differences compared with the spherical explosive nucleosynthesis. For SNIa, new models of delayed detonation and results of the explosive nucleosynthesis was proposed by [9]. These new models are expected to compensate the previous models in many observational points.

INTRODUCTION

The elements heavier than carbon are believed to be synthesized in stars. In particular, supernovae must play an important role from the early epoch of the universe. Then it is crucial to calculate the supernova nucleosynthesis to be 'consistent' with the stellar evolution models. The calculations performed according to the following procedure are called 'realistic'. First, performed is the stellar evolutionary calculation of massive stars to the beginning of the iron core collapse. Second, the explosion models are constructed with an assumption of the explosion mechanism. Finally, the explosive nucleosynthesis using these explosion models is calculated [13]. As the result, it has been proved that global agreement compared with the solar system abundances of $A < 60$ is fairly well except for some elements. We should note that the abundances of $A \gtrsim 60$ are also produced in massive stars: the weak components of the s-process, the p-process elements, and the r-process elements [11] [12] [1].

On the other hand, recent observations of supernova remnants have revealed difficulty to explain the origin of some elements using the available results: for example, ^{44}Ti in Cas A [4]. This would be closely related to the mechanism of the supernova explosion which has not been fully understood yet.

So far, supernova nucleosynthesis has been investigated on the assumption of the spherically symmetric explosion. However, some observations show that the explosion would have been asymmetric (e.g., [2]). Recently, a simulation of explosive nucleosynthesis in massive stars on the assumption of the axisymmeric explosion has been performed by [16] [6]. They used as initial models of explosion, presupernova models developed by [15], [8] and [3] (see Fig. 2 and Fig. 3). Certainly, qualitatively new results are obtained as is shown in the following section.

In this review, we compare the available results of supernova nucleosynthesis. Then, we show the results of the asymmetric explosive nucleosynthesis in the explosion of massive stars evolved to the beginning of the iron core collapse. The combined results of SNIa and SNII are presented with the use of the latest models for spherical explosion and the axisymmetric explosion.

EXPLOSIVE NUCLEOSYNTHESIS UNDER SPHERICAL EXPLOSION

In the present stage, we have two kinds of models of nucleosynthesis in supernovae. First, the type Ia (SNIa) model has been believed to be the explosion of accreting carbon–oxygen white dwarf [9]. In the present paper, we consider a deflagration model W7 and a delayed detonation model WDD2 which is an improved version of the deflagration model consistent with results of recent multi-dimensional calculations [9]. These models show very nice agreement with the observations of SNIa: shape of the light curves, synthesis spectra, and the velocity distribution of the abundances. However, it should be noted that disagreement with some observations are found recently; we will take into account another models in the future.

Second, the type II (SNII) model is the explosion of massive stars (core collapse driven supernovae) but it has been revealed that there are sub-types of this supernova – type Ib, Ic, IIp, IIL. Though the mechanism of the core collapse driven supernova has not yet fully understood, SN1987A has shown clearly that the scenario is right; many observations were explained and theoretical predictions were confirmed: detection of neutrinos, fit to the light curve, the amount of ^{56}Ni and ^{57}Ni. The results of supernova nucleosynthesis in massive stars are reviewed by [3]. Based on the success of both the model of SNIa and SNII, comparison of the supernova yields with the solar system abundances for some selected elements was tried [14]. It is found that agreement is fairly well as far as the 'representative' elements are concerned.

Though the explosive nucleosynthesis in massive stars has rather firm foundation, there are some discrepancies between available results. In Fig. 3, we have compared the results of two groups: Ha95 [3] and WW95 [15]. Obviously, production of some elements are inconsistent among each other. Combined results of SNIa + SNII are displayed in Fig. 4 for W7 + Ha95 and W7 + WW95, WDD2 + Ha95 and WDD2 + WW95, respectively. Despite the differences of the stellar models, noted are the differences of the initial abundances. While Ha95 evolved the helium stars with the initial abundances of ^{4}He and ^{14}N only, WW95 included all the solar system abundances in their post processing network from the main sequence stages. Since neutron capture process like s-process may change some initial abundances significantly but may not change some of them, it is not clear whether the amount of neutron rich isotopes are reproduced correctly (see Table 1). We should note that these models are constructed on the assumption of spherical explosion.

EXPLOSIVE NUCLEOSYNTHEIS UNDER AXISYMMERIC EXPLOSION

There exist observational evidences that the shape of the supernova remnant is non-spherical [2]. This indicates that the explosion would be asymmetric. Since any 3-dimensional calculation is rather difficult due to the limitation of the computer power, 2-dimensional calculations have been performed on the assumption of the axisymmetric explosion [6]. This axisymmeric explosion could be plausible if the progenitor core rotates around a fixed axis. However, no one knows the angular momentum distribution of the progenitor. Therefore, the calculation was done for some parameters which represent the strength of axisymmetric explosion. Then nucleosynthesis has been calculated using the explosion model, so called post processing. Results are shown in Fig. 5 and Fig. 6. It is found that strong alpha-rich freeze-out phenomena after the passage of the shock wave has occurred along the polar axis region. As the result, a lot of nuclei which have not been produced much in models of spherical explosion are produced significantly during the stage of alpha-rich freeze-out. As for a 20 $M_\odot$ star, which is the case of SN1987A, from the observation, there are three severe constraints for the explosive nucleosynthesis; the amount of ^{56}Ni is $\simeq 0.07 M_\odot$, the ratio of ^{57}Co to the solar value is $\simeq 1.5$, and the ratio of ^{58}Ni to the solar value is $\simeq 1$ [5]. Axisymmeric nucleosynthesis can pass all of these constraints. It is remarkable to find that the amount of ^{44}Ti which is more than 6 times compared with the amount in the spherical explosion could explain the late time behavior of the light curve. On the other hand, it is very difficult to estimate the amount of ^{44}Ti from the observation because the life time is estimated to be from 40 to 80 years. It should be noted that spherical models have failed to produce enough amount of ^{44}Ti compared to the solar value as seen in Fig. 4.

EFFECTS OF AXISYMMETRIC EXPLOSIVE NUCLEOSYNTHESIS ON THE SOLAR SYSTEM ABUNDANCES

Axisymmetric explosive nucleosynthesis of massive stars from 10 to 50 $M_\odot$ has been performed [7]. The basic assumption is the same as [6]. IMF (initial mass function) averaged results are shown in Fig. 7 and Fig. 8. In Fig. 7, compared are the abundances relative to the spherical case. It is clear that the elements of $44 \leq A \leq 65$ which are produced from the silicon rich layer and around the bottom of the oxygen rich layer are overproduced up to by a factor of hundred. Therefore, we can say that results for the $20 M_\odot$ star are qualitatively applicable to other massive stars. Integrated abundances are compared with the solar values in Fig. 8, where spherical model (S1) and axisymmeric model (A3) are compared each other. We can see that the model A3 fits some elements to the solar values better than S1. Noticeable is the amount of ^{44}Ti like the case of 20 $M_\odot$ star. Furthermore, isotopes of $Z \geq 44$ are overproduced compared to the model S1. We would stress that since these elements are produced around the mass cut, we are obliged to say that the production depends on the location and/or the form of the mass cut. These results demonstrate that the result of S1 compensates with that of A3 though the more detailed analysis should be necessary. We should remember that there are a lot of elements which are produced through the neutron capture process. During the core helium burning stage, neutron capture nucleosynthesis occurs very efficiently [10]. In Table 1, IMF averaged overproduction factors through the s-process are shown for the elements which are under or overproduced as is seen in Fig. 8. If we compare the products of these neutron rich elements with the solar values, we should take into account of the results of the full s-process calculation which depends on the distribution of seed. In our models, even if we consider the s-process during the core helium burning stage, ^{33}S and ^{47}Ti are underproduced.

DISCUSSION

It may be said that the field of the nucleosynthesis confronts a new frontier: the supernova nucleosynthesis has entered a 2nd generation beyond the spherically symmetric era. Observations will show the importance of the study of the nucleosynthesis in an asymmetric model as we have taken into account. Then, new explosion models of SNIa and SNII will present a new site for production of heavy nuclei through the s-, p- and r-process. Since supernova models become much more complicated from now on, unambiguous nuclear data like the cross sections, β-decay rates, mass excesses, and level densities should be needed. Let us conclude that new developments of models would motivate us to recognize importance of basic physical data like nuclear data.

REFERENCES

1. Arnould, M., Rayet, M., and Hashimoto, M., in this volume.
2. Catchpole R. et. al., *MNRAS* **229**, 15 (1987).
3. Hashimoto, M., *Prog. Theor. Phys.* **94**, 663 (1995).
4. Iyudin A.F. et al. *Proc. of the 2nd INTEGRAL Workshop*, ESA, SP-382 1996, p.37.
5. Kumagai, S. Shigeyama, T., Nomoto, K., Hashimoto, M., and Itoh, S., *A&A*, 273 (1993).
6. Nagataki S., Hashimoto, M., Sato, K., and Yamada, S., *ApJ.* **486**, 1026 (1997).
7. Nagataki S., Hashimoto, M., and Sato, K., in preparation.
8. Nomoto K., and Hashimoto, M., *Phys. Rep.* **163**, 13 (1988).
9. Nomoto, K., et al., astro-ph/9706025 (1997).
10. Prantzos, N., Hashimoto, M., and Nomoto, K., *A&A*, **234**, 211 (1990).
11. Rayet M., Arnould, M., Hashimoto, M., Prantzos, N., and Nomoto,K., *A&A*, **298** 517 (1995)
12. Rayet, M., Hashimoto, M., in this volume.
13. Thielemann, F.-K., Nomoto, K., and Hashimoto, M., *ApJ*, **460**, 408 (1996).
14. Tsujimoto, T., Nomoto, K., Yoshii, Y., Hashimoto, M., Yanagida, S., Thielemann, F.-K., *MNRAS.* **277**, 945 (1995).
15. Woosley, S.E, and Weaver, T.A., *ApJS.* **101**, 181 (1997).
16. Yamada, K., and Sato, K., *ApJ* **434**, 268 (1994).

TABLE 1. IMF averaged overproduction factors over 10 - 50 $M_\odot$ stars obtained from the full s-process network calculations.

^{37}Cl	^{39}K	^{41}K	^{46}Ti	^{49}Ti	^{54}Cr	^{59}Co	^{62}Ni	^{63}Cu	^{65}Cu	^{64}Zn	^{66}Zn
65	3	5	4	6	16	31	66	90	278	54	164

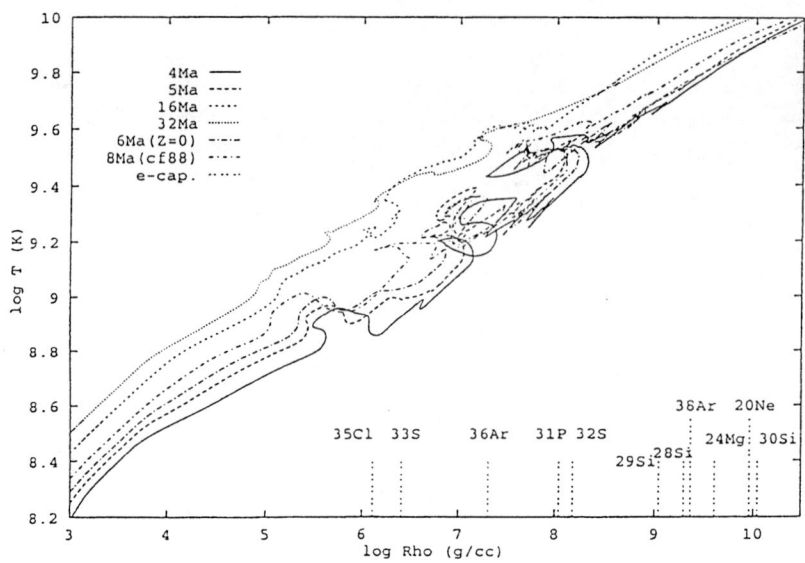

FIGURE 1. The evolution of the central density and temperature for massive stars.

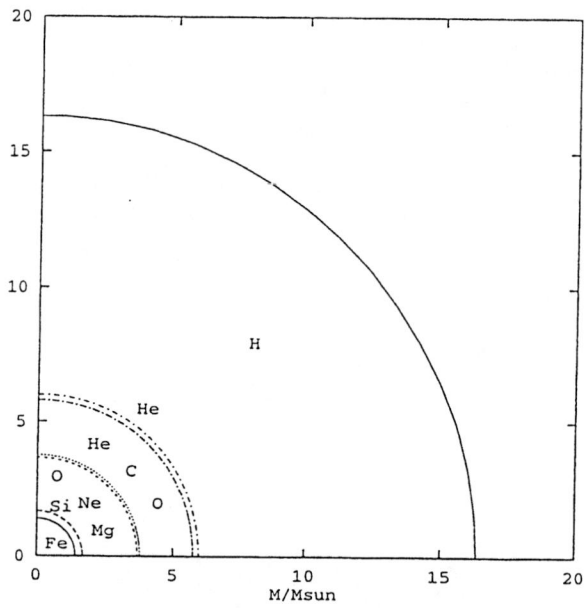

FIGURE 2. Sturcure of 20 $M_\odot$ at the beginning of the core collapse.

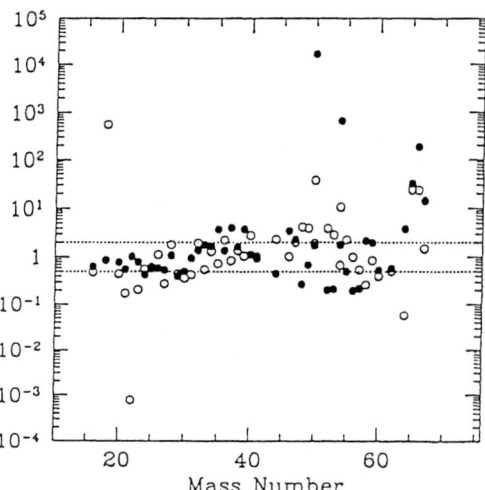

FIGURE 3. Comparison of the abundances for two groups. Open circles represent WDD2/W7 and dots WW95/Ha95.

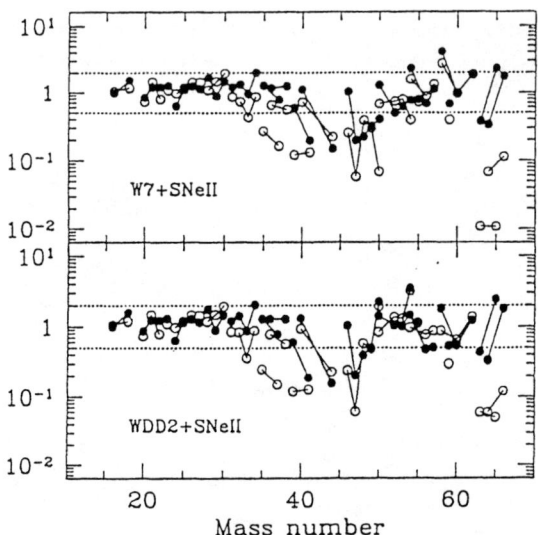

FIGURE 4. The integrated abundances of the ejecta relative to the solar values. As for SNIa, W7 (upper figure) and WDD2 (lower figure), respectively. Open circles:Ha+SNIa. Dots:WW95+SNIa.

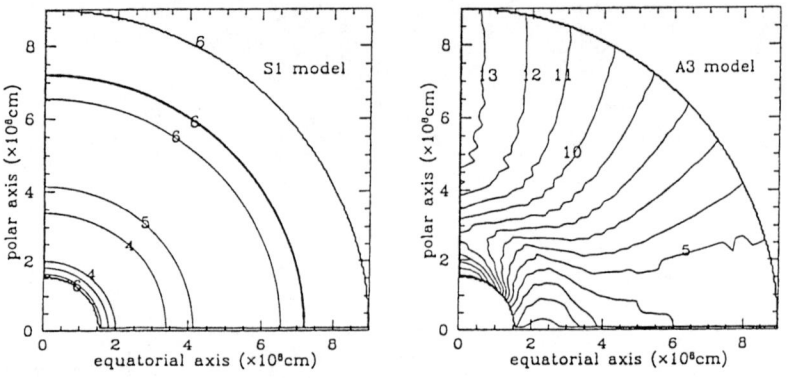

FIGURE 5. Distributions of the entropy for the two cases (left : S1 model, right : A3 model). Entropy is normalized by k_B. Contours are drawn for the initial position of test particles.

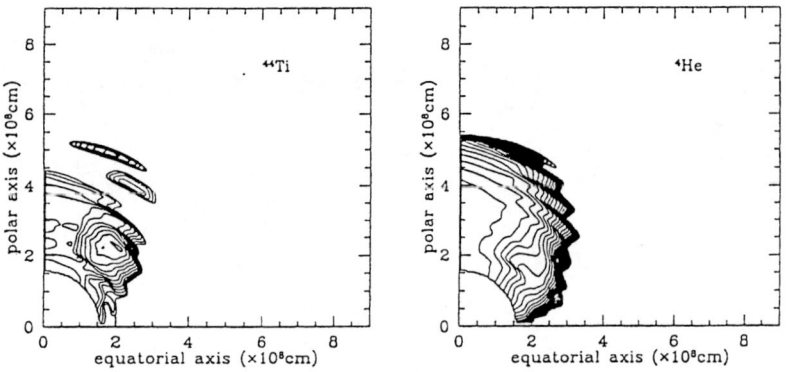

FIGURE 6. Left: Contour of the mass fraction of ^{44}Ti in the A3 case. The maximum value of the mass fraction of ^{44}Ti is 1.3×10^{-2}. Right : Same as left but for ^{4}He. The maximum value is 4.0×10^{-1}. Contours are drawn for the initial position of test particles.

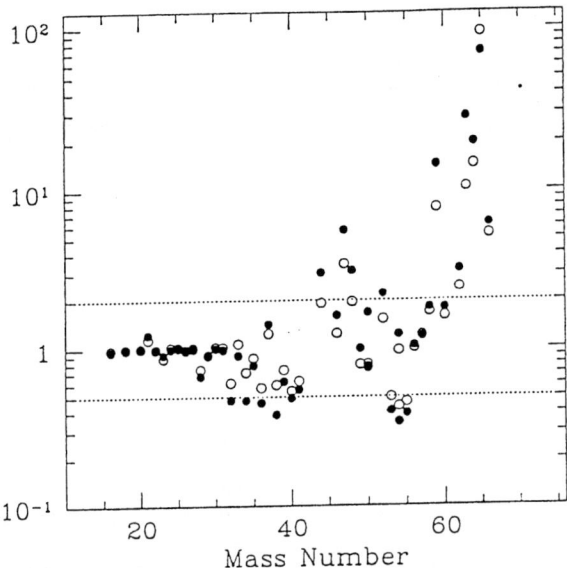

FIGURE 7. Comparison of the abundances relative to a spherical model. Open circles:A1/S1. Dots:A3/S1.

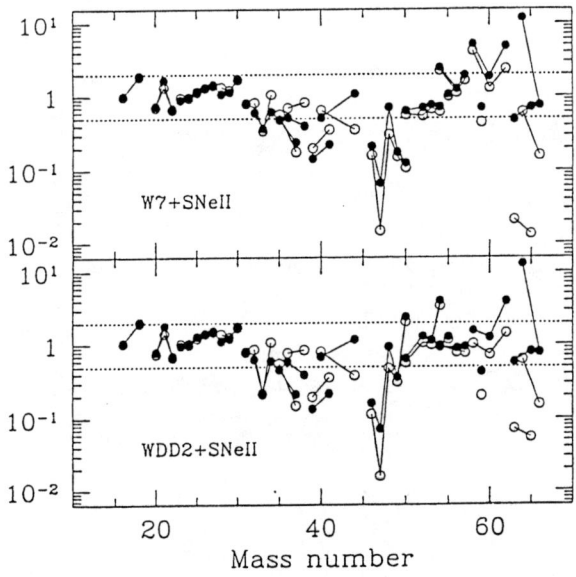

FIGURE 8. Integrated abundances as Fig. 4. Open circles: SNIa+S1. Dots: SNIa+A3.

THE PRE-SUPERNOVA EVOLUTION OF MASSIVE STARS AND CONCOMITANT NUCLEOSYNTHESIS

Georges Meynet

Geneva Observatory, CH-1290 Sauverny, Switzerland

Abstract. After a recall of the main features of massive star evolution, we briefly describe the different nuclear burning phases through which these stars evolve. We discuss determinations of stellar surface abundances which provide some clues to the nuclear processes occuring in massive stars. Finally, we emphasize the role played by mass loss in the process of enrichment of the interstellar medium in newly synthesized elements.

I THE MASSIVE STARS

Hereafter we shall call massive stars, the stars that go through the whole series of nuclear burning phases (H, He, C, Ne, O and Si-burning) until the formation of an iron core. The minimum initial mass for a star to have such an evolution is estimated to be 8 $M_\odot$ by Maeder & Meynet (1989) based on grids of stellar models with mass loss and core overshooting and 10 $M_\odot$ by Becker and Iben (1979) and Nomoto (1984) from model calculations not considering these effects. Massive stars are the progenitors of type II supernovae and probably of a part of type Ib/c (in which case they would result from the explosion of a Wolf-Rayet star, see the review by Wheeler & Swartz 1994). Their final fate is either a neutron star or a black hole, depending on the mass of their iron core: for core masses below the Oppenheimer-Volkoff's mass (around 2 $M_\odot$), the final product is a neutron star, otherwise the pressure exerted by the degenerate neutrons is insufficient to prevent core collapse and a black hole forms.

Let us add here that the most massive stars ($M > 100$ $M_\odot$), having a mass of the helium core superior to 35 $M_\odot$, encounter the pair instability during the oxygen burning phase. This instability occurs because, at temperatures around $2\ 10^9$ K, a large part of the energy released by gravitational contraction goes into the creation of pairs of electrons and positrons (Fowler & Hoyle 1964). This process subtracts energy that would otherwise provide pressure support, and therefore may trigger a dynamical collapse. Numerical models describing the final fate of these stars are presented for instance in Woosley (1986, see references therein). Let us note that this instability will never occur if, as an effect of strong mass loss by stellar winds, the mass of the He core is decreased below a value equal to ~ 35 $M_\odot$. This may well be the case at solar metallicity, where all stars more massive than 40 $M_\odot$ end their stellar life with final masses between 5 and 10 $M_\odot$ (cf. Schaller et al. 1992). In that case, the pair instability would be encountered only by very massive stars at low metallicity (in metal poor regions, the stellar winds are believed to be much weaker, see section IV).

In the following, we restrain the discussion to the case of single stars. Let us mention however that in a close binary, the evolutionary scenarios can be radically different from those followed by single stars. In a close binary system, mass can be stripped from either star in a Roche-lobe overflow event and either transferred to the other star or lost from the system or both. The reader will find a comprehensive introduction to the field of interacting binaries in van den Heuvel (1994).

How does the nucleosynthesis occuring in massive stars intervene in the more general context of the chemical evolution of galaxies ? As a consequence of their very short lifetimes (a few 10^6–10^7 years) and their ability to synthesize heavy elements, massive stars are among the most important contributors to the chemical evolution of galaxies. It is especially true during the first phases of the chemical evolution of a galaxy or of a starburst region. Indeed, at the beginning, only massive stars contribute, since the other sources haver longer evolutionary timescales.

Later, other nucleosynthetic sources become also important contributors to the chemical enrichment of the interstellar medium. Let us mention the two most important ones :

1) The low and intermediate mass stars (see the review by N. Mowlavi in this volume): the nucleosynthetic active stars originate from the initial mass range between 1 and 8–10 $M_\odot$. These stars begin to enter the game of the chemical evolution only after a period of $\sim 10^8$–10^9 years. They contribute to the synthesis of helium, carbon and nitrogen (primary and secondary). Numerous "s-process" elements are also produced by this type of stars.

2) The type Ia supernovae: these supernovae occur when a white dwarf in a binary system, as a result of accretion or of merging, acquire a mass superior to the Chandrasekhar mass (see Hashimoto et al. in this volume). Carbon or helium ignition occurs then in highly degenerate conditions leading to the complete destruction of the star. Type Ia supernovae are important contributors to the enrichment in iron of the galactic matter. According to Timmes et al. (1995), type II supernovae produce about 2/3's of the solar system iron

abundance, while type Ia are responsible for the rest. Let us note that the evolutionary timescale to produce a type Ia supernova is not known precisely. Indeed to the lifetime of an intermediate mass star (between 10^8 to 10^9 years), one must add the uncertain time needed for the accretion or merging process to occur.

Other sources may also contribute to the synthesis of some particular isotopes. Novae for instance may contribute to the synthesis of ^{15}N (see the review by Jose & Hernanz in this volume).

II NUCLEAR BURNING PHASES

A CNO-cycle

The CNO cycle is the main H-burning process in main sequence stars with central temperatures superior to $18\ 10^6$ K (initial mass $>\sim 1.2$ M$_\odot$). CNO elements act as catalysers of the proton fusion reactions. The sum of their abundances remains constant during this process, but due to the very small rate of ^{14}N$(p,\gamma)^{15}$O, most of the ^{12}C and ^{16}O are transformed into ^{14}N. As a numerical example, the Solar System number ratios of C/N and O/N amount to 4 and 9, respectively (see Anders & Grevesse 1989). When CNO has reached equilibrium, C/N and O/N take values as low as 0.03 and 0.02, respectively. Let us recall that in cosmic material, CNO represent about 80% of the heavy elements mass fraction, oxygen alone contributing to about half of the heavy elements mass fraction.

In addition to the CNO cycle, two cycles/chains are active: the Ne-Na cycle and the Mg-Al chain. These processes are responsible for the synthesis of some sodium and aluminium (in particular the isotope ^{26}Al which is important for γ-ray line astronomy, see the review by Prantzos and Diehl 1996 and section D below).

B He-burning phase

At the beginning of the He-burning phase, ^{12}C is synthesized by the 3α reaction. At the end, part of that ^{12}C is transformed into ^{16}O by ^{12}C$(\alpha,\gamma)^{16}$O. This last reaction plays an important role in stellar evolution. It determines the chemical composition of the core at the end of the He-burning phase (typically at the end of the He-burning phase of a 20 M$_\odot$, one obtains 78% of ^{16}O and 19% of ^{12}C in mass fraction, see Schaller et al. 1992) fixing thereby the quantity of available fuel for the carbon burning phase and the efficiency of the He-burning phase for producing ^{16}O [1].

---

[1] ^{16}O is synthesized from a single well defined source: hydrostatic core He-burning phase in presupernova stars.

The rate of this reaction is likely enhanced by a factor between 1.5 and 3 compared with the one given in Caughlan & Fowler (1988, see for instance the discussion in Schaller et al. 1992). The higher the rate, the more ^{16}O is synthesized at the end of the He-burning phase, and the less ^{12}C is available for the carbon burning phase. As a numerical example, the value of the C/O ratio at the end of the He-burning phase in a 15 $M_\odot$ model calculated with the rate of Fowler et al. (1975[2]) is 1.65 in mass fraction. For the same initial mass, this ratio is lowered to 0.22 when calculated with the rate enhanced by a factor 3 (see the discussion in Maeder & Meynet 1987). According to Woosley (1986), the use of a high rate for ^{12}C$(\alpha,\gamma)^{16}$O in massive star calculations favors the formation of big iron cores and thus of black holes: for instance, a 25 $M_\odot$ model calculated with the Fowler et al (1975) rate yields an iron core mass of 1.35 $M_\odot$, while the same model using a rate enhanced by about a factor 3, gives an iron core of 2.1 $M_\odot$.

Other important chains of reactions accompany the He-burning process. In particular, the ^{14}N left by the previous H-burning phase is rapidly destroyed at the beginning of the He-burning phase and transformed, through a chain of reactions, into ^{22}Ne. Some fluorine is also produced at the beginning of this phase (see Goriely et al. 1989 and section D). The ^{13}C, produced by CNO processing in the previous nuclear phase, is destroyed at the beginning of the He-burning phase by the ^{13}C$(\alpha,n)^{16}$O reaction. The neutrons emitted in this way can be captured by heavier elements and produce some "s-process" elements (see the review by Rayet & Hashimoto in the present volume). A more efficient source of neutrons intervenes at the end of the He-burning phase when ^{22}Ne is partly destroyed by ^{22}Ne$(\alpha,n)^{25}$Mg. A short study of the impact of recent determinations of the rates of ^{13}C$(\alpha,n)^{16}$O and ^{22}Ne$(\alpha,n)^{25}$Mg on the neutron production may be found in Meynet & Arnould (1993).

C Advanced evolutionary phases in massive stars

After the core He-burning phase, the star goes through the carbon burning, the neon-photodisintegration, and the oxygen and silicon burning stages (see Clayton 1968). The hydrostatic burning processes occuring during the pre-supernova evolution of massive stars produce essentially isotopes lighter than calcium. Among these isotopes, the so called α-elements which are formed through series of (α, γ) reactions from ^{12}C (such as ^{20}Ne, ^{24}Mg, ^{28}Si ...) are produced in large quantities as a consequence of their great stability. In the Solar System material, these α-elements are, as expected, the most abundant isotopes among the elements between carbon and calcium (cf Anders & Grevesse 1989). Isotopes lighter than calcium are little affected by the explosive nucleosynthesis, contrarily to heavier ones which are sensitive to the uncertain modeling of the explosion mechanism (see Woosley & Weaver 1995).

---

[2] This rate is similar to the rate proposed by Caughlan & Fowler (1988).

III SURFACE ABUNDANCES AND NUCLEAR PROCESSES

Wolf-Rayet (WR) stars give us the opportunity to look almost directly to the H and He-burning products, allowing us to test part of our ideas concerning the nucleosynthesis processes in massive stars. Let us recall that WR stars appear as a normal evolutionary stage of massive stars, with initial mass greater than $\sim 25 - 30$ M$_\odot$ for population I stars. This mass limit depends on the initial metallicity (Maeder 1991; Maeder & Meynet 1994). Due to strong stellar winds, the original stellar envelope is removed and nuclear matter which was initially in the CNO core (WN stars) and then in the 3α-burning core (WC stars) appears at the surface.

The CNO abundances observed at the surface of WN stars are well accounted for by the stellar models (Nugis 1991, Willis 1996). In that case, the surface abundance predictions are not very sensitive to the various physical uncertainties affecting stellar models (such as mass loss rates, mixing), since they are the mare result of the CNO cycles at equilibrium. The good agreement between observation and theory may thus be taken as an indication of the general correctness of our understanding of the CNO cycles and of the relevant nuclear data (Maeder 1983). For what concerns WC stars, the situation is more complicated and such a statement cannot be made. However, comparisons between predicted and observed abundances do confirm the He-burning origin of the matter seen at the surface of these stars. In that respect, let us mention the recent result obtained by Willis et al. (1997) with the ISO (Infra Red Observatory) satellite. These authors succeeded in measuring the neon abundance at the surface of a WC star (WR 146) and obtained a value in good agreement with the Ne-enrichment predicted by the models. This result is in contrast with a previous observation of another WC star (γ^2Vel) which showed no Ne enrichment (Barlow et al. 1988). In this last case, however, uncertainties concerning the intensity of the mass loss rate hampers the interpretation of the result.

The observations by Willis et al. (1997), in addition to provide a nice check of a theoretical prediction, has other interesting consequences. Indeed, by confirming the high level of neon abundance (believed to be under the form of ^{22}Ne), 1) it supports the view according to which this isotope could be an important neutron source at the end of the He-burning phase, 2) it makes more plausible the role WC stars might play in building the non solar ^{22}Ne/^{20}Ne ratio observed in the Galactic Cosmic Rays (see Webber et al. 1996 for a recent determination of the isotopic composition in Cosmic Rays and Meynet & Maeder 1997 for a comparison with the predictions of a model accounting for the effects of injection in the Cosmic Rays of material originating from WR stars [3]).

---

[3] A detailed description of this model is given in Maeder & Meynet 1993.

IV MASS LOSS AND CHEMICAL EVOLUTION

In the mass domain superior to 30-40 $M_\odot$ the mass loss by stellar winds is clearly the dominant factor affecting almost all the outputs of the stellar models (see the review by Maeder & Conti 1994).

The physical mechanisms by which mass is ejected from stars are still subject to debate. For luminous early type stars, the most developed of all the models available is the radiation-driven wind theory (Lucy & Solomon 1970; Castor et al. 1975). In the frame of this theory, it is possible to show (see e.g. Kudritzki et al. 1988, p. 174 and followings) that the mass loss rate depends on the metallicity. This dependence comes out as a consequence of the increase of the opacity in the outer layers of the star, when the metallicity increases (more numerous bound-bound and bound-free transitions). Greater opacities favor a more efficient transfer of the momentum transported by the photons to the matter. Abbott (1982) finds an almost linear dependence between mass loss and metal abundance, while Kudritzki et al. (1988) find that the mass loss rate is proportional to approximately the square root of the metallicity.

Maeder (1992) studied the metallicity dependence of the stellar yields implied mainly by the dependence of the mass loss rates on the metallicity. By stellar yields, we mean here the quantity of an element which has been synthesized and ejected into the interstellar medium by a given star during its whole life. As seen above, when the metallicity increases, the proportion of matter expelled by stellar winds increases. Thus at low metallicities, one expects that massive stars eject most of their yields at the time of the supernova explosion, while at high Z (Z is the heavy elements mass fraction), the ejection occurs both during the pre-supernova stage, through stellar winds, and during the supernova explosion itself. These two situations are not equivalent. Indeed, matter expelled by stellar winds has a chemical composition characteristic of a less advanced evolutionary stage than matter expelled during the supernova explosion. Thus its chemical composition is different. To understand this point, let us consider first an He-burning core just at the beginning of the He-burning phase. If the core is naked and suffers mass loss (typically a WC star) the matter expelled by stellar winds is rich in ^{4}He and ^{12}C. At the end of the He-burning phase, a small O-rich core forms. Now if we consider the evolution of the same core but with no mass loss, then the whole core matter would be transformed into essentially oxygen. Nearly no carbon and absolutely no helium would be ejected from the former He-burning core at the time of the supernova explosion. In this case, one expects that evolution with mass loss produces ejecta richer in elements characteristics of the beginning of the He-burning phase (^{4}He and ^{12}C), while evolution without mass loss would produce ejecta richer in elements characteristics of the end of the He-burning phase (^{16}O). The stellar yields from massive stars are thus quite different depending on the intensity of the stellar winds and therefore on the initial metallicity.

The example sketched above shows that mass loss by stellar winds enables to "save" some elements from further destruction. The abundances of the ejected material are frozen at the state reached at the time of ejection and are no longer modified by nuclear processing. This fact may play a role in the process of enrichment of the interstellar medium in ^{26}Al and ^{19}F by massive stars. Indeed, in massive stars, ^{26}Al is synthesized in the H-burning core and destroyed in the core He-burning phase. Part of the ^{26}Al produced during the main sequence may escape destruction by being ejected through stellar winds during the WN phase (see Fig. 1). Fluorine is built up at the beginning of the He-burning phase and destroyed at the end of this phase. Again, stellar winds during the WC phase enable to remove from the star part of the synthesized ^{19}F before it is destroyed. According to our stellar models, it appears that in both cases, these types of injections of ^{26}Al and ^{19}F play an important role, if not dominant, in the building of their observed abundances in the interstellar medium (Meynet et al. 1997; Arnould & Meynet 1997; Meynet & Arnould 1997).

We also analysed the importance of various physical ingredients of stellar models on the contribution of WR stars to the ^{26}Al synthesis. We concluded that the lack of a precise knowledge of the mass loss rates and the limited reliability of the mixing prescriptions are the main factors limiting the accuracy of the WR ^{26}Al yield predictions. Each of these sources of uncertainties may lead to variations in the calculated production of that radionuclide that can amount to a factor of the order of 2 to 3. In contrast, more limited uncertainties arise from a purely nuclear physics origin.

To end this section, let us add that, according to Maeder (1992), yields might depend on the metallicity also through another mechanism. Let us suppose that when a black hole forms, all the matter is swallowed by the black hole. In that case, the progenitors of black holes contribute to the enrichment of the interstellar medium only through their stellar winds. It is reasonable to believe that the initial mass range for black hole formation depends on the metallicity. Indeed at low metallicity, the most massive stars arrive at the end of their evolution with still a great part of their initial mass and thus will certainly give birth to a black hole. In contrast at high metallicity, the star nearly evaporates under the effects of the stellar winds and thus ends its stellar life with a very small final mass which produces a neutron star or may be sometimes a white dwarf (see Meynet et al. 1994). If black hole formation does not lead to a supernova explosion, one expects that at low Z, yields are poor in heavy elements (locked into the black hole). This leads to values of $\Delta Y/\Delta Z$ (helium over heavy elements enrichments) as high as 4-5 in agreement with the determinations by Pagel et al. (1992). Let us note that He-enrichment is less affected by black hole formation, since He is produced in great quantities by intermediate mass stars, massive stars contributing also through their winds. At high Z, lower values of $\Delta Y/\Delta Z$ are expected, since strong winds from massive stars are carrying away a lot of heavy elements and

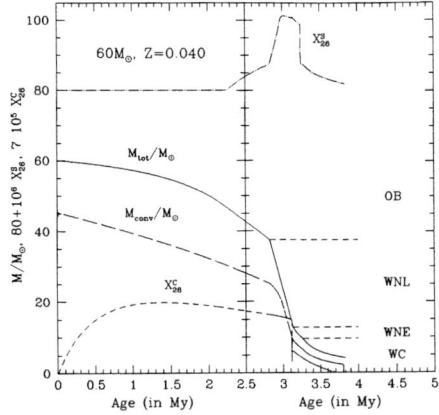

FIGURE 1. Evolution of the total mass M_{tot}, of the mass of the convective core M_{conv}, and of the central (X_{26}^c) and surface (X_{26}^s) ^{26}Al mass fractions for a 60 $M_\odot$ model star with metallicity $Z = 0.04$. The various spectroscopic types encountered during the evolution are indicated on the right of the figure: OB for OB main sequence stars, WNL, WNE and WC for the different classes of WR stars. Note the different ordinate scales defined on the left of the figure. From Meynet et al. 1997.

the process of black hole formation is expected to be less frequent as a result of smaller final masses.

V CONCLUSION

Massive stars are quite "generous" stars. They are responsible for the synthesis of the bulk of the heavy elements and are strong injectors of light, mass, momentum and energy into the interstellar medium. We recall two important consequences of their heavy stellar winds: 1) they permit to look at bare cores and in this way to test part of our ideas concerning massive star nucleosynthesis, 2) they save from further destruction some elements synthesized in the course of the evolution and are responsible for a great part of the metallicity dependence of the stellar yields.

Mass loss by stellar winds is not the only important process occuring during massive star evolution. Mixing processes and in particular those induced by rotation may also play a key role. In that respect, it is worthwhile to emphasize that the uncertainties affecting the nuclear reaction rates may be completely blurred out by other unknown physical ingredients. This is the reason why it is generally not possible from stellar models and still less from models for the chemical evolution of the Galaxy, to derive constraints on

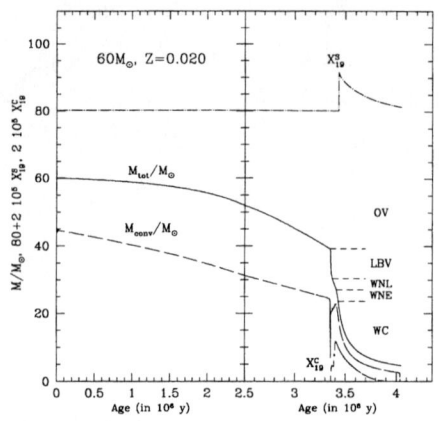

FIGURE 2. Same as Fig. 1 for ^{19}F and a 60 M$_\odot$ model star with metallicity $Z = 0.02$. OV means O-type main sequence stars, LBV, Luminous Blue Variable. From Meynet & Arnould 1997.

nuclear aspects. However numerical simulations can give indications on the consequences brought about by the change of the rate of some key nuclear reaction rates, helping us to identify the most important ones for the energetic and nucleosynthetic aspects.

ACKNOWLEDGMENTS

I am very grateful to Nami Mowlavi for helpful comments and suggestions which improved the manuscript.

REFERENCES

1. Abbott D.C., *ApJ* **259**, 282 (1982).
2. Anders E., Grevesse N., *Geochim. Cosmochim. Acta* **53**, 197 (1989).
3. Arnould M., Meynet G., in *Proceedings 2nd INTEGRAL Workshop 'The Transparent Universe'*, Eds. Winkler C. et al., ESA SP-382, p. 33 (1997).
4. Barlow M.J., Roche P.F., Aitken D.K., *MNRAS* **232**, 821 (1988).
5. Becker S.A., Iben I., *ApJ* **232**, 831 (1979).
6. Castor J., Abbott D.C., Klein R., *ApJ* **195**, 157 (1975).
7. Caughlan G.R., Fowler W.A., *Atom. Data Nucl. Data Tables* **40**, 283 (1988).
8. Clayton D.D., in *"Principles of stellar evolution and nucleosynthesis"*, McGraw-Hill, New-York (1968).
9. Fowler W.A., Caughlan G.R., Zimmermann B.A., *ARAA* **13**, 113 (1975).

10. Fowler W.A., Hoyle F., *ApJSS* **9**, 201 (1964).
11. Goriely S., Jorissen A., Arnould M. 1989, in *Proc. 5th Worshop on Nuclear Astrophys.*, eds Hillebrandt W. & Müller E., MPI Rep., p. 60.
12. Kudritzki R.P, Pauldrach A., Puls J., in *"O-stars and WR stars"*, NASA SP-497, eds. Conti P.S. & Underhill A.B., p. 173 (1988).
13. Lucy L.B., Solomon P., *ApJ* **159**, 879 (1970).
14. Maeder A., *A&A* **120**, 113 (1983).
15. Maeder A., *A&A* **242**, 93 (1991).
16. Maeder A., *A&A* **264**, 105 (1992).
17. Maeder A., Conti P., *ARAA* **32**, 227 (1994).
18. Maeder A., Meynet G., *A&A* **182**, 243 (1987).
19. Maeder A., Meynet G., *A&A* **210**, 155 (1989).
20. Maeder A., Meynet G., *A&A* **278**, 406 (1993).
21. Maeder A., Meynet G., *A&A* **287**, 803 (1994).
22. Meynet G., Arnould M., in *Nuclei in the Cosmos*, eds. Käppeler F. & Wisshak K., IOP, Bristol, p. 487 (1993).
23. Meynet G., Arnould M., in *WR Stars in the Framework of Stellar Evolution*, 33rd Liège Inter. Astrophys. Coll., Eds. Vreux J.M. et al., Université de Liège, p. 89. (1997).
24. Meynet G., Arnould M., Prantzos N., Paulus G., *A&A* **320**, 460 (1997).
25. Meynet G., Maeder A., *Adv. Space Res.* **19**, 763 (1997).
26. Meynet G., Maeder A., Schaller G., Schaerer D., Charbonnel C., *A&AS* **103**, 97 (1994).
27. Nomoto K., in *"Stellar Nucleosynthesis"*, eds. Chiosi C. & Renzini A., Reidel, Dordrecht, p. 238 (1984).
28. Nugis T., in *"Evolution of Stars: The Photospheric Abundance Connection"*, IAU Symp. **145**, eds. Michaud G. & Tutukov A., Kluwer, p. 209 (1991).
29. Pagel B.E.J., Simonson E.A., Terlevich R.J., Edmunds M.G., *MNRAS* **255**, 325 (1992).
30. Prantzos N., Diehl R., *Phys. Rep.* **267**, 1 (1996).
31. Schaller G., Schaerer D., Meynet G., Maeder A., *A&AS* **96**, 269 (1992).
32. Timmes F.X., Woosley S.E., Weaver T.A., *ApJS* **98**, 617 (1995).
33. van den Heuvel E.P.J., in *"Interacting Binaries"*, 22nd Saas-Fee Course, eds. Nussbaumer H. & Orr A., Springer-Verlag, Berlin, p. 263 (1994).
34. Webber W.R., Lukasiak A., McDonald F.B., Ferrando P., *ApJ* **457**, 435 (1996).
35. Wheeler J.C., Swartz D.A., *Space Science Reviews* **66**, 425 (1994).
36. Willis A.J., *Astrophys. and Space Science* **237**, 145 (1996).
37. Willis A.J., Dessart L., Crowther P.A., Morris P.W., Maeder A., Conti P.S., van der Hucht K.A., *MNRAS* **290**, 371 (1997).
38. Woosley S.E., in *"Nucleosynthesis and Chemical Evolution"*, 16th Saas-Fee Course, eds. Hauck B. et al., Geneva Observatory, p. 1 (1986).
39. Woosley S.E., Weaver T.A., *ApJS* **101**, 181 (1995).

BINARY STAR EVOLUTION AND THE HOT MODES OF H-BURNING

Nuclear Uncertainties and Their Role in Nova Nucleosynthesis

Jordi José[*] and Margarita Hernanz[†]

[*] *Departament de Física i Enginyeria Nuclear,*
Universitat Politècnica de Catalunya, EUPVG,
Avda. Víctor Balaguer s/n, 08800 Vilanova i la Geltrú (Barcelona), SPAIN
and
Institut d'Estudis Espacials de Catalunya (IEEC),
Edifici Nexus-104, C./ Gran Capità 2-4, 08034 Barcelona, SPAIN
[†] *Institut d'Estudis Espacials de Catalunya (IEEC), CSIC Research Unit,*
Edifici Nexus-104, C./ Gran Capità 2-4, 08034 Barcelona, SPAIN

Abstract. We report on 13 evolutionary sequences of nova outbursts, from the onset of accretion to the expansion and ejection stages, for a wide range of input parameters, such as the white dwarf mass (0.8 to 1.35 $M_\odot$), and the degree of mixing between the accreted envelope and the outermost shells of the underlying core (25 to 75%). Both CO and ONe white dwarfs are considered. The hydrodynamic code is linked to a complete nuclear reaction library including 370 reactions which follows the detailed evolution of 100 isotopes, ranging from ^{1}H to ^{40}Ca.

Special emphasis is focused on the mean composition of the ejecta, which provide large amounts of nuclei of astrophysical interest, such as ^{7}Be or ^{22}Na (which might be detected by future gamma-ray missions), ^{26}Al, or other species that may account for their Galactic content, like ^{13}C, ^{15}N or ^{17}O. A detailed analysis of the nuclear flows involved in the synthesis of several nuclear species and the corresponding uncertainties in the rates is also outlined.

INTRODUCTION

The cornerstone for the theoretical model of nova outbursts is the observational analysis of old novae (Walker 1954, Kraft 1963). These seminal works have settled the binary nature of a classical nova: a large, cool main sequence star that fills its Roche lobe, and a small white dwarf star. Mass transfer through the inner lagrangian point is responsible for the formation of an accretion disk that surrounds the compact white dwarf star. As a result, a fraction of the material lost by the secondary spirals and finally ends up on top of the white dwarf. The piling up of matter gradually compresses and heats the envelope, up to the point when ignition conditions to drive a thermonuclear runaway (hereafter, TNR) are reached.

Several hydrodynamic computations of nova outbursts have analyzed the chemical composition of the ejecta of classical novae, (see Starrfield et al. 1997, Kovetz & Prialnik 1997, José & Hernanz 1997, and references therein). Due to the high temperatures achieved in the envelope during the TNR, with typical peak values $\sim (2-3) \times 10^8$ K, novae eject significant amounts of nuclear processed material into the interstellar medium. This has raised the question of the potential role of classical novae in the Galactic abundances. Assuming a Galactic nova rate of 30 events yr^{-1}, a Galaxy's lifetime of 10^{10} yr and an average ejected mass per nova outburst of $\sim 2 \times 10^{-5}$ M$_\odot$, one can estimate the total amount of mass ejected by novae in the Galaxy's history. The estimated value, $\sim 6 \times 10^6$ M$_\odot$, accounts for only $\sim 1/3000$ of the Galactic disk's gas and dust content. This suggests that despite the large occurrence rate of novae in the Galaxy, they scarcely play a role in the Galactic chemical puzzle, compared with other major sources such as supernovae or AGB stars. Nevertheless, classical novae may represent likely sites for the synthesis of individual nuclei with overproduction factors $f \equiv X_i/X_{i,\odot} \geq 1000$ (where X_i and $X_{i,\odot}$ are the mass fractions of species i in the ejected envelope and the solar value, respectively), such as ^{13}C, ^{15}N or ^{17}O.

In this paper, we reinvestigate the role played by classical nova outbursts on the synthesis of chemical species, for both CO and ONe novae, with special emphasis on the dominant nuclear flows during the TNR.

MODEL AND RESULTS

A one-dimensional, implicit, hydrodynamical code (SHIVA), in lagrangian formulation, has been developed to analyze the course of nova outbursts, from the onset of accretion up to the expansion and ejection stages, following the method described in Kutter & Sparks (1972, 1980). The code has been linked to a reaction network, which follows the detailed evolution of 100 nuclear species, ranging from ^{1}H to ^{40}Ca, through 370 nuclear reactions, with updated rates and screening factors. Details of the code can be found in José (1996) and José & Hernanz (1997).

The effect of the white dwarf mass has been tested through a number of simulations involving both CO white dwarfs ($M_{wd} = 0.8, 1.0$ and 1.15 M$_\odot$) and ONe ones ($M_{wd} = 1.0, 1.15, 1.25$ and 1.35 M$_\odot$). The overlapping between both intervals is due to the uncertain exact upper limit for CO degenerate cores. Abundances at the edge of a 1.20 M$_\odot$ degenerate core, resulting from the evolution of a 10 M$_\odot$ Population I star (Ritossa, García–Berro & Iben 1996) have been adopted for the ONe models, whereas a composition of X(^{12}C)=0.495, X(^{16}O)=0.495 and X(^{22}Ne)=0.01, has been adopted for the CO ones. A mass accretion rate of 2×10^{-10} M$_\odot$ yr^{-1} and an initial luminosity of 10^{-2} L$_\odot$, rather typical values, have been adopted in order to limit the parameter space.

The main characteristics of the evolutionary sequences presented in this paper are given in Tables 1 and 2: the initial white dwarf mass, M_{wd}, and the adopted mixing level between core and envelope are input parameters; ΔM_{env} is the envelope's mass at the end of the accretion stage; t_{acc} is the duration of the accretion phase; t_{rise} is the time required for a temperature rise from $T_{bs} = 3 \times 10^7$ K to 10^8 K, at the burning shell; $\varepsilon_{nuc,max}$ and T_{max} are peak nuclear energy generation rate and maximum temperature at the burning shell, respectively; t_{max} is the time required to reach peak temperature from $T_{bs} = 10^8$ K; ΔM_{ejec}, v_{ejec} and K represent the total mass, the mean velocity and the mean kinetic energy of the ejected envelope. The mean composition of the ejecta is given in Tables 3 and 4, in mass fractions.

As a framework for the analysis, we will describe the evolution of Model ONe3: a 1.15 $M_\odot$ ONe white dwarf with a 50% mixing with core material and accreting mass at a rate of 2×10^{-10} $M_\odot$ yr^{-1}.

The early accretion phase is dominated by p-p chains (mainly ^{1}H(p,e$^+\nu_e$)^{2}H). As the temperature rises, contribution from the CNO-cycle reaction ^{12}C(p,γ)^{13}N, followed by its β^+-decay into ^{13}C, become also important. The accretion stage lasts for $\sim 2 \times 10^5$ yr, up to the point when the nuclear timescale becomes shorter than the accretion timescale (which is achieved for $T_{bs} = 3 \times 10^7$ K). The mass of the accreted envelope is 3.2×10^{-5} $M_\odot$.

The beginning of the TNR is accompanied by the development of a convective zone just above the burning shell, which rapidly expands towards the outer envelope. When T_{bs} reaches 7×10^7 K, convection extends already through a region of 170 km above the ignition shell, with a mean convective velocity of 10^5 cm s^{-1}. The release of nuclear energy is dominated by the cold CNO-cycle, mainly through ^{12}C(p,γ)^{13}N(β^+)^{13}C(p,γ)^{14}N, ^{14}N(p,γ)^{15}O and ^{16}O(p,γ)^{17}F(β^+)^{17}O(p,α)^{14}N, plus a contribution from some reactions of the NeNa and MgAl-cycles, such as ^{21}Ne(p,γ)^{22}Na, followed by ^{25}Mg(p,γ)^{26}Al^g. The rise of temperature is accompanied by a shift in the dominant nuclear flows towards higher Z nuclei. At $T_{bs} = 10^8$ K, convection has already extended through the whole envelope. The Model has spent $t_{rise} \sim 1.3 \times 10^7$ s to rise from $T_{bs} = 3 \times 10^7$ K to 10^8 K. When temperature reaches 2.1×10^8 K, the star achieves a maximum rate of nuclear energy generation of $\varepsilon_{nuc,max} = 7.6 \times 10^{15}$ erg g^{-1} s^{-1}. At this time, significant energy production comes from the hot CNO-cycle, mainly through ^{12}C(p,γ)^{13}N(p,γ)^{14}O, ^{14}N(p,γ)^{15}O and ^{16}O(p,γ)^{17}F. Leakage from the MgAl-cycle through proton captures on ^{27}Si and ^{27}Al becomes progressively important. A peak temperature of $T_{max} = 2.2 \times 10^8$ K is attained 540 s after the ignition shell reached 10^8 K. Main nuclear flows at peak temperature are shown in Figure 1. As a result of the violent TNR, 1.9×10^{-5} $M_\odot$ are ejected (60% of the formerly accreted envelope), with a mean velocity of ~ 2400 km s^{-1}.

In order to check the effect of the white dwarf core composition, we have evolved Model CO5, a CO white dwarf with the same input parameters than

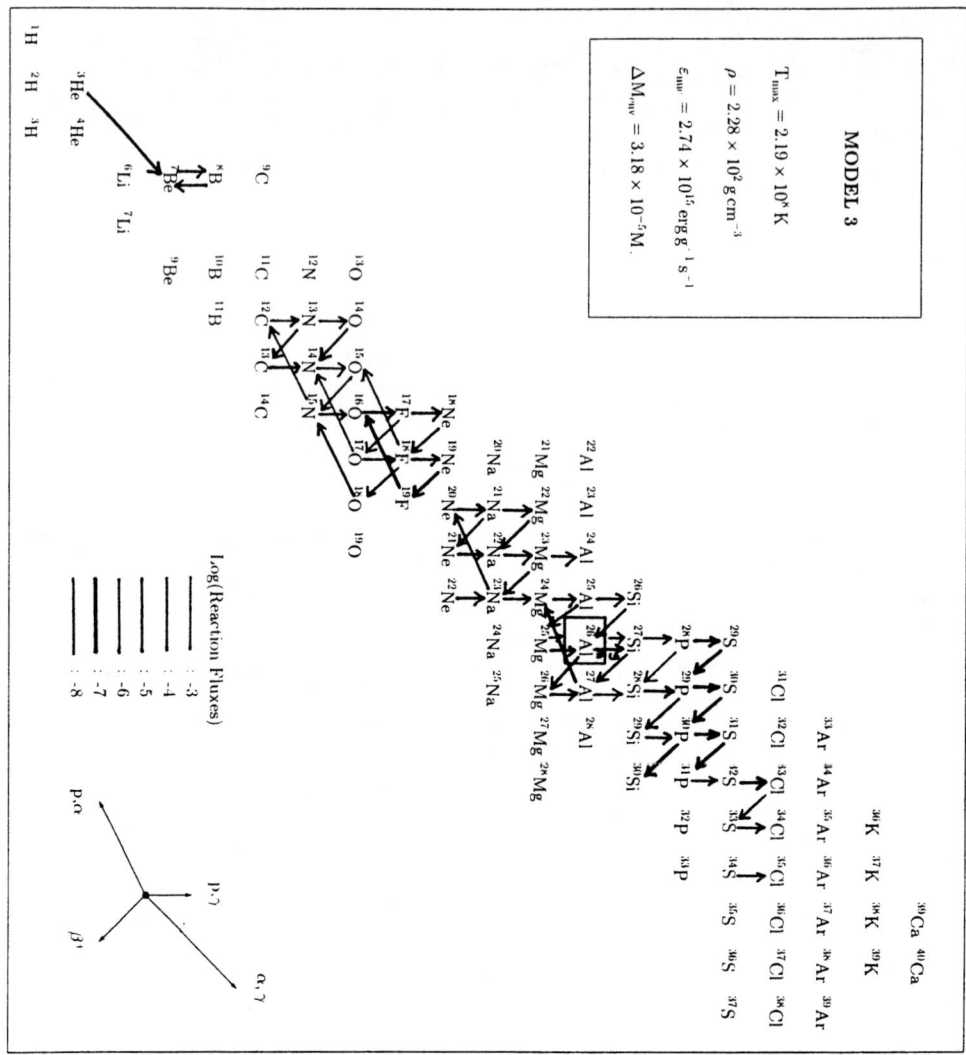

FIGURE 1. Main nuclear flows at peak temperature for Model ONe3 (a 1.15 $M_\odot$ ONe white dwarf, with 50% mixing).

TABLE 1. Input parameters and results of the ONe nova models

	ONe1	ONe2	ONe3	ONe4	ONe5	ONe6	ONe7
M_{wd} (M$_\odot$)	1.00	1.15	1.15	1.15	1.25	1.35	1.35
% Mixing	50	25	50	75	50	50	75
ΔM_{env} (10^{-5} M$_\odot$)	6.4	3.2	3.2	3.5	2.2	0.54	0.58
t_{acc} (10^5 yr)	3.3	1.9	1.9	2.1	1.3	0.31	0.33
t_{rise} (10^6 s)	20	46	13	11	6.8	2.5	2.1
$\varepsilon_{nuc,max}$ (10^{16} erg g^{-1} s^{-1})	0.29	0.36	0.76	2.4	2.1	19	14
T_{max} (10^8 K)	1.98	2.21	2.19	2.48	2.44	3.24	3.32
t_{max} (s)	768	828	540	305	380	150	108
ΔM_{ejec} (10^{-5} M$_\odot$)	4.7	2.3	1.9	2.6	1.4	0.44	0.34
v_{ejec} (km s^{-1})	1600	2100	2400	2500	3100	4100	6000
K (10^{45} erg)	1.3	1.1	1.2	1.9	1.4	0.9	1.3

Model ONe3. The higher amount of ^{12}C present in its envelope increases the role played by the CNO-cycle and more nuclear energy is released at the same temperature. Therefore, Model CO5 accretes a less massive envelope before the TNR begins (1.8×10^{-5} M$_\odot$). Since the ignition density (and, hence, the degeneracy) is also lower, a slightly lower peak temperature is attained ($T_{max} = 2.1 \times 10^8$ K). The net effect is that Model ONe3 shows an extension of the nuclear activity towards high Z nuclei, both because of the different peak temperature and the different chemical composition of the envelope. A second feature, which turns out to be crucial, is the different timescales of the TNR: Model CO5 requires $t_{rise} \sim 1.1 \times 10^6$ s to increase the temperature at the burning shell from $T_{bs} = 3 \times 10^7$ K to 10^8 K, plus $t_{max} \sim 65$ s to reach peak value. These shorter times deeply influence the final yields (see Tables 3 and 4).

In order to mimic the uncertain process of mixing between the solar-like accreted material and the outermost layers of the underlying CO or ONe white dwarf, we have adopted different degrees of mixing ranging from 25 to 75%. Computations with 1.15 M$_\odot$ ONe white dwarfs (i.e., Models ONe2, ONe3 and ONe4) show that a more massive envelope is accreted when a higher degree of mixing is adopted, leading to a more violent outburst. We have also performed several computations involving 1.15 M$_\odot$ CO white dwarfs (i.e., Models CO4, CO5 and CO6, with 25, 50 and 75% mixing, respectively), as well as 0.8 M$_\odot$ CO white dwarfs (Models CO1 and CO2, with 25 and 50%, respectively). Contrary to the ONe Models, the most massive envelopes are accreted on top of white dwarfs with 25% mixing, with a minimum mass around 50% mixing. However, the strength of the explosion, as indicated by a higher peak temperature and a higher mean kinetic energy, increases with the mixing level.

As shown also in Tables 1 and 2, massive white dwarfs develop a TNR after a shorter accretion phase (and hence, accreting less mass) as compared to lighter white dwarfs, because of the higher surface gravity. Also the evolution towards peak temperature takes place in a shorter timescale. The most

TABLE 2. Input parameters and results of the CO nova models

	CO1	CO2	CO3	CO4	CO5	CO6
M_{wd} (M$_\odot$)	0.8	0.8	1.0	1.15	1.15	1.15
% Mixing	25	50	50	25	50	75
ΔM_{env} (10^{-5} M$_\odot$)	9.7	8.8	3.9	2.1	1.8	1.8
t_{acc} (10^5 yr)	3.1	2.6	1.7	1.2	1.0	1.0
t_{rise} (10^6 s)	2.8	1.8	1.1	0.43	0.72	0.48
$\varepsilon_{nuc,max}$ (10^{16} erg g^{-1} s^{-1})	0.05	0.1	0.3	0.5	1.1	1.9
T_{max} (10^8 K)	1.45	1.51	1.70	2.03	2.05	2.08
t_{max} (s)	454	199	152	147	65	51
ΔM_{ejec} (10^{-5} M$_\odot$)	7.0	6.4	2.3	1.5	1.3	1.3
v_{ejec} (km s^{-1})	800	1200	1900	2200	2700	2900
K (10^{45} erg)	0.6	1.1	0.9	0.8	1.0	1.3

relevant outcome is the increase in the peak temperature attained during the TNR as the mass of the white dwarf increases.

Tables 3 and 4 list the mean chemical composition in the ejecta, few days after the explosion, resulting from our evolutionary sequences of ONe and CO novae, respectively. Overproduction factors, relative to solar abundances, for Models ONe3 and ONe6, are displayed in Figure 2. The elemental yields obtained in our grid of nova models fit fairly well the spectroscopic abundance determinations of some nova systems such as V693 CrA 1981, V1370 Aql 1982, QU Vul 1984, PW Vul 1984 and V1688 Cyg 1978.

A simple analysis of the mean composition of the ejecta reveals that classical novae are likely sites for the synthesis of most of the Galactic ^{13}C and ^{17}O, and may also contribute significantly to the abundance of ^{15}N, though an extra source seems to be required (José & Hernanz 1997). This conclusion does not rely on nuclear uncertainties, since most of the CNO-cycle reactions are well determined. In fact, new measurements of several reactions, such as ^{17}O(p,γ)^{18}F, ^{17}F(p,γ)^{18}Ne, or ^{18}F(p,α)^{15}O (Coc 1997), have little incidence on the resulting yields.

CO Models synthesize significant amounts of ^{7}Be (Hernanz et al. 1996), large enough to be detected from nearby novae through its γ-ray emission, providing a potential observational clue of the presence of a CO white dwarf.

The ejecta of ONe Models show significant overproduction of ^{22}Na (Gómez–Gomar et al. 1997), which strongly depends on a good determination of some crucial rates like ^{21}Na(p,γ)^{22}Mg, and to some extend, ^{22}Na(p,γ)^{23}Mg and ^{23}Na(p,γ)^{24}Mg. Its γ-ray emission might be detected by future space missions according to the mass fractions obtained in our calculations which, in turn, are in good agreement with the upper limits derived from COMPTEL observations.

Two key reactions play a significant role in the ^{26}Al synthesis: ^{26}Al(p,γ)^{27}Si and ^{27}Al(p,α)^{24}Mg. In fact, the use of recent determinations in nova calculations have significantly reduced the final amounts of ^{26}Al in the ejecta, in

TABLE 3. Mean chemical composition in the ejecta of the ONe models

Nucleus	ONe1	ONe2	ONe3	ONe4	ONe5	ONe6	ONe7
^{1}H	3.2E-1	4.7E-1	3.0E-1	1.2E-1	2.8E-1	2.4E-1	7.3E-2
^{3}He	7.1E-8	2.1E-9	4.3E-8	1.7E-7	2.8E-8	2.9E-8	9.7E-8
^{4}He	1.8E-1	2.8E-1	2.0E-1	1.3E-1	2.2E-1	2.4E-1	1.7E-1
^{7}Be	2.3E-7	4.6E-8	6.0E-7	1.2E-6	6.9E-7	1.3E-6	2.4E-6
^{12}C	1.3E-2	1.8E-2	2.3E-2	2.2E-2	2.8E-2	2.1E-2	2.6E-2
^{13}C	1.7E-2	2.3E-2	2.8E-2	2.7E-2	3.2E-2	1.5E-2	2.5E-2
^{14}N	2.6E-2	3.0E-2	2.2E-2	2.7E-2	3.2E-2	4.6E-2	3.5E-2
^{15}N	7.7E-3	1.7E-2	2.3E-2	2.4E-2	4.2E-2	1.2E-1	1.4E-1
^{16}O	1.7E-1	2.4E-2	1.2E-1	2.3E-1	7.1E-2	2.2E-2	9.1E-2
^{17}O	1.8E-2	1.1E-2	2.8E-2	4.1E-2	3.9E-2	1.7E-2	5.1E-2
^{18}O	8.2E-3	2.4E-3	6.0E-3	7.3E-3	4.2E-3	9.8E-4	1.8E-3
^{19}F	8.5E-6	4.7E-6	8.9E-6	1.2E-5	1.3E-5	2.2E-5	4.0E-5
^{20}Ne	1.8E-1	9.0E-2	1.8E-1	2.6E-1	1.8E-1	1.5E-1	2.4E-1
^{21}Ne	1.9E-5	1.3E-5	3.0E-5	4.0E-5	3.5E-5	5.1E-5	8.4E-5
^{22}Ne	2.0E-3	5.9E-4	1.7E-3	2.5E-3	1.0E-3	1.5E-4	4.2E-4
^{22}Na	4.8E-5	3.1E-5	5.3E-5	1.5E-4	9.6E-5	6.0E-4	6.5E-4
^{23}Na	1.2E-3	3.6E-4	7.5E-4	3.6E-3	1.4E-3	6.6E-3	7.9E-3
^{24}Mg	2.5E-4	1.6E-5	1.0E-4	1.5E-3	2.0E-4	3.6E-4	1.2E-3
^{25}Mg	1.0E-2	7.8E-4	2.9E-3	7.4E-3	2.4E-3	4.2E-3	6.6E-3
^{26}Mg	9.4E-4	7.8E-5	3.4E-4	1.0E-3	2.8E-4	5.9E-4	1.0E-3
^{26}Al	2.7E-3	1.8E-4	9.3E-4	2.0E-3	5.4E-4	7.2E-4	1.5E-3
^{27}Al	1.2E-2	7.6E-4	4.5E-3	9.2E-3	2.0E-3	1.8E-3	4.5E-3
^{28}Si	3.4E-2	3.0E-2	5.4E-2	7.3E-2	5.6E-2	3.5E-2	5.8E-2
^{29}Si	8.7E-5	3.1E-4	4.2E-4	7.8E-4	8.8E-4	1.7E-3	2.7E-3
^{30}Si	4.3E-5	1.4E-3	6.9E-4	1.7E-3	4.8E-3	1.1E-2	1.7E-2
^{31}P	4.5E-6	2.6E-4	5.9E-5	1.9E-4	1.3E-3	7.6E-3	1.2E-2
^{32}S	2.0E-4	3.6E-4	2.0E-4	1.2E-4	8.3E-4	2.3E-2	1.9E-2
^{33}S	4.7E-6	4.3E-5	1.2E-5	7.0E-6	7.7E-5	9.1E-3	4.4E-3
^{34}S	9.2E-6	1.8E-5	9.2E-6	4.7E-6	1.9E-5	6.4E-3	1.8E-3
^{35}Cl	1.5E-6	6.2E-6	2.2E-6	1.2E-6	6.1E-6	7.0E-3	8.7E-4
^{36}S	4.6E-8	5.4E-8	4.2E-8	2.1E-8	3.2E-8	5.4E-9	5.7E-9
^{36}Ar	3.9E-5	5.8E-5	3.9E-5	1.9E-5	3.8E-5	3.9E-3	1.9E-4
^{37}Cl	4.8E-7	1.4E-6	6.2E-7	3.4E-7	1.2E-6	2.8E-4	7.2E-6
^{38}Ar	7.7E-6	1.1E-5	7.6E-6	3.8E-6	7.4E-6	5.1E-5	3.7E-6
^{39}K	1.8E-6	2.9E-6	1.8E-6	9.1E-7	2.0E-6	6.5E-6	1.8E-6

TABLE 4. Mean chemical composition in the ejecta of the CO models

Nucleus	CO1	CO2	CO3	CO4	CO5	CO6
^{1}H	5.1E-1	3.3E-1	3.2E-1	4.7E-1	3.0E-1	1.2E-1
^{3}He	7.0E-6	9.2E-6	6.1E-6	1.5E-6	4.1E-6	2.8E-6
^{4}He	2.1E-1	1.4E-1	1.5E-1	2.5E-1	1.6E-1	9.0E-2
^{7}Be	4.4E-7	9.6E-7	3.1E-6	6.0E-6	8.1E-6	4.0E-6
^{12}C	1.4E-2	5.3E-2	3.6E-2	2.9E-2	4.8E-2	6.8E-2
^{13}C	3.4E-2	1.1E-1	1.3E-1	4.4E-2	9.6E-2	1.9E-1
^{14}N	9.5E-2	1.1E-1	1.1E-1	7.1E-2	1.1E-1	1.4E-1
^{15}N	9.9E-4	9.3E-4	6.2E-3	2.3E-2	4.0E-2	2.9E-2
^{16}O	1.3E-1	2.5E-1	2.4E-1	8.6E-2	2.1E-1	3.4E-1
^{17}O	3.3E-3	4.4E-3	8.0E-3	1.2E-2	2.1E-2	1.9E-2
^{18}O	8.4E-4	5.6E-4	2.2E-3	4.4E-3	3.8E-3	3.7E-3
^{19}F	8.5E-7	4.4E-7	9.9E-7	5.0E-6	3.4E-6	1.8E-6
^{20}Ne	1.2E-3	8.2E-4	8.5E-4	1.4E-3	9.7E-4	5.2E-4
^{21}Ne	2.9E-8	4.0E-8	5.6E-8	1.9E-7	1.7E-7	7.2E-8
^{22}Ne	2.6E-3	5.0E-3	5.0E-3	2.2E-3	4.8E-3	7.3E-3
^{22}Na	3.4E-7	3.0E-7	1.6E-7	3.8E-7	2.9E-7	1.1E-7
^{23}Na	3.6E-5	3.6E-5	3.4E-5	1.6E-5	2.0E-5	2.4E-5
^{24}Mg	5.7E-5	6.3E-5	1.6E-5	4.4E-6	1.8E-5	1.0E-5
^{25}Mg	3.8E-4	2.4E-4	2.8E-4	1.1E-4	1.6E-4	1.3E-4
^{26}Mg	5.5E-5	3.7E-5	3.0E-5	1.1E-5	1.5E-5	9.4E-6
^{26}Al	8.1E-6	3.4E-6	1.6E-5	3.1E-5	4.7E-5	3.3E-5
^{27}Al	4.8E-5	2.6E-5	4.3E-5	1.3E-4	1.3E-4	5.9E-5
^{28}Si	4.9E-4	3.3E-4	3.3E-4	9.4E-4	4.5E-4	1.9E-4
^{29}Si	2.6E-5	1.7E-5	1.7E-5	1.6E-5	1.3E-5	7.3E-6
^{30}Si	1.8E-5	1.2E-5	1.2E-5	3.2E-5	1.7E-5	7.5E-6
^{31}P	6.1E-6	4.1E-6	4.1E-6	6.2E-6	4.0E-6	2.0E-6
^{32}S	3.0E-4	2.0E-4	2.0E-4	2.9E-4	2.0E-4	9.8E-5
^{33}S	2.5E-6	1.7E-6	1.8E-6	8.6E-6	3.3E-6	1.3E-6
^{34}S	1.4E-6	9.3E-6	9.3E-6	1.4E-5	9.2E-6	4.6E-6
^{35}Cl	1.9E-6	1.3E-6	1.3E-6	2.4E-6	1.4E-6	6.7E-7
^{36}S	7.0E-8	4.7E-8	4.7E-8	6.8E-8	4.6E-8	2.3E-8
^{36}Ar	5.8E-5	3.9E-5	3.9E-5	5.8E-5	3.9E-5	1.9E-5
^{37}Cl	6.4E-7	4.3E-7	4.3E-7	7.4E-7	4.6E-7	2.2E-7
^{38}Ar	1.2E-5	7.7E-6	7.7E-6	1.2E-5	7.7E-6	3.8E-6
^{39}K	2.6E-6	1.7E-6	1.7E-6	2.6E-6	1.7E-6	8.7E-7

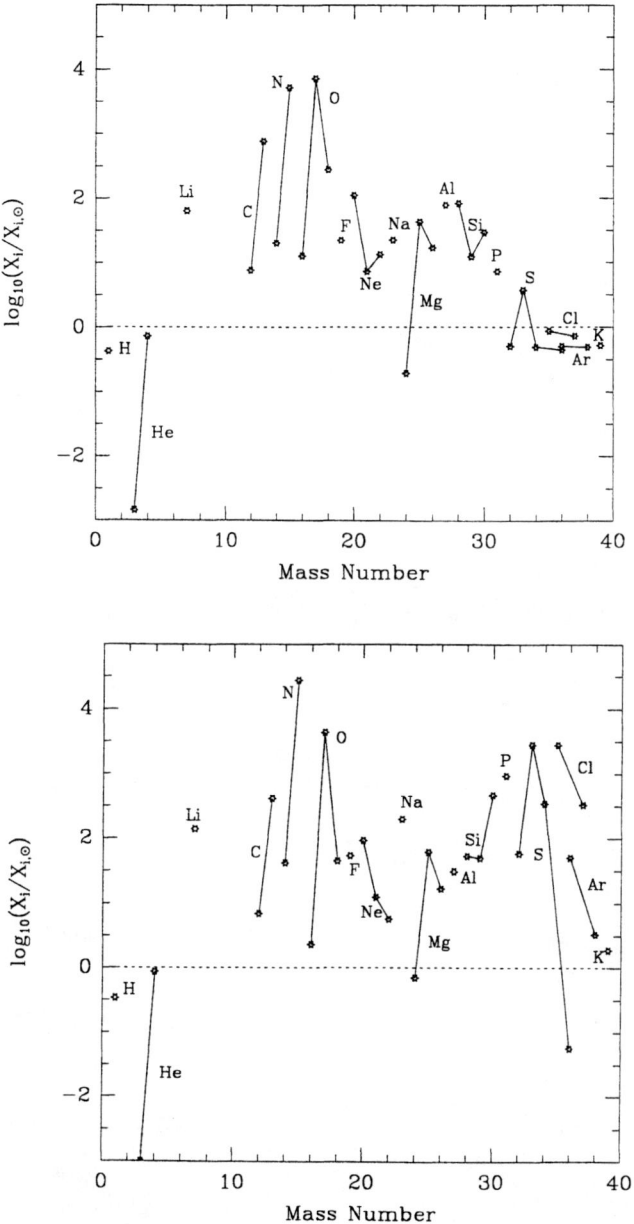

FIGURE 2. Overproduction factors, relative to solar abundances, versus mass number, for Models ONe3 (upper panel) and ONe6 (lower panel).

better agreement with the current prediction of young progenitors, based on the observed distribution of ^{26}Al in the Milky Way. Classical novae, as analyzed by José, Hernanz & Coc (1997), may only account for less than 10% of the estimated Galactic ^{26}Al.

Our results show also that massive ONe white dwarfs are characterized by low O/N and C/N ratios. Significant detection of ^{19}F, ^{35}Cl, and even ^{31}P and ^{33}S, may reveal the presence of a massive ONe white dwarf. Despite new determinations of crucial reactions like ^{31}P(p,γ)^{32}S, ^{31}P(p,α)^{28}Si, or ^{32}S(p,γ)^{33}Cl are available, more accurate rates are still needed in order to elucidate the role played by classical nova in the synthesis of isotopes in the range Si-Ca.

acknowledgments

This research has been partially supported by the DGICYT (PB94-0827-C02-02), by the CICYT (ESP95-0091), and by the CIRIT (GRQ94-8001).

REFERENCES

1. Coc, A. *private comm.* (1997).
2. Gómez–Gomar, J., Hernanz, M., José, J. and Isern, J. *MNRAS*, submitted.
3. Hernanz, M., José, J., Coc, A. and Isern, J. *ApJ* **465**, L27 (1996).
4. José, J. *PhD Thesis*, University of Barcelona (1996).
5. José, J. and Hernanz, M. *ApJ*, submitted.
6. José, J., Hernanz, M. and Coc, A. *ApJ* **479**, L55 (1997).
7. Kovetz, A. and Prialnik, D. *ApJ* **477**, 356 (1997).
8. Kraft, R.P. *Adv. Astron. Astrophys.* **2**, 43 (1963).
9. Kutter, G.S. and Sparks, W.M. *ApJ* **175**, 407 (1972).
10. Kutter, G.S. and Sparks, W.M. *ApJ* **239**, 988 (1980).
11. Ritossa, C, García–Berro, E. and Iben, I. *ApJ* **460**, 489 (1996).
12. Starrfield, S., Truran, J.W., Wiescher, M.C. and Sparks, W.M. *MNRAS*, in press.
13. Walker, M.F. *PASP* **66**, 230 (1954).

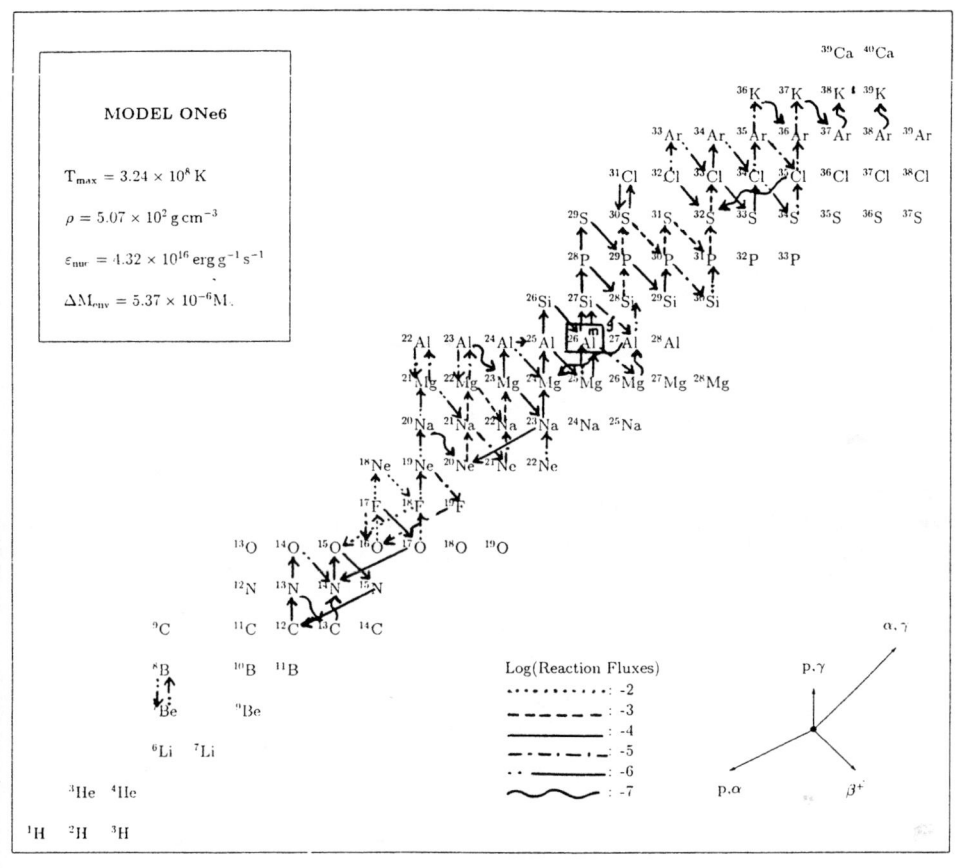

FIGURE 1. Main nuclear flows at peak temperature for Model ONe3 (a 1.15 $M_\odot$ ONe white dwarf, with 50% mixing).

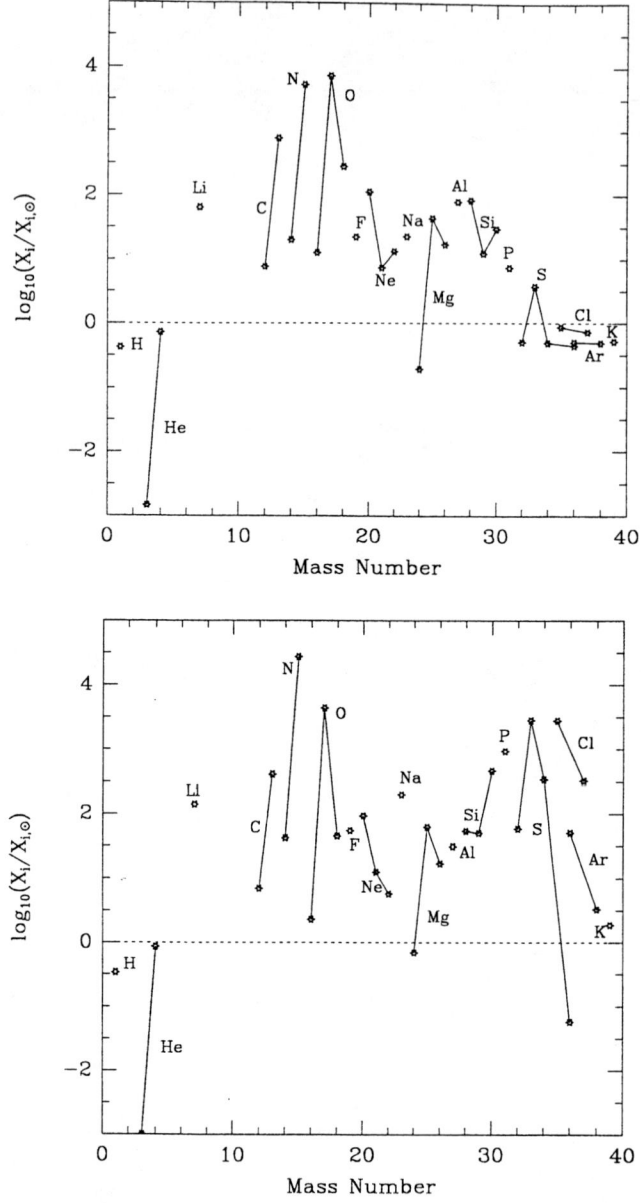

FIGURE 2. Overproduction factors, relative to solar abundances, versus mass number, for Models ONe3 (upper panel) and ONe6 (lower panel).

X-Ray Bursts

Ronald E. Taam

*Department of Physics & Astronomy,
Northwestern University,
Evanston, IL 60208*

Abstract. The observations and progress in the theoretical understanding of X-ray burst sources are reviewed. Of about 150 known low mass X-ray binary systems, X-ray bursts have been observed from about 50 sources. X-ray bursts exhibit a wide variety of profiles; rise times vary from less than a second to about 10 s and decay times are in the range of 10 s to minutes with the decays much shorter at high energies than at low energies. The bursts are characterized by peak luminosities greater than 10,000 times that of the sun, peak temperatures of about 2 keV, and recur at intervals of several hours. Study of the radiation from these sources demonstrates that these sources are binary systems in which a neutron star is a member. Theoretical studies reveal that the X-ray burst results from either a combination of hydrogen and helium burning or pure helium burning in thermonuclear flashes in the envelope of an accreting neutron star.

INTRODUCTION

The last two decades have witnessed unprecedented discoveries in astrophysics. Among the newly discovered classes of objects which have attracted great attention are compact X-ray sources. The brightest of these sources are close binary systems in which mass is transferred from a companion star to a neutron star. The bright X-ray sources are intrinsically fascinating, showing intense X-ray bursts, as well as a wealth of periodic behavior. Detailed analysis of these sources as a group has yielded fundamental understanding of the intrinsic properties of the underlying neutron star and has provided insight into the diversity of the mass transfer and accretion processes in binary star systems containing a compact object. In this paper we review the current status and progress in our understanding of the x-ray burst phenomenon based on direct confrontation of theory with observation.

I OBSERVATIONAL DATA

X-ray bursts were first discovered [9] more than twenty years ago using data obtained from the ANS satellite. Subsequently, these sources were observed intensively by a number of groups using other X-ray satellites. An extensive review of burst sources and their properties is given in [20].

Briefly, the x-ray burst phenomenon is characteristized by a number of timescales; e.g., the light curves for a majority of the bursts are typically described by a rise timescale $\sim 0.1 - 1$ s and decay timescale $\sim 10 - 100$ s. In some sources the rise timescale can be as long as 20 s and the decay timescale may be greater than 1000 s. The bursts recur on timescales of hours, although not strictly periodically. The emission from these sources is very energetic with peak luminosities $\sim 10^{38}$ erg s^{-1}. For a typical source, the peak luminosity can vary by a factor of 3 to 10, and the corresponding energy emitted during the burst is in the range of $\sim 10^{39}$ to 10^{40} ergs. During the non burst state, a persistent source is seen with a luminosity level of $\leq 10^{37}$ erg s^{-1}. The ratio of this persistent emission to the time averaged burst emission lies in the range of $\sim 30 - 250$. For some sources, this ratio can be as low as 2, or as high as 1000. Finally, the spectra of these bursts harden during the rise and soften during the decline. The spectra can be fit by a blackbody with a temperature at the peak of the outburst corresponding to 2 - 3 keV.

II MECHANISM OF THE OUTBURST

X-ray burst sources are binary systems as revealed by the presence of eclipses or X-ray dips in the light curve with orbital periods typically in the range of several hours to several days. Information on the nature of the bursting object is provided by the burst spectra itself. In particular, the burst spectra are well fit by a blackbody [25], and the flux variations during the decay of the burst can be described in terms of an object cooling off at constant radius. For an assumed distance to the source of about 30,000 light years (comparable to the distance to the center of the Milky Way galaxy as suggested by the concentration of sources toward the galactic center), the size of the emitting region is found to be comparable to the dimensions of a neutron star (~ 10 km).

The mechanism for the outburst itself is related to the fact that the material which is accreted by the neutron star is primarily hydrogen-rich, and this matter must eventually be processed to a neutron rich composition in the neutron star interior. For a wide range of conditions (e.g., the internal temperature of the neutron star and the mass accretion rate), the conversion process is unstable in either a helium layer underlying a stable burning hydrogen rich layer or in a layer in which both hydrogen and helium co-exist. In this process the material is accreted onto the neutron star surface until a critical amount of fuel has been accumulated such that the hydrogen rich mixture ignites leading to the fusion of hydrogen into helium and (or hydrogen and helium) into heavier elements. The instability is thermal in origin and is related to the high degree of electron degeneracy of the gas (because the gas pressure is relatively insensitive to the temperature in this regime) and the high temperature sensitivity of the nuclear reaction rates. In the outburst, the energy released by the exothermic reactions leads to the heating of the neutron star surface to much higher X-ray temperatures and to the emission of a burst. The fact that the ratio of the persistent luminosity to the time averaged burst luminosity is comparable to the ratio of the gravitational potential at the neutron star surface ($\sim 10^{20}$ erg gm^{-1}) to the energy yield per gram associated with nuclear fusion ($\sim 10^{18}$ erg gm^{-1}) strongly supports such a model.

Hot Hydrogen-Helium Burning

A unique feature of nuclear processing in the envelope of an accreting neutron star is the possibility that hydrogen is not completed consumed prior to the onset of helium burning. This comes about when the matter is accreted faster than hydrogen can be processed into helium. Since the timescale for this conversion is limited by the weak interaction timescale for converting a proton into a neutron, such a possibility arises in the neutron star case where the timescale for accretion of a critical amount of fuel can be of the order of an hour. In the neutron star envelope hydrogen burning is temporarily stabilized by the β limited CNO cycle. At the typical densities and temperatures in the burning region in the neutron star envelope ($\rho \sim 10^6$ gm cm^{-3}, T $\sim 10^8$ K), helium ignites producing carbon via $3\alpha \rightarrow ^{12}C$. As a consequence, the energy generation rate due to hydrogen burning via the CNO process increases and the temperatures in the envelope rise further. Since the rate of helium burning is very temperature sensitive ($\sim T^{40}$), the 3α process soon dominates and charged particle reactions dictate the nuclear burning development to nuclei in the range of ^{56}Ni to ^{64}Ge. However, at temperatures $\sim 4-5 \times 10^8$ K, the $^{15}O(\alpha,\gamma)^{19}Ne$, and $^{14}O(\alpha,p)^{17}F$ reactions lead to break out from the CNO cycle and other (α,p), (p,γ) reactions become important. These additional reactions, thus, provide new channels for the processing of hydrogen and helium beyond the normal reaction chains discussed for massive stars [28]. Specifically, [36,12,39] the nonstandard nuclear flows which involved a series of (p,γ) reactions followed by positron decays were referred to as the rapid proton process or rp process. However, because of the high densities and temperatures (approaching 10^9 K in the outburst) typical of neutron star envelopes during an x-ray burst, (α,p) reactions compete favorably with the positron decays. These additional reactions lead to a nuclear burning development involving a series of (α,p) and (p,γ) reactions up to Ca followed by (p,γ) reactions to the formation of iron peak elements (since α captures onto heavy nuclei are inhibited due to the increasing Coulomb barrier).

The previous studies [36,12,39] made use of statistical model calculations (which are valid for a high density of excited states in the compound nucleus) to determine the thermonuclear reaction rates. Although such an approach is adequate at high temperatures and for heavy nuclei (heavier than Si), the application of the approach for light nuclei is questionable especially near the proton drip line where the reaction Q-values for proton captures can be small and the compound nucleus is created at low excitation energies. Hence, more recent studies [35,24] have focused on nuclear flows based on improved reaction rate estimates including single resonance and direct capture contributions. Although there still remain large uncertainties in the reaction rates for unstable nuclei (which can be reduced with future experiments using radioactive ion beams), it is likely that the reaction paths identified for given temperature and density histories will not change significantly.

III THEORETICAL STATUS

There is a widespread consensus that thermonuclear flashes in the surface layers of accreting neutron stars are responsible for the Type I X-ray burst phenomenon [20]. The model successfully reproduces the energetics of the observed bursts, their timescales, and the spectral softening seen during burst decay. It is this success which has provided the strongest theoretical evidence that neutron stars are involved in the phenomenon.

Although the general properties of X-ray bursts can be adequately described by the thermonuclear flash model, significant discrepancies exist between the predictions of the simple thermonuclear flash model and observations [20]. These include the erratic correlations between the persistent level of emission and burst behavior, the occurrence of weak bursts whose energetics cannot be accounted for in terms of the matter accreted between bursts, the so-called "double-bursts" separated by $\sim 300-1000$ seconds, and the irregular and lack of burst behavior for very bright sources.

The primary thrust of recent studies has been directed toward investigations of the properties of individual bursts for an entire sequence of bursts. Such studies are essential because the burst properties are sensitive to thermal inertia effects [27], reflecting the heating associated with previous outbursts, and compositional history [1,11,39]. Thus, long term studies of the neutron star envelope subject to a number of successive thermonuclear flashes are required for the advancement of the theory of the X-ray burst phenomenon beyond the simple picture originally proposed [38,14,18].

As results of such investigations [8] it has been shown that the extremely long X-ray burst tail (~ 2500 s) in the soft X-ray transient source Aql X-1 [4] can be successfully explained in terms of a prolonged phase of hydrogen burning accelerated by electron captures at high density in the accreted envelope of a neutron star. Hence, these observations provide a sensitive probe of the physical conditions in the nuclear burning regions. In particular, the observations of Aql X-1 provide a stringent constraint on the mass accreted by the neutron star and its core temperature prior to the first X-ray burst event of the outburst since burning must take place at densities $\gtrsim 1.5 \times 10^7$ gm cm^{-3} to explain the extended phase of X-ray emission. In addition, it was found that the thermal and compositional inertia effects were very important in determining the burst systematics since the properties of the subsequent bursts are distinguished by a lack of a long X-ray tail reflecting the much smaller accumulated masses. The low initial core temperatures required ($\sim 10^7$ K) are consistent with the thermal relaxation of the neutron star envelope during the quiescent state ($\sim 12-16$ months) of Aql X-1 and with the time averaged mass accretion rates ($\sim 10^{-9} M_\odot yr^{-1}$) inferred from the transient outburst itself.

A more extensive study of the evolution of a neutron star undergoing a series of thermonuclear flashes in its accreted hydrogen-rich envelope has also been carried out [30]. A large number of thermonuclear flashes (typically greater than 12) were numerically followed to determine the effects of the history of the neutron star's thermal and compositional structure on the properties of the emitted X-ray bursts. It was found that the bursts exhibited erratic recurrent behavior (recurrence timescales varying by one to two orders of magnitude), especially for cool neutron star cores ($T \lesssim 10^8$ K) and low metal abundances ($Z \lesssim 0.001$) in the accreted matter. For parameters leading to erratic behavior a convective mixing mechanism was identified which operated during the quiescent phase to facilitate the initiation of the next outburst. The resulting bursts were weak and were characterized by short recurrence timescales ($\sim 1-2$ hr) and low peak luminosities ($\sim 10^{37}$ erg s^{-1}). These properties are not predicted by the simple thermonuclear flash model [7] and result from the incomplete burning of nuclear fuel in the previous outbursts coupled with the inward mixing of helium to high densities. In particular, the spatial variation of the composition in the burning layer led to the mean molecular weight increasing outwards, which was initially stabilized by the inverted temperature gradient due to the spatial variation of the energy generation rate associated with (α, γ) reactions on intermediate mass nuclei. However, when the inverted temperature gradient disappeared as a result of radiative cooling from the neutron star surface, the region became unstable and mixing of matter then occurred. The results indicate that the mixing mechanism leads to unstable burning in an irregular manner, very reminiscent of the irregularity manifested as erratic behavior seen in the X-ray transient sources Ser X-1 [21], 1735-44 [19], 1608-22 [22] and EXO 0748-67 [10].

More recently [29], the properties of successive individual X-ray bursts have been explored in regimes that have been identified where especially interesting or relevant burst behavior occurs. For example, the behavior of neutron stars accreting at rates $\sim 10^{-9} - 10^{-8} M_\odot yr^{-1}$ were explored to provide understanding of sources which exhibit very irregular burst behavior such as Cyg X-2 and GX 17+2 [15–17] and sources which show a lack of detectable bursts such as Sco X-1, GX 5-1, GX 349+2, and GX 340-0 at high persistent luminosities [34]. In this mass accretion rate regime it is well known that the nuclear processing is complicated by the presence of hydrogen and that significant hydrogen remains after the burst [1,6,39]. The burning of this residual hydrogen as accelerated by electron captures at high densities ($\sim 1.5 \times 10^7$ gm cm^{-3}) led to the heating of the neutron star envelope. The inclusion of this additional heating led to X-ray bursts that were weaker and to bursts that recurred more frequently than in studies where its affect was neglected. The bursting behavior was found to extend, at least, to $10^{-8} M_\odot yr^{-1}$. The existence of bursts (identified to be of thermonuclear origin) from the very bright low mass X-ray binary systems, GX 17+2 [31,26,16] provides support for this theoretical result. However, the lack of bursting behavior in the very bright X-ray sources - Sco X-1, GX 5-1, and GX 340-0 suggests that either mass accretion rate variations on timescales of minutes mask these variations or that the physical conditions under which nuclear burning takes place may be modified by multi-dimensional effects.

IV FUTURE WORK

Theoretical studies have now advanced to the point where multi-dimensional investigations of the thermonuclear instability should now be considered. It has been suggested [2,3] following earlier studies [5,23] that the ignition and propagation of burning fronts may take place in a nonspherical manner. Such burning is expected to occur in circumstances when the propagation time of the burning wave around the neutron star is longer than the thermal diffusion timescale to its surface. For neutron stars accreting at high rates, such conditions may be satisfied. In this case, the bursts are weak due to the low degree of electron degeneracy of the nuclear burning region and the burning front is described as a slow moving deflagration front with a speed much less than the local speed of sound. For weak bursts, the nuclear energy is transported by radiation so the influence of convection on the propagation speed of the burning front is unimportant. The relaxation of the assumption of spherical symmetry can lead to small luminosity fluctuations on timescales of minutes which may be related to variability detected in the brightest low mass X-ray binary systems [2,3]. (For observational details, see [13,32,33]). Physically, the burning front can be extinguished after a diffusion wave breaks through the surface since energy can be radiated to the surface before energy can be conducted laterally to heat adjacent unburned fuel.

Such studies will reveal a rich variety of solutions that may provide important insights for the interpretation of the variability observed in X-ray burst sources. With the determination of the propagation speeds in the conductive and convective burning regimes, it should be possible to determine the critical mass accretion rate delineating the region where multi-dimensional effects influence the X-ray light curve. For example, as the mass accretion rate is decreased the nuclear outbursts are stronger (due to the higher electron degeneracy in the burning region) and the timescale for propagation of convective burning fronts around the surface is likely to decrease. For a sufficiently low mass accretion rate, the propagation speed of the convective burning fronts will be increased to the extent that the timescale for propagation around the entire surface is shorter than the energy transport timescale to the surface. Thus, the transition level between strong bursting behavior and the absence of such behavior offers the potential for a new diagnostic that can be used to place limits on the mass accretion rate in the system. In addition to the applicability to the irregular bursting behavior described above, the results can be used to determine the viability of the nonspherical burning framework to the form and intensity (with respect to mass accretion rate) of the low frequency variability observed from these systems.

With these new theoretical models in hand one will be well positioned to extract critical insights and understanding of the X-ray burst phenomenon from the wealth of observational data obtained from EXOSAT, ASCA, and the Rossi X-ray Timing Explorer satellites. It is expected that significant progress will be made in determining the relationship between the burst properties, mass accretion rate, and the state of the neutron star in X-ray burst sources. These insights will provide for an understanding of the systems themselves and place important constraints on the thermal properties of their underlying neutron stars.

REFERENCES

1. Ayasli, S., & Joss, P. C. 1982, ApJ, 256, 637
2. Bildsten, L. 1994, ApJ, 418, L21
3. Bildsten, L. 1995, ApJ, 438, 852
4. Czerny, M., Czerny, B., & Grindlay, J. E. 1987, ApJ, 312, 122
5. Fryxell, B. A., & Woosley, S. E. 1982, ApJ, 261, 332 (New York: AIP), p. 302
6. Fujimoto, M. Y., Hanawa, T., Iben, I. Jr., & Richardson, M. B. 1995, in High Energy Transients in Astrophysics, ed. S. E. Woosley (New York: AIP Press), p. 302
7. Fujimoto, M. Y., Sztajno, M., Lewin, W. H. G., & Van Paradijs, J. 1987, ApJ, 319, 902
8. Fushiki, I., Taam, R. E., Woosley, S. E., & Lamb, D. Q. 1992, ApJ, 390, 634
9. Grindlay, J. E. et al. 1976, ApJ, 205, L127
10. Gottwald, M., Haberl, F., Parmar, A. N., & White, N. E. 1986, ApJ, 308, 213
11. Hanawa, T., & Fujimoto, M Y. 1982, PASJ, 36, 119
12. Hanawa, T., Sugimoto, D., & Hashimoto, M. 1983, PASJ, 35, 491

13. Hasinger, G., & van der Klis, M. 1989, A&A, 225, 79
14. Joss, P. C. 1977, Nature, 270, 310
15. Kahn, S. M., & Grindlay, J. E. 1982, ApJ, 281, 821
16. Kuulkers, E., van der Klis, M., Oosterbroek, T., van Paradijs, J., & Lewin W. 1994, in Evolution of X-Ray Binaries, ed. S. S. Holt & C. S. Day (New York: AIP), p. 539
17. Kuulkers, E., van der Klis, M., & van Paradijs, J. 1995, ApJ, 450, 748
18. Lamb, D. Q., & Lamb, F. K. 1978, ApJ, 220, 291
19. Lewin, W. H. G. et al. 1980, MNRAS, 193, 15
20. Lewin, W. H. G., van Paradijs, J., & Taam, R. E. 1993, Sp Sci Rev, 62, 223
21. Li, F., et al. 1977, ApJ, 228, 893
22. Murakami, T. et al. 1980, ApJ, 240, L143
23. Nozakura, T., Ikeuchi, S., & Fujimoto, M. Y. 1984, ApJ, 286, 281
24. Rembges, F., Freiburghaus, C., Rauscher, T., Thielemann, F. K., Schatz, H., & Wiescher, M. 1997, ApJ, 484, 412
25. Swank, J. H., et al. 1977, ApJ, 212, L73
26. Sztajno, M., van Paradijs, J., Lewin, W. H. G., Langmeier, A., Trumper, J., & Pietsch, W. 1986, MNRAS, 222, 499
27. Taam, R. E. 1980, ApJ, 241, 351
28. Taam, R. E. 1985, Ann Rev Nuc Part Sci, 35, 1
29. Taam, R. E., Woosley, S. E., & Lamb, D. Q. 1996, ApJ, 459, 271
30. Taam, R. E., Woosley, S. E., Weaver, T. A., & Lamb, D. Q. 1993, ApJ, 413, 324
31. Tawara, Y., Hirano, T., Kii, T. Matsuoka, M, & Murakami, T. 1984, PASJ, 36, 861
32. van der Klis, M. 1989, A&A, 27, 517
33. van der Klis, M. 1995, in X-Ray Binaries, ed. W. H. G. Lewin, J. van Paradijs, & E. P. J. van den Heuvel (Cambridge: Cambridge University Press), 252
34. van Paradijs, J., Penninx, W., & Lewin, W. H. G. 1988, MNRAS, 233, 437
35. Van Wormer, L., Gorres, J., Iliadis, C., Wiescher, M., Thielemann, F. K. 1994, ApJ, 432, 326
36. Wallace, R. K., & Woosley, S. E. 1981, ApJS, 45, 389
37. Wallace, R. K., & Woosley, S. E. 1985, in High Energy Transients in Astrophysics, ed. S. E. Woosley (New York: AIP Press), p. 319
38. Woosley, S. E., & Taam, R. E. 1976, Nature, 263, 101
39. Woosley, S. E., & Weaver, T. A. 1985, in High Energy Transients in Astrophysics, ed. S. E. Woosley (New York: AIP Press), p. 273

Nucleosynthesis at the proton drip line - a challenge for nuclear physics

H. Schatz[*], L. Bildsten[†], J. Görres[*], T. Rauscher[‡],
F.-K. Thielemann[‡], M. Wiescher[*]

[*] *University of Notre Dame, Dept. of Physics, Notre Dame, IN 46556, USA*
[†] *University of California, Depts. of Physics and of Astronomy, Berkeley, CA 94720, USA*
[‡] *Universität Basel, Institut für Physik, CH-4056 Basel, Switzerland*

Abstract. The rp-process in X-ray bursts is investigated using a complete and updated nuclear reaction network from H to Sn that is coupled to a one dimensional, one zone X-ray burst model. In particlular we consider 2p-capture reactions that can bridge proton unbound nuclei and therefore accelerate the reaction flow. This allows for the first time the calculation of the actual endpoint of the rp-process. We find that for a 25 s burst the reaction flow reaches already Cd. The consequences for energy production, final composition of the ashes and fuel consumption are discussed. In addition, the influence of the current uncertainties in the nuclear physics data base on the results is investigated and the parameters for which a future experimental determination is most desirable are identified.

I INTRODUCTION

Type I X-ray bursts are thermonuclear flashes on the surface of accreting neutron stars [1-3] (see also the review article [4]). Hydrogen and helium ignite when the accreted layer reaches a critical mass and burn explosively in a thermonuclear runaway. Helium is burned via the 3α-reaction and the αp-process (a sequence of (α,p) and (p,γ) reactions, [5]), which provides seed nuclei for the hydrogen burning via the rp-process [6] (rapid proton capture and β-decays). For a long time the doubly magic nucleus ^{56}Ni was considered to be the endpoint of the rp-process in X-ray bursts, since a captured proton is so weakly bound that photodisintegration can remove it efficiently. In fact most of the previously used reaction networks ended at ^{56}Ni. It was argued, however, that the rp-process might well continue beyond ^{56}Ni and that this would have interesting consequences. Most important is the long standing problem of occasionally observed extremely short burst intervals of the order of several minutes. These burst intervals are too short to accrete enough fuel to power the second burst and can therefore not be explained by the simple thermonuclear flash model. Several solutions to this problem were offered assuming that the second burst is powered by fuel left over from the first burst [5,7]. [8] speculated however, that taking into account processing beyond ^{56}Ni would consume all hydrogen in a single burst and presented therefore an alternative explanation based on mixing of unburned fuel from outer layers into the burning zone. The rp-process beyond ^{56}Ni would also dramatically change the composition of the ashes of the nuclear burning and thus the composition of the outer crust of the neutron star, which would affect neutron star seismology calculations.

The development of nuclear reaction networks beyond ^{56}Ni was hampered in the past by the lack of experimental and theoretical information about the very unstable nuclei at the proton drip line. Nevertheless a few exploratory studies were performed by several authors that either ended at Se [9] or Y [6], or used only 16 nuclei between H and Cd [10]. Given the different results of these studies and the lack of detailed nucleosynthesis calculations beyond Y it seemed to be necessary to investigate the rp-process between Ni and Sn with an updated and complete reaction network.

II NETWORK CALCULATIONS

In a first step, Van Wormer et al. 1994 [11] calculated the rp-process nucleosynthesis between H and Kr for constant temperatures and densities. For the study presented here, we extended their reaction network up to ^{100}Sn (see [12] and references therein).

The most important input parameters for the reaction network are the charged particle reaction rates (proton capture and (p,α)-reactions), the β-decay half-lives and the nuclear masses. The latter are crucial since they determine the particle separation energies and therefore photodisintegration and particle decay rates. Above Ge, the rp-process path enters a region, where for most nuclei along the reaction path none of these parameters are known experimentally and the calculations are therefore based on theoretical models. The parameters used in this study are described in detail in [12]. The charged particle reaction rates were calculated using the Hauser-Feshbach code SMOKER [13]. Uncertainties of less than a factor of two are typical for this method. β-decay half-lives were based on a QRPA code [14]. Comparison of the predicted half-lives (below 100 s) with experimental data for the proton rich nuclei between Ge and Sn yields an average error of a factor of 5.5. For some nuclei above $A = 84$ the half-live can be calculated using the shell model code OXBASH [15], which leads to an improved uncertainty of a factor of 2.7. The nuclear masses were taken from the FRDM (1992) mass model [16]. Comparison with experimental data shows a theoretical error of about 500 keV, which is calculated by decoupling the experimental errors [16]. However, it has also to be taken into account how the uncertainties behave when the mass

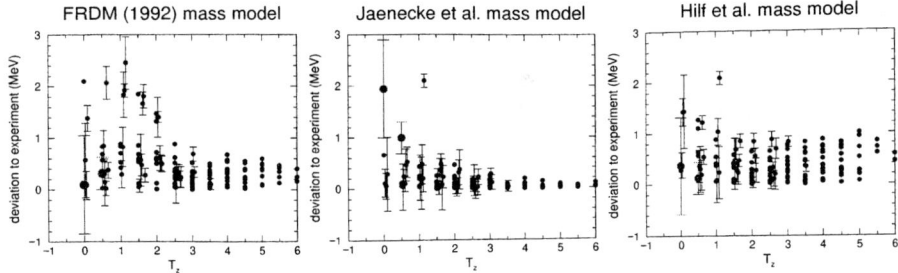

FIGURE 1. The difference of the mass model predictions to experimental nuclear masses as a function of T_z for the proton rich nuclei between Ge and Sn. The errorbars represent the experimental errors. Shown are the results for the FRDM (1992), the Jänecke et al., and the Hilf et al. mass models.

model is extrapolated towards proton rich nuclei. Fig. 1 shows the differences between mass model predictions and experimental data as a function of the z-component of the isospin $T_z = (N - Z)/2$. For comparison we also show the data for the Jänecke et al. mass model [17] and for the older Hilf et al. mass model [18]. For the FRDM (1992) mass model and for the Jaenecke et al. mass model there is a systematic increase in the deviations towards the proton drip line (around $T_z = 0$), even if the larger experimental errors are taken into account. A measure for this is the relation between the theoretical error of all proton rich nuclei in this region (σ) to the theoretical error of the most proton rich nuclei with $T_z \leq 1$ (σ_1). For the FRDM (1992) mass model σ_1 is 70% higher than σ (888 keV instead of 514 keV), while for the Jaenecke et al. mass model σ_1 is more than a factor of 3 larger than σ (286 keV instead of 89 keV). Remarkably, the only mass model that shows no increase in the uncertainty towards more proton rich nuclei but rather a slight decrease is the older Hilf et al. mass model ($\sigma = 466$ keV, $\sigma_1 = 412$ keV). For the present study the FRDM (1992) mass model was chosen, since it is based on the same description of nuclear structure as the QRPA code. It therefore allows a consistent calculation of all nuclear physics parameters in the network.

We also included the most recent experimental information on the crucial isotopes ^{65}As, ^{69}Br, and ^{73}Rb (see section III A). In addition we considered for the first time 2p-capture reactions [12] that bridge proton unbound nuclei by proton capture on an equilibrium abundance of the unbound nucleus similar to the 3α-reaction. These reactions are typically slow, but they can well compete with the slow β-decays of ^{68}Se and ^{72}Kr thus accelerating the rp-process considerably.

The reaction network was coupled to a 1 dimensional, 1 zone X-ray burst model that calculates temperature and density in the burning zone assuming constant pressure [19]. The raise of the burst is calculated selfconsistently from the generated nuclear energy, while during the burst decline the temperature curve is matched to typical observed luminosity profiles. For the present work a burst with an e-fold luminosity decline timescale of 25 s was chosen. The temperature curve is shown in Fig. 2. The density is initially around 10^6 g/cm^3 and drops down to $2.5 \cdot 10^5$ g/cm^3 when the peak temperature of 2 GK is reached. In all calculations we assumed an initial solar composition.

III RESULTS

A Reaction flow

^{56}Ni was long considered to be the endpoint of the rp-process owing to its long electron capture half-life of $2 \cdot 10^4$ s at X-ray burst conditions and to its low proton capture Q-value that prevents further proton captures at high temperatures because of photodisintegration. This can be seen in Fig. 3 that shows the lifetime of ^{56}Ni against proton capture in the rp-process as a function of temperature and for a typical density of 10^6 g/cm^3. The proton

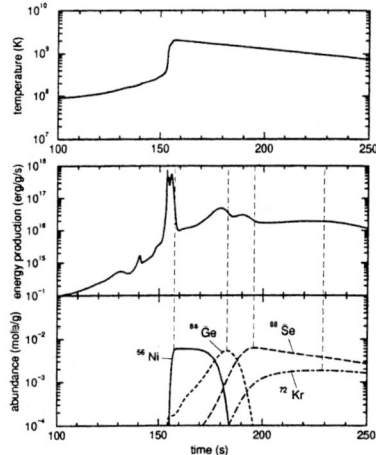

FIGURE 2. The temperature, energy production rate, and the isotopic abundances of the most important waiting points as a function of time.

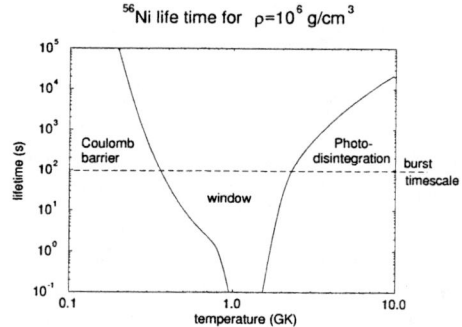

FIGURE 3. The lifetime of ^{56}Ni against proton capture as a function of temperature for a density of 10^6 g/cm^3 and a solar hydrogen abundance. Also indicated is the order of magnitude for the cooling timescale of a typical X-ray burst.

capture lifetime can be estimated quite reliably, since the proton capture Q-values of ^{56}Ni and ^{57}Cu are known experimentally. Proton capture is inhibited at low temperatures due to the Coulomb barrier and at high temperatures due to photodisintegration of ^{57}Cu. However, Fig. 3 also shows that there is a temperature window between 1 and 1.5 GK, where the lifetime of ^{56}Ni is orders of magnitude smaller than the burst timescale. At these conditions ^{56}Ni will be rapidly converted into heavier elements. Fig. 2 shows the abundances of ^{56}Ni and the heavier waiting point isotopes as a function of time. During the temperature raise, the temperature window between 1 and 1.5 GK for processing beyond ^{56}Ni is crossed before the rp-process reaction flow reaches ^{56}Ni. ^{56}Ni is therefore temporarily the endpoint for the rp-process until the temperature window is crossed again during the cooling phase. Then, as Fig. 2 shows, ^{56}Ni is rapidly depleted and the rp-process quickly reaches heavier isotopes.

The next isotopes that were considered as possible endpoints of the rp-process are the rather long lived even Z, $N = Z$ nuclei ^{64}Ge ($T_{1/2} = 64$ s), ^{68}Se ($T_{1/2} = 36$ s) and ^{72}Kr ($T_{1/2} = 17$ s). This has triggered considerable efforts in several laboratories to experimentally investigate the relevant nuclei in this region. However, while the β-decay half-lives are known with good precision, there is still no experimental proton capture Q-value for any of these nuclei. The only information available so far is a β-decay half-life measurement for ^{65}As of 190 ms [20]. In addition upper limits for the lifetimes of ^{69}Br (< 24 ns [21,22]) and ^{73}Rb (< 100 ns [23,24]) can be derived from the nonobservation of these isotopes when certain assumptions for the production cross section in spallation or projectile fragmentation reactions are made. From this it can be concluded that ^{65}As does not proton decay, while ^{69}Br and ^{73}Rb are proton unbound by at least 450 keV and 680 keV respectively. It was therefore speculated that ^{64}Ge can be bridged by fast proton captures and that ^{68}Se delays the rp-process with its full β-decay half-life representing the endpoint of the rp-process. However, the lifetime of ^{68}Se depends sensitively on the proton capture Q-value. This is demonstrated in Fig. 4 that shows the lifetime of ^{68}Se (including β-decay and proton capture) as a function of the proton capture Q-value for a typical temperature of 1.5 GK, a density of 10^6 g/cm^3 and a solar hydrogen abundance. For positive Q-values proton capture reduces the lifetime

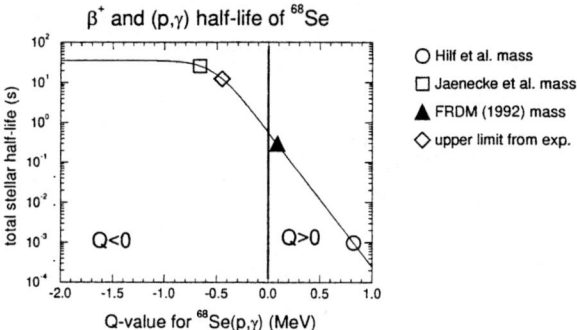

FIGURE 4. The lifetime of ^{68}Se against β-decay and proton capture as a function of the proton capture Q-value for a temperature of 1.5 GK, a density of 10^6 g/cm^3 and solar hydrogen abundance. Indicated are also the predictions of the various mass models as well as the upper limit derived from experimental results.

of ^{68}Se to less than a second and ^{68}Se is not a rp-process waiting point. On the other hand, a negative Q-value does not necessarily imply that β-decay is the only destruction mechanism. There is a range of negative Q-values between 0 and −1 MeV, where the 2p-capture reaction reduces the lifetime of ^{68}Se significantly below its β-decay value. It is therefore not sufficient to determine whether the proton capture Q-value is positive or negative, but its exact value has to be known, as long as it is above −1 MeV. For a lifetime determination of better than a factor of 2 the Q-value has to be known with an accuracy of better than 100 keV. This is about five times better than the accuracy of the FRDM (1992) mass model predictions. An experimental determination of the proton capture Q-value of ^{68}Se would therefore be crucial. The 2p-capture rate on ^{68}Se depends also linearly on the ^{69}Br(p,γ)^{70}Kr reaction rate, but as long as the proton capture Q-value of ^{68}Se is not known experimentally, the respective uncertainties are negleglible. In this study the experimental upper limit for the Q-value of −450 keV was adopted that still leads to a 50% reduction of the ^{68}Se lifetime compared to pure β-decay. For ^{72}Kr similar arguments apply. To summarize, if 2p-capture reactions are taken into account, the lifetimes of the long lived waiting points in the Ge-Kr range are much smaller than previously assumed. It also has to be taken into account that burst timescales of the order of 10 to 100 s are observed and that the 25 s burst discussed in this work has already a cooling timescale of 100 s ($L \propto T^4$, with the luminosity L and the temperature T).

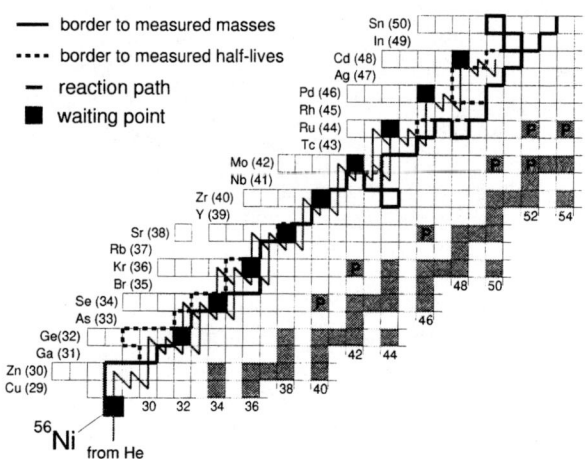

FIGURE 5. The rp-process reaction path above ^{56}Ni. Indicated are the most important waiting points as well as the borders of measuerd masses and half lives.

The complete reaction path above ^{56}Ni is shown in Fig. 5. It can be seen that the rp-process in a 25 s burst proceeds up to ^{98}Cd. The endpoint is reached when the burst timescale equals the summed lifetimes of all waiting points along the reaction path. The dominant waiting points are the even Z, $N = Z$ nuclei between Ge and Cd. Fig. 5 shows that for none of these waiting points experimental proton capture Q-values are available and for the nuclei above ^{76}Sr the β-decay half-lives are not known as well. An experimental determination of these parameters would improve the accuracy of the calculations considerably. Note that the reaction path follows the proton drip line, indicating that neither the Coulomb barrier nor fuel exhaustion are the limiting factors.

B Energy production

Fig. 2 shows the nuclear energy production rate as a function of time during the X-ray burst. A striking feature are the pronounced variations of the energy production rate that span several orders of magnitude. These variations are a consequence of the fact that the energy production drops drastically each time the reaction flow reaches a nucleus with a long lifetime (waiting point). Fig. 2 shows that the first major peak in the energy production is associated with the reaction flow reaching ^{56}Ni, while the next two peaks correlate with the waiting points ^{64}Ge and ^{68}Se. Whether these structures are related to the double burst profiles observed in some X-ray bursts (for example [25]) will have to be investigated using more complex X-ray burst models. It is also interesting to note that 43% of the total burst energy are produced by processing beyond ^{56}Ni during the tail of the burst.

C Final composition

The final composition of the ashes can be found in Fig. 6. Obviously, the endpoint of the rp-process is neither the iron group nor a specific isotope at $A = 64$ or $A = 68$ as it had been suggested before, but a range of nuclei between $A = 68$ and $A = 100$ with a relatively flat distribution. Some material is already accumulated at the end of the network at $A = 100$.

Fig. 7 shows the final overproduction factors in respect to the solar system abundances for the most proton rich isobar in each mass chain, assuming that all unstable nuclei decayed back to stability. β-delayed proton decay was neglected, which was shown to be reasonable for the $A = 96$ and 98 chains by a recent experiment [26]. More experiments of this kind would certainly be important. The distribution of the overproduction factors shows that mainly light p-nuclei are produced with comparable overproduction factors of 10^7-10^8. This is interesting since these nuclei are severely underproduced in standard p-process scenarios. It can be shown [12] that if a small fraction of the burned material (less than 0.03-0.3%) escapes the neutron star, a significant contribution to the galactic nucleosynthesis would be possible. Improved X-ray burst models will have to clarify, whether escape factors of this order of magnitude are reasonable.

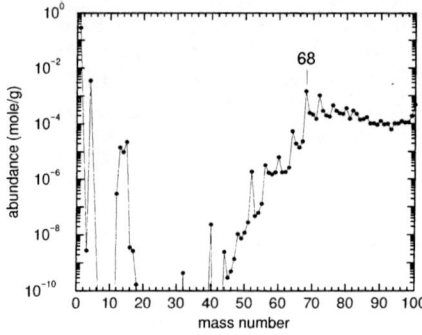
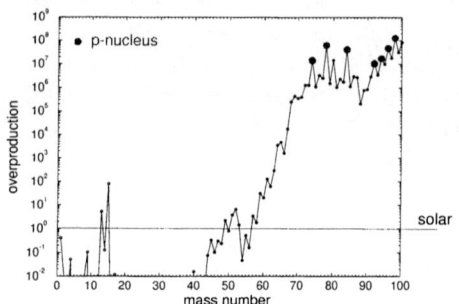

FIGURE 6. The final isotopic abundances summed in each mass chain as a function of mass number.

FIGURE 7. The final overproduction factors for the most proton rich isobar in each mass chain as a function of mass number. The overproduction factors are calculated relative to the solar abundances assuming that all unstable nuclei decayed into stable isotopes.

D Fuel consumption

In Fig. 8 we calculated the consumption of hydrogen (^{1}H) and helium (^{4}He) as a function of the luminosity timescale.

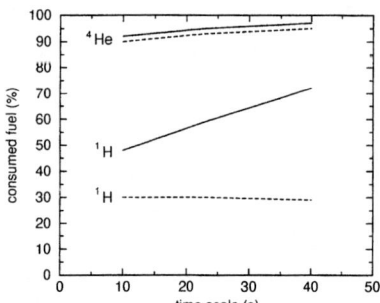

FIGURE 8. The fraction of consumed ^{1}H and ^{4}He fuel as a function of the burst (luminosity) timescale. The dashed lines indicate results from a smaller network ending at ^{56}Ni.

Since the reaction flow reaches the end of the network already for a 25 s burst, the calculations at larger timescales underestimate the fuel comsumption somewhat. Shown are the results for our full reaction network as well as for a network ending at ^{56}Ni. It can be seen that helium is consumed almost completely, independent on burst duration and network size. This is a consequence of the fact that most (70%) of the helium is rapidly burned during the few seconds in the raise of the burst, and that all α-induced reactions occur below ^{56}Ni owing to the Coulomb barrier. On the other hand, Fig. 8 also shows that the use of the full reaction network is essential for calculating the hydrogen consumption. However, even with a full reaction network significant amounts of hydrogen remain unburned (40% for a 25 s burst). This can be understood from the fact that the the transition from the αp-process into the rp-process occurs approximately at ^{41}Sc. Therefore ^{41}Sc is the seed nucleus for the rp-process. Since 10 α-particles (and one proton) are consumed to produce one ^{41}Sc nucleus, and since the ^{1}H/^{4}He number ratio is about 10 (solar value), the proton to seed ration is about 100. From this it follows that in order to consume all hydrogen, the rp-process would have to proceed on average up to $A = 141$. This is not possible, since the here considered cooling timescales of 40-160 s are much shorter than the timescale of the rp-process till $A = 141$ owing to the slow β-decays along the reaction path. Therefore, significant amounts of hydrogen remain unburned and the hydrogen consumption becomes roughly proportional to the burst timescale.

IV CONCLUSIONS

We developed the first complete and updated nuclear reaction network for rp-process calculations beyond ^{56}Ni up to ^{100}Sn. As a first step we used a simple X-ray burst model to investigate the influence of nuclear burning beyond ^{56}Ni. In contrast to [5,6,9] we find that the rp-process in typical X-ray bursts does not end at ^{56}Ni, ^{64}Ge or ^{68}Se, but reaches at least nuclei around $A = 100$. The rp-process beyond ^{56}Ni produces 43% of the burst energy that is released during the burst decline in form of pronounced energy bursts that reflect the properties of the nuclei along the reaction path. We also find that the ashes of the nuclear burning in X-ray bursts, and therefore the outer crust of an accreting neutron star does not consist of iron group nuclei as assumed previously, but of a mostly flat distribution of isotopes between $A = 68$ and $A = 100$. However, in contrast to speculations by [8] the synthesis of these heavier nuclei does not lead to a complete consumption of hydrogen. As a consequence, the models for the explanation of very short burst intervals that rely on the assumption that some hydrogen remains unburned are probably still viable. However, the amount of unburned hydrogen might be lower than assumed. The results obtained depend strongly on theoretical models for the prediction of nuclear masses and β-decay half-lives, that carry large uncertainties. Therefore, an experimental determination of proton capture Q-values and β-decay half-lives of the even Z, $N = Z$ nuclei between Ge and Sn would be most important. In order to constrain waiting point lifetimes better than a factor of two, it is especially important to determine the masses of the waiting point isotopes and the following, eventuelly proton unbound isotones with an accuracy of better than 100 keV - a challenge for theorists and experimentalists.

REFERENCES

1. S. E. Woosley and R. E. Taam, Nature **263**, 101 (1976).
2. L. Maraschi and A. Cavaliere, Highlights of Astronomy **4**, 127 (1977).
3. P. Joss, Nature **270**, 310 (1977).
4. W. H. G. Lewin, J. van Paradijs, and R. E. Taam, Space Sci. Rev. **62**, 233 (1993).
5. S. E. Woosley and T. A. Weaver, in *High Energy Transients in Astrophysics*, Vol. 115 of *AIP Conference Proceedings*, edited by S. E. Woosley (American Institute of Physics, New York, 1984), p. 273.
6. R. K. Wallace and S. E. Woosley, Ap. J. Suppl. **45**, 389 (1981).
7. R. E. Taam, S. E. Woosley, T. A. Weaver, and D. Q. Lamb, Ap. J. **413**, 324 (1993).
8. M. Y. Fujimoto, M. Sztajno, W. H. G. Lewin, and J. van Paradijs, Ap. J. **319**, 902 (1987).
9. T. Hanawa, D. Sugimoto, and M.-A. Hashimoto, Pub. Astr. Soc. Japan **35**, 491 (1983).
10. R. K. Wallace and S. E. Woosley, in *High Energy Transients in Astrophysics*, Vol. 115 of *AIP Conference Proceedings*, edited by S. E. Woosley (American Institute of Physics, New York, 1984), p. 319.
11. L. Van Wormer *et al.*, Ap. J. **432**, 326 (1994).
12. H. Schatz *et al.*, 1997, Phys. Rep., in print.
13. T. Rauscher, F.-K. Thielemann, and K.-L. Kratz, Phys. Rev. C **56**, 1613 (1997).
14. P. Möller, J. R. Nix, and K.-L. Kratz, At. Data Nucl. Data Tab. (1997), submitted.
15. H. Herndl and B. A. Brown, (1997), to be published.
16. P. Möller, J. R. Nix, W. D. Myers, and W. J. Swiatecki, At. Data Nucl. Data Tab. **59**, 185 (1995).
17. J. Jänecke and P. Masson, At. Data Nucl. Data. Tab. **39**, 265 (1988).
18. E. R. Hilf, H. von Groote, and K. Takahashi, in *Proc. 3rd Inter. Conf. Nuclei far from Stability, Cargese* (CERN, Geneva, 1976), Vol. 76-13, p. 142.
19. L. Bildsten, in *The Many Faces of Neutron Stars*, edited by A. Alpar, L. Bucceri, and J. Van Paradijs (Dordrecht, Kluwer, 1997).
20. J. Winger *et al.*, Phys. Lett. B **299**, 214 (1993).
21. B. Blank *et al.*, Phys. Rev. Lett. **74**, 4611 (1995).
22. R. Pfaff *et al.*, Phys. Rev. C **53**, 1753 (1996).
23. M. F. Mohar *et al.*, Phys. Rev. Lett. **66**, 1571 (1991).
24. A. Jokinen *et al.*, Z. Phys. A **355**, 227 (1996).
25. M. Sztajno *et al.*, Ap. J. **299**, 487 (1985).
26. M. Hellström *et al.*, Z. Phys. A **356**, 229 (1996).

CHEMICAL EVOLUTION OF THE GALAXY

Origin and Evolution of the Light Elements Li, Be, and B

Michel Cassé [1,2], Elisabeth Vangioni-Flam[1], and Sean T. Scully[1]

[1] *Institut d'Astrophysique de Paris, 98bis Boulevard Arago, 75014 Paris, France*

[2] *Service d'Astrophysique, DSM, DAPNIA, CEA, France*

Abstract. The increase in observational data in recent years concerning the light elements Li, Be, and B has led us to reexamine the standard picture of their production through galactic cosmic-ray spallation. The recent observations in the Orion star forming region of strong gamma-ray emission lines of C and O indicate that the weak shocks and turbulence found in star forming regions lead to the acceleration of these elements. Interactions between fast C and O and the ambient H and He may in fact be the primary production mechanism of B and Be and not cosmic-ray reactions as has been previously thought.

INTRODUCTION

The light elements Li, Be and B (LiBeB) are both simple and rare, contrary to the general trend observed in nature (table 1). They are rare because they are fragile. Their fragility coupled with the absence of stable nuclei of atomic mass 5 and 8 has been of major importance in the nuclear history of the universe preventing significant production of elements heavier than He in the big bang, making necessary stars to pursue further nuclear evolution. Primordial nucleosynthesis, through the process of nuclear fusion, synthesized D, ^{3}He, ^{4}He and produced a trace amount of ^{7}Li. Stellar nucleosynthesis, either quiescent or explosive, again through nuclear fusion, subsequently forms all species from C to U. But LiBeB nuclei are destroyed in stellar furnaces due to their low binding energy, with the destruction temperatures of ^{6}Li, ^{7}Li, ^{9}Be, ^{10}B and ^{11}B being 2, 2.5, 3.5, 5.3 and 5 million K respectively. Thus LiBeB are bypassed by nuclear fusion with the exception of ^{7}Li which may be produced in the AGB phase of low mass stars and/or in novae. Indeed they owe their existence to the opposite process of nuclear break up through spallation and fragmentation.

TABLE 1. Abundances of Light Isotopes in the Solar System

Li/H	$\sim 10^{-9}$
Be/H	2×10^{-11}
B/H	$3 - 6 \times 10^{-10}$
^{7}Li / ^{6}Li	12.5
^{11}B / ^{10}B	4

Since the seminal work of [1], the origin of LiBeB has been traditionally attributed to the interaction of galactic cosmic rays (GCR, constituted by fast nuclei, essentially protons and alpha particles, wandering the Galaxy) with the interstellar medium (see [2] for a review). To demonstrate the argument, consider the present Be/H ratio resulting from the nuclear break up of interstellar CNO nuclei by cosmic ray protons. The production rate is given by

$$\frac{d(Be/H)}{dt} = (CNO/H)\sigma\Phi, \qquad (1)$$

where CNO/H is the abundance of medium heavy elements. σ is the mean cross section of the reactions p + CNO -> Be, averaged on the GCR energy spectrum (5 mb) and Φ is the proton flux (~ 10 cm^{-2} s^{-1}). Since CNO/H $\sim 10^{-2}$ in the solar system, the production rate ($\sim 5 \times 10^{-28}$ s^{-1}) integrated over the Galactic lifetime ($\sim 3 \times 10^{17}$s) yields Be/H $= 1.5 \times 10^{-11}$ which is in good agreement with the observations to within an order of magnitude (table 1). The abundances of ^{6}Li and ^{10}B may be estimated similarly. This calculation based only on the observed values at the present time in our Galactic vicinity is rough since it does not take into account the evolutionary effects of CNO and of Φ. But stellar observations show that after an abrupt rise lasting about 1 Gyr, the metallicity has leveled off. Thus if Φ has evolved similarly, the above calculation makes sense if restricted to the late Galactic evolution. In retrospect, the claim that GCR spallation explains the formation of LiBeB seems however slightly exaggerated since only 3 isotopes (6, 9, 10) out of 5 are accounted for satisfactorily. Note that no excess production of any isotope is encountered which allows us to invoke other processes to fill in the gap where necessary (*e.g.* big-bang and stellar nucleosynthesis of ^{7}Li [2] and playing with the low energy threshold of the p + 14N -> ^{11}B reaction [3]).

Thus, until recently, the conviction that the primary mechanism for the production of LiBeB is GCR spallation was strong. But the situation has changed dramatically in light of new observations in recent years (for details see [4]). Indeed, the discovery of a strong gamma-ray line emission of C and O from the Orion complex [5,6] and the measurement of Be and B abundances in very metal poor stars ([7] and references therein) has prompted cosmic ray physicists to reassess the problem of the production and evolution of LiBeB [8-13,4]. The gamma-ray data indicate that C and O may be copiously accelerated by a mechanism involving presumably weak (reflected) shocks and turbulence within the Orion cloud or in its vicinity and, in all likelihood, in other active star forming regions [14].

What is the source of these fast C and O nuclei in Orion and other OB associations, and what is the contribution of the interaction of these particles with the ambient interstellar He and H to the production of light elements in the Galaxy? Two (reverse) processes (at least) compete including a fast p and α + CNO -> LiBeB (process I, GCR) as well as a fast CO + H, He -> LiBeB (process II, Orion-like). Process I is "secondary" by definition, since it is dependent on the metallicity of the target medium whereas process II is "primary" since it is independent of the metallicity. Process II is strongly predominant, at least in early Galactic evolution. Indeed the data show a proportionality between Be and Fe, and B and Fe (figure 1). Since Fe is produced by massive stars (SNII) in the early Galactic evolution, one concludes that Be and B are also co-produced by some related spallative mechanism(s).

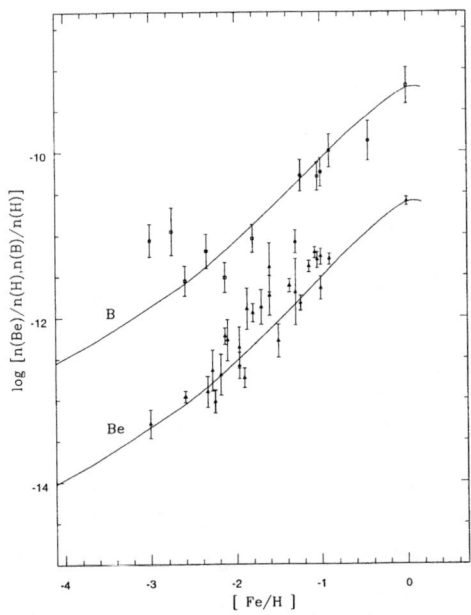

FIGURE 1. Beryllium and Boron evolution, Be Data points are from [15-17]. B Data points from [18,19].

PRODUCTION OF LIBEB BY FAST C AND O FROM MASSIVE STARS

In Process II LiBeB production, as well as the gamma-ray line emission, depend on the composition and energy spectrum of the accelerated beam, and less sensitively on the target composition. Carbon rich Wolf Rayet stars (WC) as astrophysical sources of fast nuclei have the advantages as discussed by [11] of an absence of H, a richness in C and O, moderate He content and a relatively low energy injection requirement in order to produce the gamma-ray lines (see the accompanying paper on Orion). Different source compositions including solar system-like, supernovae-like and that of Wolf Rayet stars have been tested using simple injection spectra parameterized as power laws with a variable cut off, E_0, [12]. The different elemental and isotopic ratios vs E_0 associated to various source compositions are shown in figures 2, 3, and 4.

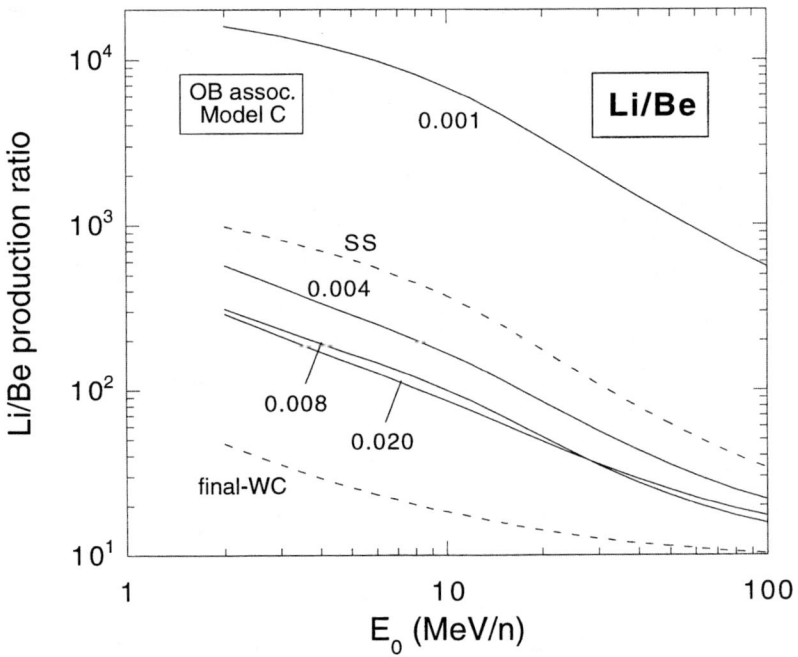

FIGURE 2. Li/Be as a function of E_0 for different wind compositions related to stars of various metallicities (10^{-3} − .02) (solid lines) grouped in OB associations compared to SS (Solar System)[20] and final-WC (dashed lines)[13].

Energetic constraints suggest a proton poor beam and a rather high E_0, of the order of 30 MeV/n, still consistent with the width of the C and O gamma-ray lines [12]. Extensive studies by [21,14] indicate that a WC composition [22] is quite satisfying. The WC composition has the additional merit of avoiding

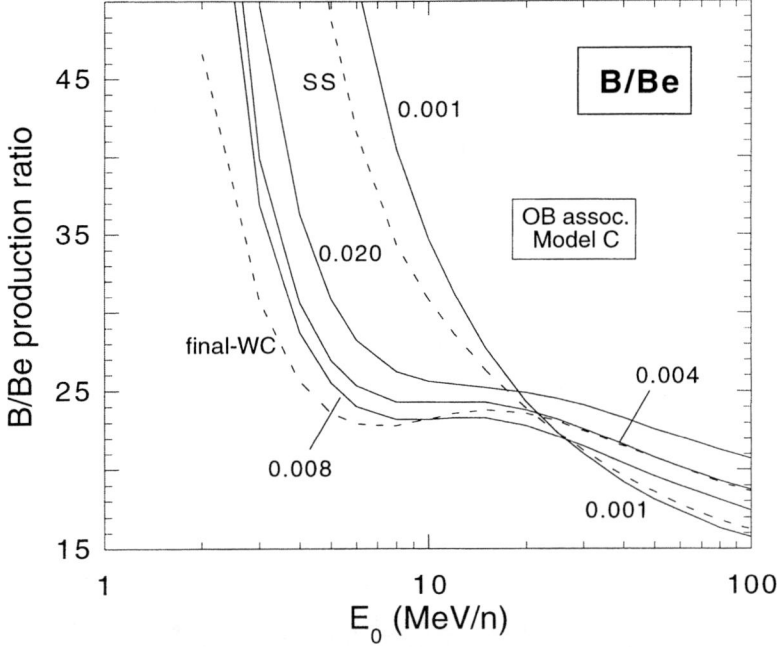

FIGURE 3. Same as figure 2 but for the B/Be ratio[20].

the production of gamma ray lines in the undesired (1-3 MeV) spectral band. The production ratios of the different isotopes and the Be production rate at $Z = 0$ and $Z = Z_\odot$ normalized to the observed C and O gamma-ray lines are presented in table 2. The adopted E_0 is 30 MeV/n.

TABLE 2. Production ratios and rates for 2 metallicities

	$Z = 0$	$Z = Z_\odot$
^{7}Li/^{6}Li	1.5	1.5
^{11}B/^{10}B	4.9	4.4
Li/Be	15.9	15.5
B/Be	26.3	22.7
dBe/dt(s^{-1})	9.3×10^{37}	2×10^{37}

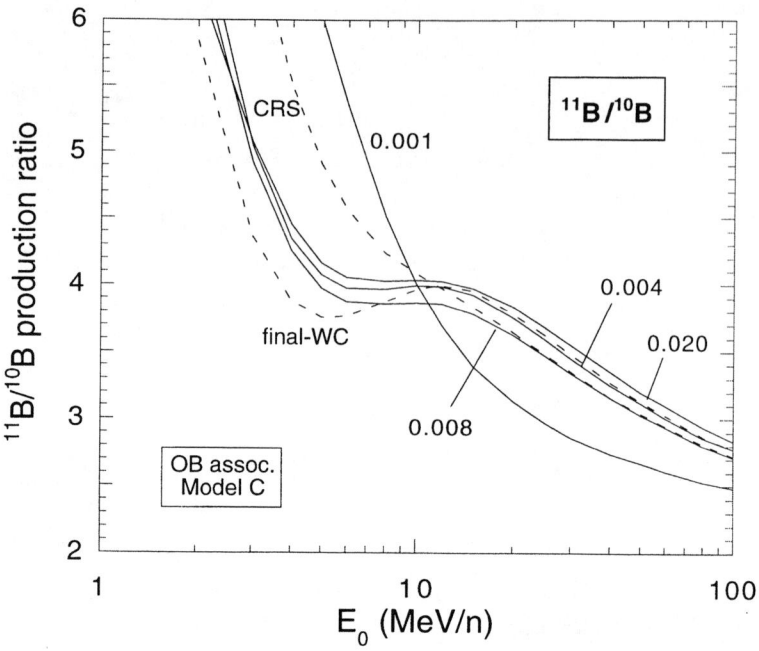

FIGURE 4. Same as figure 2 but for the $^{11}B/^{10}B$ ratio[20].

GALACTIC EVOLUTION OF LIGHT ELEMENTS

The essence of the model is that freshly synthesized C and O nuclei injected/accelerated in WR winds and supernova ejecta lead to a primary production of Be and B independent of the ISM metallicity by fragmentation of these species on the surrounding H and He nuclei, as required by observations of light element abundances in halo stars down to [Fe/H] $\sim$ -4 ([7] and references therein). In the context of the early LiBeB evolution, two strong constraints should be obeyed [23]: i) avoiding overproduction of Li at [Fe/H] < -1 to preserve the lithium plateau (e.g. [24]), ii) getting B/Be from 10 to 30 as observed in Fe poor halo stars [7]. An additional constraint was considered by [9], namely getting $^{11}B/^{10}B$ > 4, globally in the past, to obtain $\sim$4 after mixing with GCR spallation products ($\sim$2.5) In conjunction with this last constraint, one must also to take into account the possible source of ^{11}B through neutrino spallation in SNII explosions [25,23].

Galactic chemical evolution models are used to follow the behaviour of LiBeB isotopes, including stellar destruction of these fragile species (for details see [9,23]). In addition to satisfying the classical Galactic evolutionary constraints, a model is forced to respect the various constraints mentioned above on Li, Be and B. Using simple closed box galactic chemical evolution models, the irradiation time (i.e. the total number of Be nuclei produced per Orion-like region) is adjusted to fit the early Be and B evolution and results in $\tau = 0.5 - 1 \times 10^5$ yr (figure 1).

If the fast particle composition reflects that of the ejecta of massive stars, relatively He poor, overproduction of Li is avoided at halo metallicities, and thus the Spite plateau (*e.g.* [24]) is preserved (see figure 1 of [8]). For [Fe/H] > -1 which corresponds to the beginning of the disk evolution, spallation, both at low and high energy, falls short of fitting the solar abundances and the meteoritic lithium isotopic ratio ($^7Li/^6Li = 12.5$) . Thus a pure ^7Li component of stellar origin is required. Low mass AGB stars make good candidates to produce the remaining lithium. Due to their relatively long life time, they would eject their lithium with the required delay in the early Galactic evolution and thus the Spite plateau would be once again respected. With a reasonable ^7Li yield of $1 - 1.5\ 10^{-8}$ $M_\odot$ (*e.g.* [26]), the observed meteoritic ratio is reproduced.

In the case of B and Be, the low energy component (LEC) dominates their early evolution whereas GCR contribute but only marginally afterwards [9,4]. This calculation is adjusted to reproduce the observed log(Be/H) and log(B/H) vs [Fe/H]. It is additionally required to reproduce the observed $^{11}B/^{10}B$ ratio (~ 4) in the solar system, the Be/H and B/H values observed in the present and local Galactic environment (Be/H $= 1.4 \times 10^{-11}$ and B/H $= 2 \times 10^{-10}$ [27]) and the observed relationship log(Li/H) vs [Fe/H], while avoiding overproduction at [Fe/H] < -1 from the $\alpha + \alpha$ reaction.

The $^{11}B/^{10}B$ ratio is initially higher than observed value of ~ 4 but is diluted by the GCR component (which contains a ratio of ~ 2.5) It should be noted that neutrino spallation could also affect this ratio. It is worth mentioning that the boron meteoritic value is more consistent with our evolutionary scheme than the photospheric value. The B/Be ratio evolves from about 25 at the beginning of the Galactic evolution to about 18 at the present-day epoch which falls within the observed range. Fortunately, the uncertainty of the Fe yield of massive stars has little influence on these results due to the steepness of the IMF used.

WHAT TO DO NEXT: LOW METALLICITY WR AND SN AND LIBEB EVOLUTION

The situation present in Orion can be generalized to all star forming regions. We can assume that the nuclei propelled to MeV/n by winds and explosions of massive stars are further accelerated up to tens of MeV/n by ambient shocks and/or turbulence in superbubbles maintained by the high concentration of active objects grouped in OB associations. They may then propagate within the parent molecular cloud at the border of or within the superbubble produced by supersonic stellar winds and supernova explosions. In this dense medium they suffer dramatic energy losses and occasionally fertile nuclear collisions with H and He nuclei producing LiBeB and gamma-ray lines. At the low energies considered ionization losses dominate other losses (escape of the system and nuclear destruction). Specific details of these models can be found in [14].

The results which have been discussed so far have relied solely on WR models of solar composition. WR models at solar and super-solar metallicities ($Z = 0.02$ and 0.04 [22]) could be utilized to study possible variation of the emission (C, O lines and, Li-Be feature) with Galactic longitude due to the increase of the metallicity toward the center of the Galaxy [14]. WR stars are generous sources of fast C and O nuclei able to produce, in turn, specific gamma-ray lines and LiBeB nuclei by their interaction with the surrounding medium. The strength and composition of the wind, however, is heavily metallicity dependent [22]. The main effects of metallicity on light element production by massive stars in the early (low Z) Galaxy should therefore be estimated. Thus low metallicity massive stars ($Z < 10^{-3}$) could be used to explore the early production of LiBeB in the Galaxy.

In order to estimate this effect, we adopt a source spectrum of the form as above of $n(E)dE = k\ E^{-1.5}\ e^{-E/E_o}$ [11,12] where the break energy, E_0 is treated as a free parameter. The wind of massive stars of very low Z are very poor in C, N and O but rich in He so that $\alpha + \alpha$ reactions play a leading role. As a consequence, the low Z massive star model exceeds the Li/Be limit but without consequence on the Galactic evolution since the amount of matter dispersed by the stellar wind is small compared to the mass of material ejected by the explosion (1 $M_\odot$ as opposed to 10 $M_\odot$ typically for higher Z). None of the constraints on LiBeB evolution is violated by the SN model U40B of [28] associated with a break energy of $E_0 > 20$ MeV/n. The results of the calculation are quite sensitive to the adopted SN explosion energy which implies that the early evolution of LiBeB could be a test of SN models. It remains now to insert the calculated yields in a Galactic evolutionary model to follow the change of the various elemental and isotopic ratio and compare them to observations.

CONCLUSION

The material expelled by very massive stars, of typically 60 $M_\odot$ ZAMS, seems to have the right composition to explain both the gamma-ray line emission of Orion and the evolution of Be and B in the early Galaxy. The early Galactic (halo) evolution of Be and B seems to be governed by low energy nuclei of the kind that produce gamma-ray lines in Orion (see however [29]). The contribution of this component to the subsequent (disk) evolution is far from being negligible. The uncertainties on the solar abundance of Be and B preclude definite conclusion, but it seems that GCR play only a marginal role in the overall LiBeB story. Spallation is not sufficient to explain the whole

Li behaviour. The present ^{7}Li abundance can however be explained assuming that low mass (1 – 2 M$_\odot$), low metallicity stars are able to produce about 10^{-8} M$_\odot$ of this isotope. This yield furthermore settles an important constraint on the evolution of low mass stars, especially complex. The problem of the ^{11}B/^{10}B ratio is alleviated if not solved. Yet, a slight contribution from neutrino spallation in SNII explosions cannot be excluded but is severely restricted. More detailed estimates, including refined cosmic ray spallation and metallicity dependent yields can be found in [4].

ACKNOWLEDGMENTS

We would like to thank Etienne Parizot and Roland Lehoucq for their numerical expertise and Yvette Oberto for helping in the preparation of this manuscript. This work was supported by PICS, n° 319, "Gamma-ray Line Astrophysics and Associated Nucleosynthesis", CNRS.

REFERENCES

1. Meneguzzi M., Audouze J., & Reeves H., 1971, A& A, 15, 337.
2. Reeves, H., 1994, Rev. Mod. Phys., 66, 193.
3. Meneguzzi, M., & Reeves, H., 1975, A&A. 40, 99.
4. Lemoine, M., Vangioni-Flam, E.,& Cassé M., 1997, ApJ, in press.
5. Bloemen, H., et al. , 1994, A&A 281, L5.
6. Bloemen, H., et al. , 1997, ApJ, 475, L25.
7. Duncan, D., et al. , 1997, ApJ, in press.
8. Vangioni-Flam, E., Cassé, M., & Ramaty, R., 1997, in "The 2nd INTEGRAL Workshop: The Transparent Universe," ESASP 382, St. Malo, p. 123.
9. Vangioni-Flam, E., Cassé, M., 1996, in "Cosmic Abundances," Maryland, eds. Holt & Sonnebunn, ASPC Series, Vol. 99, p. 366.
10. Cassé, M., Lehoucq, R.,& Vangioni–Flam, E., 1995, Nature, 373, 318.
11. Ramaty, R., Kozlovsky, B., & Lingenfelter, R.E., 1995, ApJ. 438, L21.
12. Ramaty, R., Kozlovsky, B., & Lingenfelter, R.E., 1996, ApJ 456, 525.
13. Ramaty, R., Kozlovsky, B., & Lingenfelter, R.E., 1997, in "The Transparent Universe," ed. C. Winkler, T.J.-L. Courvoisier and Ph. Durouchoux, ESA Publications Division, p. 75.
14. Parizot, E.M.G., 1997, A&A, submitted.
15. Ryan, S., Bessel, M., Sutherland, R., & Norris, J., 1990, ApJ, 348, 157.
16. Gilmore, G., Gustafsson, B., Edvardsson, B., & Nissen, P.E., 1992, Nature 357, 379.
17. Ryan, S., Norris, I., Bessel, M., & Deliyannis, C., 1994, ApJ. 388, 184.
18. Boesgaard, A., & King, J.R., 1993, A.J., 106, 2309.
19. Duncan, D., Lambert ,D., & Lemke, M., 1992, ApJ. 401, 584.
20. Parizot, E.M.G., 1997, Ph.D. thesis.

21. Ramaty, R., 1997, A& A, in press.
22. Meynet, G., *et al.* 1991, AAS, 103,97.
23. Vangioni-Flam, E., Cassé, M., Olive, K., Fields, B., 1996, ApJ, 468, 199.
24. Spite, M., 1997, in "Fundamental Stellar Properties: The Interaction between Observations and Theory," IAU, p. 185.
25. Woosley, S.E., Hartmann, D., Hoffman, R., Haxton, W., 1990, ApJ., 356, 272.
26. Abia, C., Isern, Canal, R., 1993, A&A, 275, 96.
27. Arnould, M., & Forestini, M., 1989, Nuclear astrophysics: Proceedings of the third international Summer School, Larabida, Ed.: M. Lozano, M.I. Gallardo, J.M. Arias, p. 48
28. Woosley, S.E., & Weaver, T.A., 1995, ApJ Supp, 101, 181.
29. Ramaty, R., Kozlovsky, B., Lingenfelter, R.E., & Reeves, H., 1997, ApJ, submitted.

The "stellar yields - galactic chemical evolution" connection: a re-analysis.

S. Sandrelli[*,1], A. Visco[*] and M. Tosi[†,2]

Dipartimento di Astronomia, Università di Bologna, Italy
†Osservatorio Astronomico di Bologna, Italy

Abstract.
By means of galactic chemical evolution models we analyze the consequences of different upgraded stellar yields on the gradients along the Galactic Disk of ^{14}N, ^{16}O and of the main isotopic ratios (^{12}C/^{13}C, ^{14}N/^{15}N, ^{16}O/^{17}O, ^{16}O/^{18}O, ^{18}O/^{17}O). The evolution in the solar neighbourhood of the same ratios is also examined.

The best results are obtained by using solar composition yields during the whole galactic evolution, which provide a good fit to the data for the current gradient and the evolution except for ^{18}O/^{17}O. On the contrary, by using metallicity dependent yields, we don't reproduce satisfactorily the ^{12}C/^{13}C gradient and the time behaviour of all the isotopic ratios. Altering some of the basic assumption of the galactic model doesn't change this result.

A simple algorithm to mimic the time delayed contribution of novae to the interstellar enrichment of ^{15}N is introduced, leading to a significant improvement of the comparison with the observed nitrogen isotopic ratio. Hints to stellar nucleosynthesis can be traced from these results.

INTRODUCTION

Models for the chemical evolution of the interstellar medium (ISM) work thanks to a toolkit gathered by picking up instruments from various branches of physics and astronomy, both from the observational point of view and the theoretical one. In a complex system as our galactic disk, diffuse gas can collapse under gravitational forces and shock waves to give birth to clumpy and dense molecular clouds, which are supposed to act like actual stellar incubators. Once the stars are born, they start to synthetize and destroy isotopes by nuclear reactions and to change the ISM chemical composition by mass loss mechanisms, along typical timescales related to the stellar lifetimes. Besides,

---
[1] sandrelli@astbo4.bo.astro.it
[2] tosi@astbo3.bo.astro.it

some external matter can rain down onto the disk, inducing further evolution. It's clear that such a model rises as a complex machine, which takes into account nuclear physics, plasma physics, pure hydrodynamical effects, neutrino and photon interactions with baryonic matter, gravitational theory of collapse, thermodynamics, energy transfer, chaotic effects and so on. On the other side, these models are requested to reproduce a number of observations, so that they can be well constrained. In this case, one can demonstrate that they depend on a handy number of free parameters (see Tosi [1]). Furthermore, one finds that possibly there is a subset of parameters playing a dominant role for each specific issue. For instance, the model predictions on isotopic ratios depend mostly on the initial mass function (IMF) and stellar nucleosynthesis. In this sense, chemical evolution models become powerful tools to investigate *if* and *how* findings of other theories give consistent results when they are put together.

The disk of the Galaxy stands as one of the best candidate to be studied extensively from quite every point of view. Particularly, over the years a great effort was spent to understand the galactic evolution of CNO isotopes, since they are rather abundant and observable in a number of different locations, as the solar system, red giant stars (RGB and AGB stars), ISM, planetary nebulae (PN). In the last years, new observational improvements have taken place and new theoretical work was done so that it is useful to try to reassess properly some questions. We'll make use of the model developed by Tosi [1], shortly described in section 2. Referring to section 2.1 for the summary of new theoretical results, we cite briefly the main observational data used in this work. Shaver et al. [2] measured ^{14}N and ^{16}O abundances in several HII regions spanning a large range in galactocentric distance, $R_{gc} \sim 3 - 13.7$ kpc. In 1991 Fich & Silkey [3] extended this data set to the outer regions of the galactic disk, finding O/H and N/H[3] values very similar to the solar ones. Isotope ratios are usually inferred by measuring the abundances of molecules differing from one another just for a single isotope. Gardner & Whiteoak [4] (1979), Henkel et al. [5–7] determined the values of ^{12}C/^{13}C through the measure of H_2CO and $H_2^{13}CO$ radio spectra of a number of molecular clouds. Langer & Penzias [8,9] obtains similar results by using the molecular ratio $^{12}C^{18}O$ and $^{13}C^{18}O$. Dahmen et al. [10], Henkel et al. [7] and Wilson & Rood [11] measured the abundance ratio of $HC^{15}N$ and $H^{13}CN$ in several molecular clouds. Gardner & Whiteoak [12] applied the same method to $H_2C^{18}O/H_2^{13}CO$ ratio measured in molecular clouds to have evaluations on $^{16}O/^{18}O$. Differently, Penzias [13] got the $^{18}O/^{17}O$ directly from measures of $C^{17}O$ and $C^{18}O$ in giants molecular clouds. The standard solar abundances are taken from Anders & Grevesse [14]; the present values in the solar ring from Wilson & Rood [11], Henkel et al. [15], Kahane [16].

---

[3] ratios by number

THE MODEL'S INGREDIENTS

Basic assumptions

We examine the galactic disk between 3 and 10 kpc from the center, dividing it in rings 1 kpc wide. At $t = 0$ Gyr each ring has a non zero gas mass and a null stellar content. We assume the disk age to be 13 Gyr and the solar system to be located at a distance of 8 kpc from the galactic center. We refer to Tosi [1] for a detailed description of the models. From now on we will refer to the best of her models 1 as to the standard reference case. In this model the IMF is Tinsley's [17] and the SFR is assumed to be an exponentially deacreasing function of time, with e-folding time τ. Our standard prescription is $\tau \sim 15$ Gyr, that means a nearly constant $\psi_i(t)$ at a given galactocentric distance R. Besides, it steeply decreases outwards because of the term ψ_{0i} which is proportional to the present total and gas mass of the ring i (see Table 1).

Finally, the disk is not isolated: observed phenomena suggest both a coupling with the galactic halo by gas fall from the latter to the disk and gas flows from the intergalactic medium to the disk. For the infall rate we adopt here an analytical approximation: in Table 1, $R_{max,i}$ ($R_{min,i}$) is the outer (inner) radius of the ring i, while F_i is the infall density, assumed $4 \cdot 10^{-3} \frac{M_\odot}{kpc^2 yr}$ in the standard model 1. A low value of Θ approximates the halo infall, while a higher value is related to an extragalactic infall: we use $\Theta \to \infty$. Tosi [18] showed that the metal content of the infall can't be in excess of $\sim 0.3 - 0.4$ $Z_\odot$ in order to be consistent with the observed chemical features of the disk. This is consistent with the data of West et al. [19] and other authors. So we are allowed to assume $Z_f = 0.2 \cdot Z_\odot$ and this will be our standard choice.

The basic assumptions of our model are summarized in Table 1.

TABLE 1. Basic assumpions of our model. See text for details.

Disk Age	$R_\odot$	SFR	IMF	Infall
13 Gyr	8 Kpc	$\psi_{0i}e^{-t/\tau}$	Tinsley80	$F_i\pi(R_{max}^2 - R_{min}^2)e^{-t/\Theta}$

Stellar yields

The last basic assumption of evolutionary chemical models of the ISM are the stellar yields. In the following, we briefly describe the "source" of stellar yields we used to compute our models.

Boothroyd & Sackmann [20] (henceforth BS): it's a large set of evolutionary tracks of stellar models referring to stars from 0.85 to 9 $M_\odot$ with an initial metallicity from $Z = 0.001$ to $Z = 0.02$. Stars are followed up to the beginning of the AGB phase, which is not included. In the low mass range, the BS's code includes the so called "cool bottom processing" (see Charbonnel [21], Wasserburg et al. [22]) which has been demonstrated to fit nicely some oxygen isotope ratios measured in metheoritic grains. On the RGB and the AGB, the mass loss is included with a Reimers' law, where the scaling factor η depends on the initial metallicity of the stars. Nuclear reaction rates from Caughlan & Fowler [23] (hereinafter CF88) were used except for $^{17}O(p,\alpha)^{14}N$ and $^{17}O(p,\gamma)^{18}F$. These latter come from Landré et al. [24].

TABLE 2. Standard models calculated by varying the stellar yields: A solar symbol besides the names indicates that only solar yields were used; otherwise yields at different metallicities were assumed. The acronyms' meaning is illustrated in the text. *Int.* means that an interpolation procedure was exploited.

Model	0.8-2.75[a]	4.5	6	M_{up}[b]	11	13	40
1-m1	BS$_\odot$	FC$_\odot$		int.		WW$_\odot$	
1-m2	BS	FC		int.		WW	
1-m3	BS	FC		int.		WW$_\odot$	
1-m4	BS	FC$_\odot$		int.		WW	
1-m5	BS	FC[c]		int.		WW	
1-m6	BS$_\odot$	FC		int.		WW	
1-m1b	BS$_\odot$				int.	WW$_\odot$	
1-m2b	BS				int.	WW	
1-m3b	BS	BSM[d]			int.	WW	
1-G22	vHG$_\odot$				int.	WW$_\odot$	
1-G12	vHG				int.	WW	
1-lan	BS$_\odot$	FC$_\odot$		int.		WW$_\odot$	LH$_\odot$
1-mar22	MBC$_\odot$	BS$_\odot$		int.		WW$_\odot$	
1-mar12	MBC	BS		int.		WW	

[a] unit of $M/M_\odot$. In the following columns we indicate only the upper limit of the mass range.
[b] $M_{up} \sim 8 - 9 M_\odot$
[c] Mass loss rate parameter η halved with respect to mod2.
[d] BS modified. See text.

Forestini & Charbonnel [25] (CF): they calculate stellar models ranging from 3 to 7 $M_\odot$ at metallicities $Z = 0.005$ and $Z = 0.02$ up to around a dozen of thermal pulses. Then extrapolation formulas bring the stellar models up to the planetary ejection. Overshoot is not taken into accout and the classical Schwarzschild criterion for convection borders is used. The mass loss phenomenon is described by a Reimers' law, where η depends both on the stellar evolutionary phase and on the stellar initial mass. Stellar yields are obtained also for different values of η. CF88 provide most of the nuclear rates, but some of them are taken from other authors, for instance $^{17}O(p,\alpha)^{14}N$ and $^{17}O(p,\gamma)^{18}F$ are taken from Landré et al. [24].

Marigo et al. [26] (MBC): semi-analytical model based on stellar models calculated by the Padua group. The analytical procedure starts from the end of E-AGB phase. They take into account a moderate core overshoot. They provide yields for stars of mass included in the range 3-7 $M_\odot$, at $Z = 0.02$ and

$Z = 0.008$. The mass loss rate formula is related to the $\dot{M}$ - period relation, determined by observations of long period variables, like Miras and OH/IR pulsating stars. MBC always use CF88.

van den Hoek & Groenewegen [27] (vHG): semi-analytical model based on stellar tracks calculated by the Geneva group with moderate core overshoot. As in MBC, they adopt an analytical procedure from the end of the E-AGB phase on. Along the AGB phase, the mass loss is a Reimers' law. The isotopes ^{15}N, ^{17}O and ^{18}O are not included in their calculations. Stellar yields are computed for stars from 0.8 to 8 $M_\odot$, at $Z = 0.001, 0.004, 0.008, 0.02, 0.04$.

Woosley & Weaver [28] (WW): calculations of explosive nucleosynthesis in stars of initial mass running from 11 to 40 $M_\odot$, at metallicity $Z = 0, 0.0001, 0.01$ and 1 times solar. The Ledoux criterion for convection with further modifications for semiconvection is preferred to the Schwarzschild criterion giving smaller convective shell during the explosion. All the models are evaluated with no mass loss. For the rate of the reaction $^{12}C(\alpha, \gamma)^{16}O$ they use CF88's value multiplied by 1.7.

Langer & Henkel [29] (LH): yields of stars from 15 to 50 $M_\odot$. They adopt both a time-dependent convection and a time-dependent semiconvection at solar metallicity. The criterion of Ledoux is used as in WW, while they make use of the mass loss formula of Nieuwenhuijzen & de Jager [30] and Langer [31] for WR stars.

We tried to collect these data in the most suitable way to obtain informations about the stellar yields - ISM chemical evolution connection. Table 2 summarizes our choices: for each row, the name of the model is shown; the reference of the yields we adopt for that mass range is indicated in the column. For each author and whenever possible, we prepare two sets of stellar yields: at solar metallicity $Z_\odot$ and at a lower Z_1, the exact value of which depends on the data we are working with. In a galactic model with Z−dependent yields, we employ the latter one until $Z < Z_\odot$, and the former one for $Z \geq Z_\odot$. Model 1-m3b is a numerical experiment: we tried to isolate the contribution of the AGB evolutionary phase in intermediate mass stars from FC's yields and to correct BS's data by this amount.

RESULTS

Figure 1-a and 1-b show the observed distribution of ^{16}O/H and ^{14}N/H by number compared with some selected theoretical curves. In both of them, the observational data suggest a quite steep decrease of the N and O abundances from 3 to 11 kpc. On the contrary, the outer data are similar to the solar values. All the models reproduce fairly well this trend (we remind that the model applies in the range $3 - 10$ kpc) but they all seem to overestimate the mean observed curve. The small differences between them are easily understood. Panel a) deals with oxygen gradient: ^{16}O is produced by massive stars and its yields depends on the size of He-core and on the rate adopted for

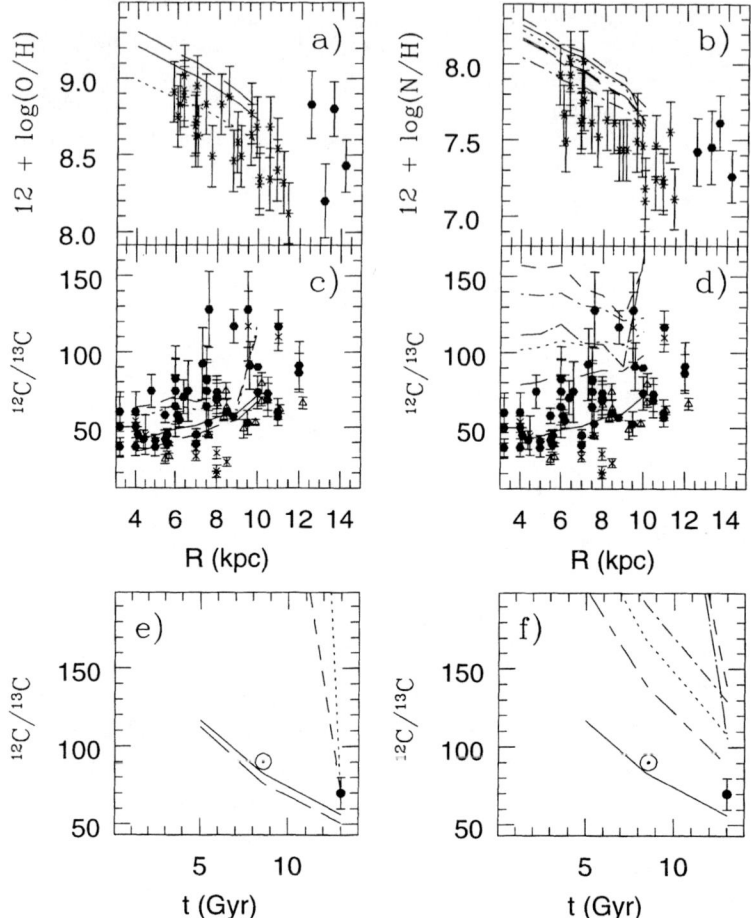

FIGURE 1. Panels a) and b): radial distribution of ^{16}O and ^{14}N at the present epoch. Data from Shaver et al. ($\times$) and from Fich & Silkey ($\bullet$). c) and d): radial distribution of the carbon isotopic ratio at the present epoch. Data from Henkel et al. ($\bullet$), Langer & Penzias ($\triangle$) and Gardner & Whiteoak ($\times$). e), f): time behaviour of $^{12}C/^{13}C$ in the solar ring. Data from Anders & Grevesse ($\odot$), Henkel et al. ($\bullet$). a): long dash = 1-m1b, dot = 1-lan. b): dot = 1-m2, dot-dash = 1-m5, long dash = 1-m1b, dot-short dash = 1-G22, short dash - long dash = 1-mar22. c) and e): dot = 1-m2, dot dash = 1-m5, long dash = 1-m1b. d) and f): dot-long dash= 1-mar12, short dash-long dash= 1-mar22, dot-short dash= 1-G22, dot dash= 1-G12, dot = 1-lan. In every panel, the solid line is the prediction of the model 1-m1.

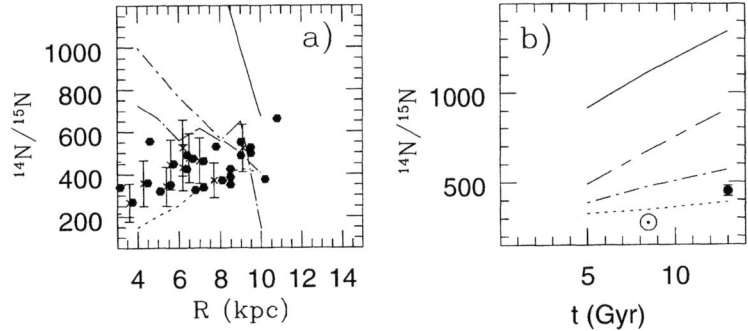

FIGURE 2. Panel a): radial distribution of $^{14}N/^{15}N$ at the present epoch. Data from Henkel et al. (•) and Dahamen et al. (×); b): time behaviour of $^{14}N/^{15}N$ in the solar ring. Data from Anders & Grevesse (⊙) and Henkel et al. (•). a) and b): short dash-long dash = 1-mar22, dot-short dash = 1-G22, dot = 1-novae, a special model with the contribution on novae (see text). In every panel, the solid line is the prediction of the model 1-m1.

$^{12}C(\alpha,\gamma)^{16}O$. In 1-lan one uses yields coming from massive stars with smaller cores than Woosley & Weaver's [28]. The gap between 1-m1 and 1-m1b signs the contribution of AGB stars. ^{14}N is thought to be originated in all mass stars as a secondary element[4] in every H-burning episodes lead by CNO cycles reaction. It's supposed to have also a primary production in TP-AGB stars in conjunction with envelope burning after third dredge-up events. Panel b) shows the most significant models. The comparison between 1-m1b and 1-m1 gives an idea of the AGB contribution to the ^{14}N enrichment.

Panels c)-f) in figure 1 deals with the spatial distribution and evolution of $^{12}C/^{13}C$. Two different sets of models are shown: in c), all the curves reproduce the data, but the general trend is met just by 1-m1 and 1-m1b, while the others seem to be too flat. And in e), the difference it's striking as 1-m2 differs from the solar abundance by a factor of 20. Inset d) it's a clear example of how results can differ from one another just assuming yields from different authors: again the trend, when not the absolute value, is reproduced by 1-mar22 and 1-lan, while the other curves show an opposite slope: all of them decrease. It's also noticeable that all of them calculate intermediate mass yields by using a semi-analytical method. In inset f), the only model which is consistent with the observed values is 1-mod1. Models with metal dependent yields never give good results. Now, ^{13}C is supposed to have both a primary and a secondary origin, the former one due to TP-AGB phase, as

---

[4]) An isotope has a *primary* origin when it's built from initial H and ^{4}He content; it's a *secondary* element when it origins from seed nuclei heavier than H and ^{4}He.

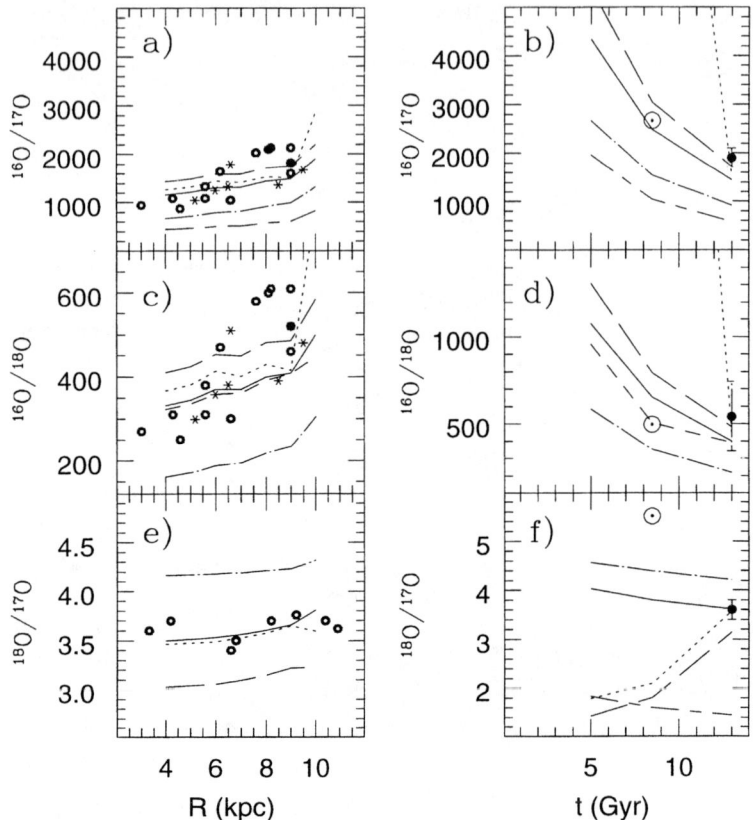

FIGURE 3. Panels a), c) e): radial distribution of the oxygen isotopic ratios at the present epoch. Data from Kahane (o) and Wilson & Rood; b), d): Anders & Grevesse (⊙), Wilson & Rood (•); f) Anders & Grevesse (⊙), and Kahane (•). a) and b): dot = 1-m2, long dash = 1-m1b, short dash-long dash= 1-mar22, dot-long dash= 1-lan. c) and d): dot = 1-m2, long dash = 1-m1b, dot dash= 1-m3, dot-long dash= 1-lan. e) and f): dot = 1-m2, long dash= 1-m3b, short dash-long dash= 1-mar22 (in c) it's always under the value of 2), dot-long dash= 1-lan. In every panel, the solid line is the prediction of the model 1-m1. Panels b), d) and f) show the models' time behaviour in the solar ring.

described for nitrogen. Non solar yields suffer from two effects: Z decreases so that there's less initial CNO abundances; inner T increases so CNO cycling is more efficient and ^{13}C easily reaches its equilibrium value, which can explain some differences.

The observed values of ^{14}N/^{15}N are consistent with a positive gradient along the galactic disk. One the other hand, one has to remark that the data suggest opposite conclusions: the spatial data are consistent with a primary/secondary ratio, while the solar value is lower or similar to the present one, which is not consistent with the previous expectation. Besides, one should remember that nitrogen is mostly produced as a secondary element. From figure 2 it is clearly seen that no model among the ones listed in table 2 meets the trend. Model 1-G22 is the only model to give a nice time pattern, but it's based on intermediate mass stars' calculations which do not take into account ^{15}N. Besides, even if all the shown models reproduce the derivative of the time evolution, they miss completely the spatial one. A possible solution (see Visco [32]) is investigated by the dotted model in figure 2, by assuming a primary production of ^{15}N in novae. As novae have long timescales, the ejection of ^{15}N is delayed with respect to a typical primary isotope as ^{16}O: this feature implies a higher abundance in the inner enriched rings of the galactic disk, giving a positive spatial gradient. To mimic novae contribution, we assumed an association between novae frequency and binary systems, as in D'Antona & Matteucci [33]. Besides, one has to sort out the evolution of the ratio: if the measured solar value makes sense, the stars should not eject ^{15}N too late. All of these hints are satisfied by model 1-novae. The problem with this model is that one needs to include timescales of novae explosion as short as the lifetime of a 4 $M_\odot$: there is no way to reproduce both the trends without such an assumption. Since theoretical data on novae yields are very rare, we are obliged to parametrize it. Obviously, 1-novae is just a model created *ad hoc* and much more effort is requested to novae's experts in order to provide estimations of their stellar yields.

Let's turn to the oxygen isotopes ratios. They are reported in figure 3. The radial observed distributions of both the ^{16}O/^{17}O and ^{16}O/^{18}O is steeper than the theoretical one. In a) 1-m1, 1-m1b and 1-m2 are quite identical, even if the difference between the first and the second shows the AGB contribution. A similar trend is shown by 1-lan, but the absolute value is lower of a factor 2-3. Again a problem rises with the solar value: the best results are obtained by 1-m1, while the non solar yields model 1-m2 is far from beeing acceptable. Since ^{17}O is thought to be a purely secondary element, its appearance in the ISM is delayed with respect to ^{16}O. Solar yields don't avoid the delay, but they are much more abundant than low metallicity yields and that's why 1-m1 or 1-m1b get the best prediction. We have opposite needs from the spatial distribution and the evolution of the ratio: more secondary production to have a steeper gradient and a primary production or a quick secondary production by intermediate or massive stars to get a lower solar value. The problem with this isotope seems to be rather open, also for the uncertainties related to its destruction rates (see CF88, Landré et al. [24]). Note, for instance, in figure 3 the model 1-mar22: it uses CF88 destruction rates which are lower than Landré et al.'s and as a consequence it underestimates the ratio by a factor 2-3. The isotope ^{18}O is produced by massive stars as a result of He-burning, coming out from the so called nitrogen-chain. So it's a purely secondary element and it follows nitrogen in its ejection into in the ISM. H burning phases destroy it in a non negligible way, as it's testified by the difference between 1m1b and 1-m1 in panel c) and d). The solar time value, in panel d), is well reproduced by 1-m3. The difference between it and 1-m1 is that the latter makes use of solar intermediate mass stars' yields, and TP-AGB stars of high Z are drastic destroyers of ^{18}O. Model 1-m2 differs from the solar value by a factor 15. Here the situation is similar to the nitrogen problem discussed before. In figure 3-e) there's a fair agreement between measured and theretical results: the absolute value does not show a large scatter from one model to the other and even 1-m2 has a good attitude with the data confirming the secondary/secondary nature of this ratio. But in panel f), just 1-m1 and 1-lan make some effort to reproduce the solar value, succeding in obtainig a negative derivative; the other lines have the opposit trend. An astrophysical solution was suggested by Wielen & Wilson: since the Sun shows chemical abundances higher than other local objects, they syggest it to be born around two kpc closer to the galactic center than usually believed. If this is the case, one must compare the sun data with the present observed ones in the ISM not local but where it was born. In this case the other isotopic ratios are qualitatively unaffected to this variation, but the present $^{16}O/^{18}O$ value in the ISM would be around 450 (instead of the local value around 540) to be compared with the solar measure 490. Evolution would have brought a decrease of the ratio, as expected from normal secondary/secondary behaviours.

DISCUSSION AND CONCLUSIONS

$\underline{^{14}N}$ seems to be overestimated. Since this effect is independent of the adopted yields, it could be due to a rough assumption of nitrogen equilibrium in stellar models.

$\underline{^{12}C/^{13}C}$: the problems with non solar yields seem to be an underestimate of ^{13}C production. Its primary yields should not increase too much, since it would mean more flatness in the spatial distribution, that's not a situation to pursue. Observational data show some scatter, but if one selects homogeneous data from the same author then the trend is rather well defined. So, there's a hint of a higher production of ^{13}C as a secondary element in massive and intermediate mass stars of low metallicities: this production is obviously limited by the CNO initial content.

$\underline{^{14}N/^{15}N}$: No model assuming standard stellar nucleosynthesis succeeds in reproducing the observed values. A possible solution could be taking into account the delayed contribution of novae to ^{15}N production. The clue requires wide investigations.

$^{16}O/^{17}O$, $^{16}O/^{18}O$: A primary origin of ^{17}O is requested to account for the solar value or a quick secondary origin. The behaviour of ^{18}O still remains a puzzle. Wide space is left for astrophysical and nuclear physics inquiries.

REFERENCES

1. Tosi, M.: *A.&A.* **197**,33, 1988
2. Shaver, P.A., McGee, R.X., Newton, L.M., Danks, A.C., Pottasch, S.R.: *M.N.R.A.S.* **204**,53, 1983
3. Fich, M., Silkey, M.: *Ap.J.* **366**,107, 1991
4. Gardner, F.F., Whiteoak, J.B.: *M.N.R.A.S.* **188**,331, 1979
5. Henkel, C., Walmsley, C.M., Wilson, T.L.: *A.&A.* **82**,41, 1980
6. Henkel, C., Wilson, T.L., Bieging, L.: *A.&A.* **109**,344, 1982
7. Henkel, C., Güsten, R., Gardner, F.F.: *M.N.R.A.S.* **143**,148, 1985
8. Langer, W.D., Penzias, A.A.: *Ap.J.* **357**,477, 1990
9. Langer, W.D., Penzias, A.A.: *Ap.J.* **408**,539, 1993
10. Dahmen, G., Wilson, T.L., Matteucci, F.: *A.&A.* **295**,194, 1995
11. Wilson, T.L., Rood, R.: *A.R.A.A.* **32**,191, 1994
12. Gardner, F.F., Whiteoak, J.B.: *M.N.R.A.S.* **194**,37, 1981
13. Penzias, A.A.: *Ap.J.* **249**,518, 1981
14. Anders, E., Grevesse, N.: *Geochim. Cosmochim. Acta* **53**, 197, 1989
15. Henkel, C., Wilson, T.L., Langer, N., Chin, Y.N., Mauersberger, R.: *The Structure and Content of Molecular Clouds*, 1995
16. Kahane, C.: *Nuclei in the Cosmos III. Third international symposium on nuclear astrophysics*, 1995
17. Tinsley, B.M.: *Fund. Cosmic Phys.* **5**, 287, 1980
18. Tosi, M.: *A.&A.* **197**,47, 1988
19. West, K.A., Pettini, M., Penston, M.V., Blades, J.C., Morton, D.C.: *M.N.R.A.S.* **215**,481, 1985
20. Boothroyd, A.I., Sackmann, I.J.: *private communication*, 1996
21. Charbonnel, C.: *A.&A.* **282**,811, 1994
22. Wasserburg, G.J.,Boothroyd, A.I., Sackmann, I.J.: *Ap.J.* **447**,L37, 1995
23. Caughlan, G.R.,Fowler, W.A.: *Atom. Data Nucl. Data Tables* **40**, 283, 1988 [CF88].
24. Landré, V., Prantzos, N., Aguer, P., Bogaert, G., LefLbvre, A., Thibaud, J.P.: *A.&A.* **240**,85, 1990
25. Forestini, M., Charbonnel, C.: *A.&A.S.S* **123**,241, 1997
26. Marigo, P., Bressan, A., Chiosi, C.: *A.&A.* **313**,545, 1996
27. van den Hoek, L.B., Groenewegen, M.A.T.: *A.&A.S.S* **123**,305, 1997
28. Woosley, S.E., Weaver, T.A.: *Ap.J.S.S.* **101**,181, 1995
29. Langer, N., Henkel, C.: 1995, *Space Science Reviews*, Proc. of the "Gamow Seminar held on Sept. 12 to 14", St. Petersburg, Russia, A.M. Bykov
30. Nieuwenhuijzen, H. de Jager, C.: *A.&A.* **231**,134, 1990
31. Langer, N.: *A.&A.* **220**,135, 1989
32. Visco, A.: *Tesi di Laurea*, Dip. Astronomia, Univ. Bologna, 1997
33. D'Antona, F., Matteucci, F.: *A.&A.* **248**,62, 1991

Heavy Elements Abundances in Metal-Poor Stars[1]

P. Magain[2], E. Jehin, C. Neuforge[3] and A. Noels

Institut d'Astrophysique et de Géophysique
Université de Liège
5, avenue de Cointe, B-4000 Liège, Belgium

Abstract. A sample of 21 metal-poor stars have been analysed on the basis of high resolution and high signal-to-noise spectra. Correlations between relative abundances of 16 elements have been studied, with a special emphasis on the neutron-capture ones.

This analysis reveals the existence of two sub-populations of field halo stars, namely Pop IIa and Pop IIb. They differ by the behaviour of the s-process elements versus the α and r-process elements.

We suggest a scenario of formation of these stars, which closely relates the field halo stars to the evolution of globular clusters. The two sub-populations would have evaporated the clusters during two different stages of their chemical evolution.

INTRODUCTION

Traditional abundance analyses of metal-poor stars aim at determining abundance ratios of some chemical elements as a function of the overall metallicity, usually measured by the iron abundance [Fe/H][4]. These trends are then compared to predictions from models of nucleosynthesis and chemical evolution of the Galaxy, in order to provide constraints on the sites and mechanisms for element synthesis.

However, many of these abundance ratios show rather considerable star-to-star scatter, so that they provide only weak constraints on the models.

With the improvement of observing and spectroscopic analysis techniques, it is now possible to reduce considerably the observational uncertainties in the

---

[1] Based on observations carried out at the European Southern Observatory (La Silla, Chile)
[2] Maître de Recherches au Fonds National Belge de la Recherche Scientifique
[3] Chargé de Recherches au Fonds National Belge de la Recherche Scientifique
[4] We use the traditional spectroscopic notation $[E/M] = \log(N_E/N_M)_\star - \log(N_E/N_M)_\odot$, where E and M are any chemical elements.

abundance determinations. In a first step, this allows to decrease the scatter in the abundance ratios, but only down to a certain point, since there is a genuine cosmic scatter which can now be measured and analysed if the data are of sufficient quality.

With these high quality data, it is thus possible to investigate the cosmic scatter in relative abundances at a given metallicity, and to identify abundance correlations between several elements. These correlations should give better constraints on the sites of formation of these elements, and on the nucleosynthetic mechanisms responsible for their formation.

In contrast to traditional spectroscopic analyses, for which the main result is the variation of a mean abundance ratio as a function of overall metallicity, what we seek are the correlations between deviations from the mean trend in different abundance ratios. In other words, we want to determine if, for example, all the stars for which the individual [Ba/Fe] is systematically higher than the mean [Ba/Fe] for the whole sample, also show the same trend in [Ti/Fe] or in [Zr/Fe], or any other relative abundance.

We can then deduce that the elements which are strongly correlated together, have very likely been synthesized by the same nucleosynthetic processes in the same kinds of objects. This is a new tool which should be quite efficient for the identification of the sites and mechanisms of element synthesis at different stages of the galactic evolution.

In the following, we present the results of such an analysis for a sample of moderately metal-poor stars and we suggest a scenario explaining the observed trends.

OBSERVATIONAL DATA AND ABUNDANCE ANALYSIS

We have analysed a sample of 21 dwarf and subgiant stars with [Fe/H] ~ -1, i.e. one tenth of the solar metallicity. This metallicity range is generally assumed to correspond to the most metal-rich part of the halo of our Galaxy.

The spectra were obtained with the Coudé Echelle Spectrometer (CES) fed by the 1.4m Coudé Auxiliary Telescope (CAT) at the European Southern Observatory (La Silla, Chile). Four spectral regions were observed, basically chosen to contain lines of neutron-capture elements.

The spectral resolution is of the order of 65 000 and the signal-to-noise ratio in the continuum is ~ 250 for each spectrum.

In order to reduce the analysis uncertainties, the lines were chosen, whenever possible, to have similar dependences on the stellar atmospheric parameters (effective temperature, surface gravity, microturbulence velocity, overall metallicity). Moreover, the analysis was carried out differentially inside the sample, i.e., each star was compared to all other stars in the sample. The zero point

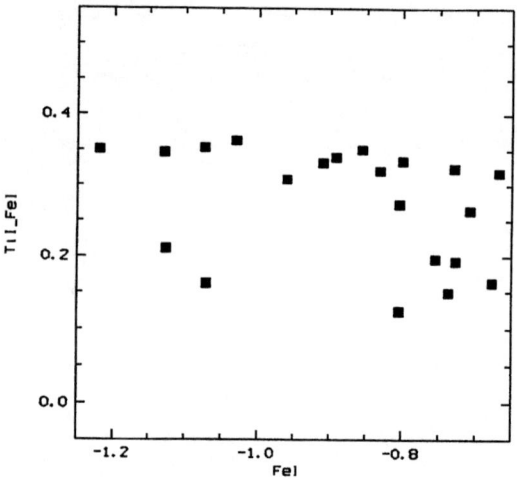

FIGURE 1. Traditional plot: [Ti/Fe] versus [Fe/H] for the 21 stars analyzed.

was then fixed by analyzing one of the stars with respect to the Sun (the precise value of this zero point is relatively unimportant as it does not affect the abundance correlations).

We have determined the abundances of 16 elements, namely Mg, Ca, Sc, Ti, V, Cr, Fe, Ni, Sr, Y, Zr, Ba, La, Ce, Sm and Eu.

If we had made a typical abundance analysis, we would have presented the results as in Fig. 1, which shows the abundance of Ti relative to that of Fe, plotted as a function of [Fe/H]. All these abundances were obtained from the analysis of neutral lines. We note a roughly constant overabundance of Ti relative to Fe, with a 1σ scatter amounting to 0.08 dex (20%). Now we address the following point: is this scatter real or is it due to observational and/or analysis uncertainties? To answer this crucial question, we compare the values of [Ti/Fe] deduced from neutral lines with the same quantities deduced from lines of the singly ionized species. This comparison is displayed in Fig. 2, which shows a very nice correlation, with a scatter of 0.026 dex (6%) only. As the neutral and ionized lines have different dependences on the stellar atmospheric parameters, the analysis uncertainties should not exceed 6%, and most of the scatter in the abundance of Ti relative to Fe should be real cosmic scatter.

We can thus conclude that our data are of sufficient quality to investigate the cosmic scatter in the relative abundances of the chemical elements, and proceed in the analysis of these correlations.

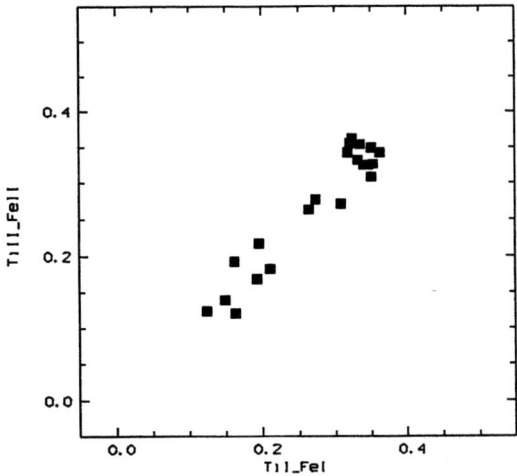

FIGURE 2. Comparison of the values of [Ti/Fe] deduced from neutral lines with those deduced from ionized lines.

ABUNDANCE CORRELATIONS

Figure 3 displays the abundance of Ca relative to Fe, [Ca/Fe], plotted as a function of [Ti/Fe], and shows that the two elements Ca and Ti are closely correlated. The same is true for Mg. We can thus conclude, and this is certainly no surprise, that the so-called α-elements were synthesized by the same process in the same objects, but that there is an important source of Fe which is unrelated with the nucleosynthesis of the α-elements.

The abundances of Cr and Fe relative to Ti also show a remarkable correlation, indicating a common origin for the two iron-peak elements Cr and Fe. The same holds true for Ni, with two exceptions: the stars HD193901 and HD194598 appear to be somewhat depleted in Ni (in these two objects, [Ni/Fe] is ~ 0.11 dex (24%) below the mean of the other stars). These two stars, which present other abundance anomalies, will be identified by open symbols in all subsequent figures.

The main aim of this work is to carry out the same kind of analysis for the neutron-capture elements, in order to identify the sites and mechanisms for the synthesis of these elements, in a relatively early phase of the galactic evolution.

A first hint was put forward by Zhao and Magain [1] who found that the elements Y and Zr are better correlated with Ti than with Fe, and suggested that this indicates that massive stars played a dominant role in the early nucleosynthesis of Y and Zr.

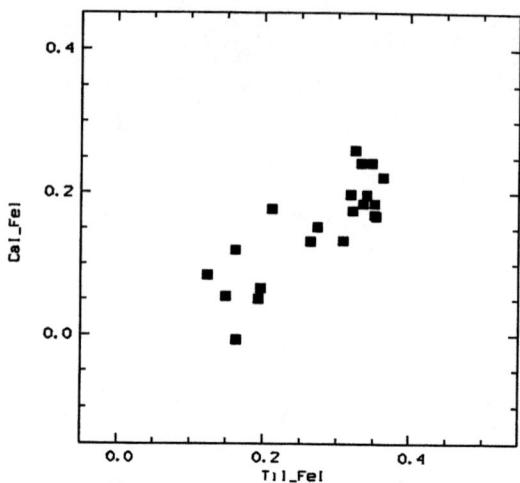

FIGURE 3. Plot of [Ca/Fe] versus [Ti/Fe].

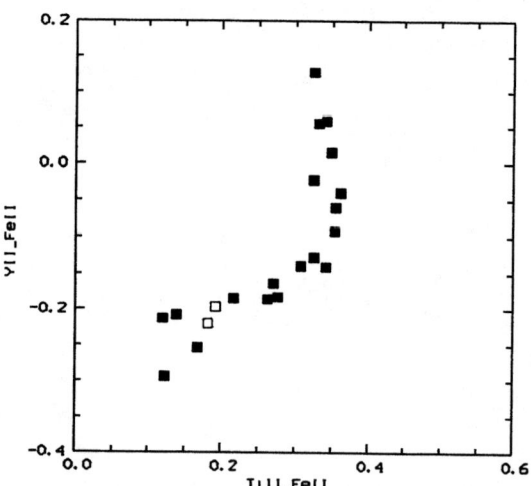

FIGURE 4. Plot of [Y/Fe] versus [Ti/Fe].

Our results confirm the findings of Zhao and Magain: for example, while the scatter of our [Y/Fe] determinations amounts to 0.12 dex (30%), the scatter of [Y/Ti] is 0.07 dex only (18%). These figures are in agreement with the results of Zhao and Magain [1]. However, the present data allow to go further than just compare the scatters, and Fig. 4 gives the star-by-star comparison of [Y/Fe] with [Ti/Fe]. We see that [Y/Fe] is indeed correlated with [Ti/Fe], but the correlation is not simple: there is a sample of stars with constant (and maximum) [Ti/Fe] for which [Y/Fe] spans a certain range, and a second sample of stars with lower (albeit positive) [Ti/Fe], for which the values of [Y/Fe] are also lower, and correlated with [Ti/Fe]. Similar results are found when any of the elements Sr, Y and Zr is compared to any of the α-elements.

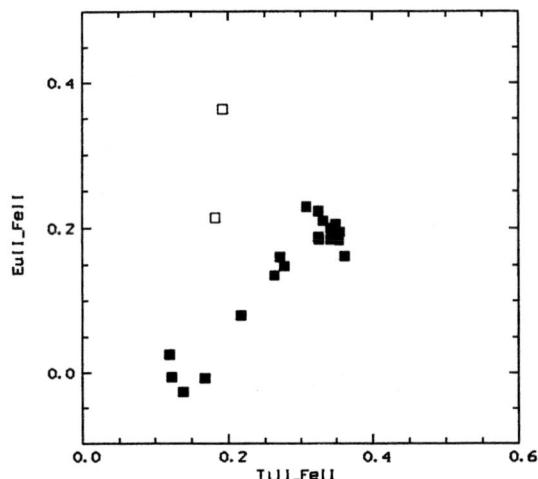

FIGURE 5. Plot of [Eu/Fe] versus [Ti/Fe].

A very clean result is presented in Fig. 5, where the abundance of the prototypical r-process element Eu is compared to the Ti abundance. The correlation is nearly perfect, except for the same two stars which show a Ni depletion. Now, they stand up as relatively enriched in Eu. The nice correlation allows us to conclude that, in general, the r-process element Eu is synthesized in the same objects as the α-elements, i.e. most probably in the supernova explosion of massive stars, which confirms the generally accepted scenario.

A more complex situation appears when one examines the heavier s-process elements (Ba, La, Ce). The abundances of these three elements behave in a similar way and, to reduce the dispersion, the mean of the abundances of these elements relative to Fe is compared to [Ti/Fe] in Fig. 7. (We take the mean

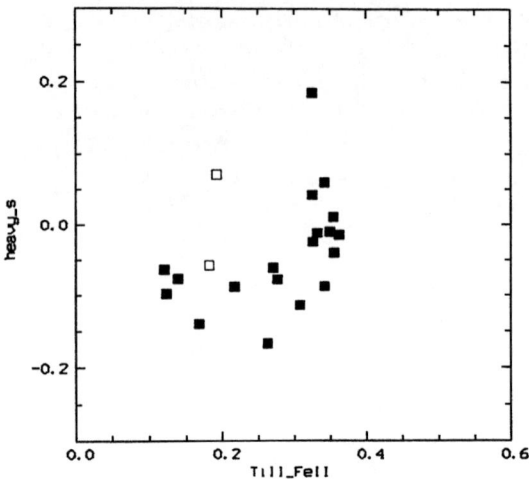

FIGURE 6. Plot of the mean of [Ba/Fe], [La/Fe] and [Ce/Fe] versus [Ti/Fe].

because, for these elements, we only have few lines of slightly lower quality). Similarly to the light s-process elements, the group of stars with constant (and maximum) [Ti/Fe] show a range of heavy s-process abundances. However, all the stars with lower [Ti/Fe] display essentially the same (and minimum) heavy s-process abundances. Once more, the star HD193901 appears enhanced in s-process elements relative to this mean trend, but HD194598 seems 'normal'.

TENTATIVE SCENARIO

Upon analysis of the correlations between the relative abundances of a number of elements for stars with chemical composition characteristic of the metal-rich limit of the halo stars, we can draw the following conclusions.

(1) A small fraction ($\sim 10\%$?) of the stars display anomalous abundances: Ni is depleted, while Eu (and possibly other heavy neutron capture elements) are enhanced.

(2) Apart from these 'anomalous' stars, one distinguishes two stellar populations: (a) roughly 50% of the stars have a range of moderate overabundances of the α-elements and either a constant or slowly varying abundance of the s-process elements relative to the iron peak, and (b) the other 50% show a constant (and maximum) overabundance of the α-elements relative to the iron-peak elements, and a range in s-process abundances. This behaviour should be related to nucleosynthesis processes.

The first interpretation which comes to mind is to relate one of these populations to the most metal-rich part of the halo and the other to the most metal-poor part of the disk. However, upon examination of the available kinematical data, there is no clear distinction between these populations on this basis alone, both populations containing high velocity stars, typical of halo kinematics.

This leads us to propose an alternative interpretation, in which all these stars would belong to the halo and form two sub-classes of Population II, namely Pop IIa and Pop IIb. In the following, we propose a temporal scenario.

General picture

At the beginning was a burst of star formation with at least a fraction of massive stars. As they evolved, more and more supernova (SN) explosions occured and α-elements together with r-process ones were formed and ejected in the surrounding environment, still rich in interstellar matter (ISM). As time passed, new stars more and more enriched in these elements were formed. These stars would correspond to Pop IIa, with $[\alpha/Fe]$ and $[r/Fe]$ increasing with time.

The slope in $[Y/Fe]$ versus $[Ti/Fe]$ for Pop IIa would indicate an overproduction of Y relative to Fe in massive stars, which requires, of course, an r-process contribution to the synthesis of this element. Our results show the same tendency for Sr and Zr.

On the other hand, the constant value obtained for the heavy s elements (Ba, La, Ce) suggests a roughly similar production of these elements and iron-peak elements during the outburst phase.

Let us assume that, after this burst, no more massive stars are formed. The lower mass stars are either still reaching the main sequence or in a more evolved phase, maybe already processing s-elements. These elements will later be ejected through stellar wind or superwind events and will contaminate the surronding ISM, already enriched in α and r-process elements and thus showing a unique $[\alpha/Fe]$ and $[r/Fe]$. This matter will condense in new stars with a constant $[\alpha/Fe]$ and an increasing value of $[s/Fe]$ in the course of time. These stars are the Pop IIb stars located on the vertical s-process feature in Fig. 4.

For a typical r element like Eu, the behaviour of $[Eu/Fe]$ versus $[\alpha/Fe]$ is, of course, completely different, showing a perfect correlation in Pop IIa stars, as expected, and an absence of the vertical s-process feature, replaced by a clumping of the points representative of Pop IIb stars at the maximum value of $[\alpha/Fe]$ and $[r/Fe]$, i.e. at the end of the massive stars outburst. This shows that, if produced by lower mass stars, it must be in the same proportions as Fe.

Globular cluster scenario

In the general picture, we have not specified the astrophysical environment in which these phases would occur.

Here we will be more specific and suggest that the formation of the field halo stars takes place in the globular clusters (GCs). This requires two reasonable assumptions. The first one is that the evaporation of low mass stars from GCs happens since the early phases of the evolution of the cluster and accounts for the field Pop II stars. The second one is that the matter ejected by SNe and stellar winds, although generally assumed to be mostly expelled from the cluster, nevertheless contributes to the enrichment of the lower mass stars, first by mixing with the ISM and then by accretion at the surface of already formed stars. Such a possibility of self-enrichment by SNe has been discussed by Smith [2,3] and Morgan and Lake [4].

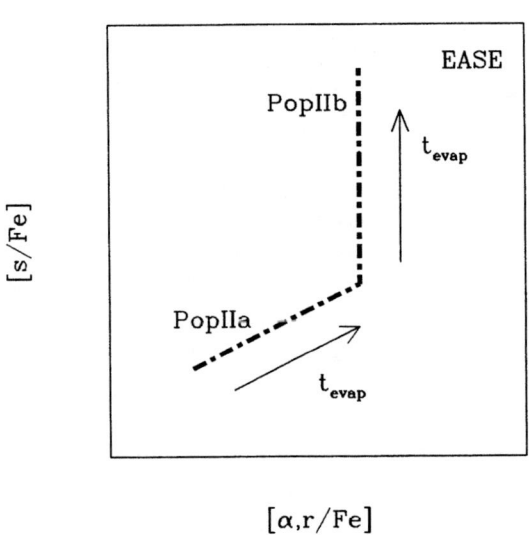

FIGURE 7. EASE scenario.

In the early phase of the GC evolution, massive stars will form SNe until all stars more massive than about 8 $M_\odot$ have completed their evolution. This fixes the end of the α and r elements synthesis and the maximum value of $[\alpha/Fe]$ observed in Fig. 4. The following evolution will lead to a relative enrichment of s elements only.

This explains nicely the features observed in Fig. 4. Pop IIa stars are evaporated during the massive stars outburst, $[\alpha/Fe]$ increasing with the time at which evaporation occurs t_{evap}, and the Pop IIb escape later in the evolution of the cluster, after the end of the SN phase. The stars located at the top of the vertical branch are those which have escaped the cluster in the most advanced phases of its evolution. A schematical illustration is given in Fig. 7.

The range of metallicity at a given location in Fig. 4 corresponds to stars evaporated from clusters of various global enrichments (due to different initial mass functions) and, thus, different present metallicity. As the evolution time for a star of a given mass is about the same in all GCs, it is no surprise that the abundance ratios, contrary to the metallicity, do not depend on the cluster from which the halo stars have evaporated.

The similar proportion of Pop IIa and Pop IIb stars suggests that either the evaporation was much more efficient in the early phases of the GC evolution or that a large fraction of Pop IIa stars originate from GCs which have disrupted during the massive stars outburst. This is in agreement with the view that GCs with a flat mass function are weakly bound [5].

This EASE (Evaporation/Accretion/Self-Enrichment) scenario also nicely explains the larger metallicity range covered by the field halo stars, extending to much lower metallicities than the GCs. The very metal-poor stars would be evaporated from the GCs at a very early stage of the outburst phase, when the self-enrichment of the cluster is still very small.

ACKNOWLEDGMENTS

This work has been supported by contracts ARC 94/99-178 "Action de Recherche Concertée de la Communauté Française de Belgique and Pôle d'Attraction Interuniversitaire P4/05 (SSTC, Belgium).

REFERENCES

1. Zhao G., Magain P., *AA* **244**, 425 (1991).
2. Smith G., *PASP* **99**, 67 (1986).
3. Smith G., *ApJ* **306**, 565 (1987).
4. Morgan S., Lake G., *ApJ* **339**, 171 (1989).
5. Meylan G., Heggie D.C., *Astron. Astrophys. Rev.* **8**, 1 (1997).

THE SYNTHESIS OF THE A>56 NUCLIDES IN REALISTIC MODEL STARS

The s-Process Efficiency in Massive Stars

Marc Rayet[1] and Masaaki Hashimoto[2]

[1] Institut d'Astronomie et d'Astrophysique
Université Libre de Bruxelles
Campus Plaine - CP 226, B-1050 Brussels - Belgium
[2] Faculty of Science, Department of Physics
Kyushu University, Fukuoka 810 - Japan

Abstract. The s-process is studied in the He burning core of a $M_\alpha = 8\,M_\odot$ He star using temperature and density profiles obtained in standard stellar evolution models and an updated nuclear network. We discuss in detail the s-process efficiency for metallicities $z/z_\odot$ ranging between 1 and 10^{-3} in the light of recent nuclear data. We show that ^{16}O acts as a powerful neutron poison at low z although its effect on the s-process yields is still subject to large uncertainties due to our imperfect knowledge of some nuclear reaction rates.

I INTRODUCTION

About half the abundance of nuclei with $A \gtrsim 70$ present in the solar system (and presumably in many other regions of the universe with similar metallicity) is believed to be produced in stars by slow neutron captures (s-process) on $A \approx 56$ (iron-peak) nuclei which are abundantly produced in thermonuclear burning processes. The rest of the $A \gtrsim 70$ isotopic content of the solar like material is accounted for by rapid neutron captures (r-process) [1], with the exception of the few and little abundant neutron deficient nuclei produced by the p-process [2]. Although both s- and r-processes contribute to the synthesis of most $A \gtrsim 70$ nuclei (called s-r nuclei), some of them can be produced in the s-process only, because they are shielded by a stable isobar from β-decay of the very neutron-rich nuclei produced during the r-process. Those "s-only" nuclei are especially important from a conceptual point of view since if an s-process model can fit their solar abundances it is also expected to give a reasonable prediction for the s-process contributions to the solar abundances of the s-r nuclei and, by substraction, for the corresponding r-process contributions.

It has early been recognized that the solar system abundance distribution of the s-only isotopes can be accounted for by a very simple phenomenological model known as the "classical s-process" [3]. This model is based on the observation that if the β-decay rates of the unstable isotopes formed during the s-process are much faster than neutron capture rates on those isotopes, the abundances obtained at the end of the s-process depend only on the time integrated neutron flux (or "neutron exposure") $\tau = \int_0^t v_T n_n(t) dt$, n_n being the neutron number density and v_T the neutron thermal velocity, and not on $n_n(t)$ or t separately (it must also be assumed that the capture rates, i.e. the temperature, are constant with time). Since the first suggestion by Seeger et al. [4] that the s-only solar distribution resulted from the mixing of matter exposed to an exponentially decreasing distribution of τ (defined by a mean exposure τ_0), the accumulation of high quality data on neutron capture cross sections led to the conclusion that at least 2 distributions were needed: a *main* component, where a relatively small number of seed nuclei are exposed to a large neutron exposure ($\tau_0 \approx 0.3 \, \mathrm{mb}^{-1}$), to account for the production of the s-nuclei with $A \gtrsim 90$, and a *weak* component to produce the more abundant $A \lesssim 90$ nuclei with a smaller exposure ($\tau_0 \approx 0.07 \, \mathrm{mb}^{-1}$) [5].

In several cases, neutron captures actually compete with β-decay rates and the products resulting from such situations, known as "branchings", depend on the neutron density itself as well as on the temperature, the weak interaction rates being frequently temperature dependent. In the classical model the branchings act as constraints on n_n and T (which are taken constant with time). With a detailed consideration of the different branchings and a very small number of parameters the classical model has achieved a remarkable fit to the solar distribution of s-only nuclei (see e.g. [5] for a review). On the other hand it has been developed with very little relation to actual astrophysical situations.

It is now generally admitted that the main component of the s-process takes place during the helium shell burning (AGB) phase of low mass stars, a scenario which offers a natural explanation for an exponential-like distribution of neutron exposures and is strongly supported by observations. However, the presence of ^{13}C, the most probable neutron source in that scenario, in the helium burning shell of an AGB star depends on complicated and badly known mixing mechanisms and it has not been possible up to now, except in one single case, to obtain self-consistently a sufficient amount of this nucleus to account for the main component s-process in AGB star models (note that an interesting progress in the solution of this problem has been reported recently [6]).

It has long been known that the central helium burning phase in massive stars can release neutrons by the activation of the ^{22}Ne$(\alpha,\mathrm{n})^{25}$Mg reaction [7]. However the neutron irradiation is limited in this scenario by the secondary character of the neutron source, as dicussed in Sect. II A, as well as by the poisoning effect of neutron captures on light nuclei, principally on ^{25}Mg itself, as already recognized by Lamb et al. [8]. Massive stars are therefore generally considered to be at the origin of the *weak* component s-process. Contrary to the situation of the ^{13}C driven s-process in AGB stars, the presence of the neutron source in the helium burning cores of massive stars is quite natural and does not depend on the details of the stellar model. This will be briefly discussed, in relation with the nuclear physics involved, in Sect. II. The problem of the neutron poisons and of the s-process efficiency at different metallicities will be addressed in Sect. III in the light of recent measurements of neutron capture rates. It must be mentioned here that the contributions of further burning phases, like central carbon and shell helium burning, are expected to be rather weak but are still a matter of debate. They will not be considered in the present report.

II THE S-PROCESS IN A 8 $M_\odot$ HELIUM STAR

A Astrophysical Input

We present in this paper various results concerning the s-process taking place during the central He burning phase of a massive star. The evolution of helium stars with masses M_α ranging between 3.3 and 32 $M_\odot$ has been described in full detail by Nomoto and Hashimoto [9] and by Hashimoto [10]. We concentrate here on the case of a helium star of mass $M_\alpha = 8M_\odot$, which corresponds to the helium core developed, at the end of the hydrogen burning phase, by a star of initial ("main sequence") mass M_{ms} between 19 and 26 $M_\odot$. This large estimated uncertainty on M_{ms} accounts essentially for uncertainties in the treatment of convection in the H burning phase, but is of no importance for our present purpose. On the other hand, the evolution of an isolated He star, as considered here, would be hardly affected by the presence of a stellar envelope.

As helium starts burning, a convective core develops from the stellar centre, reaching a maximum extension of 5.7 $M_\odot$ (use is made of the Schwarzschild criterium for convective stability) and then recesses, on a total time of $\sim 766\,10^3$ years. The stellar evolution calculation uses a limited nuclear reaction network to compute the stellar energy generation and provides temperatures and densities as a function of time and of the stellar mass coordinate. Those temperature and density profiles are then used to calculate the s-process nucleosynthesis with the full network described below. This "post-processing" procedure has been proved [20, hereafter PHN] to be a very good approximation to the more consistent but prohibitive (in term of computing time) method of coupling stellar evolution with the complete nucleosynthesis.

The initial composition of the helium star, consisting of more than 98% (by mass) of helium, can be chosen somewhat arbitrarily for the purpose of s-process calculations since the stellar evolution is almost independent of its details. The main neutron source in the central He burning scenario being the reaction ^{22}Ne$(\alpha, n)\,^{25}$Mg, a special attention must be given to the abundance of ^{14}N which is the progenitor of ^{22}Ne through the chain of reactions ^{14}N$(\alpha, \gamma)^{18}$F$(\beta^+)^{18}$O$(\alpha, \gamma)^{22}$Ne$(\alpha, n)^{25}$Mg. Here we assume that all the CNO nuclei have been transformed into ^{14}N during the H burning phase so that, for solar metallicity, we take $X(^{14}N) = 0.0137$, a value which is somewhat overestimated and maximizes the s-process efficiency. Since the abundances of nuclides with $A \geq 20$ are not modified during the H burning phase we decide, for those nuclides, to take a solar distribution. All other initial abundances are set to zero, with the exception of $X(^{13}C)$, for which we take, again arbitrarily, its solar value in order to check a possible contribution of the ^{13}C$(\alpha, n)\,^{16}$O neutron source to the s-process. We refer to Sect. III for a discussion of initial abundances in the case of other metallicities.

B Nuclear Physics Input

The nuclear reaction network used in this work contains 472 nuclei with $Z = 0$ up to 84 (with the exception of $3 \leq Z \leq 5$) linked by 834 reaction. The reaction rates involving charged projectiles are generally taken from [11, hereafter CF88] while for most of the radiative neutron capture rates we use the compilation of Beer et al. [12]. The temperature and density dependent rates for β decays and electron captures are given by Takahashi and Yokoi [14].

In several cases however more recent determinations of reaction rates have been used. Let us first mention the main neutron producing reaction ^{22}Ne$(\alpha, n)\,^{25}$Mg whose rate has been recently re-evaluated by Drotleff et al. [15] and found to differ sensitively from CF88, being smaller than CF88 by a factor 0.2 at $T_6 = 200$ ($T_6 = T/10^6$ K) and larger by a factor 3 at $T_6 = 350$, i.e. when the reaction is the most effective. Also important is the reaction ^{22}Ne$(\alpha, \gamma)\,^{26}$Mg whose rate again differs from CF88 by a factor which increases from 0.56 to 18 in the same temperature range [16][1]. The consequences of those changes will be discussed in the next section.

We also adopt recent results by Y. Nagai and co-workers for the neutron capture rates on the light nuclei ^{12}C [18] and ^{16}O [19]. The importance of those reactions for the neutron economy at different values of the metallicity will be discussed in detail in Sect. III. In particular for ^{16}O$(n, \gamma)\,^{17}$O we compare the s-process yields obtained with the value $\sigma_{16} = 0.2\mu$b quoted by Bao and Käppeler [13] and with the large value $\sigma_{16} = 34\mu$b reported by Igashira et al. [19] for the maxwellian averaged cross-section at 30 keV. In the next section however we choose the low value of [13] for the sake of comparison with the

---

[1] The latter, unpublished, reaction rate has been revised in the recent European Nuclear Astrophysics Compilation of Reaction Rates, NACRE (to be published in ADNDT; see [17] for a preliminary report), the ratio of the new values to CF88 ranging now from 1.0 to 9.2 between 200 and 350 10^6 K

results of PHN. Note that neutron captures on ^{12}C have no influence on the neutron economy, even at low metallicities (see Sect. III).

C Some Results for Solar Metallicity

The results of the calculations performed in the conditions described in the above sections are plotted in Fig. 1 which shows the overabundances $X/X_\odot$ of the main products of the s-process nucleosynthesis in the $56 < A \lesssim 90$ range. The nucleus ^{89}Y is the last one to be significantly overproduced, the neutron irradiation obtained during the core He burning phase being too weak to enhance significantly the nuclear abundances beyond that range[2]. Figure 1 also compares our results with 2 previous calculations [20,21] in order to test the influence of astrophysical and nuclear physics conditions on the s-process yields. It is seen that in all 3 calculations the s-only nuclei (there are 6 of them, shown by an asterisk, in the considered mass range) are co-produced in comparable quantities, with the exception of ^{80}Kr which is produced twice as much as the others. This is regarded as a difficulty of the s-process in massive stars (see e.g. PHN) since it means that if ^{80}Kr *is* produced there, 50% of the other 5 nuclei must be made somewhere else independently of ^{80}Kr.

We also observe very similar abundance patterns in all 3 calculations (with few exceptions, for $A < 65$), as expected if the attained neutron densities are of the same order of magnitude and if the weak interaction rates are treated in the same way (which is the case for the 3 calculations we consider). However somewhat different values are obtained for the overabundances, reflecting differences in the neutron irradiations: $\tau = 0.18$ [this work], 0.21 [21], 0.33 [20], increasing values of τ corresponding to increasing overabundances.

The stellar model used in PHN is almost identical to the one described above while Raiteri et al [21] evolve a 25 $M_\odot$ star on the main sequence, ending up at the end of core He burning with a CO core of $M = 6.3\,M_\odot$ (5.7 in our case and in PHN). The origin of the (relatively small) differences between the results of PHN and [21] are difficult to trace because the two calculations also differ by their nuclear physics inputs. In contrast the differences between PHN and our calculations can be attributed essentially to nuclear physics effects. Important neutron capture cross-sections on light nuclei (neutron poisons) are chosen to be the same in both calculations, in particular σ_{16} for which the small Bao and Käppeler [13] value is used. On the other hand PHN adopt the CF88 rates for the neutron producing reaction ^{22}Ne$(\alpha,n)^{25}$Mg as well as for the

---

[2] Note that only two nuclides are enhanced by more than a factor 10 beyond that mass, ^{152}Gd ($X/X_\odot = 32.3$), also reported in previous calculations [20,21], and ^{158}Dy ($X/X_\odot = 24.9$), which is actually a p-nucleus but gets here a small contribution due to the β^- instability of ^{157}Gd at high temperature ($T_6 > 150$). This nuclide is made in PHN with a much smaller abundance, its production being quite sensitive to the details of the (T,ρ) conditions during the core He burning (see Table 2 of PHN)

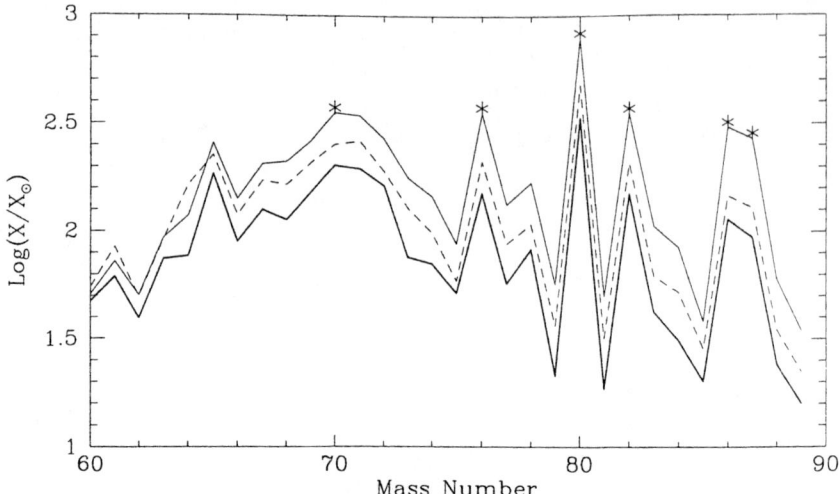

FIGURE 1. Overabundances of the main products of s-process nucleosynthesis at the end of core helium burning in 3 similar calculations (see text): this work (thick solid line), PHN [20] (thin solid line) and Raiteri et al. [21] (dashed line). The s-only isotopes are indicated by an asterisk

competing reaction channel ^{22}Ne$(\alpha,\gamma)\,^{26}$Mg while we use the revised rates of [15] and [16] respectively (see Sect. II B). Although the rate we use for the (α,n) channel is larger than the CF88 values at high temperatures $(T_6 \lesssim 350)$ our yields are smaller because of the large rate for the (α,γ) channel reported in [16] at those temperatures. At $T_6 = 350$ for example the branching ratio $(\alpha,n)/(\alpha,\gamma)$, which is 7.8 in favour of the (α,n) channel in CF88, is reduced to only 1.3 in our case. Our calculations show that increasing this ratio by reducing the (α,γ) channel brings our results in complete agreement with PHN. Let us note here that ^{13}C is burnt very early in the formation of the convective core and, as expected, plays no role in the ensuing nucleosynthesis.

III THE NEUTRON ECONOMY AT VARIOUS METALLICITIES

The s-process efficiency, i.e. the enhancement of heavy element abundances during the s-process *with respect to their initial abundances* depends on the operation of 3 components: the neutron sources (in our case essentially ^{22}Ne), the seed nuclei (with $A \gtrsim 56$) and the neutron poisons, i.e. light nuclei capturing neutrons which are lost for the s-process nucleosynthesis. Potential poisons can be products of the helium burning (like ^{12}C, ^{16}O or ^{20}Ne) in which case their abundance does not depend on the initial metallicity of the star (they are called for this reason "primary"), or result from the transformation of a nucleus whose initial abundance in the star scales with metallicity. This is the case for ^{22}Ne or ^{25}Mg whose progenitor is ^{14}N (these poisons are called "secondary").

Experimental values of the neutron capture rates on light nuclei have been subject to many important revisions during the last decade (some examples are discussed by Y. Nagai [22]). Some have decreased and nuclei which were considered as effective neutron poisons play now a negligible role, like ^{20}Ne (σ_{20} decreasing from 1.43 [13] to 0.119 mb [12]) or ^{22}Ne (σ_{22} decreasing from 720 [13] to 60 μb [12]). On the other hand, a very large value has been reported recently for σ_{16} [19](see Sect. II B) and the role of ^{16}O as a neutron poison has to be re-evaluated.

It has sometimes been suggested that ^{12}C could have a poisoning effect due to its relatively large capture cross-section, $\sigma_{12} \approx 20\,\mu$b (see e.g. PHN). However it turns out that neutrons captured by ^{12}C are instantaneously re-emitted in the reaction ^{13}C$(\alpha,n)\,^{16}$O so that the presence of ^{12}C does not modify the neutron economy. In the following we shall concentrate on the role of ^{16}O for which a similar neutron "recycling" operates but not as completely as with ^{12}C so that the net poisoning effect of ^{16}O remains significant, at least for small metallicities.

S-process calculations have been performed using the astrophysical and nuclear physics inputs described previously but for different values of the initial metallicity. The metallicity expresses the enhancement of "metals" (i.e. any element with $Z > 2$) abundances with respect to hydrogen. Observations of elemental abundances in the Galaxy suggest that all "metals" did not evolve in the same way during the galactic history, in particular that oxygen has evolved faster than iron (we refer to PHN for a brief discussion of this question and for references). We consider in the following a rather extreme scenario (case B in PHN) in which, when going back in galactic time, the abundance of iron decreases faster than the abundance of oxygen. If the metallicity z denotes the abundance of oxygen at a given galactic time (and $z_\odot$ its solar value) the abundance of Fe is given in that scenario by $X(\text{Fe})/X_\odot(\text{Fe}) = (z/z_\odot)^{1.42}$ (e.g. for $z/z_\odot = 10^{-3}$, $X(\text{Fe})/X_\odot(\text{Fe}) = 5.5\,10^{-5}$).

If we assume that the initial abundance of ^{14}N in our model is a product of a previous CNO cycle, then it should scale with $z/z_\odot$. So we decide (somewhat arbitrarily) to apply, for all the $4 < A \leq 30$ nuclides, the same scaling factor to the initial abundances which were adopted in the solar metallicity case (Sect. II A). On the other hand, for $A > 30$ we scale the initial abundances by the factor $(z/z_\odot)^{1.42}$. Therefore with this prescription we test 2 effects: 1) the increasing importance of primary poisons when the abundance of the neutron source decreases (Sect. III A) and 2) the increased s-process efficiency for larger source/seed ratios (Sect. III B).

A The Poisoning Effect of ^{16}O versus Metallicity

In order to test the role of ^{16}O as a neutron poison, we have calculated the s-process yields for values of $z/z_\odot$ ranging from 1 to 0.001 and for each of them with the 2 extreme values $\sigma_{16} = 0.2\,\mu$b [13] and $34\,\mu$b [19]. For solar metallicity ($z/z_\odot = 1$), the overabundances obtained in the 2 cases (not shown here by lack of space), differ by generally less than 20% , indicating that the poisoning effect of ^{16}O is very limited even for $\sigma_{16} = 34\,\mu$b. This is explained by the recycling effect already mentioned for ^{12}C, most of the captured neutrons being in this case re-emitted through the reaction ^{17}O$(\alpha,n)^{20}$Ne. The same conclusion is reached in a recent study by C. Travaglio et al. [23] on neutron recycling at solar metallicity. However, the captured neutrons are not 100% recycled so that going to low metallicities, $X(^{16}$O$)$ staying constant and $X(^{22}$Ne$)$ decreasing with $z/z_\odot$, the number of lost neutrons becomes comparable to the number of neutrons emitted by ^{22}Ne, and ^{16}O can act as an efficient neutron poison. This is shown quantitatively in Fig. 2 which displays the overabundances of all nuclei with $A \leq 209$ at the end of core He burning for $z/z_\odot = 10^{-3}$. It appears that the overproduction factors *with respect to the initial abundances of seed nuclei* ($X_{\text{seed}}/X_\odot = 5.5\,10^{-5}$, see the horizontal dashed line in Fig. 2) are quite similar to the $z = z_\odot$ case (with the usual sharp decrease at $A \gtrsim 90$) when $\sigma_{16} = 0.2\mu$b, but they are almost negligible with $\sigma_{16} = 34\mu$b.

One reason for the imperfect recycling of neutrons is that the ^{17}O$(\alpha,\gamma)^{21}$Ne reaction competes with the neutron emitting (α,n) channel. In the above calculations both rates are taken from CF88, with a branching ratio $(\alpha,\gamma)/(\alpha,n) \approx 0.1$ in the temperature range of interest. This ratio however is much larger than the value $(\alpha,\gamma)/(\alpha,n) \approx 10^{-4}$ suggested by the microscopic calculations of P. Descouvemont [24] (a significant reduction of the rate for the (α,n) channel is also predicted in [24] but is not confirmed by NACRE, see footnote 1). The status of those reactions being still uncertain we have re-calculated the overproduction obtained with $\sigma_{16} = 34\mu$b and the ^{17}O$(\alpha,n)^{20}$Ne rate from CF88, but with a branching ratio $(\alpha,\gamma)/(\alpha,n) = 10^{-4}$. The result of this calculation, also plotted in Fig. 2, shows a significant increase of the s-process yields with respect to those obtained with a branching ratio of ≈ 0.1, which indicates a larger (though still uncomplete) recycling of the neutrons trapped by ^{16}O.

B The s-Process Efficiency

The s-process efficiency in stars of different metallicities is an important parameter for the chemical evolution of the Galaxy. On the one hand it is expected to decrease, with the source/poison ratio, when the metallicity decreases. On the other hand it may increase whith decreasing metallicity if

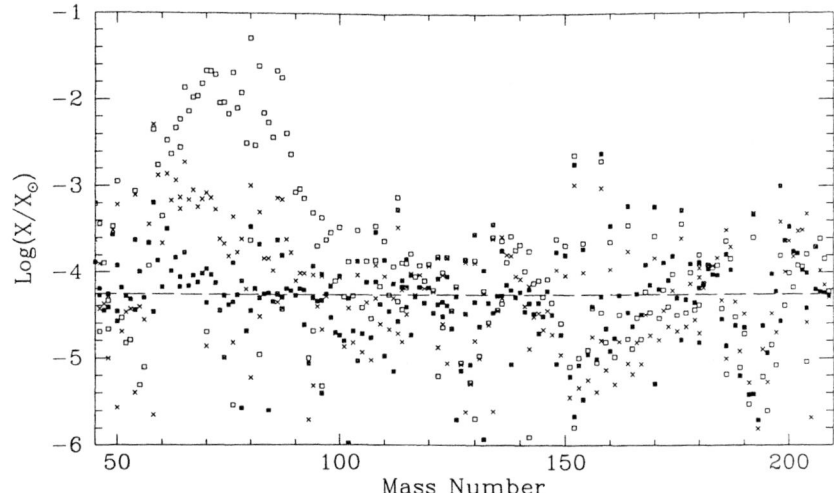

FIGURE 2. Overabundances obtained at the end of core He burning for a metallicity $z/z_\odot = 10^{-3}$. The initial abundance level of seed nuclei (relative to solar abundances) is shown by the dashed line. Open squares: $\sigma_{16} = 0.2\,\mu$b [13]; solid squares: $\sigma_{16} = 34\,\mu$b [19]; crosses: idem with reduced ^{17}O$(\alpha,\gamma)^{21}$Ne rate

the source/seed ratio increases, i.e. if we consider, as before, a scenario where the seed nuclei abundances decrease with metallicity faster than does the abundance of ^{14}N (assuming that this nucleus evolves like oxygen). In order to disentangle those two effects we have first considered a scenario where the source/seed ratio is independant of metallicity (corresponding to case A of PHN) and calculated the s-process yields for $z/z_\odot = 1, 10^{-1}, 10^{-2}, 10^{-3}$.

As before, calculations have been made with the 2 extreme values for σ_{16}. With the small value [13] the efficiency, measured by the ratio X/X_{seed} for the 6 s-only nuclei considered above, is almost constant for $z/z_\odot$ between 1 and 10^{-2}. The effect of the ^{16}O poison eventually shows up at $z/z_\odot = 10^{-3}$, bringing down the efficiency by nearly one order of magnitude. On the contrary, with $\sigma_{16} = 34\mu$b the poisoning effect is already significant for $z/z_\odot = 10^{-1}$ and decreases the efficiency by more than 2 decades at $z/z_\odot = 10^{-3}$. When the source/seed ratio increases whith decreasing metallicity as in Sect. III A, the situation, depicted in Fig. 3, is rather more complex. For $\sigma_{16} = 0.2\,\mu$b the efficiency starts to increase when metallicity decreases because the source/seed ratio gets larger, but this effect is obliterated by neutron poisoning when $z/z_\odot$ reaches 10^{-3} (although the efficiency remains larger than for $z/z_\odot = 1$). With $\sigma_{16} = 34\,\mu$b, after a very slight increase of efficiency for $z/z_\odot = 10^{-1}$, the poisoning effect already dominates the neutron economy at $z/z_\odot = 10^{-2}$, reducing the efficiency almost by a factor 100 for $z/z_\odot = 10^{-3}$.

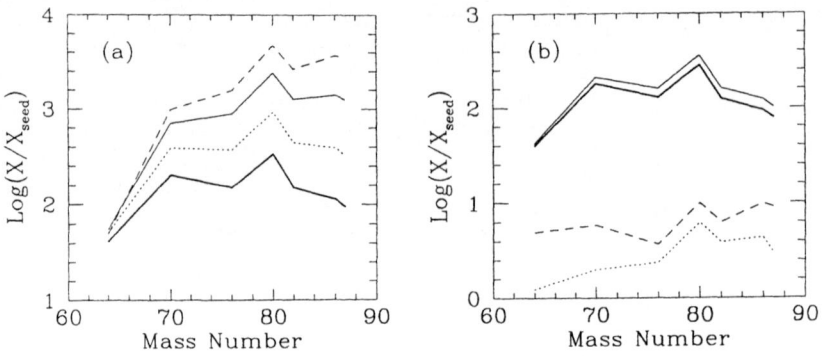

FIGURE 3. The s-process efficiency, X/X_{seed} for 6 s-only nuclei in the $70 \leq A < 90$ range: (a) with $\sigma_{16} = 0.2\,\mu\text{b}$ [13], (b) with $\sigma_{16} = 34\,\mu\text{b}$ [19], and for different metallicities: $z/z_\odot = 1$ (thick solid line), $z/z_\odot = 10^{-1}$ (thin solid line), $z/z_\odot = 10^{-2}$ (dashed line), $z/z_\odot = 10^{-3}$ (dotted line)

IV CONCLUSIONS

Although the s-process in the central He burning of massive stars is well understood in terms of the stellar models involved, contrary to the s-process in AGB stars, relatively large uncertainties in some key nuclear reactions still preclude precise estimates of the s-process yields. We have shown in particular that ^{16}O progressively emerges as a very effective neutron poison (even supplanting ^{25}Mg) as metallicity decreases, but that the poisoning effect depends dramatically on the very uncertain branching ratio between the (α,n) and (α,γ) reactions on ^{17}O. On the other hand, the efficiency of the neutron source is also conditioned by a branching ratio involving a radiative process, namely the ratio of the (α,n) to the (α,γ) reactions on ^{22}Ne which again is still subject to large uncertainties.

As already stressed in PHN another important cause of uncertainty comes from the source/seed ratio which is not precisely determined when one goes back in the Galaxy's past, i.e. at low metallicities. Clearly all those uncertainties have to be faced when studying the evolution of s-nuclei in the galactic history.

Similar calculations are now being performed for different stellar masses, covering the range $3.3 \leq M_\alpha/M_\odot \leq 32$, in order to reach more quantitative conclusions. Possible contributions of the shell He and central C burning phases to the synthesis of s-nuclei will also be re-examined.

REFERENCES

1. Takahashi K., 1997, this conference
2. Arnould M., Rayet M., Hashimoto M., 1997, this conference
3. Cameron A. G. W., 1959, ApJ 130, 452
4. Seeger P. A., Fowler W. A., Clayton D. D., 1965, ApJS 11, 121
5. Käppeler F., Beer H., Wisshak, K., 1989, Rep. Progr. Phys. 52, 945
6. Herwig F., Blöcker D., Schönberner, El Eid M., 1997, A&A 324, L81
7. Peters J. G., 1968, ApJ 154, 225
8. Lamb S. A., Howard W. M., 1977, Truran J. W., Iben I. Jr., 1977, ApJ 217, 213
9. Nomoto K., Hashimoto M., 1988, Phys. Rep. 163, 13
10. Hashimoto M., Progr. Theor. Phys. 94, 663
11. Caughlan G. R., Fowler W. A., 1988, ADNDT 40, 283
12. Beer H., Voß F., Winters R. R., 1992, ApJS 80, 403
13. Bao Z. Y., Käppeler F., 1987, ADNDT 36, 411
14. Takahashi K., Yokoi K., 1987, ADNDT 36, 375
15. Drotleff H. W., Denker A., Knee H., Soiné M., Wolf G., Hammer J. W., Greife U., Rolfs C., Trautvetter H. P., 1993, ApJ 414, 735
16. Drotleff H. W., 1992, private communication
17. Angulo, C., 1997, Nucl. Phys. A597, 231
18. Ohsaki T., Nagai Y., Igashira M., Shima T., Seino S., Irie T., 1994, ApJ 422, 912
19. Igashira M., Nagai Y., Masuda K., Ohsaki T., Kitasawa H., 1995, ApJ 441, L89
20. Prantzos N., Hashimoto M., Nomoto K., 1990, A&A 234, 211
21. Raiteri C. M., Busso M., Gallino R., Picchio G., Pulone L., 1991, ApJ 367, 228
22. Nagai Y., 1997, this conference
23. Travaglio C., Gallino R., Arlandini C., Busso M., 1997, preprint
24. Descouvemont P., 1993, Phys. Rev. C48, 2746

The ^{187}Re - ^{187}Os Cosmochronometry — the Latest Developments

Kohji Takahashi

Max-Planck-Institut für Astrophysik, D-85740 Garching[1]

Abstract. We review the progress made recently towards a reliable evaluation of the age of the Galaxy through the ^{187}Re - ^{187}Os cosmochronometry. With some exemplifying results, we assert that nucleo-cosmochronology may finally be going to take a worthy share in chronological studies of the Universe.

INTRODUCTION

Nucleo-cosmochronology aims at setting a stringent lower bound for the age of the Universe with the use of the observed abundances of radioactive nuclides. It primarily deals with the bulk solar abundances measured in meteorites with a tacit understanding that they reflect the long history of Galactic chemical evolution prior to the formation of the solar system. As such, the main task is to figure out how nucleosynthetic processes had contributed to the accumulation of chronometric pair nuclides in the interstellar medium [1–4] (see also [5–7] for reviews).

The most classical chronometry using ^{232}Th - ^{238}U and ^{235}U - ^{238}U [8] relies on the estimated ratios of their yields by the r-process nucleosynthesis. Unfortunately, no convincing scenarios on the astrophysical sites for the occurrence of the r-process are available to date.[2] Combined with the lack of naturally-occurring elements beyond ^{209}Bi but the chronometric nuclides themselves, that makes it next to impossible to estimate the yield ratios within the required accuracy even if all the necessary data on the very neutron-rich and heavy precursor nuclei were known. The bottom line here is that that chronometry will remain most unreliable for a long time to come.

---

[1] also at Physik-Department E12, Technische Universität München, Garching, and at National Astronomical Observatory, Mitaka, Tokyo

[2] Speaking of the "synthesis of the $A > 56$ nuclides in realistic model stars", it appears likely that this volume contains reviews on the s-process (by M. Rayet) and the p-process (by M. Arnould), but none on the r-process. In a minimal effort to secure the completeness, we add here that the recent furor over the possibility of the r-process in the high-entropy environment (the "hot bubble") in type-II supernovae has subsided along with the latest works seemingly overruling the optimism (see e.g. [9–12] and the references therein)

First proposed by Clayton [13], the ^{187}Re-^{187}Os chronometry makes use of the cosmogonical decay of ^{187}Re with a half-life of about 42 Gyr to ^{187}Os. It has the distinct advantage of being independent of r-process models for the post-r-process cascades produce ^{187}Re, but not ^{187}Os directly. Now, ^{187}Os is produced also by the s-process, so that the chronometry requires an estimate of the s-process component in the observed ^{187}Os abundance, the residual being the cosmogonical one. It is fortunate that because of its non-explosive nature the s-process is easier to take hold of than the r-process, and indeed its most likely astrophysical sites are known.

Realistic models of the s-process as well as semi-empirical analyses have shown that s-process abundance ratios in a limited range of mass number are well described within a steady-flow approximation as far as only the neutron captures are involved. Taking the advantage that ^{186}Os is a pure s-process element, we therefore consider the s-process production ratio of ^{187}Os and ^{186}Os, which is in number

$$N_s(^{187}\text{Os})/N_s(^{186}\text{Os}) = F_b \bar{\sigma}_s(^{186}\text{Os})/\bar{\sigma}_s(^{187}\text{Os}) = F_b F_\sigma \bar{\sigma}_{\text{lab}}(^{186}\text{Os})/\bar{\sigma}_{\text{lab}}(^{187}\text{Os}), \quad (1)$$

and the yield ratio in mass is

$$y_{s,\text{os}187}/y_{s,\text{os}186} = (187/186) N_s(^{187}\text{Os})/N_s(^{186}\text{Os}). \quad (2)$$

In the above, $\bar{\sigma}_s$ represents the Maxwellian-averaged radiative neutron capture cross section under s-process conditions, and F_b deviates from unity [14] when the steady-flow approximation fails as the result of s-process path-branchings such as ^{187}Os electron captures [15]. The factor F_σ was introduced in order to take account of the difference between the $\bar{\sigma}_s(^{186}\text{Os})/\bar{\sigma}_s(^{187}\text{Os})$ ratio and the laboratory counterpart, $\bar{\sigma}_{\text{lab}}(^{186}\text{Os})/\bar{\sigma}_{\text{lab}}(^{187}\text{Os})$. The main cause of F_σ deviating from unity would be the thermal population at s-process temperatures of the 9.75 keV first excited state of ^{187}Os (denoted by ^{187}Os* in the following) [16,17].

The transmutation of ^{187}Re and ^{187}Os by weak interaction has in fact turned out to have much more profound significance in the chronometry. Takahashi and Yokoi [18] pointed out that ^{187}Re β^- decay would be enormously enhanced at high temperatures, primarily because of the bound-state β-decay process to ^{187}Os*. This implies that a part of ^{187}Re embedded in a newly-born star would exist as ^{187}Os in the ejecta, and therefore the laboratory half-life would not apply to the net cosmogonical ^{187}Re decay, albeit the relatively short time-scales involved in stellar evolution. Yokoi, Takahashi and Arnould [3] studied the effects of such "astration" on the chronometry in the light of a chemical evolution model that was constrained by various observational data in the solar neighborhood.

A breakthrough came with the recent experimental determination of the β^- decay half-life of fully-ionized ^{187}Re to be 32.9 ± 2.0 yr [19], more than one *billion* times shorter than that of the neutral ^{187}Re. Its good agreement with the predicted value of 14 yr [20], and the consequent derivation of log ft-value of 7.87 ± 0.03 for the key transition to ^{187}Os*, have set a solid foundation for reliable evaluations of ^{187}Re β^- decay and ^{187}Os electron capture rates in stellar interiors with the use of the basic formalism [18] and of thermodynamic conditions (temperature, density and composition) given by realistic stellar evolution models.

The present talk does not aim at presenting a precisely determined age of the Galaxy, but is rather intended for outlining the due process of establishing the ^{187}Re - ^{187}Os chronometry so as to accomplish that ultimate aim. As such, we begin with a short summary of the principal ingredients of a modeling of the Galactic chemical evolution in relation to the chronometry. We then present some results of a quantitative study of astration of the concerned nuclides in realistic model stars [21,22]. In addition, we depict a subsequent attempt at determining the age of the Galaxy that uses the chemical evolution model of Yokoi et al. [3]. With those exemplifying results, and through discussions on the remaining problems to be solved, we assert a favorable outlook for the ^{187}Re - ^{187}Os chronometry.

CHEMICAL EVOLUTION IN THE SOLAR NEIGHBORHOOD

In order to discuss chemical evolution in connection with chronometries using the solar abundances, it would be appropriate to consider the constituents in a cylinder perpendicular to the Sun in the Galactic plane that covers the disk thickness.[3]

The total mass in that cylinder (or in practice the column mass density), m_{tot}, consists of that of the interstellar medium (ISM), m_{gas}, and of the stars (dead or alive), m_*:

$$m_{tot} = m_{gas} + m_*. \qquad (3)$$

The Basic Equations

Total Mass: Allowing for the possible transfer of gas between that cylindrical volume and the neighboring zones, we have the time variation of m_{tot} in terms of the rates of the mass inflow (f_{in}) and outflow (f_{out}):

$$\frac{dm_{tot}}{dt} = +f_{in} - f_{out}. \qquad (4)$$

Gas Mass: The corresponding equation for m_{gas} is

$$\frac{dm_{gas}}{dt} = -\frac{dm_*}{dt} + \frac{dm_{tot}}{dt} = -\Psi + [R\Psi]_- + f_{in} - f_{out}, \qquad (5)$$

where $\Psi(t)$ is the star formation rate in mass, and $[R\Psi]_-$ represents the rate in mass with which the stellar material is returning to the ISM from previously-born stars that are just dying at time t.

---

[3] One may extend the cylinder to the halo region, and consider a coupling of the evolutions in the two zones. The halo stars, along with Globular clusters, are a primordial component of the Galaxy. It is thought that the massive disk was formed by the infall from the halo. As such, the one-zone model has its limitations of describing the disk evolution at early times at least. In this work, we do not deal with the halo stars, although the equations may formally apply for the halo as well. [Nor do we explicitly consider the binary stars]

Under the conventional assumption that the initial mass function (IMF), i.e., the relative birth rates of stars with different masses, is independent of time, $\Psi(t)$ means the stellar birth rate integrated over the IMF. Denoting the lifetime and the remnant mass of stars with the initial mass M by τ_M and $M_{\rm rem}$, respectively, and the normalized IMF by $\Phi(M)$, we then have[4]

$$[R\,\Psi]_-(t) = \int_{M(t=\tau_M)}^{M_{\max}} [M - M_{\rm rem}] \Phi(M) \Psi(t - \tau_M) \mathrm{d}M. \qquad (6)$$

Stable Elements: The evolution of a stable element i can be expressed in terms of its mass fraction X_i ($\equiv m_i/m_{\rm gas}$) as

$$\frac{\mathrm{d}}{\mathrm{d}t}(X_i\, m_{\rm gas}) = -X_i\, \Psi + [R_i^{\rm a} X_i\, \Psi]_- + [y_i^{\rm new}\, \Psi]_- - X_i\, f_{\rm out} + X_i^{\rm in}\, f_{\rm in}. \qquad (7)$$

In Eq. (7), $[R_i^{\rm a} X_i\, \Psi]_-$ is the rate with which that element is returning to ISM after having survived astration:

$$[R_i^{\rm a} X_i\, \Psi]_-(t) = \int_{M(t=\tau_M)}^{M_{\max}} R_i^{\rm a}(M) X_i(t - \tau_M)\, M\Phi(M) \Psi(t - \tau_M) \mathrm{d}M, \qquad (8)$$

where $R_i^{\rm a}(M) \leq 1 - M_{\rm rem}/M$, whereas $[y_i^{\rm new}\, \Psi]_-$ is the fresh yield in the ejecta:

$$[y_i^{\rm new}\, \Psi]_-(t) = \int_{M(t=\tau_M)}^{M_{\max}} y_i^{\rm new}(M) \Phi(M) \Psi(t - \tau_M) \mathrm{d}M, \qquad (9)$$

with $y_i^{\rm new}(M)$ being the yield in mass by nucleosynthesis.

In the above equations, the quantities such as $R_i^{\rm a}(M), y_i^{\rm new}(M), \tau_M(M)$ and $M_{\rm rem}(M)$ depend primarily on the initial mass of the star but to some degree also on the initial metallicity $Z(t - \tau_M)$ (or more generally on the initial composition).[5]

The Chronometric Equations in Question

Mass Fraction of ^{187}Re: The ^{187}Re mass fraction in the ISM, $X_{\rm re187}$, obeys

$$\frac{\mathrm{d}}{\mathrm{d}t}(X_{\rm re187}\, m_{\rm gas}) = -\lambda_\beta\, X_{\rm re187}\, m_{\rm gas} - X_{\rm re187}\, \Psi + [R_{\rm re187 \to re187}\, X_{\rm re187}\, \Psi]_-$$
$$+ [R_{\rm os187 \to re187}\, X_{\rm os187}\, \Psi]_- + [y_{\rm r,re187}\, \Psi]_- - X_{\rm re187}\, f_{\rm out} + X_{\rm re187}^{\rm in}\, f_{\rm in}. \qquad (10)$$

The negative terms are the consumptions of ^{187}Re by: i) the β-decay with the rate λ_β equal to the laboratory value; ii) the star formation; and iii) the outflow. The positive terms are the accumulations by: i) recycled ^{187}Re that survived astration, with $R_{\rm re187 \to re187}(M)$ replacing $R_i^{\rm a}(M)$ in Eq. (8); ii) a portion of the injected ^{187}Os that have changed to ^{187}Re during stellar lifetime, with $R_{\rm os187 \to re187}(M)$ replacing $R_i^{\rm a}(M)$ in Eq. (8); iii) the r-process yield, with $y_{\rm r,re187}(M)$ replacing $y_i^{\rm new}(M)$ in Eq. (9); and iv) the amount in the inflow material, $X_{\rm re187}^{\rm in}\, f_{\rm in}$.

---

[4] Note that τ_M decreases with increasing M
[5] In the astrophysical context, the "metallicity" stands for the total mass fraction of all the elements but H and He. The solar metallicity, $Z_\odot$, is about 2%

Mass Fraction of A = 187 Nuclides Since the weak interaction processes occur between nuclei with the same mass number, it is more convenient to consider the sum

$$X_{187} \equiv X_{\text{re}187} + X_{\text{os}187}, \qquad (11)$$

rather than $X_{\text{os}187}$ as the second entity. Nonetheless, we have to consider astration because pre-existing heavy nuclei will certainly be destroyed when they encounter a high flux of neutrons. Under the plausible assumption of no selective destructions by neutrons of the concerned nuclei, we express the astration fraction by $R_n(M)$ [replacing $R_i^a(M)$ in Eq. (8)]. A portion which was produced from lighter mass nuclei is considered to be included in the fresh s-process yield $y_{\text{s, os}187}(M)$ [replacing $y_i^{\text{new}}(M)$ in Eq. (9)]. Consequently, we have

$$\frac{d}{dt}(X_{187}\, m_{\text{gas}}) = -X_{187}\,\Psi + [R_n X_{187}\Psi]_- + [(y_{\text{r, re}187} + y_{\text{s, os}187})\Psi]_-$$
$$-X_{187} f_{\text{out}} + X_{187}^{\text{in}}\, f_{\text{in}}. \qquad (12)$$

Mass Fraction of ^{186}Os: Similarly,

$$\frac{d}{dt}(X_{\text{os}186}\, m_{\text{gas}}) = -X_{\text{os}186}\,\Psi + [R_n X_{\text{os}186}\Psi]_- + [y_{\text{s, os}186}\,\Psi]_-$$
$$-X_{\text{os}186} f_{\text{out}} + X_{\text{os}186}^{\text{in}}\, f_{\text{in}}. \qquad (13)$$

The Input Quantities

The principal input data for the determination of the age of the Galaxy, T_G, are the solar abundances [23] of the concerned elements at the time of the formation of the solar system, $T_\odot \equiv T_G - (4.55\ \text{Gyr})$: $X_{\text{re}187}^\odot$, $X_{\text{os}187}^\odot$ and $X_{\text{os}186}^\odot$. Let us consider a simple case in which Eq. (9) is separable in M and t. The computed $X_{\text{os}186}(T_\odot)$ is matched with $X_{\text{os}186}^\odot$ to obtain a (mass-integrated) value of $y_{\text{s, os}186}$. With $y_{\text{s, os}187}$ evaluated from Eq. (2), we equate $X_{187}(T_\odot)$ to $X_{\text{re}187}^\odot + X_{\text{os}187}^\odot$ to have $y_{\text{r, re}187}$. Finally the comparison of $X_{\text{re}187}(T_\odot)$ and $X_{\text{re}187}^\odot$ leads to $T_\odot$ and thus to T_G.

To solve the coupled equations of chemical evolution requires many other unknown quantities. Some of them are calculable with the use of stellar evolution models and nuclear physics rather reliably and independently of chemical evolution models, whereas others are not and have to be given in due consideration of constrains from astronomical observations in the solar neighborhood [1-4].

In Eq. (1), the ratio $\bar{\sigma}_{\text{lab}}(^{186}\text{Os})/\bar{\sigma}_{\text{lab}}(^{187}\text{Os})$ can be determined by experiments with neutron energies corresponding to s-process temperatures ($= 0.478 \pm 0.022$ at 30 keV [24]). The correction factor F_σ has to be estimated theoretically but with the help of experimental data particularly on the inelastic neutron capture cross section on ^{187}Os [17]. In contrast, the evaluation of the branching correction factor F_b requires s-process models in addition to various nuclear input data in the $A = 184 - 187$ range [14]. In our numerical calculations presented later in the text, we will take $F_\sigma = 0.81$ [17], and assume $F_b = 1$ for simplicity.

Stellar evolution models give τ_M and $M_{\rm rem}$ in Eq. (6). They also provide us with stellar conditions in quest for the evaluations of the quantities describing the effects of astration: $R_{\rm n}(M)$, $R_{\rm re187\to re187}(M)$, and $R_{\rm os187\to re187}(M)$ in Eqs. (10,12-13).

On the other hand, functional forms of the stellar birth rate $\Psi(t)$, the IMF $\Phi(M)$, the inflow rate $f_{\rm in}(t)$, and the outflow rate $f_{\rm out}(t)$ are more troublesome to give. One often assumes $\Psi \sim m_{\rm gas}^n$ with an integer n (see, however, [25]). Alternatively, Ψ may be given as an external function. In either case, Φ must be consequently set so as to reproduce the number distribution (in $M \leq M_{\rm max}$) of stars alive presently (the "present-day mass function": PDMF) [26]. The physical mechanisms leading to gaseous in- and out-flows are certainly too complicated to be dealt with by simple models of chemical evolution such as the one being presented here. For this reason, the net rate of such mass transfer is often assumed to be proportional to Ψ, or is given as an external function.

ASTRATION FACTORS

Here, we improve the estimates by Yokoi et al. [3] of the astration effects in two major aspects: i) the use of realistic stellar evolution models instead of the schematic one used previously, and ii) the improvement of weak interaction rates in consideration of the experimentally determined ft-values for the transitions between the ground-state ^{187}Re and ^{187}Os* [19].

The Model Stars

We have adopted the recent models of Wagenhuber [21] for low- and intermediate-mass stars (through the He-shell thermal pulse phase), and of Langer and Henkel [22] for massive stars (through the phase of Si burning). Among other choices, we pick the model stars with the solar metallicity and the initial mass $M = 1, 1.5, 2, 3, 5, 7 M_\odot$ and $M = 10, 15, 20, 25, 30, 40, 50 M_\odot$ from the respective works, which undergo mass loss from their surfaces during evolution. It may be worth noting here that we are not interested in the ^{187}Re/^{187}Os abundance ratio in material that would be eventually enclosed in the stellar remnant. Nor do we need to follow further the time variations of the abundances of concerned elements once they are exposed to, and destroyed by, a high flux of neutrons as is expected in core or shell He-burning phases. Consequently and because of their short duration, highly-advanced burning phases of stellar evolution are not as much important for the net effect of astration as for the delimitation of the ejecta and the remnant.

We extract from those models the radial profiles of temperatures, density and composition of the major elements (particularly, H and He), and their variations in time τ. In addition, we need information on wether the energy transport at each segment of a star at a given time is radiative or convective. If a convective zone appears, nuclear reaction rates are averaged, and the abundances made uniform, throughout the zone.

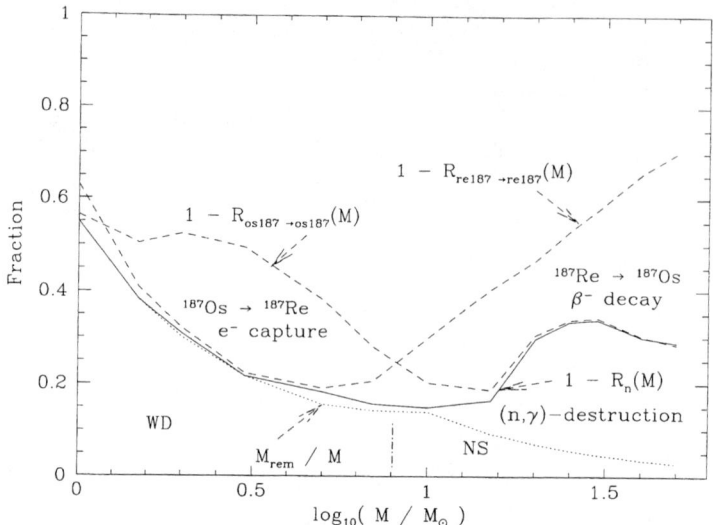

FIGURE 1. Astration fractions in stars with the initial mass range of $M = 1 \sim 50 M_\odot$. Of the unit amount of ^{187}Re embedded, the fraction $R_{\text{re}187 \to \text{re}187}$ survives astration while $R_{\text{re}187 \to \text{os}187} = R_n - R_{\text{re}187 \to \text{re}187}$ is found as ^{187}Os in the ejecta. Similarly, $R_{\text{os}187 \to \text{os}187}$ survives while $R_{\text{os}187 \to \text{re}187} = R_n - R_{\text{os}187 \to \text{os}187}$ becomes ^{187}Re

The Fate of ^{187}Re and ^{186,187}Os in Solar-Metallicity Stars

We consider a tiny, unit amount of ^{187}Re, ^{187}Os or ^{186}Os embedded in the model stars (at "zero-age main-sequence"), and subject it to the stellar conditions in the subsequent evolutionary phases. Figure 1 depicts the results.[6]

Destructions by Neutrons: The trace elements are destroyed by neutrons when they feel temperatures in excess of about 10^8 K in He-rich layers where ^{22}Ne$(\alpha,n)^{25}$Mg and/or ^{13}C$(\alpha,n)^{16}$O reactions produce neutrons.[7]

The solid curve in Fig. 1 indicates the fraction destroyed by neutrons. In low-mass stars, most of those layers will later be enclosed in white dwarfs anyway. In massive stars, they will undergo further burning stages to be partly in the ejecta with the rest enclosed in the remnants.

---

[6] Further results will appear somewhere else (Takahashi, Langer and Wagenhuber, in preparation). The white dwarf (WD) masses are taken from an external source [27]. The maximum M of stars leading to WD is tentatively taken to be 8 $M_\odot$. A neutron star (NS) of $1.4 M_\odot$ is assumed as the remnants for stars with $M \geq 10 M_\odot$, despite the possible formation of black holes in most massive stars. Less massive neutron stars are assumed as remnants of electron-capture induced supernovae in the $8 \leq M/M_\odot \leq 10$ range [28]

[7] We do not inquire here whether a strong s-processing indeed develops there. For instance, even a relatively weak flux of neutrons from the ^{22}Ne source would be sufficient for the destruction of the trace elements

Transmutations via Weak Interaction Both ^{187}Re β^- decays and free e^- captures on ^{187}Os are fastest in the innermost part of matter surviving (n,γ) destructions, where temperatures are still high ($\sim 10^8$ K). At a fixed temperature, the efficiency of the ^{187}Re $-^{187}$Os transmutation depends on the baryon density. If it is relatively high, ^{187}Re would not be ionized sufficiently for the bound-state β-decay to occur to ^{187}Os*. Instead, ^{187}Os* state would capture free electrons having correspondingly high chemical potential. This is the case of low-mass stars. In massive stars, the densities near the periphery of the extended He-burning core are much lower, so that the β-decays are much enhanced whereas the e^- captures hardly occur. [See the dashed curves in Fig. 1.]

AGE DETERMINATION BY A SIMPLE MODEL

With the above results at hand, we revisit the one-zone model of Yokoi et al. [3]. The working assumptions are: i) the "instantaneous recycling approximation," with which $\Psi(t-\tau_M)$ is replaced by $\Psi(t)$ and $M(t=\tau_M)$ by $M_{\min}$ (set equal to 1 $M_\odot$) in Eq. (6); ii) all the quantities related to stellar evolution and nucleosynthesis do not depend on time t, leading to constant R, R_i^a and y_i^{new}; iii) the rate of the infall from the halo (assumed to be void of Re and Os) $f_{\text{in}} = \gamma(1-R)\Psi$, and no outflow. [If $\gamma = 1$, the refill is complete.]; and iv) The normalized stellar birth rate $b(t) = c/t_0$ for $t \le t_0$ and $= (1-c)/(T_G - t_0)$ otherwise, with $t_0 = T_G - (6 \text{ Gyr})$ [3] (see also [29]). [If $c = t_0/T_G$, Ψ is a constant.]

The Internal Errors and the Age of the Galaxy

The adjustable parameters, γ, c and $m_{\text{gas}}(0) = m_{\text{tot}}(0)$, are varied in certain mesh sizes. For a given T_G, we pick solutions of Eqs. (4-5,12-13) that satisfy various observational constraints, and distribute them according to the respective solutions of Eq. (10), $X_{\text{re187}}(T_\odot)$, in comparison with $X_{\text{re187}}^\odot$. The constraints imposed concern: i) PDMF [26]; ii) $m_{\text{tot}}(T_G), m_{\text{gas}}(T_G), 1/b(T_G)$ and $f_{\text{in}}(T_G)$ [1]; and iii) some global aspects of age-metallicity relation [29,30] and metallicity distribution [31].

Figure 2 shows the results for $T_G = 13, 15$ and 17 Gyr. Within the adopted model of chemical evolution and the uncertainties in the parameter values, the most likely value of T_G is found to be 15 ± 1 Gyr.

EXTERNAL ERRORS AND AN OUTLOOK

The possible errors associated with atomic, nuclear and stellar physics involved in the evaluation of the astration fractions are generally small, although more model stars with varied metallicity (Z) have yet to be tested.

Much larger "external" errors stem from the uncertainties in the Re/Os abundance ratio observed in meteorites, and in the evaluation of the right-hand-side of Eq. (1).

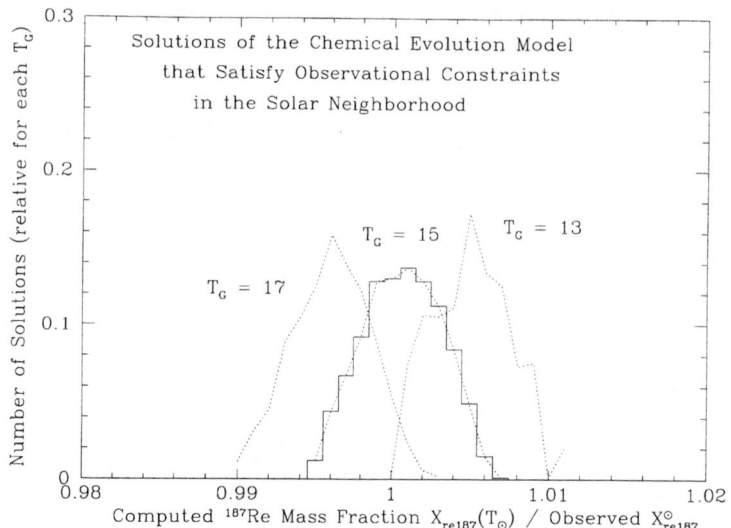

FIGURE 2. Number distribution functions of the satisfactory solutions of Eqs. (4-5,12-13) in terms of the solutions of Eq. (10) for selected T_G-values in Gyr. The widths represent the "internal" errors only. Typical values of the parameters are: $c \sim 0.8$, $\gamma \sim 0.7$ and $m_{\rm gas}(0) \sim 30 M_\odot {\rm pc}^{-2} {\rm Gyr}^{-1}$. The astration fractions are: $R_n \approx 0.94R$, $R_{\rm re187 \to re187} \approx 0.90 R_n$ and $R_{\rm os187 \to re187} \approx 0.18 R_n$, where R (≈ 0.30–0.35) is the returning fraction

Being even slightly optimistic, we estimate the combined uncertainty to be ± 10 % at present, which still makes T_G uncertain by about ± 4 Gyr. However, Anders and Grevesse [23] state that "the variation in the Re/Os ratios seems to reflect mainly differences among laboratories or analysts, not meteoritic classes." If that is the case, we have a good reason for being even more optimistic.

So far we have discussed about the possible errors in T_G derived from a one-zone model of chemical evolution. As mentioned earlier,[3] it is highly desirable to consider the halo and disk simultaneously, particularly because of the importance of the infall during the early evolution of the disk. It will be much more profitable, though how difficult it may be, to construct a model unifying the halo and the disk than to try to improve one-zone models by removing or altering the working assumptions thereof. To begin with, such a two-zone model would solve the classical "G-dwarf problem," i.e., the deficiency of very metal-poor stars in the disk, with ease and most naturally. In the last years, there has been a tremendous progress in astronomical observations of the elemental abundances in metal-deficient (i.e. the halo) stars. Those data can certainly be used as additional constrains in the modeling of chemical evolution in the early Galaxy. To set up a model that consistently describes the abundance patterns, particularly those of heavy elements which are often quite whimsical (see [32,33] for reviews), is of no small difficulty. But the efforts would not be defied.

Is there good hope that the ^{187}Re$-^{187}$Os chronometry will turn out well ? Positive.

ACKNOWLEDGMENTS

Our special thanks go to T. Faestermann, P. Kienle, F. Bosch, N. Langer and J. Wagenhuber for close collaborations. The work has been supported in part by the Deutsche Forschungsgemeinschaft under the "Sonderforschungsbereich 375 - 95 für Astro-Teilchenphysik," and in Japan by the Heiwa Nakajima Foundation.

REFERENCES

1. Tinsley, B. M., *ApJ* **216**, 548–59 (1977).
2. Tinsley, B. M., *Fundam. Cosmic Phys.* **5**, 287–388 (1980).
3. Yokoi, K., Takahashi, K., and Arnould, M., *A & A* **117**, 65–82 (1983).
4. Clayton, D. D., *MNRAS* **234**, 1–36 (1988).
5. Arnould, M., and Takahashi, K., in *Astrophysical Ages and Dating Methods*, ed. E. Vangioni-Flam et al., Gif-s-Yvette: Editions Frontières, 1990, pp. 325–48.
6. Schramm, D. N., *ibid.*, pp. 365–83.
7. Cowan, J.J., Thielemann, F.-K., and Truran, J.W., *ARA&A* **29**, 447–97 (1991).
8. Fowler, W. A., and Hoyle, F., *Ann. Phys.* **10**, 280–302 (1960).
9. Qian, Y. - Z., and Woosley, S. E., *ApJ* **471**, 331–51 (1996).
10. Hoffman, R. D., Woosley, S. E., and Qian, Y. - Z., *ApJ* **482**, 951–62 (1997).
11. Takahashi, K., and Janka, H. - Th., in *Origin of Matter and Evolution of Galaxies*, ed. T. Kajino et al., Singapore: World Scientific, 1997, pp. 213–27.
12. Cardall, C. Y., and Fuller, G. M., *preprint*, astro-ph/9701178 (1997).
13. Clayton, D. D., *ApJ* **139**, 637–63 (1964).
14. Arnould, M., Takahashi, K., and Yokoi, K., *A & A* **137**, 51–7 (1984).
15. Arnould, M., *A & A* **21**, 401–12 (1972).
16. Woosley, S. E., and Fowler, W. A., *ApJ* **233**, 411–7 (1979).
17. Winters, R. R. et al., *Phys. Rev.* **C34**, 840–9 (1986).
18. Takahashi, K., and Yokoi, K., *Nucl. Phys.* **404**, 578–98 (1983).
19. Bosch, F. et al., *Phys. Rev. Lett.* **77**, 5190–3 (1996).
20. Takahashi, K. et al., *Phys. Rev.* **C36**, 1522–8 (1987).
21. Wagenhuber, J., *Astron. Astrophys.*, to be submitted (1997).
22. Langer, N., and Henkel, C., *SSR* **74**, 343-52 (1995).
23. Anders, E., and Grevesse, N., *Geochim. Cosmochim. Acta* **53**, 197–214 (1989).
24. Winters, R. R., and Macklin, R. L., *Phys. Rev.* **C25**, 208–12 (1982).
25. Vásquez, E. C., and Scalo, J. M., *ApJ* **343**, 644–658 (1989).
26. Miller, G. E., and Scalo, J. M., *ApJS* **41**, 513–47 (1979).
27. Blöcker, T., *A & A* **297**, 727–738 (1995).
28. Nomoto, K., *ApJ* **277**, 791–805 (1984).
29. Twarog, B. A., *ApJ* **242**, 242–59 (1980).
30. Wheeler, J. C., Sneden, C., and Truran, J. W., *ARA & A* **27**, 279-349 (1989).
31. Pagel, B. E. J., and Patchett, B. E., *MNRAS* **172**, 13–40 (1975).
32. Ryan, S. G., in *Origin of Matter and Evolution of Galaxies*, op. cit., pp. 3–11.
33. Beers, T. C., *ibid.*, pp. 12–28.

The p-Process in Exploding Massive Stars

Marcel Arnould[1], Marc Rayet[1] and Masaaki Hashimoto[2]

[1]*Institut d'Astronomie et d'Astrophysique*
Université Libre de Bruxelles
Campus Plaine - CP 226, B-1050 Brussels – Belgium
[2]*Faculty of Science, Department of Physics*
Kyushu University, Fukuoka 810 – Japan

Abstract. The p-process has been studied by now in a quite large variety of astrophysical situations. This review concerns the p-nuclidic yields from the explosion of massive stars either as SNII, or as pair-creation supernovae. The nucleosynthetic predictions rely on detailed stellar models, on quantitative p-process seed abundances, as well as on extended nuclear reaction networks. The impact of some key nuclear physics and astrophysics uncertainties on the p-process yields is also investigated. In addition, a first quantitative evaluation of the metallicity dependence of the amount of ejected p-nuclides is provided by the study of SN1987A. This represents the first step in the construction of a reliable model for the evolution of the galactic content of the p-nuclides.

Our predictions are confronted with the bulk solar system composition, as well as with isotopic anomalies attributed to p-nuclides observed in various meteoritic materials. The possibility of developing a p-process chronometry on the short-lived radionuclides ^{92}Nb and ^{146}Sm that are inferred to have been present in the early solar system is also briefly discussed.

I INTRODUCTION

The stable neutron-deficient isotopes of the elements with charge number $Z \geq 34$ are classically referred to as the p-nuclei. Those nuclides are observed only in the solar system, where they represent no more than 1% to 0.1% (with increasing Z) of the bulk $Z \geq 34$ elemental abundances, made predominantly of the more neutron rich s- and r-nuclei. Isotopic anomalies involving p-nuclei are also found in some primitive meteorites.

It seems now astrophysically plausible that the p-nuclides originate from the oxygen/neon layers of highly evolved massive stars during their pre-supernova phase [2], or during their explosion either as Type II supernovae (SNII) [25], or even as pair-creation supernovae [24]. At the temperatures of about 2 to

3 billion degrees that can be reached in those layers, the p-nuclei may be synthesized by the (γ, n) photodisintegrations of pre-existing more neutron-rich species (especially s-nuclei), possibly followed by cascades of (γ, p) and/or (γ, α) reactions. It has also been proposed that those nuclear transformations could take place in the C-rich zone of Type Ia supernovae (SNIa) as well [12], [13].

This review summarizes some of the p-process calculations that have been conducted recently on grounds of detailed models of various supernova types that may represent the end stage of the evolution of massive stars. The most extended such calculations concern the SNII explosion of stars with masses in a large ($13 \leq M \leq 25$ $M_\odot$) range, and with metallicities $Z = Z_\odot$, where $Z_\odot$ is the solar metallicity. The possible influence of metallicity on these yields remains to be scrutinized in full detail. At this point, the only information on the Z dependence of the p-process yields is provided by the specific examination of SN1987A, viewed as a 20 $M_\odot$ star with $Z \approx 0.3$ $Z_\odot$ [21]. These extended calculations represent the first step towards a quantitative build-up of a model for the evolution of the galactic p-nuclidic content.

Sections II and III are concerned with the p-process yields from SNII and pair-creation SN, respectively, and with their comparison with the bulk solar system composition. Section IV deals briefly with the p-process anomalies, and in particular with the production of the long-lived p-nuclides ^{92}Nb and ^{146}Sm. Some conclusions are drawn in Sect. V.

II THE P-PROCESS IN SNII

A The input physics

Recent calculations demonstrate that the p-process can develop in the O/Ne layers of massive stars explosively heated to peak temperatures $T_{m,9}$ (in 10^9 K) in the approximate $1.8 \lesssim T_{m,9} \lesssim 3.3$ range [25]. These zones are referred to in the following as the p-process layers (PPL's). The identification of these zones and their physical conditions derive from the use of models for the pre-SN and SN explosion of $Z = Z_\odot$ stars with masses M ranging from 13 to 25 $M_\odot$ [9], [10], [11], [18]. The adopted SN1987A models are those of [18], [26]. In all cases, the PPL's are found to be located close to the bottom of the O/Ne-rich stellar zone, and are thus far enough from the SNII mass cut for their precise location not to affect the predicted p-process yields in any significant way. In contrast, the precise extent and mass of the PPL's, and consequently the predicted p-process abundances, may be sensitive to the still somewhat uncertain $^{12}C(\alpha,\gamma)^{16}O$ reaction rate, as well as to the assumed final kinetic energy of the ejecta. The nominal values adopted for these quantities are the reaction rate proposed by [6] (see [25] for a justification of this choice), and a kinetic energy of 10^{51} erg. The impact of a change of the values of these two quantities on the SNII p-process yield predictions has been evaluated by [25].

The adopted nuclear reaction network is thoroughly described in [23]. It includes approximately 1050 nuclei and 11000 reactions induced by neutrons, protons, and α-particles, as well as photodisintegrations. The 3α reaction and the various $^{12}C + ^{12}C$, $^{12}C + ^{16}O$ and $^{16}O + ^{16}O$ channels are also included, so that all light particle abundances are calculated self-consistently. This is especially the case for the neutrons, the captures of which are shown by [23] to play a non-negligible role in some p-process conditions, at variance with the assumption made in early simplified calculations [31].

As for the reaction rates, the Hauser-Feshbach model of [28] is used to calculate the rates on targets heavier than Si. For lighter nuclei, the compilations of [7] for charged particle reactions [except for the nominal $^{12}C(\alpha,\gamma)^{16}O$ rate] and of [4] for neutron captures are used.

The slow neutron capture nucleosynthesis (s-process) that develops pre-explosively in the stellar mass zones that will make the PPL's provides the seeds for the p-process. The s-process accompanying core He burning in massive stars is known to strongly enhance the $60 \lesssim A \lesssim 90$ s-nuclei, while the abundances of the heavier nuclei are only very weakly increased. For the model stars considered in this work, the heavy element abundances at the end of He burning are taken from the s-process calculations of [20]. As discussed in [25], these abundances are expected to represent quite closely the p-process seed abundances, except in some specific cases for which corrections are imposed by the astrophysical conditions encountered between the end of He-burning and the start of the p-process.

In contrast to the $A > 40$ nuclides, the lighter species can have their concentrations substantially modified between core He exhaustion and the onset of the explosion. Their PPL pre-explosion abundances are assumed to be equal to the presupernova values calculated by [18] and by [10] (see [25]).

B Some results

The calculations performed on grounds of the input physics just described indicate that the synthesis of the heavy p-nuclei requires lower temperatures than the production of the lighter ones, which are more resistant to photodisintegrations. As a consequence, the light ($N \lesssim 50$), intermediate-mass ($50 \lesssim N \lesssim 82$) and heavy ($N \gtrsim 82$) p-nuclides are essentially produced in the temperature ranges $T_{m,9} \gtrsim 3$, $3 \gtrsim T_{m,9} \gtrsim 2.7$ and $T_{m,9} \lesssim 2.5$, respectively (see [25] for details). In fact, each p-nucleus is substantially produced in a very narrow range of temperatures (see Fig. 3 of [21]). This extreme sensitivity of the results to T_m makes the use of realistic stellar models mandatory.

For the sake of presentation, let us define the mean overproduction factor $\langle F_i \rangle(M)$ of the p-nuclide i in a star of mass M as the total mass of this

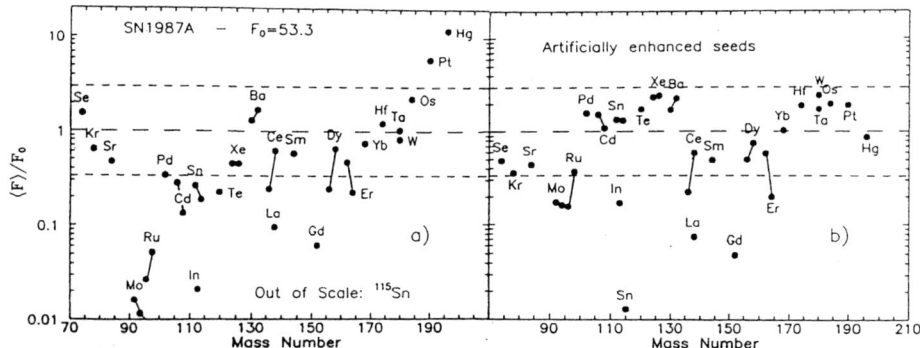

FIGURE 1. Normalized p-nuclei overproductions obtained in SN1987A. Solid lines join different p-isotopes of the same element. Fig. 1a: self-consistent seed abundances; Fig. 1b: artificial, solar-like, abundances

p-nuclide produced in the PPL's divided by the corresponding mass if the PPL's had a solar composition. We then define the normalized overproduction as the ratio $\langle F_i \rangle(M)/F_0(M)$, where $F_0(M)$ is the mean overproduction factor averaged over the 35 p-nuclei. With such definitions, all the normalized overproductions would be equal to unity if the derived abundance pattern were solar. Figure 1a displays the normalized overproductions derived for SN1987A from the nominal model ingredients defined in Sect. II.A.

The normalized overproductions predicted on the same grounds for the $Z = Z_\odot$ $13 \leq M \leq 25$ M$_\odot$ stars lie in the ranges indicated in Fig. 2 by the vertical bars associated with each p-nucleus. Roughly speaking, these overproduction factors are seen to depend only relatively weakly on the star masses, this being less true, however, for the lightest p-nuclei ^{74}Se, ^{78}Kr and ^{84}Sr. This result is discussed in [25].

Following test calculations performed for the $Z = Z_\odot$ 25 M$_\odot$ star [25], it appears that the impact on the p-process yields of a change in the $^{12}C(\alpha,\gamma)^{16}O$ rate or in the explosion energy is roughly of the same magnitude as the changes implied by the consideration of models of different masses calculated with the same input physics (see Fig. 2). The abundances of ^{74}Se, ^{78}Kr and ^{84}Sr are again the most sensitive to variations in the input physics (see [25] for a discussion of this feature).

The SNII yields for individual stars have been used by [25] in order to evaluate for the first time the evolution of the galactic content of the p-nuclides. As a zeroth-order approximation to the build-up of a full galactic chemical evolution model, the p-process yields have been averaged over an Initial Mass Function (IMF), leading to the IMF-averaged normalized overproduction factors displayed in Fig. 2.

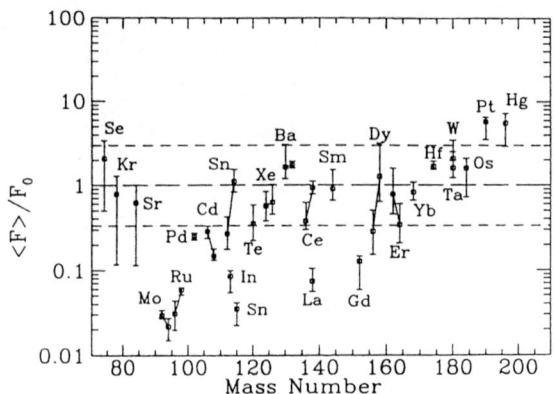

FIGURE 2. Ranges of variations of the normalized p-nuclei overproductions predicted for the individual SNII explosions of $Z = Z_\odot$ stars with masses in the range from 13 to 25 $M_\odot$. For the sake of simplicity, these ranges are just schematized by vertical bars, the results for individual model stars not being represented (Precise data for each of the considered stars can be found in Table 2 of [25]). Open squares indicate the values of the normalized overproductions obtained by integration over the IMF proposed by [14]. Solid lines join different p-isotopes of the same element

Of course, the build-up of a model for the temporal evolution of the galactic content of the p-nuclides not only requires the knowledge of the stellar mass dependence of their yields, but also their variations with the metallicity. This question has been first addressed qualitatively by [25], who have just evaluated the role of metallicity from a comparison between the results obtained for the 20 $M_\odot$ star with $Z = Z_\odot$ and the predictions by [21] for the 20 $M_\odot$ SN1987A progenitor with the LMC metallicity $Z \approx 0.3\ Z_\odot$. In the considered Z range, the 20 $M_\odot$ PPL characteristics are expected to be essentially the same, so that a metallicity change mostly affects the efficiency of the s-process responsible for the p-process seeds (see Rayet & Hashimoto, this volume). In fact, it appears that the p-process is about twice less efficient in SN1987A than in its solar metallicity counterpart. This conclusion is likely to be safely extrapolated to the other considered model stars.

From the IMF-averaged overproduction factors displayed in Fig. 2, it is concluded that about 60% of the produced p-nuclei fit the solar system composition within a factor of 3. Some discrepancies are also apparent. In particular, the Mo and Ru p-isotopes are predicted to be severely underproduced. The origin of this problem is not fully identified yet. One cannot exclude some failure in the nuclear physics input of relevance for the production or destruction of these isotopes. However, even the complete photodisintegration of all the available $A \gtrsim 100$ seeds would fall short in accounting for the especially abundant Mo and Ru p-isotopes [23]. In other words, the puzzle cannot be remedied by a mere change of the photodisintegration flux. A natural alternative calls for a significant transformation of the $A < 90$ nuclides by proton captures. The very low PPL proton concentrations obtained in our calculations make such a transformation inefficient, unless some of the predicted proton capture rates turn out to be grossly underestimated.

In order to examine the Mo-Ru question further, test calculations have been performed by enhancing artificially the abundances of the $A \gtrsim 90$ seed nuclei above the predicted core He burning s-process values. This indeed alleviates to a large extent the SNII underproduction of the Mo and Ru p-isotopes [3], as shown in Fig. 1b, based on a solar-like seed distribution used by [12] for their p-process calculations in SNIa explosions. At this point, it is not possible, however, to provide any realistic justification for such increased PPL seed abundances. However, one notes that the s-process calculations fall short to account for the Ba observed in SN1987A by a factor of about 10 [19]. Could it be that the puzzle of the Mo and Ru p-isotopes lies in our inability to model the s-process in massive stars correctly ? Alternatively, other complementary sources of these isotopes might be required. At this point, no truly plausible candidate has been identified.

The nuclides ^{113}In, ^{115}Sn, ^{138}La and ^{152}Gd are also underproduced in our SNII calculations. No other source has been identified yet for the former three species. In particular, neutron capture processes do not seem to be able to account for the solar abundances of ^{113}In and ^{115}Sn [17]. In contrast, ^{152}Gd can be produced in substantial amounts by the s-process. Finally, ^{180}Ta^m can emerge in large quantities from the PPL's in all the considered SNII models. The s-process could also produce this nuclide [16]. The relative contribution of these various mechanisms to this interesting nuclide remains to be unravelled.

In order for SNII to be potentially important contributors to the galactic p-nuclides, one has also to make sure that they do not produce too much of other species, and in particular oxygen, the abundance of which is classically attributed to SNII explosions. This question has been examined in some detail by [25], who conclude that the p-nuclides are globally underproduced in the $Z = Z_\odot$ 25 $M_\odot$ star by a factor of about 4 ± 2 relative to oxygen when all the abundances are normalized to the bulk solar values. The factor 4 is obtained with the nominal input physics sketched in Sect. II.A, and the mentioned uncertainty relates to changes in the $^{12}C(\alpha,\gamma)^{16}O$ rate and in the explosion energy referred to above. The problem of the oxygen overproduction may be eased in lower mass SNII explosions, while it is worsened with decreasing metallicity [25]. It may thus be that the p-process enrichment of the Galaxy has been slower than the oxygen enrichment. There is at present no observational test of this prediction.

At this point, it must be kept in mind that uncertainties on stellar convection introduce also a large uncertainty (a factor of 2–3) in the oxygen yield of massive stars [30].

III THE P-PROCESS IN PAIR-CREATION SN

The evolution of a star with an initial mass of 140 $M_\odot$ and solar metallicity has been computed from the main sequence. This very massive star is predicted to explode as a pair-creation SN after brief phases of carbon, neon and oxygen burning, and to leave a black hole remnant of mass $M_{bh} = 19$ $M_\odot$ (see [24] for details of the model). A zone of the massive oxygen core that develops prior to the explosion, as well a series of its exploding layers are found to be well suited for the development of the p-process. This nucleosynthesis is followed with the help of the reaction network already described in Sect. II.A. The seed abundances are derived from a full s-process calculation, the concentrations of the species lighter than Fe being obtained directly, as in Sect. II.A, through the stellar evolution computations.

The resulting normalized overabundances of the p-nuclides (defined as in Sect. II.B) are displayed in Fig. 3. Figure 3(a) shows a marked deficiency of the p-nuclides lighter than Ba relative to the heavy ones. As analyzed by [24], this mainly results from the trapping in the $M_{bh} = 19$ $M_\odot$ remnant of the layers that are hot enough for producing significant amounts of the light p-nuclei, while not contributing to the synthesis of the heavy ones (see Sect. II.B). It has to be emphasized that the M_{bh} value cannot be accurately predicted. Even the absence of a remnant cannot be excluded. As a consequence, the level of light p-nuclei depletion is quite uncertain as well. In order to evaluate the extent of this uncertainty, Fig. 3(b) displays the post-explosion overabundances predicted when $M_{bh} = 0$. In this extreme situation, the global overabundance pattern qualitatively resembles the SNII one displayed in Figs. 1 and 2. One noticeable exception is ^{180}Ta^m, which is found to be underproduced as long as $M_{bh} \lesssim 20$ $M_\odot$.

Finally, we want to stress that, as demonstrated in Fig. 3(b), the pair-creation SN p-process yields result to a very large extent from the quasi-hydrostatic evolutionary phase prior to the explosion. More specifically, the heavy p-nuclides, which emerge from relatively cool regions, appear to be almost unaffected by the explosion. In contrast, the explosion produces a substantial fraction of the three lightest ones, the synthesis of which requires high temperatures (and which would be locked in the remnant for $M_{bh} \gtrsim 15$ $M_\odot$). In the intermediate-mass range, the p-nuclides are either slightly produced or destroyed during the explosion, the largest destruction occuring for low-abundance species, like ^{113}In, ^{115}Sn or ^{152}Gd. These results thus substantiate for the first time on quantitative grounds the early claim by [2] that hydrostatic O/Ne burning in massive stars could be an adequate site for the p-process. It remains to be seen if this very same conclusion also holds for at least some stars exploding as SNII.

At this point, the question arises of the possible contribution of pair-creation SN to the galactic content of p-nuclides. It cannot be answered quantitatively at this stage, especially because the frequency of occurence and yields of these

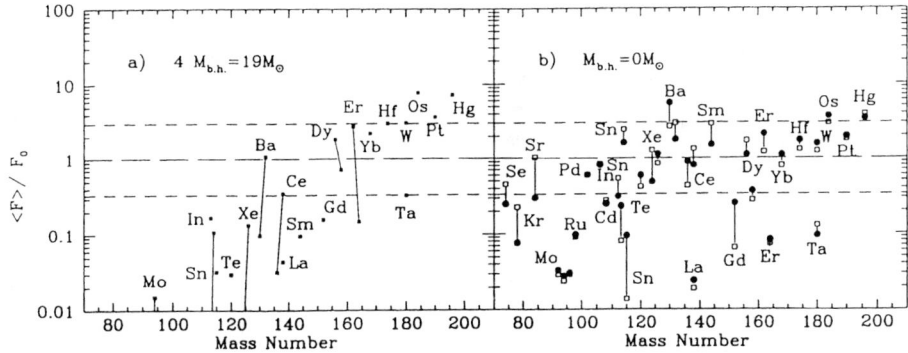

FIGURE 3. (a) Normalized p-nuclei overproductions obtained following the pair-creation SN explosion of a 140 $M_\odot$ star of solar metallicity, leaving a black hole remnant of mass $M_{bh} = 19\ M_\odot$. Solid lines join different p-isotopes of the same element; (b) Same as (a), but under the assumption that no black hole remnant is left. The post-explosion overproductions are represented by open squares, while the values prior to the explosion are shown as black dots

events versus metallicity is unknown. Obviously, such stellar explosions cannot be very frequent, except perhaps in the very young Galaxy. Despite the very large mass of processed material ejected per single event, their net yields may thus turn out not to be sufficient to account for the p-nuclei at the solar level if it is truly representative of the whole galactic disk, which is not known at the present time.

IV THE P-PROCESS ISOTOPIC ANOMALIES AND CHRONOMETRY

There is now observational evidence for the existence of isotopic anomalies due to the p-isotopes of Kr, Sr, Xe, Ba and Sm in various meteoritic materials, including the so-called FUN inclusions and various classes of grains of possible circumstellar origin (e.g. [15], [5], [1]). These anomalies manifest themselves as excesses or deficits of the abundances of the p-nuclides with respect to the more neutron-rich isotopes, when comparison is made with the bulk solar mix.

Among these discovered p-process anomalies, one of the most puzzling ones concerns the so-called Xe-L, which is characterized by excesses of ^{124}Xe and ^{126}Xe and, to a lesser extent, of ^{128}Xe with respect to solar. Remarkably enough, this special Xe is found to be correlated in diamonds of presumed circumstellar origin with Xe-H, the distinguishing characteristic of which is an excess of the r-process isotopes ^{134}Xe and ^{136}Xe. The very origin of the resulting Xe-HL unique pattern is not quantitatively understood yet. Our SNII p-process calculations, which reproduce very nicely the bulk solar isotopic pattern of the Xe p-isotopes, cannot fully account for the Xe-L isotopic composition [25]. Some changes in selected reaction rates might help, but no fully satisfactory solution has been found yet.

Meteoritic studies also provide strong evidence for the existence in the early solar system of the two p-process radionuclides ^{92}Nb^g ($t_{1/2} = 3.6\,10^7$ y) and ^{146}Sm ($t_{1/2} = 1.03\,10^8$ y) ([8], and references therein). Our SNII calculations predict ^{92}Nb/^{92}Mo production ratios ranging from about 0.003 to 0.03, depending upon the model star. These levels of production could well account for the value (^{92}Nb/^{92}Mo)$_0$ = $(2.9 \pm 0.6)\,10^{-5}$ at the start of condensation in the solar system (subscript 0) inferred by [8]. However, the remaining uncertainties in the levels of production of ^{92}Nb and ^{92}Mo by various stars, combined with a possible spallative production of these two nuclides [8], make rather unreliable at this time the development of a ^{92}Nb-based p-process chronometry.

As far as samarium is concerned, we obtain SNII ^{146}Sm/^{144}Sm production ratios in the approximate range from 0.4 to 1.5, which is compatible with the value (^{146}Sm/^{144}Sm)$_0$ = 0.008 ± 0.001 reported by [8]. We note that our predictions are significantly discrepant from the SNIa value of 0.04 reported by [13]. The nuclear physics uncertainties affecting the calculated yields of ^{144}Sm and ^{146}Sm (in particular of its progenitor ^{146}Gd) have been discussed by [32] and [22]. A remarkable experimental effort has been conducted recently on ^{144}Sm$(\alpha,\gamma)^{148}$Gd, identified as one of the key reactions involved in the calculation of the ^{144}Sm/^{146}Sm production ratio. The impact of these data on the predictions of this ratio, and the reliability of a ^{146}Sm-^{144}Sm chronometry are discussed in [27].

V CONCLUSIONS

This contribution reviews the unprecedented effort to model the p-process developing during SNII or pair-creation explosions of massive stars. More specifically, it presents the yields of the p-nuclides emerging from the explosion of $Z = Z_\odot$ stars with masses in a large ($13 \leq M \leq 25$ M$_\odot$) range, as well as from the especially interesting SN1987A supernova. The calculations make use of an extended nuclear reaction network, and are performed in the framework of detailed models for the explosion of realistic pre-supernova structures. Special attention is also paid to the evaluation of the abundances of the s-process seeds, which are of crucial importance for a quantitative description of the p-process. The variety of considered stellar cases makes possible for the first time the calculation of IMF-averaged abundances of the p-nuclides.

From the study of the SNII supernovae, it is concluded that
(i) the p-process abundance patterns do not depend drastically on the stellar mass or on other conditions, like the SN shock energy, or the dependence of

the pre-SN composition on the $^{12}C(\alpha,\gamma)^{16}O$ rate;
(ii) in all the considered cases, about 60 % of the p-nuclides are nicely co-produced (with respect to solar). The relative underproduction of the Mo and Ru p-isotopes found in previous studies is confirmed, and is shown to pervade the whole considered stellar mass range;
(iii) when normalized to the solar system values, the p-nuclides are globally underproduced with respect to oxygen by a factor of about 4. After the puzzling Mo and Ru underproduction, this prediction may represent the second most serious embarassment for the SNII p-process model. However, it is claimed at this point that various uncertainties of astrophysics or nuclear physics nature that can affect both the p-process and oxygen yields prevent to rule out the SNII model presented in this work. Time will tell if a reduction of those uncertainties can help alleviating or worsen the problem of the overproduction of oxygen relative to the p-nuclei!

Our SNII predictions are also confronted with anomalies attributed to p-isotopes found in various meteoritic materials. It is found in particular that our calculated Xe isotopic composition fails to reproduce the special pattern of Xe-L loading interstellar diamonds. On the other hand, the radionuclides ^{92}Nb and ^{146}Sm that are inferred to have been present in the early solar system are also synthesized in the SNII p-process. However, the predicted levels of production are considered to be somewhat too uncertain to allow at present the development of a reliable p-process chronometry.

Finally, the distribution of p-process yields from pair-creation supernovae is found to resemble qualitatively the SNII yields if the explosion leaves no remnant. In contrast, the p-isotopes of the elements lighter than Ba would be largely locked up in a black hole remnant if its mass exceeds about 10 $M_\odot$. These calculations also demonstrate that the p-process yields result to a very large extent from the evolutionary phase prior to the explosion.

The contribution to the p-process from the pre-supernova stage of massive stars, as well as from various SNI explosions, is currently under study, and is the necessary complement to the calculations reported here.

REFERENCES

1. Anders E, Zinner E., 1993, Meteoritics 28, 490
2. Arnould M., 1976, A&A, 46, 117
3. Arnould M., Rayet M., Hashimoto M. 1992, in: Kubuno S., Kajino T. (eds.) Unstable Nuclei in Astrophysics. World Scientific, Singapure, p. 23
4. Bao Z. Y., Käppeler F., 1987, ADNDT 36, 411
5. Begemann F., 1993, in: Prantzos N., Vangioni-Flam E., Cassé M. (eds.) Origin and Evolution of the Elements. Cambridge Univ. Press, Cambridge, p. 518
6. Caughlan G. R., Fowler W. A., Harris M. J., Zimmerman B. A., 1985, ADNDT 32, 197

7. Caughlan G. R., Fowler W. A., 1988, ADNDT, 40, 283
8. Harper C. L., 1996, ApJ. 466, 437
9. Hashimoto M., Nomoto, K., Shigeyama, T., 1989, A&A 210, L5
10. Hashimoto M., Nomoto K., Tsujimoto T., Thielemann F.-K., 1993, in: Käppeler F., Wisshak K. (eds.) Nuclei in the Cosmos. Institute of Physics Publishing, Bristol, p. 587
11. Hashimoto M., Nomoto K., Tsujimoto T., Thielemann F.-K., 1996, in: McCray R., Wang Zhenru. (eds.) Proc. IAU Coll. 145, Supernovae and Supernova Remnants. Cambridge Univ. Press, Cambridge, p. 157
12. Howard W. M., Meyer B. S., Woosley S. E., 1991, ApJ 373, L5
13. Howard W. M., Meyer B. S., 1993, in: Käppeler F., Wisshak K. (eds.) Nuclei in the Cosmos. Institute of Physics Publishing, Bristol, p. 575
14. Kroupa P., Tout C. A., Gilmore G., 1993, MNRAS 262, 545
15. Lee T., 1988, in: Kerridge J. F., Matthews M. S. (eds.) Meteorites and the Early Solar System. Univ. of Arizona Press, Tucson, p. 1063
16. Németh, Zs., Käppeler F., Reffo G., 1992, ApJ 392, 277
17. Németh Zs., Belgya T., Yates S. W., Theis Ch., Käppeler F., 1993, in: Käppeler F., Wisshak K. (eds.) Nuclei in the Cosmos. Institute of Physics Publishing, Bristol, p. 219
18. Nomoto K., Hashimoto M., 1988, Phys. Rep. 163, 13
19. Prantzos N., Arnould M., Cassé M, 1988, ApJ 331, L15
20. Prantzos N., Hashimoto M., Nomoto K., 1990, A&A 234, 211
21. Prantzos N., Hashimoto M., Rayet M., Arnould M., 1990, A&A 238, 455
22. Rayet M., Arnould, M., 1992, in: Delbar Th. (ed.) Radioactive Nuclear Beams 1991. Adam Hilger, Bristol, p. 347
23. Rayet M., Prantzos N., Arnould M., 1990, A&A 227, 271
24. Rayet M., El Eid M., Arnould M., 1993, in: Käppeler F., Wisshak K. (eds.) Nuclei in the Cosmos. Institute of Physics Publishing, Bristol, p. 613
25. Rayet M., Arnould, M., Hashimoto M., Prantzos N., Nomoto K., 1995, A&A 298, 517
26. Shigeyama T., Nomoto K., Hashimoto M., 1988, A&A 196, 141
27. Somorjai E. et al., 1997, to be published
28. Thielemann F.-K., Arnould M., Truran J. W., 1986, in: Vangioni-Flam E., Audouze J., Cassé M., Chièze J.-P., Tran Thanh Van J. (eds.) Advances in Nuclear Astrophysics, Editions Frontières, Gif-sur-Yvette, p. 525
29. Thielemann F.-K., Nomoto K., Hashimoto M., 1993, in: Prantzos N., Vangioni-Flam E., Cassé M. (eds.) Origin and Evolution of the Elements. Cambridge Univ. Press, Cambridge, p. 297
30. Weaver T. A., Woosley S. E., 1993, Phys. Rep. 227, 65
31. Woosley S. E., Howard W. M., 1978, ApJS 36, 285
32. Woosley, S. E., Howard, W. M., 1990, ApJ 354, L21

Theoretical Astronuclear Physics in Tours: A Brief Summary

Marcel Arnould

Institut d'Astronomie et d'Astrophysique
Université Libre de Bruxelles
Campus Plaine - CP 226, B-1050 Brussels, Belgium

Since the fifties, astrophysics has advanced at a remarkable pace, and has achieved an impressive record of success. Factors contributing to these rapid developments include paramount progress in ground-based and space astronomical observations, as well as in complex astrophysical modellings. These modellings not only take advantage of the rapidly accumulating observations, but also of major advances in computer technologies, as well as of spectacular breakthroughs in experimental and theoretical nuclear physics.

The special interplay between astrophysics and nuclear physics is embodied into a field commonly referred to as *Nuclear Astrophysics* or, equivalently, as *Astronuclear Physics*, which is one of the most fascinating examples of a truly interdisciplinary field of research. In fact, nuclear physics and astrophysics bring their share to the common adventure of understanding the structure and composition of the Universe and of its various constituents. The close relationship between these two major disciplines comes about because of the clear demonstration that the structure, evolution and composition of a large variety of cosmic objects, including the Solar System, bear strong imprints of the properties of atomic nuclei, as well as of their interactions. In such conditions, careful and dedicated experimental and theoretical studies of a large variety of nuclear processes are indispensable tools for the modelling of ultra-macroscopic systems such as stars.

The Tours Symposium on Nuclear Physics III has devoted nine of its sessions to a large variety of astronuclear physics questions. A rich panel of internationally recognized experts have presented high-quality reviews of, and their latest results on a broad range of questions of topical interest. Time has also been allotted to short oral presentations by young researchers. A most stimulating atmosphere has prevailed throughout all the sessions. Constructive interferences have been maintained at a high level among and between the senior attendees and young researchers, irrespective of their very diverse scientific backgrounds (observational and theoretical astrophysics, experimental and theoretical nuclear physics).

One of the highlights of the Symposium has been the identification of the input physics, and more particularly of the nuclear data, that most significantly influence astrophysical predictions concerning the evolution of non-exploding or exploding stars of various masses and ages, and their concomitant nuclidic production. More specifically, interesting reviews have been devoted to the status of the stellar modelling and of the nucleosynthesis predictions for massive stars during their pre-supernova and Type II supernova phases, as well as for low- and intermediate-mass stars. In addition, much attention has been paid to a variety of explosive events that are expected to develop in binary stars (Type I supernovae, novae, X-ray bursts). These various accounts clearly demonstrate that many uncertainties of astrophysical nature and of nuclear origin still affect the predictions. The nuclear questions of relevance for these stellar modellings have been discussed at length in various sessions, both from an experimental and from a theoretical viewpoint. The experimental contributions are reviewed by C. Rolfs in an accompanying summary. We largely limit ourselves here to some considerations about more theoretical aspects developed during the Symposium.

Quite naturally, much emphasis has been put on the experimental and theoretical problems raised by thermonuclear reactions in astrophysical plasmas. These processes are indeed pivotal for the energy budget and composition of stars, and have clear signatures on manifold quantities accessible to the astronomical observation. The thermonuclear reactions involved in astrophysical scenarios concern mainly the capture of protons and neutrons, as well as of α-particles by stable, neutron-deficient or neutron-rich nuclides in the whole mass range. As reviewed at length by many contributors, the rates of a wealth of those reactions are still more or less uncertain, in spite of much dedicated experimental and theoretical efforts.

In non-explosive conditions, corresponding in particular to the quiescent phases of stellar evolution which take place at relatively low temperatures, most of the reactions of interest concern stable nuclides. The major difficulty comes from the fact that the energies of astrophysical interest for charged-particle induced reactions are much lower than the Coulomb barrier energies. As a consequence, the corresponding cross sections can dive into the nanobarn to picobarn abysses. The state of the art in the extremely difficult measurements of such cross sections has been nicely discussed at the Symposium, and some inovative approaches have been proposed. Progress made in the experimental determination of neutron capture rates at energies of astrophysical interest has also been of concern (see the summary by C. Rolfs).

In stellar (nova, supernova, X-ray burst) or non-stellar (Big Bang) explosions, the energies of astrophysical interest are typically larger than in the non-explosive situations, and can be of the order of the Coulomb barrier. In such conditions, the relevant cross sections are also larger, and may range between the micro- and the millibarn. Unfortunately, there is a very high price to pay in order to enter this cross section range. In the explosive astrophysical sites, the thermal neutron, proton and α-particle bath is indeed able to drive the nuclear flows to either very neutron-deficient or very neutron-rich nuclides. The precise description of these flows requires the knowledge of the rates of captures of neutrons, protons or α-particles by more or less highly β-unstable nuclei. Except in some specific cases that have been reviewed at the Symposium, those reactions have not lent themselves yet to a direct experimental scrutiny (which is presently considered as a new frontier in laboratory nuclear astrophysics), so that their rates have to be evaluated theoretically.

The present status of the thermonuclear reaction models of astrophysical relevance and their attached uncertainties have been nicely discussed during the Symposium, along with some promising approaches that have to be explored further. From these discussions, and from the described lively experimental and theoretical efforts from a community of highly motivated and dedicated scientists, one can feel optimistic about future significant progress in our knowledge of thermonuclear reaction rates for use in stellar and nucleosynthesis models.

The Symposium has also produced the clearest proofs that thermonuclear reaction rates are by far not the only nuclear quantities entering in a crucial way the astrophysical models. In particular, a whole session has been devoted to interesting presentations of recent developments in the fields of nuclear mass and β-decay predictions that are of particular importance in the modelling of stellar explosions and of the concomitant nucleosynthesis. Remaining difficulties have been identified, and possible solutions have been sketched. Here again, the reported advances carry the promises of future developments of great interest.

On the other hand, non-thermal ("spallation") reactions also enter the realm of astrophysics. They can act in low temperature and density media through the interaction of non-thermally accelerated particles with the interstellar or interplanetary medium, as well as with the material (gas or grains) at stellar surfaces or in circumstellar shells. In fact, the interest of these reactions is clearly not limited to astrophysics. They are also of importance in many other fields of sciences, such as planetology, geophysics, geochemistry, climatology, glaciology, hydrology, space and aviation technology, as well as radiation protection in atmospheric and space flights. Quite unusually in a symposium concerned with astronuclear physics, room has been made for a very interesting review of the role of spallation reactions in producing so-called cosmogenic nuclides in extraterrestrial matter. A further proof that nuclear astrophysics has something to do with nuclear reactions at energies in excess of the Coulomb barrier has been produced by a discussion of the recent observations of nuclear γ-ray lines from the Orion molecular cloud. A model interpreting these lines as resulting from nuclear deexcitations following spallation reactions induced by particles accelerated by nearby massive stars (supernovae) has been presented. Interestingly enough, it points to the need for more nuclear data regarding in particular spallation reactions involving α-particles. It has to be emphasized that these reactions are also of substantial interest in reaction theory, as no general and truly reliable model is available for such reactions at energies (about 10 to 100 MeV per nucleon) that are typical of stellar (solar) cosmic rays. Last but not least, the discussion of the role of spallation reactions in astrophysics could not have been put to an end without an overview of the interesting question of the non-thermal production of Li, Be and B, this problem being revisited in the light of the observations and modellings of the above-mentioned Orion γ-ray lines.

The discussions concerning the non-explosive or explosive yields from a large variety of single or binary stars of various masses and compositions complemented with those about the Li, Be and B synthesis have quite naturally opened the way to a closer look into the chemical evolution of our Galaxy. This is a highly challenging task for theoreticians and observers as well. As made clear once more during this Symposium, these models suffer from a wealth of specific uncertainties coming on top of those of astrophysics or nuclear origin affecting the yields from individual stars. By continuously providing more and higher-quality data, particularly concerning the abundances of light and heavy nuclides at the surface of older and older stars, observers keep challenging the theoreticians' views, and contrive their models more and more. The Symposium has provided an interesting illustration of this exciting situation. Room has also been given to a careful revisitation of a special aspect of the modeling of the chemical evolution of the Galaxy that has to do with the use of radionuclides in order to estimate the galactic age. It has been clearly reiterated that any chronometry relying on the classical ^{232}Th-^{238}U and ^{235}U-^{238}U pairs will remain most unreliable for a long time to come. In contrast, the view has been expressed that there is a reasonable hope for the ^{187}Re-^{187}Os chronometry to turn out well. This optimism relies in particular on the recent laboratory demonstration that the β^--decay half-life of the fully ionized ^{187}Re is more than one billion times shorter than that of the neutral Re. This experiment provides a real breakthrough in nucleo-cosmochronomogical models, and is truly emblematic of the very intimate interplay between the microphysics of the atomic nucleus and the macrophysics of the stars and galaxies.

In fact, this interplay has been constantly at the forefront of all the astronuclear physics sessions of the Tours Symposium on Nuclear Physics III. This has made of this meeting a most scientifically fruitful and humanly enjoyable event. This success is clearly due to the high scientific standards of all the participants and of the great quality of the reviews. But not only. In fact, the Symposium would not have been what it has been without the constant involvement and dedication of the Konan University Nuclear Physics Group, and in particular of Prof. H. Ohta, Chairman of the Symposium, and of Profs. H. Utsunomiya and T. Wada, Scientific Secretaries of the Symposium. All the participants are very grateful to them, and look very much forward to coming back to Tours for the Symposium on Nuclear Physics IV.

Tours Symposium on Nuclear Physics III
Sept. 2 – 5, 1997, Tours, France

SCHEDULE

	Sept. 2 (Tue.)	Sept. 3 (Wed.)		Sept. 4 (Thu.)	Sept. 5 (Fri.)
09:00 - 10:30	Opening SHE I	FFD II	ANP I	ANP III	ANP VI
11:00 - 12:30	SHE II	FFD III	ANP II	ANP IV	ANP VII
14:00 - 15:30	SHE III	Excursion		ANP V	ANP VIII
16:00 - 17:30	FFD I			PEN I	ANP IX Closing
18:00 - 19:30	SBF I	Gala Dinner		PEN II	

SHE : Super-Heavy Elements   ANP : Astro-Nuclear Physics
FFD : Fusion-Fission Dynamics   PEN : Physics with Exotic Nuclei
SBF : Sub-Barrier Fusion

Chairparsons

SHE I	Y.T. OGANESSIAN (JINR, Russia)
SHE II	Y. ABE (Kyoto, Japan)
SHE III	G. MUENZENBERG (GSI, Germany)
FFD I	A. IWAMOTO (JAERI, Japan)
FFD II	N. TAKAHASHI (Osaka, Japan)
FFD III	T. WADA (Konan, Japan)
SBF	B. REMAUD (IRESTE, France)
PEN I	D. GUERREAU (GANIL, France)
PEN II	M. LEWITOWICZ (GANIL, France)
ANP I	J. KING (Toronto, Canada)
ANP II	Y. NAGAI (TIT, Japan)
ANP III	H. UTSUNOMIYA (Konan, Japan)
ANP IV	C. ROLFS (Bohum, Germany)
ANP V	K. TAKAHASHI (MPA, Germany)
ANP VI	M. CASSE (Saclay, France)
ANP VII	S. KUBONO (Tokyo, Japan)
ANP VIII	M. ARNOULD (Brussels, Belgium)
ANP IX	F. KAEPPELER (KfK, Germany)

SCIENTIFIC PROGRAM

Sept. 2 (Tue.)

OPENING SESSION
9:00 – 9:10   OHTA (Konan, Japan)

SHE I: EXPARIMENTAL DEVELOPMENTS   OGANESSIAN (JINR, Russia)
9:10 – 9:45   HESSBERGER (GSI, Germany)
   GSI experiments on the synthesis of superheavy elements
9:45 – 10:20   YEREMIN (JINR, Russia)
   Fusion reaction and experimental approach to the synthesis of superheavy nuclei
10:20 – 10:50   NOMURA (KEK, Japan)
   The RNB project in Japanese Hadron Facility and possible use of neutron-rich beam for the study of superheavy nuclei

SHE II: THEORETICAL DEVELOPMENTS   ABE (Kyoto, Japan)
11:20 – 11:45   CHEREPANOV (JINR, Russia)
   Fusion of massive nuclei and synthesis of superheavy elements in the framework of the DNS concept
11:45 – 12:10   ANTONIENKO (Giessen, Germany)
   Competition between complete fusion and quasi-fission in reactions with heavy nuclei
12:10 – 12:40   ARITOMO (Konan, Japan)
   Diffusion model for the synthesis of superheavy elements

SHE III: DECAY OF SUPERHEAVY ELEMENTS   MUENZENBERG (GSI, Germany)
14:00 – 14:25   MOLLER (USA)
   Stability and production of superheavy nuclei
14:25 – 14:50   BENDER (Frankfurt, Germany)
   Superheavy nuclei in self-consistent calculations
14:50 – 15:15   SMOLANCZUK (Warsaw, Poland)
   Decay properties of superheavy nuclei

FFD I: EXPERIMENTAL ADVANCES   IWAMOTO (JAERI, Japan)
15:45 – 16:25   SCHMIDT (GSI, Germany)
   Fission of exotic nuclei
16:25 – 16:50   ENGELMANN (GSI, Germany)
   Production cross-section of very neutron-rich nuclei in relativistic projectile fission of ^{238}U
16:50 – 17:15   MORJEAN (GANIL, France)
   Evolution of fission lifetime with temperature: A straightforward measurement by the blocking technique
17:15 – 17:40   YUASA-NAKAGAWA (RIKEN, Japan)
   Angular momentum dependence of prescission particle multiplicity in medium mass systems

SBF I: SUB-BARRIER FUSION   REMAUD (Nantes, France)
18:10 – 18:50   HINDE (ANU, Australia)
   Fusion barrier distributions and fission fragment anisotropies
18:50 – 19:15   IKEZOE (JAERI, Japan)
   Fusion reactions of deformed nuclei near Coulomb barrier
19:15 – 19:40   HAGINO (Tohoku, Japan)
   Anharmonic phonon excitations in subbarrier fusion reactions

Sept. 3 (Wed.)

PARALELL SESSION (ROOM 1)
FFD II: FUSION OF MASSIVE NUCLEI　　　　TAKAHASHI (Osaka, Japan)
9:00 – 9:25　RUDOLF (Strasbourg, France)
　　　　Investigation of the extra-extra push by pre-scission neutron measurements with DEMON
9:25 – 9:45　KOSENKO (JINR, Russia)
　　　　Fusion dynamics of massive nuclei
9:45 – 10:05　TOKUDA (Konan, Japan)
　　　　Multi-dimensional Langevin approach to the fusion of massive nuclei
10:05 – 10:15　GOVERDOVSKI (Obninsk, Russia)
　　　　Time scale of fission process of cold heavy nuclei

FFD III: FISSION MODES　　　　WADA (Konan, Japan)
10:45 – 11:15　CHUBARIAN (Texas A&M, USA)
　　　　Observation of fission modes in heavy ion induced reactions
11:15 – 11:40　DANIEL (JINR, Russia)
　　　　New mode of the ^{252}Cf spontaneous fission obtained with modern HPGe detectors arrays
11:40 – 12:05　YOKOYAMA (Osaka, Japan)
　　　　Fission mode study for low-energy fission of light actinide elements
12:05 – 12:30　IWAMOTO (JAERI, Japan)
　　　　Vlasov treatment of spontaneous fission and subbarrier fusion

PARALELL SESSION (ROOM 2)
ANP I: THERMONUCLEAR REACTIONS I　　　　KING (Toronto, Canada)
9:00 – 9:30　ROLFS (Bochum, Germany)
　　　　New experimental approaches to quests in static burning
9:30 – 10:00　GALSTER (Louvain-la-neuve, Belgium)
　　　　Reactions with radioactive beams: Direct measurements
10:00 – 10:10　SAGARA (Kyushu, Japan)
　　　　A plan for ^{4}He(^{12}C, ^{16}O)γ experiment
10:10 – 10:20　ROLFS (Bochum, Germany)
　　　　Laboratory electron screening
10:20 – 10:30　UTSUNOMIYA (Konan, Japan)
　　　　Time scale for non-resonant breakup of ^{7}Li over the Gamov energy region

ANP II: THERMONUCLEAR REACTIONS II　　　　NAGAI (TIT, Japan)
11:00 – 11:30　KUBONO (Tokyo, Japan)
　　　　New experiments for the breakout off the HCNO cycle
11:30 – 12:00　MOTOBAYASHI (Rikkyo, Japan)
　　　　Nuclear astrophysics with intermediate-energy RI beams
12:00 – 12:30　KING (Toronto, Canada)
　　　　The beta-delayed proton decay of ^{17}Ne and the ^{12}C(α,γ)^{16}O reaction cross section
12:30 – 12:40　IWASA (RIKEN, Japan)
　　　　Coulomb dissociation of ^{8}B at 254 MeV/u
12:40 – 12:50　TYPEL (Munchen, Germany)
　　　　Astronuclear physics with Coulomb dissociation

14:00 –　　　　EXCURSION and GALA DINNER

Sept. 4 (Thu.)

ANP III: THERMONUCLEAR REACTIONS III UTSUNOMIYA (Konan, Japan)
9:00 – 9:30 NAGAI (TIT, Japan)
  Pulsed keV neutrons for nuclear astrophysics and recent results of (n,γ) reaction of light nuclei
9:30 – 10:00 KAEPPELER (KfK, Germany)
  Constraints for s-process scenarios from neutron capture studies in the lanthanide region
10:00 – 10:30 DESCOUVEMONT (Brussels, Belgium)
  Recent progress in theoretical nuclear astrophysics
10:30 – 10:40 HERNDL (Vienna, Austria)
  One-nucleon capture rates for unstable nuclei in nuclear astrophysics
10:40 – 10:50 GORIELY (Brussels, Belgium)
  Radioactive neutron capture by exotic nuclei

ANP IV: NON-THERMAL REACTIONS AND NEUTRINO ASTROPHYSICS
  ROLFS (Bohum, Germany)
11:20 – 11:50 MICHEL (Hanover, Germany)
  Spallation reactions in extraterrestrial matter - Experiments and applications
11:50 – 12:00 COC (CSNSM, France)
  Nuclear data for gamma ray astronomy and for the cosmochemistry of isotopic anomalies
12:00 – 12:10 CASSE (Saclay, France)
  Nuclear gamma-ray lines from the Orion molecular cloud

ANP V: OTHER BASIC NUCLEAR DATA FOR ASTROPHYSICS
  TAKAHASHI (MPA, Germany)
14:00 – 14:30 PEARSON (Montreal, Canada)
  The ETFSI nuclear mass formula - Recent developments
14:30 – 15:00 BORZOV (Obninsk, Russia)
  Beta-decay rates - A self-consistent approach
15:00 – 15:30 TACHIBANA (Waseda, Japan)
  Beta-decay rates - The semi-gross theory

PEN I: PHYSICS WITH EXOTIC NUCLEI GUERREAU (GANIL, France)
16:00 – 16:30 MITTIG (GANIL, France)
  Mass and nuclear moment measurements with high and low energy RNBs
16:30 – 17:00 GELLETLY (Surrey, UK)
  Gamma spectroscopy with SISSI and SPIRAL
17:00 – 17:30 GUILLEMAUD-MUELLER (Orsay, France)
  Study of deformations in light neutron-rich nuclei

PEN II: PHYSICS WITH EXOTIC NUCLEI LEWITOWICZ (GANIL, France)
18:00 – 18:30 ALAMANOS (GANIL, France)
  Status and future development at SPIRAL
18:30 – 19:00 MARECHAL (Orsay, France)
  Inelastic proton scattering of unstable nuclei
19:00 – 19:30 SUZUKI (Fukui, Japan)
  Double giant resonance

Sept. 5 (Fri.)

ANP VI: STELLAR EVOLUITON AND NUCLEOSYNTHESIS
                                        CASSE (Saclay, France)
9:00 – 9:30   MOWLAVI (Geneva, Switzerland)
               Nucleosynthesis in low- and intermediate-mass stars
9:30 – 10:00   HASHIMOTO (Kyushu, Japan)
               Supernova nucleosynthesis
10:00 – 10:30 MEYNET (Geneva, Switzerland)
               The pre-supernova evolution of massive stars and concomitant nucleosynthesis

ANP VII: BINARY STAR EVOLUTION AND THE HOT MODES OF H-BURNING
                                        KUBONO (Tokyo, Japan)
11:00 – 11:30 JOSE (Barcelona, Spain)
               Nuclear uncertainties and their role in nova nucleosynthesis
11:30 – 12:00 TAAM (Northwestern, USA)
               X-ray bursts
12:00 – 12:30 SCHATZ (Notre-Dame, USA)
               Nucleosynthesis at the proton-drip line - A challenge for nuclear physics

ANP VIII: CHEMICAL EVOLUTION OF THE GALAXY   ARNOULD (Brussels, Belgium)
14:00 – 14:30 CASSE (Saclay, France)
               Origin and evolution of lithium, beryllium and boron
14:30 – 15:00 SANDRELLI (Bologna, Italy)
               The "stellar yields - galactic chemical evolution" connection: A re-analysis
15:00 – 15:30 MAGAIN (Liege, Belgium)
               Heavy elements abundances in metal-poor stars

ANP IX: THE SYNTHESIS OF THE A>56 NUCLIDES IN REALISTIC MODEL STARS
                                        KAEPPELER (KfK, Germany)
16:00 – 16:30 RAYET (Brussels, Belgium)
               The s-process in realistic model stars
16:30 – 17:00 TAKAHASHI (Munich, Germany)
               A progress in the $^{187}Re - {}^{187}Os$ cosmochronometry
17:00 – 17:30 ARNOULD (Brussels, Belgium)
               The p-process in realistic model stars
17:30 – 17:45 ARNOULD (Brussels, Belgium)
               Summary

List of Participants

Abe, Yasuhisa

YITP
Kyoto University
Oiwake-cho, Kitashirakawa
Kyoto 606-01
Japan

abey@yukawa.kyoto-u.ac.jp

Alamanos, Nicolai

CE- Saclay
DAPNIA/SPRIV Bat 703
Orme de Merisiers
F-91191 Gif-sur-Yvette Cedex
France

alaman@phnx7.saclay.cea.fr

Angulo, Carmen

Institut d'Astronomie et d'Astrophysique
Universite Libre de Bruxelles
CP 226, Bvd. du Triomphe
B-1050 Bruxelles
Belgium

angulo@astro.ulb.ac.be

Antonenko, Nikolai

Institut fur Theoretische Physik
Justus-Liebig-Universitat, Giessen
Heinrich-Buff-Ring 16
D-35392 Giessen
Germany

anton@professor.physik.uni-giessen.de

Arimoto, Yasushi

Research Center for Nuclear Physics
Osaka University
10-1 Mihogaoka, Ibaraki
Osaka 567
Japan

arimoto@rcnp.osaka-u.ac.jp

Aritomo, Yoshihiro

Department of Physics
Konan University
Okamoto 8-9-1, Higashinada
Kobe 658
Japan

e71a044@center.konan-u.ac.jp

Arnould, Marcel

Institut d'Astronomie et d'Astrophysique
Universite Libre de Bruxelles
CP 226, Bvd. du Triomphe
B-1050 Bruxelles
Belgium

marnould@astro.ulb.ac.be

Bender, Michael

Institut fur Theoretische Physik
J.W.Goethe-Universitat, Frankfurt/Main
Postfach 11 19 32
D-60054 Frankfurt am Main
Germany

bender@th.physik.uni-frankfurt.de

Borzov, Ivan Nikolayevich

Institute of Physics and Power Engineering
Bondarenko
249020 Obninsk
Russia

iborzov@astro.ulb.ac.be

Casse, Michel

Service d'Astrophysique
CEA Saclay
Orme de Merisiers
F-91191 Gif-sur-Yvette Cedex
France

cassse@iap.fr

Cherepanov, Evgeni

Flerov Laboratory of Nuclear Reactions
Joint Institute for Nuclear Research
Dubna 141980
Moscow region
Russia

cher@ljar.jinr.dubna.su

Chubaryan, Grigor

Cyclotron Institute
Texas A&M University
College station
TX 77843-3366
USA

chubarian@comp.tamu.edu

Coc, Alain

C.S.N.S.M.
Bat. 104
F-91405 Orsay Campus
France

coc@csnsm.in2p3.fr

Daniel, Andrei

Flerov Laboratory of Nuclear Reactions
Joint Institute for Nuclear Research
Dubna 141980
Moscow region
Russia

daniel@suntimpx.jinr.dubna.su

Descouvemont, Pierre

Physique Nucleaire Theorique
Universite Libre de Bruxelles
CP 229, Bvd. du Triomphe
B-1050 Bruxelles
Belgium

pdesc@ulb.ac.be

Engelmann, Christian

GSI
Planckstrasse 1
D-64291 Darmstadt
Germany

c.engelmann@gsi.de

Fukushima, Akira

Department of Physics
Konan University
Okamoto 8-9-1, Higashinada
Kobe 658
Japan

e71a354@center.konan-u.ac.jp

Galin, Joel

GANIL
BP 5027
F-14076 Caen Cedex
France

galin@ganil.fr

Galster, Wilfried

Department de Physique
Institut Physique Nucleaire
Universite Catholique de Louvain
Chemin du Cyclotron 2
B-1348 Louvain-la-Neuve
Belgium

galster@fynu.ucl.ac.be

Gelletly, William

Physics Department
University of Surrey
Guildford
Surrey GU2 5XH
UK

w.gelletly@surrey.ac.uk

Goriely, Stephane

Institut d'Astronomie et d'Astrophysique
Universite Libre de Bruxelles
CP 226, Bvd. du Triomphe
B-1050 Bruxelles
Belgium

sgoriely@astro.ulb.ac.be

Goverdovski , Andrei

Experimental Nuclear Physics Division
Institute of Physics and Power
Engineering
249020 Bondarenko
Obninsk
Russia

gaa@ippe.rssi.su

Guerreau, Daniel

GANIL
BP 5027
F-14076 Caen Cedex
France

guerreau@ganil.fr

Guillemaud-Mueller, Dominique

Institut de Physique Nucleaire
F-91406 Orsay Cedex
France

guillema@in2p3.fr

Hagino, Kouichi

Department of Physics
Tohoku University
Sendai, 980-77
Japan

hagino@nucl.phys.tohoku.ac.jp

Hashimoto, Masaaki

Department of Physics
Kyushu University
4-2-1 Ropponmatsu, Chuo-ku
Fukuoka 810
Japan

hashi@gemini.rc.kyushu-u.ac.jp

Hatogai , Koichi

Education and Research Center for
Information Science
Konan University
Okamoto 8-9-1, Higashinada
Kobe 658
Japan

hatogai@konan-u.ac.jp

Herndl, Harald

Institut fur Kernphysik
TU Wien
Wiedner Hauptstr. 8-10
A-1040 Wien
Austria

herndl@pp2.kph.tuwien.ac.at

Hessberger, Fritz Peter

GSI, Abt. KP2
Planckstrasse 1
D-64291 Darmstadt
Germany

hess@vzvi6f.gsi.de

Hinde, David J.

Department of Nuclear Physics
The Australian National University
Canberra, ACT2601
Australia

djh103@nuc.anu.edu.au

Hirose, Hanako
Department of Physics
Konan University
Okamoto 8-9-1, Higashinada
Kobe 658
Japan

e71a353@center.konan-u.ac.jp

Ikezoe, Hiroshi
Advanced Science Research Center
Japan Atomic Energy Research Institute
Tokai, Naka
Ibaraki 319-11
Japan

ikezoe@tdmalph1.tokai.jaeri.go.jp

Iwamoto, Akira
Advanced Science Research Center
Japan Atomic Energy Research Institute
Tokai, Naka
Ibaraki, 319-11
Japan

iwamoto@hadron01.tokai.jaeri.go.jp

Iwasa, Naohito
Cyclotron Laboratory
RIKEN
2-1 Hirosawa
Saitama 351-01
Japan

iwasa@valk.phys.wani.osaka-u.ac.jp

Jehin, Emmanuel
Institut d'Astrophysique et Geophysique
Universite de Liege
5 Avenue de Cointe
B-4000 Liege
Belgium

e.jehin@ulg.ac.be

Jose, Jordi
Dept. Fisica i Enginyeria Nuclear
Universitat Politecnica de Catalunya
Avda. Victor Balaguer s/n
08800 Vilanova i la Geltru
Barcelona
Spain

jjose@pro.eupvg.upc.es

Kaeppeler, Franz
Forschungszentrum Karlsruhe
Institut fur Kernphysik
Postfach 3640
D-76021 Karlsruhe
Germany

kaepp@ik3iris.fzk.de

Kato, Hisayuki
Department of Physics
Konan University
Okamoto 8-9-1, Higashinada
Kobe 658
Japan

e71a260@center.konan-u.ac.jp

King, James D.
Physics Department
University of Toronto
Toronto
Ontario M5S 1A7
Canada

king@physics.utoronto.ca

Kosenko, Grigoriy
Flerov Laboratory of Nuclear Reactions
Joint Institute for Nuclear Research
141980 Dubna
Moscow region
Russia

kosenko@nrsun.jinr.dubna.su

Kubono, Shigeru

Center for Nuclear Study
University of Tokyo
3-2-1 Midori-cho, Tanashi
Tokyo 188
Japan

kubono@cns.s.u-tokyo.ac.jp

Lewitowicz, Marek

GANIL
BP 5027
F-14076 Caen Cedex
France

lewito@ganac4.in2p3.fr

Magain, Pierre

Institut d'Astrophysique et Geophysique
Universite de Liege
5 Avenue de Cointe
B-4000 Liege
Belgium

magain@astro.ulg.ac.be

Marechal, Francois

Institut de Physique Nucleaire
F-91406 Orsay Cedex
France

marechal@ipno.in2p3.fr

Meynet, Georges

Geneva Observatory
CH-1290 Sauverng
Switzerland

georges.meynet@obs.unige.ch

Michel, Rolf

Zentrum fur Strahlenschutz und
Radiooekologie
Universitat Hannover
Am Kleinen Felde 30
D-30167 Hannover
Germany

michel@mbox.zsr.uni-hannover.DE

Mittig, Wolfgang

GANIL
BP 5027
F-14076 Caen Cedex
France

mittig@ganac4.in2p3.fr

Mizuno, Masato

Department of Physics
Konan University
Okamoto 8-9-1
Higashinada
Kobe 658
Japan

masato@base2.ipc.konan-u.ac.jp

Moller, Peter

P.Moller's Scientific Computing and
Graphics Inc.
P.O. Box 1440, Los Alamos
NM 87544
USA

71052.670@compuserve.com

Morikawa, Ryuji

Department of Physics
Konan University
Okamoto 8-9-1
Higashinada
Kobe 658
Japan

e71a290@center.konan-u.ac.jp

Morjean, Maurice

GANIL
BP 5027
F-14076 Caen Cedex
France

morjean@ganil.fr

Motobayashi, Tohru

Deptartment of Physics
Rikkyo University
3 Nishi-Ikebukuro, Toshima
Tokyo 171
Japan

motobaya@rikkyo.ac.jp

Mowlavi, Nami

Observatoire de Geneve
CH-1290 Versoix
Switzerland

nami.mowlavi@obs.unige.ch

Muenzenberg, Gottfried H.

GSI
Planckstrasse 1
D-64291 Darmstadt
Germany

g.muenzenberg@gsi.de

Nagai, Yasuki

Department of Applied Physics
Tokyo Institute of Technology
O-okayama 2-12-1, Meguro
Tokyo 152
Japan

nagai@atlas.nucl.ap.titech.ac.jp

Nakanishi, Norihiko

Konan University
Okamoto 8-9-1, Higashinada
Kobe 658
Japan

Neuforge, Corinne

Institut d'Astrophysique et Geophysique
Universite de Liege
5 Avenue de Cointe
B-4000 Liege
Belgium

Corinne.Neuforge@ulg.ac.be

Nomura, Toru

High Energy Accelerator Research
Organization (KEK)
Tanashi Branch
3-2-1 Midori-cho, Tanashi
Tokyo 188
Japan

toru.nomura@kek.jp

Oganessian, Yuri Ts.

Flerov Laboratory of Nuclear Reactions
Joint Institute for Nuclear Research
141980 Dubna
Moscow Region
Russia

oganessian@sungraph.jinr.dubna.su

Ohta, Masahisa

Department of Physics
Konan University
Okamoto 8-9-1, Higashinada
Kobe, 658
Japan

masaota@konan-u.ac.jp

Okazaki, Kenichiro

Department of Physics
Konan University
Okamoto 8-9-1, Higashinada
Kobe 658
Japan

e71a176@center.konan-u.ac.jp

Osada, Katsuyuki

Department of Physics
Konan University
Okamoto 8-9-1, Higashinada
Kobe 658
Japan

e71a259@center.konan-u.ac.jp

Pearson, J. Michael

Laboratoire de Physique Nucleaire
Department de Physique
Universite de Montreal
Montreal H3C 3J7
Canada

pearson@lps.umontreal.ca

Rayet, Marc Jean

Institut d'Astronomie et d'Astrophysique
Universite Libre de Bruxelles
CP 226, Bvd. du Triomphe
B-1050, Bruxelles
Belgium

mrayet@astro.ulb.ac.be

Remaud, Bernard

IRESTE
ATLANPOLE
La Chantrerie, CP 3003
F-40097 Nantes Cedex
France

bremaud@ireste.fr

Rolfs, Claus

Experimentalphysik III
Ruhr-Universitat, Bochum
D-44780 Bochum
Germany

rolfs@ep3.ruhr-uni-bochum.de

Rudolf, Gerard

IRES Strasbourg
BP 28
F-67037 Strasbourg Cedex
France

gerard.rudolf@ires.in2p3.fr

Sagara, Kenshi

Department of Physics
Kyushu University
Hakozaki, Higashi-ku
Fukuoka 812-81
Japan

sagara@kutl.kyushu-u.ac.jp

Sandrelli, Stefano

Dipartimento di Astronomia
Universita di Bologna
Via Zamboni 32
I-40126 Bologna
Italy

sandrelli@astbo4.bo.astro.it

Schatz, Hendrik

Forschungszentrum Karlsruhe IK3
Postfach 3640
D-76021 Karlsruhe
Germany

h.schatz@ik3.fzk.de

Schmidt, Karl-Heinz

GSI
Planckstrasse 1
D-64291 Darmstadt
Germany

k.h.schmidt@gsi.de

Smolanczuk, Robert

Soltan Institute for Nuclear Studies
Hoza 69
PL-00-681 Warsaw
Poland

smolan@fuw.edu.pl

Suzuki, Toshio

Department of Applied Physics
Fukui University
Bunkyo 3-9-1
Fukui 910
Japan

suzuki@quantum.apphy.fukui-u.ac.jp

Taam, Ronald

Department of Physics and Astronomy
Northwestern University
2145 Sheridan Road, Evanston
IL 60208
USA

taam@ossenu.astro.nwu.edu

Tachibana, Takahiro

Advanced Research Center for Science and Engineering
Waseda University
3-4-1 Okubo, Shinjuku
Tokyo
Japan

ttachi@mn.waseda.ac.jp

Takahashi, Noriaki

Osaka University
Graduate School of Science
Machikaneyama 1-16, Toyonaka
Osaka 560
Japan

ntakahas@rcnpax.rcnp.osaka-u.ac.jp

Takahashi, Kohji

Max-Planck-Institut fur Astrophysik
Ludwig-Merk-Str. 15
D-80805 Muenchen
Germany

kjt@mpa-garching.mpg.de

Tanaka, Yoshiro

Government & Public Systems Marketing Section
Mitsubishi Electric Corporation
Kansai Branch Office
2-2 Dojima-2, Kita-ku
Osaka 530
Japan

tanaka@icg51.saib.melco.co.jp

Tokimoto, Yoshiaki

Department of Physics
Konan University
Okamoto 8-9-1, Higashinada
Kobe 658
Japan

e71a182@center.konan-u.ac.jp

Tokuda, Takuya

Department of Physics
Konan University
Okamoto 8-9-1, Higashinada
Kobe 658
Japan

e71a045@center.konan-u.ac.jp

Typel, Stefan

Sektion Physik
Ludwig-Maximilians-Universitat, Muenchen
Am Coulombwall 1
D-85748 Garching
Germany

stypel@laser.physik.uni-muenchen.de

Uegaki, Eiji

Department of Physics
Akita University
Gakuen-cho 1-1, Tegata
Akita 010
Japan

ue@uws13.phys.akita-u.ac.jp

Utsunomiya, Hiroaki

Department of Physics
Konan University
Okamoto 8-9-1, Higashinada
Kobe, 658
Japan

hiro@konan-u.ac.jp

Wada, Takahiro

Department of Physics
Konan University
Okamoto 8-9-1, Higashinada
Kobe 658
Japan

wada@konan-u.ac.jp

Yamamoto, Yoshiki

Department of Physics
Konan University
Okamoto 8-9-1, Higashinada
Kobe 658
Japan

yoshiki@base2.ipc.konan-u.ac.jp

Yeremin, Alexander

Flerov Laboratory of Nuclear Reactions
Joint Institute for Nuclear Research
141980 Dubna
Moscow region
Russia

eremin@sunvas.jinr.dubna.su

Yokoyama, Akihiko

Department of Chemistry
Graduate School of Science
Osaka University
1-1 Machikaneyama, Toyonaka
Osaka 560
Japan

yokoyama@chem.sci.osaka-u.ac.jp

Yonehara, Katsuya

Department of Physics
Konan University
Okamoto 8-9-1, Higashinada
Kobe 658
Japan

Yuasa-Nakagawa, Keiko

Cyclotron Laboratory
RIKEN
Hirosawa 2-1, Wako
Saitama 351-01
Japan

keiko@rikaxp.riken.go.jp

AUTHOR INDEX

A

Abe, Y., 61, 107, 142, 171
Adamian, G. G., 41, 51
Agodi, C., 189
Alamanos, N., 300
Ameil, F., 126
Antonenko, N. V., 41, 51
Aoki, Y., 142, 343
Aritomo, Y., 61, 107
Armbruster, P., 3, 126
Arnould, M., 626, 637
Aryaeinejad, R., 202
Asztalos, S., 202
Azhari, A., 305
Azuma, R. E., 372

B

Baba, H., 212
Baba, T., 399
Babu, B. R. S., 202
Bateman, N., 372
Baumann, T., 382
Bazin, D., 305
Bellia, G., 189
Bender, M., 85
Benlliure, J., 113
Bernas, M., 126
Bildsten, L., 559
Blank, B., 382
Blumenfeld, Y., 305
Böckstiegel, C., 113, 126
Bonasera, A., 222
Borzov, I. N., 475, 485
Boue, F., 382
Boyd, R. N., 372
Brown, J. A., 305
Buchmann, L., 372
Bürvenich, T., 85

C

Calabretta, L., 189
Cassé, M., 465, 571

Cherepanov, E. A., 41
Chevallier, M., 134
Chow, J. C., 372
Chu, S. Y., 202
Chubarian, G. G., 189
Clerc, H.-G., 113
Coc, A., 457
Cohen, C., 134
Cole, J. D., 202
Cottle, P. D., 305
Czajkowski, S., 126, 382

D

Daniel, A. V., 202
Dardenne, Y. K., 202
Dasgupta, M., 233
D'Auria, J. M., 372
Dauvergne, D., 134
Davinson, T., 372
de Jong, M., 113
Descouvemont, P., 418
Dessagne, Ph., 126
Dombsky, M., 372
Donzaud, C., 126
Driger, M., 202
Dural, J., 134

E

Engelmann, C., 126

F

Fauerbach, M., 305
Folger, H., 3
Förster, A., 382
Fuchi, Y., 355
Furutaka, K., 142
Futami, Y., 142

G

Gai, M., 382
Galin, J., 134
Galster, W., 327, 372
Geissel, H., 126, 382
Gelletly, W., 279
Gete, E., 372
Giesen, U., 372
Ginter, T., 202
Glasmacher, T., 305
Goldenbaum, F., 134
Goriely, S., 436, 485
Görres, J., 559
Goverdovski, A. A., 179
Gregorich, K. E., 202
Greiner, W., 85
Grewe, A., 113
Grosse, E., 382
Guillemaud-Mueller, D., 290

H

Hagino, K., 259
Hahn, K. I., 355
Hamada, S., 343
Hamilton, J. H., 202
Hanappe, F., 189
Hashimoto, M., 517, 605, 626
Hatogai, K., 107
Heinz, A., 113, 126
Hellström, M., 382
Hernanz, M., 539
Herndl, H., 428
Heßberger, F. P., 3
Hinde, D. J., 233
Hirota, K., 343
Hirzebruch, S., 305
Hofinger, R., 428
Hofmann, S., 3
Huck, A., 189
Hurst, B. J., 189

I

Ieki, K., 343
Igashira, M., 399
Ikezoe, H., 249

Ikuta, T., 249
Iliadis, C., 372
Itkis, M. G., 189
Ivanyuk, F. A., 164
Iwamoto, A., 222
Iwasa, N., 382
Iwata, Y., 343

J

Jackson, K. P., 372
Jacquet, D., 134
Janas, Z., 126
Jehin, E., 592
Jeong, S. C., 355
Jewell, J. K., 305
Jiang, D. X., 142
José, J., 539
Jun, L., 249
Junghans, A. R., 113

K

Käppeler, F., 408
Kasagi, J., 142
Kato, S., 355
Katori, K., 343
Kawashima, H., 355
Kelley, J. H., 305
Kemper, K. W., 305
Kii, T., 399
Kim, J. C., 355
King, J. D., 372
Kinoshita, M., 399
Kirsch, R., 134
Kliman, J., 202
Koczon, P., 382
Kohlmeyer, B., 382
Kondratiev, N. A., 189
Kondratyev, V., 222
Kormicki, J., 202
Kosenko, G. I., 164
Kozhuharov, C., 126
Kozulin, E. M., 189
Kubono, S., 355
Kulessa, R., 382
Kurasawa, H., 315
Kurokawa, M., 355

Kuyucak, S., 259
Kuzumaki, T., 249

L

Laue, F., 382
Lavrentev, A., 3
Lee, C. H., 355
Lee, C. S., 355
Lee, I. Y., 202
Lee, J. H., 355
Lee, S. M., 142
Leino, M. E., 3
Liatard, E., 189
Lienard, E., 134
Liu, X., 142, 355
Lott, B., 134
Lui, Y.-W., 343
Lukashin, K., 189

M

Ma, W.-C., 202
Macchiavelli, A. O., 202
Magain, P., 592
Maiolino, C., 189
Mamdouh, A., 475
Mantica, P. F., 305
Marchand, C., 382
Maréchal, F., 305
Maruhn, J. A., 85
Matsuda, K., 142
Meynet, G., 526
Michel, R., 447
Miehé, Ch., 126
Minemura, T., 355
Mitsuoka, S., 249
Mittig, W., 271
Miyachi, T., 355
Mohar, M. F., 202
Möller, P., 75
Morhac, M., 202
Morissey, D. J., 305
Morjean, M., 134
Morton, A. C., 372
Motobayashi, T., 355, 362, 382
Mowlavi, N., 507
Müller, J., 113

Münzenberg, G., 3, 126

N

Nagai, Y., 399
Nagame, Y., 249
Nagataki, S., 517
Naito, S., 399
Nakagawa, T., 142
Nakata, H., 495
Nayak, R. C., 475
Neuforge, C., 592
Ninov, V., 3
Nishinaka, I., 249
Nix, J. R., 75
Nobuhara, Y., 399
Noels, A., 592
Nomura, T., 29

O

Oberhummer, H., 428
Oeschler, H., 382
Oganessian, Yu. Ts., 16, 164, 189, 202
Ohsaki, T., 399
Ohta, M., 61, 107, 171, 343
Okazaki, K., 61, 107, 171
O'Kelly, D., 189
Osada, K., 343
Ostapenko, Yu. B., 179
Ottini, S., 305
Ozawa, A., 382

P

Park, S. H., 355
Parker, P. D., 355
Pashkevich, V. V., 164, 189
Pearson, J. M., 475, 485
Péghaire, A., 134
Périer, Y., 134
Pfützner, M., 113, 126
Poizat, J. C., 134
Pokrovsky, I. V., 189
Popeko, A. G., 3
Popeko, G. S., 202
Porquet, M.-G., 457

Pravikoff, M. S., 382
Prevot, G., 134
Prussin, S. G., 202

R

Ramayya, A. V., 202
Rasmussen, J., 202
Rauscher, T., 559
Rayet, M., 475, 605, 626
Reinhard, P. G., 85
Rémillieux, J., 134
Riley, L. A., 305
Roy, G., 372
Rudolf, G., for the DEMON Collaboration, 155
Rusanov, A. Ya., 189
Rutz, K., 85

S

Sagara, K., 337
Salamatin, V. S., 189
Sandrelli, S., 581
Saro, S., 3
Sato, K., 517
Scarpaci, J. A., 305
Schatz, H., 559
Scheid, W., 51
Schilling, T., 85
Schmaus, D., 134
Schmidt, K.-H., 113
Schmitt, R. P., 189, 343
Schwab, E., 382
Schwab, W., 126, 382
Scully, S. T., 465, 571
Senger, P., 382
Shen, W. Q., 142
Shima, T., 399
Shimoda, T., 355
Shoppa, T., 372
Shotter, A., 372
Smith, M., 355
Smolańczuk, R., 97
Speer, J., 382
Steiner, M., 305
Steinhäuser, S., 113
Stéphan, C., 126

Stodel, Ch., 3
Stoyer, M. A., 202
Strasser, P., 355
Sturm, C., 382
Stuttgé, L., 189
Sümmerer, K., 126, 382
Suomijärvi, T., 142, 305
Surowiec, A., 382
Surowka, G., 382
Suzuki, T., 315

T

Taam, R. E., 551
Tachibana, T., 495
Takahashi, K., 616
Takahashi, T., 399
Takaoka, K., 399
Takigawa, N., 259
Tanaka, M. H., 355
Tassan-Got, L., 126
Ter-Akopian, G. M., 202
Teranishi, T., 382
Thielemann, F.-K., 559
Tokimoto, Y., 343
Tokuda, T., 107, 171
Tomyo, A., 399
Tondeur, F., 475
Tosi, M., 581
Toulemonde, M., 134
Tsukada, K., 249
Typel, S., 389

U

Uhlig, F., 382
Utsunomiya, H., 343, 355
Utyonkov, V. K., 16

V

Vangioni-Flam, E., 465, 571
Visco, A., 581
Volkov, V. V., 41, 51
Voss, B., 126

W

Wada, T., 61, 107, 142, 171
Wagner, A., 382
Walus, W., 382
Wiescher, M., 559
Wisshak, K., 408

Y

Yamada, M., 495
Yamada, S., 517
Yamagata, T., 343
Yamaji, S., 142
Yasue, M., 355
Yeremin, A. V., 3, 16
Yoshida, K., 142
Yuasa-Nakagawa, K., 142

Z

Zhu, S. J., 202